PLANETARY, LUNAR, AND SOLAR POSITIONS
NEW AND FULL MOONS, A.D. 1650–1805

PLANETARY, LUNAR, AND SOLAR POSITIONS
NEW AND FULL MOONS, A.D. 1650–1805

OWEN GINGERICH
and
BARBARA L. WELTHER
Harvard-Smithsonian Center for Astrophysics

The American Philosophical Society
Independence Square ● Philadelphia
1983

Printed in the U.S.A. by Science Press, Ephrata, Pennsylvania

Library of Congress Catalog Card No. 83-71297
International Standard Book Number: 0-87169-590-1
US ISSN: 0065-9738

CONTENTS

PREFACE

The computation of astronomical ephemerides for past epochs was among the earliest proposed applications of high-speed computers. The first of these appeared quite independently in the early 1960s: *Solar and Planetary Longitudes for Years −2500 to +2000 by 10-Day Intervals*,[1] giving geocentric longitudes with an accuracy of 1°; and the Tuckerman *Planetary, Lunar, and Solar Positions 601 B.C. to A.D. 1 at Five-day and Ten-day Intervals*,[2] giving both geocentric longitudes and latitudes with an accuracy approaching $0°.01$ for the sun and planets and $0°.1$ for the moon. In 1964 Tuckerman extended his tables in a second volume[3] to A.D. 1649, and in 1973 Goldstine furnished tables of new and full moons[4] for the range 1001 B.C. to A.D. 1651.

For certain historical problems, however, the cutoff of A.D. 1649 for the more accurate positions proved to be a troublesome limitation, and consequently we here provide an extension to A.D. 1805. Our computer procedures are similar to those used by Tuckerman and Goldstine, but because the original IBM program was unavailable, we have used an independent program slightly modified from one supplied by Professor Peter Huber. Details of the computational procedure and expected accuracy will be found in Section II.

---

[1] William D. Stahlman and Owen Gingerich, (Madison, 1963).
[2] Bryant Tuckerman, *Memoirs* of the American Philosophical Society, Vol. 56 (1962).
[3] Bryant Tuckerman, *Memoirs* of the American Philosophical Society, Vol. 59 (1964).
[4] Herman H. Goldstine, *New and Full Moons 1001 B.C. to A.D. 1651. Memoirs* of the American Philosophical Society, Vol. 94 (1973).

I. THE ACCURACY OF HISTORICAL EPHEMERIDES.

A prime historical problem that can be examined by means of these tables concerns the increasing accuracy of ephemerides and almanacs in the seventeenth and eighteenth centuries. This interval bridges the time from the quantum jump in positional accuracy provided by Kepler's work, through the development of Newtonian gravitational theory, to the initiation of the great national nautical almanacs. By 1800 the accuracy of the best almanacs was comparable to our tables, that is, better than a minute of arc. Perhaps the most surprising result of our analysis is how little immediate and direct impact Newton's work had on the computation of astronomical positions. Second, we can see how far most common almanacs lagged behind the state of the art during these centuries.

The computing of ephemerides in the seventeenth and eighteenth centuries rested in the hands of relatively few astronomers or astrologers, and at even fewer centers. One sequence, begun by Johannes Kepler (1571–1630), may be considered an unbroken series from 1617 to the present day. When Kepler's own ephemeris expired in 1637, Lorenz Eichstadt (1596–1660) in Gdansk produced a sequel in two segments; and following him, Johann Hecker (d. 1675) calculated the positions from 1666 through 1680.

As time marched on with no obvious Gdansk successor to Hecker in sight, the astronomers at the newly established Paris Observatory decided that Hecker's widely used ephemerides, which had been reprinted in Paris, should be explicitly continued. Consequently, the *Connoissance des temps* was founded, and within a few years the annual volumes of this almanac managed to include almost as much tabular data for the year as a typical ephemeris volume. Twenty years later Gabriel Philippe de La Hire (1677–1719), eldest son of the astronomer Philippe de La Hire (1640–1718), brought out a competing almanac based on his father's tables. In an ensuing battle of words Jean Lefèbvre (1650–1706), the editor of the *Connoissance des temps,* lost his rights to it. They were transferred by the French government from Lefèbvre to the Académie des Sciences. Lalande remarked that this was a loss for astronomy, since Lefèbvre calculated eclipses better than La Hire.[5] Deprived of both his editorship and his Academy membership, Lefèbvre seems to have issued under a pseudonym[6] a competing volume, the *Éphémérides des mouvemens célestes* for the years 1702–1714. Apparently there was a demand for ephemerides with a dozen years of data rather than the single year of the annual almanacs, so the *Éphémérides* were subsequently continued in a numbered series. The positions in these volumes were computed independently of the *Connoissance des temps* with different levels of accuracy. The computers themselves were from the same community, and Joseph-Jérôme Lefrançais de Lalande (1732–1807), for example, edited both the *Connoissance des temps* (from

1760–1775 and again from 1795 to 1807) and the *Éphémérides des mouvemens célestes* (1775–1800).

In any event, throughout the eighteenth century, Paris was a major center for ephemeris computation, perhaps reflecting the remarkable mathematical climate in France with such luminaries as Clairaut, Lagrange, and Laplace.

In contrast, in England, the home of Newton, Flamsteed, and Halley, no sustained independent calculation of ephemerides continued into the eighteenth century, despite the flourishing output of numerous small astrological almanacs. Earlier Vincent Wing (1619–1668) and John Gadbury (1627–1704) calculated or sponsored a series throughout the latter part of the seventeenth century; but it was, as we shall show, of rather uneven quality.

The other major center of ephemeris computation during this period was Bologna, beginning with a long sequence from 1621 through 1700 by Andrea Argoli (1570–1657), and subsequently paralleled by those of Francisco Montebruni (fl. 1600s) and his successors. Initiated by Manfredi in 1715, another competing series developed in Bologna, duplicating the years through 1757, and then continuing unrivaled into the nineteenth century.

The major sequences of ephemerides are shown on Figure 1. In addition, we have indicated the publication of the principal underlying tables used for the preparation of ephemerides. Bibliographical details for both the ephemerides and the tables are listed separately at the end of this section. In many instances the computers specified which tables formed the basis for their positions, but often the situation was mixed. For example, one table might have been used for the sun, another for the moon, and a third for the planets. Occasionally, the ephemeris makers used their own private tables or corrections, and, as we have discovered, they sometimes copied their positions wholesale from other workers. Finally, we should remark that just because an ephemeris maker claimed to use a certain set of tables, this does not guarantee that he knew how to use them correctly. And even if the computer did his work correctly, occasional typographical errors appear in the printed ephemerides. The most conspicuous errors are easily detected, and we have corrected them in our difference plots; smaller typographical errors are generally indistinguishable from actual errors of computation and hence go uncorrected.

We have sampled the accuracy of the published ephemerides around four representative intervals: 1651–1665, 1682–1701, 1715–1728, and 1767–1778. These selected dates were not always available in the authors we wished to examine, so we chose a few other runs including some just at the end of the eighteenth century. First we shall discuss the accuracy of the planetary positions for each of the four intervals, presenting our principal results graphically in Figures 2 through 12, and mentioning some of the many other works also analyzed. Then, we shall describe the errors for the solar and lunar positions in Figures 13 and 14.

[5] J. de Lalande, *Bibliographie astronomique,* p. 344 (Paris, 1803).
[6] See Robert M. McKeon, "Le Fèvre, Jean," *Dictionary of Scientific Biography,* Vol. 8, pp. 131–132 (New York, 1973).

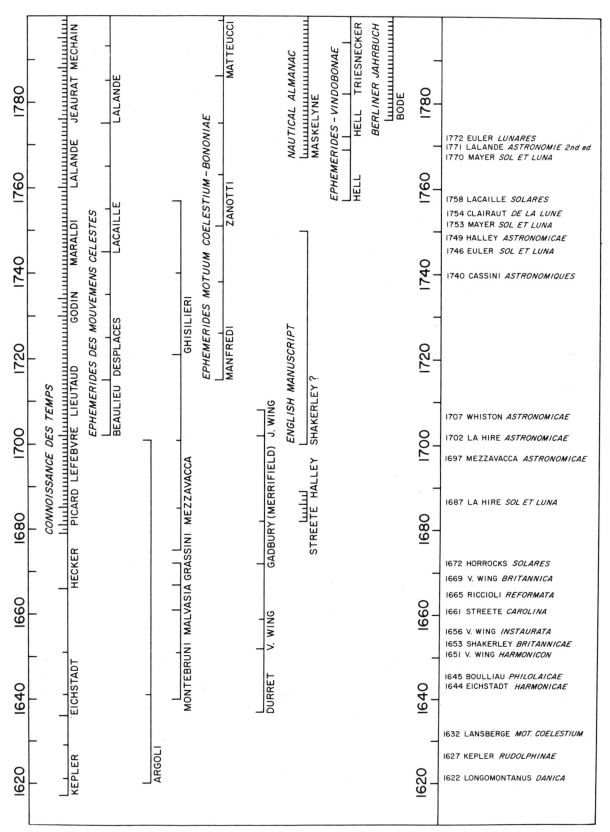

Figure 1

PLANETARY POSITIONS 1651–1665

The planet Mars has always proved a challenge for celestial mechanicians because of its close approaches to the earth and the comparatively high eccentricity of its orbit. In the work of both Ptolemy and Copernicus, the predicted Martian positions had errors about five times greater than for Saturn or Jupiter, that is upwards of 3° in longitude. Because Kepler concentrated so intensively on this planet, his accuracy was actually better for Mars than for Saturn or Jupiter. The accuracy of Kepler's basic theory is well reflected in the ephemerides of Argoli, illustrated in Figure 2. Argoli originally adopted the Copernican *Prutenicae tabulae coelestium motuum* (Tübingen, 1551) as the foundation of his *Ephemerides ab anno MDCXXI ad MDCXL* (Rome, 1621); in fact, he apparently calculated the entire run from 1600 through 1660, but he published only the two decades. Subsequently, he recalculated these years and beyond to 1700 "according to the Tychonic hypotheses." These were based on his own version of Kepler's *Tabulae Rudolphinae* (1627), which he published as *Secundorum mobilium tabulae juxta Tychonis Brahe et novas e coelo deductas observationes* (Padua, 1634). A preliminary examination of these tables suggests that Argoli retained circular orbits, leaving the appearance of his tables closer to the *Prutenicae tabulae* than to Kepler's, but he did handle the varying heliocentric distances in a far more sophisticated fashion than Copernicus attempted.

Following the publication of Kepler's *Tabulae Rudolphinae,* several astronomers adopted the elliptical orbits but sought to circumvent the indirect law-of-areas procedure with alternative geometric schemes. Before Newton's *Principia* (1687), there was little reason to demand Kepler's specific procedure over simpler constructions yielding similar results. Such a plan was offered by Ismaël Boulliau (1605–1694) in his *Astronomia Philolaica* (Paris, 1645), and another by Vincent Wing in his *Harmonicon coeleste* (London, 1651). In Figure 2, the error curve labeled "Wing" actually comes from two separate sequences.[7] The early part, to 1658, presumably derives from the theory and tables of his *Harmonicon coeleste,* and shows slightly inferior results to those of Argoli. In 1656 Wing published an improved procedure in his *Astronomia instaurata* (London), and the next installment of his *Ephemerides* (London, 1658) showed a marked improvement in the Mars' positions.[8]

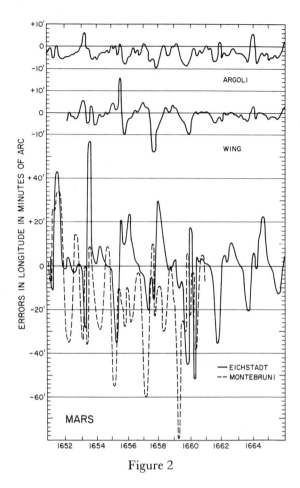

Figure 2

Meanwhile, as Figure 2 clearly reveals, two other major ephemerides for this same period show substantially larger errors for Mars. The principal deviations fall every two years, at the Martian oppositions, when any errors in the planet's heliocentric positions are greatly magnified by the earth's close approach. Eichstadt, the lineal successor to Kepler, in principle should have followed the *Tabulae Rudolphinae,* but if so, he sometimes used them incorrectly.[9] Montebruni, at Bologna, claimed that he had used the tables published in 1632 by Philip van Lansberge (1561–1632); and our check using those tables for 14 June 1651 (Julian) confirms Montebruni's calculation. Our analysis of Montebruni's ephemerides demonstrates that Lansberge's tables for Mars are definitely inferior to Kepler's.

For Jupiter and Saturn, Argoli and Eichstadt have large errors so closely identical that they have not been plotted separately (Figure 3); this is rather surprising considering Argoli's comparatively greater accuracy for Mars. The graph for 1651–1666 shows the entire 12-year error cycle for the orbital period of Jupiter, and approximately half of the 29-year cycle for Saturn. Superimposed on these broad

---

[7] See the Bibliography of Ephemerides at the end of Section I.

[8] Apparently, this improvement results from Wing's recognition that the so-called "empty focus" approximation is by no means good enough; as he wrote in his 1658 *Ephemerides,* p. 140: "Although the said motion is generally related to the supposed focus of the Ellipsis, yet the angle made at the imaginary place of the Planet (by the first halfe of the Bisected Excentricity) is not equal to the true Aequation of the Circle, neither is the equated Anomaly depending thereon of sufficient exactnesse, to determine the optique Aequation." Presumably, Wing picked up this information from Jeremy Shakerley's *Anatomy of Urania Practica,* London, 1649, and Shakerley must have found this in Horrock's manuscripts. We are indebted to Ian Lowe for this note.

[9] Our check for 14 June 1651 (Julian) yields a prediction from the *Tabulae Rudolphinae* differing only a few minutes from our computed ephemeris, whereas Eichstadt's value differs by 43'; however, the position in question falls at a particularly awkward spot for interpolation in Kepler's so-called "Tabula anguli."

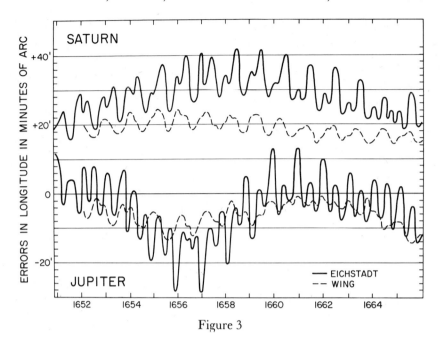

Figure 3

sweeps are the annual variations produced by errors in establishing the earth's orbit. For these two outer planets Montebruni has such large errors that we have not plotted them here: ±20′ for Saturn, and +40′ to −60′ for Jupiter. Wing on the other hand, is conspicuously more successful than the others, apparently because of a better solar theory, but as we shall see (Figure 13), his solar accuracy is no better than his contemporaries.

PLANETARY POSITIONS 1682–1701

The inferior planets proved in general more nettlesome for ephemeris computers. The problems apparently arose

because neither Venus nor Mercury could be observed during their conjunctions with the sun, and consequently the positions show enormous errors where the predictions could not be observationally controlled, particularly at inferior conjunction. (Note the change in scale required in Figure 4 for Venus.) Argoli, whose ephemeris for Mars was rather good, has substantial errors for Venus, and in 1651 errors of +5° to −9° for Mercury (not plotted here). Montebruni, whose positions are generally poor compared to the standards of the time, also has large errors for Venus, and in 1651 errors of +7° to −6° for Mercury. In contrast, Wing and Eichstadt are considerably better for both inferior

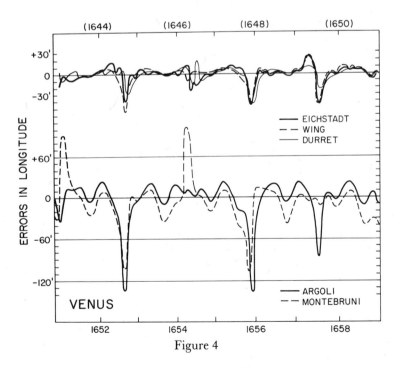

Figure 4

planets. Figure 4, for Venus, exhibits identical error patterns for Wing and Eichstadt beginning in 1656. Since Wing's *Ephemerides* were published in 1652, but Eichstadt's in 1644, we assume that Wing plagiarized from Eichstadt and not vice versa.[10]

On Figure 4 we have superimposed the error pattern from one earlier ephemeris, that of Noël Durret (1590–ca. 1650); this run from 1643 through 1650 serves to show the notable stability of the Venusian error cycle, in phase with the well-known eight-year repetition of Venus's positions. Durret, like others of this period, used his own tables[11] based on the *Tabulae Rudolphinae* but with a modification that avoided Kepler's indirect law-of-areas procedure. It is rather remarkable that Durret's positions match so well those of Eichstadt and Wing, which were each being computed from different derivative versions of Kepler's basic work.

PLANETARY POSITIONS 1715–1728

In the ensuing decades of the seventeenth century, there were few major advances in the computation of planetary ephemerides. In 1666 Hecker carried on the sequence begun by Kepler and Eichstadt, and in 1672 Gadbury followed Wing in the sequence of London imprints. Perhaps the most interesting detail turned up by our analysis is the fact that the astrologer Gadbury plagiarized his ephemerides from Hecker, and when Hecker's work ran out at the end of 1680, Gadbury hired John Merrifield (fl. 1680s) to compute for him. Merrifield did a wretched job for Jupiter, a tolerably bad job for Saturn, and he borrowed Mars from Argoli—not such a bad choice considering that Mars was Argoli's forte. Nor was Merrifield alone in borrowing; apparently Flaminio Mezzavacca (d. 1704) also copied Martian positions from the long-dead Argoli, whose ephemerides ran through 1700.

By the late 1600s small English almanacs were in their heyday in England;[12] most of these are too imprecise for our purposes, as they give the positions only to the nearest degree. There are a few exceptions: William Lilly (1602–1681), who along with Gadbury was perhaps the best known astrologer of the period, borrowed all his positions from Argoli. Thomas Streete (1622–1689), on the other hand, had earned a reputation for his planetary tables, *Astronomia Carolina* (London, 1661), and his short run of annual almanacs generally had the best positions then available. Streete had access to the as yet unpublished work of Jeremiah Horrocks (1618–1641), and as a result he adopted Horrocks's value of 0.0173 for the eccentricity of the solar orbit (= the earth's orbit), compared to the

Tychonic value of 0.018.[13] (A more accurate value for 1600 is 0.0169.) Because this parameter affects all the planetary predictions, and because it was the most poorly established key parameter in Kepler's *Tabulae Rudolphinae,* this single change produced a significant increase in accuracy of predictions in Streete's work.

John Gadbury promptly attacked Streete's solar theory in *A Bar to stop Thomas Street from his impudent Attempts and mad clambering up to Astronomy; to which is demonstrated, that his Tabula Corolina* [sic] *is all false,* an anonymous polemic appended to John Heydon's *The Holy Guide* (London, 1662). Vincent Wing, another of the leading table and almanac makers, also objected to some of Streete's results on the grounds that he had mishandled atmospheric refraction. Streete took up the defense in *An appendix to Astronomia Carolina containing ... A monitum to Mr. Vincent Wing* (London, 1664), saying "For a warning to himself and to remove some of his stumbling blocks out of the way of *Tyroes,* I propose this brief and plain demonstration ..." Wing counterattacked with his *Examen Astronomiae Carolinae T.S. or a short mathematicall discourse containing some animadversions upon ... T. Streete's Astronomicall Tables ... Wherein his errours ... are ... detected* (London, 1665), whereupon Streete systematically responded to the charges in *Examen examinatum: or, Wing's examination of Astronomia Carolina examined* (London, 1667). In general Wing was the more polemical, with frequent appeal to numerous authorities such as Tycho; Streete, in one of the responses concerning eclipses allowed that Tycho's were the best observations before 1600, but pointed out that there existed newer data on 21 post-telescopic eclipses. While it was difficult at this time, before the founding of the Royal Observatory at Greenwich, to obtain reliable runs of planetary positions for checking the comparative accuracy of ephemerides, Streete did suggest the forthcoming transit of Mercury in 1677 as a test case—a phenomenon that he attempted in vain to observe from the chambers of Robert Hooke at Gresham College. In 1710 Edmond Halley (1656–1742), a persistent admirer of Streete's, reedited the *Astronomia Carolina* (London, 1710), including as an appendix a long series of sextant observations made in 1682–84 at Halley's home in Islington, adding as his closing words that his novel daytime observations of Mercury "abundantly evince the certainty of Mr. Streete's Mercurial Astronomy."

Unfortunately, Streete's series of almanacs continued only for the four years 1682–1685.[14] In this same period the *Connoissance des temps* began its unbroken annual publication in Paris. Not until 1687, eight years after its founding,

---

[10] The error patterns for Mercury in 1656 are also virtually identical for Eichstadt and Wing, with a scatter of around ±3′ between their values; the Mercury errors themselves go between +52′ and −36′.

[11] See the Bibliography of Tables at the end of Section I.

[12] Bernard Capp, *English Almanacs, 1500–1800,* (Ithaca, 1979).

[13] Curtis Wilson, "Horrocks, Harmonies and the Exactitude of Kepler's Third Law," pp. 234–259, in *Studia Copernicana XVI, Science and History, Studies in Honor of Edward Rosen* (Wroclaw, 1978).

[14] Earlier Streete published *A Double Ephemeris ... Geocentricall and Heliocentricall* (London, 1653) which we have not seen; and possibly a second volume in 1654, not located.

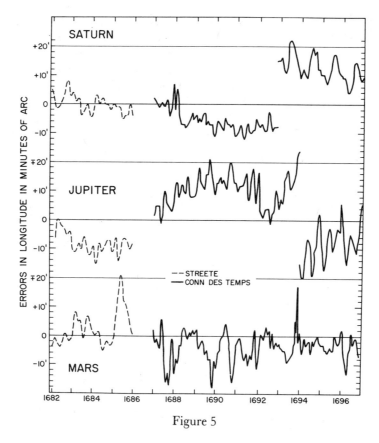

Figure 5

did the *Connoissance des temps* include planetary positions to minutes of arc. Figures 5–7 show the error patterns for Streete and for the *Connoissance des temps*. Streete's superiority is obvious, especially for Venus and, in 1683, for Mercury. The *Connoissance des temps,* on the other hand, had clearly not stabilized its computational techniques (as the discontinuities on Figures 5 and 7 attest), nor was its editor able to determine whether changes were for the better or worse. The case of Mercury (Figure 7) is especially disconcerting, where in 1683 the previously tolerable errors

in the *Connoissance des temps* were exchanged for a disastrous new procedure. The accuracy of both inferior planets is particularly sensitive to the adopted eccentricity of the solar orbit, and apparently Cassini's new and smaller value of 0.017 did not win prompt acceptance.[15]

---

[15] Curtis Wilson informs us that Cassini's value of the eccentricity had been published as early as 1662 by Cornelio Malvasia in his *Ephemerides novissimae,* (Modena), p. 155. Cassini himself apparently did not publish this value until his belated memoir of 1684 on the results of the Cayenne expedition.

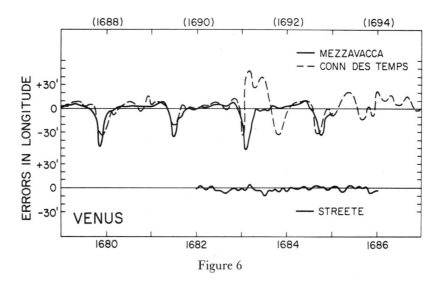

Figure 6

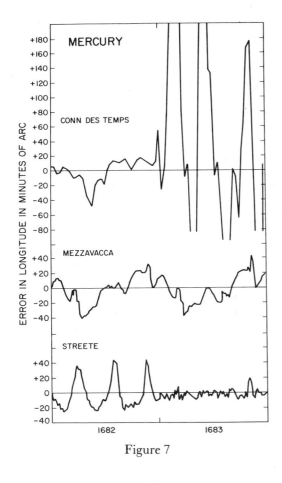

Figure 7

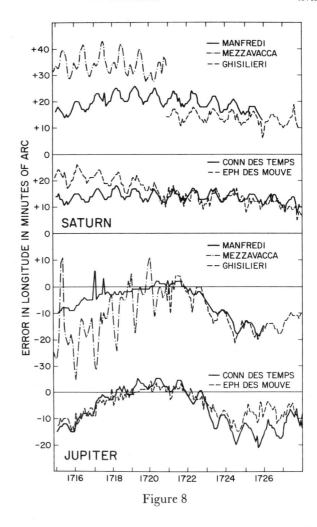

Figure 8

PLANETARY POSITIONS 1767–1778 AND LATER

Although Newton's *Principia* (London, 1687) pointed the way to the theoretical improvement of planetary positions through the concept of mutual planetary perturbations, it was many decades before the ephemerides actually benefited from his achievement. The effect of the perturbations is most conspicuous in Saturn and Jupiter, where the periods oscillate in a cycle of just over 900 years. Apparently, Horrocks first suspected an interaction and Halley (without knowing the period or even its cyclic nature) included it empirically in his *Tabulae astronomicae* (London, 1749). Unfortunately, the long delay before eventual posthumous publication of Halley's tables prevented early application of the concept. In fact, the effect was just the opposite of that in Halley's tables, as J. H. Lambert showed in 1773. Hence our plots for the 1715–1727 interval (Figure 8) show errors for Saturn and Jupiter comparable to those of 1682–1697 (Figure 5), despite the fact that the positions for the other planets are considerably improved.

In Figure 8 we have presented the error patterns for Saturn and Jupiter from five ephemerides: three Italian works in the upper band for each planet and two French works in the lower band. Up to this time Mezzavacca's ephemerides had held an unchallenged supremacy in Bologna, but following his death in 1704 it became clear that

improvement was possible (as shown in Figure 9 as well as Figure 8). Thus in 1716 Eustachio Manfredi (1674–1759) began a series that he would eventually extend to 1750. This created some animosity from Antonio Ghisilieri (1685–1734), who apparently considered himself heir to Mezzavacca's series, and who in 1731 published nearly a hundred pages of errors in the first two volumes of Manfredi's ephemerides,[16] despite the fact that Manfredi's positions were usually better than his own. The similarity between the error patterns of Saturn for the *Connoissance des temps* and Manfredi suggests that the Italian ephemerides served as a control in Paris, if they were not copied outright. Similarly, for Jupiter the *Connoissance des temps* positions somewhat resemble those of Manfredi through 1720, but beginning in 1721 they are suspiciously close to Ghisilieri's.

A comparison of Figure 9 with Figures 5 and 6 shows the appreciable improvement for Mars and Venus in the two decades between 1696 and 1715, probably largely attributable to the improved parameters for the solar theory men-

---

[16] Antonio Ghisilieri, *Ephemeridum coelestium motuum Manfredij errata insigniora* (Venice, 1731).

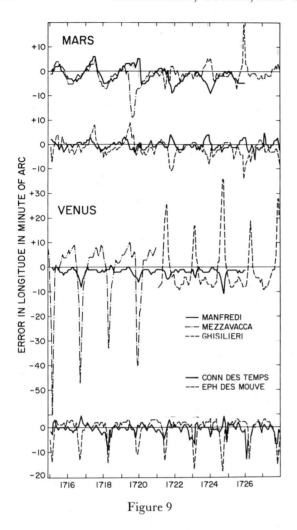

Figure 9

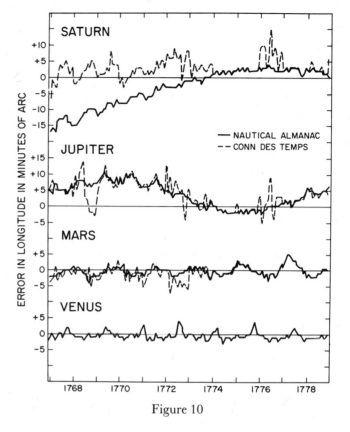

Figure 10

tioned earlier. (Note the change in scale for Venus between the two graphs.) We have again plotted the three Italian and the two French works for the 1715–1727 interval. Manfredi in Bologna and the *Connoissance des temps* in Paris were clearly the most successful for these planets.

By 1708 the long succession of English ephemerides had died, not to be revived until the founding of the *The Nautical Almanac* in 1767. Small English almanacs abounded, but without giving planetary positions more precisely than to the nearest degree. For this they could use either continental sources, or possibly an unpublished ephemeris once in the hands of Gadbury and attributed to Jeremy Shakerley (fl. 1640–1653).[17] Because the positions given in these almanacs

are so rough it is difficult to determine just what sources were actually being used in England.

The establishment of *The Nautical Almanac* did not bring with it an instant dramatic improvement (see Figure 10), but nevertheless the positions tabulated both by it and by its venerable French counterpart, the *Connoissance des temps,* show considerably better accuracy than the ephemerides half a century earlier, especially with respect to Venus.

---

[17] Edward Bernard, *Catalogi librorum manuscriptorum Angliae et Hiberniae in unum collecti* (Oxford, 1697), Vol. 2, p. 221, lists as item 7129 under Joannis Gadbury "Ephemerides for the Planetary Motions, Eclipses, Conjunctions, and Aspects for 50. years to come, calculated with all possible exactness, the Sun's place to seconds, and the Planets places to minutes, from the Brittish Tables, composed first by the Reverend Mr. Horrox, and first published by Mr. Jeremy Shackerly: Commencing A.D. 1700. when the leap day is omitted by the new Style, and ending A.D. 1749. Hereunto are added the Types of all visible Eclipses; together with the

Figures of all the vernal, and the exact time of all the other ingresses, referred to the Meridian of London."

Systematic inquiries in London, Oxford, and Cambridge have failed to locate this manuscript. However, in the Swann Galleries (New York) auction sale 1154, September 27, 1979, we purchased an anonymous manuscript (item 51) entitled *A Compleat Ephemerides of Celestial Motion for 50 Years, beginning att 1700 and ending with 1749; with Types of al the Visible Eclipses of both Sun and Moon that will happen in that time.* This same manuscript was sold in 1927 by Henry Sotheran & Co. in catalog 804, the library of J. L. E. Dreyer, where the probable connection with Horrocks and Shakerley was explicitly noted. While this manuscript indeed gives the solar positions to seconds and the lunar places to minutes, the planets are given only to degrees. Presumably Bernard, Savilian professor of astronomy at Oxford, had got the information for his entry from Gadbury, who may have described the manuscript and its known origin from Shakerley without quoting directly from a title page.

Job Gadbury (d. 1715) carried on his cousin John's almanacs until his death; the almanac for 1723 under the name of Job Gadbury was apparently pseudonymous, but its positions agree precisely with those of our manuscript. The "John Gadbury" almanac of 1746, in contrast, no longer uses these positions. The error plot for the solar longitudes (see Figure 13) shows errors typical of Gadbury of the period around 1650–70. If ours is not the original manuscript in Gadbury's possession, it must be a close copy.

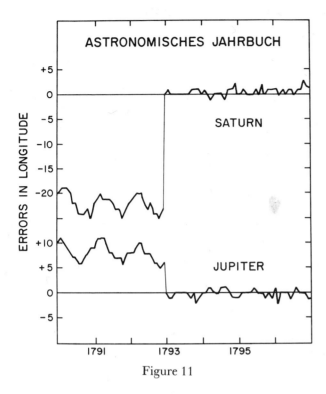

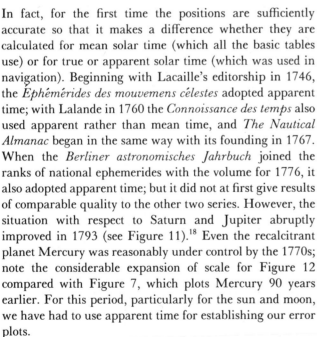

Figure 11

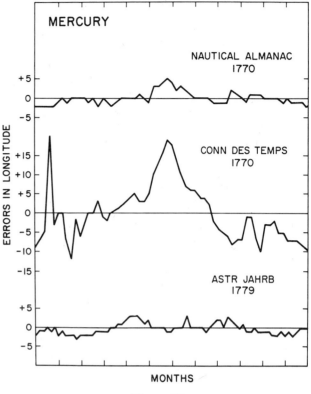

MONTHS

Figure 12

In fact, for the first time the positions are sufficiently accurate so that it makes a difference whether they are calculated for mean solar time (which all the basic tables use) or for true or apparent solar time (which was used in navigation). Beginning with Lacaille's editorship in 1746, the *Éphémérides des mouvemens célestes* adopted apparent time; with Lalande in 1760 the *Connoissance des temps* also used apparent rather than mean time, and *The Nautical Almanac* began in the same way with its founding in 1767. When the *Berliner astronomisches Jahrbuch* joined the ranks of national ephemerides with the volume for 1776, it also adopted apparent time; but it did not at first give results of comparable quality to the other two series. However, the situation with respect to Saturn and Jupiter abruptly improved in 1793 (see Figure 11).[18] Even the recalcitrant planet Mercury was reasonably under control by the 1770s; note the considerable expansion of scale for Figure 12 compared with Figure 7, which plots Mercury 90 years earlier. For this period, particularly for the sun and moon, we have had to use apparent time for establishing our error plots.

THE ACCURACY OF SOLAR POSITIONS

Fundamental to any successful planetary theory is an accurate solar theory. We have chosen to illustrate the improvement in the solar positions with a single set of

graphs for the entire period under discussion (Figure 13). The sinusoidal error patterns arise primarily from erroneous eccentricities. In the 1650s virtually all astronomers still used the value of 0.018 adopted by Kepler from Tycho's solar theory; Tycho's error arose largely through the confusion of effects of refraction and parallax.[19] Around 1640 Horrocks obtained an improved eccentricity (mentioned earlier) from his observations of Venus. Horrocks's value became available to Streete before 1661, and is reflected in the diminished errors in his solar positions plotted here for 1682–83. At around this time the calculator of the *Connoissance des temps* may have had available the new value of 0.017 derived by Cassini from his meridian line in the San Petronio Cathedral in Bologna and reconfirmed from Richer's expedition to Cayenne in 1672–73. In Italy Mezzavacca may have benefited from Streete's *Astronomia Carolina* or from Cassini's results. Note that the *Connoissance des temps* seems plagued with computational errors; there is no other way that a simple solar theory can produce such noisy results.

---

[18] Curtis Wilson points out that the remarkable improvement for Jupiter and Saturn probably resulted from Laplace's classic memoir on their long-term periodic inequalities, read in 1786, and the tables published in turn by Lalande in the third edition of his *Astronomie* (Paris, 1792).

[19] Y. Maeyama, "The Historical Development of Solar Theories in the Late-Sixteenth and Seventeenth Centuries," *Vistas in Astronomy*, ed. Arthur Beer, Vol. 16, pp. 35–60 (Oxford, 1974). Tycho's solar parallax was too large and his refraction for the meridian altitude of the equator at Hven was too small, so that the corrected altitude of the sun at either equinox was too large. Thus from the winter solstice the sun rose too rapidly to the March equinox (and vice versa in the summer), leading to an erroneously large eccentricity.

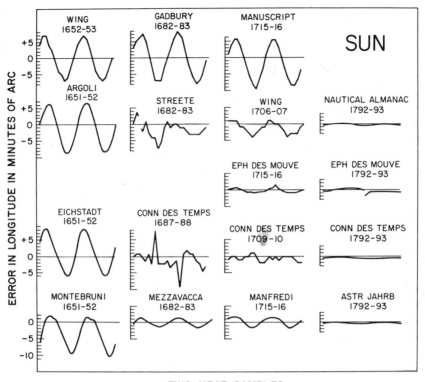

Figure 13

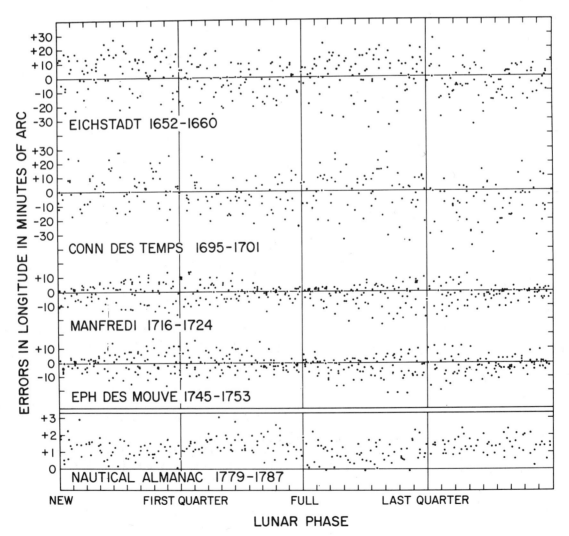

Figure 14

Meanwhile, Gadbury's ephemeris still reflected the solar eccentricity of Kepler's *Tabulae Rudolphinae,* and our manuscript ephemeris for 1700 to 1749 shows a similar error pattern, not surprisingly considering that it was calculated late in the seventeenth century. By 1715 the *Éphémérides des mouvemens célestes* had improved perceptibly beyond the best results of even the previous decade. Whether this was due to further advances in clockwork, in micrometers, or in the analysis of the accumulating data remains to be established by future research. And by the end of the eighteenth century the accuracy of a few tens of seconds of arc approaches the limits of our own computational scheme—in other words, the effects of planetary perturbations were for the first time being included in the ephemerides.

THE ACCURACY OF LUNAR POSITIONS

Because of the complexity of the moon's motion, it is comparatively difficult to represent the error patterns in a meaningful way with a small sampling of points. We have chosen to plot the errors as a function of phase (Figure 14), and we have taken 9-day steps to minimize commensurabilities with either the sidereal or synodic periods of the moon. For earlier periods, when the lunar theory was at a less sophisticated state of development, such phase plots show clearly basic periodic terms that are inadequately accounted for, but for our interval it is not easy to see any such phase effects except possibly that Eichstadt was slightly more successful at new and full moon than at the quarters. Our Figure 14 depicts the gradual improvement of lunar positions, which became increasingly important as astronomers sought to use the moon's motion as an aid for solving the longitude problem. Note the great improvement and consequent change in plotting scale for *The Nautical Almanac* in the 1779–1787 interval. The moon, despite the difficulty of its theory, ironically had some of the most accurate ephemerides because of the immense labor expended on its vagaries.

BIBLIOGRAPHY OF TABLES

Boulliau, Ismaël, *Tabulae Philolaicae*. Paris, 1645.

Cassini, Jacques, *Tables astronomiques*. Paris, 1740.

Clairaut, Alexis-Claude, *Tables de la lune, calculées suivant la théorie de la gravitation universelle*. Paris, 1754.

Eichstadt, Lorenz, *Tabulae harmonicae coelestium motuum*. Stettin, 1644.

Euler, Leonhard, "Novae tabulae astronomicae motuum solis et lunae" in *Opuscula*. Berlin, 1746.

Euler, Leonhard, *Novae tabulae lunares*. Petropoli, 1772.

Halley, Edmond, *Tabulae astronomicae*. Ed. John Bevis. London, 1749.

Horrocks, Jeremiah, *Astronomia Kepleriana defensa et promota, . . . tabulae solares, . . .* Ed. John Wallis. London, 1672.

Kepler, Johannes, *Tabulae Rudolphinae*. Ulm, 1627.

Lacaille, Nicolas-Louis de, *Tabulae solares*. Paris, 1758.

La Hire, Philippe de, *Tabularum astronomicarum pars prior, de motibus solis et lunae*. Paris, 1687.

La Hire, Philippe de, *Tabulae astronomicae*. Paris, 1702.

Lalande, Joseph-Jérôme Lefrançais de, *Astronomie*, 2nd ed., Paris, 1771.

Lansberge, Philip van, *Tabulae motuum coelestium perpetuae*. Middleburg, 1632.

Longomontanus, Christian [Severin], *Astronomia Danica*. Amsterdam, 1622.

Mayer, Johann Tobias, "Novae tabulae motuum solis et lunae," *Commentaria societatis regiae scientiarum Gottingensis*, Vol. 2, pp. 383–430, 1752. Göttingen, 1753.

Mayer, Johann Tobias, *Tabulae motuum solis et lunae novae et correctae*. Ed. Nevil Maskelyne. London, 1770.

Mezzavacca, Flaminio, *Tabulae astronomicae*. Bologna, 1697.

Riccioli, Giambattista, *Astronomia reformata*. Bologna, 1665.

Shakerley, Jeremy, *Tabulae Britannicae*. London, 1653.

Streete, Thomas, *Astronomia Carolina*. London, 1661.

Whiston, William, *Praelectiones astronomicae . . . tabulae plurimae astronomicae . . .* Cambridge, 1707.

Wing, Vincent, *Harmonicon coeleste*. London, 1651.

Wing, Vincent, *Astronomia instaurata*. London, 1656.

Wing, Vincent, *Astronomia Britannica*. London, 1669.

BIBLIOGRAPHY OF EPHEMERIDES

Argoli, Andrea, *Ephemerides ad longitudinem almae urbis Romae ab anno 1621 ad annum 1640 ex Prutenicis tabulis supputatae.* Rome, 1621.

Argoli, Andrea, *Novae caelestium motuum ephemerides ad longitudinem almae urbis ab anno 1620 ad annum 1640 ex ejusdem auctoris tabulae supputatae . . .* Rome, 1629.

Argoli, Andreae, *Ephemerides exactissimae caelestium motuum ad longitudinem almae urbis . . . ab anno 1641 ad annum 1700 . . .* Lyon, 1659.

Berliner astronomisches Jahrbuch. Johann Elert Bode, 1776-1829. Berlin, 1774-1826.

Connoissance des temps. Paris. Jean Picard, 1679-1684; Jean Lefèbvre, 1685-1701; Jacques Lieutaud, 1702-1729; Louis Godin, 1730-1733; Giovanni Domenico Maraldi, 1734-1759; Joseph-Jérôme Lefrançais de Lalande, 1760-1775; Edme-Sébastien Jeaurat, 1776-1787; Pierre-François-André Méchain, 1788-1794; Joseph-Jérôme Lefrançais de Lalande, 1795-1807.

Durret, Noël, *Novae motuum caelestium Ephemerides Richelianae annorum 15, ab anno 1637 incipientes, ubi sex anni priores e fontibus Lansbergianis, . . .* [Paris, 1641] London, 1647.

Eichstadt, Lorenz, *Ephemerides coelestium motuum ab anno 1636 ad annum 1640.* Stettin, 1634.

Eichstadt, Lorenz, *Ephemerides continuatae ab anno 1641 ad annum 1650.* Stettin, 1636.

Eichstadt, Lorenz, *Ephemerides coelestium motuum ab anno 1651 ad annum 1675.* Gdansk, 1644.

English manuscript. See footnote 17.

Éphémérides des mouvemens célestes. Jean de Beaulieu [Pseudonym], 1702-1714. Paris, 1703.

Éphémérides des mouvemens célestes. Paris. Philippe Desplaces, Vols. 1-3: 1715-1725 (1716), 1726-1734 (1727), 1735-1744 (1734); Nicolas-Louis de Lacaille, Vols. 4-6: 1745-1754 (1744), 1755-1764 (1755), 1765-1774 (1763); Joseph-Jérôme Lefrançais de Lalande, Vols. 7-9: 1775-1784 (1774), 1785-1792 (1783), 1793-1800 (1791).

Ephemerides motuum coelestium. Bologna. Eustachio Manfredi, Vols. 1-3: 1715-1725 (1715), 1726-1737 (1725), 1738-1750 (1725); Eustachio Zanotti, Vols. 4-6: 1751-1762 (1750), 1763-1774 (1762), 1775-1786 (1774); Petronio Matteucci, Vols. 7-8: 1787-1798 (1786), 1799-1810 (1798).

Ephemerides astronomicae ad meridianum Vindobonae. Vienna. Maximilian Hell, 1757-1768; Anton Pilgram, 1769-1771; M. Hell, 1772-1781; M. Hell and Franz von Paula Triesnecker, 1782-1793; F. v. P. Triesnecker and Johann Tobias Bürg, 1794-1806.

Gadbury, John, *Ephemerides of the celestial motions for X years 1672-1681.* London, 1671.

Gadbury, John, *Ephemerides of the celestial motions for XX years 1682-1701.* London, 1680.

Grassini, Girolamo, *Ephemeris Felsinea ad annos quinque contracta juxta hypotheses Lansbergii.* Bologna, 1666.

Ghisilieri, Antonio, *Ephemerides motuum coelestium ab anno 1721 ad annum 1740 e tabulis Hirii, Streetii et Flamstedii ad meridianum Bononiae supputatae.* Bologna, 1720.

Ghisilieri, Antonio, *Ephemerides motuum coelestium continuatae ab anno 1739 ad annum 1756.* Bologna, 1725.

[Halley, Edmond], *Ephemeris ad annum 1686 [1687, 1688].* London, [1685-1687].

Hecker, Joannis, *Ephemerides motuum coelestium ab anno 1666 ad annum 1680, ex observationibus correctis Tychonis Brahe, et Jo. Kepleri hypothesibus physicis, Tabulisque Rudolphinis ad meridianum Uraniburgicum.* Gdansk, 1662.

Kepler, Johannes, *Ephemerides novae motuum coelestium ab anno 1617, [1618, 1619, 1620]. Ex observationibus potissimum Tychonis Brahei, hypothesibus physicis, & tabulis Rudolphinis.* Linz, 1616-1619.

Kepler, Johannes, *Tomi primi. Ephemeridum . . . pars secunda, ab anno 1621 ad annum 1628.* Sagan, 1630.

Kepler, Johannes, and Jacob Bartsch, *Tomi primi. Ephemeridum . . . pars tertia, complexa annos à 1629 in 1636.* Sagan, 1630.

Malvasia, Cornelio, *Ephemerides novissimae motuum coelestium ad longitudinem urbis Mutinae, ex Lansbergii hypothesibus exactissime supputatae ab anno 1661 ad annum 1666.* Modena, 1662.

Mezzavacca, Flaminio, *Ephemerides Felsineae recentiores ad longitudinem urbis Bononiae ab anno 1675 ad annum 1684.* Bologna, 1675.

Mezzavacca, Flaminio, *Otia sive ephemerides Felsineae recentiores ab anno 1684 ad annum 1712.* Bologna, 1686.

Mezzavacca, Flaminio, *Otia sive ephemerides Felsineae recentiores ab anno 1701 ad annum 1720.* Bologna, 1701.

Montebruni, Francisco, *Ephemerides novissimae motuum coelestium . . . ex Lansbergii tabulis. 1640-1645.* Bologna, 1640.

Montebruni, Francisco, *Ephemerides novissimae motuum coelestium . . . Pt. 1: 1641-1650; Pt. 2: 1651-1660.* Bologna, 1650.

Nautical Almanac and Astronomical Ephemeris, The. Nevil Maskelyne, 1767–1813. London, 1765–1811.

Streete, Thomas, *A compleat ephemeris for 1682, [1683, 1684, 1685].* London, 1681–1684.

Wing, John, *Ephemerides of the coelestial motions for six years; beginning anno 1702, and ending anno 1707.* London, 1702.

Wing, Vincent, *An ephemerides of the coelestiall motions for VII years, beginning anno 1652, ending anno 1658.* London, 1652.

Wing, Vincent, *An ephemerides of the coelestial motions for XIII years. Beginning anno 1659. ending anno 1671.* London, 1658.

II. THE COMPUTED EPHEMERIDES.

The following tables give geocentric positions (tropical celestial longitudes and latitudes) of the sun, moon, and naked-eye planets to an accuracy and spacing suitable for historical purposes, and dates of new and full moons (conjunction and opposition in longitude) for the period A.D. 1650 through 1805. The planetary positions are computed for 12 hours U.T. (noon) Julian calendar. As in the Tuckerman tables, positions are given at five-day intervals for Mercury, Venus and the moon, and ten-day intervals for the sun, Mars, Jupiter, and Saturn. The central column gives a pair of Julian calendar dates, the one on the left corresponding to tabulated positions of Mars, Jupiter, and Saturn and to the ancillary column of Gregorian calendar dates shown at the far left of each page.

In the tables of new and full moons, we have used the Gregorian calendar with U.T.—in contradistinction to the Goldstine tables, which use the Julian calendar and local time in a location near Babylon (arbitrarily defined as 3 hours east of Greenwich). Instead of the arbitrary lunation count used by Goldstine, we give the Julian day number for each event.

As in the Tuckerman and Goldstine tables, the ephemerides are based upon the theories of Leverrier for the sun and inner planets, Gaillot for Jupiter and Saturn, and Hansen for the moon. Although better tables are available in some instances, these were originally chosen to match the *Tafeln zur astronomischen Chronologie* of P. V. Neugebauer. Because the computer coding originally written at IBM was unavailable, a new program was prepared by Professor Peter Huber at the Eidgenössische Technische Hochschule in Zurich. Professor Huber graciously placed his program at our disposal; we made certain modifications to increase its efficiency without fundamentally altering its accuracy, and we wrote a new control program to calculate the times of new and full moons. Professor Huber's program differs from Tuckerman's by including: 1. for the sun the largest perturbation in longitude by the moon, 2. for the moon, Hansen's perturbations in longitude and latitude with amplitudes greater than 10 seconds of arc, and 3. for Jupiter and Saturn, perturbations from Uranus. As a consequence, our lunar phenomena differ slightly from Goldstine's; for 1650 the difference is 2 or 3 minutes of time, but we are confident that the present scheme is the more accurate one. For 1950 our dates for full moon agree on the average with the *American Ephemeris,* with a maximum deviation in one case of 2 minutes of time.

In tests against 37 equi-spaced dates from the *Connaissance des temps* for 1958, Professor Huber found the following maximal differences:

	$\triangle$ Long.	$\triangle$ Lat.
Moon	0°.0244	0°.0103
Sun	0.0005	0.0002
Mercury	0.0042	0.0003
Venus	0.0053	0.0005

	$\triangle$ Long.	$\triangle$ Lat.
Mars	0.0037	0.0003
Jupiter	0.0020	0.0012
Saturn	0.0081	0.0015

Recently, F. R. Stephenson and M. A. Houlden ["The accuracy of Tuckerman's solar and planetary tables," *Journal for the History of Astronomy,* vol. 12, pp. 133–138, 1981] have examined the Tuckerman tables in light of the Integrated Ephemeris of the Jet Propulsion Laboratory using 200 randomly selected dates over the interval from -600 to $+1649$. For all of the objects they found the latitudes almost invariably accurate to the precision quoted, and for all except Mars the agreement in longitude was satisfactory. For Mars the principal discrepancy arose at perihelic oppositions, and after about 1200 A.D. the agreement was satisfactory. Hence, we expect the accuracy of our ephemeris longitudes to be as good or better than the error limits quoted by Tuckerman:

	Long.
Sun	$0°.00_3$
Mercury	0.01_0
Venus	0.01_5
Mars	0.01_5
Jupiter	0.01_2
Saturn	0.08_4
Moon	0.2_1

Their error limits in latitude are bounded "by $0°.00_5$ for the Sun and planets, and $0°.0_7$ for the Moon."

We give, finally, the longitudes for other meridians, particularly for the ephemerides and tables cited in the first part of our paper.

City	Long.	Time Correction	Local Mean Time
Philadelphia	75°10′W	-5^h01^m	6:59 a.m.
Cambridge	71 05 W	$-4\ 44$	7:16 a.m.
Toledo	4 02 W	$-0\ 16$	11:44 a.m.
Oxford	1 15 W	$-0\ 05$	11:55 a.m.
London	0 10 W	$-0\ 01$	11:59 a.m.
Paris	2 20 E	$+0\ 09$	12:09 p.m.
Lyons	4 50 E	$+0\ 19$	12.19 p.m.
Bologna	11 20 E	$+0\ 45$	12:45 p.m.
Padua	11 53 E	$+0\ 48$	12:48 p.m.
Leipzig	12 20 E	$+0\ 49$	12:49 p.m.
Venice	12 20 E	$+0\ 49$	12:49 p.m.
Rome	12 30 E	$+0\ 50$	12:50 p.m.
Copenhagen	12 34 E	$+0\ 50$	12:50 p.m.
Hven	12 45 E	$+0\ 51$	12:51 p.m.
Berlin	13 25 E	$+0\ 54$	12:54 p.m.
Gdansk	18 41 E	$+1\ 15$	1:15 p.m.
Frombork	19 40 E	$+1\ 19$	1:19 p.m.
Cracow	19 55 E	$+1\ 20$	1:20 p.m.

ACKNOWLEDGMENTS

We wish to thank Professor Peter Huber for generously supplying his computer program, and the Smithsonian Astrophysical Observatory for providing the computer time required for this study and these tables. For a grant to procure the microfilms of early ephemerides, we thank the Publications Committee of the American Philosophical Society, and for making the microfilms, the Observatoire de Paris (which, incidentally, has probably the world's best collection of early ephemerides). We also heartily thank John Hamwey for his care and patience in drafting the figures. We are grateful for the comments and criticism on our earlier drafts given by Curtis Wilson, Ian Lowe, and Brian Marsden.

REFERENCES FOR OUR COMPUTED EPHEMERIDES

SUN

Leverrier, Urbain Jean Joseph, *Annales de l'Observatoire impérial de Paris,* Vol. 4, pp. 34–37. Paris, 1858.
The program uses terms for perturbations in longitude with coefficients greater than 0″.1 and terms in the radius vector with coefficients greater than 0″.3.

MOON

Hansen, Peter Andreas, *Tables de la lune construites d'après le principe newtonien de la gravitation universelle.* London, 1857
Program includes terms for perturbations in longitude and latitude with amplitudes greater than 10″.

MERCURY

Leverrier, Urbain Jean Joseph, *Annales de l'Observatoire impérial de Paris,* Vol. 5, pp. 12–16. Paris, 1859.
Program uses terms for perturbations in longitude with coefficients greater than 0″.4; there are no terms in the radius vector greater than 0″.425.

VENUS

Leverrier, Urbain Jean Joseph, *Annales de l'Observatoire impérial de Paris,* Vol. 6, pp. 17–19. Paris, 1861.
Program uses terms for perturbations in longitude with coefficients greater than 0″.1 and terms in the radius vector with coefficients greater than 0″.3.

MARS

Leverrier, Urbain Jean Joseph, *Annales de l'Observatoire impérial de Paris,* Vol. 6, pp. 206–211. Paris, 1861.
Program uses terms for perturbations in longitude with coefficients greater than 0″.1 and terms in the radius vector with coefficients greater than 0″.3.

JUPITER

Gaillot, Aimable Jean-Baptiste, *Annales de l'Observatoire de Paris,* Vol. 31, pp. 113–130. Paris. 1913.
The program uses terms with coefficients greater than 0″.4 for perturbations from Saturn in longitude, eccentricity, and perihelion, and greater than 10″.0 for perturbations in semi-major axis. Contrary to the Tuckerman tables, perturbations from Uranus are included.

SATURN

Gaillot, Aimable Jean-Baptiste, *Annales de l'Observatoire de Paris,* Vol. 24, pp. 183–212. Paris, 1904.
The program uses terms with coefficients greater than 4″ for perturbations from Jupiter in longitude, eccentricity, and perihelion, and greater than 100″ for perturbations in semi-major axis. Contrary to the Tuckerman tables, perturbations from Uranus are included.

JULIAN 12 UT	SATURN LONG.	SATURN LAT.	JUPITER LONG.	JUPITER LAT.	MARS LONG.	MARS LAT.	SUN LONG.
1649							
Dec 31	85.01	-1.08	216.73	1.20	310.93	-1.12	290.55
Jan 10	84.37	-1.05	217.83	1.23	318.84	-1.08	300.73
Jan 20	83.86	-1.02	218.68	1.26	326.74	-1.04	310.88
Jan 30	83.53	-0.99	219.25	1.29	334.63	-0.98	321.00
Feb 9	83.37	-0.96	219.52	1.33	342.49	-0.91	331.08
Feb 19	83.40	-0.93	219.47	1.36	350.31	-0.84	341.11
Mar 1	83.62	-0.89	219.12	1.39	358.08	-0.75	351.09
Mar 11	84.02	-0.86	218.47	1.41	5.79	-0.66	1.01
Mar 21	84.59	-0.83	217.56	1.43	13.43	-0.57	10.88
Mar 31	85.31	-0.80	216.46	1.44	21.00	-0.47	20.69
Apr 10	86.16	-0.78	215.22	1.44	28.49	-0.37	30.44
Apr 20	87.14	-0.75	213.95	1.43	35.89	-0.26	40.15
Apr 30	88.21	-0.73	212.73	1.41	43.20	-0.16	49.80
May 10	89.37	-0.71	211.63	1.38	50.43	-0.05	59.42
May 20	90.59	-0.69	210.73	1.35	57.56	0.06	69.00
May 30	91.86	-0.67	210.09	1.31	64.61	0.16	78.56
Jun 9	93.15	-0.65	209.73	1.26	71.57	0.27	88.10
Jun 19	94.45	-0.64	209.67	1.22	78.45	0.37	97.63
Jun 29	95.75	-0.63	209.90	1.17	85.24	0.48	107.16
Jul 9	97.02	-0.62	210.43	1.13	91.95	0.58	116.70
Jul 19	98.26	-0.61	211.22	1.08	98.58	0.68	126.27
Jul 29	99.43	-0.60	212.27	1.04	105.13	0.78	135.86
Aug 8	100.52	-0.59	213.54	1.01	111.60	0.88	145.47
Aug 18	101.52	-0.58	215.01	0.97	117.99	0.98	155.14
Aug 28	102.41	-0.57	216.66	0.94	124.30	1.07	164.85
Sep 7	103.16	-0.57	218.45	0.91	130.52	1.17	174.61
Sep 17	103.77	-0.56	220.37	0.89	136.66	1.26	184.43
Sep 27	104.21	-0.56	222.40	0.87	142.70	1.36	194.31
Oct 7	104.47	-0.55	224.50	0.85	148.64	1.46	204.24
Oct 17	104.55	-0.54	226.67	0.83	154.47	1.55	214.23
Oct 27	104.44	-0.54	228.87	0.82	160.16	1.66	224.27
Nov 6	104.15	-0.53	231.09	0.81	165.70	1.76	234.35
Nov 16	103.69	-0.52	233.30	0.80	171.07	1.87	244.48
Nov 26	103.08	-0.51	235.49	0.80	176.22	1.98	254.64
Dec 6	102.36	-0.49	237.62	0.79	181.12	2.10	264.82
Dec 16	101.56	-0.48	239.68	0.79	185.71	2.22	275.01
1650							
Dec 26	100.73	-0.46	241.63	0.80	189.92	2.35	285.21
Jan 5	99.92	-0.44	243.46	0.80	193.67	2.48	295.39
Jan 15	99.18	-0.42	245.13	0.81	196.83	2.62	305.56
Jan 25	98.55	-0.40	246.62	0.81	199.27	2.75	315.70
Feb 4	98.06	-0.37	247.90	0.82	200.82	2.88	325.80
Feb 14	97.73	-0.35	248.93	0.83	201.29	2.99	335.86
Feb 24	97.59	-0.33	249.70	0.84	200.53	3.04	345.87
Mar 6	97.63	-0.30	250.19	0.85	198.50	3.02	355.82
Mar 16	97.85	-0.28	250.36	0.86	195.37	2.89	5.71
Mar 26	98.26	-0.26	250.23	0.87	191.62	2.63	15.55
Apr 5	98.83	-0.24	249.78	0.88	187.95	2.26	25.33
Apr 15	99.55	-0.22	249.06	0.88	185.01	1.83	35.07
Apr 25	100.40	-0.20	248.09	0.88	183.26	1.38	44.75
May 5	101.37	-0.18	246.95	0.87	182.83	0.95	54.38
May 15	102.45	-0.17	245.70	0.86	183.66	0.57	63.99
May 25	103.60	-0.15	244.43	0.83	185.59	0.23	73.55
Jun 4	104.81	-0.14	243.23	0.81	188.45	-0.06	83.10
Jun 14	106.07	-0.12	242.19	0.78	192.07	-0.32	92.64
Jun 24	107.36	-0.10	241.36	0.74	196.32	-0.53	102.17
Jul 4	108.66	-0.09	240.80	0.71	201.09	-0.72	111.71
Jul 14	109.96	-0.08	240.53	0.67	206.30	-0.89	121.26
Jul 24	111.23	-0.06	240.57	0.63	211.89	-1.03	130.83
Aug 3	112.45	-0.05	240.92	0.59	217.80	-1.15	140.43
Aug 13	113.62	-0.03	241.55	0.56	223.99	-1.25	150.08
Aug 23	114.71	-0.02	242.46	0.53	230.43	-1.33	159.76
Sep 2	115.70	0.00	243.61	0.50	237.10	-1.40	169.49
Sep 12	116.58	0.01	244.99	0.47	243.97	-1.45	179.28
Sep 22	117.33	0.03	246.57	0.44	251.02	-1.48	189.13
Oct 2	117.92	0.05	248.32	0.41	258.22	-1.50	199.03
Oct 12	118.35	0.06	250.22	0.39	265.57	-1.50	208.99
Oct 22	118.60	0.08	252.24	0.37	273.04	-1.48	219.00
Nov 1	118.67	0.10	254.36	0.35	280.61	-1.45	229.06
Nov 11	118.55	0.12	256.56	0.33	288.27	-1.41	239.17
Nov 21	118.24	0.14	258.81	0.31	296.00	-1.36	249.31
Dec 1	117.77	0.16	261.09	0.30	303.77	-1.29	259.48
Dec 11	117.15	0.18	263.37	0.28	311.58	-1.21	269.67
Dec 21	116.42	0.20	265.64	0.27	319.41	-1.12	279.86
1651							

GREGORIAN 12 UT	MOON LONG.	MOON LONG.	MOON LAT.	MOON LAT.	VENUS LONG.	VENUS LONG.	VENUS LAT.	VENUS LAT.	MERCURY LONG.	MERCURY LONG.	MERCURY LAT.	MERCURY LAT.
1650												1650
Jan 5		326.1		-5.1		257.39		1.01		265.91		0.25
10 Jan 15	29.9	89.1	-2.2	2.8	263.61	269.83	0.79	0.57	273.04	280.47	-0.39	-0.94
20 Jan 25	151.9	219.5	5.0	1.3	276.05	282.28	0.34	0.11	288.15	296.10	-1.41	-1.77
30 J-F 4	291.4	0.6	-4.3	-4.0	288.51	294.74	-0.11	-0.33	304.33	312.88	-2.00	-2.08
9 Feb 14	61.5	122.0	0.8	4.7	300.97	307.20	-0.53	-0.72	321.76	330.96	-1.98	-1.66
19 Feb 24	188.6	258.6	3.4	-2.4	313.42	319.65	-0.89	-1.05	340.35	349.57	-1.09	-0.26
1 Mar 6	328.7	33.5	-5.0	-1.4	325.87	332.09	-1.18	-1.29	357.91	4.34	0.79	1.92
11 Mar 16	93.0	156.3	3.5	4.9	338.31	344.52	-1.37	-1.42	7.84	7.93	2.91	3.47
21 Mar 26	226.8	297.6	0.1	-4.8	350.72	356.92	-1.45	-1.45	5.10	0.96	3.36	2.54
31 M-A 5	4.2	65.4	-3.6	1.6	3.12	9.31	-1.43	-1.38	357.49	355.96	1.29	-0.03
10 Apr 15	125.3	192.3	5.1	3.0	15.50	21.68	-1.30	-1.20	356.62	359.23	-1.17	-2.02
20 Apr 25	265.7	335.1	-3.3	-5.0	27.86	34.03	-1.08	-0.94	3.39	8.80	-2.59	-2.87
30 Apr	38.0		-0.9		40.19		-0.78		15.25		-2.90	
May 5		97.4		4.0		46.36		-0.60		22.62		-2.67
10 May 15	159.1	229.9	4.9	-0.2	52.51	58.66	-0.42	-0.23	30.86	39.95	-2.22	-1.57
20 May 25	304.3	10.5	-5.1	-3.2	64.81	70.96	-0.03	0.17	49.88	60.50	-0.76	0.12
30 M-J 4	70.7	130.2	2.1	5.2	77.10	83.24	0.37	0.56	71.48	82.27	0.95	1.59
9 Jun 14	194.8	268.6	2.8	-3.4	89.37	95.50	0.75	0.92	92.43	101.69	1.95	2.00
19 Jun 24	341.3	44.0	-4.7	-0.3	101.63	107.75	1.08	1.22	109.98	117.28	1.75	1.25
29 Jun	103.2		4.3		113.88		1.34		123.52		0.51	
Jul 4		164.6		4.5		120.00		1.43		128.58		-0.42
9 Jul 14	232.3	307.1	-0.6	-5.0	126.11	132.22	1.50	1.54	132.26	134.28	-1.51	-2.67
19 Jul 24	16.1	76.4	-2.4	2.7	138.33	144.43	1.56	1.54	134.36	132.42	-3.78	-4.62
29 J-A 3	136.6	200.8	5.0	1.9	150.52	156.61	1.49	1.42	128.96	125.28	-4.91	-4.46
8 Aug 13	271.2	344.3	-3.8	-4.2	162.69	168.76	1.31	1.17	123.07	123.54	-3.38	-1.98
18 Aug 23	49.1	108.5	0.6	4.6	174.82	180.88	1.01	0.82	127.07	133.38	-0.57	0.59
28 Aug	171.3		3.8		186.92		0.61		141.69		1.37	
Sep 2		238.9		-1.6		192.95		0.37		151.03		1.75
7 Sep 12	310.2	19.4	-5.1	-1.8	198.97	204.97	0.12	-0.15	160.61	169.99	1.78	1.55
17 Sep 22	80.9	141.2	3.4	5.1	210.95	216.93	-0.43	-0.71	178.99	187.60	1.16	0.66
27 S-O 2	207.6	278.1	1.0	-4.5	222.88	228.81	-1.00	-1.29	195.83	203.71	0.09	-0.49
7 Oct 12	347.9	52.6	-4.0	1.3	234.72	240.61	-1.57	-1.83	211.26	218.50	-1.08	-1.63
17 Oct 22	112.4	175.0	5.1	3.6	246.46	252.29	-2.08	-2.31	225.39	231.82	-2.13	-2.53
27 Oct	245.4		-2.4		258.07		-2.51		237.58		-2.79	
Nov 1		317.2		-5.2		263.82		-2.67		242.21		-2.81
6 Nov 11	23.6	84.9	-1.4	3.8	269.51	275.14	-2.80	-2.88	244.89	244.34	-2.47	-1.59
16 Nov 21	144.6	210.2	5.1	0.8	280.69	286.17	-2.91	-2.89	239.71	233.11	-0.09	1.54
26 N-D 1	284.3	354.7	-4.8	-3.6	291.55	296.80	-2.81	-2.66	229.21	229.81	2.49	2.65
6 Dec 11	57.5	117.1	1.7	5.1	301.91	306.84	-2.44	-2.14	233.70	239.40	2.32	1.76
16 Dec 21	178.0	247.2	3.3	-2.6	311.57	316.04	-1.76	-1.29	246.01	253.10	1.12	0.47
26 Dec 31	322.8	30.1	-4.9	-0.6	320.21	323.99	-0.72	-0.06	260.47	268.03	-0.15	-0.71
1651												1651
5 Jan 10	90.3	150.0	4.1	4.5	327.31	330.07	0.72	1.60	275.76	283.67	-1.21	-1.61
15 Jan 20	213.2	285.5	0.1	-4.7	332.14	333.40	2.60	3.69	291.80	300.16	-1.90	-2.06
25 Jan 30	359.7	63.5	-2.8	2.5	333.72	333.00	4.85	6.01	308.78	317.63	-2.05	-1.84
4 Feb 9	122.9	184.2	5.0	2.3	331.26	328.67	7.09	7.95	326.60	335.34	-1.38	-0.63
14 Feb 19	250.5	324.2	-3.2	-4.7	325.61	322.60	8.49	8.63	343.13	348.80	0.40	1.62
24 Feb	34.5		0.3		320.11		8.39		351.13		2.81	
Mar 1		95.7		4.6		318.47		7.85		349.61		3.58
6 Mar 11	155.8	220.0	4.2	-0.9	317.85	318.23	7.11	6.26	345.32	340.71	3.59	2.82
16 Mar 21	289.5	1.7	-5.1	-2.4	319.52	321.58	5.37	4.49	337.88	337.53	1.64	0.40
26 Mar 31	67.5	127.5	3.2	5.2	324.30	327.54	3.64	2.83	339.42	343.06	-0.68	-1.53
5 Apr 10	189.7	257.9	1.8	-4.0	331.22	335.26	2.08	1.40	348.03	354.04	-2.14	-2.52
15 Apr 20	328.8	37.3	-4.6	0.7	339.58	344.16	0.77	0.20	0.93	8.59	-2.66	-2.58
25 Apr 30	99.6	159.6	5.0	4.0	348.93	353.86	-0.30	-0.75	17.00	26.15	-2.27	-1.76
5 May 10	225.2	297.1	-1.5	-5.2	358.93	4.13	-1.13	-1.46	36.05	46.59	-1.05	-0.21
15 May 20	6.8	71.3	-2.0	3.5	9.42	14.80	-1.73	-1.94	57.50	68.28	0.66	1.42
25 May 30	131.5	192.8	5.1	1.4	20.26	25.78	-2.10	-2.22	78.40	87.54	1.93	2.14
4 Jun 9	262.7	336.3	-4.2	-4.0	31.36	36.98	-2.28	-2.30	95.56	102.40	2.03	1.61
14 Jun 19	42.9	104.2	1.3	4.9	42.66	48.37	-2.27	-2.21	107.98	112.16	0.91	-0.06
24 Jun 29	163.8	227.7	3.4	-1.8	54.13	59.91	-2.11	-1.98	114.73	115.45	-1.24	-2.52
4 Jul 9	301.7	13.9	-5.0	-1.0	65.73	71.58	-1.82	-1.63	114.27	111.55	-3.74	-4.62
14 Jul 19	77.1	136.9	3.9	4.6	77.46	83.36	-1.43	-1.21	108.26	105.80	-4.91	-4.51
24 Jul 29	197.2	264.8	0.6	-4.4	89.30	95.26	-0.97	-0.73	105.29	107.30	-3.56	-2.31
3 Aug 8	340.5	49.4	-3.4	2.3	101.24	107.25	-0.49	-0.24	111.92	118.87	-1.00	0.18
13 Aug 18	110.2	169.8	5.0	2.7	113.28	119.34	0.00	0.23	127.61	137.34	1.07	1.60
23 Aug 28	232.4	303.5	-2.7	-5.0	125.41	131.51	0.45	0.65	147.29	157.01	1.77	1.66
2 Sep 7	17.8	83.0	-0.3	4.6	137.63	143.77	0.84	1.01	166.28	175.07	1.34	0.87
12 Sep 17	142.6	203.7	4.4	-0.3	149.93	156.11	1.15	1.27	183.40	191.30	0.32	-0.28
22 Sep 27	269.8	342.4	-4.9	-3.2	162.31	168.52	1.36	1.43	198.79	205.86	-0.91	-1.52
2 Oct 7	53.1	115.3	2.9	5.3	174.74	180.97	1.47	1.47	212.47	218.49	-2.09	-2.59
12 Oct 17	175.2	239.4	2.2	-3.4	187.22	193.47	1.45	1.41	223.67	227.54	-2.96	-3.12
22 Oct 27	308.9	20.1	-5.0	0.0	199.73	206.00	1.33	1.23	229.26	227.71	-2.92	-2.16
1 Nov 6	86.6	146.9	4.8	4.2	212.28	218.56	1.11	0.97	222.52	216.27	-0.73	0.94
11 Nov 16	208.5	277.2	-0.8	-5.1	224.84	231.13	0.81	0.64	213.29	214.84	2.07	2.44
21 Nov 26	348.1	56.1	-2.7	3.1	237.42	243.71	0.46	0.26	219.53	225.87	2.28	1.85
1 Dec 6	119.0	178.6	5.0	1.7	250.00	256.29	0.07	-0.13	232.96	240.40	1.29	0.68
11 Dec 16	243.5	316.2	-3.6	-4.4	262.58	268.87	-0.33	-0.52	248.00	255.70	0.08	-0.49
21 Dec 26	26.1	90.6	0.8	4.8	275.16	281.45	-0.71	-0.88	263.48	271.36	-1.01	-1.45
31 Dec	151.0		3.5		287.73		-1.03		279.38		-1.79	
1651												1651

JULIAN 12 UT	SATURN LONG.	SATURN LAT.	JUPITER LONG.	JUPITER LAT.	MARS LONG.	MARS LAT.	SUN LONG.
1651							
Dec 31	115.62	0.21	267.88	0.26	327.23	-1.03	290.05
Jan 10	114.80	0.23	270.05	0.24	335.04	-0.92	300.23
Jan 20	114.00	0.25	272.14	0.23	342.81	-0.81	310.39
Jan 30	113.27	0.27	274.11	0.22	350.55	-0.70	320.51
Feb 9	112.64	0.28	275.96	0.21	358.23	-0.59	330.59
Feb 19	112.17	0.29	277.63	0.19	5.85	-0.47	340.62
Feb 29	111.85	0.31	279.12	0.18	13.40	-0.36	350.61
Mar 10	111.71	0.32	280.39	0.17	20.88	-0.24	0.54
Mar 20	111.76	0.33	281.42	0.15	28.27	-0.13	10.40
Mar 30	111.99	0.34	282.17	0.14	35.59	-0.02	20.22
Apr 9	112.40	0.35	282.64	0.12	42.82	0.09	29.97
Apr 19	112.97	0.36	282.79	0.11	49.97	0.19	39.68
Apr 29	113.69	0.37	282.63	0.09	57.04	0.30	49.34
May 9	114.54	0.38	282.16	0.07	64.03	0.39	58.96
May 19	115.51	0.39	281.41	0.04	70.94	0.48	68.54
May 29	116.57	0.40	280.42	0.02	77.78	0.57	78.10
Jun 8	117.72	0.41	279.26	0.00	84.55	0.66	87.64
Jun 18	118.92	0.42	277.99	-0.03	91.25	0.74	97.17
Jun 28	120.18	0.43	276.72	-0.05	97.89	0.81	106.71
Jul 8	121.46	0.45	275.53	-0.07	104.47	0.88	116.25
Jul 18	122.75	0.46	274.50	-0.09	110.99	0.95	125.81
Jul 28	124.03	0.47	273.70	-0.12	117.46	1.01	135.40
Aug 7	125.29	0.49	273.18	-0.13	123.89	1.07	145.02
Aug 17	126.51	0.51	272.96	-0.15	130.26	1.13	154.67
Aug 27	127.67	0.53	273.06	-0.17	136.60	1.18	164.39
Sep 6	128.75	0.55	273.48	-0.18	142.89	1.23	174.14
Sep 16	129.73	0.57	274.20	-0.20	149.13	1.28	183.96
Sep 26	130.60	0.59	275.20	-0.21	155.34	1.32	193.83
Oct 6	131.33	0.62	276.45	-0.22	161.49	1.36	203.76
Oct 16	131.91	0.65	277.93	-0.24	167.59	1.40	213.74
Oct 26	132.32	0.67	279.61	-0.25	173.65	1.44	223.78
Nov 5	132.55	0.70	281.47	-0.26	179.63	1.47	233.86
Nov 15	132.60	0.73	283.46	-0.27	185.55	1.50	243.98
Nov 25	132.46	0.76	285.58	-0.28	191.38	1.52	254.14
Dec 5	132.13	0.79	287.80	-0.30	197.12	1.54	264.32
Dec 15	131.64	0.82	290.08	-0.31	202.73	1.56	274.51
1652							
Dec 25	131.02	0.84	292.41	-0.32	208.21	1.57	284.71
Jan 4	130.28	0.86	294.77	-0.34	213.52	1.57	294.90
Jan 14	129.48	0.88	297.13	-0.36	218.62	1.57	305.07
Jan 24	128.67	0.90	299.47	-0.37	223.47	1.55	315.21
Feb 3	127.88	0.91	301.76	-0.39	228.01	1.51	325.31
Feb 13	127.16	0.92	303.99	-0.41	232.16	1.46	335.37
Feb 23	126.55	0.93	306.13	-0.44	235.82	1.38	345.39
Mar 5	126.09	0.93	308.15	-0.46	238.87	1.26	355.34
Mar 15	125.79	0.94	310.04	-0.49	241.17	1.10	5.24
Mar 25	125.66	0.94	311.77	-0.52	242.51	0.88	15.09
Apr 4	125.72	0.93	313.31	-0.55	242.75	0.58	24.87
Apr 14	125.96	0.93	314.63	-0.58	241.74	0.21	34.60
Apr 24	126.37	0.93	315.71	-0.62	239.51	-0.25	44.29
May 4	126.94	0.93	316.52	-0.65	236.37	-0.76	53.92
May 14	127.65	0.93	317.04	-0.69	232.89	-1.27	63.52
May 24	128.50	0.93	317.24	-0.74	229.81	-1.73	73.10
Jun 3	129.46	0.93	317.13	-0.78	227.75	-2.10	82.64
Jun 13	130.51	0.93	316.71	-0.82	227.02	-2.37	92.18
Jun 23	131.65	0.94	315.98	-0.86	227.65	-2.56	101.71
Jul 3	132.85	0.94	315.01	-0.90	229.54	-2.68	111.24
Jul 13	134.09	0.95	313.85	-0.93	232.49	-2.74	120.79
Jul 23	135.36	0.96	312.57	-0.95	236.33	-2.76	130.37
Aug 2	136.64	0.98	311.27	-0.97	240.90	-2.74	139.96
Aug 12	137.92	0.99	310.05	-0.98	246.07	-2.70	149.60
Aug 22	139.17	1.01	308.98	-0.99	251.73	-2.63	159.29
Sep 1	140.37	1.03	308.16	-0.98	257.79	-2.55	169.02
Sep 11	141.52	1.05	307.62	-0.98	264.19	-2.44	178.80
Sep 21	142.58	1.08	307.41	-0.97	270.86	-2.32	188.65
Oct 1	143.55	1.10	307.52	-0.95	277.76	-2.19	198.54
Oct 11	144.39	1.13	307.96	-0.94	284.85	-2.04	208.50
Oct 21	145.10	1.17	308.72	-0.92	292.08	-1.89	218.51
Oct 31	145.66	1.20	309.76	-0.91	299.43	-1.73	228.57
Nov 10	146.04	1.24	311.07	-0.90	306.86	-1.56	238.67
Nov 20	146.25	1.27	312.62	-0.89	314.36	-1.39	248.81
Nov 30	146.27	1.31	314.36	-0.88	321.89	-1.22	258.98
Dec 10	146.10	1.35	316.28	-0.87	329.43	-1.04	269.17
Dec 20	145.76	1.38	318.35	-0.87	336.98	-0.88	279.36
1653							

GREGORIAN 12 UT	MOON LONG.		MOON LAT.		VENUS LONG.		VENUS LAT.		MERCURY LONG.		MERCURY LAT.	
1652												1652
Jan 5		211.3		-1.4		294.01		-1.17		287.55		-2.02
10 Jan 15	280.4	355.0	-5.0	-1.7	300.29	306.57	-1.29	-1.38	295.91	304.41	-2.10	-1.99
20 Jan 25	62.3	123.9	3.8	4.7	312.84	319.10	-1.45	-1.49	312.94	321.15	-1.64	-1.00
30 J-F 4	183.2	245.7	0.8	-4.1	325.36	331.60	-1.51	-1.50	328.33	333.20	-0.02	1.25
9 Feb 14	318.6	32.4	-4.1	1.8	337.84	344.07	-1.46	-1.39	334.33	331.31	2.59	3.56
19 Feb 24	96.7	156.4	5.1	2.9	350.29	356.49	-1.29	-1.17	326.01	321.52	3.68	2.99
29 Feb	216.5		-2.3		2.68		-1.02		319.65		1.90	
Mar 5		282.4		-5.2		8.86		-0.85		320.49		0.75
10 Mar 15	357.1	67.9	-1.2	4.4	15.02	21.16	-0.66	-0.45	323.42	327.87	-0.26	-1.10
20 Mar 25	129.7	189.1	4.6	0.2	27.28	33.38	-0.23	0.01	333.44	339.89	-1.74	-2.18
30 M-A 4	251.7	320.9	-4.6	-4.0	39.46	45.52	0.25	0.50	347.06	354.90	-2.42	-2.46
9 Apr 14	34.6	101.7	2.1	5.3	51.55	57.56	0.75	1.00	3.39	12.53	-2.28	-1.89
19 Apr 24	161.7	222.7	2.6	-2.8	63.54	69.49	1.23	1.46	22.33	32.73	-1.30	-0.53
29 Apr	289.1		-5.2		75.40		1.66		43.51		0.34	
May 4		360.0		-0.9		81.28		1.85		54.18		1.19
9 May 14	70.7	134.1	4.5	4.3	87.13	92.93	2.01	2.14	64.18	73.09	1.86	2.24
19 May 24	193.7	258.1	-0.4	-4.7	98.69	104.40	2.24	2.30	80.69	86.88	2.30	2.00
29 M-J 3	328.0	38.4	-3.3	2.5	110.05	115.64	2.31	2.29	91.54	94.53	1.35	0.37
8 Jun 13	105.2	165.8	5.0	1.9	121.17	126.61	2.21	2.09	95.68	94.97	-0.87	-2.26
18 Jun 23	226.7	295.5	-3.2	-4.7	131.97	137.23	1.91	1.68	92.73	89.82	-3.54	-4.43
28 Jun	7.3		0.1		142.37		1.38		87.43		-4.71	
Jul 3		75.2		4.7		147.38		1.03		86.59		-4.40
8 Jul 13	138.2	197.6	3.7	-1.1	152.23	156.90	0.62	0.14	87.83	91.28	-3.61	-2.55
18 Jul 23	261.4	334.3	-4.9	-2.5	161.36	165.55	-0.41	-1.02	96.88	104.41	-1.37	-0.22
28 J-A 2	45.5	110.3	3.4	4.9	169.44	172.96	-1.69	-2.44	113.51	123.56	0.74	1.41
7 Aug 12	170.5	230.3	1.1	-3.8	176.04	178.57	-3.25	-4.11	133.85	143.87	1.73	1.74
17 Aug 22	298.1	13.1	-4.7	1.0	180.44	181.53	-5.02	-5.95	153.40	162.37	1.50	1.09
27 Aug	82.1		5.2		181.73		-6.86		170.79		0.56	
Sep 1		143.7		3.2		180.96		-7.67		178.70		-0.06
6 Sep 11	202.8	264.9	-1.9	-5.2	179.23	176.69	-8.31	-8.65	186.12	193.01	-0.71	-1.39
16 Sep 21	336.2	50.9	-2.3	4.0	173.69	170.68	-8.63	-8.23	199.32	204.89	-2.04	-2.64
26 S-O 1	116.6	176.2	4.8	0.5	168.17	166.48	-7.50	-6.53	209.44	212.46	-3.13	-3.43
6 Oct 11	236.2	301.4	-4.3	-4.6	165.79	166.09	-5.46	-4.37	213.16	210.66	-3.37	-2.72
16 Oct 21	14.7	86.9	1.2	5.2	167.31	169.33	-3.31	-2.33	205.14	199.52	-1.37	0.32
26 Oct 31	149.4	208.6	2.8	-2.4	172.03	175.29	-1.43	-0.63	197.56	200.02	1.61	2.18
5 Nov 10	271.4	339.7	-5.2	-1.9	179.03	183.15	0.07	0.68	205.46	212.40	2.21	1.91
15 Nov 20	52.6	120.9	4.0	4.4	187.58	192.29	1.20	1.64	219.97	227.76	1.43	0.87
25 Nov 30	181.1	242.0	-0.1	-4.5	197.22	202.33	1.99	2.28	235.59	243.43	0.29	-0.28
5 Dec 10	308.6	18.6	-3.8	1.8	207.61	213.02	2.48	2.63	251.27	259.14	-0.81	-1.28
15 Dec 20	89.2	153.4	5.0	2.0	218.53	224.15	2.71	2.73	267.05	275.04	-1.68	-1.97
25 Dec 30	212.8	276.9	-3.0	-4.9	229.85	235.61	2.70	2.62	283.13	291.28	-2.13	-2.12
1653												1653
4 Jan 9	347.2	57.3	-0.7	4.5	241.44	247.31	2.50	2.34	299.35	307.01	-1.88	-1.35
14 Jan 19	124.0	184.9	3.9	-1.0	253.23	259.18	2.15	1.93	313.52	317.54	-0.46	0.81
24 Jan 29	245.2	313.5	-4.9	-3.3	265.16	271.17	1.69	1.43	317.50	313.20	2.28	3.41
3 Feb 8	26.1	94.5	2.8	5.1	277.20	283.24	1.16	0.88	307.34	303.40	3.66	3.07
13 Feb 18	157.3	216.7	1.4	-3.7	289.31	295.38	0.60	0.32	302.63	304.54	2.08	1.04
23 Feb 28	279.3	351.3	-5.0	-0.1	301.47	307.56	0.05	-0.20	308.34	313.44	0.09	-0.72
5 Mar 10	64.5	129.7	5.0	3.6	313.67	319.77	-0.45	-0.67	319.48	326.23	-1.38	-1.87
15 Mar 20	189.8	249.5	-1.6	-5.2	325.88	332.00	-0.88	-1.06	333.60	341.53	-2.18	-2.32
25 Mar 30	315.1	29.8	-3.2	3.2	338.12	344.24	-1.22	-1.34	350.02	359.09	-2.26	-1.99
4 Apr 9	101.1	163.1	5.0	0.9	350.36	356.48	-1.44	-1.51	8.75	18.95	-1.51	-0.83
14 Apr 19	222.3	283.9	-4.0	-4.8	2.60	8.72	-1.55	-1.57	29.51	40.00	0.01	0.91
24 Apr 29	352.8	67.9	0.1	5.0	14.85	20.97	-1.55	-1.50	49.81	58.38	1.72	2.29
4 May 9	135.6	195.4	3.0	-2.1	27.09	33.21	-1.43	-1.33	65.40	70.69	2.53	2.39
14 May 19	255.7	320.4	-5.0	-2.5	39.33	45.45	-1.21	-1.07	74.12	75.59	1.83	0.87
24 May 29	31.7	104.6	3.4	4.5	51.57	57.70	-0.91	-0.74	75.12	73.09	-0.42	-1.86
3 Jun 8	168.3	227.8	0.0	-4.3	63.82	69.95	-0.56	-0.36	70.35	68.03	-3.15	-4.02
13 Jun 18	290.8	358.6	-4.2	1.0	76.08	82.21	-0.17	0.03	67.09	67.99	-4.35	-4.16
23 Jun 28	70.5	139.3	5.0	2.3	88.35	94.49	0.23	0.42	70.84	75.55	-3.57	-2.70
3 Jul 8	200.1	260.9	-2.9	-5.1	100.64	106.79	0.61	0.78	81.98	90.02	-1.67	-0.60
13 Jul 18	327.6	37.9	-1.5	4.1	112.94	119.10	0.94	1.08	99.42	109.71	0.39	1.17
23 Jul 28	107.9	172.2	4.3	-0.7	125.26	131.43	1.20	1.30	120.27	130.57	1.63	1.78
2 Aug 7	231.8	295.3	-4.8	-3.9	137.60	143.77	1.37	1.42	140.33	149.46	1.65	1.31
12 Aug 17	5.9	76.9	2.0	5.2	149.95	156.12	1.44	1.43	157.96	165.87	0.81	0.19
22 Aug 27	143.3	204.2	1.8	-3.5	162.31	168.49	1.40	1.34	173.18	179.86	-0.50	-1.23
1 Sep 6	264.3	331.4	-5.2	-1.1	174.67	180.85	1.25	1.14	185.82	190.87	-1.97	-2.68
11 Sep 16	45.1	114.3	4.6	4.0	187.03	193.21	1.00	0.84	194.71	196.81	-3.30	-3.73
21 Sep 26	176.8	236.3	-1.2	-5.0	199.39	205.57	0.66	0.47	196.44	193.08	-3.80	-3.26
1 Oct 6	298.0	9.0	-3.7	2.3	211.74	217.91	0.25	0.03	187.53	182.82	-1.97	-0.28
11 Oct 16	83.6	149.5	5.1	1.2	224.08	230.24	-0.20	-0.43	181.95	185.29	1.11	1.88
21 Oct 26	209.5	269.2	-3.7	-4.9	236.40	242.56	-0.66	-0.89	191.44	198.96	2.10	1.94
31 Oct	333.5		-0.7		248.71		-1.11		206.99		1.55	
Nov 5		47.5		4.6		254.85		-1.31		215.12		1.05
10 Nov 15	120.4	182.8	3.3	-1.9	260.98	267.11	-1.50	-1.67	223.20	231.20	0.50	-0.07
20 Nov 25	242.2	303.5	-4.9	-2.9	273.23	279.33	-1.82	-1.93	239.11	246.96	-0.61	-1.12
30 N-D 5	10.9	85.8	2.7	4.8	285.42	291.50	-2.02	-2.07	254.78	262.60	-1.56	-1.91
10 Dec 15	154.8	215.1	0.3	-4.3	297.55	303.58	-2.08	-2.06	270.43	278.21	-2.15	-2.23
20 Dec 25	275.4	339.7	-4.4	0.3	309.58	315.55	-1.99	-1.88	285.81	292.88	-2.10	-1.69
30 Dec	49.8		4.9		321.48		-1.73		298.67		-0.91	
1653												1653

JULIAN 12 UT	SATURN		JUPITER		MARS		SUN	GREGORIAN 12 UT	MOON				VENUS				MERCURY			
	LONG.	LAT.	LONG.	LAT.	LONG.	LAT.	LONG.		LONG.		LAT.		LONG.		LAT.		LONG.		LAT.	
1653								1654												1654
								Jan 4		122.6		2.8		327.37		-1.53		301.81		0.32
Dec 30	145.26	1.42	320.53	-0.86	344.50	-0.71	289.55	9 Jan 14	187.5	247.1	-2.8	-5.2	333.20	338.97	-1.29	-1.01	300.66	295.33	1.86	3.15
Jan 9	144.62	1.45	322.81	-0.86	351.99	-0.55	299.74	19 Jan 24	309.9	17.7	-2.2	3.6	344.68	350.30	-0.68	-0.31	289.34	286.21	3.54	3.07
Jan 19	143.88	1.47	325.16	-0.87	359.44	-0.40	309.89	29 J-F 3	88.8	157.5	4.8	-0.3	355.82	1.23	0.09	0.54	286.53	289.41	2.21	1.27
Jan 29	143.08	1.49	327.54	-0.87	6.84	-0.25	320.02	8 Feb 13	219.2	279.5	-4.8	-4.3	6.52	11.66	1.01	1.52	293.95	299.58	0.39	-0.40
Feb 8	142.28	1.50	329.95	-0.88	14.17	-0.11	330.10	18 Feb 23	345.8	57.0	1.1	5.2	16.62	21.37	2.05	2.60	305.98	312.97	-1.05	-1.58
Feb 18	141.50	1.51	332.36	-0.89	21.45	0.02	340.14	28 Feb	126.4		2.5		25.87		3.17		320.46		-1.95	
								Mar 5		190.8		-3.2		30.08		3.74		328.42		-2.16
Feb 28	140.80	1.51	334.75	-0.91	28.65	0.14	350.12	10 Mar 15	250.8	313.0	-5.2	-1.9	33.93	37.35	4.32	4.88	336.87	345.82	-2.20	-2.05
Mar 10	140.21	1.51	337.10	-0.92	35.78	0.26	0.06	20 Mar 25	23.6	96.0	4.0	4.4	40.25	42.52	5.42	5.91	355.29	5.25	-1.69	-1.11
Mar 20	139.76	1.51	339.38	-0.95	42.84	0.36	9.93	30 M-A 4	162.1	223.4	-0.7	-4.8	44.05	44.69	6.32	6.62	15.55	25.77	-0.32	0.59
Mar 30	139.48	1.50	341.57	-0.97	49.83	0.46	19.74	9 Apr 14	283.1	348.2	-3.9	1.3	44.35	42.99	6.74	6.64	35.29	43.42	1.51	2.27
Apr 9	139.37	1.48	343.66	-1.00	56.75	0.55	29.50	19 Apr 24	62.5	133.4	5.1	1.6	40.72	37.80	6.25	5.56	49.69	53.84	2.72	2.76
Apr 19	139.44	1.47	345.61	-1.03	63.61	0.64	39.21	29 Apr	196.1		-3.5		34.66		4.59		55.74		2.32	
								May 4		255.8		-4.8		31.82		3.45		55.45		1.41
Apr 29	139.68	1.46	347.41	-1.06	70.40	0.71	48.87	9 May 14	316.6	25.5	-1.2	4.1	29.67	28.48	2.26	1.11	53.40	50.55	0.10	-1.35
May 9	140.10	1.45	349.03	-1.10	77.13	0.78	58.49	19 May 24	101.2	168.7	3.8	-1.7	28.28	29.03	0.08	-0.80	48.09	46.99	-2.62	-3.48
May 19	140.67	1.43	350.44	-1.14	83.80	0.85	68.08	29 M-J 3	229.1	288.8	-4.9	-3.2	30.60	32.88	-1.53	-2.12	47.69	50.22	-3.87	-3.83
May 29	141.38	1.42	351.61	-1.19	90.42	0.90	77.64	8 Jun 13	352.1	64.4	2.1	5.0	35.75	39.11	-2.58	-2.92	54.42	60.14	-3.44	-2.78
Jun 8	142.22	1.41	352.52	-1.23	96.99	0.95	87.18	18 Jun 23	138.2	202.2	0.8	-4.2	42.86	46.94	-3.16	-3.31	67.27	75.71	-1.92	-0.94
Jun 18	143.18	1.41	353.14	-1.28	103.52	1.00	96.71	28 Jun	261.7		-4.6		51.29		-3.38		85.34		0.04	
								Jul 3		323.0		-0.2		55.87		-3.38		95.81		0.89
Jun 28	144.23	1.40	353.45	-1.33	110.01	1.04	106.24	8 Jul 13	29.7	103.3	4.7	3.5	60.64	65.57	-3.32	-3.20	106.58	117.11	1.49	1.78
Jul 8	145.35	1.40	353.44	-1.39	116.47	1.07	115.79	18 Jul 23	173.1	234.5	-2.4	-5.2	70.65	75.84	-3.04	-2.84	127.06	136.32	1.78	1.52
Jul 18	146.54	1.40	353.10	-1.43	122.90	1.10	125.35	28 J-A 2	294.7	359.1	-2.7	3.0	81.14	86.54	-2.61	-2.35	144.87	152.73	1.07	0.46
Jul 28	147.78	1.41	352.45	-1.48	129.30	1.13	134.93	7 Aug 12	68.9	140.7	5.1	0.4	92.02	97.57	-2.06	-1.76	159.88	166.29	-0.25	-1.04
Aug 7	149.04	1.42	351.53	-1.52	135.68	1.15	144.55	17 Aug 22	206.4	266.4	-4.6	-4.5	103.20	108.89	-1.46	-1.14	171.83	176.28	-1.87	-2.70
Aug 17	150.30	1.43	350.38	-1.55	142.05	1.17	154.21	27 Aug	328.7		0.3		114.63		-0.82		179.30		-3.46	
								Sep 1		37.2		5.0		120.43		-0.51		180.36		-4.03
Aug 27	151.56	1.44	349.10	-1.57	148.40	1.18	163.91	6 Sep 11	108.1	176.3	3.2	-2.7	126.28	132.18	-0.21	0.08	178.93	174.91	-4.21	-3.74
Sep 6	152.80	1.46	347.77	-1.58	154.75	1.19	173.67	16 Sep 21	238.5	298.4	-5.2	-2.3	138.12	144.09	0.36	0.61	169.71	166.16	-2.51	-0.86
Sep 16	153.99	1.48	346.50	-1.57	161.08	1.19	183.48	26 S-O 1	4.4	76.5	3.3	4.8	150.11	156.16	0.84	1.05	166.45	170.63	0.60	1.54
Sep 26	155.11	1.51	345.37	-1.55	167.41	1.19	193.35	6 Oct 11	145.9	210.2	-0.2	-4.7	162.24	168.35	1.22	1.37	177.42	185.49	1.94	1.93
Oct 6	156.16	1.53	344.48	-1.53	173.73	1.18	203.28	16 Oct 21	270.3	331.5	-4.0	0.8	174.49	180.65	1.48	1.57	193.97	202.46	1.65	1.22
Oct 16	157.10	1.57	343.88	-1.49	180.05	1.17	213.26	26 Oct 31	42.0	115.5	4.9	2.2	186.83	193.03	1.62	1.64	210.80	218.96	0.70	0.14
Oct 26	157.92	1.60	343.60	-1.45	186.36	1.15	223.29	5 Nov 10	181.8	243.2	-3.2	-4.9	199.25	205.48	1.63	1.58	226.95	234.80	-0.41	-0.94
Nov 5	158.60	1.64	343.67	-1.41	192.67	1.13	233.38	15 Nov 20	302.5	6.4	-1.5	3.7	211.73	217.98	1.51	1.41	242.55	250.20	-1.43	-1.84
Nov 15	159.12	1.68	344.08	-1.37	198.97	1.10	243.50	25 Nov 30	80.6	152.8	4.3	-1.3	224.24	230.51	1.29	1.15	257.77	265.19	-2.15	-2.33
Nov 25	159.48	1.72	344.81	-1.33	205.26	1.07	253.65	5 Dec 10	216.0	275.7	-4.9	-3.4	236.79	243.07	0.98	0.80	272.31	278.77	-2.31	-2.02
Dec 5	159.65	1.77	345.85	-1.29	211.54	1.03	263.83	15 Dec 20	335.8	43.3	1.7	5.1	249.35	255.63	0.61	0.41	283.79	286.01	-1.36	-0.21
Dec 15	159.64	1.81	347.15	-1.26	217.80	0.97	274.02	25 Dec 30	119.0	188.1	1.5	-4.1	261.92	268.20	0.21	0.01	283.82	277.76	1.36	2.78
1654								1655												1655
Dec 25	159.45	1.85	348.70	-1.23	224.03	0.91	284.21	4 Jan 9	248.9	308.4	-4.8	-0.7	274.49	280.77	-0.20	-0.39	271.97	269.78	3.32	3.02
Jan 4	159.08	1.89	350.45	-1.20	230.23	0.84	294.40	14 Jan 19	11.1	81.7	4.3	4.3	287.06	293.34	-0.58	-0.76	271.15	274.88	2.30	1.46
Jan 14	158.56	1.93	352.38	-1.17	236.40	0.75	304.57	24 Jan 29	155.9	221.5	-1.8	-5.3	299.62	305.89	-0.92	-1.06	280.06	286.14	0.64	-0.11
Jan 24	157.91	1.96	354.45	-1.15	242.51	0.65	314.71	3 Feb 8	281.1	342.3	-3.0	2.4	312.16	318.43	-1.18	-1.28	292.84	300.01	-0.76	-1.30
Feb 3	157.17	1.98	356.64	-1.13	248.56	0.53	324.82	13 Feb 18	48.6	120.4	5.3	1.4	324.70	330.95	-1.36	-1.41	307.59	315.56	-1.73	-2.01
Feb 13	156.38	2.00	358.93	-1.12	254.54	0.39	334.88	23 Feb 28	191.1	253.7	-4.3	-4.6	337.20	343.45	-1.43	-1.43	323.92	332.72	-2.13	-2.08
Feb 23	155.58	2.01	1.28	-1.11	260.42	0.23	344.89	5 Mar 10	313.4	18.0	-0.2	4.6	349.69	355.92	-1.40	-1.35	341.97	351.65	-1.83	-1.36
Mar 5	154.82	2.01	3.67	-1.10	266.18	0.04	354.85	15 Mar 20	87.7	158.3	3.8	-2.0	2.14	8.35	-1.27	-1.16	1.63	11.52	-0.65	0.25
Mar 15	154.14	2.01	6.09	-1.10	271.79	-0.19	4.75	25 Mar 30	224.9	285.2	-5.1	-2.5	14.55	20.75	-1.04	-0.89	20.67	28.26	1.25	2.17
Mar 25	153.57	2.00	8.50	-1.10	277.21	-0.45	14.60	4 Apr 9	346.7	55.7	2.7	4.9	26.93	33.10	-0.72	-0.54	33.63	36.40	2.84	3.09
Apr 4	153.15	1.98	10.90	-1.11	282.39	-0.76	24.39	14 Apr 19	126.8	194.7	0.5	-4.4	39.27	45.42	-0.34	-0.14	36.54	34.48	2.80	1.93
Apr 14	152.88	1.96	13.26	-1.11	287.27	-1.12	34.12	24 Apr 29	257.5	317.0	-4.1	0.5	51.56	57.68	0.07	0.29	31.29	28.37	0.64	-0.78
Apr 24	152.79	1.94	15.56	-1.12	291.75	-1.55	43.81	4 May 9	21.8	94.7	4.7	2.9	63.80	69.91	0.50	0.71	26.87	27.25	-2.01	-2.88
May 4	152.87	1.92	17.79	-1.14	295.72	-2.04	53.46	14 May 19	164.9	229.4	-2.9	-4.9	76.00	82.08	0.91	1.10	29.48	33.35	-3.35	-3.47
May 14	153.13	1.89	19.91	-1.16	299.04	-2.60	63.06	24 May 29	289.6	349.9	-1.7	3.4	88.14	94.19	1.27	1.43	38.61	45.11	-3.26	-2.79
May 24	153.55	1.87	21.91	-1.18	301.55	-3.25	72.63	3 Jun 8	58.9	133.5	4.8	-0.5	100.23	106.25	1.56	1.67	52.76	61.52	-2.09	-1.24
Jun 3	154.12	1.85	23.76	-1.20	303.04	-3.97	82.18	13 Jun 18	201.2	262.7	-4.9	-3.6	112.26	118.24	1.75	1.80	71.30	81.88	-0.31	0.58
Jun 13	154.84	1.83	25.44	-1.23	303.36	-4.72	91.72	23 Jun 28	321.9	24.8	1.4	5.1	124.21	130.16	1.81	1.79	92.79	103.48	1.30	1.74
Jun 23	155.68	1.81	26.93	-1.26	302.42	-5.46	101.25	3 Jul 8	97.4	171.1	2.6	-3.7	136.08	141.98	1.74	1.65	113.58	122.92	1.87	1.72
Jul 3	156.63	1.80	28.18	-1.30	300.40	-6.07	110.79	13 Jul 18	235.7	295.3	-5.0	-1.0	147.86	153.70	1.52	1.35	131.46	139.21	1.34	0.76
Jul 13	157.67	1.79	29.19	-1.33	297.76	-6.44	120.33	23 Jul 28	355.4	61.7	4.0	4.8	159.51	165.29	1.15	0.91	146.13	152.17	0.03	-0.82
Jul 23	158.79	1.78	29.90	-1.37	295.19	-6.52	129.90	2 Aug 7	136.1	206.9	-0.9	-5.2	171.03	176.72	0.63	0.32	157.16	160.87	-1.74	-2.70
Aug 2	159.98	1.78	30.32	-1.41	293.39	-6.31	139.50	12 Aug 17	268.5	328.0	-3.3	1.9	182.37	187.96	-0.01	-0.38	162.93	162.87	-3.60	-4.30
Aug 12	161.20	1.78	30.40	-1.45	292.82	-5.89	149.14	22 Aug 27	30.9	100.2	5.2	2.3	193.49	198.95	-0.76	-1.17	160.44	156.12	-4.57	-4.14
Aug 22	162.45	1.79	30.16	-1.49	293.60	-5.35	158.82	1 Sep 6	173.7	240.7	-3.8	-4.7	204.34	209.63	-1.60	-2.03	151.70	149.54	-2.96	-1.39
Sep 1	163.70	1.79	29.58	-1.52	295.64	-4.77	168.55	11 Sep 16	300.5	1.8	-0.5	4.3	214.82	219.88	-2.47	-2.91	151.02	156.00	0.10	1.17
Sep 11	164.95	1.81	28.72	-1.55	298.78	-4.19	178.33	21 Sep 26	68.4	139.1	4.3	-1.2	224.80	229.55	-3.34	-3.75	163.39	171.97	1.74	1.89
Sep 21	166.16	1.83	27.61	-1.57	302.79	-3.63	188.17	1 Oct 6	209.8	273.0	-5.0	-2.6	234.10	238.40	-4.15	-4.51	180.89	189.74	1.73	1.37
Oct 1	167.33	1.85	26.33	-1.57	307.51	-3.11	198.07	11 Oct 16	332.5	37.4	2.4	5.0	242.39	246.02	-4.82	-5.07	198.35	206.68	0.89	0.35
Oct 11	168.43	1.87	24.98	-1.57	312.76	-2.63	208.02	21 Oct 26	107.2	177.4	1.3	-4.1	249.19	251.82	-5.25	-5.33	214.76	222.62	-0.21	-0.76
Oct 21	169.45	1.90	23.66	-1.55	318.44	-2.19	218.03	31 Oct	244.1		-4.2		253.78		-5.28		230.30		-1.29	
								Nov 5		304.5		0.3		254.95		-5.08		237.80		-1.76
Oct 31	170.36	1.94	22.48	-1.52	324.43	-1.79	228.08	10 Nov 15	5.5	74.7	4.6	3.6	255.18	254.41	-4.67	-4.03	245.11	252.17	-2.14	-2.41
Nov 10	171.15	1.98	21.51	-1.48	330.67	-1.43	238.18	20 Nov 25	146.3	214.1	-2.3	-5.1	252.65	250.09	-3.15	-2.03	258.79	264.60	-2.50	-2.34
Nov 20	171.80	2.02	20.83	-1.43	337.09	-1.11	248.32	30 N-D 5	276.9	336.1	-1.9	3.2	247.12	244.21	-0.77	0.52	268.82	270.08	-1.81	-0.76
Nov 30	172.28	2.06	20.48	-1.38	343.64	-0.81	258.49	10 Dec 15	40.0	113.0	5.1	0.4	241.85	240.37	1.71	2.72	266.94	260.44	0.79	2.31
Dec 10	172.60	2.11	20.47	-1.33	350.28	-0.55	268.67	20 Dec 25	184.5	249.0	-4.8	-3.9	239.92	240.48	3.51	4.07	255.13	253.93	3.04	2.91
Dec 20	172.74	2.16	20.81	-1.28	356.98	-0.31	278.87	30 Dec	309.0		1.1		241.95		4.45		256.26		2.34	
1655								1655												1655

JULIAN	SATURN		JUPITER		MARS		SUN	GREGORIAN	MOON				VENUS				MERCURY			
12 UT	LONG.	LAT.	LONG.	LAT.	LONG.	LAT.	LONG.	12 UT	LONG.		LAT.		LONG.		LAT.		LONG.		LAT.	
1655								1656												1656
								Jan 4		8.9		5.0		244.20		4.66		260.78		1.62
Dec 30	172.69	2.20	21.48	-1.22	3.73	-0.11	289.06	9 Jan 14	76.3	151.5	3.5	-3.0	247.08	250.51	4.73	4.69	266.53	273.02	0.87	0.15
Jan 9	172.47	2.25	22.45	-1.18	10.49	0.08	299.24	19 Jan 24	220.9	282.3	-5.2	-1.4	254.36	258.58	4.56	4.36	279.98	287.30	-0.49	-1.05
Jan 19	172.07	2.29	23.70	-1.13	17.25	0.24	309.40	29 J-F 3	341.4	43.4	3.7	5.1	263.09	267.83	4.10	3.79	294.94	302.89	-1.51	-1.85
Jan 29	171.53	2.33	25.19	-1.09	24.01	0.39	319.53	8 Feb 13	114.0	189.1	0.3	-5.0	272.77	277.87	3.45	3.08	311.16	319.79	-2.05	-2.09
Feb 8	170.87	2.35	26.88	-1.05	30.75	0.52	329.61	18 Feb 23	255.2	314.6	-3.5	1.6	283.11	288.47	2.68	2.28	328.80	338.17	-1.94	-1.57
Feb 18	170.13	2.38	28.76	-1.02	37.46	0.63	339.65	28 Feb	14.9		5.0		293.92		1.87		347.78		-0.96	
								Mar 4		80.0		3.1		299.45		1.47		357.28		-0.11
Feb 28	169.34	2.39	30.78	-0.99	44.14	0.73	349.64	9 Mar 14	152.5	225.1	-3.0	-4.8	305.05	310.71	1.07	0.69	5.99	12.96	0.92	1.99
Mar 9	168.56	2.40	32.91	-0.96	50.79	0.81	359.57	19 Mar 24	287.8	347.1	-0.7	4.0	316.42	322.17	0.32	-0.03	17.29	18.52	2.88	3.35
Mar 19	167.82	2.39	35.14	-0.94	57.40	0.89	9.45	29 M-A 3	50.2	118.3	4.5	-0.4	327.97	333.79	-0.35	-0.65	16.84	13.32	3.20	2.39
Mar 29	167.16	2.38	37.44	-0.92	63.98	0.95	19.27	8 Apr 13	190.8	259.2	-4.9	-2.8	339.64	345.52	-0.91	-1.15	9.68	7.42	1.12	-0.24
Apr 8	166.62	2.36	39.78	-0.91	70.52	1.00	29.03	18 Apr 23	319.5	20.5	2.3	5.0	351.41	357.33	-1.35	-1.51	7.20	8.98	-1.42	-2.29
Apr 18	166.22	2.34	42.16	-0.89	77.03	1.05	38.74	28 Apr	87.3		2.0		3.26		-1.64		12.47		-2.84	
								May 3		157.5		-3.7		9.21		-1.73		17.36		-3.08
Apr 28	165.98	2.31	44.53	-0.88	83.51	1.08	48.41	8 May 13	227.8	291.7	-4.5	0.2	15.16	21.14	-1.78	-1.80	23.41	30.47	-3.04	-2.74
May 8	165.91	2.28	46.89	-0.87	89.95	1.11	58.03	18 May 23	351.3	55.4	4.5	4.1	27.12	33.11	-1.79	-1.75	38.50	47.47	-2.21	-1.49
May 18	166.01	2.25	49.22	-0.87	96.37	1.14	67.62	28 M-J 2	125.8	196.3	-1.5	-5.2	39.11	45.13	-1.67	-1.57	57.33	67.94	-0.64	0.26
May 28	166.28	2.21	51.50	-0.87	102.77	1.15	77.18	7 Jun 12	262.8	323.5	-2.3	3.0	51.15	57.18	-1.44	-1.29	78.91	89.70	1.06	1.64
Jun 7	166.71	2.18	53.70	-0.87	109.14	1.16	86.72	17 Jun 22	24.0	92.1	5.2	1.5	63.22	69.27	-1.12	-0.94	99.89	109.24	1.92	1.91
Jun 17	167.29	2.15	55.81	-0.87	115.51	1.17	96.26	27 Jun	165.0		-4.4		75.33		-0.74		117.69		1.62	
								Jul 2		233.6		-4.3		81.40		-0.54		125.23		1.09
Jun 27	168.01	2.13	57.81	-0.87	121.86	1.17	105.79	7 Jul 12	296.1	355.5	0.7	4.9	87.49	93.58	-0.33	-0.11	131.81	137.33	0.35	-0.54
Jul 7	168.85	2.11	59.67	-0.88	128.20	1.17	115.33	17 Jul 22	58.2	130.2	4.1	-2.0	99.69	105.80	0.10	0.30	141.62	144.40	-1.57	-2.65
Jul 17	169.80	2.09	61.37	-0.89	134.55	1.16	124.89	27 J-A 1	203.5	268.6	-5.2	-1.8	111.93	118.08	0.50	0.68	145.32	144.12	-3.69	-4.51
Jul 27	170.84	2.07	62.88	-0.90	140.90	1.14	134.47	6 Aug 11	328.6	28.5	3.4	5.1	124.23	130.39	0.85	1.00	140.94	136.83	-4.83	-4.42
Aug 6	171.96	2.06	64.16	-0.91	147.25	1.13	144.08	16 Aug 21	94.2	169.0	1.2	-4.5	136.57	142.75	1.13	1.24	133.62	132.99	-3.32	-1.84
Aug 16	173.13	2.05	65.20	-0.92	153.62	1.10	153.74	26 Aug 31	240.2	301.9	-3.8	1.3	148.95	155.15	1.32	1.38	135.67	141.40	-0.39	0.77
Aug 26	174.35	2.05	65.96	-0.94	160.00	1.08	163.44	5 Sep 10	1.3	63.0	4.9	3.5	161.36	167.58	1.41	1.41	149.33	158.38	1.50	1.80
Sep 5	175.58	2.05	66.41	-0.95	166.40	1.05	173.20	15 Sep 20	132.0	207.2	-2.2	-4.9	173.81	180.04	1.39	1.34	167.73	176.93	1.77	1.50
Sep 15	176.82	2.06	66.54	-0.96	172.83	1.01	183.01	25 Sep 30	274.5	334.4	-0.9	3.8	186.28	192.52	1.27	1.17	185.80	194.33	1.08	0.57
Sep 25	178.05	2.07	66.34	-0.97	179.28	0.97	192.87	5 Oct 10	34.9	99.4	4.7	0.3	198.77	205.01	1.04	0.90	202.51	210.39	0.00	-0.57
Oct 5	179.25	2.09	65.81	-0.98	185.75	0.93	202.79	15 Oct 20	170.8	243.6	-4.6	-3.1	211.26	217.52	0.74	0.56	218.00	225.35	-1.14	-1.66
Oct 15	180.39	2.11	64.97	-0.98	192.26	0.88	212.77	25 Oct 30	307.1	6.7	2.2	5.0	223.77	230.02	0.37	0.17	232.41	239.10	-2.12	-2.48
Oct 25	181.47	2.14	63.88	-0.98	198.80	0.82	222.80	4 Nov 9	69.8	137.6	2.6	-3.1	236.28	242.53	-0.03	-0.24	245.21	250.34	-2.68	-2.65
Nov 4	182.45	2.17	62.61	-0.97	205.37	0.76	232.88	14 Nov 19	209.4	277.9	-4.8	-0.2	248.78	255.03	-0.45	-0.65	253.69	253.96	-2.25	-1.33
Nov 14	183.33	2.21	61.25	-0.95	211.98	0.69	243.00	24 Nov 29	338.8	39.6	4.5	4.5	261.29	267.53	-0.84	-1.02	249.99	243.34	0.17	1.78
Nov 24	184.08	2.25	59.92	-0.93	218.63	0.61	253.15	4 Dec 9	106.5	177.0	-0.7	-5.1	273.78	280.02	-1.19	-1.34	238.71	238.49	2.69	2.76
Dec 4	184.68	2.29	58.70	-0.89	225.31	0.53	263.33	14 Dec 19	246.5	310.7	-2.8	2.8	286.26	292.49	-1.46	-1.56	241.73	246.96	2.35	1.74
Dec 14	185.13	2.33	57.70	-0.86	232.02	0.44	273.52	24 Dec 29	10.4	73.7	5.3	2.3	298.72	304.94	-1.64	-1.68	253.24	260.10	1.06	0.39
1656								1657												1657
Dec 24	185.40	2.38	56.98	-0.82	238.77	0.34	283.72	3 Jan 8	144.7	215.9	-3.8	-4.7	311.15	317.34	-1.70	-1.68	267.31	274.78	-0.25	-0.81
Jan 3	185.50	2.43	56.58	-0.77	245.55	0.23	293.91	13 Jan 18	281.7	342.7	0.2	4.7	323.53	329.70	-1.64	-1.56	282.47	290.39	-1.30	-1.68
Jan 13	185.41	2.48	56.51	-0.73	252.37	0.11	304.08	23 Jan 28	42.6	109.3	4.4	-0.9	335.85	341.98	-1.44	-1.30	298.56	307.01	-1.95	-2.07
Jan 23	185.15	2.52	56.79	-0.69	259.22	-0.02	314.22	2 Feb 7	183.7	253.0	-5.2	-2.2	348.09	354.18	-1.12	-0.91	315.76	324.80	-2.02	-1.76
Feb 2	184.73	2.56	57.40	-0.65	266.10	-0.16	324.33	12 Feb 17	315.3	14.9	3.1	5.1	0.23	6.25	-0.68	-0.42	334.01	343.06	-1.26	-0.48
Feb 12	184.17	2.60	58.30	-0.61	273.00	-0.31	334.40	22 Feb 27	76.2	146.8	1.8	-3.9	12.24	18.18	-0.14	0.16	351.26	357.53	0.55	1.73
Feb 22	183.50	2.62	59.47	-0.57	279.93	-0.48	344.41	4 Mar 9	222.0	288.1	-4.1	1.0	24.08	29.93	0.47	0.79	0.75	0.37	2.82	3.51
Mar 4	182.75	2.64	60.88	-0.54	286.88	-0.66	354.37	14 Mar 19	348.1	47.9	4.8	3.7	35.73	41.47	1.12	1.45	357.01	352.58	3.50	2.74
Mar 14	181.97	2.65	62.49	-0.51	293.84	-0.85	4.28	24 Mar 29	111.8	185.5	-1.5	-5.0	47.13	52.72	1.78	2.09	349.21	348.02	1.53	0.24
Mar 24	181.20	2.65	64.28	-0.48	300.80	-1.06	14.12	3 Apr 8	258.6	321.4	-1.3	3.8	58.22	63.63	2.39	2.67	349.10	352.07	-0.89	-1.76
Apr 3	180.48	2.64	66.21	-0.46	307.75	-1.28	23.92	13 Apr 18	20.8	82.2	4.8	0.8	68.92	74.09	2.93	3.14	356.53	2.16	-2.37	-2.71
Apr 13	179.85	2.62	68.27	-0.43	314.68	-1.51	33.66	23 Apr 28	149.6	224.0	-4.3	-3.7	79.10	83.95	3.32	3.45	8.76	16.22	-2.80	-2.65
Apr 23	179.34	2.60	70.41	-0.41	321.57	-1.75	43.34	3 May 8	293.1	353.7	1.9	5.1	88.59	92.99	3.52	3.53	24.49	33.57	-2.28	-1.70
May 3	178.96	2.56	72.62	-0.39	328.40	-2.00	52.99	13 May 18	54.0	118.5	3.1	-2.5	97.11	100.89	3.46	3.30	43.44	54.01	-0.94	-0.07
May 13	178.75	2.53	74.89	-0.37	335.13	-2.25	62.60	23 May 28	188.8	261.1	-5.2	-0.8	104.25	107.12	3.05	2.68	64.96	75.78	0.78	1.49
May 23	178.70	2.49	77.18	-0.36	341.74	-2.52	72.17	2 Jun 7	326.0	25.8	4.4	4.7	109.38	110.93	2.18	1.54	85.98	95.26	1.93	2.06
Jun 2	178.83	2.45	79.48	-0.34	348.17	-2.78	81.72	12 Jun 17	88.4	156.7	0.1	-4.9	111.65	111.42	0.74	-0.22	103.52	110.72	1.89	1.44
Jun 12	179.11	2.42	81.77	-0.33	354.36	-3.05	91.26	22 Jun 27	227.9	296.2	-3.5	2.3	110.20	108.05	-1.31	-2.49	116.78	121.58	0.72	-0.21
Jun 22	179.56	2.38	84.04	-0.31	0.25	-3.31	100.79	2 Jul 7	358.0	58.2	5.3	2.8	105.22	102.12	-3.65	-4.68	124.92	126.54	-1.33	-2.54
Jul 2	180.15	2.34	86.25	-0.30	5.74	-3.57	110.33	12 Jul 17	124.6	196.2	-3.0	-5.0	99.22	96.96	-5.48	-6.00	126.21	123.98	-3.71	-4.60
Jul 12	180.88	2.31	88.40	-0.29	10.71	-3.81	119.87	22 Jul 27	265.5	329.7	-0.4	4.5	95.58	95.19	-6.25	-6.28	120.54	117.25	-4.93	-4.55
Jul 22	181.73	2.29	90.47	-0.27	15.03	-4.03	129.44	1 Aug 6	29.7	91.8	4.4	-0.3	95.75	97.17	-6.13	-5.84	115.60	116.56	-3.55	-2.21
Aug 1	182.68	2.26	92.42	-0.26	18.51	-4.23	139.04	11 Aug 16	162.8	235.2	-4.9	-2.8	99.32	102.09	-5.47	-5.04	120.39	126.84	-0.84	0.35
Aug 11	183.72	2.24	94.24	-0.25	20.94	-4.37	148.67	21 Aug 26	301.1	2.4	2.8	5.0	105.37	109.07	-4.56	-4.06	135.24	144.71	1.21	1.67
Aug 21	184.84	2.23	95.90	-0.24	22.11	-4.43	158.35	31 Aug	62.0		2.1		113.13		-3.54		154.45		1.78	
								Sep 5		127.3		-3.3		117.49		-3.02		164.00		1.62
Aug 31	186.01	2.22	97.36	-0.23	21.86	-4.37	168.08	10 Sep 15	201.9	272.5	-4.5	0.6	122.10	126.92	-2.51	-2.00	173.15	181.86	1.26	0.78
Sep 10	187.21	2.22	98.61	-0.21	20.21	-4.14	177.86	20 Sep 25	335.2	34.8	4.7	3.8	131.92	137.08	-1.51	-1.03	190.16	198.06	0.22	-0.37
Sep 20	188.43	2.22	99.62	-0.20	17.47	-3.72	187.69	30 S-O 5	95.5	164.8	-1.0	-5.0	142.38	147.80	-0.59	-0.17	205.61	212.81	-0.97	-1.55
Sep 30	189.66	2.22	100.34	-0.19	14.29	-3.13	197.59	10 Oct 15	240.4	307.7	-1.9	3.6	153.33	158.95	0.22	0.58	219.61	225.91	-2.09	-2.53
Oct 10	190.87	2.23	100.76	-0.17	11.49	-2.43	207.54	20 Oct 25	8.1	67.7	4.9	1.2	164.65	170.43	0.90	1.17	231.48	235.88	-2.85	-2.95
Oct 20	192.04	2.25	100.85	-0.15	9.71	-1.73	217.54	30 Oct	131.0		-3.9		176.27		1.41		238.32		-2.70	
								Nov 4		203.4		-4.4		182.17		1.61		237.58		-1.90
Oct 30	193.16	2.27	100.61	-0.14	9.23	-1.09	227.60	9 Nov 14	277.1	341.1	1.4	5.1	188.12	194.11	1.76	1.87	232.91	226.40	-0.46	1.20
Nov 9	194.20	2.30	100.04	-0.12	10.06	-0.54	237.69	19 Nov 24	40.6	101.9	3.4	-1.9	200.15	206.21	1.94	1.97	222.59	223.34	2.28	2.57
Nov 19	195.15	2.33	99.18	-0.10	12.04	-0.10	247.83	29 N-D 4	168.7	241.8	-5.3	-1.7	212.31	218.44	1.97	1.92	227.43	233.32	2.33	1.84
Nov 29	195.99	2.36	98.08	-0.08	14.95	0.26	258.00	9 Dec 14	311.7	13.2	4.1	4.9	224.58	230.75	1.84	1.73	240.11	247.33	1.23	0.60
Dec 9	196.70	2.40	96.80	-0.05	18.60	0.55	268.18	19 Dec 24	73.3	138.0	0.7	-4.4	236.93	243.12	1.60	1.43	254.79	262.39	-0.01	-0.59
Dec 19	197.25	2.44	95.45	-0.03	22.84	0.77	278.37	29 Dec	207.7		-4.2		249.32		1.25		270.13		-1.09	
1657								1657												1657

JULIAN	SATURN		JUPITER		MARS		SUN	GREGORIAN	MOON				VENUS				MERCURY			
12 UT	LONG.	LAT.	LONG.	LAT.	LONG.	LAT.	LONG.	12 UT	LONG.		LAT.		LONG.		LAT.		LONG.		LAT.	
1657								1658												1658
								Jan 3		278.9		1.7		255.53		1.05		278.03		-1.52
Dec 29	197.65	2.49	94.13	0.00	27.51	0.95	288.57	8 Jan 13	344.9	44.8	5.2	3.0	261.75	267.98	0.84	0.62	286.09	294.36	-1.84	-2.04
Jan 8	197.88	2.54	92.94	0.02	32.53	1.08	298.75	18 Jan 23	107.0	176.1	-2.3	-5.1	274.20	280.43	0.39	0.16	302.85	311.54	-2.08	-1.92
Jan 18	197.92	2.58	91.96	0.04	37.81	1.19	308.91	28 J-F 2	246.7	314.6	-1.2	4.2	286.67	292.90	-0.06	-0.28	320.33	328.88	-1.53	-0.84
Jan 28	197.80	2.63	91.26	0.06	43.30	1.27	319.04	7 Feb 12	17.0	76.6	4.5	0.2	299.13	305.36	-0.49	-0.68	336.51	342.03	0.15	1.39
Feb 7	197.50	2.67	90.87	0.08	48.94	1.33	329.13	17 Feb 22	142.5	215.2	-4.5	-3.4	311.59	317.82	-0.86	-1.01	344.09	342.14	2.66	3.55
Feb 17	197.05	2.71	90.80	0.10	54.69	1.37	339.17	27 Feb	284.5		2.3		324.04		-1.15		337.45		3.66	
								Mar 4		348.7		5.0		330.27		-1.26		332.75		2.97
Feb 27	196.46	2.74	91.07	0.12	60.55	1.40	349.16	9 Mar 14	48.8	109.7	2.3	-2.8	336.49	342.70	-1.35	-1.41	330.14	330.15	1.83	0.63
Mar 9	195.78	2.76	91.64	0.13	66.47	1.42	359.09	19 Mar 24	180.0	253.9	-4.8	0.0	348.91	355.11	-1.45	-1.45	332.38	336.30	-0.43	-1.29
Mar 19	195.03	2.77	92.50	0.15	72.44	1.43	8.97	29 M-A 3	320.5	21.8	4.6	4.0	1.31	7.51	-1.43	-1.39	341.48	347.64	-1.93	-2.35
Mar 29	194.26	2.78	93.63	0.16	78.46	1.43	18.79	8 Apr 13	81.0	144.7	-0.7	-4.9	13.70	19.88	-1.32	-1.22	354.62	2.32	-2.55	-2.54
Apr 8	193.51	2.77	94.97	0.18	84.50	1.43	28.56	18 Apr 23	218.6	291.2	-2.9	3.3	26.06	32.24	-1.10	-0.97	10.73	19.84	-2.30	-1.85
Apr 18	192.81	2.75	96.52	0.19	90.58	1.42	38.27	28 Apr	354.8		5.1		38.40		-0.81		29.66		-1.21	
								May 3		54.3		1.6		44.57		-0.64		40.12		-0.40
Apr 28	192.21	2.73	98.24	0.20	96.68	1.40	47.94	8 May 13	114.5	181.9	-3.6	-4.9	50.73	56.88	-0.46	-0.27	50.97	61.74	0.47	1.28
May 8	191.73	2.70	100.09	0.21	102.79	1.38	57.56	18 May 23	257.1	326.6	0.5	5.1	63.03	69.17	-0.07	0.13	71.87	80.99	1.88	2.18
May 18	191.39	2.66	102.06	0.23	108.93	1.35	67.15	28 M-J 2	27.6	87.1	3.7	-1.4	75.32	81.45	0.33	0.52	88.93	95.60	2.16	1.81
May 28	191.20	2.62	104.13	0.24	115.08	1.32	76.71	7 Jun 12	150.2	220.6	-5.2	-2.7	87.59	93.72	0.71	0.89	100.92	104.73	1.14	0.18
Jun 7	191.19	2.58	106.26	0.25	121.26	1.28	86.26	17 Jun 22	294.4	0.2	3.6	5.0	99.85	105.97	1.05	1.19	106.84	107.07	-1.01	-2.34
Jun 17	191.33	2.54	108.45	0.27	127.45	1.25	95.79	27 Jun	59.8		1.1		112.09		1.32		105.50		-3.61	
								Jul 2		121.1		-4.0		118.21		1.42		102.66		-4.53
Jun 27	191.65	2.50	110.67	0.28	133.67	1.20	105.32	7 Jul 12	187.9	259.5	-4.6	0.8	124.33	130.44	1.49	1.54	99.64	97.72	-4.85	-4.52
Jul 7	192.11	2.46	112.90	0.30	139.91	1.16	114.86	17 Jul 22	330.0	32.5	5.1	3.2	136.54	142.64	1.56	1.55	97.82	100.31	-3.66	-2.50
Jul 17	192.72	2.42	115.12	0.31	146.19	1.11	124.42	27 J-A 1	92.2	157.1	-1.9	-5.1	148.73	154.82	1.51	1.44	105.22	112.31	-1.24	-0.06
Jul 27	193.46	2.38	117.31	0.33	152.49	1.06	134.00	6 Aug 11	226.9	297.5	-1.9	3.8	160.89	166.96	1.34	1.21	121.13	130.96	0.89	1.50
Aug 6	194.32	2.35	119.46	0.35	158.83	1.01	143.61	16 Aug 21	3.9	64.0	4.5	0.3	173.02	179.07	1.05	0.87	141.06	150.93	1.75	1.71
Aug 16	195.28	2.33	121.54	0.37	165.22	0.95	153.27	26 Aug 31	125.7	195.1	-4.3	-3.9	185.11	191.13	0.66	0.43	160.35	169.25	1.43	1.00
Aug 26	196.33	2.31	123.54	0.39	171.65	0.89	162.97	5 Sep 10	266.0	333.9	1.7	5.0	197.15	203.14	0.18	-0.09	177.65	185.59	0.45	-0.15
Sep 5	197.44	2.29	125.42	0.41	178.12	0.82	172.72	15 Sep 20	36.4	95.8	2.4	-2.6	209.13	215.09	-0.37	-0.66	193.07	200.10	-0.79	-1.43
Sep 15	198.60	2.28	127.17	0.44	184.64	0.75	182.53	25 Sep 30	161.0	233.9	-5.0	-0.9	221.04	226.97	-0.95	-1.24	206.62	212.51	-2.04	-2.58
Sep 25	199.80	2.27	128.74	0.46	191.22	0.68	192.39	5 Oct 10	304.1	8.4	4.5	4.2	232.87	238.75	-1.52	-1.79	217.50	221.12	-3.02	-3.25
Oct 5	201.01	2.27	130.13	0.49	197.85	0.61	202.31	15 Oct 20	68.4	128.8	-0.5	-4.8	244.59	250.41	-2.05	-2.29	222.58	220.84	-3.14	-2.47
Oct 15	202.23	2.27	131.29	0.52	204.55	0.53	212.29	25 Oct 30	198.0	272.6	-3.7	2.6	256.19	261.92	-2.50	-2.67	215.65	209.55	-1.11	0.58
Oct 25	203.41	2.28	132.19	0.56	211.30	0.44	222.32	4 Nov 9	340.3	41.5	5.2	1.9	267.60	273.22	-2.81	-2.91	206.70	208.39	1.82	2.33
Nov 4	204.56	2.30	132.81	0.59	218.12	0.35	232.39	14 Nov 19	100.7	163.6	-3.3	-5.2	278.76	284.23	-2.95	-2.94	213.27	219.79	2.27	1.91
Nov 14	205.65	2.32	133.12	0.63	225.00	0.26	242.51	24 Nov 29	236.0	310.0	-0.6	4.9	289.59	294.83	-2.88	-2.74	227.07	234.64	1.38	0.80
Nov 24	206.66	2.34	133.11	0.67	231.94	0.16	252.66	4 Dec 9	14.6	73.9	3.9	-1.0	299.92	304.83	-2.54	-2.25	242.34	250.09	0.21	-0.37
Dec 4	207.57	2.37	132.77	0.70	238.95	0.06	262.84	14 Dec 19	134.3	200.3	-5.0	-3.5	309.54	313.98	-1.89	-1.43	257.89	265.76	-0.89	-1.35
Dec 14	208.36	2.41	132.11	0.74	246.03	-0.05	273.03	24 Dec 29	274.3	345.6	2.8	5.0	318.12	321.87	-0.87	-0.21	273.72	281.81	-1.73	-1.99
1658								1659												1659
Dec 24	209.02	2.44	131.18	0.77	253.17	-0.16	283.23	3 Jan 8	47.1	106.6	1.3	-3.6	325.15	327.86	0.56	1.44	290.04	298.39	-2.11	-2.06
Jan 3	209.53	2.48	130.04	0.81	260.37	-0.27	293.41	13 Jan 18	169.8	238.5	-4.8	-0.2	329.87	331.05	2.44	3.53	306.73	314.75	-1.77	-1.20
Jan 13	209.88	2.53	128.75	0.83	267.63	-0.39	303.59	23 Jan 28	311.9	19.2	4.8	3.3	331.28	330.48	4.70	5.88	321.75	326.46	-0.28	0.98
Jan 23	210.05	2.57	127.42	0.85	274.95	-0.52	313.73	2 Feb 7	78.9	140.3	-1.7	-5.0	328.66	326.02	6.98	7.86	327.36	323.99	2.39	3.47
Feb 2	210.05	2.61	126.15	0.86	282.32	-0.64	323.84	12 Feb 17	207.2	277.5	-2.6	3.2	322.95	319.96	8.42	8.58	318.41	313.96	3.70	3.09
Feb 12	209.87	2.65	125.02	0.87	289.75	-0.77	333.91	22 Feb 27	348.0	51.3	4.7	0.5	317.54	316.00	8.36	7.84	312.38	313.53	2.05	0.95
Feb 22	209.54	2.69	124.11	0.87	297.22	-0.90	343.93	4 Mar 9	110.8	175.7	-4.1	-4.4	315.47	315.93	7.12	6.30	316.75	321.42	-0.05	-0.89
Mar 4	209.06	2.72	123.48	0.87	304.72	-1.02	353.89	14 Mar 19	245.9	316.0	1.0	5.0	317.30	319.43	5.43	4.57	327.14	333.68	-1.55	-2.02
Mar 14	208.45	2.75	123.16	0.86	312.26	-1.15	3.80	24 Mar 29	22.4	82.8	2.7	-2.5	322.20	325.49	3.73	2.94	340.91	348.76	-2.31	-2.40
Mar 24	207.76	2.77	123.15	0.85	319.81	-1.27	13.65	3 Apr 8	143.8	212.8	-5.2	-1.8	329.21	333.28	2.20	1.51	357.22	6.30	-2.29	-1.97
Apr 3	207.01	2.77	123.45	0.84	327.38	-1.38	23.44	13 Apr 18	285.0	352.9	4.1	4.6	337.63	342.23	0.88	0.32	16.00	26.30	-1.44	-0.71
Apr 13	206.25	2.77	124.05	0.83	334.94	-1.49	33.18	23 Apr 28	55.3	114.7	-0.2	-4.6	347.02	351.97	-0.19	-0.64	36.98	47.61	0.15	1.02
Apr 23	205.52	2.76	124.92	0.82	342.48	-1.59	42.88	3 May 8	178.4	251.1	-4.3	1.6	357.06	2.27	-1.04	-1.37	57.59	66.45	1.77	2.25
May 3	204.85	2.74	126.03	0.81	349.98	-1.67	52.52	13 May 18	323.3	27.8	5.3	2.3	7.57	12.97	-1.65	-1.88	73.91	79.85	2.40	2.19
May 13	204.27	2.71	127.37	0.80	357.44	-1.74	62.13	23 May 28	87.6	147.7	-3.0	-5.3	18.43	23.96	-2.05	-2.17	84.14	86.64	1.60	0.65
May 23	203.83	2.67	128.89	0.80	4.82	-1.80	71.71	2 Jun 7	214.8	289.9	-1.7	4.4	29.55	35.18	-2.25	-2.27	87.23	86.02	-0.61	-2.02
Jun 2	203.52	2.63	130.57	0.79	12.10	-1.84	81.26	12 Jun 17	359.7	61.0	4.2	-0.7	40.86	46.58	-2.26	-2.21	83.53	80.76	-3.33	-4.24
Jun 12	203.37	2.59	132.38	0.79	19.28	-1.85	90.80	22 Jun 27	120.2	182.4	-4.8	-4.0	52.33	58.12	-2.12	-1.99	78.86	78.64	-4.58	-4.34
Jun 22	203.39	2.55	134.31	0.79	26.30	-1.85	100.33	2 Jul 7	252.8	327.9	1.8	5.1	63.95	69.80	-1.84	-1.66	80.45	84.34	-3.66	-2.69
Jul 2	203.57	2.50	136.33	0.79	33.15	-1.83	109.86	12 Jul 17	34.0	93.5	1.6	-3.4	75.67	81.58	-1.46	-1.25	90.19	97.85	-1.57	-0.45
Jul 12	203.91	2.46	138.42	0.79	39.80	-1.79	119.41	22 Jul 27	154.0	219.1	-4.9	-1.0	87.51	93.47	-1.02	-0.78	107.01	117.13	0.54	1.28
Jul 22	204.40	2.42	140.56	0.80	46.20	-1.72	128.98	1 Aug 6	291.8	4.1	4.4	3.6	99.46	105.46	-0.54	-0.30	127.55	137.72	1.68	1.76
Aug 1	205.03	2.38	142.73	0.81	52.31	-1.62	138.57	11 Aug 16	66.5	126.1	-1.5	-4.9	111.49	117.55	-0.06	0.18	147.38	156.46	1.59	1.21
Aug 11	205.79	2.34	144.91	0.82	58.06	-1.50	148.21	21 Aug 26	189.3	257.5	-3.1	2.5	123.62	129.72	0.40	0.61	164.96	172.90	0.70	0.09
Aug 21	206.66	2.31	147.08	0.83	63.38	-1.35	157.89	31 Aug	330.2		4.9		135.84		0.80		180.31		-0.58	
								Sep 5		38.1		0.7		141.98		0.97		187.15		-1.28
Aug 31	207.63	2.28	149.22	0.85	68.18	-1.17	167.61	10 Sep 15	98.3	159.6	-4.0	-4.7	148.13	154.31	1.12	1.24	193.36	198.79	-1.97	-2.62
Sep 10	208.68	2.26	151.31	0.87	72.34	-0.95	177.39	20 Sep 25	226.4	296.8	0.1	4.9	160.50	166.70	1.34	1.41	203.13	205.89	-3.18	-3.54
Sep 20	209.80	2.24	153.33	0.89	75.71	-0.68	187.22	30 S-O 5	6.9	70.5	3.2	-2.3	172.92	179.15	1.46	1.47	206.32	203.63	-3.57	-3.01
Sep 30	210.96	2.23	155.26	0.91	78.09	-0.36	197.11	10 Oct 15	130.2	194.5	-5.2	-2.6	185.40	191.65	1.46	1.42	198.16	192.77	-1.73	-0.04
Oct 10	212.16	2.22	157.06	0.94	79.29	0.02	207.06	20 Oct 25	265.1	335.6	3.5	4.9	197.91	204.18	1.35	1.26	190.97	193.58	1.34	2.04
Oct 20	213.36	2.22	158.72	0.97	79.12	0.46	217.06	30 Oct	41.5		0.2		210.45		1.14		199.19		2.17	
								Nov 4		102.3		-4.5		216.73		1.00		206.31		1.95
Oct 30	214.55	2.22	160.19	1.00	77.49	0.95	227.11	9 Nov 14	162.7	230.8	-4.6	0.6	223.01	229.29	0.85	0.68	214.06	221.99	1.52	0.98
Nov 9	215.72	2.23	161.47	1.04	74.55	1.45	237.21	19 Nov 24	304.3	12.5	5.2	2.8	235.58	241.87	0.50	0.31	229.93	237.84	0.42	-0.15
Nov 19	216.85	2.24	162.50	1.08	70.83	1.91	247.34	29 N-D 4	74.7	134.2	-2.7	-5.2	248.16	254.45	0.11	-0.09	245.71	253.56	-0.69	-1.19
Nov 29	217.90	2.26	163.27	1.13	67.09	2.27	257.50	9 Dec 14	196.7	268.7	-2.4	3.7	260.74	267.03	-0.29	-0.48	261.43	269.35	-1.61	-1.93
Dec 9	218.87	2.29	163.74	1.17	64.09	2.51	267.69	19 Dec 24	342.6	47.3	4.5	-0.4	273.32	279.61	-0.67	-0.84	277.32	285.31	-2.13	-2.17
Dec 19	219.74	2.31	163.90	1.22	62.32	2.63	277.88	29 Dec	107.3		-4.6		285.89		-1.00		293.20		-2.00	
1659								1659												1659

JULIAN 12 UT	SATURN LONG.	SATURN LAT.	JUPITER LONG.	JUPITER LAT.	MARS LONG.	MARS LAT.	SUN LONG.
1659							
Dec 29	220.49	2.34	163.74	1.27	61.90	2.66	288.07
Jan 8	221.10	2.38	163.27	1.31	62.73	2.63	298.26
Jan 18	221.55	2.41	162.50	1.35	64.63	2.57	308.41
Jan 28	221.84	2.45	161.49	1.39	67.39	2.48	318.54
Feb 7	221.96	2.49	160.29	1.41	70.82	2.39	328.64
Feb 17	221.91	2.53	159.00	1.43	74.78	2.30	338.68
Feb 27	221.69	2.56	157.71	1.43	79.16	2.20	348.67
Mar 8	221.32	2.60	156.50	1.43	83.86	2.10	358.61
Mar 18	220.81	2.62	155.47	1.42	88.82	2.01	8.49
Mar 28	220.18	2.64	154.67	1.40	93.98	1.92	18.31
Apr 7	219.48	2.65	154.15	1.37	99.32	1.82	28.09
Apr 17	218.73	2.65	153.92	1.34	104.79	1.74	37.80
Apr 27	217.98	2.64	154.01	1.31	110.39	1.65	47.47
May 7	217.27	2.63	154.39	1.28	116.08	1.56	57.10
May 17	216.63	2.60	155.05	1.25	121.86	1.47	66.69
May 27	216.09	2.57	155.98	1.22	127.72	1.39	76.25
Jun 6	215.67	2.53	157.13	1.19	133.66	1.30	85.80
Jun 16	215.41	2.49	158.49	1.17	139.66	1.22	95.33
Jun 26	215.30	2.45	160.03	1.15	145.73	1.13	104.86
Jul 6	215.35	2.40	161.72	1.13	151.87	1.05	114.41
Jul 16	215.56	2.36	163.55	1.11	158.07	0.96	123.96
Jul 26	215.93	2.32	165.48	1.10	164.34	0.87	133.54
Aug 5	216.45	2.27	167.50	1.09	170.69	0.78	143.15
Aug 15	217.11	2.24	169.59	1.09	177.10	0.69	152.80
Aug 25	217.89	2.20	171.72	1.09	183.58	0.60	162.50
Sep 4	218.78	2.17	173.87	1.09	190.14	0.51	172.25
Sep 14	219.77	2.14	176.04	1.09	196.78	0.42	182.05
Sep 24	220.83	2.12	178.18	1.10	203.50	0.32	191.92
Oct 4	221.95	2.10	180.29	1.11	210.30	0.22	201.84
Oct 14	223.12	2.09	182.35	1.13	217.18	0.13	211.80
Oct 24	224.30	2.08	184.32	1.15	224.14	0.03	221.83
Nov 3	225.50	2.08	186.18	1.17	231.19	-0.07	231.91
Nov 13	226.68	2.08	187.91	1.20	238.32	-0.18	242.02
Nov 23	227.82	2.09	189.47	1.23	245.54	-0.28	252.17
Dec 3	228.92	2.10	190.85	1.26	252.83	-0.38	262.35
Dec 13	229.94	2.11	192.00	1.30	260.21	-0.48	272.54
1660							
Dec 23	230.87	2.13	192.89	1.34	267.65	-0.58	282.73
Jan 2	231.69	2.15	193.51	1.38	275.17	-0.68	292.92
Jan 12	232.39	2.18	193.83	1.43	282.75	-0.77	303.09
Jan 22	232.94	2.21	193.84	1.47	290.39	-0.86	313.24
Feb 1	233.34	2.24	193.53	1.51	298.07	-0.95	323.36
Feb 11	233.58	2.27	192.92	1.55	305.80	-1.02	333.42
Feb 21	233.64	2.31	192.04	1.58	313.56	-1.09	343.45
Mar 3	233.54	2.34	190.95	1.60	321.33	-1.15	353.41
Mar 13	233.28	2.37	189.72	1.62	329.11	-1.20	3.32
Mar 23	232.86	2.39	188.44	1.62	336.89	-1.24	13.18
Apr 2	232.32	2.41	187.19	1.61	344.65	-1.27	22.98
Apr 12	231.68	2.42	186.07	1.59	352.39	-1.28	32.72
Apr 22	230.97	2.42	185.14	1.56	0.08	-1.28	42.41
May 2	230.22	2.42	184.45	1.52	7.71	-1.26	52.06
May 12	229.49	2.41	184.05	1.48	15.28	-1.23	61.67
May 22	228.79	2.39	183.95	1.44	22.76	-1.18	71.25
Jun 1	228.18	2.36	184.15	1.39	30.15	-1.12	80.80
Jun 11	227.67	2.33	184.64	1.35	37.44	-1.04	90.33
Jun 21	227.30	2.29	185.39	1.31	44.61	-0.95	99.87
Jul 1	227.07	2.25	186.40	1.27	51.64	-0.85	109.40
Jul 11	227.00	2.21	187.63	1.23	58.53	-0.73	118.94
Jul 21	227.09	2.16	189.06	1.20	65.27	-0.60	128.51
Jul 31	227.34	2.12	190.66	1.17	71.83	-0.46	138.11
Aug 10	227.75	2.08	192.41	1.15	78.19	-0.30	147.73
Aug 20	228.30	2.04	194.28	1.13	84.33	-0.12	157.41
Aug 30	228.99	2.00	196.26	1.11	90.21	0.07	167.13
Sep 9	229.80	1.97	198.32	1.09	95.80	0.28	176.90
Sep 19	230.71	1.94	200.44	1.08	101.04	0.51	186.74
Sep 29	231.71	1.91	202.60	1.07	105.87	0.77	196.62
Oct 9	232.79	1.89	204.78	1.07	110.20	1.05	206.57
Oct 19	233.91	1.88	206.96	1.07	113.91	1.37	216.57
Oct 29	235.08	1.86	209.11	1.07	116.88	1.73	226.62
Nov 8	236.26	1.85	211.21	1.07	118.91	2.13	236.71
Nov 18	237.45	1.85	213.25	1.08	119.83	2.56	246.85
Nov 28	238.61	1.85	215.19	1.09	119.44	3.03	257.01
Dec 8	239.73	1.85	217.00	1.11	117.67	3.48	267.19
Dec 18	240.79	1.86	218.67	1.13	114.64	3.87	277.39
1661							

GREGORIAN 12 UT	MOON LONG.	MOON LONG.	MOON LAT.	MOON LAT.	VENUS LONG.	VENUS LONG.	VENUS LAT.	VENUS LAT.	MERCURY LONG.	MERCURY LONG.	MERCURY LAT.	MERCURY LAT.
1660												1660
Jan 3		167.2		-4.1		292.17		-1.14		300.65		-1.54
8 Jan 13	232.4	307.3	0.9	5.0	298.45	304.73	-1.26	-1.36	306.97	310.83	-0.71	0.52
18 Jan 23	18.8	80.5	1.9	-3.2	311.00	317.26	-1.44	-1.49	310.61	306.04	2.02	3.26
28 J-F 2	140.0	201.6	-4.9	-1.5	323.52	329.77	-1.51	-1.51	300.02	296.20	3.63	3.13
7 Feb 12	270.1	345.3	3.9	4.0	336.01	342.24	-1.47	-1.41	295.71	297.91	2.21	1.20
17 Feb 22	52.9	112.9	-1.3	-4.9	348.46	354.67	-1.32	-1.20	301.95	307.22	0.27	-0.53
27 Feb	173.5		-3.4		0.86		-1.06		313.37		-1.20	
Mar 3		238.1		1.9		7.04		-0.89		320.19		-1.71
8 Mar 13	309.2	21.6	5.1	1.2	13.20	19.34	-0.71	-0.50	327.58	335.51	-2.07	-2.25
18 Mar 23	85.5	145.1	-4.0	-4.9	25.47	31.57	-0.28	-0.04	343.95	352.94	-2.24	-2.04
28 M-A 2	208.3	276.5	-0.6	4.6	37.65	43.71	0.20	0.45	2.48	12.56	-1.63	-1.00
7 Apr 12	348.0	56.1	3.8	-1.9	49.75	55.76	0.70	0.95	23.00	33.41	-0.19	0.72
17 Apr 22	117.2	177.9	-5.2	-3.2	61.74	67.69	1.19	1.42	43.17	51.66	1.59	2.26
27 Apr	244.8		2.6		73.60		1.63		58.49		2.61	
May 2		316.0		5.2		79.48		1.83		63.44		2.56
7 May 12	25.2	89.1	0.9	-4.3	85.33	91.13	1.99	2.13	66.39	67.26	2.08	1.16
17 May 22	149.0	212.0	-4.8	-0.2	96.89	102.59	2.24	2.31	66.20	63.79	-0.12	-1.57
27 M-J 1	283.2	354.9	4.8	3.3	108.24	113.83	2.34	2.32	61.05	59.11	-2.87	-3.77
6 Jun 11	60.5	121.5	-2.3	-5.1	119.35	124.79	2.26	2.15	58.76	60.27	-4.15	-4.05
16 Jun 21	181.4	247.9	-2.7	2.9	130.15	135.39	1.98	1.76	63.62	68.68	-3.56	-2.80
26 Jun	322.5		4.8		140.53		1.47		75.33		-1.85	
Jul 1		31.9		0.0		145.52		1.13		83.46		-0.81
6 Jul 11	94.2	153.8	-4.4	-4.2	150.36	155.02	0.72	0.25	92.90	103.24	0.19	1.01
16 Jul 21	215.2	285.9	0.4	4.8	159.46	163.63	-0.29	-0.89	113.90	124.33	1.56	1.78
26 Jul 31	1.0	67.0	2.4	-3.1	167.49	170.98	-1.57	-2.32	134.22	143.45	1.72	1.42
5 Aug 10	127.2	186.8	-5.0	-1.8	174.01	176.49	-3.13	-4.01	152.02	159.95	0.95	0.35
15 Aug 20	251.0	324.9	3.5	4.5	178.31	179.33	-4.93	-5.88	167.24	173.86	-0.35	-1.11
25 Aug 30	37.5	100.3	-0.9	-4.9	179.45	178.60	-6.81	-7.64	179.70	184.58	-1.88	-2.65
4 Sep 9	159.8	221.2	-3.7	1.4	176.79	174.20	-8.30	-8.66	188.19	189.99	-3.33	-3.84
14 Sep 19	288.9	3.2	5.2	2.0	171.18	168.20	-8.67	-8.28	189.34	185.83	-3.99	-3.53
24 Sep 29	72.0	132.7	-3.8	-5.0	165.73	164.11	-7.56	-6.62	180.42	176.00	-2.30	-0.64
4 Oct 9	193.0	257.6	-1.2	4.2	163.49	163.86	-5.57	-4.49	175.35	178.83	0.82	1.71
14 Oct 19	328.0	39.9	4.5	-1.3	165.14	167.21	-3.45	-2.47	185.13	192.83	2.03	1.95
24 Oct 29	104.8	164.5	-5.2	-3.5	169.95	173.25	-1.58	-0.78	201.04	209.33	1.62	1.16
3 Nov 8	227.3	296.1	1.9	5.2	177.01	181.15	-0.07	0.54	217.53	225.60	0.62	0.06
13 Nov 18	6.8	74.8	1.7	-4.0	185.61	190.33	1.07	1.52	233.55	241.40	-0.49	-1.01
23 Nov 28	136.7	196.7	-4.8	-0.7	195.27	200.39	1.89	2.18	249.19	256.94	-1.48	-1.86
3 Dec 8	263.5	335.4	4.4	3.9	205.68	211.09	2.41	2.56	264.66	272.30	-2.14	-2.27
13 Dec 18	44.2	108.4	-1.7	-5.0	216.62	222.24	2.66	2.69	279.72	286.58	-2.20	-1.87
23 Dec 28	168.2	230.1	-2.9	2.3	227.94	233.71	2.68	2.61	292.17	295.14	-1.16	0.01
1661												1661
2 Jan 7	301.5	14.1	4.9	0.6	239.54	245.42	2.50	2.35	293.85	288.34	1.56	2.94
12 Jan 17	79.8	141.0	-4.3	-4.3	251.34	257.29	2.17	1.96	282.28	279.30	3.46	3.10
22 Jan 27	200.4	265.4	0.1	4.6	263.28	269.29	1.73	1.47	279.88	283.01	2.31	1.41
1 Feb 6	340.2	51.1	3.1	-2.8	275.33	281.38	1.21	0.93	287.76	293.54	0.55	-0.22
11 Feb 16	113.9	173.4	-5.1	-2.1	287.45	293.53	0.66	0.38	300.03	307.07	-0.89	-1.43
21 Feb 26	234.0	302.8	3.1	5.0	299.62	305.72	0.11	-0.15	314.57	322.51	-1.83	-2.09
3 Mar 8	18.3	86.2	-0.1	-4.9	311.82	317.94	-0.39	-0.62	330.90	339.76	-2.18	-2.08
13 Mar 18	146.7	206.3	-4.0	0.9	324.05	330.17	-0.83	-1.02	349.11	358.93	-1.78	-1.26
23 Mar 28	269.7	341.4	5.1	3.0	336.29	342.42	-1.18	-1.31	9.08	19.19	-0.52	0.39
2 Apr 7	55.0	119.6	-3.3	-5.2	348.55	354.67	-1.42	-1.49	28.64	36.66	1.35	2.19
12 Apr 17	179.0	240.5	-1.6	3.7	0.80	6.92	-1.54	-1.56	42.69	46.43	2.76	2.91
22 Apr 27	307.6	20.0	4.9	-0.3	13.05	19.17	-1.55	-1.51	47.75	46.82	2.57	1.70
2 May 7	90.0	151.7	-5.0	-3.7	25.30	31.42	-1.44	-1.35	44.28	41.29	0.42	-1.02
12 May 17	211.6	276.8	1.4	5.1	37.54	43.67	-1.24	-1.10	39.09	38.49	-2.29	-3.19
22 May 27	346.7	57.6	2.5	-3.4	49.79	55.92	-0.95	-0.78	39.76	42.79	-3.64	-3.69
1 Jun 6	123.4	183.3	-4.8	-1.0	62.04	68.17	-0.60	-0.41	47.39	53.39	-3.40	-2.83
11 Jun 16	245.6	315.1	4.0	4.3	74.30	80.44	-0.21	-0.01	60.67	69.18	-2.05	-1.13
21 Jun 26	25.8	93.5	-1.1	-4.9	86.57	92.71	0.19	0.38	78.82	89.31	-0.17	0.71
1 Jul 6	155.7	215.3	-3.0	2.0	98.86	105.01	0.57	0.74	100.15	110.79	1.38	1.76
11 Jul 16	281.5	354.2	5.0	1.4	111.16	117.31	0.90	1.05	120.85	130.19	1.83	1.63
21 Jul 26	63.6	127.8	-4.1	-4.4	123.47	129.64	1.17	1.28	138.79	146.66	1.21	0.63
31 J-A 5	187.6	248.7	-0.1	4.4	135.81	141.98	1.36	1.41	153.78	160.09	-0.09	-0.90
10 Aug 15	319.2	33.0	3.9	-2.2	148.15	154.33	1.44	1.44	165.48	169.71	-1.77	-2.65
20 Aug 25	99.7	160.7	-5.2	-2.4	160.51	166.69	1.41	1.36	172.45	173.20	-3.47	-4.12
30 Aug	219.9		2.8		172.87		1.28		171.48		-4.37	
Sep 4		283.9		5.2		179.05		1.17		167.40		-3.97
9 Sep 14	357.8	70.3	0.9	-4.7	185.22	191.40	1.04	0.88	162.44	159.26	-2.81	-1.19
19 Sep 24	133.8	193.1	-4.3	0.5	197.58	203.75	0.70	0.51	159.81	164.13	0.31	1.34
29 S-O 4	253.6	321.1	4.9	3.8	209.92	216.09	0.30	0.08	171.07	179.31	1.85	1.92
9 Oct 14	36.1	105.7	-2.5	-5.2	222.25	228.41	-0.15	-0.38	187.97	196.62	1.71	1.31
19 Oct 24	166.5	226.0	-1.9	3.3	234.57	240.72	-0.61	-0.84	205.09	213.34	0.81	0.27
29 Oct	289.3		5.1		246.87		-1.07		221.38		-0.29	
Nov 3		359.4		0.7		253.01		-1.28		229.24		-0.84
8 Nov 13	73.3	139.1	-4.7	-3.8	259.14	265.26	-1.47	-1.65	236.96	244.56	-1.34	-1.78
18 Nov 23	198.5	260.1	1.1	4.9	271.37	277.48	-1.80	-1.92	252.03	259.31	-2.13	-2.35
28 N-D 3	326.9	38.1	3.2	-2.8	283.56	289.63	-2.02	-2.08	266.25	272.50	-2.40	-2.18
8 Dec 13	108.7	171.0	-4.9	-1.1	295.68	301.71	-2.10	-2.09	277.31	279.35	-1.60	-0.53
18 Dec 23	230.7	296.1	3.7	4.6	307.71	313.68	-2.03	-1.93	277.05	270.89	1.02	2.52
28 Dec	5.8		-0.3		319.60		-1.79		265.11		3.20	
1661												1661

JULIAN 12 UT	SATURN LONG.	SATURN LAT.	JUPITER LONG.	JUPITER LAT.	MARS LONG.	MARS LAT.	SUN LONG.
1661							
Dec 28	241.78	1.87	220.16	1.15	110.81	4.13	287.58
Jan 7	242.67	1.89	221.44	1.17	106.89	4.23	297.76
Jan 17	243.45	1.91	222.49	1.20	103.62	4.15	307.93
Jan 27	244.09	1.93	223.27	1.23	101.50	3.96	318.06
Feb 6	244.59	1.95	223.76	1.26	100.70	3.70	328.15
Feb 16	244.93	1.98	223.95	1.29	101.15	3.42	338.20
Feb 26	245.11	2.00	223.82	1.32	102.68	3.14	348.19
Mar 8	245.12	2.03	223.39	1.34	105.10	2.87	358.13
Mar 18	244.97	2.05	222.67	1.36	108.23	2.62	8.02
Mar 28	244.66	2.07	221.71	1.38	111.93	2.40	17.84
Apr 7	244.21	2.09	220.56	1.38	116.07	2.19	27.61
Apr 17	243.63	2.10	219.31	1.38	120.59	1.99	37.34
Apr 27	242.97	2.10	218.04	1.37	125.40	1.81	47.00
May 7	242.25	2.10	216.84	1.34	130.46	1.65	56.63
May 17	241.51	2.09	215.79	1.31	135.73	1.49	66.23
May 27	240.79	2.08	214.96	1.27	141.18	1.34	75.79
Jun 6	240.12	2.05	214.38	1.23	146.78	1.20	85.34
Jun 16	239.54	2.03	214.10	1.19	152.53	1.06	94.87
Jun 26	239.07	1.99	214.11	1.14	158.41	0.93	104.40
Jul 6	238.73	1.96	214.43	1.10	164.41	0.80	113.94
Jul 16	238.55	1.92	215.03	1.05	170.53	0.68	123.50
Jul 26	238.52	1.88	215.89	1.01	176.75	0.56	133.07
Aug 5	238.66	1.84	217.00	0.97	183.09	0.44	142.68
Aug 15	238.95	1.80	218.33	0.94	189.53	0.33	152.34
Aug 25	239.40	1.76	219.86	0.91	196.08	0.21	162.03
Sep 4	239.99	1.73	221.55	0.88	202.74	0.10	171.78
Sep 14	240.71	1.69	223.39	0.85	209.50	-0.01	181.58
Sep 24	241.55	1.66	225.35	0.82	216.36	-0.11	191.43
Oct 4	242.48	1.64	227.40	0.80	223.33	-0.22	201.35
Oct 14	243.51	1.61	229.54	0.79	230.40	-0.32	211.32
Oct 24	244.59	1.59	231.72	0.77	237.57	-0.42	221.34
Nov 3	245.73	1.58	233.94	0.76	244.84	-0.51	231.42
Nov 13	246.90	1.56	236.17	0.75	252.21	-0.60	241.53
Nov 23	248.08	1.55	238.39	0.74	259.66	-0.69	251.68
Dec 3	249.25	1.55	240.57	0.73	267.20	-0.77	261.86
Dec 13	250.40	1.55	242.69	0.73	274.81	-0.84	272.05
1662							
Dec 23	251.49	1.55	244.73	0.73	282.50	-0.90	282.24
Jan 2	252.53	1.55	246.67	0.73	290.25	-0.96	292.43
Jan 12	253.48	1.56	248.46	0.73	298.04	-1.01	302.60
Jan 22	254.32	1.57	250.10	0.73	305.87	-1.05	312.75
Feb 1	255.05	1.59	251.55	0.74	313.73	-1.08	322.87
Feb 11	255.64	1.60	252.77	0.74	321.61	-1.09	332.94
Feb 21	256.09	1.62	253.75	0.75	329.49	-1.10	342.96
Mar 3	256.37	1.63	254.46	0.76	337.35	-1.09	352.93
Mar 13	256.50	1.65	254.87	0.77	345.20	-1.07	2.84
Mar 23	256.46	1.67	254.98	0.77	353.00	-1.04	12.70
Apr 2	256.25	1.68	254.77	0.78	0.77	-1.00	22.50
Apr 12	255.90	1.69	254.26	0.78	8.47	-0.94	32.24
Apr 22	255.41	1.70	253.48	0.78	16.11	-0.88	41.94
May 2	254.81	1.71	252.46	0.77	23.67	-0.80	51.59
May 12	254.13	1.70	251.29	0.76	31.15	-0.72	61.20
May 22	253.40	1.70	250.02	0.74	38.54	-0.62	70.78
Jun 1	252.67	1.68	248.76	0.72	45.83	-0.52	80.33
Jun 11	251.96	1.67	247.60	0.69	53.02	-0.41	89.87
Jun 21	251.32	1.64	246.60	0.66	60.11	-0.29	99.40
Jul 1	250.77	1.62	245.83	0.62	67.08	-0.17	108.94
Jul 11	250.34	1.59	245.33	0.59	73.94	-0.04	118.48
Jul 21	250.04	1.55	245.14	0.55	80.68	0.09	128.05
Jul 31	249.91	1.52	245.25	0.52	87.30	0.23	137.64
Aug 10	249.93	1.48	245.67	0.49	93.78	0.37	147.27
Aug 20	250.11	1.45	246.37	0.45	100.12	0.52	156.94
Aug 30	250.45	1.41	247.35	0.42	106.31	0.68	166.66
Sep 9	250.94	1.38	248.57	0.39	112.32	0.84	176.43
Sep 19	251.57	1.35	250.01	0.37	118.15	1.01	186.26
Sep 29	252.33	1.32	251.64	0.34	123.76	1.20	196.15
Oct 9	253.20	1.29	253.44	0.32	129.12	1.39	206.08
Oct 19	254.16	1.27	255.38	0.30	134.19	1.60	216.08
Oct 29	255.21	1.25	257.44	0.28	138.91	1.83	226.13
Nov 8	256.31	1.23	259.59	0.26	143.20	2.08	236.22
Nov 18	257.46	1.22	261.81	0.24	146.97	2.35	246.35
Nov 28	258.63	1.20	264.08	0.22	150.10	2.65	256.52
Dec 8	259.80	1.20	266.37	0.21	152.44	2.98	266.70
Dec 18	260.97	1.19	268.67	0.19	153.81	3.34	276.89
1663							

GREGORIAN 12 UT	MOON LONG.		MOON LAT.		VENUS LONG.		VENUS LAT.		MERCURY LONG.		MERCURY LAT.	
1662												1662
Jan 2		76.1		-4.8		325.49		-1.60		263.08		3.01
7 Jan 12	142.4	202.4	-3.1	1.9	331.32	337.08	-1.36	-1.09	264.67	268.64	2.37	1.58
17 Jan 22	264.1	333.7	5.0	2.2	342.78	348.40	-0.76	-0.40	274.00	280.22	0.79	0.05
27 J-F 1	44.9	112.4	-3.7	-4.7	353.92	359.33	0.00	0.45	287.01	294.22	-0.61	-1.16
6 Feb 11	174.8	234.2	-0.3	4.3	4.61	9.74	0.92	1.43	301.80	309.74	-1.61	-1.92
16 Feb 21	299.1	12.2	4.5	-1.3	14.69	19.43	1.97	2.53	318.05	326.75	-2.09	-2.09
26 Feb	82.9		-5.2		23.93		3.11		335.87		-1.90	
Mar 3		147.0		-2.7		28.11		3.69		345.40		-1.49
8 Mar 13	206.8	267.2	2.5	5.3	31.94	35.33	4.29	4.87	355.21	4.97	-0.84	0.04
18 Mar 23	335.8	50.8	2.0	-4.2	38.20	40.43	5.43	5.95	14.01	21.48	1.05	2.05
28 M-A 2	119.1	180.0	-4.5	0.2	41.91	42.49	6.39	6.72	26.58	28.89	2.83	3.20
7 Apr 12	239.3	302.1	4.6	4.3	42.08	40.66	6.88	6.80	28.41	25.77	3.01	2.20
17 Apr 22	13.9	88.3	-1.4	-5.2	38.33	35.38	6.44	5.76	22.29	19.51	0.94	-0.46
27 Apr	153.1		-2.2		32.25		4.81		18.41		-1.69	
May 2		212.5		3.0		29.44		3.67		19.28		-2.58
7 May 12	273.1	339.0	5.1	1.6	27.36	26.23	2.48	1.33	21.98	26.23	-3.11	-3.30
17 May 22	52.6	124.0	-4.2	-4.0	26.10	26.91	0.29	-0.61	31.78	38.48	-3.19	-2.81
27 M-J 1	185.7	245.3	0.9	4.7	28.55	30.87	-1.36	-1.96	46.25	55.04	-2.20	-1.41
6 Jun 11	308.7	17.5	3.6	-2.0	33.78	37.17	-2.44	-2.80	64.80	75.36	-0.51	0.39
16 Jun 21	90.7	157.7	-5.0	-1.3	40.95	45.05	-3.06	-3.22	86.30	97.08	1.16	1.68
26 Jun	217.6		3.6		49.42		-3.31		107.27		1.90	
Jul 1		279.2		4.8		54.01		-3.33		116.67		1.82
6 Jul 11	346.2	56.8	0.5	-4.6	58.79	63.74	-3.28	-3.18	125.24	132.96	1.48	0.94
16 Jul 21	127.0	190.0	-3.5	1.7	68.82	74.02	-3.04	-2.85	139.82	145.72	0.21	-0.65
26 Jul 31	249.7	314.7	5.1	3.0	79.33	84.73	-2.63	-2.38	150.51	153.95	-1.61	-2.62
5 Aug 10	25.0	95.4	-3.0	-5.0	90.21	95.77	-2.10	-1.81	155.66	155.25	-3.59	-4.36
15 Aug 20	161.3	221.6	-0.7	4.2	101.40	107.09	-1.51	-1.20	152.60	148.35	-4.69	-4.32
25 Aug 30	282.8	351.9	4.8	-0.3	112.83	118.63	-0.88	-0.57	144.30	142.58	-3.21	-1.69
4 Sep 9	64.4	132.1	-5.1	-3.2	124.47	130.37	-0.27	0.02	144.35	149.47	-0.20	0.94
14 Sep 19	194.2	253.6	2.2	5.3	136.30	142.28	0.30	0.55	156.99	165.73	1.62	1.86
24 Sep 29	317.4	30.3	2.8	-3.5	148.29	154.34	0.79	1.00	174.84	183.85	1.76	1.45
4 Oct 9	102.6	166.7	-4.8	-0.2	160.42	166.53	1.18	1.33	192.59	201.02	1.00	0.48
14 Oct 19	226.5	286.6	4.4	4.6	172.66	178.82	1.45	1.54	209.16	217.05	-0.08	-0.65
24 Oct 29	353.6	69.1	-0.5	-5.1	185.00	191.20	1.60	1.63	224.71	232.16	-1.19	-1.69
3 Nov 8	138.8	199.8	-2.5	2.7	197.41	203.64	1.62	1.59	239.38	246.30	-2.11	-2.42
13 Nov 18	259.2	321.2	5.0	2.1	209.89	216.14	1.52	1.43	252.75	258.35	-2.57	-2.49
23 Nov 28	31.5	106.8	-3.6	-4.3	222.40	228.67	1.31	1.18	262.34	263.42	-2.04	-1.08
3 Dec 8	172.7	232.2	0.7	4.6	234.94	241.22	1.02	0.84	260.19	253.67	0.43	2.02
13 Dec 18	292.9	357.8	3.9	-1.3	247.51	253.79	0.65	0.46	248.40	247.36	2.87	2.87
23 Dec 28	70.3	142.6	-5.0	-1.7	260.08	266.36	0.25	0.05	249.91	254.63	2.39	1.72
1663												1663
2 Jan 7	205.1	264.8	3.5	4.9	272.65	278.94	-0.15	-0.35	260.56	267.18	1.00	0.30
12 Jan 17	328.1	36.3	1.1	-4.3	285.22	291.51	-0.54	-0.72	274.23	281.60	-0.35	-0.92
22 Jan 27	108.5	176.3	-4.1	1.5	297.79	304.06	-0.89	-1.03	289.24	297.15	-1.39	-1.76
1 Feb 6	236.8	298.0	5.1	3.5	310.34	316.61	-1.16	-1.26	305.36	313.90	-2.00	-2.08
11 Feb 16	5.1	75.6	-2.3	-5.2	322.87	329.13	-1.35	-1.40	322.78	331.99	-1.99	-1.68
21 Feb 26	145.1	208.8	-1.3	4.0	335.39	341.63	-1.43	-1.44	341.42	350.76	-1.13	-0.33
3 Mar 8	268.5	332.5	5.0	0.7	347.87	354.10	-1.41	-1.36	359.34	6.17	0.70	1.82
13 Mar 18	43.6	114.4	-4.7	-3.8	0.33	6.54	-1.29	-1.19	10.23	10.98	2.81	3.41
23 Mar 28	179.9	240.7	1.8	5.1	12.75	18.94	-1.07	-0.92	8.71	4.76	3.37	2.62
2 Apr 7	301.0	8.8	3.2	-2.5	25.13	31.31	-0.76	-0.58	1.08	359.12	1.39	0.05
12 Apr 17	82.8	151.2	-5.0	-0.7	37.47	43.62	-0.39	-0.18	359.33	1.54	-1.12	-2.02
22 Apr 27	213.3	272.9	4.2	4.6	49.76	55.89	0.03	0.24	5.38	10.53	-2.61	-2.91
2 May 7	335.0	46.9	0.2	-4.7	62.01	68.12	0.46	0.67	16.78	23.98	-2.95	-2.73
12 May 17	121.1	186.0	-2.9	2.5	74.21	80.29	0.87	1.07	32.07	41.04	-2.28	-1.63
22 May 27	246.1	306.0	5.0	2.4	86.35	92.40	1.24	1.40	50.86	61.42	-0.82	0.06
1 Jun 6	11.1	85.9	-3.0	-4.7	98.44	104.46	1.54	1.66	72.38	83.23	0.90	1.55
11 Jun 16	157.4	219.3	0.3	4.6	110.47	116.45	1.74	1.80	93.48	102.87	1.92	1.98
21 Jun 26	278.9	340.6	4.1	-0.8	122.42	128.36	1.82	1.81	111.31	118.78	1.76	1.27
1 Jul 6	49.3	124.1	-5.0	-2.4	134.29	140.18	1.77	1.68	125.24	130.56	0.55	-0.35
11 Jul 16	191.6	251.8	3.3	5.1	146.06	151.90	1.56	1.40	134.57	136.99	-1.41	-2.55
21 Jul 26	312.4	17.2	1.7	-3.8	157.70	163.47	1.20	0.97	137.51	135.97	-3.65	-4.53
31 J-A 5	88.6	160.5	-4.6	0.9	169.21	174.90	0.70	0.40	132.69	128.83	-4.89	-4.54
10 Aug 15	224.3	284.0	5.1	3.8	180.54	186.12	0.06	-0.30	126.13	125.98	-3.51	-2.10
20 Aug 25	347.2	55.9	-1.5	-5.3	191.64	197.09	-0.69	-1.10	128.96	134.85	-0.67	0.53
30 Aug	127.2		-2.1		202.46		-1.52		142.88		1.34	
Sep 4		195.0		3.7		207.74		-1.96		152.08		1.74
9 Sep 14	256.1	316.8	5.1	1.3	212.92	217.96	-2.41	-2.86	161.60	170.97	1.79	1.57
19 Sep 24	24.1	95.3	-4.2	-4.3	222.86	227.58	-3.30	-3.73	179.99	188.61	1.19	0.69
29 S-O 4	164.1	228.0	1.2	5.0	232.10	236.36	-4.14	-4.52	196.86	204.78	0.13	-0.45
9 Oct 14	287.8	350.9	3.4	-1.9	240.32	243.90	-4.86	-5.14	212.39	219.69	-1.03	-1.58
19 Oct 24	62.7	134.0	-5.1	-1.3	247.02	249.58	-5.34	-5.45	226.66	233.21	-2.08	-2.48
29 Oct	199.3		3.9		251.47		-5.44		239.15		-2.74	
Nov 3		260.4		4.6		252.54		-5.26		244.06		-2.78
8 Nov 13	320.0	26.8	0.6	-4.3	252.69	251.81	-4.89	-4.28	247.18	247.26	-2.48	-1.65
18 Nov 23	101.7	170.9	-3.5	2.2	249.96	247.34	-3.41	-2.31	243.22	236.60	-0.20	1.45
28 N-D 3	233.1	292.6	5.0	2.6	244.35	241.48	-1.05	0.24	232.06	231.99	2.48	2.69
8 Dec 13	353.7	64.5	-2.5	-5.0	239.20	237.82	1.44	2.47	235.43	240.86	2.38	1.82
18 Dec 23	139.7	205.8	-0.3	4.6	237.46	238.11	3.27	3.87	247.32	254.31	1.18	0.52
28 Dec	265.9		4.3		239.66		4.27		261.61		-0.11	
1663												1663

JULIAN 12 UT	SATURN LONG.	SATURN LAT.	JUPITER LONG.	JUPITER LAT.	MARS LONG.	MARS LAT.	SUN LONG.
1663							
Dec 28	262.09	1.19	270.94	0.18	154.02	3.70	287.09
Jan 7	263.17	1.18	273.18	0.16	152.94	4.05	297.27
Jan 17	264.17	1.18	275.34	0.15	150.55	4.33	307.43
Jan 27	265.08	1.19	277.41	0.13	147.10	4.50	317.57
Feb 6	265.88	1.19	279.37	0.12	143.17	4.49	327.66
Feb 16	266.56	1.20	281.19	0.11	139.45	4.32	337.71
Feb 26	267.11	1.21	282.83	0.09	136.59	4.01	347.71
Mar 7	267.50	1.22	284.28	0.07	134.96	3.63	357.65
Mar 17	267.73	1.22	285.51	0.06	134.62	3.23	7.54
Mar 27	267.80	1.23	286.49	0.04	135.47	2.84	17.37
Apr 6	267.70	1.24	287.19	0.02	137.33	2.48	27.15
Apr 16	267.45	1.25	287.59	0.00	140.03	2.15	36.87
Apr 26	267.05	1.25	287.69	-0.02	143.40	1.85	46.54
May 6	266.53	1.25	287.47	-0.04	147.31	1.58	56.17
May 16	265.90	1.25	286.94	-0.07	151.68	1.34	65.77
May 26	265.21	1.24	286.14	-0.09	156.42	1.12	75.34
Jun 5	264.48	1.23	285.11	-0.12	161.47	0.92	84.88
Jun 15	263.75	1.21	283.91	-0.14	166.79	0.73	94.41
Jun 25	263.06	1.20	282.64	-0.17	172.35	0.55	103.95
Jul 5	262.44	1.17	281.37	-0.19	178.11	0.39	113.48
Jul 15	261.92	1.15	280.20	-0.21	184.07	0.23	123.04
Jul 25	261.53	1.12	279.21	-0.23	190.20	0.09	132.61
Aug 4	261.29	1.09	278.47	-0.25	196.49	-0.05	142.22
Aug 14	261.19	1.06	278.01	-0.26	202.93	-0.18	151.87
Aug 24	261.27	1.03	277.86	-0.28	209.53	-0.30	161.56
Sep 3	261.50	1.00	278.03	-0.29	216.26	-0.41	171.30
Sep 13	261.89	0.98	278.52	-0.30	223.13	-0.52	181.10
Sep 23	262.42	0.95	279.30	-0.31	230.13	-0.62	190.95
Oct 3	263.10	0.92	280.37	-0.32	237.25	-0.71	200.86
Oct 13	263.89	0.90	281.68	-0.33	244.49	-0.79	210.83
Oct 23	264.80	0.88	283.21	-0.34	251.85	-0.87	220.85
Nov 2	265.79	0.86	284.94	-0.35	259.31	-0.93	230.92
Nov 12	266.86	0.84	286.84	-0.36	266.87	-0.99	241.03
Nov 22	267.98	0.83	288.88	-0.37	274.51	-1.04	251.18
Dec 2	269.14	0.81	291.03	-0.38	282.23	-1.07	261.35
Dec 12	270.32	0.80	293.27	-0.39	290.02	-1.09	271.54
1664							
Dec 22	271.49	0.79	295.57	-0.40	297.85	-1.11	281.74
Jan 1	272.64	0.78	297.92	-0.42	305.73	-1.11	291.93
Jan 11	273.75	0.78	300.29	-0.43	313.63	-1.09	302.11
Jan 21	274.80	0.77	302.65	-0.45	321.54	-1.07	312.26
Jan 31	275.77	0.77	304.99	-0.47	329.45	-1.04	322.37
Feb 10	276.64	0.76	307.28	-0.48	337.34	-0.99	332.45
Feb 20	277.41	0.76	309.50	-0.51	345.20	-0.93	342.48
Mar 2	278.04	0.76	311.62	-0.53	353.02	-0.87	352.45
Mar 12	278.53	0.76	313.63	-0.56	0.78	-0.79	2.37
Mar 22	278.87	0.76	315.50	-0.58	8.49	-0.71	12.23
Apr 1	279.05	0.76	317.20	-0.62	16.12	-0.62	22.03
Apr 11	279.06	0.76	318.71	-0.65	23.68	-0.53	31.78
Apr 21	278.92	0.76	319.99	-0.69	31.16	-0.43	41.48
May 1	278.62	0.76	321.03	-0.72	38.55	-0.33	51.13
May 11	278.18	0.75	321.79	-0.77	45.85	-0.22	60.74
May 21	277.62	0.75	322.26	-0.81	53.06	-0.11	70.32
May 31	276.98	0.74	322.41	-0.85	60.17	0.00	79.88
Jun 10	276.27	0.73	322.25	-0.90	67.20	0.11	89.42
Jun 20	275.53	0.71	321.77	-0.94	74.13	0.22	98.95
Jun 30	274.81	0.70	321.00	-0.98	80.98	0.33	108.48
Jul 10	274.14	0.68	319.98	-1.02	87.73	0.44	118.03
Jul 20	273.55	0.66	318.79	-1.05	94.39	0.55	127.59
Jul 30	273.07	0.64	317.50	-1.07	100.95	0.67	137.18
Aug 9	272.72	0.62	316.20	-1.09	107.43	0.78	146.81
Aug 19	272.52	0.59	315.00	-1.09	113.81	0.89	156.47
Aug 29	272.48	0.57	313.97	-1.09	120.10	1.00	166.19
Sep 8	272.60	0.55	313.20	-1.09	126.28	1.12	175.96
Sep 18	272.89	0.53	312.72	-1.07	132.35	1.23	185.78
Sep 28	273.33	0.51	312.56	-1.06	138.30	1.35	195.66
Oct 8	273.91	0.49	312.74	-1.04	144.11	1.48	205.60
Oct 18	274.63	0.47	313.25	-1.02	149.77	1.60	215.59
Oct 28	275.47	0.45	314.06	-1.00	155.26	1.73	225.64
Nov 7	276.41	0.43	315.17	-0.98	160.54	1.87	235.73
Nov 17	277.44	0.42	316.53	-0.97	165.58	2.02	245.86
Nov 27	278.53	0.40	318.12	-0.95	170.33	2.17	256.02
Dec 7	279.67	0.39	319.91	-0.94	174.71	2.34	266.20
Dec 17	280.84	0.38	321.86	-0.93	178.66	2.52	276.39
1665							

GREGORIAN 12 UT	MOON LONG.		MOON LAT.		VENUS LONG.		VENUS LAT.		MERCURY LONG.		MERCURY LAT.	
1664												1664
Jan 2		325.5		-0.3		241.97		4.51		269.13		-0.68
7 Jan 12	29.5	103.0	-4.8	-3.3	244.92	248.39	4.61	4.60	276.83	284.72	-1.18	-1.59
17 Jan 22	175.9	239.0	3.0	5.2	252.28	256.53	4.50	4.32	292.83	301.18	-1.89	-2.05
27 J-F 1	298.4	359.9	2.1	-3.3	261.07	265.84	4.08	3.79	309.81	318.69	-2.05	-1.85
6 Feb 11	67.5	141.1	-5.1	0.0	270.80	275.92	3.46	3.11	327.72	336.59	-1.41	-0.69
16 Feb 21	210.3	271.1	4.9	4.1	281.18	286.54	2.73	2.34	344.64	350.76	0.32	1.52
26 Feb	331.3		-0.9		292.01		1.94		353.72		2.70	
Mar 2		36.5		-5.1		297.56		1.54		352.89		3.51
7 Mar 12	106.5	177.9	-3.0	3.1	303.17	308.84	1.14	0.76	349.04	344.38	3.60	2.91
17 Mar 22	243.1	302.8	5.1	1.6	314.56	320.32	0.39	0.04	341.15	340.32	1.74	0.49
27 M-A 1	5.6	74.9	-3.6	-4.6	326.12	331.95	-0.28	-0.58	341.78	345.09	-0.63	-1.51
6 Apr 11	145.4	213.2	0.5	4.9	337.81	343.69	-0.85	-1.09	349.80	355.61	-2.15	-2.54
16 Apr 21	275.1	335.0	3.5	-1.4	349.59	355.51	-1.29	-1.46	2.34	9.87	-2.69	-2.62
26 Apr	41.9		-5.0		1.45		-1.60		18.17		-2.32	
May 1		114.1		-2.0		7.40		-1.70		27.22		-1.81
6 May 11	182.9	247.0	3.6	4.7	13.37	19.34	-1.76	-1.79	37.03	47.52	-1.11	-0.27
16 May 21	306.8	8.6	0.8	-4.0	25.33	31.32	-1.79	-1.75	58.43	69.26	0.60	1.37
26 May 31	80.0	152.8	-4.1	1.6	37.33	43.34	-1.68	-1.59	79.49	88.79	1.89	2.12
5 Jun 10	218.8	279.8	5.1	2.8	49.37	55.40	-1.46	-1.32	97.00	104.06	2.02	1.62
15 Jun 20	339.1	44.4	-2.3	-5.2	61.44	67.49	-1.16	-0.98	109.92	114.43	0.94	0.01
25 Jun 30	118.8	189.9	-1.3	4.4	73.55	79.63	-0.78	-0.58	117.39	118.56	-1.13	-2.39
5 Jul 10	252.8	312.2	4.5	0.0	85.71	91.80	-0.37	-0.16	117.80	115.33	-3.62	-4.55
15 Jul 20	13.0	82.0	-4.6	-4.1	97.91	104.02	0.05	0.26	111.99	109.14	-4.93	-4.60
25 Jul 30	157.1	225.1	2.2	5.3	110.15	116.29	0.46	0.64	108.09	109.55	-3.68	-2.43
4 Aug 9	285.6	345.2	2.4	-2.9	122.44	128.60	0.81	0.97	113.68	120.27	-1.10	0.11
14 Aug 19	49.0	120.6	-5.2	-1.1	134.77	140.96	1.10	1.21	128.76	138.35	1.03	1.58
24 Aug 29	194.0	258.5	4.6	4.2	147.15	153.35	1.30	1.37	148.26	157.97	1.77	1.67
3 Sep 8	317.8	19.7	-0.5	-4.8	159.56	165.78	1.40	1.41	167.26	176.08	1.36	0.90
13 Sep 18	86.8	159.1	-3.7	2.4	172.00	178.23	1.40	1.35	184.45	192.39	0.36	-0.24
23 Sep 28	229.1	290.5	5.1	1.8	184.47	190.71	1.28	1.19	199.94	207.09	-0.86	-1.47
3 Oct 8	350.6	56.1	-3.3	-4.8	196.95	203.19	1.07	0.93	213.81	219.97	-2.03	-2.53
13 Oct 18	125.8	196.6	-0.3	4.6	209.44	215.69	0.78	0.60	225.36	229.54	-2.90	-3.07
23 Oct 28	262.3	322.0	3.6	-1.2	221.94	228.19	0.41	0.22	231.73	230.80	-2.91	-2.21
2 Nov 7	24.6	94.3	-4.9	-2.7	234.45	240.70	0.01	-0.19	226.08	219.66	-0.84	0.84
12 Nov 17	164.8	232.4	3.1	4.8	246.95	253.20	-0.40	-0.61	215.98	216.87	2.04	2.47
22 Nov 27	294.4	353.9	0.9	-3.9	259.45	265.69	-0.80	-0.99	221.15	227.24	2.33	1.90
2 Dec 7	60.3	133.2	-4.6	0.8	271.94	278.18	-1.16	-1.31	234.20	241.55	1.34	0.73
12 Dec 17	202.6	266.5	5.1	3.1	284.41	290.65	-1.44	-1.55	249.10	256.77	0.12	-0.46
22 Dec 27	326.1	27.2	-2.0	-5.2	296.87	303.09	-1.63	-1.69	264.53	272.40	-0.98	-1.43
1665												1665
1 Jan 6	97.5	171.7	-2.3	4.0	309.30	315.50	-1.71	-1.70	280.41	288.59	-1.78	-2.01
11 Jan 16	238.5	299.2	4.8	0.4	321.68	327.85	-1.66	-1.59	296.96	305.50	-2.10	-2.00
21 Jan 26	358.5	62.2	-4.4	-4.6	334.01	340.14	-1.48	-1.34	314.10	322.47	-1.66	-1.04
31 J-F 5	135.6	208.9	1.2	5.3	346.25	352.33	-1.17	-0.97	329.92	335.28	-0.10	1.14
10 Feb 15	272.4	331.6	2.7	-2.5	358.39	4.41	-0.74	-0.48	337.11	334.77	2.48	3.49
20 Feb 25	32.3	99.2	-5.2	-2.1	10.40	16.35	-0.20	0.09	329.72	324.96	3.71	3.09
2 Mar 7	173.9	244.1	4.0	4.4	22.25	28.10	0.41	0.73	322.59	322.94	2.01	0.84
12 Mar 17	304.9	4.5	-0.3	-4.5	33.90	39.63	1.06	1.40	325.49	329.67	-0.21	-1.07
22 Mar 27	68.1	137.7	-4.1	1.5	45.30	50.89	1.73	2.05	335.04	341.33	-1.73	-2.19
1 Apr 6	211.3	277.4	5.0	1.9	56.39	61.79	2.36	2.65	348.38	356.11	-2.44	-2.48
11 Apr 16	336.9	38.8	-3.1	-4.9	67.08	72.24	2.91	3.14	4.51	13.57	-2.32	-1.94
21 Apr 26	105.7	176.7	-1.1	4.3	77.25	82.09	3.34	3.48	23.31	33.67	-1.35	-0.59
1 May 6	247.0	309.4	3.9	-1.1	86.72	91.11	3.57	3.60	44.44	55.18	0.28	1.13
11 May 16	9.3	74.7	-4.8	-3.3	95.21	98.96	3.55	3.41	65.32	74.42	1.81	2.21
21 May 26	144.7	215.0	2.5	5.0	102.30	105.14	3.18	2.83	82.25	88.73	2.27	2.00
31 M-J 5	280.9	341.0	1.2	-3.8	107.36	108.86	2.36	1.73	93.74	97.14	1.38	0.44
10 Jun 15	42.8	112.2	-4.9	-0.2	109.52	109.23	0.95	0.00	98.75	98.48	-0.77	-2.14
20 Jun 25	183.9	251.5	4.9	3.6	107.94	105.74	-1.08	-2.26	96.55	93.66	-3.44	-4.39
30 Jun	313.5		-1.8		102.88		-3.42		90.98		-4.76	
Jul 5		13.0		-5.2		99.77		-4.46		89.65		-4.49
10 Jul 15	77.9	151.0	-3.1	3.2	96.90	94.69	-5.27	-5.82	90.36	93.33	-3.73	-2.66
20 Jul 25	222.0	285.9	5.0	0.7	93.38	93.05	-6.10	-6.15	98.51	105.74	-1.46	-0.30
30 J-A 4	345.6	46.3	-4.1	-4.9	93.67	95.13	-6.03	-5.78	114.62	124.54	0.69	1.38
9 Aug 14	114.8	189.9	0.1	5.1	97.33	100.13	-5.44	-5.03	134.79	144.82	1.72	1.74
19 Aug 24	258.1	318.9	3.0	-2.2	103.44	107.16	-4.57	-4.09	154.37	163.38	1.52	1.12
29 Aug	18.3		-5.1		111.24		-3.59		171.86		0.59	
Sep 3		81.2		-2.7		115.61		-3.08		179.83		-0.01
8 Sep 13	153.1	227.6	3.3	4.6	120.23	125.05	-2.57	-2.07	187.32	194.31	-0.66	-1.33
18 Sep 23	292.0	351.4	0.0	-4.4	130.06	135.23	-1.59	-1.12	200.74	206.48	-1.97	-2.57
28 S-O 3	52.3	118.0	-4.3	0.7	140.53	145.95	-0.67	-0.25	211.28	214.65	-3.06	-3.36
8 Oct 13	191.8	263.1	5.0	2.2	151.47	157.09	0.14	0.50	215.85	213.92	-3.34	-2.76
18 Oct 23	324.5	24.3	-2.9	-5.0	162.80	168.57	0.82	1.10	208.73	202.80	-1.47	0.22
28 Oct	87.8		-1.7		174.41		1.35		200.10		1.56	
Nov 2		156.5		3.8		180.31		1.55		201.93		2.19
7 Nov 12	229.7	296.5	4.2	-0.9	186.26	192.25	1.72	1.84	206.98	213.71	2.25	1.95
17 Nov 22	356.3	58.1	-4.8	-3.8	198.28	204.35	1.92	1.96	221.15	228.87	1.48	0.92
27 N-D 2	125.2	195.8	1.8	5.2	210.45	216.57	1.96	1.92	236.67	244.48	0.33	-0.24
7 Dec 12	265.6	328.5	1.7	-3.6	222.72	228.89	1.85	1.75	252.31	260.17	-0.78	-1.26
17 Dec 22	28.3	93.3	-5.1	-1.1	235.07	241.26	1.62	1.46	268.09	276.10	-1.66	-1.95
27 Dec	164.0		4.5		247.47		1.28		284.21		-2.12	
1665												1665

JULIAN	SATURN		JUPITER		MARS		SUN	GREGORIAN	MOON				VENUS				MERCURY			
12 UT	LONG.	LAT.	LONG.	LAT.	LONG.	LAT.	LONG.	12 UT	LONG.		LAT.		LONG.		LAT.		LONG.		LAT.	
1665								1666												1666
								Jan 1		234.2		4.1		253.68		1.09		292.40		-2.12
Dec 27	282.02	0.36	323.96	-0.93	182.07	2.71	286.59	6 Jan 11	299.7	0.1	-1.4	-5.1	259.90	266.13	0.88	0.66	300.56	308.39	-1.89	-1.39
Jan 6	283.19	0.35	326.17	-0.92	184.80	2.91	296.78	16 Jan 21	61.1	130.1	-3.6	2.3	272.36	278.59	0.44	0.21	315.20	319.75	-0.53	0.71
Jan 16	284.33	0.34	328.46	-0.92	186.70	3.12	306.94	26 Jan 31	203.3	270.7	5.2	1.3	284.82	291.06	-0.01	-0.23	320.45	316.76	2.15	3.34
Jan 26	285.42	0.33	330.82	-0.93	187.58	3.32	317.08	5 Feb 10	332.6	32.1	-3.9	-4.9	297.29	303.53	-0.44	-0.64	310.96	306.57	3.70	3.17
Feb 5	286.44	0.32	333.22	-0.93	187.27	3.50	327.17	15 Feb 20	95.3	168.4	-0.8	4.6	309.76	315.99	-0.82	-0.98	305.25	306.72	2.19	1.13
Feb 15	287.39	0.32	335.63	-0.94	185.69	3.63	337.22	25 Feb	241.3		3.4		322.22		-1.12		310.19		0.15	
								Mar 2		305.2		-1.9		328.45		-1.24		315.07		-0.69
Feb 25	288.23	0.31	338.04	-0.95	182.89	3.65	347.23	7 Mar 12	5.1	65.1	-5.0	-3.0	334.67	340.89	-1.33	-1.40	320.93	327.56	-1.36	-1.86
Mar 7	288.95	0.30	340.43	-0.96	179.25	3.54	357.17	17 Mar 22	131.5	207.1	2.5	4.8	347.10	353.31	-1.44	-1.45	334.83	342.68	-2.19	-2.33
Mar 17	289.54	0.29	342.76	-0.98	175.40	3.28	7.06	27 M-A 1	277.2	338.4	0.3	-4.3	359.51	5.71	-1.44	-1.40	351.10	0.11	-2.28	-2.03
Mar 27	289.98	0.28	345.03	-1.00	172.04	2.91	16.90	6 Apr 11	37.9	99.8	-4.4	0.1	11.90	18.09	-1.33	-1.24	9.72	19.90	-1.56	-0.88
Apr 6	290.27	0.27	347.21	-1.02	169.73	2.48	26.67	16 Apr 21	169.8	244.9	4.8	2.7	24.27	30.45	-1.13	-1.00	30.47	41.03	-0.05	0.85
Apr 16	290.40	0.26	349.27	-1.05	168.70	2.03	36.40	26 Apr	311.1		-2.8		36.62		-0.85		50.99		1.66	
								May 1		10.9		-5.1		42.78		-0.68		59.79		2.24
Apr 26	290.37	0.26	351.21	-1.08	168.95	1.61	46.07	6 May 11	71.6	136.7	-2.2	3.3	48.94	55.10	-0.50	-0.31	67.10	72.76	2.49	2.37
May 6	290.17	0.25	352.98	-1.12	170.35	1.22	55.71	16 May 21	208.9	280.8	4.8	-0.4	61.25	67.39	-0.11	0.09	76.63	78.59	1.85	0.93
May 16	289.83	0.23	354.56	-1.16	172.73	0.88	65.30	26 May 31	343.7	43.4	-4.8	-4.1	73.54	79.67	0.29	0.48	78.59	76.91	-0.33	-1.75
May 26	289.35	0.22	355.93	-1.20	175.91	0.58	74.87	5 Jun 10	106.8	175.4	1.0	5.2	85.81	91.94	0.67	0.85	74.25	71.70	-3.08	-4.02
Jun 5	288.77	0.21	357.07	-1.24	179.76	0.31	84.41	15 Jun 20	247.4	314.8	2.5	-3.3	98.07	104.19	1.02	1.16	70.32	70.73	-4.40	-4.25
Jun 15	288.10	0.20	357.93	-1.29	184.15	0.07	93.95	25 Jun 30	15.5	76.6	-5.2	-1.7	110.31	116.43	1.29	1.40	73.10	77.40	-3.68	-2.81
Jun 25	287.38	0.18	358.50	-1.34	188.99	-0.14	103.48	5 Jul 10	143.9	214.9	3.9	4.6	122.54	128.65	1.48	1.54	83.51	91.28	-1.77	-0.67
Jul 5	286.64	0.17	358.76	-1.39	194.22	-0.32	113.02	15 Jul 20	284.0	347.6	-0.8	-5.0	134.75	140.85	1.57	1.56	100.49	110.67	0.34	1.13
Jul 15	285.93	0.15	358.69	-1.44	199.78	-0.49	122.57	25 Jul 30	47.3	111.3	-3.9	1.4	146.94	153.02	1.53	1.47	121.20	131.51	1.61	1.78
Jul 25	285.27	0.14	358.30	-1.49	205.63	-0.64	132.15	4 Aug 9	182.9	253.5	5.2	1.9	159.10	165.16	1.37	1.25	141.31	150.49	1.66	1.33
Aug 4	284.71	0.12	357.60	-1.53	211.74	-0.77	141.75	14 Aug 19	318.9	19.7	-3.6	-4.9	171.22	177.26	1.10	0.92	159.06	167.04	0.84	0.23
Aug 14	284.27	0.11	356.63	-1.57	218.08	-0.89	151.40	24 Aug 29	79.9	147.9	-1.2	4.1	183.30	189.32	0.71	0.48	174.44	181.24	-0.45	-1.17
Aug 24	283.96	0.09	355.46	-1.60	224.63	-0.99	161.09	3 Sep 8	222.2	290.2	3.9	-1.5	195.33	201.32	0.23	-0.03	187.35	192.61	-1.89	-2.59
Sep 3	283.81	0.08	354.15	-1.61	231.36	-1.08	170.83	13 Sep 18	352.3	51.9	-4.9	-3.2	207.30	213.26	-0.31	-0.60	196.74	199.23	-3.21	-3.65
Sep 13	283.82	0.06	352.83	-1.61	238.28	-1.16	180.62	23 Sep 28	114.0	186.3	2.0	5.0	219.20	225.12	-0.89	-1.19	199.37	196.51	-3.76	-3.29
Sep 23	284.00	0.05	351.57	-1.60	245.35	-1.22	190.48	3 Oct 8	260.3	325.0	0.8	-4.1	231.02	236.89	-1.47	-1.75	191.11	185.96	-2.07	-0.40
Oct 3	284.34	0.03	350.47	-1.58	252.57	-1.26	200.38	13 Oct 18	25.1	84.8	-4.5	-0.3	242.72	248.53	-2.02	-2.26	184.36	187.10	1.05	1.87
Oct 13	284.83	0.02	349.63	-1.55	259.93	-1.30	210.35	23 Oct 28	150.1	225.0	4.5	3.5	254.30	260.02	-2.48	-2.67	192.87	200.19	2.12	1.97
Oct 23	285.47	0.01	349.08	-1.51	267.40	-1.31	220.37	2 Nov 7	296.3	358.3	-2.5	-5.1	265.69	271.30	-2.82	-2.93	208.12	216.20	1.59	1.09
Nov 2	286.23	-0.01	348.87	-1.47	274.97	-1.32	230.43	12 Nov 17	57.8	119.5	-2.5	2.8	276.83	282.28	-2.99	-3.00	224.25	232.23	0.54	-0.03
Nov 12	287.11	-0.02	349.00	-1.43	282.64	-1.31	240.54	22 Nov 27	188.2	262.7	5.1	0.3	287.63	292.85	-2.94	-2.82	240.14	247.99	-0.58	-1.08
Nov 22	288.09	-0.03	349.47	-1.38	290.38	-1.29	250.69	2 Dec 7	330.4	30.5	-4.7	-4.4	297.93	302.83	-2.63	-2.37	255.83	263.67	-1.53	-1.88
Dec 2	289.15	-0.04	350.26	-1.34	298.17	-1.25	260.86	12 Dec 17	91.1	156.2	0.4	5.0	307.51	311.93	-2.01	-1.57	271.53	279.38	-2.13	-2.22
Dec 12	290.26	-0.05	351.34	-1.30	306.01	-1.21	271.05	22 Dec 27	227.1	298.8	3.3	-2.8	316.03	319.75	-1.03	-0.38	287.08	294.34	-2.10	-1.72
1666								1667												1667
Dec 22	291.42	-0.07	352.70	-1.26	313.87	-1.15	281.24	1 Jan 6	2.9	62.5	-5.2	-2.1	322.98	325.64	0.39	1.27	300.46	304.15	-0.96	0.22
Jan 1	292.60	-0.08	354.28	-1.23	321.74	-1.08	291.43	11 Jan 16	125.9	194.9	3.3	4.9	327.58	328.69	2.27	3.37	303.76	298.96	1.74	3.08
Jan 11	293.79	-0.09	356.06	-1.20	329.61	-1.00	301.61	21 Jan 26	265.5	333.3	0.0	-4.8	328.84	327.95	4.55	5.75	292.82	289.13	3.57	3.16
Jan 21	294.95	-0.10	358.02	-1.17	337.45	-0.91	311.76	31 J-F 5	34.6	94.9	-4.0	0.9	326.06	323.37	6.86	7.76	288.90	291.37	2.31	1.36
Jan 31	296.08	-0.12	0.11	-1.15	345.26	-0.82	321.88	10 Feb 15	162.6	234.2	5.0	2.5	320.28	317.33	8.33	8.51	295.63	301.07	0.45	-0.35
Feb 10	297.16	-0.13	2.32	-1.13	353.03	-0.72	331.95	20 Feb 25	302.6	6.5	-3.2	-4.9	314.97	313.52	8.31	7.82	307.33	314.22	-1.03	-1.56
Feb 20	298.16	-0.14	4.61	-1.12	0.74	-0.62	341.99	2 Mar 7	66.1	128.9	-1.4	3.6	313.09	313.64	7.13	6.33	321.63	329.54	-1.95	-2.17
Mar 2	299.08	-0.16	6.97	-1.11	8.39	-0.51	351.96	12 Mar 17	201.0	272.7	4.4	-1.0	315.08	317.28	5.49	4.65	337.93	346.83	-2.22	-2.08
Mar 12	299.88	-0.17	9.36	-1.10	15.96	-0.40	1.88	22 Mar 27	338.0	38.9	-4.9	-3.3	320.10	323.44	3.82	3.04	356.27	6.21	-1.72	-1.15
Mar 22	300.57	-0.19	11.78	-1.10	23.46	-0.30	11.75	1 Apr 6	98.4	164.9	1.6	5.0	327.20	331.30	2.31	1.62	16.52	26.84	-0.38	0.53
Apr 1	301.11	-0.20	14.19	-1.10	30.87	-0.19	21.55	11 Apr 16	239.8	309.5	1.6	-4.0	335.69	340.30	1.00	0.43	36.53	44.93	1.44	2.20
Apr 11	301.51	-0.22	16.57	-1.10	38.21	-0.08	31.30	21 Apr 26	11.9	71.2	-4.7	-0.6	345.11	350.08	-0.09	-0.54	51.56	56.16	2.66	2.73
Apr 21	301.76	-0.24	18.92	-1.11	45.45	0.03	41.01	1 May 6	132.4	202.8	4.2	4.2	355.19	0.41	-0.94	-1.29	58.59	58.83	2.33	1.46
May 1	301.84	-0.25	21.20	-1.12	52.62	0.13	50.66	11 May 16	277.9	344.5	-1.8	-5.2	5.73	11.13	-1.58	-1.81	57.19	54.46	0.19	-1.26
May 11	301.76	-0.27	23.40	-1.13	59.69	0.24	60.28	21 May 26	44.6	104.2	-2.8	2.4	16.61	22.14	-1.99	-2.13	51.82	50.33	-2.57	-3.49
May 21	301.52	-0.29	25.50	-1.15	66.69	0.34	69.86	31 M-J 5	168.5	241.5	5.3	1.4	27.74	33.38	-2.21	-2.25	50.55	52.63	-3.93	-3.92
May 31	301.14	-0.31	27.47	-1.17	73.60	0.43	79.41	10 Jun 15	314.3	17.8	-4.5	-4.6	39.06	44.78	-2.25	-2.20	56.45	61.84	-3.54	-2.87
Jun 10	300.63	-0.33	29.30	-1.19	80.44	0.53	88.96	20 Jun 25	77.0	138.9	0.0	4.6	50.54	56.34	-2.12	-2.01	68.69	76.92	-2.00	-1.02
Jun 20	300.01	-0.34	30.94	-1.22	87.20	0.62	98.49	30 Jun	206.7		4.0		62.16		-1.86		86.38		-0.02	
								Jul 5		279.9		-2.0		68.01		-1.69		96.76		0.84
Jun 30	299.32	-0.36	32.38	-1.25	93.89	0.70	108.02	10 Jul 15	349.0	49.9	-5.2	-2.4	73.89	79.80	-1.50	-1.29	107.50	118.05	1.46	1.77
Jul 10	298.59	-0.37	33.59	-1.28	100.52	0.79	117.57	20 Jul 25	110.1	175.6	2.9	5.1	85.73	91.69	-1.06	-0.83	128.05	137.38	1.78	1.53
Jul 20	297.85	-0.39	34.54	-1.31	107.08	0.87	127.13	30 J-A 4	245.8	317.0	0.9	-4.5	97.67	103.68	-0.59	-0.35	146.02	153.97	1.09	0.50
Jul 30	297.15	-0.40	35.21	-1.35	113.57	0.95	136.72	9 Aug 14	22.0	81.6	-4.1	0.6	109.71	115.76	-0.11	0.13	161.24	167.78	-0.20	-0.97
Aug 9	296.51	-0.41	35.56	-1.39	120.01	1.02	146.34	19 Aug 24	144.7	214.2	4.7	3.2	121.83	127.93	0.35	0.56	173.50	178.20	-1.79	-2.60
Aug 19	295.98	-0.42	35.58	-1.42	126.39	1.09	156.01	29 Aug	284.6		-2.6		134.04		0.76		181.55		-3.35	
								Sep 3		352.3		-4.9		140.18		0.93		183.05		-3.93
Aug 29	295.57	-0.43	35.28	-1.46	132.70	1.16	165.72	8 Sep 13	53.9	113.9	-1.6	3.4	146.33	152.50	1.09	1.22	182.11	178.47	-4.16	-3.77
Sep 8	295.31	-0.44	34.65	-1.49	138.97	1.23	175.49	18 Sep 23	181.2	253.4	4.7	-0.2	158.69	164.89	1.32	1.40	173.22	169.13	-2.62	-0.98
Sep 18	295.21	-0.45	33.73	-1.51	145.17	1.29	185.31	28 S-O 3	322.1	26.0	-4.8	-3.5	171.11	177.33	1.45	1.47	168.72	172.34	0.53	1.52
Sep 28	295.28	-0.45	32.58	-1.52	151.31	1.36	195.18	8 Oct 13	85.6	147.6	1.4	5.0	183.57	189.82	1.46	1.42	178.78	186.67	1.95	1.96
Oct 8	295.51	-0.46	31.28	-1.52	157.38	1.42	205.12	18 Oct 23	219.2	292.0	2.6	-3.6	196.08	202.35	1.36	1.28	195.06	203.51	1.69	1.25
Oct 18	295.90	-0.46	29.92	-1.51	163.38	1.48	215.11	28 Oct	357.8		-4.9		208.62		1.17		211.83		0.74	
								Nov 2		58.6		-0.9		214.90		1.03		219.98		0.18
Oct 28	296.45	-0.47	28.62	-1.49	169.30	1.54	225.15	7 Nov 12	117.9	183.1	4.0	4.8	221.18	227.46	0.88	0.72	227.97	235.84	-0.38	-0.91
Nov 7	297.14	-0.47	27.47	-1.45	175.12	1.59	235.24	17 Nov 22	257.7	329.0	-0.9	-5.2	233.75	240.04	0.54	0.35	243.60	251.29	-1.39	-1.81
Nov 17	297.95	-0.48	26.55	-1.41	180.84	1.65	245.37	27 N-D 2	31.7	90.9	-3.1	2.1	246.33	252.61	0.15	-0.05	258.91	266.41	-2.12	-2.30
Nov 27	298.87	-0.49	25.93	-1.36	186.43	1.70	255.52	7 Dec 12	151.8	220.4	5.3	2.5	258.90	265.19	-0.25	-0.44	273.65	280.31	-2.30	-2.03
Dec 7	299.88	-0.49	25.64	-1.31	191.88	1.76	265.71	17 Dec 22	295.7	4.0	-3.9	-4.7	271.48	277.77	-0.63	-0.81	285.69	288.49	-1.40	-0.30
Dec 17	300.97	-0.50	25.70	-1.26	197.14	1.81	275.90	27 Dec	64.2		-0.3		284.05		-0.97		287.05		1.23	
1667								1667												1667

JULIAN 12 UT	SATURN LONG.	SATURN LAT.	JUPITER LONG.	JUPITER LAT.	MARS LONG.	MARS LAT.	SUN LONG.
1667							
Dec 27	302.11	-0.51	26.10	-1.21	202.18	1.86	286.09
Jan 6	303.29	-0.52	26.83	-1.16	206.97	1.90	296.28
Jan 16	304.48	-0.53	27.85	-1.11	211.42	1.94	306.44
Jan 26	305.66	-0.55	29.14	-1.07	215.48	1.98	316.58
Feb 5	306.83	-0.56	30.67	-1.03	219.03	2.00	326.68
Feb 15	307.95	-0.57	32.39	-0.99	221.96	2.00	336.73
Feb 25	309.01	-0.59	34.28	-0.96	224.10	1.97	346.73
Mar 6	309.99	-0.61	36.32	-0.93	225.28	1.90	356.69
Mar 16	310.88	-0.63	38.47	-0.90	225.33	1.78	6.58
Mar 26	311.65	-0.65	40.70	-0.88	224.13	1.58	16.41
Apr 5	312.30	-0.67	43.00	-0.86	221.69	1.29	26.20
Apr 15	312.81	-0.70	45.34	-0.84	218.36	0.92	35.92
Apr 25	313.17	-0.72	47.70	-0.83	214.71	0.48	45.60
May 5	313.37	-0.75	50.06	-0.82	211.47	0.02	55.24
May 15	313.41	-0.77	52.41	-0.81	209.24	-0.42	64.84
May 25	313.29	-0.80	54.72	-0.80	208.32	-0.79	74.41
Jun 4	313.01	-0.82	56.97	-0.80	208.75	-1.11	83.96
Jun 14	312.59	-0.85	59.15	-0.80	210.40	-1.36	93.49
Jun 24	312.04	-0.87	61.24	-0.80	213.11	-1.56	103.02
Jul 4	311.40	-0.89	63.20	-0.80	216.68	-1.72	112.56
Jul 14	310.69	-0.91	65.03	-0.81	220.98	-1.84	122.11
Jul 24	309.94	-0.92	66.68	-0.81	225.87	-1.92	131.68
Aug 3	309.21	-0.94	68.14	-0.82	231.25	-1.98	141.29
Aug 13	308.52	-0.94	69.37	-0.83	237.05	-2.02	150.93
Aug 23	307.90	-0.95	70.35	-0.84	243.21	-2.03	160.62
Sep 2	307.40	-0.95	71.04	-0.85	249.66	-2.03	170.36
Sep 12	307.03	-0.95	71.43	-0.86	256.37	-2.00	180.15
Sep 22	306.82	-0.95	71.48	-0.87	263.30	-1.96	190.00
Oct 2	306.77	-0.95	71.21	-0.87	270.41	-1.89	199.91
Oct 12	306.89	-0.95	70.60	-0.87	277.68	-1.82	209.86
Oct 22	307.17	-0.94	69.70	-0.87	285.08	-1.73	219.88
Nov 1	307.62	-0.94	68.57	-0.86	292.59	-1.62	229.95
Nov 11	308.22	-0.94	67.27	-0.85	300.17	-1.51	240.05
Nov 21	308.96	-0.94	65.91	-0.83	307.81	-1.39	250.19
Dec 1	309.81	-0.94	64.59	-0.80	315.48	-1.26	260.37
Dec 11	310.78	-0.94	63.42	-0.77	323.18	-1.13	270.55
Dec 21	311.83	-0.94	62.47	-0.73	330.87	-0.99	280.75
1668							
Dec 31	312.94	-0.94	61.82	-0.69	338.55	-0.85	290.94
Jan 10	314.11	-0.95	61.48	-0.65	346.19	-0.71	301.12
Jan 20	315.30	-0.96	61.49	-0.61	353.79	-0.57	311.27
Jan 30	316.50	-0.97	61.84	-0.57	1.34	-0.43	321.39
Feb 9	317.69	-0.98	62.50	-0.54	8.83	-0.30	331.47
Feb 19	318.85	-1.00	63.46	-0.50	16.25	-0.17	341.50
Mar 1	319.96	-1.02	64.68	-0.47	23.60	-0.05	351.48
Mar 11	321.01	-1.04	66.12	-0.44	30.88	0.07	1.40
Mar 21	321.98	-1.06	67.77	-0.41	38.08	0.18	11.27
Mar 31	322.84	-1.09	69.58	-0.39	45.20	0.28	21.08
Apr 10	323.59	-1.11	71.53	-0.36	52.24	0.38	30.83
Apr 20	324.21	-1.14	73.59	-0.34	59.21	0.48	40.54
Apr 30	324.69	-1.17	75.73	-0.32	66.11	0.56	50.20
May 10	325.01	-1.21	77.95	-0.30	72.93	0.65	59.81
May 20	325.17	-1.24	80.21	-0.28	79.69	0.72	69.39
May 30	325.17	-1.27	82.49	-0.27	86.40	0.79	78.95
Jun 9	325.01	-1.30	84.78	-0.25	93.04	0.86	88.49
Jun 19	324.69	-1.33	87.06	-0.23	99.63	0.91	98.02
Jun 29	324.23	-1.36	89.31	-0.22	106.17	0.97	107.56
Jul 9	323.66	-1.39	91.50	-0.21	112.67	1.02	117.10
Jul 19	322.99	-1.41	93.63	-0.19	119.13	1.06	126.66
Jul 29	322.26	-1.43	95.66	-0.18	125.56	1.10	136.25
Aug 8	321.51	-1.44	97.58	-0.16	131.95	1.14	145.87
Aug 18	320.77	-1.45	99.35	-0.15	138.32	1.17	155.54
Aug 28	320.09	-1.46	100.96	-0.14	144.66	1.19	165.25
Sep 7	319.50	-1.45	102.38	-0.12	150.98	1.22	175.01
Sep 17	319.03	-1.45	103.56	-0.10	157.28	1.24	184.83
Sep 27	318.70	-1.44	104.50	-0.09	163.56	1.25	194.70
Oct 7	318.53	-1.43	105.14	-0.07	169.82	1.26	204.63
Oct 17	318.53	-1.42	105.48	-0.05	176.05	1.27	214.62
Oct 27	318.70	-1.41	105.49	-0.03	182.27	1.27	224.66
Nov 6	319.04	-1.39	105.17	-0.01	188.46	1.27	234.74
Nov 16	319.55	-1.38	104.53	0.01	194.61	1.26	244.87
Nov 26	320.20	-1.37	103.60	0.04	200.74	1.24	255.03
Dec 6	320.98	-1.36	102.44	0.06	206.82	1.22	265.21
Dec 16	321.88	-1.36	101.14	0.08	212.85	1.19	275.41
1669							

GREGORIAN 12 UT	MOON LONG.		MOON LAT.		VENUS LONG.		VENUS LAT.		MERCURY LONG.		MERCURY LAT.	
1668												1668
Jan 1		123.9		4.3		290.34		-1.12		281.40		2.70
6 Jan 11	187.7	258.7	4.5	-1.0	296.61	302.89	-1.24	-1.35	275.29	272.46	3.35	3.10
16 Jan 21	332.4	37.1	-5.1	-2.5	309.16	315.43	-1.43	-1.48	273.29	276.67	2.39	1.54
26 Jan 31	96.3	158.5	2.5	5.0	321.69	327.94	-1.51	-1.51	281.61	287.53	0.70	-0.06
5 Feb 10	225.5	297.3	1.7	-4.0	334.18	340.41	-1.49	-1.43	294.12	301.21	-0.73	-1.28
15 Feb 20	7.3	68.8	-4.2	0.5	346.63	352.84	-1.35	-1.23	308.72	316.64	-1.72	-2.01
25 Feb	128.9		4.6		359.04		-1.10		324.97		-2.14	
Mar 1		194.8		3.7		5.22		-0.94		333.74		-2.10
6 Mar 11	264.5	335.0	-1.9	-5.0	11.38	17.53	-0.75	-0.55	342.96	352.63	-1.86	-1.39
16 Mar 21	40.6	100.3	-1.8	3.3	23.65	29.76	-0.33	-0.09	2.64	12.63	-0.70	0.18
26 Mar 31	162.8	232.7	5.0	0.6	35.85	41.91	0.15	0.40	21.97	29.88	1.17	2.09
5 Apr 10	303.6	11.0	-4.6	-3.9	47.94	53.95	0.65	0.90	35.68	39.00	2.77	3.04
15 Apr 20	72.7	132.4	1.2	5.0	59.93	65.89	1.15	1.38	39.72	38.15	2.80	1.99
25 Apr 30	198.4	271.5	3.4	-2.8	71.80	77.68	1.60	1.80	35.17	32.13	0.73	-0.70
5 May 10	341.5	45.1	-5.1	-1.3	83.53	89.33	1.98	2.12	30.25	30.17	-1.97	-2.89
15 May 20	104.7	165.8	3.8	5.1	95.09	100.79	2.24	2.32	31.97	35.46	-3.40	-3.53
25 May 30	235.7	310.3	0.3	-5.0	106.44	112.02	2.36	2.36	40.42	46.66	-3.34	-2.86
4 Jun 9	17.4	77.9	-3.5	1.7	117.54	122.97	2.30	2.20	54.10	62.68	-2.17	-1.31
14 Jun 19	137.3	201.1	5.1	3.2	128.32	133.56	2.05	1.83	72.32	82.81	-0.37	0.53
24 Jun 29	274.3	347.7	-3.0	-4.9	138.69	143.67	1.56	1.23	93.69	104.42	1.26	1.71
4 Jul 9	51.2	110.5	-0.6	4.1	148.50	153.14	0.83	0.37	114.59	124.02	1.86	1.73
14 Jul 19	171.4	238.3	4.7	-0.1	157.56	161.71	-0.17	-0.77	132.66	140.53	1.36	0.79
24 Jul 29	313.0	23.0	-4.9	-2.8	165.54	168.99	-1.45	-2.20	147.60	153.81	0.08	-0.75
3 Aug 8	83.7	143.6	2.4	5.0	171.98	174.41	-3.02	-3.90	159.03	163.04	-1.65	-2.59
13 Aug 18	207.2	277.0	2.3	-3.4	176.16	177.12	-4.84	-5.80	165.47	165.88	-3.48	-4.19
23 Aug 28	350.6	56.2	-4.5	0.3	177.16	176.22	-6.75	-7.61	163.88	159.76	-4.52	-4.18
2 Sep 7	115.8	177.9	4.5	4.1	174.34	171.70	-8.28	-8.67	155.09	152.32	-3.08	-1.51
12 Sep 17	244.9	316.1	-1.2	-5.1	168.66	165.71	-8.69	-8.32	153.15	157.62	0.01	1.12
22 Sep 27	26.1	88.2	-2.1	3.1	163.29	161.74	-7.62	-6.70	164.68	173.10	1.73	1.90
2 Oct 7	148.2	213.8	5.1	1.5	161.19	161.64	-5.67	-4.61	181.94	190.76	1.75	1.40
12 Oct 17	283.9	354.2	-4.2	-4.3	162.98	165.10	-3.58	-2.61	199.35	207.69	0.93	0.39
22 Oct 27	59.7	119.7	0.9	5.0	167.88	171.21	-1.72	-0.93	215.79	223.67	-0.17	-0.72
1 Nov 6	181.6	251.3	3.9	-1.9	174.99	179.15	-0.22	0.41	231.38	238.92	-1.25	-1.72
11 Nov 16	323.2	30.3	-5.3	-1.8	183.63	188.36	0.95	1.40	246.29	253.44	-2.10	-2.37
21 Nov 26	92.1	151.7	3.6	5.2	193.31	198.44	1.78	2.09	260.21	266.25	-2.47	-2.33
1 Dec 6	216.4	289.9	1.2	-4.5	203.73	209.16	2.33	2.50	270.84	272.70	-1.83	-0.84
11 Dec 16	1.1	64.5	-3.9	1.3	214.69	220.32	2.61	2.65	270.29	264.06	0.67	2.23
21 Dec 26	124.3	184.8	5.0	3.6	226.02	231.80	2.65	2.60	258.29	256.41	3.05	2.98
31 Dec	252.9		-2.1		237.63		2.50		258.24		2.42	
1669												1669
Jan 5		328.6		-5.0		243.52		2.36		262.43		1.69
10 Jan 15	36.9	97.5	-1.0	3.9	249.45	255.41	2.19	1.99	267.98	274.33	0.93	0.20
20 Jan 25	157.1	219.6	4.7	0.6	261.40	267.42	1.76	1.52	281.20	288.46	-0.46	-1.03
30 J-F 4	291.1	5.9	-4.6	-3.2	273.46	279.52	1.26	0.99	296.05	303.96	-1.49	-1.84
9 Feb 14	70.6	130.1	2.2	5.0	285.59	291.67	0.71	0.44	312.20	320.81	-2.05	-2.09
19 Feb 24	190.9	256.5	2.7	-2.8	297.77	303.87	0.17	-0.09	329.81	339.18	-1.96	-1.60
1 Mar 6	329.9	41.2	-4.8	-0.1	309.98	316.10	-0.34	-0.57	348.83	358.43	-1.01	-0.18
11 Mar 16	103.0	162.8	4.5	4.4	322.22	328.35	-0.78	-0.97	7.36	14.69	0.84	1.90
21 Mar 26	226.4	295.3	-0.5	-5.0	334.47	340.60	-1.14	-1.28	19.54	21.39	2.79	3.29
31 M-A 5	7.8	74.6	-2.8	2.9	346.73	352.86	-1.39	-1.47	20.30	17.12	3.20	2.46
10 Apr 15	134.8	196.4	5.2	2.1	358.99	5.12	-1.53	-1.55	13.44	10.84	1.22	-0.15
20 Apr 25	263.8	334.7	-3.6	-4.8	11.25	17.38	-1.55	-1.52	10.16	11.51	-1.38	-2.29
30 Apr	43.9		0.3		23.50		-1.46		14.64		-2.87	
May 5		106.9		4.8		29.63		-1.37		19.23		-3.13
10 May 15	166.8	231.5	4.3	-1.0	35.76	41.88	-1.26	-1.13	25.03	31.90	-3.10	-2.80
20 May 25	302.9	13.1	-5.2	-2.4	48.01	54.14	-0.98	-0.82	39.76	48.58	-2.28	-1.56
30 M-J 4	78.3	138.8	3.2	5.1	60.26	66.39	-0.64	-0.45	58.33	68.86	-0.70	0.20
9 Jun 14	199.6	268.6	1.8	-3.9	72.53	78.66	-0.25	-0.05	79.81	90.65	1.01	1.60
19 Jun 24	342.2	49.6	-4.3	0.9	84.79	90.93	0.14	0.34	100.91	110.38	1.90	1.90
29 Jun	111.5		4.8		97.08		0.53		118.96		1.62	
Jul 4		171.0		3.7		103.23		0.71		126.65		1.11
9 Jul 14	234.1	307.4	-1.4	-5.0	109.38	115.53	0.87	1.02	133.41	139.15	0.40	-0.47
19 Jul 24	20.2	84.2	-1.4	3.7	121.69	127.85	1.15	1.26	143.72	146.85	-1.47	-2.53
29 J-A 3	144.1	204.1	4.7	1.0	134.02	140.19	1.34	1.40	148.20	147.45	-3.57	-4.40
8 Aug 13	270.8	346.4	-4.2	-3.7	146.36	152.53	1.44	1.44	144.58	140.45	-4.79	-4.48
18 Aug 23	56.1	117.4	1.9	5.0	158.71	164.89	1.42	1.37	136.82	135.57	-3.44	-1.97
28 Aug	176.9		3.0		171.06		1.30		137.65		-0.49	
Sep 2		238.9		-2.3		177.24		1.20		142.93		0.71
7 Sep 12	309.2	24.0	-5.1	-0.7	183.42	189.59	1.07	0.92	150.56	159.46	1.47	1.81
17 Sep 22	90.1	149.9	4.4	4.6	195.76	201.93	0.75	0.56	168.74	177.92	1.79	1.53
27 S-O 2	210.5	275.9	0.1	-4.7	208.10	214.27	0.35	0.13	186.80	195.33	1.11	0.60
7 Oct 12	348.1	59.7	-3.6	2.5	220.43	226.58	-0.10	-0.33	203.54	211.45	0.04	-0.53
17 Oct 22	122.6	182.3	5.3	2.6	232.73	238.88	-0.57	-0.80	219.10	226.51	-1.09	-1.62
27 Oct	245.8		-3.0		245.02		-1.02		233.64		-2.07	
Nov 1		314.7		-5.1		251.16		-1.24		240.44		-2.43
6 Nov 11	26.2	93.6	-0.5	4.7	257.29	263.41	-1.44	-1.62	246.71	252.10	-2.64	-2.62
16 Nov 21	154.2	215.3	4.4	-0.4	269.51	275.61	-1.78	-1.91	255.86	256.74	-2.26	-1.39
26 N-D 1	283.1	354.0	-4.9	-3.1	281.70	287.76	-2.02	-2.09	253.43	246.90	0.06	1.69
6 Dec 11	62.6	126.2	2.7	5.1	293.81	299.83	-2.12	-2.11	241.69	240.79	2.68	2.81
16 Dec 21	185.8	249.8	2.1	-3.3	305.83	311.79	-2.07	-1.97	243.55	248.49	2.42	1.81
26 Dec 31	321.9	32.3	-4.6	0.3	317.72	323.60	-1.84	-1.66	254.60	261.35	1.12	0.43
1669												1669

JULIAN 12 UT	SATURN LONG.	SATURN LAT.	JUPITER LONG.	JUPITER LAT.	MARS LONG.	MARS LAT.	SUN LONG.
1669							
Dec 26	322.89	-1.35	99.80	0.11	218.82	1.15	285.60
Jan 5	323.97	-1.35	98.50	0.13	224.71	1.10	295.78
Jan 15	325.12	-1.35	97.36	0.15	230.52	1.04	305.95
Jan 25	326.30	-1.36	96.44	0.17	236.21	0.96	316.09
Feb 4	327.51	-1.36	95.81	0.19	241.77	0.86	326.19
Feb 14	328.72	-1.37	95.49	0.20	247.17	0.74	336.25
Feb 24	329.92	-1.39	95.51	0.22	252.36	0.59	346.26
Mar 6	331.08	-1.41	95.84	0.23	257.30	0.41	356.21
Mar 16	332.19	-1.43	96.48	0.24	261.91	0.19	6.11
Mar 26	333.22	-1.45	97.40	0.25	266.12	-0.08	15.94
Apr 5	334.17	-1.48	98.57	0.26	269.82	-0.40	25.72
Apr 15	335.02	-1.51	99.96	0.28	272.87	-0.80	35.46
Apr 25	335.75	-1.54	101.54	0.29	275.12	-1.28	45.13
May 5	336.34	-1.57	103.29	0.30	276.36	-1.85	54.77
May 15	336.79	-1.61	105.16	0.31	276.43	-2.50	64.37
May 25	337.08	-1.65	107.15	0.32	275.27	-3.20	73.94
Jun 4	337.21	-1.69	109.22	0.33	273.00	-3.88	83.49
Jun 14	337.17	-1.73	111.36	0.34	270.05	-4.47	93.03
Jun 24	336.98	-1.77	113.55	0.35	267.13	-4.86	102.56
Jul 4	336.62	-1.80	115.76	0.37	264.93	-5.05	112.09
Jul 14	336.13	-1.83	117.99	0.38	263.94	-5.03	121.65
Jul 24	335.53	-1.86	120.19	0.40	264.35	-4.87	131.22
Aug 3	334.83	-1.88	122.37	0.42	266.08	-4.63	140.82
Aug 13	334.09	-1.90	124.50	0.43	268.96	-4.33	150.47
Aug 23	333.33	-1.91	126.56	0.45	272.80	-4.01	160.15
Sep 2	332.59	-1.91	128.52	0.48	277.42	-3.68	169.89
Sep 12	331.92	-1.91	130.37	0.50	282.65	-3.34	179.68
Sep 22	331.35	-1.90	132.07	0.53	288.37	-3.01	189.52
Oct 2	330.91	-1.89	133.59	0.56	294.48	-2.69	199.42
Oct 12	330.62	-1.87	134.92	0.59	300.88	-2.37	209.38
Oct 22	330.50	-1.85	136.01	0.62	307.51	-2.07	219.39
Nov 1	330.55	-1.83	136.84	0.66	314.32	-1.78	229.46
Nov 11	330.77	-1.81	137.37	0.70	321.26	-1.50	239.56
Nov 21	331.17	-1.79	137.60	0.73	328.29	-1.24	249.70
Dec 1	331.72	-1.77	137.49	0.77	335.39	-1.00	259.87
Dec 11	332.42	-1.75	137.07	0.81	342.51	-0.77	270.06
Dec 21	333.25	-1.73	136.34	0.85	349.65	-0.56	280.25
1670							
Dec 31	334.20	-1.72	135.34	0.89	356.79	-0.36	290.45
Jan 10	335.24	-1.71	134.15	0.91	3.90	-0.18	300.63
Jan 20	336.35	-1.71	132.85	0.94	10.99	-0.02	310.78
Jan 30	337.52	-1.71	131.53	0.95	18.04	0.13	320.90
Feb 9	338.73	-1.71	130.29	0.96	25.05	0.27	330.98
Feb 19	339.95	-1.71	129.21	0.97	32.01	0.39	341.01
Mar 1	341.18	-1.72	128.38	0.96	38.92	0.50	351.00
Mar 11	342.38	-1.74	127.82	0.95	45.77	0.60	0.92
Mar 21	343.54	-1.76	127.58	0.94	52.57	0.69	10.79
Mar 31	344.65	-1.78	127.65	0.93	59.32	0.77	20.60
Apr 10	345.68	-1.81	128.03	0.92	66.01	0.84	30.36
Apr 20	346.62	-1.84	128.69	0.90	72.66	0.90	40.06
Apr 30	347.45	-1.87	129.62	0.89	79.26	0.95	49.72
May 10	348.16	-1.91	130.79	0.88	85.81	1.00	59.34
May 20	348.73	-1.95	132.17	0.87	92.33	1.04	68.93
May 30	349.15	-1.99	133.73	0.86	98.81	1.07	78.49
Jun 9	349.42	-2.03	135.44	0.85	105.26	1.10	88.03
Jun 19	349.52	-2.08	137.28	0.85	111.69	1.12	97.56
Jun 29	349.45	-2.12	139.23	0.85	118.09	1.13	107.10
Jul 9	349.22	-2.16	141.27	0.85	124.48	1.14	116.64
Jul 19	348.84	-2.20	143.37	0.85	130.85	1.15	126.20
Jul 29	348.32	-2.23	145.51	0.86	137.22	1.15	135.79
Aug 8	347.68	-2.26	147.68	0.86	143.58	1.15	145.41
Aug 18	346.97	-2.28	149.86	0.88	149.95	1.14	155.07
Aug 28	346.20	-2.30	152.02	0.89	156.32	1.13	164.78
Sep 7	345.44	-2.30	154.15	0.90	162.70	1.11	174.54
Sep 17	344.71	-2.30	156.22	0.92	169.09	1.09	184.35
Sep 27	344.05	-2.29	158.22	0.94	175.49	1.06	194.23
Oct 7	343.50	-2.27	160.11	0.97	181.91	1.03	204.16
Oct 17	343.09	-2.25	161.87	1.00	188.34	0.99	214.14
Oct 27	342.84	-2.22	163.48	1.03	194.80	0.95	224.18
Nov 6	342.76	-2.19	164.90	1.07	201.27	0.90	234.26
Nov 16	342.85	-2.16	166.11	1.11	207.77	0.85	244.38
Nov 26	343.13	-2.13	167.07	1.15	214.28	0.78	254.54
Dec 6	343.57	-2.10	167.76	1.19	220.81	0.71	264.72
Dec 16	344.17	-2.08	168.15	1.24	227.36	0.63	274.91
1671							

GREGORIAN 12 UT	MOON LONG.		MOON LAT.		VENUS LONG.		VENUS LAT.		MERCURY LONG.		MERCURY LAT.	
1670												1670
5 Jan 10	97.4	158.3	4.7	3.8	329.43	335.19	-1.44	-1.16	268.48	275.90	-0.21	-0.78
15 Jan 20	218.2	286.2	-1.0	-4.9	340.89	346.50	-0.85	-0.49	283.55	291.44	-1.28	-1.67
25 Jan 30	0.8	69.0	-2.1	3.5	352.02	357.43	-0.09	0.35	299.60	308.04	-1.94	-2.07
4 Feb 9	131.0	190.4	4.8	1.2	2.70	7.82	0.83	1.34	316.79	325.84	-2.03	-1.78
14 Feb 19	252.2	324.2	-3.8	-4.4	12.76	17.50	1.88	2.45	335.10	344.27	-1.29	-0.54
24 Feb	38.6		1.4		21.97		3.04		352.71		0.47	
Mar 1		103.7		5.1		26.15		3.64		359.39		1.63
6 Mar 11	163.7	223.4	3.2	-1.9	29.95	33.32	4.25	4.85	3.20	3.49	2.72	3.44
16 Mar 21	288.5	2.8	-5.2	-1.7	36.15	38.34	5.44	5.98	0.66	356.35	3.50	2.82
26 Mar 31	74.5	136.9	4.2	4.8	39.76	40.29	6.45	6.81	352.70	351.06	1.63	0.32
5 Apr 10	196.2	258.2	0.5	-4.4	39.81	38.32	7.00	6.96	351.68	354.27	-0.84	-1.75
15 Apr 20	326.8	40.7	-4.3	1.7	35.94	32.96	6.62	5.96	358.43	3.83	-2.38	-2.74
25 Apr 30	108.7	169.0	5.3	2.9	29.84	27.07	5.02	3.89	10.24	17.55	-2.84	-2.70
5 May 10	229.5	295.2	-2.4	-5.2	25.05	23.98	2.69	1.54	25.68	34.65	-2.33	-1.76
15 May 20	5.9	77.2	-1.4	4.3	23.92	24.80	0.49	-0.42	44.43	54.93	-1.00	-0.13
25 May 30	141.4	200.9	4.5	0.0	26.49	28.87	-1.18	-1.80	65.86	76.73	0.73	1.44
4 Jun 9	264.5	333.9	-4.6	-3.7	31.81	35.23	-2.29	-2.67	87.03	96.45	1.90	2.05
14 Jun 19	44.5	112.0	2.1	5.0	39.03	43.16	-2.95	-3.14	104.88	112.27	1.89	1.45
24 Jun 29	173.2	233.6	2.2	-2.9	47.55	52.16	-3.24	-3.28	118.56	123.65	0.76	-0.14
4 Jul 9	301.5	13.2	-4.9	-0.4	56.95	61.91	-3.25	-3.16	127.34	129.37	-1.23	-2.42
14 Jul 19	81.7	145.4	4.5	3.9	67.00	72.21	-3.03	-2.86	129.48	127.61	-3.59	-4.52
24 Jul 29	204.9	267.9	-0.8	-4.8	77.52	82.93	-2.65	-2.40	124.29	120.74	-4.93	-4.63
3 Aug 8	340.1	51.8	-2.9	3.1	88.41	93.97	-2.14	-1.86	118.58	118.94	-3.67	-2.34
13 Aug 18	117.2	177.8	5.0	1.4	99.60	105.28	-1.56	-1.25	122.24	128.29	-0.94	0.28
23 Aug 28	237.3	304.2	-3.6	-4.9	111.03	116.83	-0.94	-0.64	136.42	145.75	1.17	1.66
2 Sep 7	19.0	88.7	0.5	5.1	122.67	128.56	-0.33	-0.04	155.44	164.97	1.79	1.64
12 Sep 17	150.9	210.0	3.5	-1.6	134.49	140.47	0.23	0.49	174.13	182.87	1.29	0.81
22 Sep 27	271.5	341.9	-5.2	-2.7	146.48	152.52	0.73	0.95	191.20	199.15	0.26	-0.33
2 Oct 7	57.0	123.6	3.7	5.0	158.60	164.70	1.13	1.29	206.75	214.01	-0.92	-1.50
12 Oct 17	183.4	243.2	0.9	-4.1	170.83	176.99	1.42	1.52	220.90	227.33	-2.03	-2.48
22 Oct 27	307.6	20.4	-4.8	0.7	183.17	189.36	1.59	1.62	233.09	237.78	-2.80	-2.91
1 Nov 6	93.4	156.6	5.2	3.1	195.57	201.80	1.62	1.59	240.66	240.54	-2.69	-1.95
11 Nov 16	215.8	278.0	-2.0	-5.1	208.04	214.30	1.53	1.45	236.43	229.88	-0.57	1.10
21 Nov 26	345.5	58.5	-2.3	3.7	220.56	226.83	1.34	1.20	225.42	225.50	2.26	2.60
1 Dec 6	127.8	188.4	4.6	0.2	233.10	239.38	1.05	0.88	229.13	234.76	2.39	1.89
11 Dec 16	248.9	314.7	-4.3	-4.1	245.66	251.95	0.69	0.50	241.39	248.52	1.29	0.65
21 Dec 26	24.5	95.5	1.3	5.0	258.23	264.52	0.30	0.09	255.91	263.48	0.03	-0.55
31 Dec	160.5		2.3		270.81		-0.11		271.19		-1.07	
1671												1671
Jan 5		220.0		-2.7		277.09		-0.31		279.07		-1.50
10 Jan 15	283.3	353.0	-5.0	-1.2	283.38	289.67	-0.50	-0.68	287.12	295.39	-1.83	-2.03
20 Jan 25	63.3	130.8	4.3	4.1	295.95	302.23	-0.85	-1.00	303.89	312.61	-2.08	-1.93
30 J-F 4	192.3	252.2	-0.6	-4.7	308.50	314.77	-1.13	-1.24	321.46	330.15	-1.55	-0.89
9 Feb 14	319.4	32.0	-3.6	2.4	321.04	327.30	-1.33	-1.39	338.03	344.00	0.07	1.29
19 Feb 24	100.9	164.4	5.2	1.7	333.56	339.81	-1.43	-1.44	346.72	345.46	2.55	3.48
1 Mar 6	223.9	285.8	-3.4	-5.2	346.05	352.29	-1.42	-1.38	341.17	336.34	3.68	3.06
11 Mar 16	357.0	70.5	-0.6	4.9	358.51	4.73	-1.31	-1.22	333.29	332.81	1.94	0.72
21 Mar 26	136.5	197.0	3.9	-1.2	10.94	17.14	-1.10	-0.96	334.63	338.23	-0.38	-1.27
31 M-A 5	256.5	321.3	-5.1	-3.6	23.33	29.50	-0.80	-0.62	343.17	349.15	-1.93	-2.37
10 Apr 15	35.5	107.6	2.8	5.1	35.67	41.83	-0.43	-0.23	355.99	3.57	-2.58	-2.57
20 Apr 25	170.2	229.5	1.2	-3.7	47.97	54.10	-0.02	0.20	11.88	20.90	-2.35	-1.90
30 Apr	290.6		-4.9		60.22		0.41		30.64		-1.26	
May 5		358.7		-0.4		66.33		0.63		41.05		-0.46
10 May 15	73.9	142.5	4.9	3.3	72.42	78.50	0.83	1.03	51.89	62.71	0.41	1.23
20 May 25	202.7	262.7	-1.8	-5.0	84.57	90.62	1.21	1.38	72.97	82.25	1.84	2.15
30 M-J 4	326.7	37.5	-2.9	3.0	96.65	102.67	1.52	1.64	90.40	97.32	2.14	1.81
9 Jun 14	111.0	175.6	4.7	0.4	108.68	114.66	1.74	1.80	102.94	107.11	1.17	0.25
19 Jun 24	235.0	297.4	-4.1	-4.4	120.63	126.57	1.83	1.83	109.64	110.33	-0.91	-2.22
29 Jun	4.6		0.6		132.49		1.79		109.15		-3.49	
Jul 4		76.4		4.9		138.39		1.72		106.49		-4.47
9 Jul 14	146.1	207.5	2.6	-2.6	144.25	150.09	1.60	1.45	103.33	100.99	-4.88	-4.61
19 Jul 24	267.9	333.8	-5.1	-1.9	155.90	161.66	1.26	1.03	100.54	102.50	-3.78	-2.62
29 J-A 3	43.8	114.3	3.8	4.5	167.39	173.07	0.77	0.47	106.95	113.69	-1.34	-0.13
8 Aug 13	179.4	239.1	-0.4	-4.7	178.71	184.28	0.14	-0.22	122.26	131.96	0.84	1.47
18 Aug 23	301.9	11.8	-4.2	1.5	189.79	195.23	-0.61	-1.02	142.02	151.89	1.75	1.72
28 Aug	83.0		5.3		200.59		-1.45		161.32		1.45	
Sep 2		150.0		2.2		205.86		-1.89		170.26		1.02
7 Sep 12	211.5	271.3	-3.2	-5.3	211.02	216.04	-2.35	-2.81	178.71	186.69	0.49	-0.11
17 Sep 22	337.5	50.8	-1.6	4.4	220.92	225.62	-3.26	-3.71	194.24	201.35	-0.74	-1.37
27 S-O 2	120.7	183.9	4.3	-0.8	230.10	234.33	-4.13	-4.53	207.99	214.02	-1.98	-2.52
7 Oct 12	243.6	304.7	-4.9	-4.0	238.25	241.79	-4.89	-5.20	219.23	223.18	-2.95	-3.19
17 Oct 22	14.7	89.6	1.8	5.2	244.85	247.34	-5.43	-5.57	225.11	223.97	-3.11	-2.51
27 Oct	156.3		1.6		249.15		-5.59		219.22		-1.21	
Nov 1		216.7		-3.5		250.13		-5.45		212.93		0.48
6 Nov 11	276.3	339.8	-5.0	-1.2	250.18	249.21	-5.10	-4.52	209.38	210.41	1.79	2.34
16 Nov 21	53.1	126.8	4.4	3.6	247.27	244.59	-3.67	-2.58	214.86	221.15	2.31	1.96
26 N-D 1	190.0	249.3	-1.6	-4.9	241.59	238.75	-1.33	-0.04	228.28	235.78	1.43	0.84
6 Dec 11	310.2	16.8	-3.2	2.3	236.55	235.27	1.17	2.21	243.43	251.15	0.25	-0.33
16 Dec 21	91.6	161.6	4.9	0.6	235.01	235.74	3.04	3.66	258.93	266.79	-0.86	-1.33
26 Dec 31	222.4	282.4	-4.1	-4.6	237.36	239.75	4.09	4.36	274.76	282.86	-1.71	-1.97
1671												1671

JULIAN 12 UT	SATURN LONG.	SATURN LAT.	JUPITER LONG.	JUPITER LAT.	MARS LONG.	MARS LAT.	SUN LONG.
1671							
Dec 26	344.92	-2.05	168.22	1.29	233.93	0.54	285.11
Jan 5	345.80	-2.03	167.97	1.33	240.51	0.44	295.29
Jan 15	346.78	-2.01	167.41	1.38	247.10	0.33	305.46
Jan 25	347.85	-2.00	166.57	1.42	253.70	0.21	315.60
Feb 4	349.00	-1.99	165.50	1.45	260.31	0.07	325.70
Feb 14	350.19	-1.99	164.28	1.47	266.91	-0.09	335.76
Feb 24	351.42	-1.99	162.98	1.48	273.51	-0.26	345.78
Mar 5	352.66	-1.99	161.70	1.48	280.10	-0.46	355.73
Mar 15	353.89	-2.00	160.54	1.47	286.66	-0.67	5.63
Mar 25	355.10	-2.02	159.56	1.46	293.18	-0.91	15.47
Apr 4	356.27	-2.04	158.84	1.43	299.65	-1.17	25.26
Apr 14	357.37	-2.06	158.39	1.40	306.04	-1.46	34.99
Apr 24	358.40	-2.09	158.25	1.37	312.33	-1.77	44.68
May 4	359.33	-2.12	158.42	1.34	318.47	-2.11	54.31
May 14	0.15	-2.15	158.88	1.30	324.42	-2.48	63.91
May 24	0.85	-2.19	159.61	1.27	330.11	-2.88	73.49
Jun 3	1.41	-2.24	160.59	1.24	335.47	-3.31	83.03
Jun 13	1.81	-2.28	161.80	1.21	340.37	-3.76	92.57
Jun 23	2.06	-2.33	163.21	1.19	344.69	-4.24	102.10
Jul 3	2.13	-2.37	164.79	1.17	348.27	-4.73	111.64
Jul 13	2.04	-2.42	166.52	1.15	350.89	-5.21	121.19
Jul 23	1.78	-2.46	168.38	1.13	352.34	-5.66	130.76
Aug 2	1.37	-2.50	170.33	1.12	352.48	-6.00	140.36
Aug 12	0.82	-2.54	172.37	1.11	351.27	-6.17	150.00
Aug 22	0.16	-2.56	174.48	1.11	348.98	-6.08	159.68
Sep 1	359.42	-2.58	176.62	1.10	346.22	-5.71	169.41
Sep 11	358.65	-2.59	178.78	1.10	343.74	-5.08	179.20
Sep 21	357.88	-2.59	180.94	1.11	342.20	-4.31	189.04
Oct 1	357.15	-2.58	183.08	1.12	341.93	-3.50	198.94
Oct 11	356.50	-2.56	185.18	1.13	342.95	-2.74	208.90
Oct 21	355.97	-2.54	187.21	1.14	345.13	-2.07	218.91
Oct 31	355.59	-2.51	189.16	1.16	348.26	-1.49	228.96
Nov 10	355.38	-2.47	190.99	1.19	352.16	-1.00	239.07
Nov 20	355.34	-2.44	192.68	1.21	356.64	-0.58	249.21
Nov 30	355.48	-2.40	194.19	1.24	1.59	-0.24	259.38
Dec 10	355.80	-2.36	195.51	1.28	6.87	0.05	269.57
Dec 20	356.29	-2.32	196.59	1.31	12.43	0.29	279.76
1672							
Dec 30	356.94	-2.29	197.42	1.35	18.18	0.49	289.95
Jan 9	357.73	-2.26	197.96	1.40	24.09	0.66	300.13
Jan 19	358.64	-2.24	198.19	1.44	30.11	0.79	310.29
Jan 29	359.66	-2.21	198.11	1.48	36.21	0.91	320.41
Feb 8	0.76	-2.20	197.72	1.52	42.37	1.00	330.50
Feb 18	1.94	-2.18	197.03	1.56	48.57	1.08	340.53
Feb 28	3.15	-2.18	196.09	1.58	54.80	1.14	350.51
Mar 10	4.40	-2.18	194.96	1.60	61.05	1.18	0.45
Mar 20	5.65	-2.18	193.71	1.61	67.30	1.22	10.32
Mar 30	6.90	-2.19	192.43	1.61	73.56	1.25	20.13
Apr 9	8.11	-2.20	191.21	1.59	79.82	1.27	29.89
Apr 19	9.28	-2.22	190.13	1.57	86.08	1.28	39.60
Apr 29	10.39	-2.24	189.26	1.53	92.34	1.28	49.26
May 9	11.42	-2.27	188.65	1.50	98.59	1.28	58.89
May 19	12.35	-2.30	188.33	1.45	104.84	1.28	68.47
May 29	13.17	-2.34	188.31	1.41	111.09	1.26	78.03
Jun 8	13.86	-2.38	188.58	1.36	117.35	1.25	87.57
Jun 18	14.40	-2.42	189.14	1.32	123.61	1.23	97.10
Jun 28	14.80	-2.47	189.97	1.28	129.88	1.20	106.63
Jul 8	15.02	-2.51	191.04	1.24	136.17	1.17	116.18
Jul 18	15.08	-2.56	192.32	1.21	142.47	1.14	125.74
Jul 28	14.97	-2.61	193.80	1.17	148.79	1.10	135.32
Aug 7	14.68	-2.65	195.45	1.14	155.14	1.06	144.94
Aug 17	14.25	-2.69	197.24	1.12	161.52	1.01	154.60
Aug 27	13.67	-2.72	199.15	1.10	167.93	0.97	164.30
Sep 6	12.99	-2.74	201.15	1.08	174.38	0.91	174.06
Sep 16	12.24	-2.76	203.24	1.06	180.86	0.86	183.87
Sep 26	11.45	-2.76	205.37	1.05	187.39	0.80	193.74
Oct 6	10.67	-2.76	207.55	1.04	193.97	0.73	203.67
Oct 16	9.94	-2.74	209.73	1.04	200.59	0.66	213.65
Oct 26	9.31	-2.72	211.91	1.04	207.27	0.58	223.69
Nov 5	8.80	-2.69	214.06	1.04	214.00	0.50	233.77
Nov 15	8.45	-2.65	216.15	1.04	220.78	0.42	243.89
Nov 25	8.26	-2.61	218.17	1.05	227.61	0.33	254.05
Dec 5	8.26	-2.56	220.08	1.06	234.51	0.23	264.23
Dec 15	8.44	-2.52	221.87	1.08	241.45	0.12	274.42
1673							

GREGORIAN 12 UT	MOON LONG.		MOON LAT.		VENUS LONG.		VENUS LAT.		MERCURY LONG.		MERCURY LAT.	
1672												1672
5 Jan 10	346.0	55.5	-0.1	4.7	242.75	246.27	4.49	4.51	291.11	299.49	-2.10	-2.06
15 Jan 20	128.8	194.6	3.1	-2.4	250.20	254.48	4.43	4.28	307.91	316.08	-1.79	-1.24
25 Jan 30	254.3	316.5	-5.2	-2.5	259.04	263.83	4.06	3.79	323.36	328.56	-0.35	0.88
4 Feb 9	23.7	94.6	3.2	4.9	268.82	273.96	3.48	3.14	330.17	327.48	2.27	3.39
14 Feb 19	164.2	226.5	0.1	-4.6	279.23	284.61	2.77	2.39	322.11	317.32	3.73	3.19
24 Feb 29	286.5	352.0	-4.5	0.6	290.09	295.65	2.00	1.60	315.21	315.89	2.16	1.04
5 Mar 10	62.8	132.7	5.2	2.9	301.28	306.96	1.21	0.83	318.75	323.15	0.01	-0.85
15 Mar 20	197.9	258.2	-2.8	-5.3	312.68	318.46	0.46	0.11	328.69	335.08	-1.53	-2.02
25 Mar 30	319.7	29.4	-2.3	3.6	324.26	330.10	-0.21	-0.52	342.20	349.96	-2.32	-2.42
4 Apr 9	102.0	168.8	4.7	-0.3	335.97	341.86	-0.79	-1.03	358.33	7.34	-2.32	-2.01
14 Apr 19	230.6	290.2	-4.7	-4.2	347.77	353.69	-1.24	-1.42	16.99	27.24	-1.48	-0.77
24 Apr 29	354.5	68.2	0.9	5.1	359.64	5.59	-1.56	-1.67	37.92	48.61	0.09	0.96
4 May 9	139.8	203.2	2.0	-3.2	11.56	17.54	-1.74	-1.78	58.74	67.80	1.71	2.21
14 May 19	263.1	323.5	-4.9	-1.6	23.53	29.53	-1.78	-1.75	75.52	81.77	2.37	2.18
24 May 29	31.4	107.1	3.8	4.1	35.54	41.56	-1.69	-1.60	86.44	89.38	1.62	0.71
3 Jun 8	175.5	236.3	-1.3	-4.9	47.59	53.62	-1.49	-1.35	90.46	89.67	-0.51	-1.90
13 Jun 18	295.9	358.5	-3.4	1.7	59.66	65.71	-1.19	-1.01	87.41	84.57	-3.24	-4.21
23 Jun 28	70.1	144.5	5.0	1.2	71.78	77.85	-0.82	-0.62	82.32	81.60	-4.62	-4.43
3 Jul 8	209.3	269.0	-4.0	-4.7	83.93	90.03	-0.41	-0.20	82.91	86.33	-3.77	-2.80
13 Jul 18	329.8	35.8	-0.6	4.4	96.13	102.24	0.01	0.21	91.80	99.16	-1.67	-0.53
23 Jul 28	109.1	179.9	3.8	-2.1	108.37	114.51	0.41	0.60	108.11	118.11	0.49	1.24
2 Aug 7	241.8	301.7	-5.2	-3.0	120.66	126.81	0.78	0.93	128.48	138.66	1.66	1.76
12 Aug 17	5.5	74.7	2.6	5.2	132.98	139.16	1.07	1.19	148.35	157.48	1.60	1.23
22 Aug 27	146.9	213.5	0.9	-4.4	145.35	151.55	1.28	1.35	166.03	174.05	0.73	0.13
1 Sop 6	273.7	335.4	-4.7	-0.1	157.76	163.97	1.40	1.41	181.53	188.47	-0.53	-1.22
11 Sep 16	43.2	114.0	4.8	3.5	170.20	176.42	1.40	1.37	194.82	200.42	-1.90	-2.54
21 Sep 26	182.9	245.8	-2.4	-5.2	182.66	188.89	1.30	1.21	205.01	208.14	-3.09	-3.47
1 Oct 6	305.6	10.6	-2.6	3.0	195.14	201.38	1.10	0.97	209.06	206.93	-3.53	-3.04
11 Oct 16	82.3	152.1	4.9	0.3	207.62	213.87	0.81	0.64	201.75	196.01	-1.83	-0.15
21 Oct 26	217.2	277.6	-4.5	-4.2	220.12	226.37	0.46	0.26	193.49	195.47	1.28	2.04
31 Oct	338.3		0.4		232.62		0.06		200.69		2.20	
Nov 5		47.7		4.8		238.87		-0.15		207.60		1.99
10 Nov 15	121.5	188.5	2.6	-2.9	245.11	251.36	-0.36	-0.56	215.22	223.08	1.56	1.03
20 Nov 25	250.3	309.6	-4.9	-1.8	257.61	263.85	-0.76	-0.95	230.99	238.88	0.45	-0.12
30 N-D 5	12.7	86.2	3.4	4.5	270.10	276.34	-1.13	-1.28	246.74	254.59	-0.66	-1.16
10 Dec 15	159.1	223.0	-0.8	-4.8	282.57	288.80	-1.42	-1.54	262.47	270.41	-1.58	-1.91
20 Dec 25	282.9	342.7	-3.6	1.3	295.03	301.24	-1.62	-1.68	278.41	286.45	-2.12	-2.16
30 Dec	49.2		5.1		307.45		-1.72		294.43		-2.00	
1673												1673
Jan 4		124.7		2.0		313.65		-1.71		302.05		-1.57
9 Jan 14	194.8	256.2	-3.8	-4.9	319.83	326.00	-1.68	-1.61	308.67	313.06	-0.77	0.42
19 Jan 24	315.5	17.6	-1.0	4.1	332.16	338.29	-1.51	-1.38	313.57	309.62	1.90	3.19
29 J-F 3	87.4	162.0	4.5	-1.4	344.40	350.49	-1.22	-1.02	303.61	299.30	3.66	3.22
8 Feb 13	228.6	288.4	-5.2	-3.3	356.54	2.57	-0.79	-0.54	298.25	300.01	2.31	1.29
18 Feb 23	349.1	54.7	2.0	5.3	8.56	14.51	-0.27	0.03	303.74	308.80	0.34	-0.49
28 Feb	126.2		1.9		20.41		0.34		314.79		-1.18	
Mar 5		197.7		-4.1		26.26		0.67		321.50		-1.71
10 Mar 15	261.0	320.5	-4.8	-0.6	32.06	37.80	1.00	1.34	328.80	336.65	-2.07	-2.26
20 Mar 25	24.4	93.5	4.4	4.1	43.46	49.05	1.68	2.01	345.03	353.97	-2.27	-2.07
30 M-A 4	164.4	231.8	-1.6	-5.1	54.55	59.95	2.32	2.62	3.47	13.52	-1.66	-1.05
9 Apr 14	292.6	353.5	-2.8	2.4	65.23	70.39	2.90	3.14	23.97	34.45	-0.25	0.66
19 Apr 24	61.7	132.8	5.0	1.0	75.40	80.23	3.35	3.51	44.37	53.10	1.53	2.20
29 Apr	201.2		-4.2		84.85		3.62		60.25		2.56	
May 4		264.8		-4.3		89.22		3.66		65.60		2.54
9 May 14	324.2	28.2	0.1	4.6	93.30	97.04	3.63	3.52	69.01	70.40	2.09	1.22
19 May 24	100.5	171.2	3.3	-2.5	100.35	103.15	3.31	2.98	69.81	67.69	-0.03	-1.47
29 M-J 3	236.3	296.9	-5.0	-2.0	105.34	106.79	2.53	1.92	64.93	62.69	-2.81	-3.77
8 Jun 13	356.9	64.9	3.1	4.9	107.39	107.03	1.16	0.23	61.87	62.89	-4.20	-4.14
18 Jun 23	139.4	207.9	0.0	-4.8	105.69	103.44	-0.85	-2.02	65.80	70.49	-3.67	-2.90
28 Jun	269.9		-3.9		100.54		-3.19		76.83		-1.94	
Jul 3		329.1		1.0		97.43		-4.23		84.71		-0.89
8 Jul 13	31.3	103.2	5.0	3.0	94.59	92.43	-5.07	-5.63	93.96	104.20	0.12	0.97
18 Jul 23	177.3	242.7	-3.3	-5.1	91.18	90.92	-5.94	-6.03	114.82	125.27	1.53	1.77
28 J-A 2	302.5	2.4	-1.4	3.7	91.60	93.11	-5.94	-5.71	135.20	144.49	1.72	1.44
7 Aug 12	67.8	141.8	5.0	-0.4	95.34	98.18	-5.39	-5.01	153.13	161.14	0.98	0.38
17 Aug 22	213.5	275.8	-5.1	-3.5	101.51	105.26	-4.58	-4.11	168.53	175.27	-0.30	-1.04
27 Aug	335.1		1.6		109.35		-3.63		181.27		-1.81	
Sep 1		37.4		5.2		113.73		-3.13		186.37		-2.56
6 Sep 11	106.0	179.8	2.8	-3.5	118.36	123.19	-2.64	-2.15	190.27	192.48	-3.23	-3.75
16 Sep 21	247.7	307.8	-4.9	-0.9	128.21	133.38	-1.67	-1.20	192.34	189.30	-3.94	-3.55
26 S-O 1	8.6	74.5	4.0	4.5	138.68	144.10	-0.76	-0.34	183.99	179.10	-2.41	-0.76
6 Oct 11	144.9	216.2	-0.8	-5.0	149.62	155.24	0.06	0.42	177.73	180.61	0.75	1.69
16 Oct 21	280.3	339.7	-2.9	2.1	160.94	166.72	0.74	1.03	186.55	194.05	2.05	1.98
26 Oct 31	43.7	113.1	5.0	1.8	172.55	178.45	1.29	1.50	202.16	210.39	1.66	1.19
5 Nov 10	183.4	250.9	-3.8	-4.4	184.40	190.39	1.67	1.80	218.57	226.63	0.66	0.09
15 Nov 20	311.8	12.3	0.0	4.4	196.43	202.49	1.89	1.93	234.57	242.44	-0.46	-0.98
25 Nov 30	80.6	152.2	3.9	-1.9	208.59	214.71	1.94	1.92	250.24	258.02	-1.45	-1.83
5 Dec 10	220.6	284.0	-5.1	-2.2	220.86	227.03	1.86	1.76	265.78	273.48	-2.11	-2.25
15 Dec 20	343.4	46.4	2.9	5.1	233.21	239.41	1.64	1.49	281.01	288.07	-2.20	-1.88
25 Dec 30	118.7	190.7	0.9	-4.6	245.61	251.83	1.32	1.13	293.98	297.50	-1.20	-0.08
1673												1673

JULIAN 12 UT	SATURN LONG.	SATURN LAT.	JUPITER LONG.	JUPITER LAT.	MARS LONG.	MARS LAT.	SUN LONG.	GREGORIAN 12 UT	MOON LONG.	MOON LONG.	MOON LAT.	MOON LAT.	VENUS LONG.	VENUS LONG.	VENUS LAT.	VENUS LAT.	MERCURY LONG.	MERCURY LONG.	MERCURY LAT.	MERCURY LAT.
1673								1674												1674
Dec 25	8.81	-2.47	223.49	1.09	248.45	0.01	284.61	4 Jan 9	255.9	316.3	-4.2	0.7	258.05	264.28	0.92	0.71	296.95	291.97	1.43	2.86
Jan 4	9.34	-2.43	224.93	1.11	255.51	-0.10	294.80	14 Jan 19	15.8	82.3	4.9	3.8	270.51	276.75	0.48	0.26	285.73	282.16	3.48	3.18
Jan 14	10.02	-2.40	226.16	1.14	262.62	-0.23	304.97	24 Jan 29	157.1	227.5	-2.5	-5.2	282.98	289.22	0.04	-0.18	282.18	284.92	2.40	1.50
Jan 24	10.85	-2.36	227.14	1.16	269.78	-0.36	315.11	3 Feb 8	289.4	348.6	-1.8	3.4	295.46	301.70	-0.39	-0.59	289.40	295.00	0.61	-0.18
Feb 3	11.80	-2.33	227.85	1.19	276.99	-0.49	325.22	13 Feb 18	50.0	119.7	5.2	0.8	307.93	314.17	-0.78	-0.94	301.36	308.30	-0.86	-1.41
Feb 13	12.85	-2.31	228.26	1.22	284.24	-0.63	335.28	23 Feb 28	195.1	262.2	-4.8	-3.8	320.40	326.62	-1.09	-1.21	315.73	323.62	-1.83	-2.09
Feb 23	13.98	-2.29	228.37	1.24	291.54	-0.78	345.29	5 Mar 10	321.9	21.9	1.2	4.9	332.85	339.07	-1.31	-1.38	331.96	340.78	-2.19	-2.10
Mar 5	15.17	-2.27	228.16	1.27	298.88	-0.93	355.25	15 Mar 20	86.2	158.2	3.4	-2.6	345.29	351.50	-1.43	-1.45	350.10	359.90	-1.81	-1.30
Mar 15	16.41	-2.26	227.66	1.29	306.24	-1.08	5.15	25 Mar 30	231.5	295.0	-4.9	-1.0	357.70	3.90	-1.44	-1.41	10.07	20.27	-0.57	0.32
Mar 25	17.67	-2.26	226.87	1.31	313.62	-1.24	14.99	4 Apr 9	354.3	56.7	3.7	4.7	10.10	16.29	-1.35	-1.27	29.88	38.19	1.28	2.12
Apr 4	18.94	-2.26	225.86	1.32	321.02	-1.39	24.78	14 Apr 19	124.2	196.8	0.1	-4.8	22.48	28.65	-1.16	-1.03	44.62	48.84	2.69	2.87
Apr 14	20.20	-2.26	224.68	1.32	328.41	-1.54	34.52	24 Apr 29	266.1	326.8	-3.1	2.0	34.83	40.99	-0.88	-0.72	50.71	50.30	2.57	1.75
Apr 24	21.42	-2.27	223.42	1.31	335.78	-1.69	44.20	4 May 9	27.4	93.4	5.0	2.4	47.16	53.31	-0.54	-0.35	48.12	45.17	0.51	-0.93
May 4	22.60	-2.29	222.16	1.29	343.12	-1.83	53.85	14 May 19	163.4	234.2	-3.3	-4.7	59.47	65.61	-0.16	0.04	42.71	41.68	-2.25	-3.19
May 14	23.71	-2.31	220.99	1.27	350.40	-1.96	63.45	24 May 29	299.0	358.6	-0.2	4.3	71.76	77.89	0.24	0.44	42.48	45.09	-3.69	-3.76
May 24	24.74	-2.34	219.99	1.23	357.61	-2.07	73.02	3 Jun 8	61.9	131.7	4.4	-1.0	84.03	90.16	0.63	0.81	49.33	55.03	-3.48	-2.92
Jun 3	25.68	-2.37	219.21	1.19	4.70	-2.18	82.57	13 Jun 18	202.4	269.6	-5.1	-2.6	96.29	102.41	0.98	1.14	62.07	70.38	-2.14	-1.21
Jun 13	26.49	-2.40	218.71	1.15	11.66	-2.26	92.10	23 Jun 28	330.9	31.0	2.7	5.2	108.53	114.65	1.27	1.38	79.86	90.25	-0.23	0.66
Jun 23	27.18	-2.44	218.50	1.11	18.45	-2.33	101.64	3 Jul 8	98.2	170.9	1.9	-4.1	120.76	126.86	1.47	1.53	101.06	111.72	1.35	1.74
Jul 3	27.72	-2.48	218.60	1.06	25.01	-2.38	111.17	13 Jul 18	240.1	303.2	-4.6	0.3	132.96	139.06	1.57	1.57	121.84	131.26	1.82	1.64
Jul 13	28.10	-2.53	218.99	1.02	31.30	-2.40	120.72	23 Jul 28	2.8	64.9	4.7	4.3	145.14	151.23	1.55	1.49	139.95	147.92	1.23	0.66
Jul 23	28.32	-2.57	219.66	0.98	37.26	-2.39	130.29	2 Aug 7	136.0	209.7	-1.5	-5.3	157.30	163.36	1.40	1.29	155.16	161.63	-0.04	-0.83
Aug 2	28.36	-2.62	220.59	0.94	42.81	-2.36	139.89	12 Aug 17	275.5	335.9	-2.2	3.1	169.41	175.46	1.14	0.96	167.22	171.71	-1.68	-2.55
Aug 12	28.23	-2.66	221.77	0.90	47.84	-2.29	149.53	22 Aug 27	35.5	100.4	5.2	1.7	181.49	187.50	0.76	0.54	174.79	175.98	-3.36	-4.02
Aug 22	27.93	-2.70	223.15	0.86	52.23	-2.19	159.21	1 Sep 6	174.7	246.7	-4.3	-4.1	193.51	199.50	0.29	0.03	174.74	170.99	-4.32	-4.01
Sep 1	27.47	-2.74	224.73	0.83	55.83	-2.04	168.94	11 Sep 16	309.1	8.5	0.9	4.8	205.47	211.42	-0.25	-0.54	165.93	162.20	-2.92	-1.32
Sep 11	26.88	-2.76	226.47	0.80	58.45	-1.84	178.72	21 Sep 26	69.7	137.8	3.8	-1.8	217.36	223.27	-0.84	-1.13	162.05	165.82	0.22	1.30
Sep 21	26.17	-2.78	228.35	0.78	59.88	-1.56	188.56	1 Oct 6	213.2	281.5	-5.0	-1.3	229.16	235.03	-1.43	-1.71	172.41	180.47	1.85	1.94
Oct 1	25.40	-2.79	230.34	0.76	59.91	-1.21	198.46	11 Oct 16	341.6	41.8	3.6	4.8	240.86	246.66	-1.98	-2.24	189.05	197.66	1.74	1.35
Oct 11	24.60	-2.79	232.43	0.73	58.47	-0.78	208.41	21 Oct 26	105.7	176.5	0.8	-4.4	252.41	258.13	-2.46	-2.66	206.11	214.35	0.85	0.30
Oct 21	23.82	-2.78	234.58	0.72	55.71	-0.27	218.42	31 Oct	250.0		-3.5		263.79		-2.83		222.40		-0.26	
								Nov 5		314.4		1.8		269.39		-2.95		230.28		-0.80
Oct 31	23.10	-2.76	236.79	0.70	52.16	0.26	228.48	10 Nov 15	13.9	76.4	5.0	3.0	274.91	280.34	-3.02	-3.05	238.03	245.66	-1.30	-1.75
Nov 10	22.47	-2.73	239.02	0.69	48.59	0.76	238.58	20 Nov 25	143.6	215.3	-2.7	-5.0	285.68	290.88	-3.01	-2.91	253.18	260.54	-2.10	-2.32
Nov 20	21.98	-2.69	241.26	0.68	45.78	1.18	248.72	30 N-D 5	284.7	346.1	-0.6	4.3	295.94	300.82	-2.73	-2.48	267.61	274.07	-2.38	-2.18
Nov 30	21.65	-2.65	243.47	0.67	44.20	1.49	258.89	10 Dec 15	46.5	112.6	4.7	-0.2	305.48	309.88	-2.14	-1.71	279.23	281.85	-1.63	-0.61
Dec 10	21.50	-2.60	245.65	0.66	43.98	1.70	269.07	20 Dec 25	182.8	252.8	-5.0	-3.2	313.95	317.62	-1.18	-0.54	280.28	274.53	0.90	2.44
Dec 20	21.53	-2.56	247.76	0.66	45.01	1.84	279.27	30 Dec	317.8		2.4		320.82		0.22		268.39		3.22	
1674								1675												1675
								Jan 4		17.7		5.3		323.42		1.10		265.71		3.08
Dec 30	21.75	-2.51	249.79	0.65	47.10	1.92	289.46	9 Jan 14	80.2	150.4	2.7	-3.4	325.31	326.34	2.10	3.21	266.77	270.38	2.45	1.66
Jan 9	22.14	-2.46	251.69	0.65	50.04	1.96	299.64	19 Jan 24	221.9	288.5	-4.9	-0.2	326.40	325.42	4.40	5.61	275.52	281.58	0.85	0.10
Jan 19	22.71	-2.42	253.46	0.65	53.64	1.97	309.80	29 J-F 3	350.0	49.7	4.5	4.6	323.46	320.72	6.73	7.65	288.27	295.40	-0.57	-1.14
Jan 29	23.43	-2.37	255.06	0.65	57.77	1.97	319.92	8 Feb 13	115.4	189.3	-0.5	-5.1	317.63	314.71	8.24	8.44	302.93	310.82	-1.60	-1.92
Feb 8	24.29	-2.34	256.46	0.66	62.29	1.94	330.01	18 Feb 23	259.5	322.4	-2.6	2.8	312.42	311.06	8.26	7.80	319.09	327.77	-2.10	-2.10
Feb 18	25.26	-2.30	257.63	0.66	67.13	1.91	340.05	28 Feb	22.2		5.1		310.72		7.13		336.87		-1.92	
								Mar 5		83.0		2.2		311.36		6.36		346.39		-1.52
Feb 28	26.34	-2.28	258.55	0.66	72.22	1.87	350.03	10 Mar 15	152.5	228.0	-3.5	-4.4	312.88	315.14	5.54	4.72	356.23	6.08	-0.88	-0.03
Mar 10	27.49	-2.25	259.19	0.67	77.50	1.83	359.96	20 Mar 25	294.9	355.4	0.6	4.6	318.01	321.40	3.91	3.14	15.33	23.12	0.97	1.96
Mar 20	28.70	-2.24	259.53	0.67	82.94	1.78	9.84	30 M-A 4	54.9	118.2	3.9	-1.1	325.20	329.33	2.41	1.73	28.69	31.57	2.75	3.15
Mar 30	29.96	-2.22	259.57	0.67	88.50	1.72	19.66	9 Apr 14	191.1	265.0	-5.0	-1.7	333.74	338.38	1.11	0.54	31.68	29.50	3.01	2.26
Apr 9	31.23	-2.22	259.29	0.67	94.16	1.67	29.42	19 Apr 24	328.6	28.0	3.5	4.9	343.21	348.20	0.02	-0.44	26.15	23.16	1.03	-0.38
Apr 19	32.51	-2.21	258.72	0.67	99.90	1.61	39.13	29 Apr	89.0		1.2		353.32		-0.85		21.65		-1.65	
								May 4		155.5		-4.0		358.55		-1.20		22.06		-2.59
Apr 29	33.78	-2.22	257.87	0.66	105.72	1.55	48.80	9 May 14	229.9	299.9	-4.1	1.6	3.88	9.29	-1.50	-1.74	24.35	28.25	-3.15	-3.36
May 9	35.02	-2.23	256.82	0.65	111.60	1.49	58.42	19 May 24	1.0	61.0	5.0	3.4	14.78	20.32	-1.94	-2.08	33.53	40.00	-3.25	-2.88
May 19	36.20	-2.24	255.61	0.64	117.54	1.43	68.01	29 M-J 3	124.8	194.6	-2.1	-5.2	25.92	31.57	-2.17	-2.22	47.57	56.19	-2.27	-1.48
May 29	37.32	-2.26	254.34	0.62	123.53	1.37	77.57	8 Jun 13	267.4	333.1	-1.3	4.1	37.26	42.99	-2.23	-2.20	65.83	76.30	-0.57	0.34
Jun 8	38.36	-2.28	253.09	0.59	129.57	1.30	87.11	18 Jun 23	33.0	95.1	4.9	0.6	48.75	54.55	-2.12	-2.02	87.21	98.02	1.12	1.65
Jun 18	39.29	-2.31	251.96	0.57	135.66	1.23	96.65	28 Jun	162.7		-4.6		60.38		-1.88		108.28		1.88	
								Jul 3		233.9		-3.9		66.23		-1.72		117.78		1.82
Jun 28	40.11	-2.34	251.01	0.54	141.79	1.17	106.18	8 Jul 13	302.9	5.3	2.0	5.2	72.11	78.02	-1.53	-1.33	126.46	134.32	1.50	0.97
Jul 8	40.80	-2.37	250.30	0.50	147.98	1.10	115.72	18 Jul 23	65.3	130.9	3.1	-2.6	83.95	89.91	-1.11	-0.88	141.33	147.42	0.26	-0.58
Jul 18	41.34	-2.41	249.87	0.47	154.22	1.02	125.28	28 J-A 2	202.1	271.8	-5.1	-0.8	95.89	101.90	-0.64	-0.40	152.46	156.21	-1.52	-2.51
Jul 28	41.72	-2.45	249.75	0.44	160.51	0.95	134.86	7 Aug 12	336.8	37.0	4.3	4.6	107.92	113.97	-0.16	0.08	158.32	158.37	-3.47	-4.25
Aug 7	41.93	-2.49	249.94	0.40	166.85	0.87	144.48	17 Aug 22	98.6	168.6	0.2	-4.8	120.04	126.13	0.30	0.52	156.12	152.00	-4.65	-4.37
Aug 17	41.96	-2.53	250.43	0.37	173.26	0.80	154.14	27 Aug	241.3		-3.2		132.25		0.71		147.66		-3.33	
								Sep 1		307.9		2.4		138.38		0.89		145.33		-1.82
Aug 27	41.81	-2.57	251.21	0.34	179.72	0.71	163.84	6 Sep 11	9.6	69.1	5.0	2.5	144.53	150.69	1.05	1.19	146.45	151.08	-0.29	0.89
Sep 6	41.49	-2.60	252.25	0.32	186.25	0.63	173.59	16 Sep 21	133.5	207.6	-2.9	-4.7	156.88	163.08	1.30	1.38	158.27	166.85	1.60	1.86
Sep 16	41.02	-2.63	253.53	0.29	192.85	0.55	183.40	26 S-O 1	278.9	342.2	0.2	4.5	169.29	175.52	1.43	1.46	175.88	184.85	1.79	1.48
Sep 26	40.41	-2.66	255.02	0.27	199.52	0.46	193.27	6 Oct 11	42.1	102.3	4.0	-0.6	181.76	188.00	1.46	1.43	193.59	202.03	1.04	0.51
Oct 6	39.69	-2.67	256.71	0.24	206.26	0.37	203.19	16 Oct 21	170.5	246.3	-4.9	-2.4	194.26	200.52	1.38	1.30	210.19	218.10	-0.05	-0.61
Oct 16	38.90	-2.67	258.55	0.22	213.07	0.28	213.17	26 Oct 31	314.5	15.3	3.3	5.0	206.79	213.07	1.19	1.06	225.79	233.29	-1.15	-1.65
Oct 26	38.09	-2.67	260.53	0.20	219.96	0.18	223.20	5 Nov 10	74.8	137.4	1.6	-3.6	219.35	225.63	0.92	0.75	240.57	247.59	-2.07	-2.38
Nov 5	37.31	-2.65	262.62	0.18	226.93	0.08	233.28	15 Nov 20	209.0	283.3	-4.6	1.0	231.91	238.20	0.58	0.39	254.19	260.02	-2.54	-2.47
Nov 15	36.58	-2.63	264.81	0.17	233.97	-0.02	243.40	25 Nov 30	348.2	47.8	5.1	3.7	244.49	250.78	0.19	0.00	264.39	266.06	-2.06	-1.15
Nov 25	35.97	-2.59	267.05	0.15	241.09	-0.12	253.55	5 Dec 10	108.6	174.6	-1.5	-5.2	257.07	263.36	-0.20	-0.40	263.54	257.27	0.31	1.93
Dec 5	35.49	-2.55	269.34	0.13	248.29	-0.23	263.73	15 Dec 20	247.5	318.4	-2.2	3.8	269.64	275.93	-0.59	-0.77	251.52	249.79	2.88	2.93
Dec 15	35.19	-2.51	271.64	0.12	255.56	-0.33	273.92	25 Dec 30	20.5	80.3	5.0	1.1	282.22	288.50	-0.94	-1.09	251.83	256.24	2.47	1.79
1675								1675												1675

JULIAN 12 UT	SATURN LONG.	SATURN LAT.	JUPITER LONG.	JUPITER LAT.	MARS LONG.	MARS LAT.	SUN LONG.	GREGORIAN 12 UT	MOON LONG.	MOON LONG.	MOON LAT.	MOON LAT.	VENUS LONG.	VENUS LONG.	VENUS LAT.	VENUS LAT.	MERCURY LONG.	MERCURY LONG.	MERCURY LAT.	MERCURY LAT.
1675								1676												1676
Dec 25	35.06	-2.46	273.94	0.10	262.90	-0.44	284.11	4 Jan 9	144.3	213.5	-4.2	-4.5	294.78	301.06	-1.22	-1.33	261.97	268.46	1.06	0.34
Jan 4	35.12	-2.41	276.22	0.08	270.31	-0.54	294.30	14 Jan 19	285.1	351.9	1.2	5.1	307.33	313.59	-1.42	-1.48	275.42	282.73	-0.31	-0.89
Jan 14	35.36	-2.36	278.44	0.07	277.78	-0.65	304.48	24 Jan 29	52.1	113.6	3.3	-2.0	319.86	326.11	-1.51	-1.52	290.33	298.21	-1.37	-1.75
Jan 24	35.79	-2.31	280.60	0.05	285.32	-0.76	314.62	3 Feb 8	182.0	252.7	-5.1	-1.6	332.35	338.58	-1.50	-1.45	306.40	314.91	-1.99	-2.09
Feb 3	36.38	-2.26	282.65	0.04	292.90	-0.86	324.72	13 Feb 18	321.1	24.2	3.9	4.6	344.81	351.02	-1.37	-1.27	323.79	333.01	-2.00	-1.71
Feb 13	37.13	-2.22	284.59	0.02	300.54	-0.96	334.79	23 Feb 28	83.8	148.7	0.5	-4.3	357.22	3.40	-1.13	-0.98	342.48	351.93	-1.17	-0.39
Feb 23	38.01	-2.18	286.37	0.00	308.20	-1.05	344.80	4 Mar 9	221.0	290.8	-3.7	1.9	9.56	15.71	-0.80	-0.60	0.72	7.92	0.62	1.73
Mar 4	39.01	-2.15	287.99	-0.01	315.90	-1.14	354.77	14 Mar 19	355.7	56.1	5.0	2.6	21.84	27.95	-0.38	-0.15	12.52	13.93	2.72	3.35
Mar 14	40.10	-2.12	289.40	-0.03	323.62	-1.21	4.67	24 Mar 29	116.5	185.8	-2.5	-4.9	34.04	40.10	0.10	0.35	12.24	8.57	3.37	2.69
Mar 24	41.27	-2.10	290.58	-0.05	331.33	-1.28	14.52	3 Apr 8	259.9	327.2	-0.5	4.5	46.14	52.15	0.60	0.85	4.77	2.41	1.49	0.14
Apr 3	42.50	-2.08	291.50	-0.07	339.05	-1.34	24.31	13 Apr 18	29.0	88.2	4.2	-0.4	58.13	64.09	1.10	1.34	2.16	3.95	-1.07	-2.01
Apr 13	43.77	-2.06	292.14	-0.10	346.74	-1.38	34.05	23 Apr 28	151.1	224.3	-4.7	-3.3	70.00	75.89	1.57	1.77	7.45	12.33	-2.63	-2.95
Apr 23	45.05	-2.05	292.49	-0.12	354.40	-1.41	43.74	3 May 8	297.5	1.8	2.9	5.1	81.73	87.53	1.96	2.11	18.36	25.38	-3.00	-2.79
May 3	46.34	-2.05	292.52	-0.15	2.01	-1.43	53.38	13 May 18	61.5	121.4	1.9	-3.3	93.28	98.99	2.24	2.33	33.32	42.16	-2.34	-1.70
May 13	47.62	-2.05	292.24	-0.17	9.56	-1.43	62.99	23 May 28	187.9	262.9	-5.1	0.0	104.63	110.22	2.38	2.39	51.87	62.36	-0.89	0.00
May 23	48.86	-2.05	291.65	-0.20	17.03	-1.41	72.56	2 Jun 7	333.4	34.9	5.0	3.9	115.73	121.24	2.35	2.26	73.30	84.18	0.85	1.51
Jun 2	50.06	-2.06	290.80	-0.23	24.40	-1.38	82.11	12 Jun 17	94.2	156.6	-1.1	-5.1	126.50	131.73	2.11	1.91	94.52	104.03	1.89	1.97
Jun 12	51.19	-2.07	289.73	-0.26	31.67	-1.33	91.65	22 Jun 27	226.4	300.6	-3.1	3.2	136.85	141.82	1.65	1.33	112.62	120.25	1.76	1.29
Jun 22	52.23	-2.09	288.51	-0.28	38.81	-1.26	101.18	2 Jul 7	7.3	67.1	5.1	1.4	146.63	151.26	0.94	0.48	126.90	132.47	0.60	-0.28
Jul 2	53.17	-2.11	287.22	-0.31	45.81	-1.17	110.72	12 Jul 17	127.9	194.0	-3.7	-4.8	155.66	159.79	-0.05	-0.65	136.78	139.58	-1.31	-2.42
Jul 12	54.00	-2.14	285.97	-0.33	52.65	-1.07	120.27	22 Jul 27	265.4	336.6	0.3	5.0	163.59	167.01	-1.32	-2.07	140.54	139.42	-3.52	-4.42
Jul 22	54.68	-2.17	284.83	-0.35	59.30	-0.95	129.83	1 Aug 6	39.8	99.3	3.4	-1.6	169.95	172.33	-2.90	-3.79	136.39	132.44	-4.87	-4.60
Aug 1	55.22	-2.20	283.88	-0.36	65.74	-0.81	139.43	11 Aug 16	163.5	232.8	-5.0	-2.4	174.03	174.91	-4.74	-5.72	129.29	128.53	-3.63	-2.23
Aug 11	55.60	-2.23	283.19	-0.38	71.93	-0.65	149.07	21 Aug 26	303.6	10.9	3.5	4.7	174.88	173.86	-6.68	-7.56	130.94	136.37	-0.77	0.46
Aug 21	55.80	-2.26	282.80	-0.39	77.85	-0.48	158.74	31 Aug	71.4		0.7		171.90		-8.25		144.11		1.31	
								Sep 5		132.5		-4.0		169.21		-8.66		153.15		1.74
Aug 31	55.83	-2.29	282.72	-0.40	83.44	-0.28	168.47	10 Sep 15	201.0	272.0	-4.2	1.2	166.17	163.24	-8.70	-8.35	162.61	171.95	1.80	1.60
Sep 10	55.67	-2.33	282.97	-0.41	88.64	-0.05	178.25	20 Sep 25	340.4	43.7	5.0	2.7	160.88	159.40	-7.67	-6.78	180.97	189.62	1.22	0.72
Sep 20	55.34	-2.35	283.52	-0.41	93.37	0.20	188.08	30 S-O 5	103.0	167.3	-2.3	-5.1	158.92	159.43	-5.76	-4.72	197.90	205.85	0.17	-0.41
Sep 30	54.85	-2.37	284.38	-0.42	97.52	0.49	197.98	10 Oct 15	239.7	310.4	-1.4	4.2	160.83	163.00	-3.70	-2.75	213.49	220.85	-0.99	-1.54
Oct 10	54.22	-2.39	285.50	-0.43	100.98	0.82	207.93	20 Oct 25	15.4	75.7	4.4	-0.1	165.82	169.18	-1.86	-1.07	227.91	234.57	-2.03	-2.43
Oct 20	53.49	-2.40	286.87	-0.43	103.56	1.19	217.93	30 Oct	135.7		-4.6		172.98		-0.35		240.68		-2.70	
								Nov 4		203.8		-4.0		177.16		0.27		245.85		-2.75
Oct 30	52.69	-2.40	288.46	-0.44	105.09	1.61	227.99	9 Nov 14	278.5	347.0	2.2	5.2	181.65	186.40	0.82	1.29	249.38	250.07	-2.47	-1.70
Nov 9	51.87	-2.39	290.24	-0.44	105.36	2.07	238.09	19 Nov 24	48.7	107.9	2.2	-3.0	191.36	196.50	1.68	2.00	246.67	240.16	-0.31	1.35
Nov 19	51.08	-2.37	292.18	-0.45	104.24	2.56	248.22	29 N-D 4	170.0	241.6	-5.3	-1.1	201.80	207.23	2.25	2.43	235.02	234.26	2.47	2.74
Nov 29	50.36	-2.34	294.25	-0.46	101.74	3.03	258.39	9 Dec 14	316.2	21.6	4.7	4.2	212.77	218.40	2.55	2.61	237.22	242.36	2.44	1.89
Dec 9	49.76	-2.30	296.43	-0.47	98.19	3.42	268.58	19 Dec 24	81.1	141.2	-0.7	-4.9	224.11	229.89	2.62	2.58	248.65	255.53	1.23	0.57
Dec 19	49.30	-2.26	298.70	-0.48	94.24	3.67	278.77	29 Dec	206.3		-3.9		235.73		2.50		262.76		-0.07	
1676								1677												1677
								Jan 3		279.9		2.3		241.62		2.37		270.23		-0.65
Dec 29	49.00	-2.21	301.03	-0.49	90.66	3.76	288.97	8 Jan 13	352.2	54.4	5.1	1.7	247.56	253.52	2.21	2.02	277.90	285.77	-1.16	-1.57
Jan 8	48.90	-2.17	303.39	-0.50	88.09	3.71	299.15	18 Jan 23	113.7	176.3	-3.4	-5.0	259.52	265.54	1.80	1.56	293.86	302.21	-1.88	-2.05
Jan 18	48.98	-2.12	305.76	-0.52	86.82	3.57	309.30	28 J-F 2	244.4	317.8	-0.7	4.6	271.59	277.65	1.30	1.04	310.84	319.74	-2.06	-1.87
Jan 28	49.25	-2.07	308.13	-0.53	86.86	3.37	319.44	7 Feb 12	26.1	86.2	3.6	-1.3	283.73	289.82	0.77	0.49	328.82	337.81	-1.44	-0.74
Feb 7	49.69	-2.02	310.47	-0.55	88.06	3.16	329.52	17 Feb 22	147.1	213.3	-4.9	-3.0	295.92	302.03	0.22	-0.04	346.10	352.63	0.24	1.42
Feb 17	50.31	-1.98	312.75	-0.57	90.23	2.94	339.56	27 Feb	283.4		2.8		308.14		-0.29		356.20		2.59	
								Mar 4		354.5		4.8		314.27		-0.52		356.06		3.44
Feb 27	51.07	-1.94	314.96	-0.60	93.16	2.74	349.56	9 Mar 14	58.5	118.0	0.8	-3.9	320.39	326.52	-0.74	-0.93	352.72	348.11	3.61	3.00
Mar 9	51.97	-1.90	317.07	-0.62	96.71	2.54	359.49	19 Mar 24	182.1	251.8	-4.6	0.5	332.65	338.79	-1.10	-1.25	344.54	343.22	1.85	0.58
Mar 19	52.98	-1.87	319.06	-0.65	100.74	2.36	9.37	29 M-A 3	322.1	29.3	4.9	3.0	344.92	351.06	-1.36	-1.45	344.24	347.19	-0.57	-1.49
Mar 29	54.09	-1.84	320.90	-0.68	105.15	2.19	19.19	8 Apr 13	90.2	150.7	-2.2	-5.1	357.19	3.32	-1.51	-1.55	351.62	357.23	-2.15	-2.56
Apr 8	55.27	-1.81	322.57	-0.71	109.88	2.03	28.95	18 Apr 23	218.7	290.9	-2.3	3.8	9.45	15.58	-1.55	-1.52	3.78	11.17	-2.73	-2.66
Apr 18	56.51	-1.79	324.04	-0.75	114.85	1.89	38.67	28 Apr	359.5		4.7		21.72		-1.47		19.35		-2.37	
								May 3		62.5		0.2		27.85		-1.39		28.30		-1.86
Apr 28	57.78	-1.78	325.28	-0.79	120.04	1.75	48.34	8 May 13	121.9	184.9	-4.4	-4.6	33.98	40.10	-1.28	-1.16	38.02	48.45	-1.16	-0.33
May 8	59.08	-1.76	326.27	-0.83	125.40	1.61	57.96	18 May 23	256.8	329.5	1.1	5.3	46.23	52.36	-1.01	-0.85	59.34	70.22	0.54	1.32
May 18	60.38	-1.76	326.99	-0.87	130.91	1.48	67.55	28 M-J 2	34.7	94.9	2.6	-2.7	58.49	64.62	-0.68	-0.49	80.56	90.00	1.85	2.09
May 28	61.66	-1.75	327.40	-0.92	136.55	1.36	77.11	7 Jun 12	154.7	220.9	-5.3	-2.1	70.75	76.88	-0.30	-0.10	98.39	105.67	2.01	1.64
Jun 7	62.91	-1.75	327.50	-0.96	142.31	1.24	86.65	17 Jun 22	295.7	6.4	4.1	4.4	83.02	89.16	0.10	0.30	111.78	116.60	0.98	0.08
Jun 17	64.12	-1.76	327.28	-1.01	148.18	1.13	96.18	27 Jun	68.2		-0.4		95.30		0.49		119.94		-1.03	
								Jul 2		127.4		-4.6		101.45		0.67		121.54		-2.27
Jun 27	65.25	-1.76	326.74	-1.05	154.16	1.01	105.72	7 Jul 12	189.1	258.6	-4.2	1.3	107.60	113.75	0.84	0.99	121.22	119.06	-3.49	-4.47
Jul 7	66.30	-1.78	325.92	-1.09	160.23	0.90	115.25	17 Jul 22	334.0	41.0	5.1	1.9	119.91	126.07	1.12	1.24	115.74	112.58	-4.93	-4.68
Jul 17	67.25	-1.79	324.87	-1.13	166.39	0.79	124.81	27 J-A 1	100.7	160.9	-3.1	-5.0	132.23	138.40	1.33	1.39	110.99	111.88	-3.80	-2.55
Jul 27	68.07	-1.81	323.65	-1.16	172.65	0.69	134.40	6 Aug 11	225.3	297.5	-1.4	4.2	144.57	150.74	1.43	1.45	115.51	121.71	-1.20	0.04
Aug 6	68.77	-1.83	322.34	-1.18	178.99	0.58	144.01	16 Aug 21	10.6	73.8	3.8	-1.2	156.91	163.09	1.43	1.39	129.94	139.38	0.99	1.57
Aug 16	69.31	-1.85	321.06	-1.19	185.43	0.47	153.66	26 Aug 31	133.3	195.9	-4.8	-3.4	169.26	175.44	1.32	1.22	149.23	158.93	1.78	1.69
Aug 26	69.68	-1.87	319.88	-1.19	191.96	0.37	163.37	5 Sep 10	263.4	336.2	2.1	5.0	181.61	187.78	1.10	0.96	168.23	177.08	1.38	0.93
Sep 5	69.88	-1.89	318.89	-1.19	198.59	0.26	173.11	15 Sep 20	45.0	105.7	1.1	-3.8	193.95	200.12	0.79	0.60	185.49	193.48	0.39	-0.20
Sep 15	69.90	-1.92	318.17	-1.18	205.31	0.16	182.92	25 Sep 30	166.5	232.6	-4.8	-0.3	206.28	212.45	0.40	0.18	201.08	208.31	-0.81	-1.41
Sep 25	69.73	-1.94	317.75	-1.16	212.12	0.05	192.79	5 Oct 10	302.6	13.3	4.7	3.5	218.60	224.76	-0.05	-0.28	215.12	221.42	-1.98	-2.47
Oct 5	69.39	-1.96	317.66	-1.14	219.03	-0.05	202.70	15 Oct 20	77.7	137.4	-2.0	-5.2	230.90	237.05	-0.52	-0.75	227.00	231.47	-2.84	-3.03
Oct 15	68.88	-1.97	317.91	-1.12	226.04	-0.15	212.68	25 Oct 30	200.9	270.9	-3.0	3.1	243.18	249.32	-0.98	-1.20	234.11	233.78	-2.89	-2.25
Oct 25	68.24	-1.98	318.48	-1.09	233.14	-0.25	222.71	4 Nov 9	341.6	48.4	5.1	0.6	255.44	261.56	-1.41	-1.60	229.61	223.14	-0.95	0.74
Nov 4	67.50	-1.98	319.36	-1.07	240.33	-0.35	232.78	14 Nov 19	109.6	169.7	-4.3	-4.8	267.66	273.75	-1.76	-1.90	218.79	219.00	2.01	2.49
Nov 14	66.69	-1.97	320.52	-1.05	247.62	-0.45	242.90	24 Nov 29	236.8	310.1	0.1	5.1	279.83	285.90	-2.01	-2.09	222.82	228.65	2.38	1.96
Nov 24	65.87	-1.96	321.94	-1.03	254.99	-0.54	253.06	4 Dec 9	19.0	81.8	3.1	-2.3	291.94	297.96	-2.13	-2.14	235.46	242.73	1.39	0.77
Dec 4	65.08	-1.93	323.57	-1.01	262.45	-0.63	263.23	14 Dec 19	141.5	203.2	-5.2	-2.7	303.95	309.91	-2.10	-2.02	250.22	257.84	0.16	-0.42
Dec 14	64.37	-1.90	325.40	-1.00	269.98	-0.72	273.42	24 Dec 29	274.3	348.6	3.3	4.7	315.83	321.71	-1.89	-1.72	265.58	273.44	-0.95	-1.40
1677								1677												1677

JULIAN 12 UT	SATURN LONG.	SATURN LAT.	JUPITER LONG.	JUPITER LAT.	MARS LONG.	MARS LAT.	SUN LONG.
1677							
Dec 24	63.77	-1.87	327.39	-0.99	277.59	-0.80	283.62
Jan 3	63.32	-1.83	329.52	-0.98	285.26	-0.87	293.80
Jan 13	63.04	-1.79	331.75	-0.98	292.99	-0.94	303.98
Jan 23	62.95	-1.74	334.07	-0.97	300.77	-1.00	314.13
Feb 2	63.05	-1.70	336.44	-0.97	308.58	-1.05	324.23
Feb 12	63.34	-1.66	338.84	-0.98	316.41	-1.09	334.30
Feb 22	63.80	-1.61	341.26	-0.99	324.26	-1.12	344.32
Mar 4	64.43	-1.57	343.67	-1.00	332.11	-1.14	354.28
Mar 14	65.20	-1.54	346.05	-1.01	339.95	-1.14	4.19
Mar 24	66.11	-1.50	348.37	-1.03	347.76	-1.13	14.04
Apr 3	67.13	-1.47	350.63	-1.05	355.54	-1.11	23.84
Apr 13	68.24	-1.44	352.79	-1.07	3.27	-1.08	33.58
Apr 23	69.43	-1.42	354.84	-1.10	10.94	-1.03	43.27
May 3	70.67	-1.40	356.74	-1.13	18.54	-0.98	52.91
May 13	71.96	-1.38	358.49	-1.16	26.06	-0.90	62.52
May 23	73.26	-1.37	0.04	-1.20	33.49	-0.82	72.10
Jun 2	74.56	-1.36	1.37	-1.24	40.82	-0.73	81.65
Jun 12	75.85	-1.35	2.46	-1.29	48.06	-0.63	91.19
Jun 22	77.11	-1.35	3.28	-1.34	55.18	-0.52	100.72
Jul 2	78.31	-1.34	3.80	-1.39	62.18	-0.40	110.25
Jul 12	79.45	-1.34	4.00	-1.44	69.06	-0.27	119.80
Jul 22	80.51	-1.35	3.88	-1.48	75.80	-0.13	129.37
Aug 1	81.46	-1.35	3.43	-1.53	82.40	0.01	138.96
Aug 11	82.29	-1.36	2.68	-1.57	88.84	0.17	148.60
Aug 21	82.98	-1.37	1.67	-1.61	95.12	0.33	158.28
Aug 31	83.52	-1.38	0.46	-1.63	101.20	0.50	168.00
Sep 10	83.90	-1.39	359.14	-1.64	107.08	0.69	177.78
Sep 20	84.09	-1.41	357.82	-1.64	112.72	0.89	187.61
Sep 30	84.10	-1.42	356.58	-1.62	118.08	1.10	197.50
Oct 10	83.93	-1.43	355.52	-1.60	123.11	1.33	207.45
Oct 20	83.57	-1.43	354.73	-1.56	127.76	1.58	217.45
Oct 30	83.06	-1.44	354.24	-1.52	131.92	1.86	227.50
Nov 9	82.41	-1.44	354.09	-1.47	135.51	2.16	237.60
Nov 19	81.66	-1.43	354.28	-1.43	138.37	2.50	247.73
Nov 29	80.85	-1.42	354.82	-1.38	140.35	2.87	257.90
Dec 9	80.02	-1.40	355.67	-1.34	141.26	3.27	268.08
Dec 19	79.23	-1.38	356.80	-1.29	140.92	3.68	278.27
1678							
Dec 29	78.52	-1.35	358.20	-1.26	139.23	4.07	288.47
Jan 8	77.94	-1.32	359.83	-1.22	136.29	4.37	298.65
Jan 18	77.50	-1.29	1.65	-1.19	132.52	4.53	308.81
Jan 28	77.24	-1.25	3.63	-1.16	128.58	4.52	318.94
Feb 7	77.16	-1.21	5.74	-1.14	125.19	4.34	329.03
Feb 17	77.27	-1.17	7.96	-1.12	122.89	4.04	339.07
Feb 27	77.56	-1.14	10.27	-1.11	121.88	3.69	349.06
Mar 9	78.04	-1.10	12.63	-1.10	122.12	3.32	359.00
Mar 19	78.67	-1.07	15.02	-1.09	123.47	2.97	8.88
Mar 29	79.46	-1.04	17.43	-1.08	125.74	2.64	18.70
Apr 8	80.37	-1.01	19.84	-1.08	128.76	2.35	28.48
Apr 18	81.39	-0.98	22.22	-1.08	132.37	2.08	38.19
Apr 28	82.50	-0.96	24.55	-1.09	136.47	1.83	47.86
May 8	83.69	-0.94	26.82	-1.10	140.96	1.61	57.49
May 18	84.94	-0.92	29.00	-1.11	145.78	1.40	67.08
May 28	86.22	-0.90	31.07	-1.12	150.88	1.21	76.64
Jun 7	87.52	-0.89	33.02	-1.14	156.21	1.03	86.19
Jun 17	88.83	-0.87	34.81	-1.16	161.75	0.87	95.72
Jun 27	90.12	-0.86	36.42	-1.19	167.48	0.71	105.25
Jul 7	91.38	-0.85	37.82	-1.22	173.37	0.56	114.80
Jul 17	92.58	-0.85	38.98	-1.25	179.41	0.42	124.35
Jul 27	93.72	-0.84	39.88	-1.28	185.60	0.28	133.93
Aug 6	94.78	-0.84	40.49	-1.31	191.94	0.15	143.55
Aug 16	95.73	-0.83	40.78	-1.35	198.40	0.03	153.19
Aug 26	96.57	-0.83	40.75	-1.38	204.99	-0.09	162.90
Sep 5	97.26	-0.83	40.38	-1.41	211.71	-0.21	172.65
Sep 15	97.80	-0.83	39.69	-1.43	218.55	-0.32	182.45
Sep 25	98.17	-0.83	38.72	-1.45	225.51	-0.42	192.31
Oct 5	98.35	-0.83	37.53	-1.46	232.59	-0.52	202.23
Oct 15	98.36	-0.83	36.21	-1.45	239.78	-0.61	212.20
Oct 25	98.17	-0.82	34.86	-1.44	247.07	-0.70	222.23
Nov 4	97.81	-0.82	33.57	-1.41	254.47	-0.78	232.30
Nov 14	97.29	-0.81	32.46	-1.38	261.97	-0.85	242.41
Nov 24	96.63	-0.80	31.59	-1.33	269.56	-0.91	252.57
Dec 4	95.87	-0.79	31.03	-1.28	277.22	-0.97	262.74
Dec 14	95.06	-0.78	30.81	-1.23	284.95	-1.01	272.93
1679							

GREGORIAN 12 UT	MOON LONG.		MOON LAT.		VENUS LONG.		VENUS LAT.		MERCURY LONG.		MERCURY LAT.	
1678												
3 Jan 8	54.2	114.5	0.0	-4.4	327.53	333.30	-1.51	-1.24	281.44	289.63	-1.76	-2.00
13 Jan 18	174.2	238.5	-4.3	0.5	338.99	344.60	-0.93	-0.58	298.01	306.57	-2.09	-2.00
23 Jan 28	312.9	25.3	5.0	2.3	350.11	355.52	-0.18	0.26	315.24	323.74	-1.68	-1.09
2 Feb 7	87.7	147.2	-2.9	-5.0	0.79	5.90	0.73	1.25	331.46	337.27	-0.17	1.04
12 Feb 17	208.3	275.9	-1.9	3.6	10.84	15.55	1.79	2.37	339.76	338.13	2.36	3.41
22 Feb 27	351.2	59.8	4.3	-1.0	20.02	24.18	2.97	3.58	333.43	328.48	3.73	3.18
4 Mar 9	120.2	180.4	-4.8	-3.7	27.96	31.30	4.21	4.83	325.63	325.49	2.12	0.93
14 Mar 19	244.4	314.9	1.5	5.1	34.10	36.24	5.44	6.01	327.65	331.53	-0.14	-1.04
24 Mar 29	28.0	92.7	1.7	-3.7	37.61	38.08	6.51	6.91	336.67	342.79	-1.72	-2.20
3 Apr 8	152.3	214.8	-5.0	-1.0	37.54	35.98	7.13	7.11	349.71	357.34	-2.46	-2.51
13 Apr 18	282.4	353.9	4.4	4.2	33.54	30.54	6.80	6.15	5.64	14.62	-2.36	-1.98
23 Apr 28	62.9	124.6	-1.5	-5.2	27.43	24.71	5.22	4.10	24.29	34.60	-1.40	-0.64
3 May 8	184.9	251.0	-3.5	2.2	22.75	21.75	2.91	1.75	45.37	56.16	0.22	1.07
13 May 18	321.8	31.6	5.3	1.3	21.76	22.70	0.69	-0.23	66.42	75.70	1.76	2.17
23 May 28	96.2	156.3	-4.0	-4.9	24.44	26.87	-1.00	-1.64	83.76	90.51	2.25	1.99
2 Jun 7	218.6	289.1	-0.7	4.6	29.85	33.30	-2.15	-2.55	95.85	99.64	1.40	0.50
12 Jun 17	1.0	67.3	3.7	-1.9	37.13	41.28	-2.84	-3.05	101.70	101.88	-0.67	-2.02
22 Jun 27	128.8	188.5	-5.1	-3.0	45.69	50.31	-3.17	-3.22	100.30	97.51	-3.33	-4.34
2 Jul 7	254.1	328.3	2.5	4.9	55.12	60.08	-3.21	-3.14	94.62	92.83	-4.79	-4.59
12 Jul 17	38.5	101.3	0.5	-4.3	65.18	70.40	-3.02	-2.86	92.99	95.45	-3.85	-2.78
22 Jul 27	161.1	221.9	-4.4	0.0	75.72	81.12	-2.66	-2.43	100.22	107.11	-1.56	-0.37
1 Aug 6	291.7	7.1	4.7	2.8	86.61	92.17	-2.17	-1.90	115.76	125.54	0.64	1.35
11 Aug 16	73.9	134.4	-2.7	-5.0	97.80	103.49	-1.61	-1.31	135.74	145.76	1.71	1.75
21 Aug 26	193.9	257.3	-2.2	3.1	109.23	115.03	-1.00	-0.70	155.35	164.40	1.54	1.14
31 Aug	330.5		4.7		120.87		-0.40		172.92		0.63	
Sep 5		44.0		-0.5		126.76		-0.10		180.95		0.03
10 Sep 15	107.5	167.0	-4.8	-4.0	132.69	138.66	0.17	0.44	188.51	195.58	-0.61	-1.27
20 Sep 25	228.0	294.8	1.0	5.1	144.67	150.71	0.68	0.90	202.13	208.04	-1.91	-2.50
30 S-O 5	9.1	78.8	2.5	-3.5	156.78	162.88	1.09	1.25	213.06	216.77	-2.99	-3.30
10 Oct 15	140.0	199.9	-5.1	-1.5	169.01	175.16	1.39	1.49	218.44	217.09	-3.30	-2.79
20 Oct 25	263.9	333.8	3.9	4.7	181.34	187.53	1.57	1.61	212.31	206.17	-1.58	0.11
30 Oct	46.2		-0.9		193.74		1.62		202.76		1.51	
Nov 4		112.0		-5.1		199.97		1.59		203.93		2.20
9 Nov 14	171.8	233.9	-3.7	1.5	206.21	212.46	1.54	1.46	208.56	215.04	2.28	2.00
19 Nov 24	302.0	12.7	5.2	2.1	218.72	224.99	1.36	1.23	222.36	230.00	1.52	0.96
29 N-D 4	81.5	144.0	-3.7	-4.9	231.26	237.54	1.08	0.91	237.75	245.54	0.37	-0.21
9 Dec 14	203.8	269.6	-1.1	4.1	243.82	250.11	0.73	0.54	253.35	261.20	-0.75	-1.23
19 Dec 24	341.2	50.5	4.2	-1.3	256.39	262.68	0.34	0.14	269.13	277.15	-1.63	-1.93
29 Dec	115.4		-5.0		268.97		-0.07		285.28		-2.11	
1679												
Jan 3		175.6		-3.2		275.26		-0.27		293.51		-2.11
8 Jan 13	236.8	307.2	2.0	5.0	281.54	287.83	-0.46	-0.64	301.75	309.73	-1.90	-1.42
18 Jan 23	20.1	86.5	1.1	-4.1	294.11	300.39	-0.81	-0.97	316.82	321.86	-0.59	0.60
28 J-F 2	148.2	207.5	-4.5	-0.3	306.67	312.94	-1.11	-1.22	323.27	320.26	2.03	3.26
7 Feb 12	271.6	345.8	4.4	3.5	319.21	325.48	-1.31	-1.38	314.61	309.85	3.72	3.26
17 Feb 22	57.5	120.9	-2.4	-5.1	331.74	337.99	-1.42	-1.44	307.98	308.98	2.30	1.22
27 Feb	180.6		-2.4		344.23		-1.43		312.11		0.21	
Mar 4		240.8		2.8		350.47		-1.39		316.74		-0.65
9 Mar 14	308.6	24.2	5.1	0.3	356.70	2.92	-1.33	-1.24	322.42	328.92	-1.34	-1.86
19 Mar 24	93.0	153.9	-4.7	-4.2	9.13	15.33	-1.13	-0.99	336.08	343.85	-2.20	-2.35
29 M-A 3	213.3	276.1	0.5	4.9	21.52	27.70	-0.84	-0.66	352.20	1.14	-2.31	-2.06
8 Apr 13	347.1	61.3	3.4	-2.9	33.87	40.03	-0.48	-0.28	10.70	20.84	-1.60	-0.94
18 Apr 23	126.7	186.2	-5.2	-1.9	46.18	52.31	-0.07	0.15	31.41	42.04	-0.11	0.78
28 Apr	247.3		3.4		58.43		0.37		52.13		1.60	
May 3		313.6		5.1		64.54		0.58		61.15		2.18
8 May 13	25.9	96.7	0.2	-4.9	70.63	76.72	0.79	0.99	68.74	74.75	2.45	2.35
18 May 23	159.0	218.7	-3.9	1.0	82.78	88.83	1.18	1.35	79.03	81.46	1.86	0.98
28 M-J 2	283.1	352.6	5.0	2.9	94.87	100.89	1.50	1.63	81.96	80.66	-0.23	-1.65
7 Jun 12	63.8	130.5	-3.1	-4.9	106.89	112.88	1.73	1.80	78.15	75.45	-3.00	-4.00
17 Jun 22	190.7	252.3	-1.3	3.7	118.84	124.78	1.84	1.85	73.67	73.58	-4.45	-4.34
27 Jun	321.0		4.5		130.70		1.82		75.47		-3.78	
Jul 2		31.8		-0.6		136.59		1.75		79.34		-2.91
7 Jul 12	100.2	163.0	-4.9	-3.3	142.46	148.29	1.64	1.50	85.09	92.58	-1.86	-0.75
17 Jul 22	222.5	287.8	1.7	5.0	154.09	159.85	1.32	1.10	101.59	111.65	0.27	1.09
27 J-A 1	0.1	69.9	1.9	-3.8	165.57	171.25	0.84	0.54	122.13	132.45	1.59	1.77
6 Aug 11	134.8	194.9	-4.6	-0.5	176.87	182.44	0.22	-0.14	142.29	151.52	1.67	1.35
16 Aug 21	255.5	325.1	4.2	4.2	187.95	193.38	-0.53	-0.94	160.16	168.21	0.87	0.27
26 Aug 31	39.0	106.4	-1.7	-5.2	198.72	203.97	-1.37	-1.82	175.69	182.59	-0.39	-1.10
5 Sep 10	167.9	227.0	-2.7	2.5	209.11	214.12	-2.28	-2.75	188.84	194.30	-1.82	-2.51
15 Sep 20	290.3	3.5	5.2	1.4	218.98	223.65	-3.22	-3.68	198.69	201.56	-3.12	-3.57
25 Sep 30	76.7	140.9	-4.5	-4.5	228.11	232.30	-4.12	-4.54	202.19	199.86	-3.71	-3.31
5 Oct 10	200.4	260.4	0.2	4.7	236.18	239.67	-4.92	-5.26	194.70	189.19	-2.18	-0.52
15 Oct 20	327.0	41.9	4.2	-2.1	242.68	245.10	-5.52	-5.69	186.87	188.98	0.98	1.86
25 Oct 30	112.4	173.7	-5.3	-2.2	246.82	247.71	-5.74	-5.63	194.36	201.46	2.14	2.01
4 Nov 9	233.0	295.7	3.0	5.2	247.66	246.60	-5.31	-4.75	209.27	217.29	1.63	1.13
14 Nov 19	5.1	79.3	1.2	-4.5	244.58	241.84	-3.92	-2.85	225.31	233.27	0.58	0.01
24 Nov 29	146.2	205.7	-4.0	0.7	238.83	236.03	-1.61	-0.31	241.17	249.03	-0.54	-1.05
4 Dec 9	266.9	333.0	4.8	3.5	233.91	232.72	0.91	1.96	256.88	264.74	-1.50	-1.86
14 Dec 19	43.9	115.2	-2.3	-5.0	232.56	233.39	2.81	3.45	272.64	280.54	-2.11	-2.20
24 Dec 29	178.3	237.8	-1.5	3.4	235.08	237.53	3.91	4.21	288.33	295.76	-2.10	-1.73
1679												

JULIAN 12 UT	SATURN LONG.	SATURN LAT.	JUPITER LONG.	JUPITER LAT.	MARS LONG.	MARS LAT.	SUN LONG.
1679							
Dec 24	94.23	-0.76	30.93	-1.18	292.75	-1.05	283.12
Jan 3	93.45	-0.74	31.40	-1.13	300.59	-1.07	293.31
Jan 13	92.75	-0.71	32.18	-1.08	308.46	-1.08	303.48
Jan 23	92.17	-0.69	33.25	-1.03	316.36	-1.08	313.63
Feb 2	91.74	-0.66	34.59	-0.99	324.26	-1.07	323.74
Feb 12	91.49	-0.63	36.15	-0.95	332.16	-1.05	333.81
Feb 22	91.42	-0.60	37.90	-0.92	340.04	-1.01	343.83
Mar 3	91.54	-0.58	39.82	-0.89	347.89	-0.97	353.80
Mar 13	91.84	-0.55	41.87	-0.86	355.70	-0.91	3.71
Mar 23	92.32	-0.52	44.02	-0.83	3.46	-0.85	13.57
Apr 2	92.96	-0.50	46.26	-0.81	11.15	-0.77	23.36
Apr 12	93.74	-0.48	48.56	-0.79	18.77	-0.69	33.10
Apr 22	94.65	-0.45	50.90	-0.78	26.31	-0.60	42.80
May 2	95.67	-0.43	53.25	-0.76	33.77	-0.50	52.45
May 12	96.78	-0.41	55.61	-0.75	41.14	-0.40	62.06
May 22	97.97	-0.40	57.94	-0.74	48.42	-0.29	71.64
Jun 1	99.21	-0.38	60.23	-0.73	55.60	-0.18	81.19
Jun 11	100.49	-0.36	62.47	-0.73	62.69	-0.07	90.73
Jun 21	101.79	-0.35	64.63	-0.72	69.68	0.05	100.26
Jul 1	103.09	-0.33	66.68	-0.72	76.57	0.16	109.80
Jul 11	104.38	-0.32	68.62	-0.72	83.36	0.28	119.34
Jul 21	105.63	-0.31	70.40	-0.72	90.04	0.41	128.91
Jul 31	106.84	-0.30	72.02	-0.73	96.63	0.53	138.50
Aug 10	107.98	-0.28	73.43	-0.73	103.10	0.66	148.13
Aug 20	109.03	-0.27	74.61	-0.74	109.47	0.78	157.81
Aug 30	109.98	-0.26	75.53	-0.74	115.71	0.91	167.53
Sep 9	110.81	-0.25	76.16	-0.75	121.83	1.05	177.30
Sep 19	111.49	-0.24	76.47	-0.75	127.81	1.19	187.13
Sep 29	112.02	-0.22	76.46	-0.76	133.64	1.33	197.02
Oct 9	112.38	-0.21	76.11	-0.76	139.30	1.48	206.96
Oct 19	112.56	-0.20	75.44	-0.76	144.75	1.64	216.96
Oct 29	112.55	-0.19	74.48	-0.75	149.97	1.80	227.01
Nov 8	112.35	-0.17	73.30	-0.74	154.92	1.98	237.10
Nov 18	111.98	-0.16	71.98	-0.72	159.53	2.17	247.24
Nov 28	111.45	-0.14	70.63	-0.70	163.74	2.38	257.40
Dec 8	110.78	-0.12	69.33	-0.67	167.45	2.61	267.58
Dec 18	110.02	-0.11	68.20	-0.64	170.53	2.85	277.78
1680							
Dec 28	109.20	-0.09	67.31	-0.60	172.86	3.11	287.97
Jan 7	108.38	-0.07	66.72	-0.57	174.24	3.39	298.15
Jan 17	107.60	-0.05	66.46	-0.53	174.50	3.66	308.32
Jan 27	106.91	-0.03	66.54	-0.49	173.49	3.90	318.45
Feb 6	106.34	-0.02	66.95	-0.46	171.20	4.07	328.54
Feb 16	105.93	0.00	67.67	-0.42	167.85	4.12	338.59
Feb 26	105.69	0.02	68.68	-0.39	163.97	4.01	348.58
Mar 8	105.63	0.04	69.94	-0.36	160.25	3.74	358.52
Mar 18	105.76	0.05	71.42	-0.33	157.36	3.37	8.41
Mar 28	106.06	0.07	73.09	-0.31	155.67	2.94	18.23
Apr 7	106.54	0.08	74.92	-0.29	155.27	2.50	28.00
Apr 17	107.18	0.10	76.88	-0.26	156.09	2.09	37.72
Apr 27	107.96	0.11	78.95	-0.24	157.95	1.72	47.39
May 7	108.86	0.12	81.11	-0.22	160.67	1.38	57.02
May 17	109.88	0.14	83.32	-0.21	164.10	1.09	66.62
May 27	110.98	0.15	85.58	-0.19	168.11	0.82	76.18
Jun 6	112.16	0.16	87.86	-0.17	172.61	0.58	85.73
Jun 16	113.39	0.18	90.14	-0.16	177.50	0.36	95.26
Jun 26	114.66	0.19	92.40	-0.14	182.73	0.16	104.79
Jul 6	115.96	0.21	94.63	-0.12	188.25	-0.02	114.33
Jul 16	117.25	0.22	96.81	-0.11	194.04	-0.18	123.89
Jul 26	118.53	0.24	98.90	-0.09	200.06	-0.33	133.46
Aug 5	119.78	0.25	100.91	-0.08	206.29	-0.47	143.08
Aug 15	120.98	0.27	102.79	-0.06	212.71	-0.59	152.73
Aug 25	122.11	0.29	104.53	-0.05	219.31	-0.71	162.42
Sep 4	123.15	0.30	106.09	-0.03	226.08	-0.81	172.17
Sep 14	124.09	0.32	107.45	-0.01	233.01	-0.90	181.97
Sep 24	124.91	0.34	108.58	0.01	240.08	-0.98	191.83
Oct 4	125.58	0.37	109.44	0.03	247.29	-1.05	201.74
Oct 14	126.10	0.39	110.02	0.05	254.62	-1.10	211.71
Oct 24	126.45	0.41	110.27	0.07	262.07	-1.15	221.73
Nov 3	126.61	0.44	110.20	0.09	269.63	-1.18	231.81
Nov 13	126.58	0.46	109.80	0.12	277.27	-1.20	241.92
Nov 23	126.37	0.49	109.09	0.15	285.00	-1.20	252.07
Dec 3	125.98	0.51	108.10	0.17	292.79	-1.20	262.25
Dec 13	125.43	0.53	106.90	0.20	300.63	-1.18	272.44
1681							

GREGORIAN 12 UT	MOON LONG.		MOON LAT.		VENUS LONG.		VENUS LAT.		MERCURY LONG.		MERCURY LAT.	
1680												1680
3 Jan 8	302.4	11.7	4.7	0.2	240.59	244.15	4.37	4.41	302.18	306.40	-1.02	0.13
13 Jan 18	82.1	149.3	-4.7	-3.4	248.12	252.43	4.36	4.23	306.74	302.55	1.61	3.00
23 Jan 28	209.7	270.8	1.6	5.0	257.02	261.84	4.03	3.78	296.37	292.15	3.59	3.26
2 Feb 7	339.6	50.8	2.7	-3.3	266.84	272.00	3.49	3.16	291.36	293.41	2.41	1.44
12 Feb 17	119.0	182.0	-4.9	-0.7	277.29	282.69	2.81	2.44	297.37	302.60	0.51	-0.31
22 Feb 27	241.4	305.4	4.1	4.7	288.18	293.75	2.05	1.67	308.72	315.50	-1.00	-1.55
3 Mar 8	17.9	89.2	-0.8	-5.2	299.38	305.07	1.28	0.90	322.83	330.67	-1.95	-2.18
13 Mar 18	153.9	214.1	-3.1	2.2	310.81	316.59	0.54	0.19	339.00	347.86	-2.24	-2.10
23 Mar 28	274.1	341.7	5.3	2.5	322.41	328.25	-0.14	-0.45	357.25	7.17	-1.76	-1.20
2 Apr 7	56.6	125.7	-3.9	-4.7	334.13	340.02	-0.73	-0.97	17.50	27.88	-0.44	0.46
12 Apr 17	187.2	246.5	-0.2	4.4	345.94	351.87	-1.19	-1.37	37.73	46.39	1.37	2.14
22 Apr 27	308.7	19.6	4.6	-0.9	357.82	3.78	-1.52	-1.64	53.36	58.39	2.61	2.69
2 May 7	94.5	160.2	-5.2	-2.5	9.76	15.74	-1.72	-1.76	61.32	62.09	2.33	1.51
12 May 17	219.7	280.0	2.7	5.1	21.74	27.74	-1.77	-1.75	60.90	58.39	0.28	-1.16
22 May 27	345.2	58.4	2.0	-3.9	33.75	39.77	-1.70	-1.61	55.63	53.77	-2.51	-3.49
1 Jun 6	130.7	193.0	-4.2	0.5	45.80	51.84	-1.51	-1.37	53.53	55.14	-3.98	-4.00
11 Jun 16	252.4	315.2	4.6	3.9	57.88	63.94	-1.22	-1.05	58.55	63.60	-3.63	-2.96
21 Jun 26	23.4	96.8	-1.6	-5.0	70.00	76.07	-0.86	-0.66	70.17	78.16	-2.09	-1.09
1 Jul 6	164.8	224.9	-1.7	3.3	82.16	88.25	-0.46	-0.25	87.45	97.71	-0.09	0.79
11 Jul 16	286.0	352.3	4.9	0.9	94.35	100.46	-0.04	0.17	108.41	118.98	1.43	1.75
21 Jul 26	62.7	133.5	-4.4	-3.8	106.59	112.72	0.37	0.56	129.04	138.43	1.78	1.55
31 J-A 5	197.3	256.9	1.4	5.0	118.87	125.03	0.74	0.90	147.14	155.19	1.12	0.54
10 Aug 15	321.1	30.9	3.3	-2.6	131.19	137.37	1.04	1.17	162.56	169.24	-0.15	-0.91
20 Aug 25	101.5	168.2	-5.1	-1.1	143.56	149.76	1.26	1.34	175.13	180.06	-1.71	-2.51
30 Aug	229.0		3.9		155.96		1.39		183.72		-3.25	
Sep 4		289.7		5.0		162.17		1.41		185.64		-3.84
9 Sep 14	357.9	70.3	0.2	-5.0	168.39	174.61	1.41	1.38	185.19	181.98	-4.10	-3.80
19 Sep 24	138.6	201.3	-3.5	1.9	180.84	187.08	1.32	1.24	176.77	172.20	-2.72	-1.11
29 S-O 4	260.8	323.8	5.2	3.1	193.32	199.56	1.13	1.00	171.08	174.11	0.44	1.48
9 Oct 14	36.0	108.8	-3.1	-5.0	205.80	212.05	0.85	0.68	180.19	187.87	1.95	1.98
19 Oct 24	173.6	233.7	-0.6	4.2	218.29	224.54	0.50	0.31	196.16	204.56	1.72	1.29
29 Oct	293.6		4.7		230.79		0.10		212.86		0.77	
Nov 3		359.6		0.0		237.03		-0.10		221.00		0.22
8 Nov 13	74.8	145.4	-5.0	-2.8	243.28	249.53	-0.31	-0.52	229.00	236.87	-0.34	-0.87
18 Nov 23	207.0	266.3	2.4	5.0	255.77	262.01	-0.72	-0.91	244.66	252.38	-1.36	-1.77
28 N-D 3	327.8	37.1	2.5	-3.2	268.26	274.49	-1.09	-1.26	260.04	267.61	-2.09	-2.28
8 Dec 13	112.8	179.7	-4.5	0.3	280.73	286.95	-1.40	-1.52	274.96	281.81	-2.28	-2.04
18 Dec 23	239.5	299.8	4.5	4.1	293.18	299.40	-1.61	-1.68	287.51	290.86	-1.44	-0.38
28 Dec	4.0		-0.8		305.60		-1.72		290.17		1.11	
1681												1681
Jan 2		75.9		-5.0		311.80		-1.73		285.03		2.62
7 Jan 12	149.0	212.4	-2.1	3.2	317.98	324.15	-1.70	-1.64	278.71	275.26	3.37	3.18
17 Jan 22	271.9	334.6	5.0	1.5	330.31	336.44	-1.55	-1.42	275.53	278.52	2.48	1.62
27 J-F 1	42.1	114.4	-4.0	-4.3	342.56	348.64	-1.26	-1.07	283.21	288.96	0.77	-0.01
6 Feb 11	183.2	244.1	1.1	5.0	354.70	0.72	-0.85	-0.61	295.42	302.42	-0.70	-1.26
16 Feb 21	304.8	11.1	3.8	-1.8	6.71	12.66	-0.33	-0.04	309.87	317.74	-1.71	-2.01
26 Feb	81.5		-5.3		18.57		0.27		326.03		-2.15	
Mar 3		151.5		-1.8		24.42		0.60		334.76		-2.11
8 Mar 13	215.9	275.8	3.7	5.1	30.22	35.95	0.94	1.28	343.96	353.62	-1.88	-1.43
18 Mar 23	338.9	49.4	1.1	-4.5	41.62	47.21	1.62	1.96	3.64	13.71	-0.76	0.12
28 M-A 2	120.5	186.7	-4.1	1.4	52.71	58.11	2.29	2.60	23.24	31.44	1.09	2.01
7 Apr 12	248.0	308.1	5.1	3.5	63.39	68.54	2.88	3.14	37.66	41.48	2.70	2.99
17 Apr 22	14.8	88.6	-2.1	-5.1	73.54	78.36	3.36	3.54	42.78	41.74	2.79	2.04
27 Apr	157.7		-1.1		82.97		3.66		39.05		0.82	
May 2		220.4		3.9		87.33		3.73		35.96		-0.61
7 May 12	280.2	341.7	4.7	0.6	91.39	95.11	3.72	3.63	33.75	33.21	-1.92	-2.89
17 May 22	52.6	127.2	-4.5	-3.3	98.39	101.16	3.44	3.13	34.56	37.66	-3.44	-3.60
27 M-J 1	193.0	253.3	2.1	5.0	103.31	104.71	2.70	2.12	42.29	48.27	-3.41	-2.94
6 Jun 11	313.0	17.4	2.7	-2.6	105.25	104.83	1.37	0.45	55.48	63.86	-2.25	-1.39
16 Jun 21	91.6	163.9	-4.8	-0.1	103.42	101.13	-0.62	-1.78	73.36	83.75	-0.44	0.47
26 Jun	226.5		4.4		98.21		-2.95		94.60		1.22	
Jul 1		286.1		4.3		95.10		-4.01		105.35		1.69
6 Jul 11	347.3	55.2	-0.3	-4.8	92.28	90.17	-4.86	-5.45	115.58	125.09	1.85	1.73
16 Jul 21	130.1	198.6	-2.8	3.0	88.99	88.79	-5.78	-5.90	133.83	141.81	1.37	0.83
26 Jul 31	259.1	319.3	5.1	2.0	89.53	91.09	-5.83	-5.64	149.02	155.39	0.13	-0.68
5 Aug 10	23.5	94.3	-3.5	-4.8	93.36	96.23	-5.35	-4.99	160.83	165.11	-1.56	-2.48
15 Aug 20	166.9	231.5	0.5	4.9	99.59	103.36	-4.58	-4.13	167.91	168.76	-3.36	-4.09
25 Aug 30	291.3	353.8	4.1	-1.1	107.47	111.86	-3.67	-3.19	167.23	163.38	-4.46	-4.21
4 Sep 9	61.8	133.2	-5.2	-2.5	116.49	121.33	-2.70	-2.22	158.55	155.22	-3.19	-1.64
14 Sep 19	201.8	263.5	3.4	5.2	126.35	131.53	-1.74	-1.28	155.37	159.29	-0.09	1.08
24 Sep 29	323.8	30.2	1.6	-3.9	136.83	142.25	-0.84	-0.42	166.02	174.25	1.72	1.91
4 Oct 9	101.2	170.5	-4.5	0.8	147.77	153.39	-0.03	0.34	183.00	191.78	1.78	1.43
14 Oct 19	235.1	295.1	5.0	3.7	159.09	164.87	0.67	0.96	200.36	208.71	0.96	0.43
24 Oct 29	357.4	68.4	-1.5	-5.1	170.70	176.60	1.22	1.44	216.81	224.71	-0.13	-0.69
3 Nov 8	140.1	206.1	-1.7	3.7	182.54	188.53	1.62	1.76	232.45	240.03	-1.21	-1.68
13 Nov 18	267.6	327.1	4.7	1.0	194.57	200.63	1.85	1.91	247.46	254.70	-2.07	-2.34
23 Nov 28	32.8	107.4	-4.1	-3.8	206.73	212.85	1.93	1.91	261.59	267.84	-2.45	-2.32
3 Dec 8	177.4	240.2	1.8	5.0	219.00	225.17	1.86	1.77	272.79	275.22	-1.86	-0.91
13 Dec 18	299.8	0.4	2.9	-2.2	231.35	237.55	1.66	1.51	273.53	267.70	0.55	2.15
23 Dec 28	70.1	145.7	-5.0	-0.8	243.75	249.97	1.35	1.16	261.55	259.00	3.06	3.05
1681												1681

JULIAN 12 UT	SATURN LONG.	SATURN LAT.	JUPITER LONG.	JUPITER LAT.	MARS LONG.	MARS LAT.	SUN LONG.
1681							
Dec 23	124.76	0.56	105.59	0.22	308.50	-1.15	282.63
Jan 2	123.99	0.58	104.24	0.24	316.39	-1.11	292.82
Jan 12	123.18	0.60	102.98	0.26	324.29	-1.05	302.99
Jan 22	122.37	0.61	101.88	0.28	332.18	-0.99	313.14
Feb 1	121.60	0.63	101.03	0.30	340.04	-0.92	323.26
Feb 11	120.92	0.64	100.47	0.31	347.87	-0.84	333.33
Feb 21	120.37	0.65	100.23	0.32	355.65	-0.75	343.35
Mar 3	119.97	0.66	100.31	0.33	3.37	-0.66	353.32
Mar 13	119.74	0.66	100.72	0.34	11.03	-0.56	3.23
Mar 23	119.69	0.67	101.42	0.35	18.61	-0.46	13.09
Apr 2	119.82	0.67	102.39	0.36	26.11	-0.35	22.89
Apr 12	120.13	0.67	103.61	0.37	33.53	-0.24	32.63
Apr 22	120.61	0.68	105.04	0.37	40.86	-0.14	42.33
May 2	121.25	0.68	106.65	0.38	48.11	-0.03	51.98
May 12	122.02	0.68	108.41	0.39	55.27	0.07	61.59
May 22	122.92	0.69	110.31	0.40	62.35	0.18	71.16
Jun 1	123.93	0.69	112.31	0.41	69.33	0.28	80.72
Jun 11	125.03	0.70	114.39	0.42	76.24	0.39	90.26
Jun 21	126.20	0.71	116.54	0.43	83.06	0.49	99.79
Jul 1	127.42	0.72	118.72	0.44	89.80	0.58	109.33
Jul 11	128.68	0.73	120.93	0.45	96.47	0.68	118.87
Jul 21	129.96	0.74	123.15	0.47	103.06	0.77	128.44
Jul 31	131.25	0.76	125.35	0.48	109.58	0.86	138.03
Aug 10	132.52	0.77	127.51	0.50	116.03	0.95	147.66
Aug 20	133.76	0.79	129.62	0.52	122.40	1.04	157.33
Aug 30	134.94	0.81	131.65	0.54	128.71	1.13	167.05
Sep 9	136.06	0.83	133.58	0.57	134.93	1.21	176.82
Sep 19	137.09	0.86	135.39	0.59	141.08	1.29	186.65
Sep 29	138.02	0.88	137.04	0.62	147.15	1.38	196.54
Oct 9	138.82	0.91	138.52	0.65	153.13	1.46	206.48
Oct 19	139.47	0.94	139.78	0.68	159.00	1.55	216.47
Oct 29	139.97	0.97	140.80	0.72	164.76	1.63	226.52
Nov 8	140.29	1.01	141.56	0.76	170.39	1.72	236.61
Nov 18	140.43	1.04	142.01	0.80	175.87	1.80	246.75
Nov 28	140.38	1.07	142.15	0.84	181.16	1.89	256.91
Dec 8	140.15	1.11	141.96	0.88	186.24	1.99	267.09
Dec 18	139.74	1.14	141.45	0.92	191.05	2.08	277.28
1682							
Dec 28	139.18	1.17	140.65	0.96	195.55	2.18	287.48
Jan 7	138.50	1.19	139.60	0.99	199.64	2.28	297.66
Jan 17	137.74	1.22	138.37	1.02	203.24	2.38	307.82
Jan 27	136.93	1.24	137.06	1.04	206.22	2.47	317.96
Feb 6	136.13	1.25	135.75	1.05	208.44	2.55	328.05
Feb 16	135.37	1.26	134.55	1.06	209.73	2.62	338.10
Feb 26	134.71	1.26	133.53	1.06	209.89	2.65	348.10
Mar 8	134.17	1.26	132.77	1.05	208.80	2.61	358.04
Mar 18	133.79	1.26	132.29	1.04	206.46	2.49	7.93
Mar 28	133.57	1.25	132.12	1.02	203.14	2.25	17.76
Apr 7	133.53	1.24	132.27	1.01	199.38	1.91	27.53
Apr 17	133.67	1.24	132.71	0.99	195.90	1.49	37.25
Apr 27	133.99	1.23	133.44	0.97	193.32	1.04	46.93
May 7	134.47	1.22	134.43	0.96	192.02	0.61	56.55
May 17	135.10	1.21	135.64	0.94	192.03	0.21	66.15
May 27	135.88	1.21	137.06	0.93	193.28	-0.13	75.72
Jun 6	136.77	1.20	138.65	0.92	195.60	-0.42	85.26
Jun 16	137.77	1.20	140.39	0.91	198.79	-0.67	94.80
Jun 26	138.86	1.20	142.26	0.91	202.72	-0.88	104.33
Jul 6	140.02	1.20	144.23	0.90	207.25	-1.06	113.87
Jul 16	141.23	1.21	146.28	0.90	212.29	-1.21	123.43
Jul 26	142.48	1.22	148.39	0.91	217.75	-1.33	133.00
Aug 5	143.75	1.23	150.53	0.91	223.57	-1.43	142.61
Aug 15	145.03	1.24	152.70	0.92	229.71	-1.51	152.26
Aug 25	146.29	1.26	154.88	0.93	236.12	-1.57	161.96
Sep 4	147.51	1.28	157.03	0.94	242.77	-1.61	171.70
Sep 14	148.68	1.30	159.14	0.96	249.64	-1.64	181.50
Sep 24	149.78	1.32	161.19	0.98	256.69	-1.64	191.35
Oct 4	150.79	1.35	163.16	1.00	263.90	-1.63	201.26
Oct 14	151.70	1.38	165.02	1.02	271.26	-1.61	211.23
Oct 24	152.47	1.42	166.74	1.05	278.74	-1.57	221.25
Nov 3	153.10	1.45	168.30	1.09	286.32	-1.51	231.32
Nov 13	153.57	1.49	169.66	1.12	293.98	-1.44	241.43
Nov 23	153.86	1.53	170.81	1.16	301.69	-1.36	251.58
Dec 3	153.97	1.57	171.69	1.21	309.45	-1.27	261.75
Dec 13	153.90	1.61	172.30	1.25	317.24	-1.17	271.94
1683							

GREGORIAN 12 UT	MOON LONG.		MOON LAT.		VENUS LONG.		VENUS LAT.		MERCURY LONG.		MERCURY LAT.	
1682												1682
2 Jan 7	212.7	273.1	4.4	4.5	256.20	262.43	0.96	0.75	260.29	264.13	2.51	1.76
12 Jan 17	332.5	35.8	0.0	-4.6	268.66	274.90	0.53	0.31	269.45	275.66	0.99	0.25
22 Jan 27	108.6	182.3	-3.6	2.6	281.14	287.38	0.08	-0.14	282.44	289.62	-0.42	-1.00
1 Feb 6	246.2	305.6	5.2	2.4	293.62	299.86	-0.35	-0.55	297.16	305.03	-1.47	-1.83
11 Feb 16	6.6	73.4	-3.0	-5.2	306.10	312.34	-0.74	-0.91	313.25	321.83	-2.05	-2.10
21 Feb 26	146.9	217.0	-0.5	4.8	318.57	324.80	-1.06	-1.18	330.81	340.18	-1.97	-1.63
3 Mar 8	278.4	338.3	4.3	-0.5	331.03	337.25	-1.29	-1.37	349.86	359.56	-1.05	-0.24
13 Mar 18	42.7	112.4	-4.9	-3.4	343.47	349.69	-1.42	-1.45	8.69	16.35	0.76	1.81
23 Mar 28	184.1	250.3	2.8	5.1	355.90	2.10	-1.44	-1.42	21.69	24.14	2.70	3.23
2 Apr 7	310.1	12.2	1.9	-3.3	8.30	14.49	-1.36	-1.28	23.65	20.89	3.19	2.52
12 Apr 17	80.8	151.4	-4.8	0.0	20.68	26.86	-1.18	-1.06	17.26	14.37	1.32	-0.07
22 Apr 27	219.9	282.4	4.7	3.7	33.03	39.20	-0.92	-0.76	13.25	14.15	-1.33	-2.29
2 May 7	342.1	48.0	-1.1	-4.9	45.37	51.53	-0.58	-0.39	16.88	21.16	-2.90	-3.18
12 May 17	120.0	189.3	-2.4	3.3	57.68	63.83	-0.20	0.00	26.71	33.37	-3.16	-2.87
22 May 27	254.1	314.1	4.8	1.1	69.97	76.11	0.20	0.40	41.05	49.72	-2.35	-1.63
1 Jun 6	15.4	85.8	-3.8	-4.4	82.25	88.38	0.59	0.78	59.34	69.78	-0.77	0.14
11 Jun 16	158.8	225.5	1.2	5.0	94.51	100.63	0.95	1.11	80.71	91.58	0.96	1.57
21 Jun 26	287.0	346.3	3.1	-1.9	106.75	112.86	1.25	1.36	101.92	111.49	1.88	1.89
1 Jul 6	50.6	124.6	-5.1	-1.7	118.97	125.08	1.46	1.53	120.19	128.03	1.63	1.14
11 Jul 16	196.4	259.9	4.1	4.7	131.18	137.27	1.57	1.58	134.96	140.92	0.45	-0.40
21 Jul 26	319.4	19.8	0.4	-4.4	143.35	149.43	1.56	1.51	145.75	149.22	-1.38	-2.42
31 J-A 5	87.9	163.1	-4.4	1.8	155.50	161.56	1.43	1.32	150.98	150.67	-3.44	-4.30
10 Aug 15	232.0	292.8	5.3	2.7	167.61	173.65	1.18	1.01	148.17	144.10	-4.75	-4.53
20 Aug 25	352.2	55.3	-2.5	-5.3	179.68	185.69	0.81	0.59	140.13	138.27	-3.55	-2.10
30 Aug	126.3		-1.6		191.69		0.35		139.73		-0.59	
Sep 4		200.3		4.4		197.68		0.09		144.52		0.65
9 Sep 14	265.7	325.1	4.4	-0.2	203.64	209.59	-0.19	-0.48	151.84	160.56	1.45	1.81
19 Sep 24	26.4	92.9	-4.6	-4.0	215.52	221.43	-0.78	-1.08	169.76	178.91	1.81	1.55
29 S-O 4	164.9	235.8	1.9	5.1	227.31	233.17	-1.37	-1.67	187.79	196.34	1.14	0.64
9 Oct 14	297.8	357.6	2.1	-3.0	238.99	244.78	-1.94	-2.21	204.57	212.51	0.08	-0.49
19 Oct 24	62.4	131.7	-5.0	-0.8	250.53	256.23	-2.45	-2.66	220.20	227.65	-1.05	-1.57
29 Oct	202.7		4.4		261.88		-2.83		234.85		-2.03	
Nov 3		269.4		3.9		267.47		-2.97		241.75		-2.39
8 Nov 13	329.3	31.2	-0.9	-4.8	272.98	278.40	-3.06	-3.09	248.18	253.80	-2.60	-2.60
18 Nov 23	100.2	170.7	-3.1	2.8	283.72	288.91	-3.07	-2.98	257.95	259.42	-2.27	-1.45
28 N-D 3	239.0	301.6	4.9	1.3	293.95	298.81	-2.83	-2.59	256.79	250.51	-0.05	1.60
8 Dec 13	1.0	66.4	-3.6	-4.8	303.44	307.81	-2.27	-1.85	244.80	243.20	2.67	2.87
18 Dec 23	138.9	208.9	0.3	5.0	311.84	315.48	-1.33	-0.70	245.45	250.07	2.49	1.87
28 Dec	273.4		3.4		318.63		0.04		255.98		1.18	
1683												1683
Jan 2		333.4		-1.7		321.18		0.92		262.61		0.48
7 Jan 12	33.9	103.3	-5.1	-2.7	323.00	323.95	1.92	3.03	269.67	277.02	-0.17	-0.75
17 Jan 22	177.6	245.2	3.6	4.9	323.92	322.85	4.23	5.46	284.64	292.50	-1.25	-1.65
27 J-F 1	306.4	5.6	0.7	-4.1	320.81	318.03	6.60	7.53	300.63	309.06	-1.93	-2.07
6 Feb 11	68.6	141.2	-4.8	0.6	314.94	312.06	8.13	8.35	317.80	326.87	-2.04	-1.80
16 Feb 21	215.1	279.5	5.2	3.0	309.85	308.58	8.20	7.76	336.17	345.44	-1.33	-0.59
26 Feb	338.8		-2.2		308.33		7.13		354.10		0.40	
Mar 3		39.1		-5.2		309.07		6.38		1.16		1.53
8 Mar 13	105.2	179.6	-2.5	3.7	310.66	312.99	5.59	4.78	5.53	6.49	2.62	3.37
18 Mar 23	250.8	312.2	4.6	0.1	315.92	319.35	3.99	3.23	4.23	0.14	3.50	2.89
28 M-A 2	11.6	74.5	-4.3	-4.3	323.18	327.35	2.51	1.84	356.28	354.20	1.74	0.41
7 Apr 12	143.5	217.4	1.0	5.0	331.79	336.45	1.22	0.64	354.36	356.56	-0.78	-1.73
17 Apr 22	284.5	344.2	2.3	-2.8	341.30	346.30	0.12	-0.34	0.40	5.55	-2.39	-2.77
27 Apr	45.5		-5.0		351.44		-0.75		11.76		-2.89	
May 2		111.8		-1.5		356.69		-1.11		18.91		-2.75
7 May 12	182.6	253.6	4.0	4.1	2.03	7.45	-1.42	-1.67	26.91	35.75	-2.39	-1.81
17 May 22	316.7	16.4	-0.8	-4.7	12.94	18.50	-1.88	-2.03	45.43	55.86	-1.06	-0.19
27 M-J 1	81.0	150.6	-3.6	2.1	24.11	29.76	-2.14	-2.19	66.77	77.68	0.67	1.40
6 Jun 11	221.2	287.9	5.1	1.6	35.46	41.19	-2.21	-2.19	88.07	97.62	1.86	2.03
16 Jun 21	348.3	49.6	-3.5	-5.0	46.96	52.76	-2.12	-2.03	106.20	113.77	1.89	1.47
26 Jun	118.2		-0.7		58.59		-1.90		120.29		0.80	
Jul 1		189.9		4.7		64.45		-1.74		125.65		-0.08
6 Jul 11	258.1	320.7	3.9	-1.4	70.33	76.24	-1.56	-1.36	129.67	132.10	-1.13	-2.30
16 Jul 21	20.2	84.3	-5.1	-3.5	82.17	88.13	-1.15	-0.92	132.65	131.18	-3.46	-4.42
26 Jul 31	156.8	228.3	2.8	5.2	94.11	100.11	-0.69	-0.45	128.03	124.32	-4.91	-4.69
5 Aug 10	292.9	352.9	1.1	-3.9	106.14	112.18	-0.21	0.03	121.68	121.44	-3.79	-2.47
15 Aug 20	53.2	120.8	-5.0	-0.4	118.25	124.34	0.25	0.47	124.18	129.79	-1.04	0.21
25 Aug 30	195.8	264.8	4.9	3.4	130.45	136.58	0.67	0.85	137.64	146.81	1.13	1.65
4 Sep 9	326.1	25.5	-1.9	-5.1	142.73	148.89	1.02	1.15	156.43	165.94	1.80	1.65
14 Sep 19	87.7	158.8	-3.1	2.9	155.07	161.27	1.27	1.36	175.11	183.87	1.31	0.84
24 Sep 29	233.8	299.1	4.8	0.4	167.48	173.70	1.42	1.45	192.23	200.22	0.30	-0.29
4 Oct 9	358.7	59.1	-4.2	-4.5	179.94	186.18	1.46	1.44	207.87	215.19	-0.88	-1.45
14 Oct 19	124.1	197.5	0.2	4.9	192.44	198.70	1.39	1.31	222.16	228.71	-1.98	-2.43
24 Oct 29	269.7	331.8	2.6	-2.6	204.97	211.24	1.21	1.09	234.65	239.61	-2.75	-2.87
3 Nov 8	31.4	94.3	-5.0	-2.1	217.52	223.80	0.95	0.79	242.90	243.38	-2.68	-2.00
13 Nov 18	162.3	235.6	3.5	4.5	230.08	236.37	0.62	0.43	239.89	233.42	-0.68	1.00
23 Nov 28	303.5	3.6	-0.5	-4.7	242.66	248.94	0.24	0.04	228.37	227.74	2.23	2.64
3 Dec 8	64.9	131.3	-4.0	1.3	255.23	261.52	-0.16	-0.36	230.89	236.24	2.44	1.95
13 Dec 18	201.6	272.0	5.2	2.1	267.81	274.09	-0.55	-0.73	242.70	249.72	1.34	0.70
23 Dec 28	335.7	35.4	-3.3	-5.2	280.38	286.66	-0.90	-1.06	257.05	264.57	0.07	-0.52
1683												1683

JULIAN 12 UT	SATURN LONG.	SATURN LAT.	JUPITER LONG.	JUPITER LAT.	MARS LONG.	MARS LAT.	SUN LONG.
1683							
Dec 23	153.64	1.65	172.60	1.30	325.03	-1.07	282.13
Jan 2	153.22	1.69	172.59	1.35	332.81	-0.95	292.32
Jan 12	152.64	1.72	172.25	1.39	340.57	-0.83	302.50
Jan 22	151.96	1.75	171.62	1.44	348.29	-0.71	312.65
Feb 1	151.19	1.77	170.71	1.47	355.96	-0.59	322.76
Feb 11	150.39	1.79	169.59	1.50	3.58	-0.47	332.84
Feb 21	149.59	1.80	168.34	1.52	11.13	-0.35	342.86
Mar 2	148.86	1.80	167.04	1.53	18.60	-0.23	352.83
Mar 12	148.21	1.80	165.79	1.52	26.00	-0.11	2.75
Mar 22	147.70	1.79	164.67	1.51	33.32	0.00	12.61
Apr 1	147.33	1.77	163.76	1.49	40.56	0.11	22.41
Apr 11	147.13	1.76	163.10	1.46	47.72	0.22	32.17
Apr 21	147.11	1.74	162.74	1.43	54.79	0.32	41.86
May 1	147.26	1.72	162.68	1.39	61.79	0.41	51.51
May 11	147.58	1.70	162.92	1.36	68.71	0.50	61.13
May 21	148.07	1.68	163.44	1.32	75.56	0.59	70.71
May 31	148.70	1.67	164.24	1.29	82.34	0.67	80.26
Jun 10	149.47	1.65	165.28	1.26	89.05	0.75	89.80
Jun 20	150.36	1.64	166.54	1.23	95.70	0.82	99.33
Jun 30	151.36	1.63	167.99	1.20	102.30	0.88	108.87
Jul 10	152.44	1.63	169.61	1.18	108.84	0.95	118.41
Jul 20	153.59	1.62	171.37	1.16	115.33	1.01	127.98
Jul 30	154.80	1.62	173.26	1.14	121.78	1.06	137.57
Aug 9	156.04	1.63	175.24	1.13	128.19	1.11	147.20
Aug 19	157.29	1.64	177.29	1.12	134.55	1.16	156.86
Aug 29	158.55	1.65	179.41	1.12	140.88	1.20	166.58
Sep 8	159.80	1.66	181.55	1.11	147.18	1.24	176.35
Sep 18	161.00	1.68	183.72	1.11	153.44	1.27	186.17
Sep 28	162.15	1.71	185.87	1.12	159.66	1.31	196.06
Oct 8	163.23	1.73	188.01	1.13	165.85	1.33	206.00
Oct 18	164.22	1.76	190.09	1.14	171.99	1.36	215.99
Oct 28	165.10	1.80	192.10	1.15	178.10	1.38	226.03
Nov 7	165.84	1.84	194.02	1.17	184.15	1.40	236.13
Nov 17	166.44	1.88	195.81	1.20	190.15	1.41	246.25
Nov 27	166.87	1.92	197.45	1.22	196.08	1.42	256.42
Dec 7	167.13	1.97	198.92	1.25	201.94	1.43	266.60
Dec 17	167.21	2.01	200.17	1.29	207.70	1.42	276.79
1684							
Dec 27	167.10	2.06	201.18	1.32	213.35	1.41	286.99
Jan 6	166.82	2.10	201.93	1.36	218.86	1.39	297.17
Jan 16	166.37	2.14	202.39	1.40	224.22	1.36	307.33
Jan 26	165.78	2.18	202.54	1.44	229.37	1.31	317.47
Feb 5	165.08	2.20	202.37	1.48	234.28	1.25	327.57
Feb 15	164.32	2.22	201.89	1.52	238.88	1.16	337.62
Feb 25	163.52	2.23	201.13	1.55	243.11	1.04	347.62
Mar 7	162.75	2.24	200.14	1.58	246.86	0.89	357.57
Mar 17	162.03	2.23	198.97	1.59	250.01	0.69	7.45
Mar 27	161.41	2.22	197.70	1.59	252.41	0.44	17.29
Apr 6	160.91	2.20	196.43	1.58	253.89	0.11	27.06
Apr 16	160.57	2.18	195.24	1.57	254.26	-0.29	36.79
Apr 26	160.39	2.16	194.22	1.54	253.39	-0.78	46.46
May 6	160.38	2.13	193.41	1.50	251.31	-1.34	56.10
May 16	160.55	2.10	192.88	1.46	248.31	-1.91	65.69
May 26	160.89	2.07	192.64	1.42	244.96	-2.45	75.26
Jun 5	161.38	2.05	192.69	1.37	242.01	-2.88	84.80
Jun 15	162.02	2.02	193.04	1.33	240.08	-3.18	94.34
Jun 25	162.79	2.00	193.68	1.28	239.47	-3.35	103.87
Jul 5	163.68	1.98	194.57	1.24	240.26	-3.43	113.41
Jul 15	164.67	1.97	195.69	1.20	242.30	-3.43	122.96
Jul 25	165.75	1.95	197.03	1.17	245.40	-3.38	132.54
Aug 4	166.89	1.95	198.56	1.14	249.40	-3.29	142.14
Aug 14	168.09	1.94	200.25	1.11	254.13	-3.17	151.79
Aug 24	169.32	1.94	202.07	1.08	259.44	-3.03	161.48
Sep 3	170.57	1.95	204.01	1.06	265.23	-2.87	171.22
Sep 13	171.81	1.96	206.04	1.04	271.41	-2.70	181.01
Sep 23	173.04	1.97	208.14	1.03	277.90	-2.52	190.87
Oct 3	174.22	1.99	210.30	1.02	284.65	-2.33	200.78
Oct 13	175.35	2.02	212.48	1.01	291.60	-2.13	210.74
Oct 23	176.41	2.05	214.67	1.00	298.71	-1.93	220.76
Nov 2	177.36	2.08	216.84	1.00	305.94	-1.72	230.83
Nov 12	178.21	2.12	218.98	1.00	313.26	-1.52	240.94
Nov 22	178.92	2.16	221.06	1.01	320.64	-1.32	251.09
Dec 2	179.47	2.20	223.05	1.01	328.05	-1.12	261.26
Dec 12	179.87	2.25	224.93	1.02	335.48	-0.93	271.45
1685							

GREGORIAN 12 UT	MOON LONG.		MOON LAT.		VENUS LONG.		VENUS LAT.		MERCURY LONG.		MERCURY LAT.	
1684												1684
2 Jan 7	99.6	169.9	-1.5	4.3	292.94	299.22	-1.19	-1.31	272.26	280.11	-1.04	-1.48
12 Jan 17	240.3	306.5	4.4	-1.0	305.49	311.76	-1.40	-1.47	288.15	296.42	-1.81	-2.02
22 Jan 27	7.4	67.9	-5.0	-3.9	318.02	324.27	-1.51	-1.53	304.92	313.67	-2.08	-1.94
1 Feb 6	136.0	209.1	1.8	5.2	330.52	336.75	-1.51	-1.47	322.57	331.38	-1.58	-0.94
11 Feb 16	277.3	339.7	1.7	-3.6	342.98	349.19	-1.40	-1.30	339.50	345.89	0.00	1.19
21 Feb 26	39.3	101.8	-5.0	-1.2	355.39	1.57	-1.17	-1.02	349.23	348.68	2.44	3.40
2 Mar 7	174.0	247.5	4.4	3.7	7.74	13.89	-0.84	-0.64	344.86	340.00	3.69	3.15
12 Mar 17	312.2	12.3	-1.6	-4.9	20.02	26.14	-0.43	-0.20	336.56	335.58	2.05	0.81
22 Mar 27	72.1	137.6	-3.3	2.1	32.23	38.29	0.04	0.29	336.97	340.24	-0.32	-1.24
1 Apr 6	212.8	283.9	4.9	0.7	44.33	50.34	0.55	0.80	344.93	350.71	-1.93	-2.38
11 Apr 16	345.6	45.1	-4.1	-4.5	56.33	62.28	1.06	1.30	357.39	4.85	-2.61	-2.61
21 Apr 26	106.5	175.5	-0.3	4.6	68.20	74.08	1.53	1.74	13.05	21.97	-2.39	-1.95
1 May 6	251.0	318.2	3.1	-2.5	79.92	85.72	1.94	2.10	31.63	41.98	-1.32	-0.52
11 May 16	18.2	78.5	-5.1	-2.5	91.48	97.18	2.24	2.34	52.81	63.68	0.36	1.17
21 May 26	142.9	214.7	3.0	4.9	102.82	108.40	2.40	2.41	74.04	83.49	1.79	2.12
31 M-J 5	287.3	350.9	0.0	-4.7	113.91	119.33	2.39	2.31	91.83	98.98	2.13	1.82
10 Jun 15	50.6	113.3	-4.3	0.6	124.67	129.90	2.18	1.99	104.88	109.39	1.20	0.31
20 Jun 25	181.3	253.4	5.1	2.9	135.00	139.97	1.74	1.42	112.33	113.47	-0.82	-2.10
30 Jun	321.8		-3.0		144.77		1.05		112.70		-3.37	
Jul 5		22.9		-5.2		149.37		0.60		110.29		-4.40
10 Jul 15	83.6	150.1	-2.1	3.6	153.75	157.86	0.08	-0.52	107.07	104.36	-4.88	-4.69
20 Jul 25	220.8	290.5	4.8	-0.3	161.64	165.02	-1.19	-1.94	103.38	104.78	-3.90	-2.74
30 J-A 4	354.8	54.6	-4.9	-4.1	167.92	170.24	-2.77	-3.67	108.75	115.12	-1.44	-0.21
9 Aug 14	117.8	188.7	1.0	5.1	171.88	172.69	-4.63	-5.63	123.43	132.98	0.79	1.45
19 Aug 24	259.7	325.7	2.3	-3.3	172.59	171.49	-6.61	-7.51	142.98	152.84	1.75	1.73
29 Aug	27.0		-5.0		169.46		-8.22		162.30		1.47	
Sep 3		86.9		-1.6		166.72		-8.65		171.26		1.05
8 Sep 13	154.0	228.0	3.8	4.2	163.66	160.77	-8.70	-8.37	179.75	187.79	0.52	-0.07
18 Sep 23	296.8	359.5	-1.1	-4.8	158.47	157.06	-7.72	-6.84	195.40	202.59	-0.69	-1.32
28 S-O 3	59.1	120.6	-3.5	1.6	156.65	157.22	-5.85	-4.82	209.32	215.50	-1.92	-2.45
8 Oct 13	191.9	266.4	5.0	1.3	158.68	160.90	-3.82	-2.88	220.91	225.16	-2.88	-3.14
18 Oct 23	332.0	32.3	-3.9	-4.6	163.76	167.15	-2.00	-1.20	227.54	226.99	-3.08	-2.54
28 Oct	91.9		-0.7		170.98		-0.49		222.76		-1.31	
Nov 2		156.2		4.3		175.18		0.14		216.39		0.37
7 Nov 12	230.6	302.9	3.8	-2.1	179.69	184.45	0.69	1.17	212.17	212.51	1.74	2.36
17 Nov 22	5.5	64.9	-5.1	-2.8	189.42	194.57	1.57	1.90	216.51	222.54	2.36	2.01
27 N-D 2	126.1	193.9	2.5	5.2	199.87	205.31	2.16	2.36	229.53	236.94	1.48	0.89
7 Dec 12	268.6	337.3	0.8	-4.5	210.85	216.49	2.49	2.57	244.53	252.22	0.29	-0.30
17 Dec 22	37.8	98.0	-4.6	0.0	222.21	227.99	2.59	2.56	259.99	267.83	-0.83	-1.30
27 Dec	162.4		4.8		233.84		2.49		275.79		-1.68	
1685												1685
Jan 1		232.8		3.7		239.73		2.37		283.90		-1.96
6 Jan 11	305.1	10.1	-2.4	-5.3	245.67	251.64	2.22	2.04	292.16	300.58	-2.10	-2.06
16 Jan 21	69.7	132.3	-2.5	3.0	257.64	263.67	1.83	1.60	309.07	317.37	-1.81	-1.27
26 Jan 31	200.7	271.5	5.1	0.5	269.72	275.79	1.35	1.09	324.91	330.57	-0.41	0.78
5 Feb 10	340.0	41.9	-4.6	-4.2	281.87	287.96	0.82	0.55	332.85	330.87	2.15	3.31
15 Feb 20	101.9	168.6	0.5	4.8	294.07	300.18	0.28	0.02	325.81	320.78	3.74	3.28
25 Feb	240.0		3.0		306.30		-0.23		318.15		2.27	
Mar 2		309.0		-2.8		312.43		-0.47		318.34		1.13
7 Mar 12	13.6	73.4	-5.0	-1.8	318.56	324.69	-0.69	-0.89	320.82	324.95	0.07	-0.82
17 Mar 22	135.4	206.7	3.3	4.6	330.83	336.97	-1.06	-1.21	330.27	336.52	-1.52	-2.02
27 M-A 1	278.8	344.8	-0.6	-4.8	343.10	349.24	-1.34	-1.43	343.51	351.16	-2.34	-2.45
6 Apr 11	46.2	105.5	-3.6	1.3	355.38	1.52	-1.50	-1.54	359.46	8.39	-2.35	-2.05
16 Apr 21	171.0	245.5	5.0	2.1	7.65	13.79	-1.54	-1.52	17.97	28.18	-1.53	-0.82
26 Apr	316.0		-3.7		19.92		-1.48		38.85		0.03	
May 1		19.0		-4.8		26.05		-1.40		49.60		0.90
6 May 11	78.4	139.1	-0.9	4.0	32.19	38.32	-1.31	-1.18	59.85	69.10	1.66	2.16
16 May 21	208.6	283.9	4.5	-1.4	44.45	50.58	-1.04	-0.89	77.07	83.61	2.34	2.17
26 May 31	351.4	51.9	-5.2	-3.1	56.71	62.84	-0.71	-0.53	88.64	92.01	1.64	0.76
5 Jun 10	111.3	174.8	2.1	5.3	68.97	75.11	-0.34	-0.14	93.55	93.21	-0.42	-1.79
15 Jun 20	247.2	320.8	1.9	-4.2	81.24	87.38	0.06	0.26	91.25	88.41	-3.14	-4.18
25 Jun 30	25.0	84.3	-4.7	-0.4	93.52	99.67	0.45	0.63	85.86	84.66	-4.66	-4.53
5 Jul 10	145.6	212.6	4.4	4.3	105.82	111.97	0.80	0.96	85.45	88.40	-3.89	-2.91
15 Jul 20	285.8	355.9	-1.6	-5.2	118.12	124.28	1.10	1.22	93.47	100.51	-1.77	-0.61
25 Jul 30	57.2	117.1	-2.7	2.5	130.44	136.61	1.31	1.38	109.23	119.10	0.43	1.21
4 Aug 9	181.9	251.6	5.1	1.4	142.78	148.95	1.43	1.45	129.41	139.59	1.65	1.76
14 Aug 19	323.3	29.2	-4.2	-4.3	155.12	161.29	1.44	1.41	149.32	158.49	1.61	1.26
24 Aug 29	88.9	151.3	0.2	4.6	167.46	173.63	1.34	1.25	167.09	175.17	0.76	0.17
3 Sep 8	220.1	290.6	3.6	-2.2	179.80	185.97	1.13	0.99	182.73	189.77	-0.48	-1.16
13 Sep 18	359.0	61.3	-5.0	-1.9	192.14	198.30	0.83	0.65	196.24	202.00	-1.83	-2.47
23 Sep 28	121.0	187.3	3.1	4.9	204.46	210.62	0.45	0.23	206.84	210.31	-3.02	-3.40
3 Oct 8	259.2	328.5	0.2	-4.6	216.78	222.93	0.00	-0.23	211.71	210.14	-3.49	-3.07
13 Oct 18	33.1	92.9	-3.8	1.1	229.07	235.21	-0.47	-0.70	205.33	199.35	-1.93	-0.26
23 Oct 28	154.3	224.9	4.9	3.0	241.35	247.47	-0.94	-1.16	196.13	197.44	1.22	2.03
2 Nov 7	298.0	4.6	-3.2	-5.0	253.59	259.71	-1.37	-1.57	202.24	208.91	2.23	2.03
12 Nov 17	65.8	125.0	-1.2	3.8	265.81	271.89	-1.74	-1.89	216.40	224.20	1.60	1.07
22 Nov 27	189.3	263.3	5.0	-0.4	277.97	284.03	-2.01	-2.09	232.06	239.93	0.49	-0.08
2 Dec 7	335.4	38.7	-5.1	-3.4	290.07	296.09	-2.15	-2.16	247.78	255.63	-0.63	-1.13
12 Dec 17	98.1	158.6	1.7	5.2	302.07	308.03	-2.13	-2.06	263.51	271.46	-1.55	-1.89
22 Dec 27	226.2	301.4	2.9	-3.5	313.95	319.82	-1.95	-1.78	279.49	287.58	-2.10	-2.16
1685												1685

JULIAN 12 UT	SATURN		JUPITER		MARS		SUN	GREGORIAN 12 UT	MOON				VENUS				MERCURY			
	LONG.	LAT.	LONG.	LAT.	LONG.	LAT.	LONG.		LONG.		LAT.		LONG.		LAT.		LONG.		LAT.	
1685								1686												1686
Dec 22	180.09	2.29	226.68	1.04	342.91	-0.74	281.64	1 Jan 6	10.8	71.4	-4.9	-0.7	325.65	331.41	-1.58	-1.32	295.64	303.42	-2.01	-1.59
Jan 1	180.13	2.34	228.25	1.05	350.32	-0.57	291.83	11 Jan 16	130.9	194.1	4.1	4.7	337.09	342.70	-1.02	-0.67	310.32	315.19	-0.83	0.32
Jan 11	179.98	2.39	229.64	1.07	357.69	-0.40	302.01	21 Jan 26	264.4	338.6	-0.5	-5.1	348.21	353.61	-0.28	0.16	316.42	313.14	1.77	3.10
Jan 21	179.67	2.43	230.80	1.09	5.02	-0.24	312.16	31 J-F 5	44.2	103.5	-2.8	2.2	358.87	3.98	0.64	1.15	307.25	302.51	3.67	3.31
Jan 31	179.19	2.47	231.71	1.12	12.30	-0.10	322.28	10 Feb 15	165.1	231.5	5.0	2.1	8.90	13.61	1.70	2.29	300.90	302.20	2.42	1.38
Feb 10	178.59	2.50	232.34	1.14	19.53	0.04	332.35	20 Feb 25	303.1	14.0	-3.7	-4.4	18.06	22.20	2.89	3.52	305.60	310.42	0.40	-0.45
Feb 20	177.89	2.53	232.68	1.17	26.69	0.17	342.38	2 Mar 7	76.2	135.9	0.1	4.4	25.97	29.27	4.16	4.81	316.25	322.83	-1.15	-1.70
Mar 2	177.12	2.55	232.70	1.19	33.79	0.29	352.36	12 Mar 17	201.0	270.4	4.0	-1.5	32.03	34.13	5.44	6.04	330.04	337.81	-2.07	-2.27
Mar 12	176.34	2.55	232.42	1.21	40.82	0.40	2.28	22 Mar 27	341.2	47.6	-5.0	-2.1	35.45	35.85	6.57	6.99	346.12	355.00	-2.29	-2.10
Mar 22	175.58	2.55	231.83	1.23	47.78	0.50	12.14	1 Apr 6	107.6	169.5	3.0	5.0	35.25	33.63	7.25	7.26	4.45	14.47	-1.70	-1.10
Apr 1	174.88	2.54	230.98	1.24	54.68	0.59	21.94	11 Apr 16	238.6	309.6	1.1	-4.4	31.14	28.10	6.97	6.34	24.92	35.47	-0.30	0.59
Apr 11	174.28	2.52	229.92	1.24	61.52	0.67	31.69	21 Apr 26	17.6	80.0	-4.1	0.9	25.00	22.33	5.42	4.31	45.52	54.48	1.46	2.14
Apr 21	173.82	2.49	228.71	1.24	68.29	0.75	41.40	1 May 6	139.5	204.7	4.9	3.7	20.43	19.51	3.11	1.95	61.93	67.66	2.51	2.51
May 1	173.50	2.46	227.45	1.23	75.01	0.81	51.05	11 May 16	277.3	347.9	-2.4	-5.2	19.59	20.59	0.89	-0.04	71.52	73.40	2.10	1.27
May 11	173.35	2.43	226.20	1.21	81.67	0.87	60.66	21 May 26	52.1	112.0	-1.6	3.5	22.39	24.86	-0.83	-1.48	73.30	71.52	0.06	-1.36
May 21	173.36	2.40	225.07	1.18	88.28	0.93	70.25	31 M-J 5	172.7	241.6	5.2	0.8	27.88	31.36	-2.01	-2.42	68.83	66.35	-2.73	-3.75
May 31	173.54	2.36	224.13	1.15	94.85	0.98	79.80	10 Jun 15	316.2	24.2	-4.8	-3.8	35.22	39.39	-2.74	-2.96	65.09	65.62	-4.25	-4.22
Jun 10	173.89	2.33	223.42	1.11	101.37	1.02	89.34	20 Jun 25	85.2	144.5	1.4	5.0	43.81	48.46	-3.10	-3.17	68.07	72.37	-3.77	-3.00
Jun 20	174.40	2.29	222.99	1.06	107.86	1.05	98.87	30 Jun	207.5		3.5		53.28		-3.17		78.38		-2.03	
								Jul 5		280.0		-2.6		58.25		-3.11		86.00		-0.97
Jun 30	175.05	2.26	222.86	1.02	114.31	1.08	108.40	10 Jul 15	354.0	58.3	-5.0	-1.0	63.36	68.58	-3.01	-2.86	95.06	105.16	0.06	0.92
Jul 10	175.82	2.23	223.04	0.98	120.74	1.11	117.95	20 Jul 25	117.7	178.2	3.8	4.8	73.91	79.32	-2.67	-2.45	115.74	126.20	1.50	1.76
Jul 20	176.72	2.21	223.50	0.94	127.15	1.13	127.51	30 J-A 4	244.3	318.8	0.4	-4.8	84.81	90.37	-2.21	-1.94	136.17	145.52	1.73	1.46
Jul 30	177.71	2.19	224.25	0.89	133.53	1.14	137.10	9 Aug 14	29.8	91.0	-3.1	2.0	96.00	101.69	-1.66	-1.36	154.23	162.32	1.00	0.42
Aug 9	178.78	2.17	225.25	0.86	139.91	1.16	146.73	19 Aug 24	150.6	213.6	5.0	2.7	107.43	113.23	-1.06	-0.76	169.80	176.66	-0.25	-0.98
Aug 19	179.93	2.16	226.48	0.82	146.27	1.16	156.40	29 Aug	282.7		-3.0		119.07		-0.46		182.81		-1.73	
								Sep 3		356.7		-4.7		124.95		-0.17		188.11		-2.47
Aug 29	181.11	2.16	227.92	0.79	152.63	1.16	166.11	8 Sep 13	63.4	123.1	-0.1	4.3	130.88	136.85	0.11	0.38	192.29	194.88	-3.14	-3.66
Sep 8	182.34	2.16	229.55	0.76	158.98	1.16	175.88	18 Sep 23	184.6	250.9	4.3	-0.7	142.85	148.89	0.62	0.85	195.23	192.70	-3.88	-3.57
Sep 18	183.57	2.16	231.33	0.73	165.33	1.15	185.70	28 S-O 3	321.9	32.7	-5.0	-2.5	154.96	161.06	1.04	1.21	187.58	182.31	-2.51	-0.88
Sep 28	184.80	2.17	233.25	0.71	171.69	1.14	195.58	8 Oct 13	95.5	155.3	2.8	5.1	167.18	173.33	1.35	1.47	180.22	182.48	0.67	1.67
Oct 8	186.01	2.18	235.27	0.68	178.05	1.13	205.51	18 Oct 23	220.1	289.8	1.9	-3.9	179.50	185.70	1.55	1.59	188.01	195.30	2.06	2.01
Oct 18	187.17	2.20	237.39	0.66	184.41	1.10	215.51	28 Oct	0.3		-4.6		191.91		1.61		203.29		1.70	
								Nov 2		66.6		0.5		198.13		1.59		211.46		1.23
Oct 28	188.27	2.22	239.56	0.65	190.77	1.08	225.55	7 Nov 12	127.0	188.3	4.8	4.2	204.37	210.62	1.55	1.48	219.61	227.66	0.69	0.13
Nov 7	189.30	2.25	241.78	0.63	197.14	1.04	235.64	17 Nov 22	257.2	329.1	-1.5	-5.3	216.88	223.14	1.38	1.26	235.60	243.47	-0.42	-0.95
Nov 17	190.22	2.28	244.02	0.62	203.51	1.00	245.77	27 N-D 2	36.9	99.3	-2.2	3.3	229.42	235.70	1.11	0.95	251.29	259.09	-1.41	-1.80
Nov 27	191.02	2.32	246.26	0.61	209.88	0.96	255.92	7 Dec 12	158.9	222.7	5.2	1.6	241.98	248.26	0.77	0.58	266.89	274.65	-2.09	-2.23
Dec 7	191.69	2.36	248.48	0.60	216.26	0.90	266.11	17 Dec 22	295.6	7.4	-4.3	-4.2	254.55	260.84	0.38	0.18	282.28	289.50	-2.19	-1.89
Dec 17	192.20	2.41	250.65	0.59	222.62	0.84	276.30	27 Dec	71.5		1.0		267.13		-0.02		295.72		-1.25	
1686								1687												1687
								Jan 1		131.6		4.9		273.42		-0.22		299.76		-0.17
Dec 27	192.55	2.45	252.74	0.58	228.98	0.76	286.49	6 Jan 11	191.7	258.8	3.8	-1.7	279.71	285.99	-0.42	-0.60	299.94	295.56	1.31	2.78
Jan 6	192.73	2.50	254.74	0.58	235.32	0.67	296.68	16 Jan 21	334.3	43.6	-5.1	-1.4	292.28	298.56	-0.78	-0.94	289.25	285.13	3.50	3.27
Jan 16	192.72	2.55	256.62	0.58	241.65	0.57	306.84	26 Jan 31	104.7	164.2	3.6	4.8	304.84	311.12	-1.08	-1.20	284.58	286.89	2.50	1.58
Jan 26	192.54	2.59	258.35	0.57	247.95	0.46	316.98	5 Feb 10	226.1	296.7	1.0	-4.3	317.39	323.65	-1.30	-1.37	291.08	296.48	0.68	-0.13
Feb 5	192.19	2.64	259.90	0.57	254.21	0.33	327.08	15 Feb 20	12.0	77.6	-3.5	1.9	329.92	336.17	-1.42	-1.44	302.71	309.56	-0.83	-1.40
Feb 15	191.69	2.67	261.24	0.57	260.44	0.17	337.13	25 Feb	137.3		4.9		342.42		-1.44		316.91		-1.82	
								Mar 2		197.8		3.0		348.66		-1.40		324.74		-2.10
Feb 25	191.07	2.70	262.36	0.57	266.60	0.00	347.13	7 Mar 12	262.6	335.6	-2.4	-4.9	354.89	1.11	-1.35	-1.26	333.03	341.80	-2.20	-2.12
Mar 7	190.36	2.72	263.21	0.57	272.69	-0.20	357.08	17 Mar 22	47.8	110.2	-0.5	4.3	7.33	13.53	-1.16	-1.03	351.09	0.87	-1.84	-1.34
Mar 17	189.60	2.73	263.78	0.57	278.68	-0.43	6.98	27 M-A 1	169.9	232.9	4.5	-0.1	19.72	25.90	-0.87	-0.70	11.05	21.32	-0.63	0.26
Mar 27	188.82	2.74	264.05	0.57	284.55	-0.70	16.81	6 Apr 11	301.1	13.9	-4.8	-3.2	32.08	38.24	-0.52	-0.32	31.10	39.67	1.20	2.05
Apr 6	188.08	2.73	264.01	0.57	290.27	-1.00	26.59	16 Apr 21	81.6	142.1	2.6	5.2	44.38	50.52	-0.11	0.10	46.46	51.14	2.63	2.83
Apr 16	187.41	2.71	263.66	0.57	295.78	-1.34	36.32	26 Apr	203.2		2.5		56.64		0.32		53.55		2.56	
								May 1		269.9		-3.3		62.75		0.54		53.68		1.80
Apr 26	186.84	2.68	263.02	0.56	301.03	-1.73	45.99	6 May 11	340.6	50.4	-5.0	-0.1	68.85	74.93	0.75	0.95	51.90	49.08	0.59	-0.84
May 6	186.40	2.65	262.11	0.55	305.93	-2.18	55.63	16 May 21	114.2	174.0	4.7	4.5	81.00	87.05	1.15	1.32	46.43	44.99	-2.19	-3.19
May 16	186.11	2.62	261.01	0.54	310.40	-2.69	65.23	26 May 31	237.9	308.7	-0.6	-5.1	93.08	99.10	1.48	1.61	45.32	47.49	-3.74	-3.84
May 26	185.99	2.58	259.78	0.52	314.30	-3.26	74.79	5 Jun 10	19.3	85.2	-2.8	2.9	105.10	111.09	1.72	1.80	51.35	56.74	-3.57	-3.01
Jun 5	186.03	2.54	258.51	0.50	317.45	-3.90	84.34	15 Jun 20	146.1	206.6	5.2	2.2	117.05	122.99	1.85	1.86	63.52	71.61	-2.22	-1.29
Jun 15	186.23	2.50	257.28	0.47	319.69	-4.58	93.88	25 Jun 30	274.6	348.1	-3.6	-4.5	128.90	134.79	1.84	1.78	80.93	91.20	-0.30	0.61
Jun 25	186.60	2.46	256.19	0.44	320.81	-5.29	103.41	5 Jul 10	56.2	118.6	0.5	4.7	140.65	146.49	1.68	1.55	101.97	112.65	1.31	1.72
Jul 5	187.12	2.42	255.29	0.41	320.67	-5.97	112.95	15 Jul 20	178.2	240.7	3.9	-1.0	152.28	158.04	1.37	1.16	122.82	132.32	1.82	1.64
Jul 15	187.78	2.38	254.65	0.38	319.30	-6.52	122.50	25 Jul 30	313.1	26.5	-5.0	-1.9	163.75	169.42	0.91	0.62	141.09	149.16	1.26	0.70
Jul 25	188.57	2.35	254.30	0.35	317.01	-6.83	132.07	4 Aug 9	91.2	151.4	3.4	4.8	175.04	180.60	0.29	-0.06	156.52	163.13	0.01	-0.77
Aug 4	189.47	2.33	254.25	0.32	314.39	-6.82	141.68	14 Aug 19	211.1	276.8	1.3	-3.9	186.10	191.52	-0.45	-0.86	168.90	173.64	-1.60	-2.45
Aug 14	190.47	2.30	254.52	0.29	312.20	-6.50	151.32	24 Aug 29	352.1	62.8	-4.0	1.5	196.85	202.08	-1.29	-1.75	177.05	178.66	-3.25	-3.91
Aug 24	191.54	2.28	255.09	0.26	311.02	-5.94	161.01	3 Sep 8	124.6	184.1	5.0	3.3	207.21	212.20	-2.22	-2.69	177.91	174.54	-4.26	-4.02
Sep 3	192.68	2.27	255.93	0.24	311.14	-5.26	170.75	13 Sep 18	245.5	314.9	-1.9	-5.1	217.04	221.68	-3.17	-3.64	169.48	165.25	-3.02	-1.44
Sep 13	193.86	2.26	257.04	0.21	312.57	-4.55	180.54	23 Sep 28	30.1	97.1	-1.2	4.2	226.11	230.27	-4.11	-4.55	164.40	167.58	0.13	1.26
Sep 23	195.08	2.26	258.39	0.19	315.15	-3.85	190.39	3 Oct 8	157.2	217.4	4.7	0.5	234.10	237.54	-4.95	-5.31	173.80	181.66	1.84	1.96
Oct 3	196.30	2.26	259.94	0.17	318.68	-3.21	200.30	13 Oct 18	282.0	353.8	-4.5	-4.0	240.49	242.85	-5.60	-5.80	190.14	198.70	1.77	1.38
Oct 13	197.51	2.27	261.67	0.15	322.97	-2.63	210.26	23 Oct 28	66.2	129.8	2.1	5.3	244.49	245.29	-5.88	-5.80	207.13	215.37	0.89	0.34
Oct 23	198.70	2.28	263.56	0.13	327.85	-2.11	220.27	2 Nov 7	189.4	252.3	2.9	-2.6	245.13	243.97	-5.52	-4.98	223.42	231.32	-0.22	-0.76
Nov 2	199.83	2.30	265.58	0.11	333.18	-1.66	230.34	12 Nov 17	320.5	32.2	-5.2	-1.0	241.87	239.09	-4.18	-3.12	239.09	246.75	-1.27	-1.71
Nov 12	200.91	2.32	267.71	0.09	338.86	-1.25	240.44	22 Nov 27	100.4	161.5	4.5	4.6	236.07	233.32	-1.88	-0.58	254.32	261.75	-2.07	-2.30
Nov 22	201.90	2.35	269.92	0.07	344.80	-0.90	250.59	2 Dec 7	222.1	289.1	0.0	-4.8	231.27	230.19	0.64	1.71	268.94	275.59	-2.36	-2.18
Dec 2	202.78	2.38	272.18	0.06	350.94	-0.59	260.76	12 Dec 17	359.9	69.0	-3.5	2.4	230.12	231.03	2.58	3.25	281.08	284.24	-1.66	-0.68
Dec 12	203.54	2.41	274.48	0.04	357.23	-0.32	270.95	22 Dec 27	133.4	193.0	5.1	2.4	232.81	235.31	3.73	4.05	283.41	278.15	0.78	2.35
1687								1687												1687

JULIAN 12 UT	SATURN		JUPITER		MARS		SUN	GREGORIAN 12 UT	MOON				VENUS				MERCURY			
	LONG.	LAT.	LONG.	LAT.	LONG.	LAT.	LONG.		LONG.		LAT.		LONG.		LAT.		LONG.		LAT.	
1687								1688												1688
Dec 22	204.16	2.45	276.80	0.03	3.62	-0.08	281.14	1 Jan 6	256.2	327.6	-2.9	-4.8	238.43	242.03	4.24	4.31	271.77	268.45	3.22	3.16
Jan 1	204.63	2.50	279.10	0.01	10.08	0.12	291.34	11 Jan 16	38.5	104.2	-0.1	4.5	246.05	250.39	4.28	4.18	268.95	272.18	2.54	1.74
Jan 11	204.92	2.54	281.37	-0.01	16.59	0.30	301.51	21 Jan 26	165.5	225.2	4.0	-0.6	255.00	259.84	4.00	3.77	277.07	282.97	0.92	0.15
Jan 21	205.05	2.59	283.59	-0.02	23.13	0.46	311.67	31 J-F 5	292.1	6.6	-4.8	-2.5	264.87	270.05	3.49	3.18	289.54	296.60	-0.54	-1.12
Jan 31	205.00	2.63	285.73	-0.04	29.68	0.59	321.79	10 Feb 15	75.5	138.1	3.2	4.9	275.35	280.76	2.84	2.48	304.06	311.91	-1.58	-1.92
Feb 10	204.77	2.67	287.76	-0.06	36.24	0.70	331.87	20 Feb 25	197.6	258.9	1.5	-3.5	286.27	291.85	2.11	1.73	320.14	328.79	-2.10	-2.12
Feb 20	204.39	2.71	289.67	-0.08	42.78	0.80	341.90	1 Mar 6	329.9	44.6	-4.6	0.9	297.49	303.20	1.35	0.97	337.87	347.39	-1.94	-1.55
Mar 1	203.86	2.74	291.42	-0.10	49.32	0.89	351.88	11 Mar 16	110.7	170.9	5.0	3.5	308.95	314.73	0.61	0.26	357.25	7.18	-0.93	-0.09
Mar 11	203.23	2.77	292.99	-0.12	55.83	0.96	1.80	21 Mar 26	230.4	294.7	-1.5	-5.2	320.56	326.41	-0.07	-0.38	16.60	24.70	0.90	1.88
Mar 21	202.51	2.78	294.36	-0.14	62.32	1.02	11.67	31 M-A 5	8.4	81.0	-2.2	3.9	332.29	338.20	-0.66	-0.92	30.71	34.14	2.67	3.09
Mar 31	201.75	2.79	295.49	-0.16	68.79	1.07	21.48	10 Apr 15	144.1	203.4	4.9	0.9	344.12	350.06	-1.14	-1.33	34.86	33.17	3.00	2.31
Apr 10	200.99	2.78	296.35	-0.19	75.23	1.11	31.23	20 Apr 25	264.8	332.7	-4.2	-4.6	356.01	1.98	-1.48	-1.60	30.04	26.90	1.12	-0.29
Apr 20	200.27	2.77	296.93	-0.21	81.65	1.14	40.93	30 Apr	46.6		1.2		7.96		-1.69		25.01		-1.60	
								May 5		115.6		5.2		13.95		-1.74		24.97		-2.58
Apr 30	199.62	2.75	297.21	-0.24	88.05	1.16	50.59	10 May 15	176.3	236.5	3.2	-2.1	19.94	25.95	-1.76	-1.75	26.82	30.36	-3.18	-3.42
May 10	199.08	2.72	297.18	-0.27	94.42	1.18	60.20	20 May 25	301.4	11.7	-5.2	-1.9	31.97	37.99	-1.70	-1.63	35.33	41.56	-3.32	-2.95
May 20	198.67	2.68	296.82	-0.30	100.78	1.19	69.79	30 M-J 4	83.6	148.6	4.0	4.6	44.02	50.06	-1.52	-1.40	48.92	57.37	-2.34	-1.55
May 30	198.42	2.64	296.18	-0.33	107.13	1.20	79.35	9 Jun 14	208.2	271.0	0.4	-4.4	56.11	62.16	-1.25	-1.08	66.87	77.24	-0.64	0.28
Jun 9	198.32	2.60	295.27	-0.36	113.46	1.20	88.89	19 Jun 24	339.9	50.5	-4.0	1.7	68.23	74.30	-0.90	-0.71	88.12	98.95	1.07	1.62
Jun 19	198.39	2.56	294.16	-0.39	119.78	1.19	98.42	29 Jun	118.9		5.0		80.38		-0.50		109.28		1.86	
								Jul 4		180.6		2.5		86.47		-0.29		118.87		1.81
Jun 29	198.63	2.51	292.91	-0.42	126.11	1.18	107.95	9 Jul 14	240.6	307.6	-2.6	-5.0	92.58	98.69	-0.08	0.12	127.66	135.64	1.51	0.99
Jul 9	199.02	2.47	291.63	-0.44	132.43	1.17	117.49	19 Jul 24	19.1	88.1	-0.9	4.3	104.81	110.94	0.33	0.52	142.79	149.07	0.31	-0.51
Jul 19	199.56	2.43	290.39	-0.46	138.76	1.15	127.06	29 J-A 3	152.5	212.2	4.1	-0.4	117.09	123.24	0.70	0.86	154.34	158.38	-1.43	-2.40
Jul 29	200.24	2.39	289.28	-0.48	145.10	1.13	136.64	8 Aug 13	274.5	345.9	-4.7	-3.3	129.41	135.58	1.01	1.14	160.87	161.37	-3.34	-4.14
Aug 8	201.04	2.36	288.39	-0.49	151.46	1.10	146.26	18 Aug 23	57.9	124.0	2.7	5.0	141.76	147.96	1.24	1.32	159.55	155.65	-4.59	-4.40
Aug 18	201.95	2.33	287.76	-0.50	157.83	1.07	155.93	28 Aug	185.1		1.8		154.16		1.38		151.09		-3.44	
								Sep 2		244.4		-3.3		160.37		1.41		148.19		-1.94
Aug 28	202.95	2.30	287.44	-0.51	164.23	1.03	165.64	7 Sep 12	310.3	24.7	-5.0	0.0	166.59	172.81	1.41	1.39	148.65	152.74	-0.39	0.83
Sep 7	204.03	2.28	287.44	-0.51	170.65	0.99	175.40	17 Sep 22	95.2	158.0	4.9	3.8	179.03	185.27	1.33	1.26	159.59	167.98	1.58	1.87
Sep 17	205.17	2.27	287.76	-0.51	177.10	0.95	185.22	27 S-O 2	217.3	278.1	-1.2	-5.1	191.50	197.74	1.15	1.03	176.92	185.86	1.81	1.51
Sep 27	206.35	2.26	288.39	-0.52	183.58	0.90	195.10	7 Oct 12	347.7	63.0	-3.2	3.3	203.98	210.22	0.88	0.72	194.59	203.03	1.07	0.55
Oct 7	207.55	2.25	289.32	-0.52	190.09	0.85	205.02	17 Oct 22	130.5	190.7	5.1	1.2	216.47	222.71	0.54	0.35	211.21	219.14	-0.01	-0.57
Oct 17	208.76	2.25	290.51	-0.52	196.65	0.79	215.01	27 Oct	250.2		-3.8		228.96		0.15		226.86		-1.11	
								Nov 1		313.9		-5.0		235.20		-0.06		234.40		-1.61
Oct 27	209.96	2.26	291.94	-0.53	203.24	0.72	225.05	6 Nov 11	26.0	99.7	0.2	5.1	241.45	247.69	-0.27	-0.48	241.75	248.85	-2.03	-2.35
Nov 6	211.12	2.27	293.59	-0.53	209.88	0.65	235.14	16 Nov 21	163.8	223.0	3.4	-1.7	253.93	260.17	-0.68	-0.88	255.58	261.63	-2.51	-2.45
Nov 16	212.23	2.29	295.41	-0.53	216.55	0.58	245.27	26 N-D 1	284.6	351.5	-5.1	-2.7	266.41	272.65	-1.06	-1.23	266.36	268.60	-2.07	-1.21
Nov 26	213.27	2.31	297.40	-0.54	223.27	0.49	255.42	6 Dec 11	64.3	134.6	3.4	4.7	278.88	285.11	-1.38	-1.50	266.79	260.89	0.20	1.83
Dec 6	214.22	2.33	299.51	-0.54	230.03	0.40	265.60	16 Dec 21	195.8	255.8	0.6	-4.1	291.33	297.55	-1.60	-1.68	254.75	252.33	2.87	2.99
Dec 16	215.06	2.36	301.72	-0.55	236.84	0.31	275.80	26 Dec 31	320.9	30.3	-4.4	0.8	303.75	309.95	-1.72	-1.74	253.84	257.90	2.54	1.86
1688								1689												1689
Dec 26	215.78	2.40	304.01	-0.56	243.69	0.20	285.99	5 Jan 10	101.7	167.7	5.0	2.6	316.14	322.31	-1.72	-1.67	263.41	269.76	1.12	0.39
Jan 5	216.35	2.43	306.35	-0.57	250.58	0.08	296.18	15 Jan 20	227.3	289.8	-2.4	-5.0	328.46	334.60	-1.58	-1.46	276.63	283.88	-0.27	-0.86
Jan 15	216.76	2.47	308.73	-0.58	257.51	-0.04	306.35	25 Jan 30	358.9	69.3	-1.6	4.0	340.71	346.80	-1.31	-1.13	291.43	299.28	-1.35	-1.74
Jan 25	217.00	2.51	311.11	-0.60	264.49	-0.17	316.48	4 Feb 9	137.4	199.6	4.4	-0.3	352.86	358.88	-0.91	-0.67	307.44	315.93	-1.99	-2.09
Feb 4	217.08	2.55	313.48	-0.62	271.50	-0.32	326.59	14 Feb 19	259.2	325.5	-4.6	-4.0	4.87	10.83	-0.40	-0.10	324.79	334.02	-2.01	-1.73
Feb 14	216.98	2.59	315.81	-0.63	278.55	-0.47	336.65	24 Feb	37.8		2.0		16.73		0.21		343.53		-1.21	
								Mar 1		107.3		5.2		22.59		0.53		353.07		-0.44
Feb 24	216.71	2.63	318.09	-0.65	285.63	-0.63	346.65	6 Mar 11	171.5	231.2	2.1	-3.1	28.39	34.12	0.87	1.22	2.06	9.60	0.54	1.64
Mar 6	216.29	2.66	320.28	-0.68	292.73	-0.80	356.60	16 Mar 21	292.5	2.8	-5.2	-1.1	39.79	45.38	1.57	1.91	14.72	16.75	2.63	3.28
Mar 16	215.74	2.69	322.38	-0.70	299.86	-0.98	6.50	26 Mar 31	76.5	143.3	4.6	4.2	50.87	56.27	2.24	2.56	15.68	12.37	3.36	2.75
Mar 26	215.08	2.71	324.34	-0.73	307.00	-1.17	16.34	5 Apr 10	204.2	263.6	-0.9	-4.9	61.55	66.69	2.86	3.13	8.53	5.81	1.59	0.23
Apr 5	214.36	2.72	326.15	-0.76	314.14	-1.37	26.12	15 Apr 20	327.6	41.1	-3.9	2.3	71.69	76.50	3.37	3.56	5.10	6.46	-1.02	-1.99
Apr 15	213.60	2.72	327.79	-0.80	321.27	-1.57	35.85	25 Apr 30	114.0	177.4	5.2	1.6	81.10	85.45	3.71	3.79	9.60	14.20	-2.65	-2.99
Apr 25	212.86	2.71	329.22	-0.84	328.38	-1.77	45.53	5 May 10	236.7	297.4	-3.5	-5.1	89.50	93.19	3.80	3.73	19.99	26.81	-3.05	-2.85
May 5	212.16	2.69	330.42	-0.88	335.43	-1.98	55.17	15 May 20	4.7	79.7	-0.9	4.7	96.45	99.18	3.56	3.28	34.59	43.29	-2.41	-1.76
May 15	211.54	2.66	331.37	-0.92	342.41	-2.19	64.77	25 May 30	149.4	210.0	3.6	-1.4	101.29	102.65	2.87	2.31	52.89	63.29	-0.95	-0.06
May 25	211.03	2.63	332.03	-0.97	349.28	-2.39	74.34	4 Jun 9	269.7	333.1	-4.9	-3.2	103.13	102.64	1.58	0.68	74.20	85.11	0.79	1.47
Jun 4	210.66	2.59	332.38	-1.01	356.01	-2.58	83.89	14 Jun 19	43.2	117.3	2.6	4.8	101.17	98.83	-0.38	-1.54	95.53	105.16	1.87	1.96
Jun 14	210.44	2.55	332.42	-1.06	2.55	-2.77	93.42	24 Jun 29	182.8	242.2	0.7	-3.9	95.88	92.78	-2.71	-3.78	113.88	121.68	1.76	1.31
Jun 24	210.38	2.50	332.13	-1.11	8.84	-2.94	102.95	4 Jul 9	304.1	10.6	-4.6	0.1	90.00	87.94	-4.64	-5.26	128.51	134.31	0.64	-0.22
Jul 4	210.48	2.46	331.54	-1.16	14.81	-3.10	112.49	14 Jul 19	82.3	152.9	4.8	3.0	86.81	86.68	-5.62	-5.76	138.91	142.06	-1.21	-2.30
Jul 14	210.75	2.41	330.67	-1.20	20.37	-3.24	122.04	24 Jul 29	214.9	274.9	-2.3	-5.1	87.47	89.08	-5.73	-5.56	143.45	142.77	-3.39	-4.32
Jul 24	211.17	2.37	329.57	-1.23	25.42	-3.35	131.61	3 Aug 8	340.1	49.6	-2.3	3.5	91.39	94.28	-5.30	-4.96	140.05	136.08	-4.82	-4.65
Aug 3	211.73	2.33	328.32	-1.26	29.82	-3.44	141.21	13 Aug 18	120.5	186.5	4.7	0.0	97.67	101.46	-4.57	-4.15	132.54	131.18	-3.75	-2.36
Aug 13	212.43	2.29	327.01	-1.28	33.39	-3.48	150.86	23 Aug 28	246.5	308.6	-4.5	-4.4	105.58	109.99	-3.70	-3.23	132.99	137.95	-0.88	0.40
Aug 23	213.25	2.26	325.73	-1.28	35.93	-3.46	160.54	2 Sep 7	17.7	88.9	1.1	5.2	114.63	119.48	-2.76	-2.28	145.38	154.24	1.28	1.73
Sep 2	214.18	2.23	324.58	-1.28	37.23	-3.37	170.28	12 Sep 17	156.7	218.8	2.6	-2.9	124.50	129.68	-1.82	-1.36	163.61	172.94	1.81	1.62
Sep 12	215.19	2.20	323.65	-1.27	37.11	-3.17	180.07	22 Sep 27	278.5	343.7	-5.3	-2.0	134.98	140.41	-0.92	-0.50	181.96	190.62	1.24	0.76
Sep 22	216.28	2.18	322.98	-1.25	35.54	-2.84	189.91	2 Oct 7	56.6	127.1	4.1	4.5	145.93	151.55	-0.11	0.26	198.92	206.90	0.21	-0.37
Oct 2	217.42	2.17	322.63	-1.23	32.77	-2.37	199.82	12 Oct 17	191.0	250.8	-0.5	-4.8	157.25	163.02	0.59	0.89	214.59	222.01	-0.95	-1.49
Oct 12	218.60	2.16	322.61	-1.21	29.40	-1.79	209.78	22 Oct 27	311.5	20.5	-4.3	1.3	168.85	174.74	1.16	1.38	229.13	235.91	-1.98	-2.38
Oct 22	219.80	2.15	322.93	-1.18	26.27	-1.16	219.79	1 Nov 6	95.5	163.1	5.2	2.0	180.69	186.68	1.57	1.72	242.17	247.59	-2.65	-2.72
Nov 1	220.99	2.15	323.58	-1.15	24.05	-0.56	229.85	11 Nov 16	223.9	283.4	-3.2	-5.0	192.71	198.77	1.82	1.89	251.50	252.77	-2.47	-1.75
Nov 11	222.17	2.15	324.53	-1.13	23.12	-0.04	239.96	21 Nov 26	346.2	58.6	-1.6	4.1	204.87	210.99	1.92	1.91	250.04	243.75	-0.42	1.25
Nov 21	223.31	2.16	325.75	-1.10	23.54	0.39	250.10	1 Dec 6	133.0	197.1	3.9	-1.2	217.14	223.31	1.86	1.78	238.10	236.63	2.45	2.78
Dec 1	224.39	2.18	327.22	-1.08	25.14	0.72	260.27	11 Dec 16	256.5	317.1	-4.8	-3.5	229.49	235.69	1.67	1.54	239.08	243.91	2.51	1.95
Dec 11	225.40	2.20	328.91	-1.06	27.73	0.97	270.45	21 Dec 26	22.8	97.2	1.8	5.0	241.90	248.12	1.38	1.20	250.00	256.77	1.29	0.62
Dec 21	226.30	2.22	330.78	-1.05	31.11	1.16	280.65	31 Dec	168.4		1.0		254.34		1.00		263.92		-0.03	
1689								1689												1689

JULIAN 12 UT	SATURN LONG.	SATURN LAT.	JUPITER LONG.	JUPITER LAT.	MARS LONG.	MARS LAT.	SUN LONG.
1689							
Dec 31	227.10	2.24	332.81	-1.03	35.09	1.30	290.84
Jan 10	227.76	2.27	334.97	-1.03	39.55	1.40	301.02
Jan 20	228.28	2.31	337.23	-1.02	44.37	1.47	311.17
Jan 30	228.64	2.34	339.56	-1.02	49.47	1.52	321.30
Feb 9	228.83	2.38	341.95	-1.02	54.79	1.55	331.38
Feb 19	228.84	2.41	344.36	-1.02	60.28	1.57	341.41
Mar 1	228.70	2.45	346.78	-1.02	65.91	1.57	351.39
Mar 11	228.38	2.48	349.19	-1.03	71.64	1.57	1.32
Mar 21	227.93	2.50	351.56	-1.05	77.44	1.56	11.18
Mar 31	227.35	2.52	353.88	-1.06	83.32	1.54	21.00
Apr 10	226.68	2.53	356.12	-1.08	89.25	1.51	30.76
Apr 20	225.95	2.53	358.27	-1.11	95.22	1.49	40.46
Apr 30	225.20	2.53	0.29	-1.13	101.23	1.45	50.12
May 10	224.47	2.51	2.17	-1.17	107.27	1.42	59.74
May 20	223.79	2.49	3.89	-1.20	113.34	1.38	69.32
May 30	223.20	2.46	5.40	-1.24	119.44	1.33	78.88
Jun 9	222.73	2.43	6.70	-1.28	125.57	1.29	88.42
Jun 19	222.40	2.39	7.74	-1.32	131.73	1.24	97.95
Jun 29	222.22	2.35	8.51	-1.37	137.93	1.19	107.49
Jul 9	222.20	2.30	8.98	-1.42	144.15	1.13	117.03
Jul 19	222.34	2.26	9.12	-1.47	150.42	1.07	126.59
Jul 29	222.64	2.22	8.94	-1.52	156.73	1.01	136.18
Aug 8	223.09	2.17	8.43	-1.56	163.08	0.95	145.80
Aug 18	223.69	2.13	7.63	-1.60	169.48	0.88	155.46
Aug 28	224.41	2.10	6.57	-1.63	175.92	0.82	165.17
Sep 7	225.26	2.07	5.33	-1.65	182.43	0.74	174.93
Sep 17	226.20	2.04	4.00	-1.65	188.98	0.67	184.74
Sep 27	227.23	2.01	2.68	-1.65	195.60	0.59	194.62
Oct 7	228.32	1.99	1.46	-1.63	202.28	0.51	204.55
Oct 17	229.47	1.98	0.45	-1.60	209.03	0.42	214.53
Oct 27	230.65	1.97	359.71	-1.56	215.84	0.33	224.57
Nov 6	231.84	1.96	359.29	-1.51	222.72	0.24	234.65
Nov 16	233.02	1.96	359.20	-1.46	229.67	0.14	244.77
Nov 26	234.18	1.96	359.47	-1.42	236.69	0.04	254.93
Dec 6	235.29	1.97	0.07	-1.37	243.78	-0.06	265.11
Dec 16	236.34	1.98	0.98	-1.32	250.93	-0.17	275.30
1690							
Dec 26	237.31	1.99	2.18	-1.28	258.16	-0.28	285.50
Jan 5	238.17	2.01	3.62	-1.24	265.44	-0.40	295.68
Jan 15	238.92	2.03	5.29	-1.20	272.79	-0.51	305.85
Jan 25	239.53	2.06	7.15	-1.17	280.20	-0.63	315.99
Feb 4	239.99	2.09	9.16	-1.15	287.67	-0.75	326.09
Feb 14	240.29	2.11	11.30	-1.12	295.18	-0.87	336.15
Feb 24	240.43	2.14	13.54	-1.10	302.73	-0.98	346.16
Mar 6	240.39	2.17	15.85	-1.09	310.32	-1.09	356.11
Mar 16	240.19	2.20	18.22	-1.07	317.93	-1.20	6.01
Mar 26	239.84	2.22	20.62	-1.07	325.56	-1.30	15.86
Apr 5	239.35	2.23	23.03	-1.06	333.18	-1.39	25.64
Apr 15	238.75	2.25	25.43	-1.06	340.80	-1.47	35.38
Apr 25	238.06	2.25	27.80	-1.06	348.39	-1.54	45.06
May 5	237.33	2.25	30.12	-1.06	355.94	-1.60	54.70
May 15	236.58	2.24	32.37	-1.07	3.43	-1.64	64.30
May 25	235.87	2.22	34.53	-1.08	10.85	-1.66	73.88
Jun 4	235.22	2.20	36.58	-1.09	18.17	-1.66	83.42
Jun 14	234.66	2.17	38.50	-1.11	25.38	-1.65	92.96
Jun 24	234.23	2.13	40.26	-1.13	32.44	-1.62	102.50
Jul 4	233.94	2.09	41.83	-1.15	39.35	-1.56	112.03
Jul 14	233.80	2.05	43.19	-1.18	46.06	-1.49	121.58
Jul 24	233.82	2.01	44.31	-1.21	52.55	-1.39	131.15
Aug 3	234.00	1.97	45.16	-1.24	58.77	-1.27	140.75
Aug 13	234.34	1.93	45.71	-1.27	64.68	-1.13	150.39
Aug 23	234.83	1.89	45.94	-1.30	70.22	-0.96	160.08
Sep 2	235.46	1.85	45.84	-1.33	75.31	-0.75	169.81
Sep 12	236.22	1.82	45.41	-1.35	79.86	-0.52	179.60
Sep 22	237.09	1.79	44.66	-1.37	83.74	-0.24	189.44
Oct 2	238.05	1.76	43.64	-1.38	86.79	0.08	199.34
Oct 12	239.10	1.74	42.42	-1.38	88.82	0.45	209.29
Oct 22	240.21	1.72	41.08	-1.38	89.64	0.87	219.30
Nov 1	241.36	1.71	39.73	-1.36	89.06	1.35	229.36
Nov 11	242.53	1.70	38.47	-1.33	87.05	1.85	239.46
Nov 21	243.72	1.69	37.40	-1.29	83.82	2.32	249.61
Dec 1	244.89	1.69	36.59	-1.25	79.97	2.71	259.77
Dec 11	246.03	1.69	36.09	-1.20	76.27	2.97	269.96
Dec 21	247.11	1.69	35.93	-1.15	73.44	3.09	280.15
1691							

GREGORIAN 12 UT	MOON LONG.		MOON LAT.		VENUS LONG.		VENUS LAT.		MERCURY LONG.		MERCURY LAT.	
1690												1690
Jan 5		229.7		-3.8		260.57		0.79		271.34		-0.62
10 Jan 15	289.4	352.4	-4.7	-0.5	266.81	273.05	0.58	0.35	278.97	286.82	-1.13	-1.56
20 Jan 25	61.2	134.9	4.5	3.5	279.29	285.54	0.13	-0.09	294.89	303.23	-1.87	-2.04
30 J-F 4	201.7	261.6	-2.1	-5.1	291.78	298.02	-0.30	-0.51	311.86	320.77	-2.06	-1.88
9 Feb 14	323.1	29.7	-2.9	2.8	304.26	310.50	-0.69	-0.87	329.90	339.00	-1.47	-0.79
19 Feb 24	100.5	170.7	5.1	0.5	316.74	322.98	-1.02	-1.15	347.50	354.42	0.16	1.32
1 Mar 6	233.8	293.6	-4.4	-4.7	329.21	335.43	-1.26	-1.35	358.56	359.11	2.49	3.36
11 Mar 16	358.3	68.6	0.2	5.0	341.66	347.88	-1.41	-1.44	356.33	351.87	3.61	3.07
21 Mar 26	138.9	204.8	3.2	-2.5	354.09	0.30	-1.45	-1.42	348.01	346.23	1.96	0.67
31 M-A 5	265.5	326.5	-5.3	-2.6	6.50	12.69	-1.38	-1.30	346.79	349.36	-0.51	-1.46
10 Apr 15	35.3	107.9	3.3	4.9	18.88	25.07	-1.21	-1.09	353.51	358.88	-2.15	-2.58
20 Apr 25	175.5	237.8	0.1	-4.5	31.25	37.42	-0.95	-0.79	5.26	12.50	-2.76	-2.71
30 Apr	297.5		-4.4		43.58		-0.62		20.55		-2.42	
May 5		0.9		0.5		49.74		-0.44		29.39		-1.92
10 May 15	73.9	146.1	5.0	2.4	55.90	62.05	-0.24	-0.04	39.02	49.38	-1.22	-0.39
20 May 25	210.2	270.4	-2.9	-5.0	68.20	74.33	0.16	0.36	60.25	71.17	0.49	1.27
30 M-J 4	330.4	37.4	-2.0	3.5	80.47	86.60	0.55	0.74	81.60	91.18	1.82	2.07
9 Jun 14	112.9	182.2	4.3	-0.9	92.73	98.85	0.92	1.08	99.74	107.22	2.01	1.65
19 Jun 24	243.6	303.1	-4.8	-3.7	104.97	111.08	1.22	1.35	113.57	118.69	1.01	0.14
29 Jun	5.1		1.3		117.19		1.45		122.38		-0.94	
Jul 4		75.8		5.0		123.29		1.52		124.40		-2.15
9 Jul 14	150.8	216.4	1.6	-3.7	129.39	135.48	1.57	1.59	124.52	122.72	-3.37	-4.38
19 Jul 24	276.2	336.7	-4.9	-1.0	141.56	147.64	1.57	1.53	119.51	116.11	-4.91	-4.75
29 J-A 3	41.9	114.9	4.2	4.1	153.71	159.76	1.46	1.35	114.01	114.31	-3.92	-2.68
8 Aug 13	186.6	249.1	-1.7	-5.2	165.81	171.85	1.22	1.05	117.42	123.20	-1.30	-0.04
18 Aug 23	308.9	11.9	-3.3	2.2	177.87	183.88	0.86	0.65	131.15	140.43	0.94	1.55
28 Aug	80.6		5.3		189.87		0.41		150.21		1.78	
Sep 2		153.0		1.3		195.85		0.15		159.89		1.70
7 Sep 12	220.5	281.0	-4.2	-4.8	201.82	207.76	-0.13	-0.42	169.21	178.09	1.41	0.96
17 Sep 22	342.2	49.2	-0.5	4.6	213.68	219.58	-0.72	-1.02	186.52	194.56	0.43	-0.16
27 S-O 2	119.9	189.4	3.9	-2.0	225.46	231.31	-1.32	-1.62	202.21	209.50	-0.76	-1.36
7 Oct 12	253.1	312.8	-5.2	-2.9	237.13	242.91	-1.91	-2.18	216.41	222.83	-1.92	-2.41
17 Oct 22	16.9	88.1	2.6	5.0	248.64	254.34	-2.42	-2.64	228.60	233.35	-2.79	-2.98
27 Oct	158.4		0.7		259.97		-2.83		236.40		-2.87	
Nov 1		224.0		-4.3		265.55		-2.98		236.66		-2.28
6 Nov 11	284.9	345.2	-4.4	0.0	271.05	276.46	-3.08	-3.14	233.08	226.67	-1.05	0.63
16 Nov 21	53.6	127.4	4.7	3.0	281.76	286.93	-3.13	-3.06	221.71	221.22	1.97	2.51
26 N-D 1	195.1	257.4	-2.6	-5.0	291.95	296.79	-2.92	-2.70	224.56	230.11	2.43	2.01
6 Dec 11	316.9	19.1	-2.2	3.0	301.40	305.74	-2.39	-1.99	236.75	243.91	1.44	0.82
16 Dec 21	91.8	165.3	4.7	-0.4	309.74	313.34	-1.49	-0.87	251.34	258.93	0.20	-0.39
26 Dec 31	230.0	290.1	-4.7	-3.9	316.43	318.92	-0.13	0.74	266.64	274.48	-0.92	-1.38
1691												1691
5 Jan 10	349.7	55.2	0.9	5.0	320.68	321.55	1.73	2.85	282.48	290.66	-1.74	-1.99
15 Jan 20	130.3	201.4	2.5	-3.5	321.44	320.28	4.06	5.30	299.05	307.64	-2.09	-2.01
25 Jan 30	263.3	322.7	-5.0	-1.4	318.16	315.34	6.45	7.40	316.37	324.99	-1.71	-1.13
4 Feb 9	24.1	93.1	3.8	4.8	312.25	309.42	8.02	8.26	332.94	339.17	-0.24	0.94
14 Feb 19	167.9	235.6	-0.9	-5.2	307.28	306.10	8.13	7.72	342.30	341.38	2.25	3.33
24 Feb	295.7		-3.6		305.95		7.11		337.12		3.73	
Mar 1		355.9		1.6		306.78		6.40		332.08		3.27
6 Mar 11	60.8	132.0	5.2	2.3	308.45	310.84	5.63	4.84	328.79	328.14	2.23	1.03
16 Mar 21	204.1	268.2	-3.8	-4.9	313.82	317.29	4.07	3.32	329.88	333.45	-0.08	-1.00
26 Mar 31	327.7	30.8	-0.9	4.1	321.17	325.37	2.61	1.95	338.36	344.30	-1.71	-2.20
5 Apr 10	99.4	170.5	4.4	-1.2	329.83	334.52	1.32	0.75	351.08	358.59	-2.48	-2.55
15 Apr 20	238.7	300.0	-5.1	-3.1	339.39	344.41	0.23	-0.24	6.79	15.69	-2.39	-2.03
25 Apr 30	0.4	67.7	2.1	5.1	349.56	354.83	-0.66	-1.03	25.28	35.54	-1.45	-0.70
5 May 10	138.7	207.7	1.5	-4.0	0.18	5.61	-1.34	-1.60	46.29	57.13	0.16	1.01
15 May 20	271.9	331.5	-4.5	-0.2	11.12	16.68	-1.82	-1.98	67.50	76.95	1.71	2.13
25 May 30	34.6	106.3	4.4	3.7	22.29	27.96	-2.09	-2.16	85.22	92.23	2.23	1.99
4 Jun 9	177.4	243.2	-2.1	-5.0	33.66	39.40	-2.19	-2.18	97.88	102.04	1.43	0.56
14 Jun 19	304.2	3.9	-2.3	2.8	45.17	50.97	-2.12	-2.03	104.53	105.18	-0.58	-1.90
24 Jun 29	70.9	145.3	5.0	0.5	56.81	62.67	-1.91	-1.77	103.98	101.37	-3.22	-4.28
4 Jul 9	214.5	277.1	-4.6	-4.1	68.55	74.46	-1.59	-1.40	98.33	96.12	-4.80	-4.67
14 Jul 19	336.4	37.9	0.6	4.9	80.39	86.35	-1.19	-0.97	95.75	97.68	-3.97	-2.90
24 Jul 29	108.9	183.5	3.4	-3.0	92.33	98.33	-0.73	-0.50	101.99	108.53	-1.67	-0.45
3 Aug 8	249.7	309.8	-5.2	-1.7	104.36	110.40	-0.26	-0.02	116.92	126.56	0.58	1.32
13 Aug 18	9.4	74.0	3.4	5.1	116.47	122.55	0.20	0.42	136.69	146.71	1.70	1.75
23 Aug 28	147.6	220.1	0.1	-5.0	128.66	134.79	0.62	0.81	156.32	165.41	1.55	1.17
2 Sep 7	283.0	342.3	-3.8	1.2	140.93	147.09	0.98	1.12	173.98	182.06	0.66	0.07
12 Sep 17	44.0	111.9	5.1	3.2	153.27	159.46	1.24	1.34	189.68	196.84	-0.57	-1.21
22 Sep 27	185.7	254.7	-3.1	-5.0	165.67	171.89	1.41	1.45	203.49	209.55	-1.85	-2.43
2 Oct 7	315.2	15.5	-1.2	3.8	178.12	184.36	1.46	1.44	214.78	218.80	-2.92	-3.24
12 Oct 17	80.7	150.7	4.7	-0.3	190.61	196.87	1.40	1.33	220.92	220.15	-3.27	-2.81
22 Oct 27	222.6	287.5	-4.9	-3.2	203.14	209.41	1.24	1.12	215.85	209.61	-1.67	-0.01
1 Nov 6	346.9	50.2	1.8	5.0	215.69	221.97	0.98	0.83	205.54	206.02	1.45	2.20
11 Nov 16	119.0	189.4	2.2	-3.5	228.25	234.53	0.66	0.47	210.19	216.41	2.32	2.04
21 Nov 26	257.7	319.2	-4.6	-0.4	240.82	247.11	0.28	0.08	223.58	231.14	1.57	1.00
1 Dec 6	19.3	86.6	4.2	4.2	253.40	259.68	-0.12	-0.32	238.84	246.60	0.41	-0.17
11 Dec 16	158.1	227.0	-1.4	-5.1	265.97	272.25	-0.51	-0.69	254.39	262.24	-0.71	-1.20
21 Dec 26	291.1	350.6	-2.6	2.6	278.54	284.82	-0.87	-1.03	270.16	278.19	-1.61	-1.91
31 Dec	52.9		5.2		291.10		-1.17		286.35		-2.09	
1691												1691

JULIAN 12 UT	SATURN LONG.	SATURN LAT.	JUPITER LONG.	JUPITER LAT.	MARS LONG.	MARS LAT.	SUN LONG.
1691							
Dec 31	248.13	1.70	36.13	-1.09	71.89	3.09	290.34
Jan 10	249.06	1.71	36.65	-1.04	71.67	3.02	300.52
Jan 20	249.88	1.72	37.49	-1.00	72.67	2.91	310.68
Jan 30	250.58	1.74	38.62	-0.95	74.69	2.78	320.80
Feb 9	251.14	1.76	40.00	-0.91	77.52	2.64	330.88
Feb 19	251.54	1.78	41.59	-0.88	81.01	2.50	340.92
Feb 29	251.79	1.80	43.38	-0.84	85.00	2.37	350.90
Mar 10	251.86	1.82	45.31	-0.81	89.39	2.24	0.83
Mar 20	251.78	1.84	47.38	-0.78	94.10	2.12	10.70
Mar 30	251.53	1.86	49.55	-0.76	99.06	2.00	20.52
Apr 9	251.13	1.87	51.80	-0.74	104.22	1.89	30.28
Apr 19	250.61	1.88	54.10	-0.72	109.56	1.78	39.99
Apr 29	249.98	1.89	56.43	-0.70	115.04	1.67	49.65
May 9	249.28	1.88	58.78	-0.69	120.64	1.57	59.27
May 19	248.55	1.88	61.13	-0.67	126.35	1.47	68.86
May 29	247.81	1.86	63.45	-0.66	132.15	1.37	78.42
Jun 8	247.12	1.84	65.73	-0.66	138.04	1.27	87.96
Jun 18	246.49	1.82	67.94	-0.65	144.01	1.17	97.50
Jun 28	245.97	1.79	70.08	-0.64	150.05	1.08	107.03
Jul 8	245.58	1.75	72.11	-0.64	156.18	0.98	116.57
Jul 18	245.33	1.72	74.01	-0.64	162.38	0.89	126.13
Jul 28	245.23	1.68	75.76	-0.64	168.66	0.79	135.71
Aug 7	245.30	1.64	77.33	-0.64	175.01	0.70	145.34
Aug 17	245.53	1.61	78.70	-0.64	181.45	0.60	154.99
Aug 27	245.91	1.57	79.82	-0.64	187.96	0.50	164.70
Sep 6	246.44	1.54	80.68	-0.64	194.56	0.41	174.46
Sep 16	247.11	1.50	81.25	-0.64	201.24	0.31	184.27
Sep 26	247.90	1.48	81.49	-0.65	208.00	0.21	194.14
Oct 6	248.80	1.45	81.40	-0.64	214.85	0.11	204.07
Oct 16	249.79	1.42	80.98	-0.64	221.79	0.01	214.05
Oct 26	250.85	1.40	80.25	-0.64	228.82	-0.09	224.08
Nov 5	251.97	1.39	79.24	-0.62	235.93	-0.19	234.16
Nov 15	253.13	1.37	78.02	-0.61	243.13	-0.29	244.28
Nov 25	254.31	1.36	76.69	-0.59	250.42	-0.39	254.44
Dec 5	255.49	1.35	75.33	-0.56	257.79	-0.49	264.62
Dec 15	256.64	1.35	74.07	-0.54	265.23	-0.59	274.81
1692							
Dec 25	257.76	1.35	72.98	-0.50	272.75	-0.68	285.00
Jan 4	258.82	1.35	72.15	-0.47	280.34	-0.77	295.19
Jan 14	259.81	1.35	71.63	-0.44	287.99	-0.85	305.36
Jan 24	260.69	1.36	71.44	-0.40	295.69	-0.93	315.50
Feb 3	261.47	1.37	71.58	-0.37	303.43	-1.00	325.61
Feb 13	262.12	1.38	72.06	-0.34	311.21	-1.07	335.67
Feb 23	262.62	1.39	72.84	-0.31	319.01	-1.12	345.68
Mar 5	262.97	1.40	73.89	-0.28	326.82	-1.16	355.64
Mar 15	263.16	1.41	75.20	-0.25	334.62	-1.19	5.54
Mar 25	263.19	1.43	76.71	-0.23	342.42	-1.21	15.38
Apr 4	263.05	1.44	78.41	-0.21	350.19	-1.22	25.17
Apr 14	262.75	1.44	80.26	-0.18	357.92	-1.21	34.90
Apr 24	262.31	1.45	82.24	-0.16	5.60	-1.19	44.59
May 4	261.76	1.45	84.32	-0.15	13.22	-1.15	54.23
May 14	261.11	1.45	86.48	-0.13	20.76	-1.10	63.83
May 24	260.40	1.44	88.69	-0.11	28.22	-1.04	73.41
Jun 3	259.66	1.43	90.95	-0.09	35.58	-0.96	82.96
Jun 13	258.94	1.42	93.22	-0.08	42.83	-0.87	92.49
Jun 23	258.26	1.40	95.49	-0.06	49.97	-0.77	102.03
Jul 3	257.67	1.37	97.74	-0.04	56.98	-0.65	111.57
Jul 13	257.18	1.34	99.95	-0.03	63.85	-0.53	121.11
Jul 23	256.83	1.31	102.10	-0.01	70.57	-0.39	130.68
Aug 2	256.62	1.28	104.18	0.00	77.13	-0.24	140.29
Aug 12	256.58	1.25	106.15	0.02	83.50	-0.08	149.92
Aug 22	256.69	1.22	108.00	0.04	89.67	0.09	159.60
Sep 1	256.97	1.19	109.69	0.06	95.61	0.28	169.33
Sep 11	257.40	1.16	111.20	0.08	101.28	0.49	179.12
Sep 21	257.97	1.13	112.51	0.10	106.64	0.71	188.96
Oct 1	258.68	1.10	113.58	0.12	111.64	0.95	198.85
Oct 11	259.51	1.08	114.37	0.14	116.20	1.22	208.80
Oct 21	260.44	1.05	114.87	0.17	120.22	1.52	218.81
Oct 31	261.45	1.03	115.05	0.19	123.59	1.85	228.87
Nov 10	262.54	1.01	114.90	0.22	126.15	2.22	238.97
Nov 20	263.67	1.00	114.42	0.25	127.72	2.63	249.11
Nov 30	264.84	0.99	113.63	0.28	128.11	3.07	259.28
Dec 10	266.02	0.97	112.59	0.31	127.16	3.52	269.46
Dec 20	267.19	0.97	111.36	0.33	124.85	3.93	279.66
1693							

GREGORIAN 12 UT	MOON LONG.		MOON LAT.		VENUS LONG.		VENUS LAT.		MERCURY LONG.		MERCURY LAT.	
1692												1692
Jan 5		124.4		1.4		297.38		-1.29		294.62		-2.11
10 Jan 15	196.8	262.7	-4.4	-4.4	303.65	309.92	-1.39	-1.46	302.93	311.04	-1.91	-1.45
20 Jan 25	323.5	22.9	0.3	4.7	316.18	322.44	-1.51	-1.53	318.38	323.88	-0.65	0.51
30 J-F 4	88.3	162.8	4.1	-2.1	328.68	334.92	-1.52	-1.48	325.97	323.67	1.91	3.18
9 Feb 14	234.0	296.6	-5.3	-2.1	341.14	347.36	-1.42	-1.33	318.29	313.22	3.73	3.35
19 Feb 24	355.8	56.7	3.1	5.2	353.56	359.75	-1.20	-1.06	310.82	311.33	2.41	1.31
29 Feb	125.5		1.3		5.92		-0.89		314.10		0.28	
Mar 5		200.9		-4.6		12.07		-0.69		318.46		-0.60
10 Mar 15	269.1	329.2	-4.0	0.9	18.21	24.32	-0.48	-0.25	323.96	330.31	-1.32	-1.86
20 Mar 25	28.8	92.5	4.8	3.7	30.41	36.48	-0.01	0.24	337.36	345.04	-2.21	-2.37
30 M-A 4	163.9	237.9	-2.2	-5.0	42.52	48.54	0.50	0.75	353.31	2.19	-2.34	-2.10
9 Apr 14	302.3	1.5	-1.4	3.5	54.52	60.48	1.01	1.26	11.68	21.78	-1.64	-0.99
19 Apr 24	63.4	130.2	4.8	0.6	66.40	72.28	1.49	1.71	32.35	43.03	-0.17	0.72
29 Apr	202.7		-4.6		78.12		1.91		53.26		1.54	
May 4		272.9		-3.4		83.92		2.09		62.47		2.13
9 May 14	334.2	34.4	1.7	4.9	89.67	95.38	2.23	2.34	70.33	76.66	2.41	2.33
19 May 24	99.7	169.2	2.8	-3.0	101.02	106.60	2.41	2.44	81.34	84.23	1.87	1.03
29 M-J 3	240.5	306.1	-4.8	-0.6	112.10	117.52	2.42	2.36	85.21	84.34	-0.14	-1.53
8 Jun 13	5.9	68.5	4.1	4.6	122.85	128.07	2.24	2.06	82.05	79.26	-2.91	-3.97
18 Jun 23	137.7	208.5	-0.5	-5.0	133.17	138.12	1.82	1.52	77.13	76.55	-4.49	-4.43
28 Jun	276.4		-3.0		142.91		1.15		77.93		-3.89	
Jul 3		338.2		2.4		147.50		0.71		81.35		-3.02
8 Jul 13	38.1	104.3	5.2	2.4	151.86	155.94	0.20	-0.39	86.73	93.93	-1.96	-0.84
18 Jul 23	176.7	246.5	-3.8	-4.8	159.69	163.04	-1.06	-1.81	102.71	112.64	0.21	1.05
28 J-A 2	310.3	10.1	0.0	4.5	165.90	168.17	-2.64	-3.55	123.06	133.39	1.57	1.77
7 Aug 12	71.6	141.8	4.5	-1.0	169.74	170.49	-4.52	-5.53	143.26	152.54	1.68	1.37
17 Aug 22	215.7	282.4	-5.2	-2.5	170.31	169.14	-6.53	-7.44	161.23	169.35	0.90	0.31
27 Aug	343.1		2.8		167.04		-8.18		176.92		-0.35	
Sep 1		42.6		5.2		164.25		-8.62		183.92		-1.04
6 Sep 11	106.7	180.4	2.1	-4.0	161.19	158.32	-8.69	-8.38	190.30	195.94	-1.75	-2.43
16 Sep 21	253.2	316.3	-4.3	0.6	156.08	154.74	-7.75	-6.90	200.58	203.79	-3.04	-3.49
26 S-O 1	15.7	76.5	4.6	4.0	154.40	155.04	-5.93	-4.92	204.91	203.12	-3.66	-3.33
6 Oct 11	143.7	219.0	-1.3	-5.0	156.55	158.82	-3.94	-3.00	198.29	192.51	-2.27	-0.64
16 Oct 21	288.4	348.9	-1.7	3.3	161.72	165.13	-2.13	-1.34	189.50	190.94	0.91	1.85
26 Oct 31	48.8	112.1	4.9	1.2	168.99	173.20	-0.63	0.01	195.89	202.75	2.16	2.04
5 Nov 10	182.1	256.2	-4.2	-3.8	177.72	182.49	0.57	1.05	210.43	218.38	1.67	1.17
15 Nov 20	321.5	21.1	1.5	4.9	187.47	192.63	1.46	1.81	226.37	234.31	0.62	0.04
25 Nov 30	83.0	149.6	3.3	-2.3	197.94	203.38	2.08	2.29	242.20	250.06	-0.51	-1.02
5 Dec 10	221.1	291.4	-5.1	-1.0	208.93	214.58	2.44	2.52	257.92	265.80	-1.47	-1.83
15 Dec 20	353.4	53.5	4.0	4.8	220.30	226.09	2.56	2.54	273.73	281.68	-2.09	-2.19
25 Dec 30	118.8	188.6	0.3	-4.8	231.94	237.84	2.48	2.38	289.56	297.14	-2.10	-1.75
1693												1693
4 Jan 9	259.1	324.8	-3.6	2.1	243.78	249.76	2.24	2.06	303.84	308.54	-1.06	0.04
14 Jan 19	24.9	86.8	5.2	3.0	255.76	261.79	1.86	1.63	309.60	306.08	1.49	2.91
24 Jan 29	156.2	227.9	-3.0	-5.1	267.85	273.92	1.39	1.14	299.99	295.29	3.60	3.34
3 Feb 8	295.1	357.2	-0.6	4.3	280.01	286.11	0.87	0.60	293.93	295.52	2.52	1.53
13 Feb 18	56.8	121.6	4.7	0.0	292.22	298.33	0.34	0.08	299.17	304.18	0.58	-0.26
23 Feb 28	195.0	265.9	-5.0	-3.0	304.46	310.59	-0.18	-0.42	310.14	316.80	-0.97	-1.54
5 Mar 10	329.4	29.4	2.5	5.1	316.72	322.86	-0.64	-0.84	324.04	331.81	-1.95	-2.19
15 Mar 20	89.8	158.3	2.6	-3.2	329.00	335.14	-1.02	-1.18	340.09	348.89	-2.26	-2.13
25 Mar 30	233.8	301.7	-4.6	0.2	341.29	347.43	-1.31	-1.41	358.24	8.13	-1.79	-1.24
4 Apr 9	2.6	62.1	4.5	4.1	353.57	359.71	-1.48	-1.52	18.46	28.91	-0.49	0.40
14 Apr 19	124.6	196.7	-0.6	-4.9	5.85	11.99	-1.54	-1.53	38.90	47.79	1.31	2.07
24 Apr 29	271.3	335.7	-2.1	3.2	18.13	24.26	-1.49	-1.42	55.09	60.52	2.55	2.66
4 May 9	35.2	95.8	4.9	1.6	30.40	36.53	-1.33	-1.21	63.93	65.23	2.33	1.56
14 May 19	161.6	235.7	-3.7	-4.3	42.66	48.79	-1.07	-0.92	64.51	62.28	0.37	-1.06
24 May 29	306.7	8.3	1.2	5.0	54.93	61.06	-0.75	-0.57	59.49	57.32	-2.44	-3.48
3 Jun 8	68.1	131.2	3.6	-1.7	67.19	73.33	-0.38	-0.18	56.62	57.76	-4.03	-4.08
13 Jun 18	200.4	273.6	-5.2	-1.7	79.47	85.60	0.02	0.21	60.73	65.42	-3.73	-3.06
23 Jun 28	340.2	40.4	3.9	5.0	91.74	97.89	0.41	0.59	71.68	79.43	-2.18	-1.17
3 Jul 8	101.8	168.8	1.0	-4.4	104.04	110.19	0.77	0.93	88.53	98.67	-0.16	0.74
13 Jul 18	239.8	309.5	-4.2	1.6	116.34	122.50	1.07	1.19	109.33	119.90	1.40	1.74
23 Jul 28	12.6	72.5	5.2	3.4	128.66	134.82	1.29	1.37	130.00	139.46	1.78	1.56
2 Aug 7	137.2	207.9	-2.2	-5.2	140.99	147.16	1.42	1.45	148.25	156.38	1.14	0.57
12 Aug 17	278.1	343.8	-1.3	4.1	153.32	159.49	1.45	1.42	163.85	170.66	-0.10	-0.85
22 Aug 27	44.4	105.4	4.8	0.6	165.66	171.83	1.36	1.28	176.71	181.85	-1.63	-2.42
1 Sep 6	174.5	247.3	-4.6	-3.5	178.00	184.17	1.17	1.03	185.81	188.12	-3.15	-3.74
11 Sep 16	314.5	16.8	2.0	5.0	190.33	196.49	0.87	0.69	188.17	185.43	-4.04	-3.81
21 Sep 26	76.3	139.9	2.8	-2.6	202.65	208.80	0.49	0.28	180.36	175.38	-2.82	-1.23
1 Oct 6	213.3	285.3	-4.8	-0.3	214.95	221.10	0.05	-0.18	173.55	175.96	0.36	1.45
11 Oct 16	349.2	49.3	4.4	4.2	227.24	233.38	-0.42	-0.66	181.64	189.10	1.96	2.00
21 Oct 26	109.2	176.4	-0.2	-4.7	239.51	245.63	-0.89	-1.12	197.28	205.62	1.75	1.33
31 Oct	252.1		-2.8		251.75		-1.34		213.89		0.81	
Nov 5		321.2		3.0		257.85		-1.54		222.03		0.25
10 Nov 15	22.5	81.9	5.0	1.9	263.95	270.04	-1.72	-1.87	230.03	237.91	-0.30	-0.84
20 Nov 25	143.8	214.6	-3.3	-4.8	276.11	282.16	-2.00	-2.10	245.71	253.46	-1.32	-1.74
30 N-D 5	289.5	355.2	0.5	5.0	288.20	294.21	-2.16	-2.18	261.16	268.79	-2.06	-2.26
10 Dec 15	55.0	115.5	3.9	-1.1	300.19	306.14	-2.16	-2.10	276.24	283.26	-2.27	-2.04
20 Dec 25	180.7	253.2	-5.2	-2.6	312.06	317.93	-2.00	-1.84	289.27	293.13	-1.47	-0.46
30 Dec	325.0		3.5		323.75		-1.64		293.17		0.99	
1693												1693

JULIAN 12 UT	SATURN LONG.	SATURN LAT.	JUPITER LONG.	JUPITER LAT.	MARS LONG.	MARS LAT.	SUN LONG.
1693							
Dec 30	268.33	0.96	110.03	0.36	121.44	4.24	289.85
Jan 9	269.43	0.95	108.70	0.38	117.48	4.39	300.03
Jan 19	270.46	0.95	107.47	0.40	113.72	4.37	310.19
Jan 29	271.41	0.95	106.43	0.41	110.84	4.19	320.31
Feb 8	272.26	0.95	105.64	0.42	109.19	3.92	330.40
Feb 18	272.99	0.95	105.15	0.43	108.85	3.61	340.44
Feb 28	273.59	0.96	104.98	0.44	109.70	3.29	350.42
Mar 10	274.04	0.96	105.14	0.44	111.56	2.99	0.35
Mar 20	274.34	0.96	105.61	0.45	114.24	2.71	10.23
Mar 30	274.47	0.97	106.37	0.45	117.58	2.45	20.04
Apr 9	274.45	0.97	107.40	0.46	121.44	2.21	29.81
Apr 19	274.26	0.97	108.65	0.46	125.73	1.99	39.52
Apr 29	273.92	0.97	110.12	0.47	130.36	1.79	49.18
May 9	273.44	0.97	111.76	0.47	135.28	1.61	58.80
May 19	272.85	0.97	113.54	0.48	140.44	1.44	68.39
May 29	272.18	0.96	115.46	0.48	145.81	1.27	77.95
Jun 8	271.46	0.95	117.47	0.49	151.35	1.12	87.49
Jun 18	270.73	0.93	119.56	0.50	157.05	0.97	97.03
Jun 28	270.02	0.92	121.71	0.51	162.90	0.84	106.56
Jul 8	269.36	0.90	123.89	0.52	168.89	0.70	116.10
Jul 18	268.80	0.88	126.10	0.53	175.00	0.57	125.67
Jul 28	268.35	0.85	128.30	0.55	181.23	0.44	135.25
Aug 7	268.04	0.83	130.49	0.56	187.58	0.32	144.87
Aug 17	267.88	0.80	132.63	0.58	194.05	0.20	154.53
Aug 27	267.89	0.78	134.72	0.60	200.63	0.09	164.23
Sep 6	268.05	0.75	136.72	0.63	207.33	-0.02	173.98
Sep 16	268.38	0.73	138.62	0.65	214.13	-0.13	183.80
Sep 26	268.86	0.70	140.38	0.68	221.04	-0.24	193.66
Oct 6	269.48	0.68	141.99	0.71	228.07	-0.34	203.59
Oct 16	270.23	0.66	143.41	0.74	235.20	-0.44	213.57
Oct 26	271.10	0.64	144.61	0.78	242.43	-0.54	223.60
Nov 5	272.06	0.62	145.56	0.81	249.76	-0.62	233.67
Nov 15	273.11	0.60	146.24	0.85	257.19	-0.71	243.80
Nov 25	274.21	0.59	146.61	0.90	264.70	-0.79	253.95
Dec 5	275.36	0.57	146.66	0.94	272.30	-0.86	264.12
Dec 15	276.53	0.56	146.39	0.98	279.98	-0.92	274.32
1694							
Dec 25	277.71	0.55	145.80	1.02	287.71	-0.97	284.51
Jan 4	278.88	0.54	144.93	1.06	295.51	-1.02	294.70
Jan 14	280.00	0.53	143.83	1.09	303.35	-1.05	304.87
Jan 24	281.08	0.52	142.58	1.12	311.21	-1.07	315.01
Feb 3	282.09	0.52	141.26	1.14	319.10	-1.08	325.12
Feb 13	283.00	0.51	139.97	1.15	326.99	-1.08	335.18
Feb 23	283.81	0.50	138.81	1.15	334.87	-1.07	345.19
Mar 5	284.50	0.50	137.86	1.14	342.74	-1.05	355.15
Mar 15	285.05	0.50	137.16	1.13	350.57	-1.01	5.06
Mar 25	285.46	0.49	136.76	1.11	358.36	-0.97	14.90
Apr 4	285.70	0.49	136.67	1.09	6.10	-0.91	24.70
Apr 14	285.79	0.48	136.89	1.07	13.77	-0.84	34.43
Apr 24	285.71	0.47	137.40	1.05	21.37	-0.76	44.12
May 4	285.47	0.47	138.19	1.03	28.88	-0.68	53.76
May 14	285.09	0.46	139.23	1.02	36.32	-0.58	63.37
May 24	284.58	0.45	140.49	1.00	43.65	-0.48	72.94
Jun 3	283.96	0.44	141.95	0.99	50.90	-0.37	82.49
Jun 13	283.27	0.43	143.57	0.97	58.04	-0.26	92.03
Jun 23	282.55	0.41	145.34	0.96	65.07	-0.14	101.56
Jul 3	281.81	0.40	147.23	0.96	72.01	-0.02	111.10
Jul 13	281.11	0.38	149.21	0.95	78.83	0.11	120.65
Jul 23	280.48	0.37	151.27	0.95	85.53	0.24	130.22
Aug 2	279.95	0.35	153.39	0.95	92.12	0.37	139.82
Aug 12	279.54	0.33	155.54	0.96	98.59	0.51	149.46
Aug 22	279.28	0.31	157.71	0.96	104.92	0.65	159.13
Sep 1	279.17	0.29	159.87	0.97	111.11	0.80	168.86
Sep 11	279.23	0.27	162.01	0.99	117.14	0.96	178.65
Sep 21	279.45	0.26	164.11	1.00	123.01	1.12	188.48
Oct 1	279.83	0.24	166.14	1.02	128.68	1.29	198.38
Oct 11	280.36	0.22	168.07	1.05	134.12	1.47	208.33
Oct 21	281.03	0.21	169.89	1.07	139.30	1.66	218.33
Oct 31	281.82	0.19	171.57	1.10	144.18	1.87	228.39
Nov 10	282.73	0.18	173.08	1.14	148.68	2.09	238.48
Nov 20	283.72	0.16	174.38	1.18	152.72	2.34	248.62
Nov 30	284.80	0.15	175.46	1.22	156.20	2.60	258.79
Dec 10	285.92	0.14	176.27	1.26	158.98	2.90	268.97
Dec 20	287.09	0.13	176.79	1.31	160.91	3.21	279.16
1695							

GREGORIAN 12 UT	MOON LONG.		MOON LAT.		VENUS LONG.		VENUS LAT.		MERCURY LONG.		MERCURY LAT.	
1694												1694
Jan 4		27.8		5.1		329.51		-1.40		288.63		2.52
9 Jan 14	87.3	150.7	1.4	-3.9	335.19	340.79	-1.10	-0.76	282.21	278.17	3.37	3.26
19 Jan 24	219.4	291.1	-4.7	0.7	346.30	351.69	-0.37	0.06	277.86	280.43	2.58	1.71
29 J-F 3	358.8	59.4	5.1	3.6	356.95	2.05	0.54	1.06	284.85	290.41	0.84	0.04
8 Feb 13	120.4	188.0	-1.6	-5.1	6.97	11.66	1.61	2.20	296.75	303.66	-0.66	-1.24
18 Feb 23	258.6	327.6	-2.1	3.7	16.09	20.21	2.82	3.46	311.04	318.85	-1.70	-2.00
28 Feb	31.5		4.8		23.96		4.11		327.09		-2.16	
Mar 5		91.0		0.9		27.24		4.78		335.78		-2.13
10 Mar 15	155.0	226.8	-4.1	-4.0	29.96	32.01	5.43	6.06	344.95	354.60	-1.91	-1.47
20 Mar 25	297.0	2.6	1.5	5.0	33.27	33.61	6.62	7.07	4.63	14.77	-0.81	0.05
30 M-A 4	63.4	123.4	2.9	-2.2	32.94	31.26	7.36	7.40	24.46	32.94	1.02	1.93
9 Apr 14	191.7	265.8	-5.0	-1.0	28.72	25.66	7.13	6.52	39.54	43.86	2.62	2.94
19 Apr 24	333.9	36.1	4.3	4.4	22.57	19.94	5.61	4.51	45.73	45.22	2.78	2.08
29 Apr	95.4		0.0		18.11		3.32		42.89		0.91	
May 4		157.6		-4.5		17.27		2.15		39.83		-0.52
9 May 14	230.0	303.7	-3.6	2.5	17.41	18.47	1.09	0.15	37.35	36.38	-1.87	-2.89
19 May 24	8.8	68.8	5.2	2.2	20.33	22.84	-0.65	-1.32	37.26	39.95	-3.48	-3.66
29 M-J 3	128.4	194.0	-3.0	-5.2	25.91	29.42	-1.87	-2.30	44.23	49.92	-3.49	-3.03
8 Jun 13	268.7	340.0	-0.5	4.8	33.31	37.50	-2.63	-2.86	56.89	65.08	-2.33	-1.46
18 Jun 23	42.2	101.3	4.2	-0.7	41.94	46.60	-3.02	-3.11	74.41	84.69	-0.51	0.42
28 Jun	163.2		-4.9		51.43		-3.13		95.50		1.18	
Jul 3		232.2		-3.5		56.42		-3.09		106.27		1.66
8 Jul 13	306.7	14.4	2.9	5.2	61.54	66.77	-3.00	-2.86	116.56	126.15	1.84	1.73
18 Jul 23	74.4	134.8	1.8	-3.4	72.09	77.51	-2.68	-2.48	134.99	143.07	1.39	0.86
28 J-A 2	200.1	271.2	-5.0	-0.2	83.01	88.57	-2.24	-1.98	150.41	156.94	0.18	-0.61
7 Aug 12	343.1	47.1	4.8	3.7	94.20	99.89	-1.70	-1.41	162.58	167.12	-1.48	-2.38
17 Aug 22	106.5	169.9	-1.2	-5.0	105.63	111.43	-1.11	-0.81	170.26	171.55	-3.24	-3.98
27 Aug	238.7		-2.8		117.27		-0.52		170.49		-4.40	
Sep 1		309.7		3.1		123.15		-0.23		166.99		-4.23
6 Sep 11	17.8	78.7	4.8	1.0	129.08	135.04	0.05	0.32	162.09	158.23	-3.29	-1.77
16 Sep 21	139.3	207.1	-3.8	-4.4	141.04	147.08	0.57	0.79	157.69	161.05	-0.18	1.03
26 S-O 1	277.9	346.8	0.8	4.9	153.15	159.24	1.00	1.17	167.40	175.42	1.71	1.93
6 Oct 11	50.9	110.3	3.0	-2.0	165.36	171.51	1.32	1.44	184.08	192.81	1.80	1.46
16 Oct 21	173.7	245.5	-5.1	-1.8	177.68	183.87	1.52	1.58	201.38	209.72	1.00	0.46
26 Oct 31	316.5	22.2	3.9	4.6	190.07	196.30	1.60	1.59	217.84	225.76	-0.10	-0.65
5 Nov 10	83.0	142.6	0.2	-4.4	202.53	208.78	1.56	1.49	233.51	241.13	-1.17	-1.64
15 Nov 20	209.8	284.3	-4.3	1.7	215.04	221.30	1.40	1.28	248.62	255.93	-2.03	-2.31
25 Nov 30	353.6	55.8	5.3	2.6	227.58	233.86	1.14	0.98	262.94	269.39	-2.42	-2.31
5 Dec 10	115.1	176.6	-2.7	-5.3	240.14	246.42	0.81	0.62	274.67	277.63	-1.88	-0.98
15 Dec 20	247.3	322.2	-1.6	4.5	252.71	259.00	0.43	0.23	276.68	271.32	0.44	2.05
25 Dec 30	28.5	88.4	4.4	-0.3	265.29	271.58	0.02	-0.18	264.91	261.70	3.05	3.11
1695												1695
4 Jan 9	148.1	212.5	-4.7	-4.2	277.87	284.15	-0.38	-0.56	262.42	265.88	2.59	1.84
14 Jan 19	285.5	358.6	1.9	5.2	290.44	296.73	-0.74	-0.90	270.97	277.02	1.05	0.30
24 Jan 29	61.6	120.8	2.0	-3.1	303.01	309.28	-1.05	-1.17	283.69	290.80	-0.38	-0.97
3 Feb 8	182.8	250.3	-5.0	-1.2	315.55	321.82	-1.28	-1.36	298.28	306.11	-1.46	-1.82
13 Feb 18	323.7	33.0	4.4	3.8	328.09	334.34	-1.41	-1.44	314.29	322.85	-2.04	-2.11
23 Feb 28	93.5	154.0	-1.0	-4.8	340.59	346.84	-1.44	-1.42	331.81	341.18	-1.99	-1.66
5 Mar 10	219.4	289.2	-3.3	2.4	353.07	359.30	-1.36	-1.29	350.88	0.66	-1.10	-0.30
15 Mar 20	0.8	65.8	4.9	1.2	5.51	11.72	-1.18	-1.06	9.97	17.95	0.69	1.72
25 Mar 30	125.2	188.5	-3.7	-4.7	17.92	24.10	-0.91	-0.74	23.75	26.78	2.62	3.16
4 Apr 9	257.7	328.2	0.0	4.8	30.27	36.44	-0.56	-0.37	26.90	24.60	3.18	2.57
14 Apr 19	36.1	97.5	3.3	-1.8	42.59	48.72	-0.16	0.06	21.11	18.00	1.42	0.02
24 Apr 29	157.7	224.8	-5.1	-2.7	54.85	60.96	0.27	0.49	16.46	16.90	-1.27	-2.28
4 May 9	296.8	5.9	3.4	4.9	67.05	73.14	0.71	0.91	19.23	23.16	-2.92	-3.23
14 May 19	69.6	129.2	0.6	-4.2	79.21	85.26	1.11	1.29	28.45	34.88	-3.22	-2.94
24 May 29	191.5	262.6	-4.8	0.6	91.29	97.31	1.45	1.59	42.38	50.88	-2.42	-1.69
3 Jun 8	335.6	41.6	5.2	3.0	103.31	109.29	1.71	1.80	60.37	70.72	-0.83	0.08
13 Jun 18	102.2	161.8	-2.4	-5.3	115.26	121.20	1.85	1.87	81.61	92.51	0.91	1.53
23 Jun 28	227.1	301.4	-2.5	3.8	127.11	133.00	1.86	1.81	102.92	112.59	1.86	1.89
3 Jul 8	12.9	75.4	4.7	0.0	138.85	144.68	1.72	1.59	121.41	129.38	1.64	1.16
13 Jul 18	134.6	195.8	-4.5	-4.4	150.47	156.23	1.43	1.22	136.47	142.63	0.49	-0.34
23 Jul 28	264.5	340.0	0.9	5.1	161.94	167.60	0.97	0.69	147.71	151.50	-1.29	-2.30
2 Aug 7	48.0	108.0	2.3	-2.8	173.21	178.76	0.37	0.02	153.65	153.80	-3.31	-4.18
12 Aug 17	167.9	231.6	-5.0	-1.8	184.25	189.66	-0.37	-0.78	151.70	147.78	-4.69	-4.56
22 Aug 27	303.2	17.0	3.9	4.1	194.98	200.20	-1.21	-1.67	143.54	141.09	-3.67	-2.23
1 Sep 6	81.1	140.4	-0.8	-4.7	205.31	210.28	-2.15	-2.63	141.90	146.18	-0.69	0.58
11 Sep 16	202.5	269.4	-3.7	1.7	215.10	219.73	-3.12	-3.61	153.15	161.69	1.42	1.80
21 Sep 26	342.0	51.9	5.0	1.5	224.12	228.24	-4.09	-4.54	170.80	179.92	1.82	1.58
1 Oct 6	113.0	173.4	-3.6	-4.9	232.04	235.42	-4.97	-5.35	188.79	197.34	1.17	0.67
11 Oct 16	238.8	308.5	-0.8	4.5	238.32	240.61	-5.67	-5.91	205.59	213.56	0.12	-0.45
21 Oct 26	19.5	84.9	3.8	-1.6	242.17	242.87	-6.02	-5.97	221.28	228.78	-1.01	-1.53
31 Oct	144.6		-5.1		242.61		-5.72		236.05		-1.99	
Nov 5		207.4		-3.3		241.35		-5.21		243.04		-2.35
10 Nov 15	276.8	347.7	2.6	5.2	239.18	236.36	-4.43	-3.38	249.61	255.45	-2.57	-2.58
20 Nov 25	55.1	116.8	1.1	-4.1	233.34	230.64	-2.15	-0.85	259.95	261.99	-2.27	-1.51
30 N-D 5	176.7	242.9	-5.0	-0.3	228.67	227.68	0.38	1.46	260.06	254.13	-0.16	1.50
10 Dec 15	315.8	25.4	4.9	3.5	227.71	228.71	2.35	3.04	248.01	245.71	2.66	2.92
20 Dec 25	88.9	148.7	-2.0	-5.2	230.55	233.12	3.55	3.90	247.42	251.70	2.56	1.94
30 Dec	209.9		-3.1		236.28		4.11		257.40		1.24	
1695												1695

JULIAN 12 UT	SATURN LONG.	SATURN LAT.	JUPITER LONG.	JUPITER LAT.	MARS LONG.	MARS LAT.	SUN LONG.	GREGORIAN 12 UT	MOON LONG.	MOON LONG.	MOON LAT.	MOON LAT.	VENUS LONG.	VENUS LONG.	VENUS LAT.	VENUS LAT.	MERCURY LONG.	MERCURY LONG.	MERCURY LAT.	MERCURY LAT.
1695								1696												1696
								Jan 4		280.0		2.9		239.93		4.21		263.89		0.53
Dec 30	288.27	0.11	177.01	1.35	161.80	3.55	289.36	9 Jan 14	354.6	61.1	4.9	0.4	243.98	248.35	4.20	4.12	270.86	278.16	-0.13	-0.72
Jan 9	289.44	0.10	176.91	1.40	161.47	3.87	299.54	19 Jan 24	121.7	181.3	-4.2	-4.5	252.99	257.85	3.97	3.76	285.73	293.56	-1.23	-1.64
Jan 19	290.60	0.09	176.49	1.44	159.83	4.16	309.70	29 J-F 3	244.8	318.4	0.0	4.8	262.89	268.09	3.50	3.20	301.67	310.08	-1.92	-2.07
Jan 29	291.72	0.08	175.77	1.48	156.96	4.36	319.83	8 Feb 13	31.7	94.9	2.6	-2.6	273.41	278.84	2.87	2.52	318.82	327.89	-2.05	-1.82
Feb 8	292.77	0.07	174.80	1.52	153.24	4.42	329.91	18 Feb 23	154.3	215.1	-5.0	-2.2	284.35	289.94	2.16	1.79	337.23	346.60	-1.36	-0.65
Feb 18	293.75	0.06	173.65	1.54	149.32	4.30	339.95	28 Feb	281.8		3.3		295.60		1.41		355.45		0.32	
								Mar 4		357.0		4.5		301.31		1.04		2.86		1.44
Feb 28	294.64	0.04	172.38	1.56	145.90	4.03	349.94	9 Mar 14	66.7	127.4	-0.6	-4.7	307.07	312.87	0.68	0.33	7.76	9.38	2.52	3.29
Mar 9	295.41	0.03	171.09	1.56	143.54	3.65	359.88	19 Mar 24	187.4	250.7	-4.0	1.1	318.70	324.56	0.00	-0.31	7.73	3.94	3.49	2.95
Mar 19	296.06	0.02	169.86	1.55	142.45	3.24	9.75	29 M-A 3	320.6	34.2	5.1	2.1	330.45	336.36	-0.60	-0.86	359.96	357.47	1.84	0.51
Mar 29	296.57	0.01	168.80	1.54	142.62	2.83	19.58	8 Apr 13	99.8	159.6	-3.5	-5.1	342.29	348.23	-1.08	-1.28	357.17	358.95	-0.73	-1.71
Apr 8	296.92	-0.01	167.95	1.51	143.92	2.44	29.34	18 Apr 23	221.5	288.4	-1.4	4.1	354.19	0.16	-1.44	-1.57	2.45	7.33	-2.40	-2.80
Apr 18	297.12	-0.02	167.37	1.48	146.17	2.09	39.05	28 Apr	359.8		4.4		6.15		-1.66		13.34		-2.93	
								May 3		69.6		-1.1		12.14		-1.72		20.31		-2.81
Apr 28	297.16	-0.03	167.08	1.44	149.18	1.77	48.72	8 May 13	131.9	192.0	-5.1	-3.8	18.15	24.16	-1.75	-1.74	28.16	36.88	-2.45	-1.87
May 8	297.03	-0.05	167.10	1.41	152.82	1.48	58.34	18 May 23	257.2	327.7	1.8	5.3	30.17	36.20	-1.71	-1.64	46.45	56.80	-1.12	-0.25
May 18	296.75	-0.06	167.41	1.37	156.97	1.22	67.93	28 M-J 2	37.9	103.3	1.8	-3.8	42.23	48.28	-1.54	-1.42	67.68	78.62	0.62	1.35
May 28	296.33	-0.08	168.01	1.33	161.55	0.99	77.50	7 Jun 12	163.6	225.3	-5.0	-1.1	54.33	60.38	-1.28	-1.12	89.10	98.77	1.83	2.01
Jun 7	295.78	-0.09	168.87	1.30	166.48	0.77	87.04	17 Jun 22	294.9	7.0	4.4	4.0	66.45	72.52	-0.94	-0.75	107.49	115.24	1.88	1.48
Jun 17	295.14	-0.11	169.97	1.26	171.71	0.58	96.57	27 Jun	74.0		-1.5		78.60		-0.55		121.96		0.84	
								Jul 2		136.0		-5.0		84.70		-0.34		127.57		-0.01
Jun 27	294.43	-0.13	171.28	1.23	177.21	0.40	106.11	7 Jul 12	195.7	260.4	-3.3	2.1	90.80	96.91	-0.13	0.08	131.90	134.71	-1.04	-2.18
Jul 7	293.70	-0.14	172.77	1.21	182.93	0.23	115.64	17 Jul 22	334.1	44.9	5.0	0.9	103.03	109.16	0.28	0.48	135.70	134.65	-3.33	-4.31
Jul 17	292.97	-0.15	174.43	1.18	188.86	0.07	125.20	27 J-A 1	108.4	168.4	-4.0	-4.6	115.30	121.45	0.66	0.83	131.75	127.95	-4.87	-4.75
Jul 27	292.28	-0.17	176.22	1.16	194.99	-0.08	134.79	6 Aug 11	228.8	297.5	-0.4	4.5	127.62	133.79	0.98	1.11	124.87	124.04	-3.91	-2.60
Aug 6	291.68	-0.18	178.13	1.15	201.29	-0.21	144.40	16 Aug 21	13.0	80.7	3.2	-2.4	139.97	146.16	1.22	1.31	126.21	131.36	-1.15	0.14
Aug 16	291.18	-0.19	180.13	1.13	207.75	-0.34	154.05	26 Aug 31	141.6	201.0	-5.0	-2.5	152.36	158.57	1.37	1.40	138.89	147.89	1.09	1.64
Aug 26	290.81	-0.20	182.21	1.12	214.37	-0.46	163.76	5 Sep 10	263.7	336.2	2.8	4.9	164.78	171.00	1.41	1.39	157.43	166.92	1.80	1.67
Sep 5	290.59	-0.22	184.33	1.12	221.14	-0.57	173.51	15 Sep 20	50.4	114.7	-0.1	-4.6	177.23	183.45	1.35	1.28	176.09	184.87	1.34	0.88
Sep 15	290.54	-0.23	186.48	1.12	228.05	-0.67	183.31	25 Sep 30	174.3	234.8	-4.2	0.6	189.69	195.93	1.18	1.06	193.25	201.28	0.33	-0.25
Sep 25	290.65	-0.24	188.65	1.12	235.10	-0.76	193.18	5 Oct 10	300.8	14.8	4.9	2.9	202.17	208.41	0.92	0.76	208.97	216.35	-0.84	-1.40
Oct 5	290.92	-0.24	190.80	1.12	242.27	-0.84	203.10	15 Oct 20	85.6	147.2	-3.2	-5.2	214.65	220.89	0.58	0.39	223.41	230.06	-1.93	-2.37
Oct 15	291.36	-0.25	192.92	1.13	249.57	-0.92	213.08	25 Oct 30	207.0	270.3	-1.9	3.6	227.13	233.38	0.19	-0.01	236.16	241.38	-2.70	-2.83
Oct 25	291.94	-0.26	194.99	1.14	256.98	-0.98	223.11	4 Nov 9	339.5	52.3	4.9	-0.4	239.62	245.86	-0.22	-0.43	245.06	246.11	-2.67	-2.03
Nov 4	292.66	-0.27	196.98	1.16	264.49	-1.03	233.18	14 Nov 19	119.0	179.1	-5.0	-4.0	252.10	258.34	-0.64	-0.84	243.27	237.00	-0.78	0.89
Nov 14	293.50	-0.28	198.86	1.17	272.09	-1.08	243.30	24 Nov 29	240.6	307.9	1.1	5.2	264.57	270.81	-1.02	-1.20	231.42	230.08	2.20	2.67
Nov 24	294.44	-0.29	200.62	1.20	279.78	-1.11	253.45	4 Dec 9	18.6	88.0	2.6	-3.4	277.04	283.27	-1.35	-1.48	232.72	237.76	2.50	2.02
Dec 4	295.47	-0.30	202.22	1.22	287.54	-1.13	263.63	14 Dec 19	151.2	210.9	-5.0	-1.5	289.48	295.70	-1.59	-1.67	244.03	250.94	1.40	0.74
Dec 14	296.57	-0.31	203.62	1.25	295.36	-1.13	273.82	24 Dec 29	275.8	347.0	3.8	4.4	301.91	308.10	-1.73	-1.75	258.20	265.68	0.11	-0.49
1696								1697												1697
Dec 24	297.72	-0.32	204.81	1.29	303.22	-1.13	284.02	3 Jan 8	56.7	122.3	-0.9	-4.9	314.29	320.46	-1.74	-1.69	273.33	281.16	-1.01	-1.46
Jan 3	298.90	-0.33	205.76	1.32	311.11	-1.11	294.20	13 Jan 18	182.9	243.6	-3.4	1.6	326.61	332.74	-1.61	-1.50	289.19	297.45	-1.80	-2.01
Jan 13	300.08	-0.34	206.43	1.36	319.02	-1.08	304.37	23 Jan 28	313.0	26.1	5.0	1.5	338.86	344.95	-1.36	-1.18	305.96	314.72	-2.08	-1.95
Jan 23	301.26	-0.36	206.80	1.40	326.92	-1.04	314.52	2 Feb 7	93.2	155.4	-3.8	-4.6	351.01	357.03	-0.97	-0.73	323.67	332.59	-1.60	-0.98
Feb 2	302.41	-0.37	206.87	1.44	334.82	-0.99	324.63	12 Feb 17	214.7	278.0	-0.6	4.2	3.03	8.98	-0.46	-0.17	340.92	347.70	-0.07	1.09
Feb 12	303.52	-0.38	206.62	1.48	342.69	-0.93	334.70	22 Feb 27	351.4	63.8	3.9	-2.0	14.89	20.74	0.14	0.47	351.63	351.79	2.33	3.32
Feb 22	304.55	-0.40	206.06	1.51	350.52	-0.86	344.71	4 Mar 9	127.9	187.9	-5.1	-2.7	26.54	32.28	0.81	1.16	348.50	343.72	3.68	3.22
Mar 4	305.51	-0.42	205.24	1.54	358.30	-0.78	354.67	14 Mar 19	247.7	314.6	2.4	5.1	37.94	43.53	1.51	1.86	339.93	338.47	2.16	0.91
Mar 14	306.36	-0.43	204.19	1.56	6.03	-0.70	4.58	24 Mar 29	29.9	99.7	0.8	-4.5	49.03	54.42	2.20	2.53	339.42	342.33	-0.25	-1.21
Mar 24	307.10	-0.45	202.98	1.57	13.69	-0.61	14.43	3 Apr 8	161.2	220.5	-4.4	0.1	59.70	64.84	2.84	3.12	346.74	352.32	-1.92	-2.40
Apr 3	307.71	-0.47	201.71	1.56	21.27	-0.51	24.23	13 Apr 18	282.6	352.9	4.7	3.8	69.82	74.62	3.37	3.58	358.83	6.15	-2.64	-2.65
Apr 13	308.18	-0.49	200.46	1.55	28.77	-0.41	33.97	23 Apr 28	67.4	133.8	-2.5	-5.3	79.22	83.55	3.75	3.85	14.24	23.06	-2.43	-2.00
Apr 23	308.49	-0.51	199.31	1.53	36.19	-0.31	43.66	3 May 8	193.5	254.1	-2.3	3.0	87.58	91.25	3.88	3.83	32.64	42.92	-1.37	-0.58
May 3	308.64	-0.54	198.33	1.50	43.52	-0.20	53.30	13 May 18	319.7	31.7	5.2	0.7	94.48	97.18	3.69	3.43	53.73	64.63	0.30	1.12
May 13	308.63	-0.56	197.60	1.46	50.76	-0.09	62.91	23 May 28	103.3	166.3	-4.7	-4.1	99.25	100.56	3.04	2.50	75.10	84.69	1.75	2.09
May 23	308.46	-0.58	197.14	1.42	57.91	0.01	72.49	2 Jun 7	225.8	289.5	0.7	4.9	100.98	100.43	1.79	0.90	93.21	100.58	2.11	1.82
Jun 2	308.14	-0.60	196.97	1.37	64.97	0.12	82.03	12 Jun 17	358.5	69.9	3.3	-2.7	98.89	96.50	-0.14	-1.30	106.75	111.58	1.23	0.37
Jun 12	307.68	-0.62	197.11	1.33	71.94	0.23	91.57	22 Jun 27	137.5	198.0	-5.0	-1.7	93.53	90.44	-2.47	-3.54	114.91	116.49	-0.72	-1.97
Jun 22	307.10	-0.64	197.54	1.28	78.82	0.34	101.11	2 Jul 7	259.1	327.0	3.4	4.7	87.69	85.68	-4.43	-5.06	116.15	114.04	-3.25	-4.31
Jul 2	306.44	-0.66	198.24	1.24	85.62	0.45	110.64	12 Jul 17	37.7	106.7	-0.2	-4.7	84.62	84.54	-5.45	-5.62	110.84	107.82	-4.88	-4.77
Jul 12	305.72	-0.68	199.19	1.20	92.33	0.55	120.19	22 Jul 27	170.3	229.7	-3.5	1.3	85.39	87.06	-5.62	-5.48	106.31	107.16	-4.02	-2.86
Jul 22	304.97	-0.69	200.38	1.16	98.95	0.66	129.75	1 Aug 6	294.1	5.9	4.9	2.3	89.41	92.33	-5.25	-4.93	110.63	116.60	-1.55	-0.29
Aug 1	304.25	-0.71	201.77	1.13	105.49	0.76	139.35	11 Aug 16	76.2	141.7	-3.5	-4.8	95.74	99.55	-4.57	-4.16	124.63	134.02	0.74	1.42
Aug 11	303.58	-0.72	203.35	1.09	111.95	0.87	148.99	21 Aug 26	202.2	262.4	-0.8	4.0	103.70	108.12	-3.73	-3.28	143.95	153.80	1.74	1.74
Aug 21	303.00	-0.72	205.08	1.07	118.31	0.97	158.66	31 Aug	331.0		4.4		112.77		-2.81		163.27		1.49	
								Sep 5		44.9		-1.3		117.62		-2.35		172.26		1.08
Aug 31	302.53	-0.73	206.94	1.04	124.59	1.07	168.38	10 Sep 15	113.0	175.1	-5.1	-3.0	122.65	127.83	-1.89	-1.44	180.79	188.87	0.56	-0.03
Sep 10	302.21	-0.73	208.91	1.02	130.78	1.18	178.17	20 Sep 25	234.3	296.7	2.1	5.2	133.14	138.56	-1.00	-0.58	196.54	203.80	-0.64	-1.26
Sep 20	302.04	-0.73	210.96	1.00	136.86	1.28	188.00	30 S-O 5	9.1	82.9	1.9	-4.2	144.08	149.70	-0.19	0.18	210.63	216.94	-1.86	-2.39
Sep 30	302.04	-0.74	213.09	0.99	142.84	1.39	197.89	10 Oct 15	147.9	207.6	-4.7	-0.2	155.40	161.17	0.52	0.82	222.54	227.07	-2.82	-3.08
Oct 10	302.20	-0.74	215.25	0.98	148.70	1.49	207.84	20 Oct 25	267.3	333.0	4.5	4.4	167.00	172.89	1.09	1.32	229.87	229.91	-3.05	-2.56
Oct 20	302.53	-0.74	217.44	0.97	154.43	1.61	217.84	30 Oct	47.6		-1.6		178.83		1.52		226.23		-1.41	
								Nov 4		119.1		-5.3		184.82		1.67		219.90		0.26
Oct 30	303.02	-0.74	219.64	0.96	159.99	1.72	227.89	9 Nov 14	181.0	240.2	-2.6	2.7	190.85	196.92	1.79	1.86	215.07	214.71	1.69	2.37
Nov 9	303.65	-0.74	221.81	0.96	165.39	1.84	237.99	19 Nov 24	302.2	10.9	5.2	1.7	203.02	209.14	1.90	1.90	218.23	223.97	2.40	2.06
Nov 19	304.42	-0.74	223.94	0.96	170.56	1.96	248.13	29 N-D 4	85.3	153.2	-4.3	-4.3	215.28	221.45	1.86	1.79	230.79	238.11	1.53	0.94
Nov 29	305.30	-0.74	226.00	0.96	175.48	2.09	258.29	9 Dec 14	213.0	273.7	0.4	4.6	227.64	233.84	1.69	1.56	245.64	253.30	0.33	-0.26
Dec 9	306.28	-0.75	227.97	0.97	180.10	2.23	268.48	19 Dec 24	339.1	49.6	3.8	-1.9	240.05	246.27	1.41	1.23	261.04	268.87	-0.80	-1.28
Dec 19	307.35	-0.75	229.82	0.98	184.33	2.38	278.67	29 Dec	121.6		-5.0		252.49		1.04		276.83		-1.66	
1697								1697												1697

JULIAN 12 UT	SATURN LONG.	SATURN LAT.	JUPITER LONG.	JUPITER LAT.	MARS LONG.	MARS LAT.	SUN LONG.
1697							
Dec 29	308.47	-0.76	231.53	0.99	188.09	2.54	288.87
Jan 8	309.64	-0.77	233.06	1.01	191.28	2.70	299.05
Jan 18	310.83	-0.78	234.39	1.02	193.74	2.86	309.21
Jan 28	312.03	-0.79	235.50	1.04	195.32	3.03	319.34
Feb 7	313.21	-0.80	236.34	1.06	195.81	3.17	329.43
Feb 17	314.35	-0.82	236.90	1.08	195.09	3.28	339.47
Feb 27	315.44	-0.84	237.15	1.11	193.09	3.31	349.46
Mar 9	316.46	-0.86	237.10	1.13	189.97	3.23	359.40
Mar 19	317.40	-0.88	236.74	1.14	186.21	3.02	9.28
Mar 29	318.23	-0.90	236.08	1.16	182.48	2.68	19.10
Apr 8	318.94	-0.93	235.18	1.16	179.47	2.26	28.87
Apr 18	319.51	-0.95	234.07	1.16	177.63	1.81	38.58
Apr 28	319.94	-0.98	232.84	1.16	177.09	1.38	48.25
May 8	320.21	-1.01	231.57	1.14	177.82	0.98	57.88
May 18	320.32	-1.04	230.35	1.12	179.65	0.62	67.47
May 28	320.27	-1.07	229.26	1.09	182.39	0.31	77.03
Jun 7	320.06	-1.10	228.37	1.05	185.90	0.04	86.58
Jun 17	319.70	-1.13	227.73	1.01	190.04	-0.20	96.11
Jun 27	319.20	-1.15	227.38	0.97	194.70	-0.40	105.64
Jul 7	318.60	-1.18	227.32	0.93	199.80	-0.59	115.18
Jul 17	317.91	-1.20	227.57	0.89	205.26	-0.75	124.73
Jul 27	317.17	-1.22	228.11	0.85	211.05	-0.89	134.31
Aug 6	316.43	-1.23	228.93	0.81	217.12	-1.01	143.93
Aug 16	315.71	-1.24	230.00	0.77	223.45	-1.11	153.58
Aug 26	315.05	-1.24	231.29	0.74	230.00	-1.20	163.28
Sep 5	314.50	-1.24	232.79	0.71	236.75	-1.27	173.03
Sep 15	314.07	-1.24	234.46	0.68	243.68	-1.33	182.83
Sep 25	313.79	-1.23	236.29	0.65	250.78	-1.37	192.70
Oct 5	313.67	-1.23	238.24	0.63	258.03	-1.40	202.62
Oct 15	313.71	-1.22	240.30	0.61	265.41	-1.41	212.59
Oct 25	313.93	-1.21	242.44	0.59	272.91	-1.41	222.62
Nov 4	314.32	-1.20	244.64	0.57	280.51	-1.39	232.69
Nov 14	314.86	-1.19	246.88	0.55	288.19	-1.36	242.81
Nov 24	315.54	-1.19	249.13	0.54	295.94	-1.32	252.96
Dec 4	316.36	-1.18	251.38	0.53	303.74	-1.26	263.14
Dec 14	317.28	-1.18	253.59	0.52	311.57	-1.19	273.32
1698							
Dec 24	318.30	-1.18	255.75	0.51	319.42	-1.11	283.52
Jan 3	319.40	-1.18	257.83	0.50	327.27	-1.03	293.71
Jan 13	320.55	-1.18	259.81	0.50	335.10	-0.93	303.88
Jan 23	321.74	-1.19	261.66	0.49	342.90	-0.83	314.03
Feb 2	322.94	-1.20	263.35	0.49	350.66	-0.73	324.14
Feb 12	324.15	-1.21	264.86	0.48	358.37	-0.62	334.21
Feb 22	325.33	-1.22	266.16	0.48	6.02	-0.51	344.23
Mar 4	326.47	-1.24	267.22	0.48	13.60	-0.40	354.19
Mar 14	327.55	-1.26	268.02	0.47	21.10	-0.28	4.10
Mar 24	328.56	-1.29	268.52	0.47	28.53	-0.17	13.96
Apr 3	329.47	-1.31	268.72	0.47	35.87	-0.06	23.75
Apr 13	330.28	-1.34	268.61	0.46	43.13	0.05	33.50
Apr 23	330.96	-1.37	268.19	0.45	50.31	0.15	43.19
May 3	331.50	-1.41	267.49	0.44	57.40	0.25	52.84
May 13	331.90	-1.44	266.54	0.43	64.41	0.35	62.44
May 23	332.14	-1.48	265.40	0.41	71.34	0.45	72.03
Jun 2	332.21	-1.51	264.15	0.39	78.20	0.54	81.58
Jun 12	332.12	-1.55	262.88	0.37	84.98	0.63	91.11
Jun 22	331.88	-1.59	261.67	0.34	91.70	0.71	100.65
Jul 2	331.48	-1.62	260.61	0.31	98.34	0.79	110.18
Jul 12	330.95	-1.65	259.77	0.29	104.93	0.86	119.73
Jul 22	330.32	-1.67	259.19	0.26	111.46	0.94	129.30
Aug 1	329.61	-1.69	258.91	0.23	117.93	1.00	138.89
Aug 11	328.86	-1.71	258.95	0.20	124.35	1.07	148.52
Aug 21	328.11	-1.72	259.29	0.18	130.71	1.13	158.20
Aug 31	327.40	-1.72	259.93	0.15	137.03	1.19	167.92
Sep 10	326.76	-1.72	260.84	0.13	143.30	1.25	177.69
Sep 20	326.23	-1.71	262.02	0.11	149.51	1.30	187.52
Sep 30	325.83	-1.70	263.42	0.09	155.67	1.35	197.41
Oct 10	325.59	-1.69	265.03	0.07	161.78	1.40	207.35
Oct 20	325.52	-1.67	266.81	0.05	167.83	1.45	217.35
Oct 30	325.62	-1.65	268.75	0.03	173.80	1.49	227.40
Nov 9	325.89	-1.63	270.80	0.01	179.70	1.53	237.50
Nov 19	326.33	-1.62	272.96	0.00	185.51	1.57	247.63
Nov 29	326.92	-1.60	275.20	-0.02	191.22	1.61	257.79
Dec 9	327.66	-1.59	277.48	-0.04	196.80	1.64	267.98
Dec 19	328.52	-1.57	279.80	-0.05	202.23	1.67	278.17
1699							

GREGORIAN 12 UT	MOON LONG.		MOON LAT.		VENUS LONG.		VENUS LAT.		MERCURY LONG.		MERCURY LAT.	
1698												1698
Jan 3		185.6		-1.8		258.73		0.84		284.93		-1.94
8 Jan 13	244.9	308.7	3.2	4.9	264.97	271.21	0.62	0.40	293.21	301.66	-2.09	-2.06
18 Jan 23	17.5	88.2	0.6	-4.5	277.45	283.70	0.18	-0.04	310.21	318.64	-1.82	-1.31
28 J-F 2	156.1	217.1	-3.7	1.2	289.94	296.19	-0.26	-0.46	326.41	332.48	-0.48	0.68
7 Feb 12	277.7	345.5	4.9	3.0	302.43	308.68	-0.65	-0.83	335.42	334.16	2.03	3.22
17 Feb 22	56.7	125.4	-2.9	-5.0	314.92	321.15	-0.99	-1.13	329.51	324.32	3.74	3.36
27 Feb	189.2		-1.1		327.39		-1.24		321.20		2.38	
Mar 4		248.6		3.8		333.61		-1.33		320.89		1.23
9 Mar 14	311.8	23.6	4.9	-0.3	339.84	346.06	-1.40	-1.43	322.96	326.80	0.14	-0.78
19 Mar 24	95.3	160.8	-5.1	-3.4	352.28	358.49	-1.45	-1.43	331.90	337.98	-1.50	-2.03
29 M-A 3	221.4	281.1	1.9	5.2	4.69	10.89	-1.39	-1.32	344.84	352.39	-2.35	-2.47
8 Apr 13	347.8	62.3	2.9	-3.5	17.08	23.27	-1.23	-1.12	0.59	9.45	-2.39	-2.09
18 Apr 23	132.3	194.4	-4.9	-0.6	29.45	35.63	-0.98	-0.83	18.97	29.12	-1.58	-0.87
28 Apr	253.7		4.2		41.79		-0.66		39.78		-0.03	
May 3		315.3		4.8		47.96		-0.48		50.57		0.84
8 May 13	25.4	100.5	-0.4	-5.1	54.11	60.27	-0.29	-0.09	60.94	70.37	1.60	2.12
18 May 23	167.2	227.0	-2.8	2.3	66.41	72.55	0.11	0.31	78.56	85.39	2.31	2.16
28 M-J 2	287.0	351.4	5.1	2.4	78.69	84.82	0.51	0.70	90.75	94.52	1.66	0.82
7 Jun 12	64.1	137.2	-3.6	-4.5	90.95	97.07	0.88	1.05	96.52	96.64	-0.33	-1.67
17 Jun 22	200.2	259.6	0.2	4.4	103.19	109.30	1.20	1.32	95.02	92.28	-3.03	-4.12
27 Jun	321.8		4.1		115.40		1.43		89.49		-4.68	
Jul 2		29.4		-1.1		121.50		1.51		87.84		-4.61
7 Jul 12	102.8	171.7	-5.0	-2.0	127.60	133.69	1.57	1.59	88.10	90.56	-4.00	-3.03
17 Jul 22	232.2	292.9	3.0	5.0	139.77	145.84	1.59	1.55	95.21	101.91	-1.88	-0.69
27 J-A 1	358.5	68.5	1.4	-4.1	151.91	157.96	1.48	1.39	110.39	120.11	0.37	1.17
6 Aug 11	140.0	204.5	-4.1	1.0	164.01	170.04	1.26	1.10	130.36	140.54	1.63	1.77
16 Aug 21	264.1	327.6	4.9	3.6	176.06	182.07	0.91	0.70	150.29	159.50	1.62	1.28
26 Aug 31	36.9	107.6	-2.2	-5.2	188.06	194.03	0.46	0.21	168.15	176.28	0.79	0.21
5 Sep 10	175.1	236.3	-1.5	3.7	199.99	205.93	-0.07	-0.36	183.92	191.05	-0.44	-1.10
15 Sep 20	296.7	3.9	5.1	0.7	211.85	217.74	-0.66	-0.96	197.63	203.55	-1.77	-2.39
25 Sep 30	76.1	145.1	-4.8	-3.8	223.61	229.45	-1.27	-1.57	208.61	212.40	-2.94	-3.33
5 Oct 10	208.5	268.1	1.5	5.1	235.26	241.03	-1.86	-2.14	214.25	213.25	-3.44	-3.08
15 Oct 20	330.4	41.7	3.5	-2.6	246.76	252.44	-2.40	-2.63	208.88	202.78	-2.02	-0.38
25 Oct 30	114.9	180.5	-5.1	-1.0	258.07	263.63	-2.83	-2.99	198.89	199.50	1.15	2.02
4 Nov 9	241.0	300.6	4.0	4.9	269.12	274.51	-3.11	-3.18	203.84	210.26	2.26	2.07
14 Nov 19	5.7	80.4	0.5	-4.8	279.80	284.96	-3.19	-3.14	217.61	225.32	1.65	1.11
24 Nov 29	151.9	214.1	-3.2	2.1	289.96	294.77	-3.01	-2.81	233.14	240.98	0.54	-0.04
4 Dec 9	273.5	334.5	5.0	2.8	299.36	303.67	-2.52	-2.13	248.81	256.66	-0.59	-1.10
14 Dec 19	42.9	118.7	-2.8	-4.7	307.64	311.20	-1.64	-1.04	264.55	272.51	-1.53	-1.86
24 Dec 29	186.7	246.7	-0.1	4.3	314.25	316.68	-0.31	0.55	280.56	288.70	-2.08	-2.15
1699												1699
3 Jan 8	306.7	10.3	4.3	-0.4	318.36	319.16	1.55	2.67	296.83	304.75	-2.01	-1.62
13 Jan 18	81.5	155.3	-4.9	-2.5	318.96	317.71	3.88	5.13	311.90	317.22	-0.88	0.23
23 Jan 28	219.6	279.1	2.9	5.1	315.51	312.65	6.30	7.26	319.14	316.57	1.65	3.01
2 Feb 7	341.2	48.1	1.9	-3.6	309.58	306.80	7.89	8.15	310.92	305.83	3.68	3.40
12 Feb 17	120.3	190.0	-4.6	0.7	304.75	303.66	8.05	7.67	303.65	304.47	2.52	1.48
22 Feb 27	251.4	311.8	4.9	4.0	303.60	304.50	7.09	6.41	307.51	312.09	0.47	-0.40
4 Mar 9	17.3	87.3	-1.4	-5.3	306.25	308.70	5.66	4.90	317.74	324.19	-1.13	-1.69
14 Mar 19	157.8	223.0	-2.2	3.5	311.74	315.26	4.14	3.41	331.29	338.98	-2.08	-2.29
24 Mar 29	283.0	345.5	5.2	1.5	319.16	323.39	2.71	2.05	347.22	356.04	-2.31	-2.13
3 Apr 8	55.2	126.5	-4.2	-4.4	327.89	332.60	1.43	0.86	5.44	15.43	-1.74	-1.14
13 Apr 18	193.5	255.3	1.0	5.0	337.48	342.52	0.33	-0.14	25.87	36.47	-0.36	0.53
23 Apr 28	315.2	20.9	3.8	-1.7	347.69	352.96	-0.57	-0.94	46.66	55.83	1.40	2.08
3 May 8	94.4	164.2	-5.1	-1.5	358.33	3.78	-1.26	-1.53	63.56	69.64	2.47	2.49
13 May 18	227.5	287.4	3.7	4.9	9.29	14.86	-1.75	-1.93	73.93	76.30	2.11	1.31
23 May 28	348.5	58.4	1.0	-4.3	20.48	26.15	-2.05	-2.13	76.69	75.30	0.15	-1.26
2 Jun 7	133.2	199.8	-3.6	1.8	31.86	37.60	-2.17	-2.16	72.75	70.10	-2.65	-3.73
12 Jun 17	260.6	320.1	4.9	3.1	43.38	49.18	-2.12	-2.04	68.44	68.47	-4.29	-4.31
22 Jun 27	23.7	97.3	-2.2	-4.9	55.02	60.88	-1.93	-1.79	70.44	74.32	-3.87	-3.10
2 Jul 7	170.4	233.7	-0.5	4.3	66.77	72.68	-1.62	-1.43	79.99	87.34	-2.13	-1.05
12 Jul 17	293.3	354.0	4.4	0.1	78.61	84.57	-1.23	-1.01	96.18	106.15	-0.01	0.88
22 Jul 27	61.1	136.0	-4.7	-3.2	90.55	96.55	-0.78	-0.55	116.67	127.13	1.48	1.75
1 Aug 6	205.4	266.4	2.7	5.1	102.57	108.61	-0.31	-0.07	137.14	146.54	1.73	1.47
11 Aug 16	326.3	29.8	2.4	-3.1	114.68	120.76	0.15	0.37	155.31	163.48	1.03	0.46
21 Aug 26	100.1	173.2	-5.0	0.0	126.86	132.99	0.58	0.77	171.05	178.01	-0.20	-0.92
31 Aug	238.7		4.8		139.13		0.94		184.31		-1.66	
Sep 5		298.5		4.3		145.29		1.09		189.80		-2.38
10 Sep 15	0.5	67.8	-0.7	-5.1	151.46	157.65	1.22	1.32	194.24	197.19	-3.05	-3.57
20 Sep 25	139.1	208.5	-3.0	3.1	163.86	170.07	1.39	1.44	198.02	196.01	-3.82	-3.57
30 S-O 5	270.8	330.8	5.2	2.0	176.30	182.54	1.46	1.45	191.17	185.62	-2.60	-1.00
10 Oct 15	36.4	107.0	-3.6	-4.8	188.79	195.05	1.41	1.35	182.83	184.42	0.59	1.64
20 Oct 25	176.8	242.2	0.4	4.8	201.31	207.58	1.26	1.15	189.53	196.57	2.07	2.04
30 Oct	302.4		3.9		213.86		1.01		204.44		1.73	
Nov 4		4.0		-1.1		220.14		0.86		212.55		1.27
9 Nov 14	74.2	146.2	-5.0	-2.2	226.42	232.70	0.69	0.51	220.66	228.69	0.73	0.17
19 Nov 24	212.8	274.8	3.4	4.9	238.98	245.27	0.32	0.13	236.63	244.50	-0.39	-0.91
29 N-D 4	334.2	38.9	1.3	-3.8	251.56	257.85	-0.07	-0.27	252.33	260.15	-1.38	-1.78
9 Dec 14	113.1	183.8	-4.1	1.4	264.13	270.42	-0.47	-0.66	267.98	275.80	-2.06	-2.22
19 Dec 24	247.2	307.1	4.9	3.2	276.70	282.99	-0.83	-1.00	283.52	290.89	-2.19	-1.90
29 Dec	7.2		-1.8		289.27		-1.14		297.39		-1.29	
1699												1699

JULIAN 12 UT	SATURN LONG.	SATURN LAT.	JUPITER LONG.	JUPITER LAT.	MARS LONG.	MARS LAT.	SUN LONG.
1699							
Dec 29	329.49	-1.57	282.13	-0.07	207.49	1.69	288.37
Jan 8	330.54	-1.56	284.44	-0.08	212.52	1.71	298.55
Jan 18	331.67	-1.56	286.71	-0.10	217.29	1.72	308.71
Jan 28	332.84	-1.56	288.92	-0.12	221.73	1.72	318.84
Feb 7	334.05	-1.56	291.05	-0.14	225.77	1.70	328.93
Feb 17	335.27	-1.57	293.06	-0.16	229.29	1.66	338.98
Feb 27	336.48	-1.59	294.95	-0.18	232.17	1.59	348.97
Mar 8	337.66	-1.60	296.67	-0.20	234.25	1.48	358.92
Mar 18	338.80	-1.62	298.21	-0.22	235.36	1.32	8.80
Mar 28	339.88	-1.64	299.53	-0.25	235.32	1.09	18.63
Apr 7	340.87	-1.67	300.62	-0.28	234.02	0.79	28.40
Apr 17	341.77	-1.70	301.43	-0.30	231.53	0.39	38.12
Apr 27	342.56	-1.73	301.96	-0.34	228.20	-0.07	47.79
May 7	343.22	-1.77	302.18	-0.37	224.64	-0.55	57.42
May 17	343.74	-1.81	302.08	-0.40	221.60	-1.00	67.01
May 27	344.11	-1.85	301.68	-0.43	219.65	-1.40	76.57
Jun 6	344.32	-1.89	300.98	-0.47	219.03	-1.71	86.12
Jun 16	344.36	-1.93	300.02	-0.50	219.76	-1.95	95.65
Jun 26	344.24	-1.97	298.88	-0.53	221.71	-2.12	105.18
Jul 6	343.95	-2.01	297.61	-0.55	224.68	-2.24	114.72
Jul 16	343.52	-2.05	296.32	-0.57	228.52	-2.32	124.28
Jul 26	342.97	-2.08	295.10	-0.59	233.06	-2.37	133.86
Aug 5	342.31	-2.11	294.03	-0.61	238.19	-2.38	143.47
Aug 15	341.58	-2.13	293.19	-0.62	243.79	-2.37	153.12
Aug 25	340.82	-2.14	292.62	-0.62	249.80	-2.34	162.81
Sep 4	340.06	-2.14	292.36	-0.62	256.14	-2.29	172.56
Sep 14	339.36	-2.14	292.43	-0.62	262.76	-2.22	182.36
Sep 24	338.74	-2.13	292.83	-0.62	269.62	-2.13	192.22
Oct 4	338.23	-2.11	293.53	-0.62	276.68	-2.03	202.13
Oct 14	337.87	-2.09	294.52	-0.62	283.90	-1.91	212.10
Oct 24	337.67	-2.07	295.77	-0.62	291.25	-1.78	222.12
Nov 3	337.65	-2.04	297.26	-0.62	298.70	-1.65	232.20
Nov 13	337.80	-2.02	298.95	-0.62	306.23	-1.50	242.31
Nov 23	338.12	-1.99	300.82	-0.62	313.81	-1.35	252.46
Dec 3	338.61	-1.96	302.85	-0.62	321.42	-1.20	262.64
Dec 13	339.25	-1.94	304.99	-0.63	329.05	-1.04	272.83
1700							
Dec 23	340.03	-1.92	307.23	-0.63	336.66	-0.89	283.02
Jan 2	340.94	-1.90	309.55	-0.64	344.26	-0.73	293.21
Jan 12	341.94	-1.89	311.91	-0.65	351.81	-0.58	303.39
Jan 22	343.03	-1.88	314.29	-0.66	359.33	-0.44	313.54
Feb 1	344.18	-1.87	316.69	-0.68	6.78	-0.29	323.65
Feb 11	345.38	-1.87	319.06	-0.70	14.18	-0.16	333.72
Feb 21	346.61	-1.88	321.39	-0.71	21.50	-0.03	343.74
Mar 3	347.84	-1.88	323.66	-0.74	28.76	0.09	353.72
Mar 13	349.06	-1.90	325.84	-0.76	35.94	0.20	3.63
Mar 23	350.24	-1.91	327.92	-0.79	43.05	0.31	13.48
Apr 2	351.38	-1.93	329.87	-0.82	50.08	0.41	23.29
Apr 12	352.45	-1.96	331.66	-0.85	57.04	0.51	33.03
Apr 22	353.44	-1.99	333.26	-0.89	63.93	0.59	42.73
May 2	354.33	-2.02	334.66	-0.93	70.76	0.67	52.38
May 12	355.10	-2.06	335.83	-0.97	77.52	0.74	61.99
May 22	355.75	-2.10	336.73	-1.01	84.22	0.81	71.56
Jun 1	356.25	-2.14	337.34	-1.06	90.86	0.87	81.12
Jun 11	356.59	-2.18	337.65	-1.11	97.45	0.93	90.65
Jun 21	356.78	-2.23	337.64	-1.16	104.00	0.98	100.19
Jul 1	356.79	-2.27	337.30	-1.21	110.50	1.02	109.72
Jul 11	356.64	-2.32	336.66	-1.25	116.97	1.06	119.26
Jul 21	356.33	-2.36	335.74	-1.29	123.41	1.10	128.83
Jul 31	355.87	-2.40	334.61	-1.33	129.82	1.13	138.43
Aug 10	355.28	-2.43	333.34	-1.35	136.20	1.15	148.05
Aug 20	354.60	-2.45	332.02	-1.37	142.56	1.17	157.72
Aug 30	353.85	-2.47	330.76	-1.37	148.91	1.19	167.45
Sep 9	353.08	-2.48	329.64	-1.36	155.23	1.20	177.21
Sep 19	352.32	-2.47	328.74	-1.35	161.55	1.21	187.04
Sep 29	351.62	-2.46	328.13	-1.33	167.86	1.22	196.93
Oct 9	351.02	-2.45	327.84	-1.30	174.15	1.22	206.87
Oct 19	350.54	-2.42	327.88	-1.27	180.43	1.21	216.86
Oct 29	350.21	-2.39	328.27	-1.24	186.70	1.20	226.91
Nov 8	350.06	-2.36	328.97	-1.21	192.95	1.18	237.00
Nov 18	350.07	-2.33	329.98	-1.18	199.18	1.16	247.14
Nov 28	350.27	-2.29	331.26	-1.15	205.40	1.13	257.30
Dec 8	350.64	-2.26	332.78	-1.13	211.58	1.10	267.48
Dec 18	351.18	-2.23	334.51	-1.11	217.73	1.05	277.68
1701							

GREGORIAN 12 UT	MOON LONG.		MOON LAT.		VENUS LONG.		VENUS LAT.		MERCURY LONG.		MERCURY LAT.	
1700												1700
Jan 3		75.8		-5.1		295.54		-1.27		301.91		-0.25
8 Jan 13	151.5	219.5	-1.2	4.2	301.81	308.08	-1.37	-1.45	302.81	299.09	1.19	2.68
18 Jan 23	280.3	339.6	4.6	0.4	314.35	320.60	-1.51	-1.53	292.83	288.20	3.50	3.35
28 J-F 2	42.2	114.2	-4.4	-4.0	326.85	333.09	-1.53	-1.50	287.07	288.93	2.60	1.67
7 Feb 12	188.5	253.3	2.2	5.3	339.32	345.53	-1.44	-1.35	292.82	298.01	0.75	-0.08
17 Feb 22	312.8	13.4	2.7	-2.6	351.74	357.93	-1.24	-1.10	304.09	310.83	-0.80	-1.38
27 Feb	79.4		-5.3		4.10		-0.93		318.11		-1.82	
Mar 4		152.7		-1.0		10.26		-0.74		325.87		-2.10
9 Mar 14	223.7	285.7	4.6	4.5	16.39	22.51	-0.53	-0.30	334.10	342.83	-2.21	-2.14
19 Mar 24	345.3	49.0	-0.1	-4.8	28.60	34.67	-0.06	0.19	352.08	1.84	-1.87	-1.38
29 M-A 3	118.2	190.3	-3.7	2.3	40.72	46.73	0.44	0.70	12.02	22.35	-0.68	0.20
8 Apr 13	257.3	317.4	5.2	2.3	52.72	58.67	0.96	1.21	32.28	41.10	1.13	1.98
18 Apr 23	18.9	86.8	-3.0	-5.0	64.59	70.48	1.45	1.68	48.23	53.36	2.57	2.79
28 Apr	157.4		-0.5		76.32		1.89		56.28		2.56	
May 3		226.5		4.6		82.12		2.07		56.95		1.84
8 May 13	289.7	349.3	4.0	-0.7	87.87	93.57	2.22	2.34	55.63	53.00	0.68	-0.75
18 May 23	54.3	125.8	-4.8	-2.8	99.21	104.78	2.42	2.47	50.22	48.42	-2.13	-3.19
28 M-J 2	195.6	261.1	2.9	4.9	110.28	115.70	2.46	2.41	48.28	49.99	-3.78	-3.91
7 Jun 12	321.5	22.2	1.5	-3.5	121.02	126.23	2.30	2.13	53.46	58.51	-3.66	-3.10
17 Jun 22	91.7	164.8	-4.6	0.7	131.32	136.27	1.91	1.62	65.01	72.88	-2.31	-1.36
27 Jun	232.2		4.9		141.04		1.26		82.01		-0.37	
Jul 2		294.3		3.4		145.62		0.83		92.17		0.55
7 Jul 12	353.5	57.0	-1.6	-5.1	149.95	154.01	0.32	-0.26	102.88	113.57	1.27	1.70
17 Jul 22	130.3	202.7	-2.2	3.8	157.74	161.05	-0.93	-1.67	123.79	133.36	1.81	1.65
27 J-A 1	267.0	326.7	4.8	0.7	163.87	166.09	-2.51	-3.42	142.22	150.37	1.28	0.73
6 Aug 11	26.7	93.8	-4.1	-4.6	167.59	168.27	-4.40	-5.42	157.84	164.59	0.06	-0.70
16 Aug 21	168.9	238.7	1.3	5.3	168.01	166.77	-6.44	-7.37	170.53	175.51	-1.52	-2.35
26 Aug 31	300.1	359.4	3.0	-2.2	164.61	161.78	-8.13	-8.59	179.22	181.23	-3.15	-3.81
5 Sep 10	61.8	132.1	-5.3	-2.0	158.70	155.87	-8.68	-8.39	180.97	178.04	-4.19	-4.03
15 Sep 20	206.5	272.8	4.1	4.6	153.69	152.42	-7.78	-6.95	173.06	168.40	-3.11	-1.57
25 Sep 30	332.4	33.2	0.2	-4.4	152.16	152.86	-6.00	-5.02	166.85	169.42	0.04	1.22
5 Oct 10	98.9	170.7	-4.3	1.5	154.42	156.73	-4.05	-3.12	175.24	182.87	1.84	1.98
15 Oct 20	242.4	305.2	5.1	2.4	159.67	163.12	-2.26	-1.47	191.24	199.75	1.80	1.41
25 Oct 30	4.7	68.8	-2.7	-5.0	166.99	171.23	-0.76	-0.12	208.15	216.39	0.92	0.38
4 Nov 9	137.6	208.8	-1.3	4.2	175.76	180.54	0.45	0.94	224.44	232.35	-0.18	-0.73
14 Nov 19	276.4	336.7	4.1	-0.5	185.53	190.70	1.36	1.71	240.14	247.83	-1.23	-1.68
24 Nov 29	37.9	106.2	-4.6	-3.4	196.02	201.46	1.99	2.22	255.45	262.95	-2.03	-2.27
4 Dec 9	176.7	245.5	2.3	5.0	207.01	212.66	2.38	2.48	270.24	277.06	-2.34	-2.18
14 Dec 19	308.9	8.2	1.6	-3.3	218.39	224.19	2.52	2.52	282.86	286.53	-1.69	-0.76
24 Dec 29	72.7	144.7	-4.9	-0.2	230.04	235.94	2.47	2.38	286.43	281.74	0.66	2.25
1701												1701
3 Jan 8	215.1	280.3	4.8	3.7	241.89	247.87	2.25	2.08	275.24	271.30	3.22	3.23
13 Jan 18	340.7	40.8	-1.3	-5.1	253.88	259.92	1.89	1.67	271.22	274.04	2.63	1.82
23 Jan 28	109.1	183.4	-3.1	3.2	265.98	272.05	1.43	1.18	278.67	284.39	0.98	0.20
2 Feb 7	251.9	313.6	5.1	1.1	278.14	284.25	0.92	0.66	290.84	297.81	-0.50	-1.09
12 Feb 17	12.8	75.1	-3.9	-5.0	290.37	296.49	0.39	0.13	305.21	313.01	-1.57	-1.91
22 Feb 27	146.8	221.3	0.1	5.1	302.62	308.75	-0.12	-0.36	321.20	329.81	-2.10	-2.13
4 Mar 9	286.5	346.1	3.3	-1.8	314.89	321.04	-0.59	-0.79	338.86	348.37	-1.96	-1.59
14 Mar 19	46.0	111.3	-5.2	-2.9	327.18	333.33	-0.98	-1.14	358.25	8.25	-0.98	-0.15
24 Mar 29	185.3	257.4	3.4	4.8	339.47	345.62	-1.27	-1.38	17.84	26.22	0.82	1.80
3 Apr 8	319.5	18.7	0.5	-4.1	351.76	357.91	-1.46	-1.51	32.64	36.60	2.59	3.03
13 Apr 18	81.0	149.3	-4.5	0.5	4.05	10.20	-1.53	-1.53	37.90	36.75	2.98	2.35
23 Apr 28	223.4	291.5	5.0	2.6	16.34	22.47	-1.49	-1.43	33.91	30.71	1.21	-0.20
3 May 8	351.5	52.3	-2.5	-5.0	28.61	34.75	-1.34	-1.24	28.47	27.98	-1.54	-2.57
13 May 18	117.9	188.4	-1.9	3.7	40.88	47.02	-1.10	-0.95	29.40	32.55	-3.22	-3.47
23 May 28	260.1	324.0	4.3	-0.4	53.15	59.28	-0.79	-0.61	37.20	43.16	-3.39	-3.03
2 Jun 7	23.6	87.4	-4.6	-3.9	65.42	71.55	-0.42	-0.23	50.30	58.57	-2.42	-1.62
12 Jun 17	156.5	227.3	1.6	5.1	77.69	83.83	-0.03	0.17	67.92	78.19	-0.71	0.22
22 Jun 27	294.8	355.7	2.0	-3.2	89.97	96.11	0.36	0.55	89.02	99.87	1.03	1.59
2 Jul 7	56.5	124.3	-5.1	-1.1	102.26	108.41	0.73	0.89	110.26	119.94	1.85	1.81
12 Jul 17	195.8	264.6	4.5	4.1	114.56	120.72	1.04	1.17	128.83	136.92	1.52	1.02
22 Jul 27	327.9	27.5	-1.0	-5.0	126.87	133.03	1.28	1.36	144.21	150.66	0.35	-0.45
1 Aug 6	90.8	162.6	-3.8	2.3	139.20	145.36	1.42	1.45	156.14	160.47	-1.35	-2.29
11 Aug 16	234.5	299.8	5.2	1.5	151.53	157.70	1.45	1.43	163.32	164.27	-3.22	-4.03
21 Aug 26	0.2	60.2	-3.6	-5.1	163.86	170.03	1.38	1.30	162.90	159.27	-4.52	-4.42
31 Aug	126.8		-0.9		176.19		1.20		154.59		-3.54	
Sep 5		201.6		4.7		182.36		1.06		151.16		-2.08
10 Sep 15	271.5	333.3	3.7	-1.5	188.52	194.68	0.91	0.73	150.95	154.48	-0.50	0.78
20 Sep 25	32.7	94.3	-5.0	-3.4	200.83	206.98	0.54	0.33	160.96	169.14	1.56	1.87
30 S-O 5	164.5	239.9	2.5	4.9	213.13	219.27	0.10	-0.13	177.98	186.88	1.83	1.54
10 Oct 15	306.1	5.9	0.8	-4.0	225.41	231.54	-0.37	-0.61	195.59	204.04	1.10	0.58
20 Oct 25	66.0	130.3	-4.6	-0.2	237.67	243.79	-0.85	-1.08	212.23	220.18	0.03	-0.53
30 Oct	203.1		4.7		249.90		-1.30		227.93		-1.07	
Nov 4		276.2		2.9		256.00		-1.50		235.51		-1.57
9 Nov 14	339.0	38.5	-2.3	-5.0	262.10	268.18	-1.69	-1.85	242.92	250.10	-1.99	-2.31
19 Nov 24	100.8	168.2	-2.5	3.1	274.24	280.29	-1.99	-2.10	256.96	263.20	-2.48	-2.44
29 N-D 4	241.5	310.4	4.7	-0.1	286.32	292.33	-2.17	-2.20	268.26	271.03	-2.08	-1.27
9 Dec 14	10.9	71.7	-4.5	-4.3	298.31	304.26	-2.19	-2.14	269.94	264.51	0.09	1.73
19 Dec 24	137.4	207.4	0.9	5.1	310.17	316.04	-2.04	-1.90	258.08	254.98	2.85	3.05
29 Dec	278.4		2.5		321.85		-1.71		255.92		2.62	
1701												1701

JULIAN 12 UT	SATURN LONG.	LAT.	JUPITER LONG.	LAT.	MARS LONG.	LAT.	SUN LONG.	GREGORIAN 12 UT	MOON LONG.		LAT.		VENUS LONG.		LAT.		MERCURY LONG.		LAT.	
1701								1702												1702
								Jan 3		342.9		-3.1		327.61		-1.47		259.62		1.93
Dec 28	351.87	-2.20	336.42	-1.09	223.85	0.99	287.87	8 Jan 13	42.6	106.0	-5.2	-1.9	333.29	338.88	-1.19	-0.85	264.90	271.10	1.18	0.44
Jan 7	352.69	-2.17	338.48	-1.07	229.91	0.93	298.05	18 Jan 23	175.7	246.4	4.0	4.6	344.38	349.77	-0.47	-0.04	277.87	285.04	-0.23	-0.83
Jan 17	353.63	-2.15	340.66	-1.06	235.90	0.84	308.22	28 J-F 2	313.3	14.7	-0.6	-4.9	355.02	0.12	0.44	0.96	292.54	300.35	-1.33	-1.72
Jan 27	354.67	-2.13	342.94	-1.06	241.82	0.75	318.35	7 Feb 12	74.9	142.0	-4.2	1.3	5.03	9.71	1.51	2.11	308.48	316.96	-1.98	-2.09
Feb 6	355.79	-2.12	345.29	-1.05	247.65	0.63	328.44	17 Feb 22	214.9	283.8	5.2	2.1	14.13	18.23	2.73	3.39	325.80	335.02	-2.03	-1.76
Feb 16	356.97	-2.11	347.68	-1.05	253.36	0.49	338.49	27 Feb	346.8		-3.3		21.95		4.06		344.56		-1.25	
								Mar 4		46.6		-5.1		25.20		4.74		354.18		-0.50
Feb 26	358.19	-2.10	350.10	-1.05	258.92	0.32	348.49	9 Mar 14	108.4	179.6	-1.6	4.1	27.89	29.90	5.42	6.07	3.35	11.22	0.47	1.54
Mar 8	359.43	-2.11	352.52	-1.06	264.30	0.13	358.43	19 Mar 24	253.6	319.1	4.0	-1.2	31.10	31.37	6.67	7.15	16.81	19.45	2.53	3.20
Mar 18	0.68	-2.11	354.93	-1.07	269.46	-0.11	8.32	29 M-A 3	19.6	79.2	-4.9	-3.6	30.64	28.89	7.47	7.53	19.00	16.12	3.34	2.80
Mar 28	1.90	-2.12	357.30	-1.08	274.33	-0.39	18.15	8 Apr 13	143.8	218.4	1.7	5.0	26.30	23.23	7.29	6.70	12.33	9.32	1.69	0.32
Apr 7	3.10	-2.14	359.61	-1.09	278.84	-0.72	27.92	18 Apr 23	290.4	352.8	1.1	-3.8	20.15	17.56	5.80	4.70	8.15	9.06	-0.96	-1.98
Apr 17	4.24	-2.16	1.84	-1.11	282.89	-1.12	37.65	28 Apr	52.2		-4.7		15.81		3.52		11.83		-2.66	
								May 3		113.2		-0.7		15.04		2.35		16.12		-3.03
Apr 27	5.31	-2.18	3.96	-1.14	286.35	-1.59	47.32	8 May 13	181.4	256.9	4.4	3.5	15.25	16.38	1.28	0.33	21.66	28.28	-3.10	-2.91
May 7	6.29	-2.21	5.97	-1.16	289.06	-2.14	56.95	18 May 23	325.1	25.5	-2.1	-5.1	18.28	20.84	-0.48	-1.16	35.89	44.44	-2.47	-1.82
May 17	7.18	-2.25	7.83	-1.19	290.84	-2.78	66.55	28 M-J 2	85.5	149.2	-2.8	2.6	23.95	27.50	-1.72	-2.17	53.92	64.23	-1.01	-0.12
May 27	7.94	-2.29	9.51	-1.23	291.52	-3.49	76.11	7 Jun 12	220.4	293.8	5.0	0.4	31.41	35.62	-2.51	-2.77	75.10	86.04	0.74	1.43
Jun 6	8.57	-2.33	11.00	-1.26	290.96	-4.24	85.65	17 Jun 22	358.2	57.8	-4.5	-4.5	40.08	44.75	-2.94	-3.04	96.53	106.26	1.84	1.95
Jun 16	9.06	-2.37	12.26	-1.30	289.22	-4.96	95.19	27 Jun	119.9		0.2		49.60		-3.08		115.11		1.77	
								Jul 2		187.3		4.9		54.60		-3.06		123.06		1.33
Jun 26	9.39	-2.42	13.27	-1.35	286.62	-5.54	104.72	7 Jul 12	259.4	328.6	3.3	-2.7	59.72	64.96	-2.98	-2.86	130.07	136.08	0.68	-0.15
Jul 6	9.55	-2.47	13.99	-1.39	283.80	-5.89	114.26	17 Jul 22	30.2	90.5	-5.3	-2.4	70.29	75.71	-2.69	-2.49	140.95	144.45	-1.12	-2.19
Jul 16	9.54	-2.51	14.41	-1.44	281.47	-5.97	123.82	27 J-A 1	156.3	226.7	3.3	5.0	81.21	86.78	-2.27	-2.02	146.25	146.02	-3.26	-4.20
Jul 26	9.37	-2.56	14.50	-1.49	280.23	-5.81	133.39	6 Aug 11	296.9	1.9	0.1	-4.7	92.41	98.10	-1.75	-1.46	143.66	139.76	-4.77	-4.69
Aug 5	9.03	-2.60	14.26	-1.53	280.34	-5.49	143.00	16 Aug 21	61.9	124.4	-4.3	0.6	103.84	109.63	-1.17	-0.87	135.89	133.95	-3.86	-2.49
Aug 15	8.54	-2.63	13.71	-1.58	281.78	-5.07	152.65	26 Aug 31	194.5	265.8	5.0	2.7	115.47	121.35	-0.58	-0.29	135.13	139.59	-0.98	0.32
Aug 25	7.93	-2.66	12.85	-1.61	284.43	-4.60	162.35	5 Sep 10	332.5	34.3	-3.0	-5.1	127.27	133.24	-0.01	0.26	146.68	155.35	1.24	1.72
Sep 4	7.23	-2.69	11.76	-1.64	288.07	-4.13	172.09	15 Sep 20	94.0	160.1	-2.0	3.5	139.23	145.26	0.51	0.74	164.64	173.93	1.82	1.64
Sep 14	6.46	-2.70	10.49	-1.65	292.51	-3.67	181.89	25 Sep 30	233.8	303.3	4.5	-0.7	151.33	157.42	0.95	1.13	182.95	191.62	1.27	0.79
Sep 24	5.68	-2.70	9.15	-1.65	297.59	-3.23	191.74	5 Oct 10	6.5	66.4	-4.7	-3.7	163.54	169.68	1.28	1.41	199.95	207.95	0.24	-0.33
Oct 4	4.92	-2.69	7.84	-1.64	303.16	-2.81	201.65	15 Oct 20	127.3	197.6	1.2	5.0	175.85	182.03	1.50	1.56	215.68	223.15	-0.90	-1.45
Oct 14	4.23	-2.68	6.65	-1.62	309.12	-2.41	211.62	25 Oct 30	272.5	338.8	1.7	-3.7	188.24	194.46	1.59	1.59	230.35	237.22	-1.94	-2.34
Oct 24	3.64	-2.65	5.68	-1.58	315.37	-2.04	221.64	4 Nov 9	39.5	98.9	-4.8	-1.0	200.69	206.94	1.56	1.50	243.62	249.27	-2.61	-2.69
Nov 3	3.19	-2.62	4.99	-1.54	321.84	-1.70	231.71	14 Nov 19	162.5	236.2	4.0	4.1	213.20	219.46	1.42	1.30	253.54	255.37	-2.47	-1.79
Nov 13	2.89	-2.59	4.62	-1.49	328.48	-1.39	241.83	24 Nov 29	309.3	12.6	-1.7	-5.1	225.74	232.01	1.17	1.02	253.32	247.37	-0.52	1.15
Nov 23	2.77	-2.55	4.60	-1.44	335.24	-1.10	251.97	4 Dec 9	72.1	132.8	-3.1	2.1	238.30	244.58	0.85	0.66	241.28	239.11	2.42	2.82
Dec 3	2.83	-2.50	4.92	-1.39	342.08	-0.83	262.14	14 Dec 19	199.8	274.4	5.2	1.3	250.87	257.16	0.47	0.27	241.01	245.51	2.57	2.02
Dec 13	3.07	-2.46	5.58	-1.34	348.98	-0.59	272.33	24 Dec 29	344.1	45.1	-4.3	-4.7	263.45	269.74	0.07	-0.14	251.40	258.04	1.35	0.67
1702								1703												1703
Dec 23	3.49	-2.42	6.54	-1.30	355.91	-0.37	282.53	3 Jan 8	104.9	168.7	-0.4	4.6	276.03	282.32	-0.33	-0.52	265.11	272.46	0.01	-0.59
Jan 2	4.07	-2.38	7.78	-1.25	2.85	-0.17	292.71	13 Jan 18	238.6	311.3	4.1	-2.0	288.60	294.89	-0.70	-0.87	280.05	287.87	-1.11	-1.54
Jan 12	4.79	-2.35	9.27	-1.21	9.78	0.01	302.89	23 Jan 28	17.2	76.9	-5.3	-2.8	301.17	307.45	-1.02	-1.15	295.93	304.26	-1.86	-2.04
Jan 22	5.66	-2.32	10.97	-1.18	16.70	0.17	313.04	2 Feb 7	138.9	206.7	2.6	5.2	313.72	319.99	-1.26	-1.34	312.88	321.80	-2.07	-1.90
Feb 1	6.63	-2.29	12.85	-1.15	23.59	0.31	323.15	12 Feb 17	277.4	346.7	1.0	-4.4	326.26	332.52	-1.40	-1.44	330.97	340.16	-1.50	-0.84
Feb 11	7.70	-2.27	14.88	-1.12	30.45	0.44	333.23	22 Feb 27	49.2	108.9	-4.4	0.1	338.77	345.02	-1.44	-1.42	348.86	356.13	0.09	1.22
Feb 21	8.85	-2.25	17.04	-1.09	37.27	0.55	343.25	4 Mar 9	174.7	245.9	4.7	3.3	351.25	357.48	-1.38	-1.31	0.82	2.04	2.38	3.28
Mar 3	10.05	-2.24	19.29	-1.07	44.05	0.66	353.22	14 Mar 19	315.3	20.6	-2.4	-5.0	3.70	9.91	-1.21	-1.09	359.85	355.63	3.59	3.13
Mar 13	11.29	-2.24	21.61	-1.06	50.78	0.74	3.14	24 Mar 29	80.7	142.1	-2.1	3.0	16.11	22.30	-0.95	-0.78	351.58	349.36	2.07	0.77
Mar 23	12.55	-2.23	23.98	-1.04	57.47	0.82	13.00	3 Apr 8	212.5	284.8	4.8	-0.1	28.47	34.64	-0.60	-0.41	349.45	351.62	-0.45	-1.44
Apr 2	13.80	-2.24	26.38	-1.03	64.12	0.89	22.80	13 Apr 18	351.5	53.4	-4.6	-3.8	40.79	46.93	-0.20	0.01	355.46	0.59	-2.16	-2.61
Apr 12	15.04	-2.25	28.78	-1.03	70.72	0.95	32.55	23 Apr 28	112.7	177.2	0.9	4.9	53.06	59.17	0.23	0.45	6.78	13.86	-2.80	-2.75
Apr 22	16.24	-2.26	31.17	-1.02	77.29	1.00	42.25	3 May 8	251.2	322.4	2.6	-3.4	65.27	71.35	0.66	0.87	21.78	30.50	-2.47	-1.97
May 2	17.39	-2.28	33.53	-1.02	83.81	1.04	51.90	13 May 18	26.0	85.7	-4.9	-1.3	77.42	83.47	1.07	1.26	40.03	50.32	-1.28	-0.45
May 12	18.46	-2.31	35.83	-1.03	90.30	1.07	61.52	23 May 28	146.0	214.4	3.7	4.7	89.51	95.53	1.43	1.57	61.16	72.11	0.43	1.22
May 22	19.45	-2.34	38.06	-1.03	96.75	1.10	71.10	2 Jun 7	289.8	358.2	-0.9	-5.1	101.53	107.51	1.70	1.79	82.63	92.33	1.78	2.04
Jun 1	20.33	-2.37	40.20	-1.04	103.18	1.12	80.65	12 Jun 17	59.1	118.4	-3.4	1.7	113.47	119.41	1.85	1.88	101.05	108.73	2.00	1.65
Jun 11	21.09	-2.41	42.23	-1.05	109.59	1.14	90.20	22 Jun 27	181.3	253.0	5.2	2.4	125.32	131.20	1.88	1.84	115.31	120.70	1.05	0.20
Jun 21	21.71	-2.45	44.12	-1.07	115.98	1.15	99.73	2 Jul 7	327.1	32.1	-3.9	-4.9	137.06	142.88	1.76	1.64	124.73	127.16	-0.84	-2.03
Jul 1	22.19	-2.49	45.85	-1.09	122.35	1.16	109.26	12 Jul 17	91.5	152.3	-0.8	4.2	148.67	154.42	1.48	1.28	127.72	126.31	-3.24	-4.28
Jul 11	22.51	-2.54	47.38	-1.11	128.71	1.16	118.81	22 Jul 27	218.7	291.7	4.6	-1.1	160.13	165.78	1.04	0.76	123.28	119.71	-4.88	-4.81
Jul 21	22.65	-2.58	48.70	-1.13	135.07	1.15	128.37	1 Aug 6	2.6	64.5	-5.2	-3.0	171.39	176.93	0.45	0.10	117.15	116.86	-4.04	-2.81
Jul 31	22.63	-2.63	49.77	-1.15	141.43	1.15	137.96	11 Aug 16	124.2	188.2	2.2	5.1	182.41	187.81	-0.28	-0.69	119.41	124.76	-1.41	-0.12
Aug 10	22.43	-2.67	50.57	-1.18	147.79	1.13	147.59	21 Aug 26	257.5	329.5	1.9	-4.0	193.12	198.32	-1.13	-1.59	132.40	141.50	0.89	1.53
Aug 20	22.07	-2.71	51.06	-1.21	154.15	1.12	157.26	31 Aug	36.3		-4.5		203.41		-2.07		151.20		1.78	
								Sep 5		96.2		-0.1		208.37		-2.57		160.86		1.72
Aug 30	21.56	-2.74	51.23	-1.23	160.53	1.09	166.98	10 Sep 15	158.0	226.1	4.4	3.9	213.16	217.76	-3.07	-3.57	170.19	179.09	1.43	0.99
Sep 9	20.93	-2.77	51.07	-1.26	166.93	1.07	176.75	20 Sep 25	296.5	5.6	-1.8	-5.0	222.13	226.21	-4.06	-4.54	187.56	195.63	0.46	-0.12
Sep 19	20.20	-2.79	50.58	-1.28	173.34	1.04	186.57	30 S-O 5	68.6	128.1	-2.2	2.8	229.96	233.30	-4.99	-5.40	203.33	210.69	-0.72	-1.31
Sep 29	19.42	-2.80	49.77	-1.29	179.78	1.00	196.45	10 Oct 15	193.5	265.1	5.0	0.7	236.13	238.36	-5.75	-6.01	217.67	224.22	-1.87	-2.36
Oct 9	18.63	-2.79	48.71	-1.30	186.23	0.96	206.39	20 Oct 25	334.7	40.0	-4.5	-4.0	239.84	240.45	-6.15	-6.14	230.15	235.16	-2.73	-2.93
Oct 19	17.87	-2.78	47.46	-1.30	192.71	0.91	216.38	30 Oct	100.2		0.7		240.08		-5.92		238.60		-2.85	
								Nov 4		161.0		4.8		238.72		-5.44		239.43		-2.31
Oct 29	17.18	-2.76	46.11	-1.28	199.22	0.86	226.43	9 Nov 14	230.7	304.0	3.4	-2.8	236.47	233.62	-4.67	-3.64	236.47	230.25	-1.14	0.52
Nov 8	16.61	-2.73	44.77	-1.26	205.76	0.80	236.52	19 Nov 24	11.3	73.0	-5.1	-1.6	230.61	227.96	-2.41	-1.12	224.76	223.54	1.93	2.53
Nov 18	16.18	-2.69	43.54	-1.23	212.32	0.74	246.65	29 N-D 4	132.2	195.6	3.5	5.1	226.07	225.17	0.12	1.21	226.36	231.60	2.48	2.07
Nov 28	15.91	-2.65	42.51	-1.19	218.92	0.66	256.81	9 Dec 14	268.9	341.7	0.2	-5.0	225.29	226.38	2.12	2.83	238.06	245.12	1.50	0.87
Dec 8	15.82	-2.60	41.75	-1.15	225.54	0.58	266.99	19 Dec 24	45.8	105.3	-3.7	1.4	228.30	230.92	3.36	3.74	252.48	260.02	0.24	-0.35
Dec 18	15.92	-2.55	41.31	-1.10	232.19	0.50	277.18	29 Dec	165.4		5.1		234.13		3.98		267.70		-0.89	
1703								1703												1703

JULIAN 12 UT	SATURN LONG.	SATURN LAT.	JUPITER LONG.	JUPITER LAT.	MARS LONG.	MARS LAT.	SUN LONG.
1703							
Dec 28	16.20	-2.51	41.22	-1.05	238.87	0.40	287.38
Jan 7	16.65	-2.46	41.47	-1.00	245.57	0.29	297.56
Jan 17	17.27	-2.42	42.05	-0.95	252.30	0.17	307.72
Jan 27	18.03	-2.38	42.94	-0.91	259.05	0.04	317.86
Feb 6	18.93	-2.35	44.11	-0.86	265.82	-0.11	327.96
Feb 16	19.93	-2.32	45.52	-0.83	272.60	-0.27	338.00
Feb 26	21.03	-2.29	47.15	-0.79	279.40	-0.44	348.01
Mar 7	22.20	-2.27	48.95	-0.76	286.20	-0.63	357.95
Mar 17	23.42	-2.26	50.90	-0.73	292.99	-0.84	7.84
Mar 27	24.68	-2.25	52.98	-0.70	299.77	-1.06	17.68
Apr 6	25.95	-2.25	55.15	-0.68	306.53	-1.30	27.45
Apr 16	27.22	-2.25	57.40	-0.66	313.24	-1.56	37.18
Apr 26	28.46	-2.26	59.70	-0.64	319.88	-1.84	46.86
May 6	29.67	-2.27	62.02	-0.62	326.43	-2.13	56.49
May 16	30.82	-2.29	64.36	-0.61	332.84	-2.43	66.08
May 26	31.90	-2.31	66.69	-0.59	339.06	-2.75	75.65
Jun 5	32.89	-2.33	69.00	-0.58	345.04	-3.09	85.20
Jun 15	33.78	-2.37	71.25	-0.57	350.70	-3.43	94.73
Jun 25	34.54	-2.40	73.45	-0.56	355.93	-3.77	104.27
Jul 5	35.16	-2.44	75.55	-0.56	0.61	-4.12	113.80
Jul 15	35.63	-2.48	77.55	-0.55	4.58	-4.46	123.36
Jul 25	35.94	-2.52	79.42	-0.55	7.64	-4.77	132.93
Aug 4	36.07	-2.56	81.13	-0.54	9.50	-5.03	142.54
Aug 14	36.03	-2.60	82.65	-0.54	10.21	-5.20	152.18
Aug 24	35.81	-2.64	83.97	-0.54	9.42	-5.21	161.88
Sep 3	35.43	-2.68	85.03	-0.54	7.36	-5.02	171.62
Sep 13	34.91	-2.71	85.83	-0.54	4.50	-4.59	181.41
Sep 23	34.25	-2.73	86.32	-0.53	1.58	-3.96	191.26
Oct 3	33.51	-2.74	86.50	-0.53	359.38	-3.21	201.17
Oct 13	32.72	-2.75	86.34	-0.52	358.36	-2.45	211.13
Oct 23	31.92	-2.74	85.84	-0.52	358.65	-1.75	221.16
Nov 2	31.16	-2.72	85.04	-0.51	0.19	-1.15	231.22
Nov 12	30.48	-2.69	83.98	-0.49	2.77	-0.64	241.33
Nov 22	29.91	-2.66	82.73	-0.47	6.19	-0.23	251.48
Dec 2	29.50	-2.61	81.39	-0.45	10.27	0.11	261.65
Dec 12	29.26	-2.57	80.05	-0.43	14.85	0.38	271.84
1704							
Dec 22	29.20	-2.52	78.81	-0.40	19.83	0.61	282.03
Jan 1	29.33	-2.47	77.78	-0.37	25.10	0.78	292.22
Jan 11	29.64	-2.42	77.01	-0.34	30.61	0.93	302.40
Jan 21	30.13	-2.37	76.55	-0.31	36.29	1.04	312.55
Jan 31	30.78	-2.33	76.43	-0.28	42.10	1.13	322.67
Feb 10	31.57	-2.29	76.65	-0.25	48.01	1.20	332.74
Feb 20	32.49	-2.25	77.18	-0.22	54.00	1.26	342.77
Mar 2	33.52	-2.22	78.01	-0.19	60.04	1.30	352.74
Mar 12	34.64	-2.19	79.11	-0.17	66.12	1.33	2.66
Mar 22	35.83	-2.17	80.45	-0.14	72.23	1.35	12.53
Apr 1	37.07	-2.16	81.99	-0.12	78.37	1.36	22.33
Apr 11	38.34	-2.15	83.71	-0.10	84.52	1.36	32.08
Apr 21	39.62	-2.14	85.58	-0.08	90.68	1.36	41.79
May 1	40.90	-2.14	87.57	-0.07	96.85	1.35	51.44
May 11	42.16	-2.14	89.65	-0.05	103.03	1.33	61.05
May 21	43.37	-2.15	91.81	-0.03	109.23	1.31	70.64
May 31	44.54	-2.16	94.03	-0.01	115.43	1.29	80.19
Jun 10	45.62	-2.18	96.27	0.00	121.64	1.26	89.73
Jun 20	46.62	-2.20	98.53	0.02	127.87	1.23	99.27
Jun 30	47.50	-2.23	100.79	0.03	134.11	1.19	108.80
Jul 10	48.27	-2.26	103.02	0.05	140.38	1.15	118.34
Jul 20	48.89	-2.29	105.21	0.07	146.68	1.11	127.91
Jul 30	49.36	-2.33	107.34	0.08	153.00	1.07	137.50
Aug 9	49.66	-2.36	109.38	0.10	159.35	1.02	147.12
Aug 19	49.79	-2.40	111.32	0.12	165.74	0.96	156.79
Aug 29	49.74	-2.43	113.12	0.14	172.17	0.91	166.50
Sep 8	49.51	-2.47	114.77	0.16	178.64	0.84	176.27
Sep 18	49.11	-2.50	116.23	0.18	185.16	0.78	186.09
Sep 28	48.57	-2.52	117.47	0.21	191.72	0.71	195.97
Oct 8	47.90	-2.53	118.47	0.23	198.34	0.64	205.91
Oct 18	47.14	-2.54	119.19	0.26	205.02	0.56	215.90
Oct 28	46.33	-2.54	119.60	0.29	211.75	0.48	225.94
Nov 7	45.53	-2.52	119.70	0.32	218.54	0.39	236.03
Nov 17	44.77	-2.50	119.46	0.35	225.39	0.30	246.16
Nov 27	44.09	-2.47	118.90	0.38	232.30	0.20	256.31
Dec 7	43.54	-2.43	118.05	0.41	239.27	0.10	266.50
Dec 17	43.15	-2.39	116.96	0.44	246.31	0.00	276.69
1705							

GREGORIAN 12 UT	MOON LONG.		MOON LAT.		VENUS LONG.		VENUS LAT.		MERCURY LONG.		MERCURY LAT.	
1704												1704
Jan 3		232.1		3.3		237.82		4.10		275.52		-1.35
8 Jan 13	307.2	17.5	-3.1	-5.0	241.90	246.31	4.12	4.06	283.51	291.69	-1.72	-1.97
18 Jan 23	78.7	138.0	-1.1	3.8	250.97	255.85	3.93	3.74	300.09	308.71	-2.08	-2.01
28 J-F 2	200.5	270.1	4.9	0.0	260.92	266.13	3.50	3.22	317.48	326.21	-1.73	-1.17
7 Feb 12	344.6	51.3	-5.0	-3.2	271.47	276.91	2.90	2.57	334.37	341.00	-0.31	0.84
17 Feb 22	110.8	171.8	1.9	5.0	282.44	288.04	2.21	1.85	344.73	344.53	2.14	3.24
27 Feb	237.6		2.5		293.70		1.48		340.77		3.72	
Mar 3		308.9		-3.3		299.43		1.11		335.74		3.34
8 Mar 13	20.5	83.5	-4.6	-0.3	305.19	311.00	0.75	0.40	332.06	330.92	2.34	1.13
18 Mar 23	143.0	207.3	4.2	4.2	316.84	322.71	0.07	-0.24	332.22	335.44	-0.01	-0.97
28 M-A 2	276.3	347.3	-1.0	-5.0	328.60	334.52	-0.53	-0.80	340.10	345.85	-1.70	-2.21
7 Apr 12	54.6	115.0	-2.5	2.7	340.46	346.41	-1.03	-1.23	352.48	359.87	-2.50	-2.58
17 Apr 22	176.3	244.6	5.1	1.6	352.37	358.35	-1.40	-1.54	7.97	16.77	-2.43	-2.07
27 Apr	315.5		-4.1		4.34		-1.64		26.28		-1.50	
May 2		24.2		-4.4		10.34		-1.70		36.48		-0.76
7 May 12	87.3	146.8	0.5	4.7	16.35	22.36	-1.74	-1.74	47.21	58.09	0.10	0.96
17 May 22	211.1	283.1	4.0	-1.9	28.39	34.41	-1.71	-1.65	68.56	78.16	1.66	2.09
27 M-J 1	354.1	59.1	-5.3	-2.0	40.45	46.50	-1.56	-1.44	86.64	93.88	2.20	1.99
6 Jun 11	119.3	179.6	3.3	5.2	52.55	58.61	-1.31	-1.15	99.83	104.34	1.45	0.61
16 Jun 21	247.6	322.1	1.3	-4.6	64.67	70.75	-0.97	-0.79	107.25	108.35	-0.49	-1.78
26 Jun	30.8		-4.1		76.83		-0.59		107.56		-3.10	
Jul 1		92.4		1.0		82.92		-0.38		105.19		-4.20
6 Jul 11	151.7	214.1	4.9	3.8	89.02	95.13	-0.17	0.03	102.08	99.51	-4.80	-4.75
16 Jul 21	285.7	0.3	-2.2	-5.1	101.25	107.38	0.24	0.43	98.61	100.00	-4.09	-3.02
26 Jul 31	65.4	125.0	-1.4	3.6	113.52	119.67	0.62	0.79	103.84	109.99	-1.78	-0.54
5 Aug 10	185.1	250.4	4.9	0.8	125.83	132.00	0.95	1.08	118.12	127.60	0.53	1.29
15 Aug 20	324.5	36.5	-4.6	-3.4	138.18	144.37	1.20	1.29	137.65	147.66	1.69	1.76
25 Aug 30	98.3	157.7	1.7	4.9	150.56	156.77	1.36	1.40	157.28	166.41	1.57	1.19
4 Sep 9	220.2	288.5	3.0	-2.6	162.98	169.20	1.41	1.40	175.02	183.15	0.69	0.11
14 Sep 19	2.8	70.4	-4.8	-0.5	175.42	181.64	1.36	1.29	190.83	198.07	-0.52	-1.16
24 Sep 29	130.4	191.4	4.1	4.5	187.88	194.11	1.20	1.09	204.83	211.02	-1.78	-2.36
4 Oct 9	257.1	327.7	-0.3	-4.9	200.35	206.58	0.95	0.80	216.45	220.76	-2.85	-3.17
14 Oct 19	39.2	102.8	-2.9	2.5	212.82	219.06	0.62	0.44	223.31	223.11	-3.23	-2.83
24 Oct 29	162.4	226.5	5.2	2.3	225.30	231.54	0.24	0.03	219.34	213.11	-1.76	-0.12
3 Nov 8	295.7	6.4	-3.6	-4.8	237.79	244.02	-0.18	-0.39	208.42	208.19	1.39	2.20
13 Nov 18	73.5	134.4	0.1	4.7	250.26	256.50	-0.60	-0.80	211.87	217.81	2.35	2.09
23 Nov 28	195.2	263.1	4.4	-1.0	262.73	268.97	-0.99	-1.17	224.82	232.29	1.62	1.05
3 Dec 8	335.0	43.4	-5.2	-2.6	275.19	281.42	-1.32	-1.46	239.94	247.67	0.45	-0.13
13 Dec 18	106.5	166.1	3.0	5.3	287.64	293.85	-1.58	-1.67	255.44	263.27	-0.68	-1.17
23 Dec 28	229.1	301.2	2.1	-4.0	300.05	306.25	-1.73	-1.76	271.19	279.23	-1.58	-1.90
1705												1705
2 Jan 7	13.6	78.4	-4.4	0.6	312.43	318.60	-1.75	-1.72	287.40	295.71	-2.08	-2.11
12 Jan 17	138.8	198.7	4.7	4.1	324.76	330.89	-1.64	-1.54	304.08	312.32	-1.92	-1.48
22 Jan 27	264.8	340.0	-1.2	-5.1	337.01	343.10	-1.40	-1.23	319.89	325.81	-0.71	0.41
1 Feb 6	50.2	111.9	-1.7	3.4	349.16	355.19	-1.02	-0.79	328.56	326.99	1.79	3.08
11 Feb 16	171.4	232.7	4.9	1.4	1.18	7.13	-0.53	-0.24	321.99	316.70	3.72	3.43
21 Feb 26	302.4	18.0	-4.1	-3.8	13.04	18.90	0.07	0.40	313.78	313.79	2.52	1.41
3 Mar 8	84.6	144.5	1.5	4.9	24.70	30.43	0.74	1.09	316.17	320.25	0.35	-0.56
13 Mar 18	204.7	268.8	3.3	-2.0	36.10	41.68	1.45	1.80	325.54	331.74	-1.30	-1.85
23 Mar 28	341.2	54.3	-5.0	-0.9	47.18	52.57	2.16	2.49	338.67	346.25	-2.22	-2.39
2 Apr 7	117.5	177.0	4.1	4.7	57.85	62.98	2.81	3.11	354.44	3.25	-2.36	-2.13
12 Apr 17	239.4	307.0	0.4	-4.6	67.96	72.75	3.38	3.60	12.68	22.73	-1.69	-1.04
22 Apr 27	19.9	88.5	-3.6	2.2	77.33	81.66	3.78	3.90	33.29	44.01	-0.23	0.66
2 May 7	149.4	210.1	5.2	2.9	85.67	89.32	3.96	3.93	54.36	63.76	1.48	2.08
12 May 17	276.0	346.5	-2.9	-5.1	92.52	95.19	3.81	3.58	71.86	78.49	2.38	2.32
22 May 27	56.9	121.4	-0.6	4.5	97.21	98.47	3.21	2.69	83.54	86.87	1.88	1.08
1 Jun 6	181.2	244.5	4.6	-0.1	98.84	98.23	2.00	1.13	88.33	87.91	-0.05	-1.42
11 Jun 16	314.6	25.5	-5.0	-3.2	96.63	94.18	0.10	-1.05	85.90	83.11	-2.81	-3.93
21 Jun 26	92.1	153.5	2.6	5.2	91.19	88.11	-2.22	-3.31	80.68	79.62	-4.52	-4.52
1 Jul 6	213.6	280.7	2.5	-3.2	85.41	83.45	-4.21	-4.86	80.49	83.45	-4.00	-3.14
11 Jul 16	354.0	62.8	-4.8	0.1	82.45	82.43	-5.28	-5.48	88.43	95.31	-2.06	-0.92
21 Jul 26	125.8	185.5	4.6	4.2	83.34	85.05	-5.51	-5.40	103.86	113.64	0.15	1.00
31 J-A 5	247.3	318.8	-0.6	-4.9	87.44	90.40	-5.19	-4.90	124.00	134.32	1.55	1.76
10 Aug 15	32.7	98.1	-2.3	3.1	93.83	97.66	-4.56	-4.17	144.21	153.55	1.69	1.39
20 Aug 25	158.6	218.1	4.9	1.7	101.82	106.25	-3.75	-3.32	162.29	170.48	0.93	0.35
30 Aug	283.0		-3.6		110.92		-2.87		178.12		-0.30	
Sep 4		357.8		-4.3		115.78		-2.41		185.22		-0.99
9 Sep 14	69.4	131.8	1.1	4.9	120.81	125.99	-1.96	-1.51	191.72	197.53	-1.68	-2.35
19 Sep 24	191.3	252.2	3.5	-1.5	131.30	136.72	-1.08	-0.66	202.40	205.95	-2.95	-3.41
29 S-O 4	320.7	36.1	-5.1	-1.6	142.25	147.86	-0.27	0.10	207.52	206.28	-3.61	-3.33
9 Oct 14	104.1	164.4	4.0	4.9	153.56	159.32	0.44	0.75	201.86	195.93	-2.35	-0.76
19 Oct 24	224.4	288.3	0.9	-4.3	165.16	171.05	1.03	1.27	192.24	193.00	0.83	1.83
29 Oct	359.5		-4.3		176.99		1.47		197.48		2.18	
Nov 3		72.6		1.7		182.97		1.63		204.08		2.08
8 Nov 13	137.0	196.6	5.2	3.2	189.00	195.06	1.75	1.83	211.62	219.49	1.71	1.22
18 Nov 23	258.8	326.4	-2.2	-5.3	201.16	207.28	1.88	1.89	227.44	235.36	0.66	0.08
28 N-D 3	38.1	107.2	-1.4	4.2	213.43	219.59	1.86	1.80	243.24	251.09	-0.47	-0.99
8 Dec 13	168.9	229.0	4.7	0.4	225.78	231.98	1.70	1.58	258.95	266.85	-1.44	-1.81
18 Dec 23	295.2	5.7	-4.6	-3.8	238.19	244.41	1.44	1.27	274.81	282.80	-2.07	-2.18
28 Dec	75.3		1.9		250.64		1.08		290.77		-2.10	
1705												1705

JULIAN 12 UT	SATURN LONG.	SATURN LAT.	JUPITER LONG.	JUPITER LAT.	MARS LONG.	MARS LAT.	SUN LONG.
1705							
Dec 27	42.93	-2.34	115.70	0.46	253.40	-0.12	286.88
Jan 6	42.90	-2.29	114.36	0.48	260.55	-0.24	297.07
Jan 16	43.05	-2.24	113.05	0.50	267.76	-0.36	307.23
Jan 26	43.39	-2.19	111.86	0.52	275.02	-0.49	317.37
Feb 5	43.90	-2.15	110.88	0.53	282.34	-0.62	327.47
Feb 15	44.57	-2.10	110.16	0.54	289.71	-0.75	337.52
Feb 25	45.39	-2.06	109.75	0.54	297.11	-0.89	347.52
Mar 7	46.33	-2.03	109.65	0.55	304.56	-1.02	357.47
Mar 17	47.38	-2.00	109.88	0.55	312.03	-1.16	7.36
Mar 27	48.51	-1.97	110.42	0.55	319.52	-1.30	17.20
Apr 6	49.71	-1.95	111.24	0.55	327.02	-1.43	26.98
Apr 16	50.97	-1.93	112.31	0.55	334.51	-1.55	36.70
Apr 26	52.25	-1.91	113.61	0.55	341.98	-1.66	46.38
May 6	53.54	-1.91	115.11	0.55	349.42	-1.77	56.02
May 16	54.83	-1.90	116.77	0.56	356.79	-1.86	65.61
May 26	56.10	-1.90	118.58	0.56	4.09	-1.93	75.18
Jun 5	57.32	-1.90	120.51	0.56	11.29	-1.99	84.73
Jun 15	58.49	-1.91	122.53	0.57	18.35	-2.03	94.26
Jun 25	59.58	-1.92	124.63	0.58	25.25	-2.05	103.80
Jul 5	60.59	-1.94	126.78	0.59	31.95	-2.05	113.34
Jul 15	61.48	-1.96	128.96	0.60	38.41	-2.02	122.89
Jul 25	62.25	-1.98	131.16	0.61	44.58	-1.97	132.46
Aug 4	62.87	-2.01	133.35	0.63	50.40	-1.89	142.07
Aug 14	63.34	-2.03	135.52	0.64	55.78	-1.77	151.71
Aug 24	63.64	-2.06	137.64	0.66	60.63	-1.63	161.40
Sep 3	63.76	-2.09	139.70	0.68	64.83	-1.45	171.15
Sep 13	63.70	-2.11	141.67	0.70	68.22	-1.21	180.94
Sep 23	63.46	-2.14	143.53	0.73	70.61	-0.93	190.78
Oct 3	63.05	-2.16	145.26	0.76	71.79	-0.58	200.69
Oct 13	62.49	-2.17	146.81	0.79	71.58	-0.16	210.65
Oct 23	61.81	-2.18	148.17	0.82	69.90	0.32	220.67
Nov 2	61.04	-2.18	149.30	0.86	66.93	0.83	230.74
Nov 12	60.23	-2.17	150.18	0.90	63.24	1.32	240.84
Nov 22	59.41	-2.16	150.77	0.94	59.60	1.73	250.99
Dec 2	58.65	-2.13	151.05	0.99	56.76	2.03	261.16
Dec 12	57.98	-2.10	151.01	1.03	55.19	2.22	271.34
1706							
Dec 22	57.45	-2.06	150.65	1.08	54.97	2.31	281.54
Jan 1	57.07	-2.02	149.98	1.12	55.98	2.34	291.73
Jan 11	56.86	-1.98	149.04	1.15	58.05	2.33	301.90
Jan 21	56.85	-1.93	147.90	1.18	60.94	2.30	312.06
Jan 31	57.02	-1.88	146.62	1.21	64.50	2.24	322.18
Feb 10	57.38	-1.84	145.31	1.22	68.56	2.18	332.25
Feb 20	57.91	-1.79	144.05	1.23	73.03	2.11	342.28
Mar 2	58.60	-1.75	142.94	1.22	77.81	2.04	352.26
Mar 12	59.43	-1.71	142.05	1.21	82.84	1.96	2.18
Mar 22	60.38	-1.68	141.43	1.20	88.06	1.89	12.05
Apr 1	61.44	-1.65	141.11	1.18	93.45	1.81	21.86
Apr 11	62.59	-1.62	141.10	1.15	98.97	1.74	31.61
Apr 21	63.80	-1.60	141.39	1.13	104.59	1.66	41.32
May 1	65.05	-1.58	141.97	1.11	110.31	1.59	50.97
May 11	66.34	-1.57	142.83	1.09	116.11	1.51	60.59
May 21	67.64	-1.55	143.92	1.07	121.98	1.44	70.17
May 31	68.94	-1.55	145.23	1.05	127.91	1.36	79.73
Jun 10	70.21	-1.54	146.73	1.03	133.91	1.28	89.27
Jun 20	71.44	-1.54	148.39	1.02	139.96	1.20	98.81
Jun 30	72.62	-1.54	150.18	1.01	146.08	1.13	108.34
Jul 10	73.72	-1.55	152.09	1.00	152.25	1.05	117.88
Jul 20	74.73	-1.56	154.09	1.00	158.48	0.97	127.45
Jul 30	75.62	-1.57	156.16	0.99	164.78	0.88	137.04
Aug 9	76.39	-1.58	158.29	0.99	171.14	0.80	146.66
Aug 19	77.02	-1.60	160.44	1.00	177.56	0.71	156.32
Aug 29	77.49	-1.61	162.60	1.00	184.05	0.63	166.04
Sep 8	77.78	-1.63	164.76	1.01	190.61	0.54	175.80
Sep 18	77.90	-1.65	166.89	1.03	197.25	0.45	185.62
Sep 28	77.83	-1.66	168.97	1.04	203.96	0.35	195.50
Oct 8	77.58	-1.68	170.97	1.06	210.75	0.26	205.43
Oct 18	77.16	-1.69	172.88	1.09	217.62	0.16	215.42
Oct 28	76.58	-1.69	174.66	1.12	224.57	0.06	225.45
Nov 7	75.89	-1.69	176.29	1.15	231.60	-0.04	235.54
Nov 17	75.11	-1.69	177.74	1.18	238.71	-0.14	245.67
Nov 27	74.29	-1.68	178.98	1.22	245.90	-0.24	255.82
Dec 7	73.47	-1.66	179.98	1.26	253.17	-0.34	266.00
Dec 17	72.71	-1.63	180.71	1.30	260.51	-0.45	276.19
1707							

GREGORIAN 12 UT	MOON LONG.		MOON LAT.		VENUS LONG.		VENUS LAT.		MERCURY LONG.		MERCURY LAT.	
1706												1706
Jan 2		140.4		5.1		256.87		0.88		298.48		-1.77
7 Jan 12	200.3	262.7	2.7	-2.6	263.11	269.36	0.67	0.45	305.43	310.59	-1.11	-0.05
17 Jan 22	333.4	44.6	-4.9	-0.6	275.61	281.85	0.23	0.01	312.34	309.53	1.37	2.81
27 J-F 1	111.0	172.8	4.4	4.2	288.10	294.35	-0.21	-0.42	303.65	298.55	3.60	3.42
6 Feb 11	232.3	298.1	-0.3	-4.7	300.59	306.84	-0.61	-0.79	296.60	297.71	2.62	1.63
16 Feb 21	12.3	82.0	-2.9	2.8	313.08	319.32	-0.95	-1.10	301.02	305.80	0.65	-0.21
26 Feb	145.1		5.0		325.56		-1.22		311.59		-0.94	
Mar 3		204.9		1.9		331.79		-1.31		318.13		-1.52
8 Mar 13	265.6	335.6	-3.2	-4.8	338.02	344.25	-1.38	-1.43	325.28	332.97	-1.94	-2.20
18 Mar 23	50.6	117.5	0.5	4.9	350.47	356.68	-1.44	-1.43	341.18	349.93	-2.27	-2.15
28 M-A 2	178.1	237.5	3.8	-1.2	2.89	9.09	-1.40	-1.34	359.24	9.09	-1.83	-1.29
7 Apr 12	301.0	14.1	-5.1	-2.7	15.28	21.47	-1.25	-1.14	19.41	29.91	-0.55	0.33
17 Apr 22	87.4	151.3	3.6	5.1	27.66	33.83	-1.01	-0.86	40.04	49.16	1.24	2.00
27 Apr	210.6		1.3		40.00		-0.70		56.75		2.50	
May 2		271.6		-3.9		46.17		-0.52		62.57		2.62
7 May 12	338.6	52.5	-4.8	0.7	52.33	58.48	-0.33	-0.13	66.43	68.25	2.33	1.60
17 May 22	122.5	183.7	5.2	3.5	64.63	70.77	0.07	0.27	68.03	66.13	0.45	-0.96
27 M-J 1	243.5	307.7	-1.7	-5.2	76.91	83.04	0.47	0.66	63.40	60.97	-2.37	-3.46
6 Jun 11	17.6	89.9	-2.4	3.7	89.16	95.29	0.84	1.01	59.83	60.48	-4.07	-4.16
16 Jun 21	155.8	215.5	4.8	0.7	101.40	107.52	1.17	1.30	63.01	67.31	-3.82	-3.16
26 Jun	277.7		-4.1		113.62		1.41		73.26		-2.27	
Jul 1		345.8		-4.3		119.72		1.50		80.75		-1.26
6 Jul 11	56.5	125.6	1.2	5.0	125.82	131.90	1.56	1.59	89.64	99.65	-0.22	0.69
16 Jul 21	187.9	247.7	2.8	-2.2	137.98	144.05	1.60	1.57	110.25	120.83	1.36	1.72
26 Jul 31	313.8	25.0	-5.0	-1.3	150.12	156.17	1.51	1.42	130.97	140.48	1.78	1.57
5 Aug 10	94.5	159.6	4.1	4.3	162.21	168.24	1.29	1.14	149.34	157.55	1.17	0.61
15 Aug 20	219.5	281.2	-0.1	-4.5	174.26	180.26	0.96	0.75	165.12	172.05	-0.05	-0.78
25 Aug 30	351.7	64.0	-3.6	2.3	186.25	192.22	0.52	0.26	178.25	183.59	-1.56	-2.33
4 Sep 9	130.7	192.3	5.1	2.1	198.17	204.10	-0.01	-0.30	187.82	190.51	-3.05	-3.65
14 Sep 19	251.5	316.4	-3.0	-5.1	210.01	215.90	-0.60	-0.91	191.04	188.80	-3.97	-3.81
24 Sep 29	30.5	101.6	-0.5	4.8	221.76	227.59	-1.21	-1.52	183.97	178.67	-2.91	-1.36
4 Oct 9	165.1	224.5	4.0	-0.8	233.39	239.16	-1.82	-2.11	176.14	177.90	0.26	1.41
14 Oct 19	284.9	353.5	-4.9	-3.5	244.88	250.55	-2.37	-2.62	183.14	190.36	1.96	2.03
24 Oct 29	68.9	137.3	2.9	5.2	256.17	261.71	-2.83	-3.00	198.41	206.69	1.79	1.36
3 Nov 8	197.9	257.2	1.6	-3.5	267.19	272.57	-3.14	-3.22	214.93	223.06	0.85	0.29
13 Nov 18	320.2	31.6	-5.1	-0.4	277.84	282.98	-3.24	-3.21	231.05	238.94	-0.27	-0.80
23 Nov 28	106.0	171.0	4.9	3.7	287.96	292.75	-3.10	-2.92	246.76	254.53	-1.29	-1.71
3 Dec 8	230.2	291.3	-1.3	-5.0	297.31	301.59	-2.64	-2.28	262.27	269.96	-2.04	-2.23
13 Dec 18	357.4	70.0	-3.1	3.0	305.53	309.04	-1.80	-1.21	277.51	284.68	-2.26	-2.05
23 Dec 28	141.2	203.1	4.9	0.9	312.03	314.40	-0.49	0.36	290.97	295.31	-1.51	-0.54
1707												1707
2 Jan 7	262.8	327.2	-3.8	-4.6	316.01	316.73	1.35	2.48	296.06	292.17	0.88	2.42
12 Jan 17	36.1	107.8	0.4	4.9	316.44	315.10	3.70	4.95	285.77	281.19	3.37	3.33
22 Jan 27	174.7	234.6	3.0	-2.1	312.83	309.93	6.13	7.11	280.29	282.42	2.67	1.79
1 Feb 6	296.4	4.8	-5.0	-2.1	306.86	304.14	7.76	8.04	286.53	291.90	0.91	0.09
11 Feb 16	75.2	144.0	3.7	4.6	302.18	301.19	7.96	7.61	298.10	304.91	-0.63	-1.22
21 Feb 26	206.8	266.4	0.1	-4.4	301.22	302.21	7.07	6.41	312.21	319.96	-1.69	-2.00
3 Mar 8	331.7	43.6	-4.2	1.5	304.03	306.55	5.69	4.95	328.16	336.81	-2.16	-2.15
13 Mar 18	113.5	178.4	5.2	2.5	309.64	313.20	4.21	3.49	345.95	355.57	-1.93	-1.51
23 Mar 28	238.6	299.3	-2.8	-5.3	317.15	321.41	2.80	2.15	5.61	15.81	-0.86	-0.01
2 Apr 7	8.6	82.4	-1.6	4.4	325.93	330.66	1.53	0.96	25.65	34.39	0.95	1.86
12 Apr 17	150.0	211.5	4.4	-0.5	335.57	340.63	0.44	-0.04	41.36	46.15	2.55	2.89
22 Apr 27	270.8	334.1	-4.8	-4.2	345.81	351.10	-0.47	-0.85	48.56	48.60	2.77	2.12
2 May 7	46.8	120.3	1.9	5.2	356.47	1.93	-1.18	-1.46	46.68	43.74	0.99	-0.43
12 May 17	184.5	243.9	2.0	-3.2	7.45	13.03	-1.69	-1.87	41.05	39.67	-1.81	-2.88
22 May 27	304.3	10.7	-5.1	-1.3	18.66	24.33	-2.01	-2.10	40.08	42.34	-3.52	-3.73
1 Jun 6	85.5	156.1	4.5	3.9	30.05	35.80	-2.14	-2.15	46.25	51.64	-3.57	-3.11
11 Jun 16	217.3	276.8	-1.1	-4.8	41.58	47.39	-2.12	-2.04	58.36	66.33	-2.41	-1.54
21 Jun 26	339.6	49.0	-3.5	2.2	53.23	59.09	-1.94	-1.81	75.49	85.66	-0.58	0.36
1 Jul 6	123.4	189.9	4.9	1.1	64.98	70.90	-1.65	-1.47	96.42	107.20	1.13	1.64
11 Jul 16	249.5	310.8	-3.7	-4.7	76.83	82.79	-1.27	-1.05	117.54	127.20	1.83	1.74
21 Jul 26	16.7	88.2	-0.4	4.6	88.77	94.77	-0.83	-0.59	136.12	144.31	1.41	0.89
31 J-A 5	159.5	222.2	3.3	-1.9	100.79	106.83	-0.36	-0.12	151.76	158.45	0.22	-0.55
10 Aug 15	282.0	346.4	-5.0	-2.7	112.89	118.97	0.11	0.33	164.27	169.06	-1.40	-2.28
20 Aug 25	55.5	126.6	3.1	4.9	125.07	131.19	0.53	0.73	172.52	174.22	-3.13	-3.87
30 Aug	193.5		0.4		137.33		0.90		173.65		-4.32	
Sep 4		253.8		-4.3		143.49		1.06		170.54		-4.23
9 Sep 14	315.3	23.7	-4.6	0.6	149.66	155.85	1.19	1.29	165.67	161.36	-3.39	-1.90
19 Sep 24	94.9	163.3	5.1	2.9	162.05	168.26	1.37	1.43	160.13	162.87	-0.29	0.97
29 S-O 4	226.1	285.6	-2.6	-5.3	174.49	180.73	1.45	1.45	168.83	176.62	1.70	1.94
9 Oct 14	350.0	62.4	-2.4	3.7	186.98	193.23	1.42	1.36	185.17	193.85	1.83	1.50
19 Oct 24	133.4	197.9	4.7	-0.1	199.49	205.76	1.28	1.17	202.39	210.73	1.03	0.50
29 Oct	258.1		-4.6		212.03		1.04		218.86		-0.06	
Nov 3		318.4		-4.5		218.31		0.89		226.79		-0.61
8 Nov 13	26.3	101.3	0.8	5.2	224.59	230.87	0.73	0.55	234.57	242.22	-1.14	-1.61
18 Nov 23	169.8	231.1	2.3	-2.9	237.15	243.44	0.36	0.17	249.75	257.14	-2.00	-2.28
28 N-D 3	290.5	352.6	-5.1	-2.0	249.73	256.01	-0.03	-0.23	264.26	270.89	-2.40	-2.31
8 Dec 13	64.2	139.2	3.8	4.2	262.30	268.58	-0.43	-0.62	276.47	279.94	-1.90	-1.04
18 Dec 23	204.2	263.7	-0.8	-4.7	274.86	281.15	-0.80	-0.96	279.70	274.91	0.32	1.94
28 Dec	323.9		-3.8		287.43		-1.11		268.35		3.04	
1707												1707

JULIAN 12 UT	SATURN LONG.	SATURN LAT.	JUPITER LONG.	JUPITER LAT.	MARS LONG.	MARS LAT.	SUN LONG.
1707							
Dec 27	72.05	-1.60	181.15	1.35	267.93	-0.55	286.39
Jan 6	71.52	-1.57	181.27	1.40	275.42	-0.65	296.57
Jan 16	71.15	-1.53	181.08	1.44	282.96	-0.75	306.74
Jan 26	70.96	-1.49	180.57	1.49	290.57	-0.84	316.87
Feb 5	70.96	-1.45	179.79	1.52	298.22	-0.93	326.97
Feb 15	71.15	-1.41	178.76	1.55	305.92	-1.02	337.03
Feb 25	71.52	-1.37	177.56	1.57	313.65	-1.10	347.04
Mar 6	72.06	-1.33	176.28	1.58	321.39	-1.17	356.99
Mar 16	72.76	-1.29	175.00	1.58	329.15	-1.23	6.89
Mar 26	73.60	-1.26	173.82	1.57	336.91	-1.27	16.72
Apr 5	74.56	-1.23	172.81	1.55	344.65	-1.31	26.50
Apr 15	75.62	-1.20	172.03	1.52	352.36	-1.33	36.24
Apr 25	76.77	-1.18	171.53	1.49	0.03	-1.34	45.92
May 5	77.99	-1.16	171.33	1.45	7.64	-1.33	55.55
May 15	79.25	-1.14	171.42	1.41	15.19	-1.30	65.15
May 25	80.54	-1.12	171.86	1.37	22.65	-1.27	74.72
Jun 4	81.84	-1.11	172.48	1.33	30.02	-1.21	84.27
Jun 14	83.14	-1.10	173.41	1.30	37.28	-1.14	93.81
Jun 24	84.41	-1.09	174.56	1.26	44.41	-1.05	103.34
Jul 4	85.65	-1.08	175.93	1.23	51.40	-0.95	112.88
Jul 14	86.83	-1.08	177.47	1.20	58.24	-0.84	122.43
Jul 24	87.93	-1.08	179.16	1.18	64.91	-0.70	132.00
Aug 3	88.94	-1.08	180.99	1.16	71.30	-0.56	141.61
Aug 13	89.84	-1.08	182.93	1.14	77.62	-0.39	151.25
Aug 23	90.61	-1.08	184.95	1.13	83.62	-0.21	160.93
Sep 2	91.24	-1.09	187.04	1.12	89.32	-0.01	170.67
Sep 12	91.70	-1.09	189.18	1.12	94.67	0.22	180.46
Sep 22	91.99	-1.10	191.34	1.11	99.62	0.46	190.31
Oct 2	92.10	-1.10	193.51	1.11	104.06	0.74	200.21
Oct 12	92.02	-1.11	195.66	1.12	107.90	1.06	210.17
Oct 22	91.76	-1.11	197.77	1.13	110.99	1.41	220.18
Nov 1	91.32	-1.11	199.82	1.14	113.17	1.80	230.25
Nov 11	90.74	-1.11	201.78	1.15	114.22	2.24	240.35
Nov 21	90.03	-1.10	203.64	1.17	113.98	2.71	250.49
Dec 1	89.25	-1.09	205.35	1.19	112.34	3.18	260.66
Dec 11	88.42	-1.07	206.90	1.22	109.41	3.60	270.85
1708							
Dec 21	87.61	-1.05	208.25	1.25	105.62	3.91	281.05
Dec 31	86.85	-1.03	209.38	1.28	101.69	4.05	291.24
Jan 10	86.20	-1.00	210.25	1.32	98.37	4.03	301.42
Jan 20	85.68	-0.97	210.84	1.35	96.18	3.88	311.57
Jan 30	85.32	-0.94	211.13	1.39	95.31	3.66	321.69
Feb 9	85.14	-0.91	211.11	1.43	95.70	3.41	331.77
Feb 19	85.16	-0.88	210.77	1.46	97.20	3.15	341.80
Mar 1	85.35	-0.84	210.14	1.49	99.58	2.90	351.78
Mar 11	85.73	-0.81	209.25	1.51	102.69	2.67	1.71
Mar 21	86.28	-0.78	208.15	1.53	106.37	2.46	11.58
Mar 31	86.98	-0.76	206.92	1.53	110.51	2.26	21.39
Apr 10	87.83	-0.73	205.65	1.53	115.01	2.07	31.15
Apr 20	88.79	-0.70	204.41	1.51	119.81	1.90	40.85
Apr 30	89.85	-0.68	203.31	1.49	124.85	1.74	50.51
May 10	91.00	-0.66	202.39	1.45	130.09	1.59	60.13
May 20	92.21	-0.64	201.73	1.41	135.51	1.45	69.71
May 30	93.47	-0.62	201.35	1.37	141.08	1.31	79.27
Jun 9	94.76	-0.61	201.27	1.32	146.79	1.18	88.81
Jun 19	96.06	-0.59	201.48	1.28	152.63	1.06	98.34
Jun 29	97.36	-0.58	201.98	1.23	158.57	0.94	107.88
Jul 9	98.64	-0.57	202.76	1.19	164.63	0.82	117.42
Jul 19	99.87	-0.56	203.78	1.15	170.79	0.70	126.98
Jul 29	101.05	-0.55	205.03	1.11	177.06	0.58	136.57
Aug 8	102.15	-0.54	206.48	1.08	183.42	0.47	146.19
Aug 18	103.16	-0.53	208.10	1.05	189.88	0.36	155.85
Aug 28	104.06	-0.52	209.87	1.02	196.45	0.25	165.56
Sep 7	104.83	-0.52	211.77	0.99	203.11	0.14	175.32
Sep 17	105.45	-0.51	213.77	0.97	209.88	0.04	185.13
Sep 27	105.91	-0.50	215.86	0.96	216.74	-0.07	195.01
Oct 7	106.19	-0.50	218.00	0.94	223.71	-0.17	204.94
Oct 17	106.29	-0.49	220.19	0.93	230.77	-0.27	214.92
Oct 27	106.20	-0.48	222.39	0.92	237.93	-0.37	224.96
Nov 6	105.92	-0.47	224.58	0.92	245.19	-0.47	235.05
Nov 16	105.48	-0.46	226.76	0.91	252.54	-0.56	245.17
Nov 26	104.88	-0.45	228.88	0.91	259.98	-0.65	255.33
Dec 6	104.17	-0.43	230.92	0.92	267.50	-0.73	265.51
Dec 16	103.38	-0.42	232.87	0.92	275.10	-0.81	275.70
1709							

GREGORIAN 12 UT	MOON LONG.		MOON LAT.		VENUS LONG.		VENUS LAT.		MERCURY LONG.		MERCURY LAT.	
1708												1708
Jan 2		28.8		1.4		293.70		-1.24		264.50		3.18
7 Jan 12	102.8	175.0	5.0	1.4	299.98	306.24	-1.35	-1.44	264.64	267.70	2.67	1.92
17 Jan 22	236.9	296.5	-3.6	-4.8	312.51	318.76	-1.50	-1.53	272.52	278.41	1.12	0.35
27 J-F 1	358.9	67.0	-1.0	4.3	325.01	331.25	-1.54	-1.51	284.96	292.00	-0.34	-0.95
6 Feb 11	140.8	208.7	3.8	-1.7	337.48	343.70	-1.46	-1.38	299.42	307.20	-1.44	-1.81
16 Feb 21	268.9	329.9	-5.1	-3.2	349.91	356.10	-1.27	-1.13	315.34	323.87	-2.04	-2.12
26 Feb	35.8		2.4		2.27		-0.97		332.81		-2.01	
Mar 2		106.3		5.2		8.43		-0.79		342.18		-1.69
7 Mar 12	177.2	241.0	1.0	-4.2	14.57	20.69	-0.58	-0.36	351.90	1.74	-1.14	-0.36
17 Mar 22	300.8	4.7	-4.9	-0.3	26.78	32.86	-0.12	0.13	11.21	19.49	0.61	1.64
27 M-A 1	74.4	145.0	4.9	3.6	38.90	44.92	0.39	0.65	25.72	29.30	2.53	3.10
6 Apr 11	211.7	272.8	-2.1	-5.2	50.91	56.86	0.91	1.17	30.04	28.25	3.16	2.61
16 Apr 21	333.4	41.2	-3.0	2.9	62.79	68.67	1.41	1.65	24.97	21.72	1.50	0.12
26 Apr	113.8		5.0		74.52		1.86		19.79		-1.22	
May 1		182.1		0.5		80.32		2.05		19.77		-2.26
6 May 11	245.0	304.7	-4.4	-4.6	86.07	91.76	2.21	2.34	21.68	25.26	-2.95	-3.28
16 May 21	7.4	79.6	0.1	4.9	97.40	102.97	2.44	2.49	30.25	36.45	-3.28	-3.01
26 May 31	152.3	217.1	2.8	-2.6	108.47	113.88	2.50	2.45	43.74	52.08	-2.49	-1.76
5 Jun 10	277.6	337.4	-5.0	-2.3	119.19	124.40	2.36	2.20	61.43	71.68	-0.90	0.02
15 Jun 20	43.5	118.7	3.2	4.5	129.48	134.41	1.99	1.71	82.52	93.43	0.86	1.50
25 Jun 30	188.9	250.7	-0.5	-4.6	139.18	143.74	1.36	0.94	103.91	113.67	1.84	1.88
5 Jul 10	310.3	11.7	-3.9	0.9	148.06	152.09	0.45	-0.13	122.60	130.70	1.65	1.19
15 Jul 20	81.6	156.9	5.0	2.1	155.78	159.06	-0.79	-1.54	137.94	144.29	0.53	-0.28
25 Jul 30	223.4	283.5	-3.5	-4.9	161.84	164.01	-2.37	-3.29	149.60	153.69	-1.20	-2.19
4 Aug 9	343.6	48.0	-1.4	3.9	165.46	166.06	-4.28	-5.31	156.22	156.81	-3.19	-4.06
14 Aug 19	120.6	193.2	4.4	-1.3	165.73	164.41	-6.35	-7.30	155.14	151.44	-4.62	-4.58
24 Aug 29	256.4	316.0	-5.1	-3.6	162.19	159.33	-8.07	-8.55	147.01	144.02	-3.77	-2.36
3 Sep 8	18.5	86.4	1.8	5.3	156.25	153.44	-8.66	-8.39	144.18	147.90	-0.80	0.52
13 Sep 18	159.0	227.4	1.8	-4.0	151.32	150.13	-7.80	-6.99	154.51	162.83	1.39	1.80
23 Sep 28	288.3	349.0	-5.0	-0.9	149.94	150.70	-6.07	-5.10	171.85	180.92	1.84	1.61
3 Oct 8	55.3	125.7	4.4	4.2	152.32	154.67	-4.15	-3.24	189.78	198.35	1.20	0.70
13 Oct 18	195.8	260.3	-1.5	-5.2	157.64	161.11	-2.39	-1.60	206.61	214.60	0.15	-0.41
23 Oct 28	320.0	23.4	-3.2	2.2	165.02	169.27	-0.89	-0.25	222.36	229.90	-0.97	-1.49
2 Nov 7	93.9	164.5	5.1	1.2	173.81	178.60	0.32	0.82	237.23	244.30	-1.94	-2.31
12 Nov 17	230.8	292.2	-4.1	-4.6	183.60	188.78	1.25	1.61	251.00	257.05	-2.53	-2.55
22 Nov 27	352.2	59.5	-0.4	4.5	194.10	199.55	1.91	2.14	261.89	264.45	-2.28	-1.55
2 Dec 7	133.2	201.6	3.4	-2.2	205.11	210.76	2.31	2.43	263.22	257.74	-0.26	1.39
12 Dec 17	264.5	324.1	-5.0	-2.5	216.49	222.29	2.49	2.50	251.32	248.33	2.63	2.96
22 Dec 27	25.7	97.4	2.7	4.9	228.15	234.06	2.46	2.38	249.47	253.38	2.64	2.02
1709												1709
1 Jan 6	171.4	236.9	0.1	-4.6	240.00	245.99	2.26	2.10	258.85	265.20	1.30	0.58
11 Jan 16	297.3	356.9	-4.1	0.5	252.01	258.05	1.92	1.70	272.08	279.31	-0.09	-0.69
21 Jan 26	61.3	135.9	4.8	2.9	264.11	270.19	1.47	1.23	286.84	294.63	-1.21	-1.62
31 J-F 5	207.9	270.5	-3.2	-5.1	276.28	282.39	0.97	0.71	302.71	311.10	-1.92	-2.07
10 Feb 15	329.8	30.8	-1.8	3.5	288.51	294.64	0.45	0.19	319.83	328.91	-2.05	-1.84
20 Feb 25	98.9	173.8	5.0	-0.4	300.78	306.91	-0.07	-0.31	338.28	347.73	-1.40	-0.70
2 Mar 7	242.5	303.0	-5.1	-3.8	313.06	319.20	-0.54	-0.75	356.76	4.49	0.25	1.35
12 Mar 17	2.9	67.1	1.2	5.2	325.35	331.50	-0.94	-1.10	9.89	12.14	2.42	3.21
22 Mar 27	137.8	210.5	2.8	-3.5	337.65	343.80	-1.24	-1.36	11.12	7.71	3.46	3.01
1 Apr 6	275.4	334.9	-5.0	-1.3	349.95	356.10	-1.44	-1.50	3.69	0.85	1.94	0.60
11 Apr 16	37.3	105.4	3.9	4.6	2.25	8.39	-1.53	-1.53	0.08	1.43	-0.66	-1.69
21 Apr 26	176.5	245.5	-0.7	-5.0	14.54	20.68	-1.50	-1.44	4.58	9.18	-2.41	-2.83
1 May 6	307.3	7.4	-3.3	1.7	26.82	32.96	-1.36	-1.26	14.96	21.74	-2.98	-2.86
11 May 16	73.8	144.6	5.0	1.9	39.10	45.23	-1.13	-0.99	29.44	38.02	-2.51	-1.93
21 May 26	214.1	279.1	-3.7	-4.6	51.37	57.50	-0.82	-0.65	47.48	57.74	-1.18	-0.32
31 M-J 5	338.9	41.2	-0.6	4.2	63.64	69.77	-0.46	-0.27	68.58	79.55	0.56	1.30
10 Jun 15	112.1	183.5	4.0	-1.7	75.91	82.05	-0.07	0.13	90.10	99.89	1.80	1.99
20 Jun 25	250.0	311.5	-5.0	-2.6	88.19	94.33	0.32	0.51	108.75	116.66	1.88	1.50
30 Jun	11.0		2.4		100.48		0.69		123.57		0.88	
Jul 5		77.1		5.1		106.63		0.86		129.42		0.05
10 Jul 15	151.1	221.0	0.9	-4.4	112.78	118.93	1.01	1.14	134.05	137.22	-0.95	-2.06
20 Jul 25	284.2	343.7	-4.3	0.3	125.09	131.25	1.26	1.35	138.63	138.02	-3.19	-4.20
30 J-A 4	44.6	114.7	4.7	3.8	137.41	143.57	1.41	1.45	135.42	131.61	-4.82	-4.79
9 Aug 14	189.6	256.6	-2.6	-5.2	149.74	155.90	1.46	1.44	128.17	126.75	-4.02	-2.73
19 Aug 24	317.1	16.4	-2.1	3.1	162.06	168.23	1.40	1.32	128.32	132.98	-1.26	0.06
29 Aug	80.3		5.2		174.39		1.22		140.19		1.05	
Sep 3		153.3		0.7		180.55		1.10		148.99		1.62
8 Sep 13	226.5	290.2	-4.8	-4.1	186.71	192.86	0.95	0.78	158.44	167.90	1.81	1.69
18 Sep 23	349.5	50.7	0.8	5.0	199.02	205.16	0.59	0.38	177.07	185.87	1.37	0.91
28 S-O 3	117.8	191.6	3.5	-2.7	211.31	217.45	0.15	-0.08	194.27	202.33	0.37	-0.21
8 Oct 13	261.5	322.5	-5.1	-1.6	223.58	229.71	-0.32	-0.56	210.07	217.51	-0.79	-1.36
18 Oct 23	22.4	86.9	3.5	4.9	235.83	241.95	-0.80	-1.04	224.63	231.39	-1.88	-2.32
28 Oct	156.5		0.2		248.06		-1.26		237.64		-2.65	
Nov 2		228.8		-4.8		254.16		-1.47		243.09		-2.79
7 Nov 12	294.7	354.2	-3.5	1.4	260.25	266.32	-1.66	-1.84	247.14	248.74	-2.65	-2.07
17 Nov 22	56.7	125.0	4.9	2.6	272.38	278.43	-1.98	-2.09	246.56	240.61	-0.88	0.78
27 N-D 2	195.4	264.3	-3.2	-4.7	284.45	290.46	-2.17	-2.22	234.59	232.53	2.16	2.69
7 Dec 12	326.5	26.3	-0.7	4.0	296.44	302.38	-2.22	-2.18	234.63	239.32	2.56	2.08
17 Dec 22	92.6	164.0	4.4	-1.0	308.29	314.15	-2.09	-1.96	245.40	252.19	1.46	0.79
27 Dec	233.3		-5.0		319.96		-1.78		259.37		0.15	
1709												1709

JULIAN 12 UT	SATURN LONG.	SATURN LAT.	JUPITER LONG.	JUPITER LAT.	MARS LONG.	MARS LAT.	SUN LONG.	GREGORIAN 12 UT	MOON LONG.		MOON LAT.		VENUS LONG.		VENUS LAT.		MERCURY LONG.		MERCURY LAT.	
1709								1710												1710
								Jan 1		298.2		-2.9		325.71		-1.55		266.79		-0.45
Dec 26	102.55	-0.40	234.69	0.93	282.77	-0.88	285.90	6 Jan 11	358.0	59.5	2.3	5.2	331.39	336.98	-1.27	-0.94	274.41	282.21	-0.99	-1.44
Jan 5	101.74	-0.38	236.36	0.94	290.49	-0.94	296.08	16 Jan 21	130.1	202.8	1.9	-4.1	342.47	347.85	-0.56	-0.14	290.23	298.47	-1.78	-2.00
Jan 15	101.00	-0.36	237.85	0.95	298.27	-1.00	306.25	26 Jan 31	269.4	330.8	-4.6	0.0	353.10	358.18	0.34	0.86	306.99	315.77	-2.08	-1.96
Jan 25	100.35	-0.34	239.13	0.97	306.09	-1.04	316.39	5 Feb 10	30.0	94.5	4.5	4.4	3.08	7.76	1.42	2.02	324.76	333.77	-1.63	-1.03
Feb 4	99.84	-0.32	240.17	0.98	313.93	-1.08	326.49	15 Feb 20	168.4	240.3	-1.6	-5.3	12.16	16.24	2.65	3.31	342.30	349.43	-0.14	0.99
Feb 14	99.50	-0.30	240.95	1.00	321.80	-1.10	336.55	25 Feb	303.7		-2.5		19.93		4.00		353.92		2.22	
								Mar 2		3.0		2.8		23.15		4.70		354.78		3.23
Feb 24	99.33	-0.27	241.43	1.02	329.66	-1.11	346.56	7 Mar 12	63.5	131.3	5.3	1.8	25.81	27.77	5.40	6.08	352.08	347.48	3.66	3.29
Mar 6	99.35	-0.25	241.61	1.04	337.52	-1.11	356.51	17 Mar 22	206.8	275.9	-4.3	-4.3	28.91	29.11	6.70	7.22	343.41	341.47	2.27	1.01
Mar 16	99.56	-0.23	241.48	1.05	345.36	-1.10	6.41	27 M-A 1	336.4	35.9	0.5	4.7	28.30	26.49	7.57	7.66	341.96	344.49	-0.19	-1.18
Mar 26	99.94	-0.21	241.05	1.07	353.16	-1.07	16.25	6 Apr 11	98.9	169.6	4.0	-1.7	23.86	20.77	7.44	6.87	348.62	353.97	-1.92	-2.41
Apr 5	100.50	-0.19	240.32	1.08	0.92	-1.04	26.03	16 Apr 21	244.1	309.4	-5.1	-1.7	17.71	15.17	5.98	4.89	0.30	7.48	-2.67	-2.69
Apr 15	101.20	-0.17	239.36	1.08	8.63	-0.99	35.77	26 Apr	8.7		3.2		13.48		3.71		15.44		-2.48	
								May 1		70.1		4.9		12.79		2.55		24.16		-2.05
Apr 25	102.04	-0.15	238.22	1.08	16.27	-0.93	45.45	6 May 11	136.3	208.5	1.0	-4.4	13.08	14.27	1.47	0.52	33.65	43.86	-1.43	-0.64
May 5	103.00	-0.14	236.97	1.07	23.83	-0.86	55.09	16 May 21	279.6	341.6	-3.7	1.3	16.22	18.83	-0.31	-1.00	54.64	65.57	0.24	1.06
May 15	104.06	-0.12	235.70	1.05	31.31	-0.77	64.69	26 May 31	41.4	106.0	4.9	3.2	21.97	25.55	-1.58	-2.04	76.12	85.86	1.70	2.06
May 25	105.20	-0.10	234.51	1.02	38.70	-0.68	74.26	5 Jun 10	175.1	246.7	-2.6	-4.9	29.49	33.73	-2.40	-2.67	94.56	102.14	2.10	1.82
Jun 4	106.41	-0.09	233.46	0.99	45.99	-0.58	83.81	15 Jun 20	313.2	13.2	-0.9	3.9	38.20	42.89	-2.86	-2.98	108.55	113.68	1.26	0.43
Jun 14	107.67	-0.07	232.63	0.95	53.17	-0.47	93.35	25 Jun 30	75.2	143.6	4.7	-0.1	47.75	52.76	-3.03	-3.02	117.38	119.38	-0.63	-1.85
Jun 24	108.95	-0.06	232.06	0.91	60.25	-0.35	102.88	5 Jul 10	214.4	283.0	-4.9	-3.3	57.90	63.14	-2.96	-2.85	119.50	117.73	-3.12	-4.21
Jul 4	110.25	-0.04	231.78	0.87	67.22	-0.23	112.41	15 Jul 20	345.5	45.3	2.1	5.2	68.48	73.91	-2.70	-2.51	114.63	111.37	-4.86	-4.83
Jul 14	111.55	-0.03	231.81	0.83	74.06	-0.10	121.96	25 Jul 30	110.6	182.6	2.8	-3.4	79.41	84.98	-2.29	-2.05	109.37	109.64	-4.14	-2.99
Jul 24	112.82	-0.02	232.13	0.79	80.78	0.04	131.54	4 Aug 9	252.8	317.4	-4.9	-0.4	90.61	96.30	-1.79	-1.51	112.59	118.14	-1.66	-0.37
Aug 3	114.05	0.00	232.75	0.75	87.35	0.18	141.14	14 Aug 19	17.4	78.4	4.3	4.7	102.04	107.83	-1.22	-0.93	125.88	135.09	0.68	1.40
Aug 13	115.22	0.01	233.64	0.72	93.79	0.33	150.78	24 Aug 29	147.7	221.7	-0.5	-5.2	113.67	119.55	-0.64	-0.35	144.93	154.76	1.73	1.75
Aug 23	116.32	0.03	234.77	0.68	100.07	0.49	160.47	3 Sep 8	289.1	350.3	-2.9	2.5	125.47	131.43	-0.06	0.20	164.24	173.26	1.51	1.11
Sep 2	117.32	0.05	236.13	0.65	106.18	0.65	170.20	13 Sep 18	49.8	113.1	5.2	2.5	137.43	143.45	0.46	0.69	181.82	189.95	0.59	0.01
Sep 12	118.22	0.06	237.68	0.62	112.09	0.83	179.99	23 Sep 28	186.0	259.5	-3.6	-4.5	149.51	155.60	0.90	1.09	197.67	205.00	-0.60	-1.21
Sep 22	118.97	0.08	239.41	0.59	117.79	1.02	189.83	3 Oct 8	323.4	22.9	0.2	4.5	161.72	167.86	1.25	1.38	211.92	218.35	-1.80	-2.33
Oct 2	119.59	0.10	241.28	0.57	123.24	1.21	199.73	13 Oct 18	83.3	149.6	4.3	-0.8	174.02	180.21	1.47	1.54	224.13	228.92	-2.76	-3.03
Oct 12	120.03	0.11	243.28	0.54	128.40	1.43	209.69	23 Oct 28	224.7	295.2	-5.0	-2.0	186.41	192.63	1.58	1.59	232.12	232.72	-3.02	-2.58
Oct 22	120.30	0.13	245.37	0.52	133.21	1.66	219.70	2 Nov 7	356.2	55.8	3.1	4.9	198.86	205.11	1.56	1.51	229.65	223.47	-1.49	0.14
Nov 1	120.39	0.15	247.54	0.51	137.61	1.92	229.76	12 Nov 17	118.5	187.8	1.6	-3.9	211.36	217.62	1.43	1.33	218.10	217.01	1.63	2.38
Nov 11	120.28	0.17	249.76	0.49	141.49	2.19	239.86	22 Nov 27	262.3	328.7	-4.1	1.1	223.90	230.18	1.20	1.05	220.00	225.44	2.44	2.11
Nov 21	119.99	0.19	252.02	0.47	144.74	2.50	250.00	2 Dec 7	28.3	89.7	4.8	3.6	236.46	242.74	0.88	0.70	232.09	239.29	1.59	0.98
Dec 1	119.54	0.21	254.28	0.46	147.22	2.84	260.17	12 Dec 17	155.6	226.9	-1.9	-5.1	249.03	255.32	0.51	0.31	246.77	254.38	0.37	-0.22
Dec 11	118.93	0.23	256.53	0.45	148.74	3.21	270.36	22 Dec 27	298.0	0.7	-1.4	3.8	261.61	267.90	0.11	-0.09	262.10	269.92	-0.77	-1.25
1710								1711												1711
Dec 21	118.21	0.25	258.74	0.44	149.11	3.59	280.55	1 Jan 6	60.6	125.1	5.0	0.7	274.19	280.48	-0.29	-0.48	277.86	285.97	-1.64	-1.93
Dec 31	117.42	0.27	260.90	0.43	148.20	3.96	290.74	11 Jan 16	194.4	265.2	-4.6	-3.9	286.77	293.06	-0.67	-0.83	294.26	302.73	-2.08	-2.06
Jan 10	116.60	0.29	262.97	0.42	145.96	4.29	300.92	21 Jan 26	331.8	32.3	1.7	5.2	299.34	305.62	-0.99	-1.12	311.34	319.87	-1.83	-1.34
Jan 20	115.79	0.31	264.93	0.41	142.62	4.50	311.08	31 J-F 5	93.6	162.1	3.4	-2.6	311.90	318.17	-1.23	-1.33	327.85	334.32	-0.54	0.59
Jan 30	115.05	0.32	266.75	0.40	138.70	4.55	321.20	10 Feb 15	233.8	301.7	-5.2	-1.0	324.43	330.70	-1.39	-1.43	337.87	337.34	1.92	3.13
Feb 9	114.42	0.34	268.41	0.40	134.90	4.42	331.28	20 Feb 25	4.4	64.0	4.1	4.9	336.95	343.20	-1.45	-1.43	333.18	327.93	3.73	3.44
Feb 19	113.93	0.35	269.88	0.39	131.92	4.15	341.32	2 Mar 7	127.9	200.6	0.4	-4.8	349.44	355.67	-1.39	-1.33	324.37	323.55	2.49	1.33
Mar 1	113.59	0.36	271.14	0.38	130.14	3.79	351.30	12 Mar 17	272.2	336.4	-3.3	2.1	1.89	8.10	-1.24	-1.12	325.19	328.71	0.21	-0.73
Mar 11	113.44	0.37	272.14	0.38	129.64	3.40	1.23	22 Mar 27	36.7	96.8	5.1	2.9	14.30	20.49	-0.98	-0.82	333.58	339.48	-1.48	-2.03
Mar 21	113.47	0.38	272.88	0.37	130.36	3.02	11.10	1 Apr 6	164.2	239.6	-2.8	-4.8	26.67	32.84	-0.65	-0.46	346.21	353.64	-2.37	-2.50
Mar 31	113.68	0.39	273.33	0.36	132.11	2.66	20.91	11 Apr 16	308.5	9.8	-0.2	4.3	38.99	45.13	-0.25	-0.04	1.75	10.52	-2.42	-2.13
Apr 10	114.07	0.40	273.46	0.35	134.70	2.34	30.67	21 Apr 26	69.2	131.2	4.3	-0.2	51.26	57.38	0.18	0.40	19.97	30.07	-1.63	-0.93
Apr 20	114.62	0.41	273.28	0.34	137.98	2.04	40.38	1 May 6	202.4	277.5	-4.8	-2.5	63.48	69.56	0.62	0.83	40.71	51.53	-0.09	0.78
Apr 30	115.32	0.42	272.80	0.33	141.82	1.78	50.04	11 May 16	342.8	42.5	2.9	5.0	75.63	81.68	1.04	1.23	62.00	71.60	1.55	2.07
May 10	116.16	0.43	272.04	0.31	146.12	1.53	59.66	21 May 26	102.7	167.7	2.0	-3.4	87.72	93.74	1.40	1.55	80.01	87.10	2.28	2.15
May 20	117.11	0.43	271.04	0.30	150.78	1.31	69.25	31 M-J 5	241.4	313.3	-4.6	0.8	99.74	105.72	1.68	1.78	92.79	96.94	1.67	0.86
May 30	118.17	0.44	269.87	0.28	155.76	1.11	78.80	10 Jun 15	15.6	75.2	4.8	3.9	111.68	117.61	1.85	1.89	99.38	99.96	-0.24	-1.56
Jun 9	119.30	0.45	268.61	0.26	161.00	0.92	88.35	20 Jun 25	137.7	206.3	-1.2	-5.2	123.52	129.41	1.90	1.86	98.73	96.15	-2.92	-4.06
Jun 19	120.51	0.47	267.34	0.23	166.48	0.74	97.88	30 Jun	279.7		-2.1		135.26		1.79		93.21		-4.69	
								Jul 5		347.3		3.6		141.08		1.68		91.14		-4.69
Jun 29	121.75	0.48	266.16	0.21	172.16	0.58	107.41	10 Jul 15	47.7	108.7	5.1	1.4	146.86	152.61	1.53	1.34	90.87	92.81	-4.12	-3.15
Jul 9	123.03	0.49	265.14	0.18	178.02	0.42	116.96	20 Jul 25	174.8	245.7	-4.1	-4.5	158.31	163.96	1.11	0.84	97.02	103.37	-1.98	-0.78
Jul 19	124.32	0.50	264.35	0.15	184.06	0.28	126.52	30 J-A 4	316.1	19.9	1.1	5.1	169.56	175.09	0.53	0.18	111.58	121.13	0.30	1.13
Jul 29	125.61	0.52	263.84	0.13	190.25	0.14	136.10	9 Aug 14	79.7	143.6	3.7	-1.8	180.56	185.95	-0.20	-0.61	131.31	141.48	1.62	1.77
Aug 8	126.87	0.54	263.63	0.10	196.60	0.00	145.72	19 Aug 24	213.8	284.3	-5.2	-1.7	191.25	196.44	-1.05	-1.51	151.25	160.50	1.64	1.30
Aug 18	128.09	0.55	263.73	0.08	203.08	-0.12	155.38	29 Aug	350.7		3.8		201.52		-2.00		169.20		0.82	
								Sep 3		51.7		4.9		206.45		-2.50		177.39		0.25
Aug 28	129.26	0.57	264.15	0.06	209.71	-0.24	165.09	8 Sep 13	112.3	180.5	1.0	-4.3	211.22	215.79	-3.01	-3.52	185.09	192.30	-0.39	-1.05
Sep 7	130.35	0.59	264.86	0.04	216.47	-0.35	174.85	18 Sep 23	253.2	321.1	-3.9	1.6	220.13	224.18	-4.03	-4.53	198.99	205.07	-1.70	-2.32
Sep 17	131.34	0.62	265.85	0.02	223.35	-0.46	184.66	28 S-O 3	24.0	83.6	5.0	3.1	227.89	231.17	-5.00	-5.44	210.33	214.42	-2.86	-3.26
Sep 27	132.22	0.64	267.09	0.00	230.36	-0.56	194.53	8 Oct 13	146.3	219.0	-2.2	-5.0	233.95	236.10	-5.81	-6.11	216.70	216.26	-3.39	-3.08
Oct 7	132.96	0.66	268.55	-0.02	237.49	-0.65	204.46	18 Oct 23	291.5	356.1	-0.7	4.2	237.50	238.02	-6.28	-6.30	212.40	206.28	-2.10	-0.50
Oct 17	133.55	0.69	270.21	-0.04	244.74	-0.74	214.44	28 Oct	56.6		4.4		237.55		-6.11		201.77		1.08	
								Nov 2		116.2		0.1		236.09		-5.66		201.67		2.01
Oct 27	133.98	0.72	272.04	-0.05	252.09	-0.82	224.47	7 Nov 12	182.3	257.8	-4.6	-3.2	233.76	230.87	-4.91	-3.89	205.51	211.64	2.28	2.11
Nov 6	134.23	0.75	274.02	-0.07	259.55	-0.89	234.56	17 Nov 22	327.9	29.7	2.6	5.1	227.88	225.29	-2.67	-1.38	218.84	226.46	1.69	1.16
Nov 16	134.30	0.78	276.12	-0.08	267.11	-0.95	244.68	27 N-D 2	89.1	150.4	2.3	-3.0	223.48	222.67	-0.14	0.97	234.23	242.04	0.58	-0.01
Nov 26	134.17	0.81	278.31	-0.10	274.74	-1.00	254.83	7 Dec 12	220.2	295.5	-5.0	0.1	222.89	224.06	1.89	2.62	249.85	257.70	-0.56	-1.07
Dec 6	133.87	0.84	280.57	-0.12	282.46	-1.04	265.01	17 Dec 22	2.2	62.2	4.8	4.1	226.05	228.73	3.18	3.58	265.59	273.56	-1.50	-1.84
Dec 16	133.40	0.87	282.88	-0.13	290.24	-1.07	275.20	27 Dec	122.3		-0.7		231.99		3.84		281.63		-2.07	
1711								1711												1711

JULIAN 12 UT	SATURN LONG.	SATURN LAT.	JUPITER LONG.	JUPITER LAT.	MARS LONG.	MARS LAT.	SUN LONG.
1711							
Dec 26	132.78	0.89	285.21	-0.15	298.07	-1.09	285.40
Jan 5	132.06	0.92	287.55	-0.16	305.94	-1.09	295.59
Jan 15	131.26	0.94	289.86	-0.18	313.83	-1.09	305.75
Jan 25	130.44	0.95	292.13	-0.20	321.74	-1.07	315.90
Feb 4	129.65	0.97	294.34	-0.22	329.65	-1.04	326.01
Feb 14	128.93	0.98	296.45	-0.24	337.54	-1.01	336.06
Feb 24	128.31	0.98	298.46	-0.26	345.41	-0.96	346.08
Mar 5	127.83	0.99	300.32	-0.28	353.23	-0.90	356.04
Mar 15	127.51	0.99	302.02	-0.31	1.01	-0.83	5.93
Mar 25	127.37	0.98	303.52	-0.33	8.72	-0.75	15.78
Apr 4	127.41	0.98	304.81	-0.36	16.37	-0.66	25.57
Apr 14	127.62	0.98	305.85	-0.39	23.94	-0.57	35.30
Apr 24	128.02	0.98	306.62	-0.42	31.42	-0.48	44.99
May 4	128.57	0.97	307.10	-0.46	38.83	-0.37	54.63
May 14	129.27	0.97	307.27	-0.49	46.14	-0.27	64.23
May 24	130.10	0.97	307.12	-0.53	53.36	-0.16	73.80
Jun 3	131.05	0.97	306.65	-0.57	60.49	-0.05	83.35
Jun 13	132.10	0.97	305.90	-0.60	67.52	0.06	92.89
Jun 23	133.22	0.98	304.91	-0.63	74.45	0.18	102.42
Jul 3	134.42	0.98	303.73	-0.66	81.30	0.29	111.96
Jul 13	135.65	0.99	302.45	-0.69	88.04	0.41	121.50
Jul 23	136.92	1.00	301.17	-0.71	94.70	0.53	131.08
Aug 2	138.20	1.01	299.96	-0.72	101.25	0.64	140.68
Aug 12	139.48	1.03	298.93	-0.73	107.70	0.76	150.31
Aug 22	140.73	1.05	298.13	-0.74	114.06	0.88	160.00
Sep 1	141.94	1.07	297.62	-0.74	120.30	1.00	169.73
Sep 11	143.09	1.09	297.43	-0.74	126.43	1.12	179.51
Sep 21	144.16	1.11	297.56	-0.73	132.43	1.25	189.35
Oct 1	145.14	1.14	298.02	-0.73	138.30	1.38	199.25
Oct 11	145.99	1.17	298.79	-0.72	144.01	1.52	209.20
Oct 21	146.72	1.20	299.84	-0.72	149.55	1.66	219.21
Oct 31	147.29	1.24	301.15	-0.71	154.87	1.80	229.26
Nov 10	147.69	1.28	302.69	-0.71	159.96	1.96	239.36
Nov 20	147.91	1.31	304.44	-0.71	164.75	2.13	249.50
Nov 30	147.95	1.35	306.35	-0.71	169.19	2.31	259.67
Dec 10	147.81	1.39	308.41	-0.71	173.19	2.50	269.85
Dec 20	147.48	1.43	310.58	-0.71	176.65	2.71	280.05
1712							
Dec 30	146.99	1.46	312.85	-0.71	179.45	2.93	290.24
Jan 9	146.36	1.49	315.18	-0.72	181.42	3.17	300.42
Jan 19	145.63	1.51	317.56	-0.73	182.39	3.40	310.58
Jan 29	144.84	1.53	319.96	-0.74	182.18	3.62	320.71
Feb 8	144.03	1.55	322.36	-0.76	180.68	3.79	330.79
Feb 18	143.25	1.56	324.73	-0.77	177.95	3.87	340.83
Feb 28	142.54	1.56	327.06	-0.79	174.34	3.80	350.82
Mar 10	141.94	1.56	329.32	-0.81	170.46	3.59	0.75
Mar 20	141.48	1.55	331.49	-0.84	167.01	3.25	10.63
Mar 30	141.18	1.54	333.56	-0.87	164.58	2.82	20.44
Apr 9	141.05	1.53	335.48	-0.90	163.42	2.38	30.20
Apr 19	141.10	1.51	337.25	-0.93	163.54	1.95	39.92
Apr 29	141.33	1.50	338.83	-0.97	164.81	1.56	49.58
May 9	141.73	1.49	340.20	-1.01	167.07	1.21	59.20
May 19	142.28	1.47	341.33	-1.05	170.14	0.90	68.79
May 29	142.98	1.46	342.19	-1.10	173.87	0.62	78.35
Jun 8	143.81	1.45	342.76	-1.15	178.15	0.37	87.89
Jun 18	144.75	1.44	343.02	-1.20	182.89	0.15	97.42
Jun 28	145.79	1.44	342.95	-1.25	188.00	-0.05	106.95
Jul 8	146.91	1.44	342.57	-1.30	193.45	-0.23	116.49
Jul 18	148.09	1.44	341.88	-1.34	199.19	-0.39	126.06
Jul 28	149.32	1.44	340.92	-1.38	205.18	-0.54	135.64
Aug 7	150.58	1.45	339.76	-1.41	211.39	-0.67	145.25
Aug 17	151.85	1.46	338.47	-1.43	217.82	-0.79	154.91
Aug 27	153.11	1.47	337.15	-1.44	224.44	-0.89	164.62
Sep 6	154.34	1.49	335.90	-1.44	231.23	-0.98	174.37
Sep 16	155.54	1.51	334.81	-1.43	238.19	-1.06	184.19
Sep 26	156.67	1.54	333.96	-1.41	245.30	-1.13	194.05
Oct 6	157.72	1.57	333.39	-1.39	252.55	-1.18	203.97
Oct 16	158.67	1.60	333.16	-1.35	259.92	-1.22	213.96
Oct 26	159.50	1.63	333.27	-1.32	267.41	-1.25	223.98
Nov 5	160.20	1.67	333.71	-1.29	275.00	-1.26	234.06
Nov 15	160.74	1.71	334.48	-1.25	282.68	-1.26	244.19
Nov 25	161.11	1.75	335.54	-1.22	290.43	-1.25	254.34
Dec 5	161.30	1.80	336.86	-1.19	298.24	-1.22	264.52
Dec 15	161.31	1.84	338.43	-1.16	306.09	-1.19	274.71
1713							

GREGORIAN 12 UT	MOON LONG.		MOON LAT.		VENUS LONG.		VENUS LAT.		MERCURY LONG.		MERCURY LAT.	
1712												
Jan 1		186.8		-5.0		235.72		3.99		289.80		-2.14
6 Jan 11	258.9	331.4	-3.1	3.2	239.84	244.27	4.04	4.00	298.00	306.04	-2.02	-1.64
16 Jan 21	35.0	94.5	5.2	1.8	248.96	253.87	3.89	3.72	313.43	319.17	-0.94	0.14
26 Jan 31	157.1	225.2	-3.6	-4.9	258.95	264.17	3.49	3.23	321.75	319.91	1.53	2.91
5 Feb 10	297.1	5.7	0.3	5.0	269.53	274.98	2.93	2.60	314.61	309.24	3.67	3.48
15 Feb 20	66.7	127.2	3.8	-1.2	280.52	286.14	2.26	1.90	306.52	306.83	2.63	1.57
25 Feb	194.0		-5.1		291.81		1.54		309.50		0.54	
Mar 1		264.5		-2.5		297.55		1.18		313.81		-0.36
6 Mar 11	334.0	38.6	3.3	4.9	303.32	309.14	0.82	0.47	319.27	325.58	-1.10	-1.68
16 Mar 21	98.3	161.5	1.2	-3.8	314.99	320.86	0.14	-0.18	332.58	340.17	-2.08	-2.30
26 Mar 31	232.5	303.2	-4.3	1.1	326.76	332.69	-0.47	-0.74	348.34	357.09	-2.33	-2.16
5 Apr 10	9.4	70.7	4.9	3.1	338.63	344.59	-0.97	-1.18	6.44	16.38	-1.78	-1.19
15 Apr 20	130.5	197.7	-1.8	-5.0	350.56	356.54	-1.36	-1.50	26.82	37.46	-0.42	0.47
25 Apr 30	271.6	340.4	-1.4	4.0	2.53	8.54	-1.61	-1.68	47.76	57.13	1.33	2.02
5 May 10	43.3	102.7	4.5	0.4	14.55	20.57	-1.72	-1.73	65.13	71.55	2.42	2.46
15 May 20	164.3	235.7	-4.3	-4.0	26.60	32.63	-1.71	-1.65	76.24	79.08	2.11	1.35
25 May 30	309.9	15.8	2.1	5.2	38.67	44.71	-1.57	-1.46	79.97	79.00	0.23	-1.15
4 Jun 9	76.1	135.5	2.6	-2.6	50.77	56.83	-1.33	-1.18	76.66	73.92	-2.56	-3.70
14 Jun 19	200.2	274.4	-5.2	-1.0	62.90	68.97	-1.01	-0.83	71.89	71.44	-4.33	-4.39
24 Jun 29	346.6	49.4	4.6	4.4	75.05	81.14	-0.63	-0.43	72.91	76.36	-3.98	-3.21
4 Jul 9	108.6	169.8	-0.3	-4.8	87.24	93.35	-0.22	-0.01	81.66	88.71	-2.23	-1.14
14 Jul 19	238.1	312.7	-3.9	2.4	99.47	105.60	0.19	0.39	97.33	107.15	-0.07	0.83
24 Jul 29	21.3	81.7	5.2	2.1	111.74	117.89	0.58	0.76	117.60	128.06	1.45	1.74
3 Aug 8	141.7	206.3	-3.1	-5.1	124.05	130.21	0.91	1.06	138.10	147.55	1.74	1.49
13 Aug 18	277.0	349.5	-0.7	4.7	136.39	142.58	1.18	1.27	156.38	164.62	1.06	0.50
23 Aug 28	54.3	113.7	4.0	-0.8	148.77	154.97	1.35	1.39	172.27	179.34	-0.15	-0.86
2 Sep 7	176.5	244.7	-4.9	-3.2	161.18	167.39	1.41	1.41	185.77	191.44	-1.59	-2.30
12 Sep 17	315.7	24.6	2.7	4.9	173.61	179.84	1.37	1.31	196.12	199.42	-2.96	-3.49
22 Sep 27	86.1	146.3	1.4	-3.5	186.06	192.30	1.23	1.12	200.70	199.22	-3.76	-3.57
2 Oct 7	213.1	283.8	-4.6	0.3	198.53	204.77	0.98	0.83	194.75	189.01	-2.68	-1.12
12 Oct 17	353.2	58.0	4.8	3.3	211.00	217.24	0.66	0.48	185.55	186.46	0.50	1.61
22 Oct 27	117.6	180.2	-1.6	-5.0	223.48	229.72	0.28	0.08	191.10	197.88	2.08	2.07
1 Nov 6	251.3	322.6	-2.3	3.6	235.95	242.19	-0.13	-0.35	205.61	213.64	1.77	1.31
11 Nov 16	29.0	90.2	4.8	0.6	248.43	254.66	-0.55	-0.76	221.72	229.72	0.77	0.21
21 Nov 26	149.7	215.8	-4.2	-4.5	260.89	267.12	-0.95	-1.13	237.66	245.53	-0.35	-0.88
1 Dec 6	290.0	0.1	1.2	5.2	273.35	279.57	-1.30	-1.44	253.37	261.21	-1.35	-1.75
11 Dec 16	62.9	122.3	2.9	-2.3	285.79	292.00	-1.56	-1.66	269.07	276.94	-2.04	-2.20
21 Dec 26	183.2	253.0	-5.3	-2.1	298.20	304.40	-1.72	-1.76	284.73	292.26	-2.18	-1.92
31 Dec	328.2		4.2		310.58		-1.77		299.00		-1.33	
1713												
Jan 5		35.4		4.6		316.75		-1.74		303.98		-0.33
10 Jan 15	95.6	155.1	0.1	-4.5	322.91	329.04	-1.67	-1.58	305.57	302.55	1.07	2.58
20 Jan 25	218.8	291.1	-4.4	1.4	335.15	341.24	-1.44	-1.28	296.46	291.38	3.49	3.42
30 J-F 4	4.9	68.8	5.2	2.3	347.31	353.34	-1.08	-0.85	289.66	291.05	2.71	1.76
9 Feb 14	128.0	189.5	-2.8	-5.1	359.33	5.28	-0.59	-0.31	294.61	299.58	0.82	-0.03
19 Feb 24	256.2	329.5	-1.6	4.2	11.19	17.05	0.00	0.33	305.51	312.13	-0.76	-1.36
1 Mar 6	39.8	100.9	4.1	-0.7	22.86	28.59	0.67	1.03	319.32	327.01	-1.81	-2.10
11 Mar 16	160.9	225.7	-4.6	-3.6	34.26	39.84	1.39	1.75	335.19	343.87	-2.23	-2.17
21 Mar 26	295.0	7.0	2.0	5.0	45.33	50.73	2.11	2.46	353.07	2.81	-1.90	-1.42
31 M-A 5	72.9	132.5	1.5	-3.4	56.00	61.13	2.79	3.10	12.99	23.37	-0.73	0.13
10 Apr 15	195.1	263.6	-4.9	-0.5	66.10	70.89	3.38	3.62	33.43	42.48	1.06	1.91
20 Apr 25	334.2	42.8	4.6	3.6	75.45	79.76	3.81	3.96	49.94	55.47	2.50	2.74
30 Apr	104.8		-1.5		83.76		4.03		58.89		2.55	
May 5		164.7		-5.0		87.39		4.03		60.11		1.88
10 May 15	231.0	302.6	-3.1	3.0	90.56	93.19	3.93	3.72	59.26	56.90	0.76	-0.65
20 May 25	12.3	76.7	5.0	0.9	95.17	96.38	3.38	2.88	54.08	51.96	-2.06	-3.17
30 M-J 4	136.5	198.2	-4.0	-4.9	96.69	96.02	2.21	1.36	51.36	52.60	-3.82	-3.99
9 Jun 14	268.5	341.7	0.1	5.1	94.36	91.86	0.34	-0.80	55.64	60.34	-3.75	-3.19
19 Jun 24	48.4	109.5	3.3	-2.0	88.85	85.78	-1.98	-3.07	66.55	74.18	-2.39	-1.44
29 Jun	168.9		-5.2		83.12		-3.98		83.12		-0.44	
Jul 4		233.4		-2.9		81.22		-4.66		93.15		0.50
9 Jul 14	307.2	19.3	3.4	4.8	80.28	80.32	-5.10	-5.33	103.80	114.49	1.23	1.68
19 Jul 24	82.5	141.9	0.4	-4.3	81.28	83.04	-5.39	-5.31	124.75	134.38	1.80	1.66
29 J-A 3	202.6	270.4	-4.6	0.4	85.47	88.46	-5.13	-4.87	143.32	151.57	1.30	0.76
8 Aug 13	345.9	54.8	5.0	2.6	91.92	95.77	-4.54	-4.18	159.14	166.01	0.10	-0.64
18 Aug 23	115.3	174.9	-2.5	-5.0	99.94	104.38	-3.78	-3.35	172.12	177.31	-1.44	-2.26
28 Aug	238.0		-2.2		109.06		-2.92		181.31		-3.04	
Sep 2		308.9		3.6		113.93		-2.47		183.71		-3.71
7 Sep 12	23.3	88.3	4.3	-0.5	118.96	124.14	-2.03	-1.59	183.93	181.46	-4.12	-4.02
17 Sep 22	147.7	209.2	-4.5	-4.0	129.46	134.88	-1.16	-0.74	176.66	171.65	-3.20	-1.70
27 S-O 2	275.4	347.8	1.2	5.0	140.41	146.02	-0.35	0.02	169.42	171.35	-0.06	1.17
7 Oct 12	58.6	120.4	1.9	-3.3	151.71	157.48	0.37	0.68	176.73	184.12	1.83	2.00
17 Oct 22	180.4	245.1	-5.0	-1.2	163.31	169.20	0.96	1.20	192.36	200.81	1.83	1.45
27 Oct	314.3		4.2		175.14		1.41		209.19		0.96	
Nov 1		25.7		4.1		181.12		1.58		217.41		0.41
6 Nov 11	91.9	151.9	-1.2	-5.0	187.15	193.21	1.71	1.81	225.47	233.38	-0.15	-0.69
16 Nov 21	214.0	282.7	-3.6	2.2	199.30	205.42	1.86	1.88	241.19	248.91	-1.20	-1.64
26 N-D 1	353.6	61.7	5.2	1.5	211.57	217.74	1.86	1.80	256.57	264.13	-2.00	-2.24
6 Dec 11	124.1	183.8	-3.8	-5.1	223.92	230.12	1.72	1.60	271.52	278.50	-2.32	-2.18
16 Dec 21	249.1	321.6	-0.8	4.7	236.33	242.56	1.46	1.30	284.58	288.72	-1.71	-0.82
26 Dec 31	31.8	95.9	3.8	-1.6	248.79	255.02	1.12	0.92	289.33	285.29	0.55	2.14
1713												

JULIAN 12 UT	SATURN LONG.	SATURN LAT.	JUPITER LONG.	JUPITER LAT.	MARS LONG.	MARS LAT.	SUN LONG.	GREGORIAN 12 UT	MOON LONG.	MOON LONG.	MOON LAT.	MOON LAT.	VENUS LONG.	VENUS LONG.	VENUS LAT.	VENUS LAT.	MERCURY LONG.	MERCURY LONG.	MERCURY LAT.	MERCURY LAT.
1713								1714												1714
Dec 25	161.13	1.88	340.19	-1.14	313.96	-1.14	284.90	5 Jan 10	156.0	216.7	-5.1	-3.4	261.26	267.51	0.71	0.49	278.78	274.27	3.20	3.29
Jan 4	160.78	1.92	342.14	-1.12	321.85	-1.08	295.09	15 Jan 20	285.8	0.5	2.5	5.0	273.76	280.01	0.27	0.05	273.59	275.98	2.72	1.91
Jan 14	160.27	1.96	344.22	-1.11	329.73	-1.01	305.26	25 Jan 30	67.8	128.9	0.8	-4.0	286.26	292.51	-0.16	-0.37	280.31	285.85	1.05	0.25
Jan 24	159.64	1.99	346.42	-1.09	337.59	-0.93	315.40	4 Feb 9	188.4	251.2	-4.7	-0.4	298.76	305.01	-0.57	-0.75	292.16	299.04	-0.46	-1.07
Feb 3	158.90	2.02	348.72	-1.08	345.42	-0.84	325.51	14 Feb 19	324.0	38.0	4.7	3.0	311.25	317.50	-0.92	-1.06	306.37	314.11	-1.55	-1.91
Feb 13	158.11	2.04	351.08	-1.08	353.21	-0.75	335.57	24 Feb	101.9		-2.3		323.74		-1.19		322.26		-2.11	
								Mar 1		161.5		-5.0		329.97		-1.29		330.84		-2.14
Feb 23	157.32	2.05	353.48	-1.08	0.94	-0.65	345.59	6 Mar 11	221.9	287.8	-2.6	2.9	336.20	342.43	-1.37	-1.42	339.86	349.35	-1.99	-1.62
Mar 5	156.55	2.05	355.90	-1.08	8.61	-0.55	355.55	16 Mar 21	2.7	73.4	4.7	-0.2	348.65	354.87	-1.44	-1.44	359.24	9.30	-1.02	-0.21
Mar 15	155.86	2.04	358.32	-1.08	16.20	-0.44	5.45	26 Mar 31	134.7	194.4	-4.5	-4.2	1.08	7.29	-1.41	-1.35	19.04	27.68	0.75	1.72
Mar 25	155.28	2.03	0.72	-1.09	23.72	-0.34	15.30	5 Apr 10	257.2	326.4	0.7	5.0	13.48	19.68	-1.27	-1.17	34.48	38.93	2.51	2.97
Apr 4	154.84	2.02	3.08	-1.10	31.16	-0.23	25.09	15 Apr 20	40.4	106.9	2.5	-3.2	25.86	32.04	-1.04	-0.90	40.82	40.22	2.96	2.39
Apr 14	154.56	2.00	5.37	-1.11	38.51	-0.12	34.83	25 Apr 30	166.9	228.2	-5.1	-1.8	38.22	44.38	-0.74	-0.56	37.74	34.56	1.29	-0.10
Apr 24	154.45	1.97	7.59	-1.13	45.78	-0.01	44.51	5 May 10	294.5	5.7	3.8	4.7	50.54	56.70	-0.37	-0.18	32.04	31.11	-1.48	-2.56
May 4	154.52	1.95	9.70	-1.16	52.96	0.09	54.16	15 May 20	76.2	139.2	-0.7	-5.0	62.85	68.99	0.03	0.23	32.07	34.82	-3.25	-3.53
May 14	154.75	1.93	11.68	-1.18	60.06	0.20	63.76	25 May 30	199.1	263.6	-4.0	1.4	75.13	81.26	0.43	0.62	39.14	44.82	-3.47	-3.10
May 24	155.16	1.90	13.51	-1.21	67.07	0.30	73.33	4 Jun 9	333.5	44.1	5.3	2.2	87.38	93.51	0.81	0.98	51.73	59.80	-2.49	-1.70
Jun 3	155.72	1.88	15.17	-1.24	74.00	0.40	82.89	14 Jun 19	110.3	170.9	-3.5	-5.1	99.62	105.73	1.14	1.28	68.99	79.15	-0.78	0.16
Jun 13	156.42	1.86	16.62	-1.28	80.85	0.49	92.42	24 Jun 29	232.1	300.9	-1.5	4.1	111.84	117.94	1.40	1.49	89.93	100.78	0.98	1.56
Jun 23	157.25	1.84	17.84	-1.32	87.62	0.59	101.95	4 Jul 9	13.0	80.7	4.3	-1.1	124.03	130.11	1.56	1.60	111.23	120.99	1.83	1.81
Jul 3	158.19	1.83	18.81	-1.36	94.32	0.68	111.49	14 Jul 19	143.2	202.9	-5.0	-3.6	136.19	142.26	1.61	1.58	129.97	138.18	1.53	1.05
Jul 13	159.22	1.82	19.48	-1.41	100.95	0.77	121.04	24 Jul 29	266.8	339.8	1.7	5.1	148.32	154.37	1.53	1.45	145.61	152.21	0.39	-0.39
Jul 23	160.33	1.81	19.85	-1.45	107.50	0.85	130.61	3 Aug 8	51.2	115.4	1.3	-3.8	160.41	166.44	1.33	1.18	157.90	162.48	-1.26	-2.19
Aug 2	161.51	1.81	19.89	-1.50	114.00	0.93	140.21	13 Aug 18	175.6	235.7	-4.7	-0.8	172.45	178.45	1.01	0.81	165.68	167.05	-3.11	-3.92
Aug 12	162.73	1.81	19.60	-1.54	120.42	1.01	149.84	23 Aug 28	303.4	18.9	4.3	3.5	184.43	190.40	0.58	0.32	166.15	162.87	-4.44	-4.42
Aug 22	163.97	1.81	18.99	-1.58	126.78	1.09	159.52	2 Sep 7	87.4	148.8	-2.0	-5.0	196.34	202.27	0.05	-0.24	158.15	154.23	-3.63	-2.20
Sep 1	165.23	1.82	18.10	-1.61	133.08	1.17	169.25	12 Sep 17	208.2	270.2	-2.8	2.4	208.18	214.06	-0.54	-0.85	153.36	156.29	-0.60	0.71
Sep 11	166.47	1.83	16.97	-1.63	139.32	1.24	179.03	22 Sep 27	341.8	56.6	5.0	0.4	219.91	225.74	-1.16	-1.47	162.37	170.33	1.54	1.88
Sep 21	167.69	1.85	15.68	-1.64	145.48	1.31	188.87	2 Oct 7	121.7	181.5	-4.5	-4.4	231.53	237.28	-1.77	-2.07	179.05	187.91	1.85	1.57
Oct 1	168.87	1.87	14.33	-1.64	151.57	1.39	198.77	12 Oct 17	241.6	306.8	0.2	4.8	243.00	248.66	-2.34	-2.60	196.60	205.05	1.14	0.62
Oct 11	169.98	1.90	13.03	-1.62	157.58	1.46	208.72	22 Oct 27	20.5	92.2	3.3	-2.8	254.26	259.80	-2.82	-3.01	213.25	221.22	0.06	-0.50
Oct 21	171.00	1.93	11.87	-1.59	163.51	1.53	218.72	1 Nov 6	154.5	214.1	-5.2	-2.3	265.25	270.62	-3.16	-3.25	229.00	236.61	-1.03	-1.53
Oct 31	171.92	1.96	10.94	-1.55	169.34	1.59	228.78	11 Nov 16	276.8	345.3	3.2	5.1	275.88	281.00	-3.30	-3.28	244.07	251.33	-1.95	-2.27
Nov 10	172.72	2.00	10.30	-1.51	175.06	1.66	238.88	21 Nov 26	58.3	126.0	0.1	-4.9	285.96	290.72	-3.19	-3.02	258.30	264.73	-2.45	-2.42
Nov 20	173.38	2.04	9.99	-1.46	180.64	1.73	249.01	1 Dec 6	186.4	247.3	-4.2	0.7	295.26	299.51	-2.77	-2.42	270.09	273.37	-2.09	-1.33
Nov 30	173.88	2.08	10.02	-1.41	186.07	1.80	259.18	11 Dec 16	314.0	24.5	5.1	3.0	303.41	306.88	-1.96	-1.38	272.98	268.11	-0.02	1.62
Dec 10	174.22	2.13	10.41	-1.36	191.31	1.87	269.36	21 Dec 26	94.5	158.4	-3.0	-5.1	309.82	312.12	-0.68	0.17	261.50	257.75	2.83	3.10
Dec 20	174.37	2.18	11.11	-1.31	196.34	1.94	279.55	31 Dec	218.0		-1.8		313.65		1.16		258.10		2.70	
1714								1715												1715
								Jan 5		282.1		3.5		314.29		2.28		261.39		2.01
Dec 30	174.34	2.22	12.13	-1.26	201.09	2.01	289.75	10 Jan 15	352.8	62.9	4.6	-0.4	313.91	312.49	3.51	4.77	266.42	272.46	1.25	0.50
Jan 9	174.13	2.27	13.41	-1.22	205.50	2.07	299.93	20 Jan 25	129.1	190.2	-4.8	-3.7	310.15	307.21	5.96	6.95	279.12	286.22	-0.19	-0.80
Jan 19	173.76	2.31	14.93	-1.18	209.51	2.13	310.08	30 J-F 4	250.4	318.8	1.2	5.0	304.15	301.49	7.62	7.92	293.66	301.43	-1.31	-1.71
Jan 29	173.23	2.35	16.66	-1.14	212.99	2.19	320.21	9 Feb 14	32.0	99.7	2.0	-3.5	299.62	298.73	7.87	7.54	309.53	317.98	-1.98	-2.10
Feb 8	172.58	2.38	18.56	-1.11	215.84	2.23	330.30	19 Feb 24	162.5	222.0	-4.7	-1.0	298.85	299.93	7.03	6.40	326.81	336.03	-2.04	-1.78
Feb 18	171.84	2.40	20.61	-1.08	217.89	2.25	340.34	1 Mar 6	284.4	357.0	3.9	4.2	301.82	304.40	5.71	4.99	345.58	355.27	-1.29	-0.56
Feb 28	171.06	2.42	22.78	-1.06	218.96	2.24	350.33	11 Mar 16	70.0	134.8	-1.6	-5.0	307.55	311.15	4.27	3.57	4.61	12.77	0.40	1.46
Mar 10	170.27	2.42	25.03	-1.04	218.89	2.17	0.26	21 Mar 26	195.1	254.7	-3.0	2.1	315.13	319.42	2.89	2.24	18.81	22.02	2.44	3.13
Mar 20	169.52	2.42	27.36	-1.02	217.55	2.04	10.14	31 M-A 5	320.6	35.7	5.2	1.3	323.97	328.73	1.63	1.06	22.20	19.81	3.31	2.84
Mar 30	168.86	2.41	29.73	-1.00	214.99	1.81	19.96	10 Apr 15	106.3	168.4	-4.3	-4.6	333.66	338.73	0.54	0.06	16.15	12.92	1.78	0.42
Apr 9	168.30	2.39	32.11	-0.99	211.56	1.48	29.73	20 Apr 25	227.6	289.2	-0.2	4.5	343.93	349.23	-0.38	-0.76	11.33	11.78	-0.90	-1.96
Apr 19	167.89	2.36	34.51	-0.99	207.86	1.07	39.44	30 Apr	358.6		4.1		354.62		-1.10		14.14		-2.68	
								May 5		73.5		-2.1		0.09		-1.39		18.11		-3.07
Apr 29	167.63	2.33	36.89	-0.98	204.60	0.62	49.11	10 May 15	140.8	200.8	-5.3	-2.6	5.62	11.21	-1.63	-1.82	23.39	29.80	-3.16	-2.97
May 9	167.54	2.30	39.23	-0.98	202.39	0.18	58.73	20 May 25	261.0	325.9	2.7	5.2	16.84	22.52	-1.96	-2.06	37.22	45.62	-2.53	-1.89
May 19	167.63	2.27	41.51	-0.98	201.47	-0.22	68.32	30 M-J 4	37.5	109.8	1.1	-4.5	28.24	34.00	-2.12	-2.13	54.97	65.18	-1.08	-0.18
May 29	167.88	2.24	43.73	-0.99	201.89	-0.57	77.89	9 Jun 14	173.6	233.0	-4.4	0.3	39.79	45.60	-2.11	-2.05	76.01	86.96	0.69	1.39
Jun 8	168.30	2.21	45.84	-0.99	203.52	-0.85	87.43	19 Jun 24	296.0	4.4	4.7	3.7	51.44	57.31	-1.95	-1.83	97.52	107.35	1.81	1.93
Jun 18	168.86	2.18	47.84	-1.00	206.17	-1.09	96.96	29 Jun	76.0		-2.3		63.20		-1.67		116.32		1.77	
								Jul 4		144.4		-5.1		69.12		-1.50		124.41		1.35
Jun 28	169.57	2.15	49.70	-1.02	209.68	-1.29	106.50	9 Jul 14	205.4	266.0	-2.0	3.1	75.05	81.01	-1.30	-1.09	131.59	137.81	0.72	-0.09
Jul 8	170.40	2.13	51.39	-1.03	213.90	-1.44	116.04	19 Jul 24	333.1	43.7	4.9	0.3	86.99	92.99	-0.87	-0.64	142.93	146.74	-1.04	-2.08
Jul 18	171.34	2.11	52.88	-1.05	218.70	-1.57	125.59	29 J-A 3	113.2	177.5	-4.6	-3.8	99.01	105.05	-0.41	-0.17	148.94	149.16	-3.14	-4.08
Jul 28	172.37	2.09	54.15	-1.07	223.99	-1.67	135.18	8 Aug 13	237.0	300.6	1.0	4.9	111.11	117.19	0.06	0.28	147.20	143.45	-4.70	-4.71
Aug 7	173.48	2.08	55.17	-1.09	229.69	-1.74	144.79	18 Aug 23	11.8	82.4	2.7	-3.1	123.28	129.40	0.49	0.68	139.33	136.82	-3.96	-2.62
Aug 17	174.65	2.07	55.92	-1.11	235.74	-1.79	154.45	28 Aug	148.6		-4.9		135.54		0.86		137.38		-1.10	
								Sep 2		209.6		-1.2		141.69		1.02		141.31		0.25
Aug 27	175.86	2.07	56.35	-1.13	242.10	-1.83	164.15	7 Sep 12	269.4	337.0	3.7	4.7	147.86	154.04	1.16	1.27	148.03	156.49	1.21	1.72
Sep 6	177.09	2.07	56.46	-1.16	248.71	-1.84	173.90	17 Sep 22	50.8	119.6	-0.8	-5.0	160.24	166.45	1.35	1.41	165.67	174.93	1.83	1.66
Sep 16	178.34	2.08	56.23	-1.18	255.55	-1.83	183.71	27 S-O 2	182.3	241.5	-3.3	1.8	172.68	178.91	1.44	1.45	183.94	192.62	1.30	0.82
Sep 26	179.56	2.09	55.67	-1.19	262.59	-1.81	193.57	7 Oct 12	303.3	14.8	5.2	2.3	185.16	191.41	1.42	1.37	200.97	209.00	0.28	-0.29
Oct 6	180.76	2.10	54.82	-1.20	269.79	-1.76	203.49	17 Oct 22	89.0	154.8	-3.9	-4.8	197.66	203.93	1.30	1.19	216.76	224.27	-0.86	-1.40
Oct 16	181.91	2.13	53.71	-1.20	277.14	-1.71	213.47	27 Oct	214.9		-0.6		210.20		1.07		231.54		-1.89	
								Nov 1		274.3		4.2		216.48		0.93		238.50		-2.29
Oct 26	183.00	2.15	52.43	-1.20	284.61	-1.64	223.50	6 Nov 11	339.1	53.2	4.7	-1.1	222.76	229.04	0.77	0.59	245.04	250.89	-2.57	-2.66
Nov 5	183.99	2.18	51.08	-1.18	292.17	-1.55	233.57	16 Nov 21	125.6	188.2	-5.3	-2.9	235.32	241.60	0.41	0.21	255.50	257.85	-2.46	-1.83
Nov 15	184.88	2.22	49.75	-1.16	299.81	-1.46	243.69	26 N-D 1	247.3	308.8	2.3	5.2	247.89	254.17	0.01	-0.19	256.49	250.98	-0.62	1.03
Nov 25	185.64	2.26	48.55	-1.12	307.50	-1.35	253.85	6 Dec 11	16.7	91.1	2.1	-4.0	260.46	266.74	-0.38	-0.58	244.58	241.69	2.38	2.85
Dec 5	186.26	2.30	47.56	-1.08	315.23	-1.24	264.02	16 Dec 21	160.1	220.3	-4.5	0.0	273.03	279.31	-0.76	-0.93	243.02	247.15	2.64	2.09
Dec 15	186.72	2.35	46.86	-1.04	322.97	-1.12	274.21	26 Dec 31	280.5	345.3	4.4	4.1	285.59	291.87	-1.08	-1.22	252.82	259.32	1.41	0.72
1715								1715												1715

JULIAN 12 UT	SATURN LONG.	SATURN LAT.	JUPITER LONG.	JUPITER LAT.	MARS LONG.	MARS LAT.	SUN LONG.	GREGORIAN 12 UT	MOON LONG.		MOON LAT.		VENUS LONG.		VENUS LAT.		MERCURY LONG.		MERCURY LAT.	
1715								1716												1716
Dec 25	187.01	2.39	46.49	-0.99	330.71	-0.99	284.41	5 Jan 10	55.4	127.9	-1.4	-5.0	298.14	304.41	-1.34	-1.43	266.30	273.60	0.05	-0.55
Jan 4	187.12	2.44	46.46	-0.94	338.43	-0.86	294.59	15 Jan 20	192.8	252.1	-2.1	2.9	310.67	316.93	-1.49	-1.53	281.15	288.93	-1.08	-1.52
Jan 14	187.05	2.49	46.77	-0.90	346.12	-0.73	304.76	25 Jan 30	315.2	23.5	4.9	1.1	323.18	329.42	-1.54	-1.53	296.97	305.28	-1.84	-2.04
Jan 24	186.81	2.53	47.41	-0.85	353.77	-0.60	314.91	4 Feb 9	94.1	162.9	-4.3	-4.0	335.65	341.87	-1.48	-1.41	313.90	322.83	-2.07	-1.91
Feb 3	186.40	2.57	48.34	-0.81	1.37	-0.47	325.01	14 Feb 19	224.4	284.6	0.9	4.8	348.08	354.27	-1.30	-1.17	332.03	341.30	-1.53	-0.89
Feb 13	185.85	2.61	49.55	-0.77	8.90	-0.34	335.08	24 Feb 29	351.6	62.5	3.4	-2.5	0.45	6.61	-1.01	-0.83	350.18	357.78	0.02	1.13
Feb 23	185.19	2.64	51.00	-0.73	16.36	-0.22	345.10	5 Mar 10	131.8	196.3	-5.1	-1.4	12.75	18.87	-0.63	-0.41	2.98	4.86	2.28	3.19
Mar 4	184.45	2.65	52.65	-0.70	23.76	-0.10	355.06	15 Mar 20	255.9	318.4	3.6	5.0	24.97	31.04	-0.17	0.08	3.30	359.40	3.56	3.19
Mar 14	183.67	2.66	54.47	-0.67	31.07	0.02	4.97	25 Mar 30	29.4	101.4	0.2	-4.9	37.09	43.11	0.34	0.60	355.23	352.60	2.17	0.87
Mar 24	182.90	2.66	56.44	-0.64	38.31	0.13	14.82	4 Apr 9	167.6	228.6	-3.7	1.5	49.10	55.06	0.86	1.12	352.23	353.99	-0.38	-1.41
Apr 3	182.17	2.65	58.52	-0.62	45.46	0.24	24.61	14 Apr 19	288.2	353.9	5.1	3.3	60.98	66.87	1.37	1.61	357.49	2.36	-2.16	-2.63
Apr 13	181.53	2.64	60.70	-0.59	52.54	0.34	34.35	24 Apr 29	68.0	138.8	-3.1	-5.1	72.71	78.51	1.83	2.03	8.34	15.26	-2.84	-2.80
Apr 23	181.00	2.61	62.94	-0.57	59.54	0.43	44.05	4 May 9	201.5	260.9	-1.0	4.0	84.26	89.95	2.20	2.34	23.03	31.63	-2.52	-2.03
May 3	180.62	2.58	65.23	-0.55	66.47	0.52	53.69	14 May 19	322.0	31.2	4.9	0.0	95.59	101.16	2.45	2.51	41.06	51.27	-1.34	-0.51
May 13	180.39	2.54	67.55	-0.54	73.32	0.61	63.30	24 May 29	106.5	174.1	-5.0	-3.2	106.65	112.06	2.53	2.50	62.07	73.04	0.37	1.16
May 23	180.33	2.51	69.88	-0.52	80.11	0.69	72.88	3 Jun 8	234.3	294.0	2.0	5.1	117.37	122.57	2.41	2.27	83.64	93.47	1.74	2.02
Jun 2	180.43	2.47	72.19	-0.51	86.83	0.76	82.43	13 Jun 18	357.7	69.8	2.8	-3.2	127.64	132.56	2.07	1.81	102.34	110.19	1.99	1.66
Jun 12	180.71	2.43	74.48	-0.50	93.49	0.83	91.97	23 Jun 28	143.6	207.5	-4.6	-0.2	137.31	141.86	1.47	1.06	116.98	122.63	1.08	0.26
Jun 22	181.14	2.39	76.72	-0.49	100.10	0.89	101.50	3 Jul 8	266.8	328.5	4.2	4.3	146.16	150.17	0.57	0.00	126.98	129.80	-0.76	-1.91
Jul 2	181.72	2.36	78.88	-0.48	106.66	0.95	111.03	13 Jul 18	35.3	108.7	-0.6	-4.9	153.83	157.08	-0.65	-1.40	130.80	129.80	-3.10	-4.17
Jul 12	182.43	2.33	80.96	-0.47	113.17	1.00	120.58	23 Jul 28	178.6	239.6	-2.4	2.7	159.81	161.94	-2.23	-3.15	127.02	123.36	-4.83	-4.85
Jul 22	183.27	2.30	82.93	-0.46	119.63	1.05	130.15	2 Aug 7	299.9	4.8	5.0	1.8	163.32	163.85	-4.15	-5.20	120.38	119.51	-4.15	-2.94
Aug 1	184.21	2.27	84.75	-0.45	126.06	1.10	139.75	12 Aug 17	74.4	146.3	-3.8	-4.3	163.44	162.04	-6.24	-7.21	121.49	126.37	-1.53	-0.20
Aug 11	185.25	2.25	86.42	-0.45	132.45	1.14	149.38	22 Aug 27	211.7	271.4	0.7	4.7	159.77	156.88	-8.00	-8.50	133.68	142.60	0.85	1.51
Aug 21	186.35	2.24	87.90	-0.44	138.81	1.17	159.06	1 Sep 6	334.2	42.8	3.9	-1.7	153.80	151.02	-8.62	-8.37	152.20	161.83	1.78	1.73
Aug 31	187.52	2.23	89.15	-0.44	145.14	1.21	168.78	11 Sep 16	113.6	181.8	-5.2	-1.9	148.96	147.84	-7.81	-7.03	171.16	180.08	1.45	1.02
Sep 10	188.72	2.22	90.16	-0.43	151.44	1.23	178.56	21 Sep 26	243.6	303.7	3.4	5.2	147.71	148.54	-6.13	-5.19	188.58	196.69	0.50	-0.08
Sep 20	189.94	2.22	90.89	-0.42	157.72	1.26	188.39	1 Oct 6	10.1	82.0	1.2	-4.5	150.21	152.60	-4.25	-3.35	204.44	211.85	-0.68	-1.26
Sep 30	191.17	2.23	91.31	-0.42	163.96	1.28	198.28	11 Oct 16	151.5	215.5	-4.1	1.2	155.60	159.11	-2.51	-1.73	218.91	225.57	-1.82	-2.30
Oct 10	192.38	2.24	91.41	-0.41	170.19	1.30	208.23	21 Oct 26	275.4	337.0	5.0	3.8	163.03	167.30	-1.01	-0.37	231.66	236.90	-2.68	-2.89
Oct 20	193.56	2.25	91.17	-0.40	176.38	1.31	218.23	31 Oct	47.5		-2.2		171.86		0.20		240.71		-2.83	
								Nov 5		121.0		-5.2		176.66		0.71		242.09		-2.33
Oct 30	194.68	2.27	90.61	-0.39	182.53	1.32	228.28	10 Nov 15	187.3	248.2	-1.4	3.7	181.66	186.85	1.14	1.51	239.77	233.85	-1.23	0.41
Nov 9	195.73	2.30	89.74	-0.37	188.66	1.32	238.39	20 Nov 25	307.7	11.9	5.0	0.9	192.18	197.63	1.82	2.07	227.91	225.96	1.87	2.55
Nov 19	196.69	2.33	88.63	-0.36	194.74	1.32	248.52	30 N-D 5	86.0	158.4	-4.6	-3.5	203.19	208.85	2.25	2.38	228.23	233.13	2.53	2.13
Nov 29	197.54	2.36	87.36	-0.34	200.76	1.31	258.68	10 Dec 15	221.2	280.7	1.7	5.0	214.59	220.39	2.45	2.47	239.40	246.34	1.55	0.92
Dec 9	198.26	2.40	86.00	-0.31	206.73	1.30	268.87	20 Dec 25	341.3	48.7	3.2	-2.4	226.25	232.16	2.44	2.37	253.63	261.12	0.28	-0.32
Dec 19	198.83	2.44	84.68	-0.29	212.63	1.28	279.06	30 Dec	124.4		-4.8		238.11		2.26		268.77		-0.86	
1716								1717												1717
								Jan 4		193.5		-0.5		244.10		2.12		276.57		-1.33
Dec 29	199.24	2.49	83.48	-0.26	218.45	1.25	289.26	9 Jan 14	254.0	313.7	4.1	4.5	250.12	256.17	1.94	1.74	284.55	292.72	-1.70	-1.96
Jan 8	199.48	2.54	82.50	-0.23	224.16	1.21	299.44	19 Jan 24	16.6	87.1	0.0	-4.7	262.24	268.32	1.51	1.27	301.12	309.76	-2.08	-2.02
Jan 18	199.55	2.58	81.80	-0.20	229.74	1.15	309.60	29 J-F 3	161.5	226.7	-2.8	2.6	274.42	280.53	1.02	0.76	318.58	327.40	-1.74	-1.21
Jan 28	199.44	2.63	81.42	-0.18	235.18	1.08	319.73	8 Feb 13	286.3	347.8	5.1	2.3	286.66	292.79	0.50	0.24	335.76	342.74	-0.37	0.75
Feb 7	199.15	2.67	81.36	-0.15	240.42	0.99	329.82	18 Feb 23	54.1	126.1	-3.3	-4.8	298.93	305.07	-0.01	-0.26	347.05	347.56	2.02	3.14
Feb 17	198.72	2.71	81.64	-0.12	245.44	0.87	339.86	28 Feb	196.6		0.3		311.22		-0.49		344.38		3.70	
								Mar 5		258.7		4.7		317.37		-0.70		339.46		3.41
Feb 27	198.14	2.74	82.24	-0.10	250.17	0.73	349.85	10 Mar 15	318.8	23.5	4.3	-0.9	323.52	329.68	-0.89	-1.06	335.43	333.80	2.45	1.23
Mar 9	197.47	2.76	83.12	-0.08	254.54	0.55	359.79	20 Mar 25	93.2	164.0	-5.2	-2.6	335.83	341.99	-1.21	-1.33	334.65	337.52	0.06	-0.93
Mar 19	196.73	2.77	84.27	-0.06	258.45	0.32	9.67	30 M-A 4	230.1	290.4	3.2	5.2	348.14	354.29	-1.42	-1.48	341.90	347.44	-1.69	-2.22
Mar 29	195.96	2.78	85.64	-0.04	261.80	0.04	19.49	9 Apr 14	352.2	61.1	1.9	-3.9	0.44	6.59	-1.52	-1.53	353.91	1.17	-2.53	-2.61
Apr 8	195.20	2.77	87.21	-0.02	264.42	-0.31	29.26	19 Apr 24	132.5	200.1	-4.6	0.6	12.74	18.88	-1.50	-1.45	9.16	17.87	-2.47	-2.12
Apr 18	194.50	2.76	88.96	0.00	266.16	-0.74	38.97	29 Apr	262.6		4.9		25.03		-1.38		27.29		-1.56	
								May 4		322.3		4.0		31.17		-1.28		37.43		-0.81
Apr 28	193.88	2.73	90.84	0.02	266.81	-1.25	48.64	9 May 14	27.2	100.1	-1.3	-5.1	37.31	43.45	-1.16	-1.02	48.13	59.03	0.04	0.90
May 8	193.39	2.70	92.84	0.03	266.25	-1.84	58.27	19 May 24	170.6	234.5	-1.9	3.4	49.59	55.72	-0.86	-0.69	69.60	79.34	1.61	2.05
May 18	193.03	2.67	94.95	0.05	264.47	-2.49	67.86	29 M-J 3	294.7	355.3	5.0	1.4	61.86	67.99	-0.50	-0.31	88.01	95.48	2.18	1.98
May 28	192.84	2.63	97.09	0.06	261.71	-3.12	77.42	8 Jun 13	64.3	139.2	-4.0	-3.9	74.13	80.27	-0.11	0.08	101.69	106.54	1.47	0.66
Jun 7	192.80	2.58	99.30	0.08	258.52	-3.66	86.97	18 Jun 23	206.6	267.8	1.4	4.9	86.41	92.56	0.28	0.47	109.84	111.39	-0.40	-1.66
Jun 17	192.93	2.54	101.54	0.09	255.67	-4.05	96.50	28 Jun	327.3		3.3		98.70		0.65		111.03		-2.97	
								Jul 3		30.1		-1.8		104.85		0.82		108.95		-4.11
Jun 27	193.23	2.50	103.79	0.11	253.76	-4.26	106.03	8 Jul 13	103.0	176.8	-5.0	-1.0	111.00	117.15	0.98	1.12	105.86	102.98	-4.79	-4.81
Jul 7	193.68	2.46	106.03	0.13	253.17	-4.32	115.57	18 Jul 23	240.8	300.5	4.0	4.6	123.31	129.46	1.23	1.33	101.57	102.41	-4.20	-3.15
Jul 17	194.28	2.42	108.25	0.14	253.97	-4.26	125.13	28 J-A 2	0.9	67.1	0.5	-4.5	135.62	141.78	1.40	1.44	105.75	111.51	-1.89	-0.62
Jul 27	195.01	2.39	110.41	0.16	256.05	-4.12	134.71	7 Aug 12	141.9	212.2	-3.5	2.3	147.94	154.11	1.46	1.45	119.34	128.65	0.47	1.25
Aug 6	195.86	2.35	112.51	0.18	259.21	-3.93	144.32	17 Aug 22	273.7	333.4	5.1	2.7	160.27	166.43	1.41	1.35	138.62	148.61	1.68	1.76
Aug 16	196.81	2.33	114.53	0.20	263.28	-3.71	153.97	27 Aug	36.3		-2.7		172.59		1.25		158.24		1.59	
								Sep 1		105.8		-5.1		178.74		1.13		167.40		1.22
Aug 26	197.84	2.31	116.43	0.22	268.07	-3.48	163.68	6 Sep 11	179.4	245.8	-0.5	4.6	184.90	191.05	0.99	0.82	176.05	184.23	0.72	0.15
Sep 5	198.95	2.29	118.19	0.24	273.45	-3.23	173.43	16 Sep 21	305.8	7.2	4.5	-0.3	197.20	203.35	0.63	0.43	191.97	199.28	-0.47	-1.11
Sep 15	200.11	2.28	119.78	0.26	279.29	-2.97	183.23	26 S-O 1	73.8	144.9	-5.0	-3.4	209.49	215.62	0.20	-0.03	206.13	212.45	-1.73	-2.30
Sep 25	201.31	2.27	121.19	0.29	285.51	-2.71	193.09	6 Oct 11	215.1	278.1	2.7	5.3	221.75	227.88	-0.27	-0.51	218.07	222.66	-2.78	-3.11
Oct 5	202.52	2.27	122.37	0.31	292.02	-2.45	203.01	16 Oct 21	337.9	42.6	2.3	-3.2	234.00	240.11	-0.75	-0.99	225.61	225.96	-3.19	-2.83
Oct 15	203.73	2.27	123.29	0.34	298.77	-2.19	212.98	26 Oct 31	112.9	183.0	-4.9	-0.1	246.21	252.30	-1.22	-1.44	222.76	216.66	-1.84	-0.24
Oct 25	204.92	2.28	123.94	0.37	305.69	-1.94	223.01	5 Nov 10	249.1	309.7	4.7	4.2	258.39	264.46	-1.64	-1.81	211.43	210.47	1.32	2.20
Nov 4	206.08	2.29	124.27	0.40	312.75	-1.69	233.08	15 Nov 20	10.8	80.0	-0.7	-5.0	270.52	276.56	-1.97	-2.09	213.62	219.25	2.39	2.14
Nov 14	207.17	2.31	124.28	0.44	319.91	-1.45	243.20	25 Nov 30	152.1	219.4	-2.6	3.1	282.58	288.58	-2.18	-2.23	226.09	233.46	1.67	1.10
Nov 24	208.19	2.34	123.96	0.47	327.13	-1.22	253.35	5 Dec 10	282.0	341.4	4.9	1.7	294.55	300.49	-2.24	-2.21	241.05	248.74	0.50	-0.10
Dec 4	209.11	2.36	123.32	0.50	334.40	-1.00	263.53	15 Dec 20	45.2	118.7	-3.5	-4.4	306.39	312.25	-2.14	-2.02	256.49	264.31	-0.65	-1.14
Dec 14	209.91	2.40	122.41	0.53	341.68	-0.79	273.71	25 Dec 30	190.1	254.1	0.9	4.8	318.06	323.80	-1.84	-1.62	272.23	280.27	-1.56	-1.88
1717								1717												1717

JULIAN 12 UT	SATURN LONG.	SATURN LAT.	JUPITER LONG.	JUPITER LAT.	MARS LONG.	MARS LAT.	SUN LONG.	GREGORIAN 12 UT	MOON LONG.		MOON LAT.		VENUS LONG.		VENUS LAT.		MERCURY LONG.		MERCURY LAT.	
1717								1718												1718
Dec 24	210.58	2.43	121.26	0.56	348.96	-0.59	283.91	4 Jan 9	314.3	14.1	3.5	-1.4	329.48	335.06	-1.35	-1.03	288.45	296.79	-2.07	-2.10
Jan 3	211.11	2.48	119.98	0.59	356.22	-0.41	294.10	14 Jan 19	81.7	157.3	-5.0	-1.7	340.55	345.93	-0.66	-0.24	305.22	313.57	-1.93	-1.51
Jan 13	211.47	2.52	118.64	0.61	3.46	-0.24	304.27	24 Jan 29	226.2	287.5	3.9	4.8	351.16	356.24	0.24	0.75	321.36	327.66	-0.76	0.32
Jan 23	211.66	2.56	117.36	0.63	10.65	-0.08	314.42	3 Feb 8	346.8	48.6	0.8	-4.2	1.13	5.79	1.31	1.92	331.04	330.21	1.68	2.98
Feb 2	211.67	2.60	116.21	0.64	17.80	0.07	324.53	13 Feb 18	119.8	194.6	-4.3	1.7	10.18	14.24	2.56	3.24	325.67	320.25	3.70	3.51
Feb 12	211.52	2.64	115.29	0.65	24.90	0.20	334.60	23 Feb 28	260.4	320.0	5.2	3.0	17.90	21.10	3.94	4.66	316.85	316.34	2.63	1.51
Feb 22	211.19	2.68	114.65	0.65	31.94	0.33	344.62	5 Mar 10	20.2	85.4	-2.3	-5.3	23.71	25.62	5.38	6.09	318.31	322.09	0.42	-0.52
Mar 4	210.72	2.71	114.31	0.65	38.92	0.44	354.58	15 Mar 20	158.4	230.3	-1.5	4.3	26.71	26.84	6.74	7.29	327.16	333.19	-1.28	-1.85
Mar 14	210.13	2.74	114.30	0.65	45.84	0.54	4.49	25 Mar 30	293.0	352.4	4.7	0.2	25.96	24.09	7.66	7.78	340.00	347.48	-2.23	-2.41
Mar 24	209.44	2.76	114.60	0.65	52.71	0.63	14.35	4 Apr 9	55.5	124.1	-4.6	-4.1	21.41	18.31	7.58	7.03	355.59	4.31	-2.39	-2.17
Apr 3	208.70	2.76	115.20	0.64	59.51	0.71	24.14	14 Apr 19	196.4	264.3	1.9	5.2	15.27	12.78	6.16	5.07	13.68	23.68	-1.73	-1.09
Apr 13	207.94	2.76	116.07	0.64	66.25	0.79	33.88	24 Apr 29	324.8	25.7	2.6	-2.7	11.16	10.54	3.90	2.74	34.22	44.98	-0.29	0.60
Apr 23	207.20	2.75	117.20	0.64	72.95	0.85	43.58	4 May 9	92.8	163.3	-5.1	-0.9	10.91	12.16	1.66	0.70	55.43	65.00	1.42	2.03
May 3	206.52	2.73	118.54	0.64	79.59	0.91	53.22	14 May 19	233.0	297.0	4.4	4.2	14.17	16.82	-0.14	-0.84	73.34	80.26	2.34	2.30
May 13	205.94	2.70	120.07	0.64	86.18	0.96	62.83	24 May 29	356.5	60.7	-0.4	-4.7	20.00	23.61	-1.43	-1.91	85.65	89.40	1.89	1.13
May 23	205.48	2.67	121.76	0.64	92.73	1.00	72.41	3 Jun 8	131.7	201.8	-3.2	2.6	27.58	31.84	-2.29	-2.58	91.33	91.37	0.03	-1.31
Jun 2	205.16	2.63	123.59	0.64	99.24	1.04	81.96	13 Jun 18	268.0	328.8	5.0	1.8	36.33	41.04	-2.78	-2.91	89.70	86.98	-2.70	-3.87
Jun 12	204.99	2.59	125.54	0.64	105.72	1.07	91.50	23 Jun 28	29.1	97.6	-3.2	-4.8	45.91	50.93	-2.98	-2.99	84.31	82.80	-4.54	-4.60
Jun 22	204.99	2.54	127.57	0.65	112.16	1.10	101.04	3 Jul 8	170.8	238.8	0.2	4.8	56.08	61.33	-2.94	-2.84	83.16	85.63	-4.11	-3.25
Jul 2	205.16	2.50	129.67	0.65	118.58	1.12	110.57	13 Jul 18	301.5	0.8	3.7	-1.2	66.67	72.10	-2.70	-2.53	90.20	96.74	-2.17	-1.01
Jul 12	205.48	2.45	131.83	0.66	124.99	1.13	120.12	23 Jul 28	63.5	136.1	-5.0	-2.6	77.61	83.18	-2.32	-2.08	105.04	114.66	0.08	0.96
Jul 22	205.96	2.41	134.01	0.67	131.37	1.14	129.69	2 Aug 7	209.0	273.9	3.5	5.0	88.82	94.51	-1.83	-1.56	124.94	135.25	1.53	1.76
Aug 1	206.58	2.37	136.20	0.68	137.75	1.15	139.28	12 Aug 17	334.0	33.7	1.1	-3.9	100.25	106.04	-1.27	-0.98	145.17	154.55	1.70	1.41
Aug 11	207.32	2.34	138.38	0.70	144.11	1.15	148.91	22 Aug 27	99.9	174.7	-4.8	0.8	111.87	117.75	-0.69	-0.40	163.35	171.59	0.95	0.39
Aug 21	208.19	2.30	140.53	0.71	150.48	1.15	158.59	1 Sep 6	245.4	307.3	5.2	3.3	123.67	129.63	-0.12	0.15	179.31	186.50	-0.25	-0.93
Aug 31	209.15	2.28	142.64	0.73	156.84	1.14	168.31	11 Sep 16	6.5	68.3	-1.9	-5.2	135.62	141.65	0.40	0.64	193.12	199.07	-1.62	-2.28
Sep 10	210.19	2.25	144.67	0.75	163.21	1.13	178.09	21 Sep 26	137.8	212.6	-2.5	3.8	147.70	153.79	0.85	1.04	204.17	208.02	-2.87	-3.33
Sep 20	211.31	2.23	146.61	0.78	169.58	1.11	187.92	1 Oct 6	279.8	339.6	4.8	0.6	159.90	166.04	1.21	1.34	210.04	209.35	-3.55	-3.33
Sep 30	212.47	2.22	148.43	0.81	175.97	1.09	197.81	11 Oct 16	40.0	105.1	-4.2	-4.5	172.20	178.38	1.45	1.52	205.39	199.41	-2.43	-0.88
Oct 10	213.66	2.21	150.11	0.84	182.36	1.06	207.75	21 Oct 26	176.4	248.8	1.0	5.1	184.58	190.80	1.57	1.58	195.10	195.14	0.75	1.80
Oct 20	214.86	2.21	151.61	0.87	188.77	1.03	217.75	31 Oct	312.4		2.7		197.03		1.57		199.14		2.20	
								Nov 5		11.8		-2.3		203.27		1.52		205.45		2.11
Oct 30	216.06	2.21	152.91	0.90	195.18	0.99	227.80	10 Nov 15	75.2	143.4	-5.1	-1.7	209.52	215.79	1.45	1.35	212.83	220.62	1.76	1.26
Nov 9	217.23	2.22	153.97	0.94	201.61	0.95	237.90	20 Nov 25	214.8	283.2	3.9	4.3	222.06	228.33	1.22	1.08	228.51	236.41	0.70	0.12
Nov 19	218.36	2.23	154.77	0.99	208.06	0.90	248.03	30 N-D 5	344.0	44.7	-0.2	-4.5	234.62	240.90	0.92	0.74	244.28	252.13	-0.44	-0.96
Nov 29	219.42	2.25	155.27	1.03	214.51	0.84	258.19	10 Dec 15	112.2	182.6	-3.8	1.9	247.19	253.48	0.55	0.36	259.99	267.90	-1.41	-1.78
Dec 9	220.40	2.27	155.47	1.07	220.97	0.77	268.38	20 Dec 25	251.9	316.0	5.0	2.0	259.77	266.06	0.15	-0.05	275.88	283.92	-2.05	-2.17
Dec 19	221.28	2.30	155.34	1.12	227.45	0.70	278.57	30 Dec	15.5		-3.1		272.35		-0.25		291.95		-2.10	
1718								1719												1719
								Jan 4		79.0		-5.0		278.64		-0.44		299.80		-1.79
Dec 29	222.04	2.33	154.89	1.16	233.92	0.61	288.76	9 Jan 14	150.5	221.2	-0.7	4.6	284.93	291.22	-0.63	-0.80	306.98	312.55	-1.16	-0.13
Jan 8	222.66	2.36	154.15	1.20	240.40	0.51	298.94	19 Jan 24	287.2	348.0	4.0	-1.0	297.50	303.79	-0.96	-1.09	314.96	312.89	1.25	2.70
Jan 18	223.13	2.40	153.15	1.24	246.88	0.40	309.11	29 J-F 3	47.8	115.0	-4.9	-3.5	310.06	316.34	-1.21	-1.31	307.32	301.89	3.58	3.50
Jan 28	223.44	2.43	151.96	1.27	253.35	0.27	319.23	8 Feb 13	189.2	258.4	2.8	5.2	322.60	328.87	-1.38	-1.43	299.38	299.99	2.73	1.72
Feb 7	223.57	2.47	150.67	1.29	259.81	0.13	329.32	18 Feb 23	320.8	20.0	1.5	-3.6	335.13	341.38	-1.45	-1.44	302.94	307.46	0.73	-0.16
Feb 17	223.54	2.51	149.36	1.30	266.25	-0.03	339.37	28 Feb	81.6		-5.1		347.62		-1.41		313.07		-0.91	
								Mar 5		152.5		-0.4		353.85		-1.35		319.48		-1.51
Feb 27	223.33	2.54	148.14	1.30	272.66	-0.21	349.36	10 Mar 15	227.3	293.5	5.0	3.6	0.07	6.29	-1.26	-1.15	326.53	334.14	-1.94	-2.21
Mar 9	222.97	2.57	147.09	1.29	279.04	-0.41	359.30	20 Mar 25	353.3	53.0	-1.5	-5.1	12.49	18.69	-1.02	-0.86	342.29	350.98	-2.29	-2.18
Mar 19	222.47	2.60	146.26	1.28	285.36	-0.64	9.19	30 M-A 4	117.5	191.0	-3.3	2.9	24.87	31.04	-0.69	-0.50	0.23	10.05	-1.86	-1.33
Mar 29	221.86	2.62	145.72	1.26	291.60	-0.90	19.01	9 Apr 14	263.9	326.7	4.9	0.8	37.19	43.34	-0.30	-0.09	20.37	30.90	-0.60	0.27
Apr 8	221.16	2.63	145.48	1.23	297.75	-1.19	28.78	19 Apr 24	25.9	87.7	-3.9	-4.7	49.47	55.58	0.13	0.35	41.15	50.47	1.17	1.94
Apr 18	220.42	2.63	145.55	1.21	303.77	-1.51	38.50	29 Apr	155.2		0.0		61.69		0.57		58.36		2.44	
								May 4		229.4		4.9		67.77		0.79		64.53		2.59
Apr 28	219.67	2.62	145.92	1.18	309.62	-1.86	48.17	9 May 14	298.5	358.8	2.9	-2.2	73.84	79.89	1.00	1.19	68.84	71.15	2.32	1.63
May 8	218.95	2.61	146.57	1.15	315.24	-2.26	57.80	19 May 24	59.2	124.2	-5.0	-2.4	85.93	91.95	1.37	1.53	71.43	69.92	0.53	-0.86
May 18	218.30	2.58	147.48	1.13	320.57	-2.70	67.39	29 M-J 3	194.3	266.5	3.4	4.5	97.95	103.93	1.67	1.77	67.32	64.70	-2.28	-3.43
May 28	217.74	2.55	148.63	1.11	325.51	-3.18	76.96	8 Jun 13	331.3	30.8	0.0	-4.4	109.89	115.82	1.85	1.90	63.15	63.32	-4.10	-4.24
Jun 7	217.32	2.52	149.98	1.09	329.93	-3.71	86.50	18 Jun 23	94.0	162.4	-4.2	1.2	121.73	127.61	1.91	1.89	65.38	69.28	-3.92	-3.26
Jun 17	217.04	2.48	151.52	1.07	333.70	-4.27	96.04	28 Jun	233.3		5.1		133.46		1.83		74.89		-2.37	
								Jul 3		301.7		2.3		139.28		1.72		82.10		-1.34
Jun 27	216.91	2.43	153.21	1.06	336.63	-4.87	105.57	8 Jul 13	3.1	63.5	-2.9	-5.2	145.06	150.80	1.58	1.40	90.78	100.65	-0.29	0.64
Jul 7	216.95	2.39	155.04	1.05	338.50	-5.47	115.11	18 Jul 23	130.4	201.7	-1.6	4.2	156.49	162.14	1.17	0.91	111.17	121.75	1.33	1.71
Jul 17	217.15	2.34	156.97	1.04	339.14	-6.02	124.67	28 J-A 2	271.0	335.1	4.4	-0.7	167.73	173.26	0.60	0.26	131.93	141.50	1.78	1.58
Jul 27	217.50	2.30	158.99	1.03	338.47	-6.45	134.24	7 Aug 12	34.8	97.4	-4.9	-4.1	178.72	184.09	-0.12	-0.53	150.42	158.71	1.19	0.64
Aug 6	218.01	2.26	161.07	1.03	336.61	-6.67	143.86	17 Aug 22	168.4	240.7	1.9	5.3	189.38	194.56	-0.97	-1.43	166.37	173.41	-0.01	-0.73
Aug 16	218.66	2.22	163.20	1.03	334.05	-6.58	153.51	27 Aug	306.7		1.9		199.62		-1.92		179.75		-1.48	
								Sep 1		7.5		-3.4		204.54		-2.43		185.28		-2.24
Aug 26	219.43	2.18	165.36	1.03	331.51	-6.17	163.21	6 Sep 11	67.2	132.9	-5.2	-1.3	209.28	213.83	-2.95	-3.47	189.77	192.81	-2.96	-3.55
Sep 5	220.31	2.15	167.52	1.04	329.67	-5.53	172.96	16 Sep 21	207.3	278.0	4.5	4.0	218.14	222.16	-4.00	-4.51	193.81	192.08	-3.90	-3.80
Sep 15	221.29	2.13	169.67	1.05	329.02	-4.76	182.76	26 S-O 1	340.4	39.9	-1.2	-4.9	225.82	229.05	-5.01	-5.47	187.56	182.04	-2.98	-1.48
Sep 25	222.34	2.10	171.79	1.06	329.69	-3.97	192.61	6 Oct 11	101.0	170.2	-3.7	2.1	231.76	233.84	-5.87	-6.20	178.84	179.92	0.17	1.37
Oct 5	223.46	2.08	173.85	1.08	331.58	-3.22	202.53	16 Oct 21	245.8	313.1	5.0	1.1	235.16	235.59	-6.41	-6.46	184.70	191.66	1.96	2.05
Oct 15	224.62	2.07	175.82	1.10	334.51	-2.55	212.50	26 Oct 31	13.1	73.0	-3.7	-4.8	235.03	233.47	-6.30	-5.87	199.57	207.78	1.82	1.40
Oct 25	225.81	2.06	177.70	1.12	338.29	-1.96	222.52	5 Nov 10	136.6	208.7	-0.7	4.5	231.07	228.14	-5.14	-4.13	215.98	224.09	0.89	0.33
Nov 4	227.00	2.06	179.44	1.15	342.71	-1.45	232.60	15 Nov 20	282.6	346.2	3.3	-2.0	225.17	222.64	-2.92	-1.63	232.08	239.97	-0.23	-0.77
Nov 14	228.18	2.06	181.02	1.18	347.65	-1.01	242.71	25 Nov 30	45.7	107.5	-5.0	-2.8	220.92	220.20	-0.39	0.72	247.80	255.59	-1.26	-1.68
Nov 24	229.34	2.06	182.42	1.22	352.97	-0.64	252.86	5 Dec 10	174.1	247.3	2.8	4.8	220.51	221.75	1.66	2.41	263.36	271.11	-2.01	-2.21
Dec 4	230.44	2.07	183.59	1.26	358.59	-0.32	263.04	15 Dec 20	317.2	18.2	0.3	-4.4	223.81	226.56	2.99	3.42	278.74	286.07	-2.25	-2.05
Dec 14	231.47	2.09	184.52	1.30	4.43	-0.05	273.22	25 Dec 30	78.6	143.6	-4.5	0.4	229.86	233.63	3.71	3.88	292.61	297.39	-1.54	-0.61
1719								1719												1719

JULIAN 12 UT	SATURN LONG.	SATURN LAT.	JUPITER LONG.	JUPITER LAT.	MARS LONG.	MARS LAT.	SUN LONG.
1719							
Dec 24	232.41	2.11	185.17	1.34	10.44	0.18	283.42
Jan 3	233.24	2.13	185.52	1.39	16.57	0.37	293.61
Jan 13	233.95	2.16	185.56	1.43	22.79	0.54	303.78
Jan 23	234.52	2.18	185.28	1.48	29.08	0.68	313.93
Feb 2	234.93	2.21	184.69	1.52	35.40	0.80	324.04
Feb 12	235.19	2.25	183.83	1.55	41.76	0.90	334.11
Feb 22	235.27	2.28	182.76	1.58	48.13	0.98	344.13
Mar 3	235.18	2.31	181.53	1.60	54.50	1.05	354.11
Mar 13	234.93	2.34	180.24	1.60	60.87	1.10	4.02
Mar 23	234.53	2.36	178.99	1.60	67.24	1.15	13.87
Apr 2	234.00	2.38	177.84	1.58	73.59	1.18	23.68
Apr 12	233.36	2.39	176.89	1.56	79.93	1.21	33.42
Apr 22	232.66	2.39	176.19	1.52	86.26	1.23	43.11
May 2	231.92	2.39	175.76	1.49	92.58	1.24	52.77
May 12	231.18	2.38	175.64	1.45	98.88	1.24	62.37
May 22	230.48	2.36	175.81	1.40	105.18	1.24	71.95
Jun 1	229.86	2.33	176.28	1.36	111.47	1.24	81.51
Jun 11	229.34	2.30	177.02	1.32	117.76	1.22	91.04
Jun 21	228.95	2.26	178.01	1.29	124.05	1.21	100.58
Jul 1	228.71	2.22	179.22	1.25	130.35	1.19	110.11
Jul 11	228.62	2.18	180.63	1.22	136.66	1.17	119.66
Jul 21	228.70	2.14	182.22	1.20	142.97	1.14	129.22
Jul 31	228.94	2.09	183.95	1.17	149.31	1.10	138.82
Aug 10	229.33	2.05	185.81	1.15	155.66	1.07	148.44
Aug 20	229.87	2.01	187.78	1.14	162.05	1.03	158.12
Aug 30	230.55	1.98	189.83	1.12	168.46	0.98	167.84
Sep 9	231.34	1.94	191.93	1.11	174.90	0.93	177.61
Sep 19	232.25	1.91	194.08	1.11	181.38	0.88	187.44
Sep 29	233.24	1.89	196.25	1.10	187.90	0.82	197.33
Oct 9	234.31	1.87	198.42	1.10	194.45	0.76	207.26
Oct 19	235.44	1.85	200.56	1.11	201.06	0.70	217.26
Oct 29	236.60	1.84	202.66	1.12	207.71	0.62	227.31
Nov 8	237.78	1.83	204.70	1.13	214.40	0.55	237.40
Nov 18	238.97	1.82	206.64	1.14	221.15	0.46	247.53
Nov 28	240.13	1.82	208.46	1.16	227.95	0.37	257.70
Dec 8	241.26	1.82	210.13	1.18	234.79	0.27	267.88
Dec 18	242.33	1.83	211.64	1.21	241.69	0.17	278.07
1720							
Dec 28	243.33	1.84	212.93	1.24	248.63	0.06	288.27
Jan 7	244.23	1.86	213.99	1.27	255.63	-0.06	298.45
Jan 17	245.01	1.88	214.79	1.30	262.67	-0.18	308.61
Jan 27	245.67	1.90	215.30	1.34	269.76	-0.32	318.75
Feb 6	246.19	1.92	215.51	1.37	276.90	-0.46	328.84
Feb 16	246.55	1.94	215.40	1.40	284.08	-0.61	338.89
Feb 26	246.74	1.97	214.99	1.44	291.29	-0.76	348.89
Mar 8	246.77	1.99	214.29	1.46	298.54	-0.92	358.83
Mar 18	246.63	2.01	213.33	1.48	305.81	-1.09	8.71
Mar 28	246.33	2.03	212.20	1.49	313.10	-1.26	18.54
Apr 7	245.89	2.05	210.95	1.49	320.39	-1.43	28.31
Apr 17	245.33	2.06	209.68	1.48	327.68	-1.60	38.04
Apr 27	244.68	2.06	208.47	1.46	334.94	-1.77	47.71
May 7	243.96	2.06	207.41	1.43	342.16	-1.94	57.34
May 17	243.22	2.05	206.55	1.40	349.31	-2.10	66.94
May 27	242.50	2.04	205.96	1.35	356.37	-2.24	76.50
Jun 6	241.82	2.02	205.66	1.31	3.30	-2.38	86.04
Jun 16	241.23	1.99	205.66	1.26	10.06	-2.50	95.58
Jun 26	240.75	1.96	205.95	1.22	16.62	-2.60	105.11
Jul 6	240.40	1.92	206.53	1.17	22.91	-2.68	114.65
Jul 16	240.20	1.88	207.37	1.13	28.86	-2.74	124.20
Jul 26	240.16	1.84	208.46	1.09	34.39	-2.77	133.78
Aug 5	240.28	1.81	209.77	1.06	39.39	-2.77	143.39
Aug 15	240.56	1.77	211.27	1.02	43.74	-2.73	153.04
Aug 25	240.99	1.73	212.94	0.99	47.26	-2.65	162.73
Sep 4	241.57	1.69	214.75	0.96	49.76	-2.51	172.48
Sep 14	242.27	1.66	216.69	0.94	51.02	-2.29	182.28
Sep 24	243.10	1.63	218.72	0.92	50.86	-1.98	192.13
Oct 4	244.03	1.60	220.83	0.90	49.24	-1.57	202.05
Oct 14	245.05	1.58	223.00	0.89	46.37	-1.07	212.02
Oct 24	246.13	1.56	225.19	0.88	42.85	-0.51	222.04
Nov 3	247.27	1.54	227.40	0.87	39.50	0.05	232.11
Nov 13	248.43	1.53	229.60	0.86	37.02	0.53	242.22
Nov 23	249.61	1.52	231.77	0.86	35.84	0.92	252.37
Dec 3	250.79	1.51	233.88	0.86	36.00	1.22	262.54
Dec 13	251.94	1.51	235.91	0.86	37.37	1.43	272.73
1721							

GREGORIAN 12 UT	MOON LONG.		MOON LAT.		VENUS LONG.		VENUS LAT.		MERCURY LONG.		MERCURY LAT.	
1720												1720
4 Jan 9	213.2	284.6	5.0	2.9	237.78	242.24	3.95	3.93	298.83	295.64	0.76	2.31
14 Jan 19	350.1	49.9	-2.7	-5.2	246.95	251.88	3.84	3.69	289.38	284.33	3.35	3.40
24 Jan 29	112.6	181.5	-2.3	3.6	256.97	262.22	3.49	3.24	282.82	284.48	2.77	1.88
3 Feb 8	252.4	320.0	4.8	-0.1	267.58	273.05	2.95	2.64	288.28	293.43	0.98	0.14
13 Feb 18	22.0	81.9	-4.7	-4.4	278.61	284.23	2.30	1.95	299.49	306.19	-0.59	-1.20
23 Feb 28	148.0	220.7	0.9	5.2	289.92	295.66	1.60	1.24	313.41	321.10	-1.68	-2.00
4 Mar 9	290.2	353.9	2.5	-3.0	301.44	307.27	0.88	0.54	329.24	337.85	-2.17	-2.17
14 Mar 19	53.9	115.1	-5.1	-2.0	313.13	319.01	0.21	-0.11	346.95	356.55	-1.96	-1.54
24 Mar 29	185.4	259.6	3.8	4.3	324.92	330.85	-0.40	-0.67	6.59	16.85	-0.90	-0.07
3 Apr 8	325.9	26.8	-0.8	-4.8	336.79	342.76	-0.92	-1.13	26.81	35.79	0.88	1.78
13 Apr 18	86.3	150.1	-3.8	1.3	348.74	354.73	-1.31	-1.46	43.10	48.34	2.48	2.84
23 Apr 28	224.1	296.8	5.0	1.5	0.72	6.73	-1.58	-1.66	51.27	51.87	2.75	2.15
3 May 8	359.9	59.5	-3.6	-4.8	12.75	18.77	-1.71	-1.72	50.40	47.66	1.07	-0.33
13 May 18	120.0	187.3	-1.1	4.1	24.80	30.84	-1.71	-1.66	44.84	43.08	-1.74	-2.86
23 May 28	262.8	332.0	3.8	-1.8	36.88	42.93	-1.59	-1.48	43.03	44.83	-3.55	-3.80
2 Jun 7	32.7	92.5	-5.0	-3.1	48.99	55.05	-1.36	-1.21	48.35	53.42	-3.66	-3.20
12 Jun 17	155.6	226.2	2.2	5.1	61.12	67.19	-1.04	-0.86	59.87	67.62	-2.50	-1.62
22 Jun 27	300.1	5.4	0.9	-4.3	73.28	79.37	-0.67	-0.47	76.60	86.64	-0.65	0.30
2 Jul 7	65.1	126.6	-4.7	-0.2	85.47	91.58	-0.26	-0.06	97.33	108.11	1.09	1.61
12 Jul 17	193.3	265.3	4.7	3.7	97.70	103.82	0.15	0.35	118.50	128.23	1.82	1.74
22 Jul 27	335.4	37.5	-2.3	-5.3	109.96	116.10	0.54	0.72	137.24	145.52	1.42	0.92
1 Aug 6	97.6	162.5	-2.8	2.9	122.26	128.43	0.88	1.03	153.09	159.91	0.27	-0.49
11 Aug 16	232.5	303.2	5.1	0.6	134.60	140.78	1.15	1.25	165.91	170.93	-1.32	-2.18
21 Aug 26	9.0	69.2	-4.5	-4.5	146.97	153.17	1.33	1.38	174.69	176.79	-3.02	-3.76
31 Aug	131.1		0.2		159.38		1.41		176.69		-4.24	
Sep 5		200.4		4.9		165.59		1.41		174.01		-4.23
10 Sep 15	271.8	339.2	3.1	-2.6	171.81	178.03	1.38	1.33	169.26	164.57	-3.47	-2.03
20 Sep 25	41.5	101.1	-5.1	-2.3	184.25	190.48	1.25	1.14	162.66	164.77	-0.39	0.91
30 S-O 5	166.2	239.6	3.2	4.7	196.71	202.95	1.02	0.87	170.30	177.85	1.68	1.95
10 Oct 15	309.7	13.5	-0.3	-4.6	209.18	215.42	0.70	0.52	186.27	194.89	1.85	1.53
20 Oct 25	73.6	134.1	-3.9	0.8	221.65	227.89	0.33	0.12	203.42	211.75	1.07	0.54
30 Oct	203.3		4.9		234.12		-0.09		219.87		-0.02	
Nov 4		278.4		2.2		240.36		-0.30		227.82		-0.58
9 Nov 14	345.6	46.7	-3.4	-4.9	246.59	252.83	-0.51	-0.72	235.62	243.31	-1.10	-1.57
19 Nov 24	106.1	168.8	-1.4	3.7	259.06	265.28	-0.91	-1.10	250.88	258.33	-1.96	-2.25
29 N-D 4	241.7	315.6	4.4	-1.3	271.51	277.73	-1.27	-1.42	265.56	272.35	-2.38	-2.30
9 Dec 14	19.7	79.3	-5.0	-3.4	283.95	290.15	-1.55	-1.65	278.21	282.16	-1.91	-1.10
19 Dec 24	139.6	205.7	1.7	5.2	296.36	302.55	-1.72	-1.77	282.61	278.46	0.22	1.83
29 Dec	280.1		1.8		308.73		-1.78		271.87		3.01	
1721												1721
Jan 3		350.8		-4.1		314.90		-1.76		267.43		3.23
8 Jan 13	52.4	112.0	-4.9	-0.7	321.05	327.19	-1.70	-1.61	266.95	269.58	2.76	2.00
18 Jan 23	175.1	244.3	4.3	4.4	333.30	339.39	-1.49	-1.33	274.13	279.83	1.19	0.40
28 J-F 2	317.4	24.3	-1.6	-5.2	345.45	351.49	-1.13	-0.91	286.26	293.21	-0.31	-0.92
7 Feb 12	84.2	145.6	-3.1	2.3	357.48	3.44	-0.66	-0.37	300.57	308.30	-1.42	-1.80
17 Feb 22	212.7	283.3	5.2	1.5	9.34	15.20	-0.07	0.26	316.40	324.90	-2.04	-2.13
27 Feb	353.2		-4.2		21.01		0.60		333.82		-2.03	
Mar 4		56.5		-4.6		26.75		0.96		343.17		-1.72
9 Mar 14	116.1	180.9	-0.3	4.5	32.41	37.99	1.32	1.69	352.91	2.81	-1.18	-0.41
19 Mar 24	251.7	321.5	3.7	-2.0	43.48	48.87	2.06	2.41	12.43	20.97	0.54	1.55
29 M-A 3	27.5	88.1	-5.0	-2.4	54.14	59.27	2.76	3.08	27.60	31.72	2.45	3.03
8 Apr 13	148.9	218.3	2.7	4.9	64.24	69.01	3.37	3.63	33.06	31.82	3.13	2.64
18 Apr 23	290.8	358.2	0.4	-4.5	73.57	77.86	3.84	4.01	28.83	25.51	1.59	0.21
28 Apr	60.6		-4.0		81.85		4.10		23.23		-1.15	
May 3		119.9		0.6		85.45		4.12		22.77		-2.25
8 May 13	183.6	257.0	4.8	3.0	88.60	91.19	4.05	3.86	24.23	27.43	-2.97	-3.33
18 May 23	328.8	33.0	-3.0	-5.0	93.13	94.28	3.54	3.07	32.11	38.06	-3.35	-3.08
28 M-J 2	93.0	152.9	-1.6	3.4	94.54	93.81	2.43	1.59	45.14	53.30	-2.56	-1.83
7 Jun 12	220.4	295.7	4.9	-0.4	92.10	89.55	0.58	-0.55	62.51	72.64	-0.97	-0.04
17 Jun 22	5.0	66.4	-5.0	-3.7	86.51	83.45	-1.73	-2.83	83.43	94.35	0.81	1.46
27 Jun	125.5		1.3		80.83		-3.76		104.89		1.82	
Jul 2		187.7		5.1		78.99		-4.46		114.73		1.87
7 Jul 12	258.7	333.3	2.8	-3.6	78.12	78.22	-4.92	-5.18	123.77	131.99	1.66	1.21
17 Jul 22	39.2	98.8	-5.0	-1.1	79.23	81.03	-5.27	-5.22	139.37	145.89	0.57	-0.22
27 J-A 1	159.2	224.7	3.9	4.8	83.50	86.53	-5.06	-4.83	151.42	155.79	-1.12	-2.09
6 Aug 11	297.5	9.3	-0.6	-5.1	90.01	93.88	-4.53	-4.18	158.68	159.71	-3.07	-3.95
16 Aug 21	71.9	131.3	-3.3	1.8	98.06	102.52	-3.80	-3.39	158.49	155.07	-4.54	-4.58
26 Aug 31	194.7	263.3	5.1	2.3	107.20	112.08	-2.96	-2.53	150.55	147.05	-3.86	-2.49
5 Sep 10	335.5	43.3	-3.7	-4.6	117.12	122.30	-2.09	-1.66	146.55	149.70	-0.91	0.45
15 Sep 20	103.6	164.7	-0.5	4.2	127.62	133.04	-1.23	-0.82	155.90	164.01	1.36	1.80
25 Sep 30	232.1	302.4	4.2	-1.3	138.57	144.18	-0.43	-0.05	172.91	181.94	1.85	1.63
5 Oct 10	12.1	75.8	-5.0	-2.5	149.88	155.64	0.29	0.61	190.79	199.35	1.24	0.74
15 Oct 20	135.3	199.8	2.5	5.0	161.47	167.35	0.89	1.14	207.62	215.64	0.19	-0.38
25 Oct 30	270.9	340.9	1.2	-4.2	173.29	179.27	1.36	1.54	223.42	231.01	-0.93	-1.45
4 Nov 9	46.9	107.5	-4.2	0.4	185.30	191.36	1.67	1.77	238.39	245.54	-1.90	-2.27
14 Nov 19	167.8	236.6	4.6	3.7	197.45	203.57	1.84	1.86	252.36	258.60	-2.50	-2.53
24 Nov 29	309.9	17.8	-2.4	-5.2	209.72	215.88	1.85	1.80	263.75	266.81	-2.28	-1.60
4 Dec 9	80.2	139.4	-1.9	3.2	222.07	228.27	1.73	1.62	266.28	261.33	-0.36	1.28
14 Dec 19	202.0	274.5	5.2	0.7	234.48	240.71	1.49	1.33	254.72	251.06	2.60	3.00
24 Dec 29	347.9	52.8	-4.8	-4.0	246.94	253.18	1.15	0.96	251.60	255.12	2.71	2.09
1721												1721

JULIAN 12 UT	SATURN		JUPITER		MARS		SUN	GREGORIAN 12 UT	MOON		VENUS		MERCURY	
	LONG.	LAT.	LONG.	LAT.	LONG.	LAT.	LONG.		LONG.	LAT.	LONG.	LAT.	LONG.	LAT.
1721								1722						1722
Dec 23	253.04	1.51	237.84	0.86	39.76	1.57	282.93	3 Jan 8	112.6 172.3	1.0 5.0	259.42 265.67	0.75 0.54	260.35 266.54	1.37 0.64
Jan 2	254.08	1.52	239.63	0.87	42.95	1.67	293.12	13 Jan 18	238.1 312.8	3.7 -2.7	271.92 278.17	0.32 0.10	273.31 280.48	-0.05 -0.66
Jan 12	255.04	1.52	241.26	0.88	46.76	1.73	303.29	23 Jan 28	24.1 85.9	-5.1 -1.4	284.42 290.68	-0.12 -0.33	287.95 295.71	-1.19 -1.61
Jan 22	255.90	1.53	242.71	0.89	51.06	1.76	313.44	2 Feb 7	145.1 207.1	3.6 5.0	296.93 303.18	-0.52 -0.71	303.76 312.13	-1.91 -2.07
Feb 1	256.64	1.55	243.93	0.90	55.74	1.77	323.55	12 Feb 17	275.8 350.6	0.5 -4.8	309.42 315.67	-0.88 -1.03	320.85 329.93	-2.06 -1.86
Feb 11	257.24	1.56	244.91	0.92	60.72	1.77	333.63	22 Feb 27	58.3 118.0	-3.4 1.6	321.91 328.15	-1.16 -1.27	339.32 348.84	-1.43 -0.75
Feb 21	257.70	1.58	245.62	0.93	65.92	1.76	343.65	4 Mar 9	178.6 243.8	5.0 2.9	334.39 340.62	-1.35 -1.41	358.03 6.06	0.17 1.26
Mar 3	258.01	1.59	246.03	0.95	71.30	1.74	353.62	14 Mar 19	314.6 27.0	-3.0 -4.8	346.84 353.06	-1.44 -1.44	11.92 14.79	2.32 3.13
Mar 13	258.15	1.61	246.14	0.96	76.82	1.71	3.54	24 Mar 29	90.7 150.2	-0.6 4.0	359.28 5.48	-1.42 -1.37	14.41 11.44	3.43 3.05
Mar 23	258.12	1.62	245.94	0.97	82.46	1.67	13.40	3 Apr 8	213.8 282.2	4.4 -0.5	11.68 17.88	-1.29 -1.19	7.48 4.33	2.03 0.70
Apr 2	257.93	1.64	245.43	0.98	88.19	1.63	23.20	13 Apr 18	353.3 61.5	-4.9 -2.8	24.07 30.25	-1.07 -0.93	3.11 4.02	-0.60 -1.66
Apr 12	257.59	1.65	244.65	0.99	93.99	1.59	32.95	23 Apr 28	122.3 183.1	2.4 5.1	36.43 42.59	-0.77 -0.60	6.79 11.09	-2.42 -2.86
Apr 22	257.11	1.66	243.63	0.99	99.85	1.55	42.64	3 May 8	250.6 321.4	2.0 -3.8	48.76 54.91	-0.41 -0.22	16.62 23.21	-3.02 -2.91
May 2	256.52	1.66	242.46	0.98	105.77	1.50	52.29	13 May 18	30.7 94.5	-4.6 0.1	61.06 67.21	-0.02 0.18	30.74 39.19	-2.56 -2.00
May 12	255.85	1.66	241.19	0.96	111.74	1.45	61.91	23 May 28	154.0 217.5	4.6 4.3	73.35 79.48	0.39 0.58	48.52 58.69	-1.25 -0.38
May 22	255.13	1.65	239.94	0.94	117.75	1.39	71.48	2 Jun 7	288.9 0.3	-1.5 -5.3	85.60 91.73	0.77 0.95	69.49 80.47	0.50 1.26
Jun 1	254.39	1.64	238.77	0.92	123.80	1.34	81.04	12 Jun 17	66.0 126.6	-2.4 3.0	97.84 103.95	1.11 1.25	91.09 100.98	1.77 1.97
Jun 11	253.68	1.62	237.77	0.88	129.89	1.28	90.58	22 Jun 27	186.6 253.7	5.3 1.7	110.06 116.15	1.38 1.48	109.98 118.04	1.88 1.51
Jun 21	253.03	1.60	237.00	0.85	136.02	1.22	100.11	2 Jul 7	328.0 37.5	-4.3 -4.3	122.24 128.33	1.55 1.60	125.14 131.20	0.91 0.11
Jul 1	252.47	1.57	236.50	0.81	142.19	1.16	109.65	12 Jul 17	99.6 158.9	0.6 4.8	134.40 140.47	1.61 1.60	136.11 139.63	-0.86 -1.95
Jul 11	252.02	1.54	236.29	0.77	148.40	1.09	119.19	22 Jul 27	220.7 291.5	4.1 -1.7	146.53 152.58	1.55 1.48	141.46 141.28	-3.06 -4.08
Jul 21	251.72	1.51	236.40	0.73	154.66	1.03	128.75	1 Aug 6	6.4 72.4	-5.2 -1.7	158.61 164.63	1.37 1.23	139.04 135.31	-4.75 -4.81
Jul 31	251.56	1.48	236.80	0.69	160.97	0.96	138.35	11 Aug 16	132.2 192.1	3.3 5.0	170.64 176.64	1.06 0.86	131.56 129.57	-4.12 -2.86
Aug 10	251.57	1.44	237.49	0.65	167.33	0.89	147.98	21 Aug 26	256.6 330.2	1.3 -4.4	182.62 188.58	0.63 0.38	130.53 134.68	-1.38 -0.02
Aug 20	251.74	1.41	238.44	0.62	173.75	0.81	157.64	31 Aug	43.0	-3.7	194.52	0.11	141.52	1.01
								Sep 5	105.6	1.4	200.44	-0.17	150.11	1.61
Aug 30	252.06	1.37	239.64	0.59	180.22	0.74	167.36	10 Sep 15	164.9 226.8	4.8 3.3	206.34 212.22	-0.47 -0.79	159.46 168.89	1.81 1.71
Sep 9	252.54	1.34	241.06	0.56	186.75	0.66	177.13	20 Sep 25	294.4 8.7	-2.2 -4.9	218.07 223.88	-1.10 -1.42	178.06 186.86	1.39 0.94
Sep 19	253.16	1.31	242.66	0.53	193.34	0.58	186.96	30 S-O 5	77.4 137.7	-0.8 3.9	229.67 235.41	-1.73 -2.03	195.29 203.38	0.41 -0.17
Sep 29	253.90	1.28	244.44	0.50	200.00	0.49	196.84	10 Oct 15	198.3 263.3	4.7 0.2	241.11 246.77	-2.31 -2.58	211.16 218.65	-0.75 -1.31
Oct 9	254.76	1.26	246.35	0.48	206.73	0.40	206.78	20 Oct 25	333.4 45.5	-4.7 -3.2	252.36 257.88	-2.81 -3.01	225.84 232.70	-1.83 -2.27
Oct 19	255.72	1.23	248.38	0.46	213.53	0.31	216.77	30 Oct	110.0	2.2	263.32	-3.17	239.09	-2.60
								Nov 4	169.6	5.1	268.67	-3.29	244.75	-2.75
Oct 29	256.76	1.21	250.51	0.44	220.40	0.22	226.82	9 Nov 14	232.9 301.6	2.7 -3.2	273.91 279.02	-3.35 -3.35	249.13 251.26	-2.64 -2.10
Nov 8	257.86	1.19	252.71	0.42	227.35	0.12	236.91	19 Nov 24	12.4 80.3	-4.9 -0.3	283.96 288.70	-3.28 -3.13	249.75 244.22	-0.97 0.67
Nov 18	259.00	1.18	254.95	0.40	234.37	0.02	247.04	29 N-D 4	141.7 202.1	4.5 4.6	293.21 297.42	-2.89 -2.56	237.86 235.08	2.11 2.72
Nov 28	260.17	1.17	257.22	0.39	241.46	-0.08	257.20	9 Dec 14	269.1 340.8	-0.5 -5.2	301.29 304.72	-2.12 -1.56	236.60 240.94	2.62 2.14
Dec 8	261.35	1.16	259.49	0.37	248.62	-0.19	267.39	19 Dec 24	49.8 113.5	-2.9 2.6	307.61 309.84	-0.86 -0.03	246.80 253.45	1.52 0.85
Dec 18	262.52	1.15	261.75	0.36	255.86	-0.29	277.58	29 Dec	173.3	5.3	311.30	0.95	260.54	0.19
1722								1723						1723
								Jan 3	235.6	2.5	311.84	2.08	267.91	-0.42
Dec 28	263.65	1.14	263.96	0.35	263.17	-0.40	287.77	8 Jan 13	306.9 19.7	-3.6 -4.7	311.38 309.87	3.31 4.58	275.49 283.27	-0.96 -1.41
Jan 7	264.73	1.14	266.10	0.34	270.54	-0.51	297.96	18 Jan 23	85.2 146.0	0.2 4.6	307.47 304.50	5.78 6.78	291.27 299.50	-1.77 -2.00
Jan 17	265.74	1.14	268.16	0.33	277.97	-0.62	308.12	28 J-F 2	205.7 270.9	4.3 -0.8	301.46 298.86	7.47 7.79	308.01 316.80	-2.08 -1.97
Jan 27	266.66	1.15	270.10	0.32	285.47	-0.74	318.26	7 Feb 12	345.6 56.7	-5.1 -2.1	297.07 296.28	7.76 7.47	325.83 334.92	-1.65 -1.07
Feb 6	267.48	1.15	271.89	0.31	293.01	-0.84	328.35	17 Feb 22	119.0 178.5	3.1 5.0	296.51 297.66	6.99 6.39	343.63 351.09	-0.21 0.90
Feb 16	268.17	1.16	273.52	0.30	300.60	-0.95	338.40	27 Feb	239.4	1.8	299.62	5.72	356.11	2.11
								Mar 4	308.1	-3.8	302.27	5.03	357.65	3.14
Feb 26	268.73	1.16	274.95	0.29	308.23	-1.05	348.40	9 Mar 14	23.9 91.6	-4.1 1.1	305.46 309.11	4.33 3.64	355.57 351.24	3.63 3.34
Mar 8	269.13	1.17	276.16	0.28	315.89	-1.15	358.35	19 Mar 24	151.7 211.7	4.8 3.6	313.13 317.45	2.97 2.34	346.98 344.59	2.37 1.11
Mar 18	269.38	1.18	277.12	0.27	323.57	-1.23	8.23	29 M-A 3	275.1 346.9	-1.6 -5.1	322.02 326.80	1.73 1.16	344.60 346.74	-0.12 -1.14
Mar 28	269.47	1.19	277.80	0.26	331.25	-1.31	18.07	8 Apr 13	60.6 124.7	-1.3 3.9	331.75 336.84	0.64 0.16	350.56 355.67	-1.91 -2.43
Apr 7	269.39	1.19	278.18	0.25	338.93	-1.38	27.84	18 Apr 23	184.2 246.1	4.8 0.8	342.05 347.36	-0.28 -0.67	1.82 8.85	-2.70 -2.73
Apr 17	269.15	1.20	278.25	0.23	346.58	-1.44	37.57	28 Apr	313.0	-4.4	352.77	-1.02	16.68	-2.53
								May 3	25.8	-3.9	358.24	-1.31	25.29	-2.11
Apr 27	268.77	1.20	278.01	0.22	354.21	-1.48	47.24	8 May 13	95.4 156.8	1.9 5.2	3.79 9.38	-1.56 -1.76	34.68 44.82	-1.48 -0.69
May 7	268.25	1.20	277.47	0.20	1.79	-1.51	56.88	18 May 23	217.1 282.3	3.2 -2.5	15.03 20.71	-1.92 -2.03	55.55 66.50	0.18 1.01
May 17	267.64	1.20	276.65	0.18	9.30	-1.52	66.47	28 M-J 2	352.3 63.3	-5.2 -1.0	26.44 32.20	-2.09 -2.12	77.14 87.00	1.66 2.03
May 27	266.95	1.19	275.61	0.16	16.73	-1.51	76.04	7 Jun 12	128.5 188.5	4.3 4.8	37.99 43.81	-2.10 -2.05	95.87 103.65	2.08 1.82
Jun 6	266.22	1.18	274.41	0.14	24.07	-1.49	85.59	17 Jun 22	251.1 320.5	0.3 -4.8	49.66 55.53	-1.96 -1.84	110.29 115.71	1.28 0.48
Jun 16	265.49	1.16	273.14	0.12	31.29	-1.45	95.12	27 Jun	31.5	-3.6	61.42	-1.70	119.75	-0.55
								Jul 2	98.8	2.2	67.34	-1.53	122.17	-1.74
Jun 26	264.79	1.15	271.88	0.09	38.37	-1.39	104.66	7 Jul 12	160.8 220.7	5.2 2.8	73.27 79.23	-1.34 -1.13	122.73 121.36	-2.99 -4.11
Jul 6	264.17	1.12	270.72	0.07	45.30	-1.31	114.19	17 Jul 22	286.8 359.8	-2.8 -4.9	85.21 91.21	-0.92 -0.69	118.42 115.00	-4.82 -4.88
Jul 16	263.64	1.10	269.75	0.05	52.05	-1.21	123.74	27 J-A 1	69.2 132.8	-0.4 4.4	97.23 103.27	-0.46 -0.22	112.54 112.24	-4.25 -3.12
Jul 26	263.23	1.07	269.01	0.02	58.59	-1.09	133.32	6 Aug 11	192.8 254.0	4.4 -0.2	109.33 115.40	0.01 0.23	114.64 119.74	-1.77 -0.46
Aug 5	262.97	1.05	268.57	0.00	64.89	-0.95	142.93	16 Aug 21	324.6 38.8	-4.8 -2.7	121.50 127.61	0.44 0.64	127.16 136.17	0.63 1.37
Aug 15	262.87	1.02	268.43	-0.02	70.91	-0.79	152.58	26 Aug 31	104.9 165.8	2.8 5.0	133.74 139.89	0.82 0.99	145.92 155.72	1.73 1.76
Aug 25	262.92	0.99	268.60	-0.04	76.61	-0.61	162.27	5 Sep 10	225.2 289.2	2.0 -3.3	146.06 152.24	1.13 1.24	165.21 174.26	1.53 1.13
Sep 4	263.14	0.96	269.09	-0.06	81.92	-0.40	172.01	15 Sep 20	3.5 75.9	-4.5 0.7	158.44 164.64	1.34 1.40	182.85 191.01	0.63 0.05
Sep 14	263.51	0.93	269.87	-0.08	86.75	-0.16	181.80	25 Sep 30	138.9 198.5	4.8 3.8	170.86 177.10	1.44 1.45	198.78 206.18	-0.55 -1.16
Sep 24	264.04	0.91	270.92	-0.09	91.01	0.11	191.66	5 Oct 10	258.9 326.5	-1.1 -5.0	183.34 189.59	1.43 1.38	213.18 219.73	-1.75 -2.27
Oct 4	264.70	0.88	272.22	-0.11	94.58	0.43	201.56	15 Oct 20	41.9 110.9	-2.1 3.7	195.84 202.11	1.31 1.22	225.67 230.71	-2.70 -2.97
Oct 14	265.48	0.86	273.74	-0.13	97.27	0.79	211.53	25 Oct 30	171.7 231.4	5.0 1.2	208.38 214.65	1.10 0.96	234.28 235.42	-2.99 -2.59
Oct 24	266.38	0.84	275.45	-0.14	98.90	1.20	221.55	4 Nov 9	294.6 5.2	-4.0 -4.5	220.93 227.21	0.80 0.63	232.97 227.07	-1.57 0.03
Nov 3	267.37	0.82	277.33	-0.15	99.28	1.65	231.61	14 Nov 19	78.8 144.2	1.3 5.2	233.49 239.77	0.45 0.25	221.23 219.40	1.57 2.38
Nov 13	268.43	0.80	279.35	-0.17	98.24	2.14	241.72	24 Nov 29	203.8 265.5	3.5 -1.8	246.05 252.34	0.06 -0.14	221.84 226.94	2.48 2.16
Nov 23	269.55	0.78	281.48	-0.18	95.81	2.63	251.87	4 Dec 9	332.4 43.9	-5.3 -1.9	258.62 264.91	-0.34 -0.54	233.40 240.50	1.64 1.03
Dec 3	270.70	0.77	283.70	-0.20	92.31	3.05	262.04	14 Dec 19	113.9 176.2	4.0 4.9	271.19 277.47	-0.72 -0.90	247.90 255.47	0.41 -0.19
Dec 13	271.88	0.76	285.98	-0.21	88.38	3.35	272.23	24 Dec 29	236.0 301.4	0.8 -4.3	283.75 290.03	-1.05 -1.19	263.15 270.96	-0.74 -1.22
1723								1723						1723

JULIAN 12 UT	SATURN LONG.	SATURN LAT.	JUPITER LONG.	JUPITER LAT.	MARS LONG.	MARS LAT.	SUN LONG.	GREGORIAN 12 UT	MOON LONG.		MOON LAT.		VENUS LONG.		VENUS LAT.		MERCURY LONG.		MERCURY LAT.	
1723								1724												1724
Dec 23	273.05	0.75	288.31	-0.23	84.81	3.49	282.43	3 Jan 8	11.6	81.5	-4.1	1.5	296.30	302.57	-1.32	-1.41	278.89	287.00	-1.62	-1.91
Jan 2	274.21	0.74	290.65	-0.24	82.24	3.50	292.62	13 Jan 18	147.4	207.6	5.1	3.0	308.83	315.09	-1.49	-1.53	295.29	303.79	-2.07	-2.06
Jan 12	275.32	0.73	293.00	-0.26	80.98	3.40	302.79	23 Jan 28	269.3	339.2	-2.2	-5.0	321.34	327.59	-1.55	-1.54	312.44	321.08	-1.85	-1.38
Jan 22	276.38	0.73	295.31	-0.28	81.03	3.25	312.95	2 Feb 7	50.6	117.6	-1.0	4.1	333.82	340.04	-1.50	-1.43	329.25	336.08	-0.60	0.50
Feb 1	277.36	0.72	297.58	-0.29	82.25	3.07	323.06	12 Feb 17	180.0	239.4	4.4	0.1	346.25	352.44	-1.33	-1.21	340.21	340.40	1.80	3.03
Feb 11	278.25	0.72	299.78	-0.31	84.43	2.89	333.14	22 Feb 27	304.3	18.0	-4.5	-3.3	358.62	4.79	-1.06	-0.88	336.80	331.60	3.70	3.50
Feb 21	279.02	0.72	301.88	-0.34	87.39	2.71	343.17	3 Mar 8	88.4	152.1	2.5	5.0	10.93	17.05	-0.68	-0.46	327.65	326.31	2.60	1.43
Mar 2	279.67	0.71	303.87	-0.36	90.96	2.53	353.14	13 Mar 18	212.1	272.4	2.2	-2.9	23.15	29.23	-0.23	0.02	327.51	330.69	0.29	-0.69
Mar 12	280.17	0.71	305.71	-0.39	95.02	2.37	3.06	23 Mar 28	341.3	56.5	-4.9	0.0	35.28	41.30	0.28	0.55	335.31	341.02	-1.47	-2.03
Mar 22	280.53	0.71	307.38	-0.41	99.45	2.22	12.92	2 Apr 7	124.3	185.4	4.8	4.0	47.30	53.25	0.81	1.07	347.60	354.92	-2.38	-2.53
Apr 1	280.72	0.71	308.85	-0.44	104.19	2.07	22.73	12 Apr 17	244.6	307.4	-0.8	-5.0	59.18	65.06	1.33	1.57	2.93	11.61	-2.45	-2.17
Apr 11	280.76	0.71	310.10	-0.47	109.18	1.94	32.48	22 Apr 27	19.8	93.7	-3.1	3.2	70.91	76.71	1.80	2.01	20.98	31.03	-1.67	-0.98
Apr 21	280.63	0.71	311.09	-0.51	114.37	1.81	42.18	2 May 7	158.4	217.9	5.1	1.6	82.46	88.15	2.19	2.34	41.63	52.48	-0.15	0.72
May 1	280.34	0.71	311.82	-0.54	119.73	1.68	51.83	12 May 17	278.4	344.6	-3.6	-5.0	93.78	99.35	2.45	2.53	63.05	72.80	1.49	2.03
May 11	279.92	0.70	312.24	-0.58	125.23	1.56	61.45	22 May 27	58.4	129.2	0.2	5.1	104.84	110.24	2.56	2.54	81.42	88.76	2.25	2.14
May 21	279.37	0.69	312.35	-0.62	130.86	1.45	71.03	1 Jun 6	191.0	250.5	3.7	-1.3	115.55	120.74	2.47	2.34	94.75	99.26	1.68	0.91
May 31	278.73	0.68	312.15	-0.66	136.60	1.34	80.58	11 Jun 16	314.1	23.4	-5.1	-2.8	125.80	130.71	2.15	1.90	102.13	103.17	-0.16	-1.44
Jun 10	278.03	0.67	311.63	-0.70	142.45	1.23	90.13	21 Jun 26	96.1	162.9	3.4	4.9	135.45	139.98	1.57	1.18	102.35	100.00	-2.80	-3.98
Jun 20	277.29	0.66	310.83	-0.73	148.39	1.12	99.66	1 Jul 6	222.8	284.4	1.1	-3.9	144.26	148.25	0.70	0.14	96.99	94.55	-4.69	-4.76
Jun 30	276.57	0.64	309.80	-0.77	154.41	1.01	109.19	11 Jul 16	351.8	62.5	-4.5	0.7	151.88	155.09	-0.51	-1.25	93.76	95.16	-4.23	-3.27
Jul 10	275.89	0.63	308.59	-0.80	160.53	0.91	118.74	21 Jul 26	132.2	195.2	5.0	3.1	157.79	159.86	-2.09	-3.01	98.91	104.88	-2.09	-0.87
Jul 20	275.29	0.61	307.30	-0.82	166.73	0.81	128.30	31 J-A 5	254.8	320.1	-1.9	-5.0	161.19	161.65	-4.02	-5.07	112.81	122.18	0.24	1.09
Jul 30	274.80	0.59	306.02	-0.84	173.01	0.70	137.89	10 Aug 15	30.9	100.7	-1.8	3.8	161.15	159.68	-6.13	-7.12	132.28	142.42	1.60	1.77
Aug 9	274.43	0.57	304.84	-0.85	179.38	0.60	147.52	20 Aug 25	166.6	226.9	4.5	0.3	157.35	154.43	-7.93	-8.44	152.21	161.49	1.65	1.32
Aug 19	274.22	0.54	303.84	-0.85	185.84	0.50	157.18	30 Aug	287.9		-4.3		151.36		-8.58		170.24		0.85	
								Sep 4		357.6		-4.0		148.62		-8.35		178.48		0.28
Aug 29	274.16	0.52	303.09	-0.86	192.38	0.40	166.89	9 Sep 14	70.0	137.4	1.8	5.1	146.62	145.56	-7.81	-7.06	186.24	193.53	-0.34	-0.99
Sep 8	274.27	0.50	302.64	-0.85	199.02	0.30	176.66	19 Sep 24	199.6	258.7	2.4	-2.7	145.51	146.40	-6.18	-5.26	200.33	206.54	-1.64	-2.26
Sep 18	274.54	0.48	302.52	-0.84	205.74	0.19	186.49	29 S-O 4	322.7	36.2	-5.1	-1.0	148.12	150.55	-4.34	-3.46	212.00	216.36	-2.79	-3.19
Sep 28	274.96	0.46	302.72	-0.84	212.55	0.09	196.36	9 Oct 14	108.0	172.1	4.6	4.3	153.58	157.11	-2.62	-1.85	219.05	219.17	-3.34	-3.08
Oct 8	275.54	0.44	303.24	-0.83	219.45	-0.01	206.30	19 Oct 24	231.8	291.8	-0.5	-4.8	161.06	165.34	-1.14	-0.50	215.84	209.83	-2.17	-0.62
Oct 18	276.24	0.42	304.07	-0.82	226.45	-0.11	216.29	29 Oct	359.3		-3.9		169.91		0.08		204.76		1.00	
								Nov 3		74.7		2.5		174.72		0.59		203.93		2.00
Oct 28	277.07	0.40	305.18	-0.81	233.54	-0.21	226.33	8 Nov 13	144.0	205.2	5.3	2.0	179.74	184.93	1.04	1.42	207.24	213.06	2.31	2.15
Nov 7	278.00	0.39	306.55	-0.80	240.72	-0.31	236.42	18 Nov 23	264.3	326.6	-3.2	-5.2	190.26	195.72	1.73	1.99	220.09	227.62	1.74	1.20
Nov 17	279.02	0.37	308.14	-0.79	247.98	-0.41	246.55	28 N-D 3	37.2	112.1	-0.9	4.8	201.29	206.95	2.18	2.32	235.33	243.10	0.62	0.03
Nov 27	280.11	0.36	309.93	-0.79	255.34	-0.50	256.70	8 Dec 13	178.1	237.4	3.9	-1.0	212.68	218.49	2.41	2.44	250.90	258.73	-0.53	-1.04
Dec 7	281.25	0.34	311.88	-0.78	262.77	-0.60	266.89	18 Dec 23	298.1	3.5	-4.9	-3.5	224.36	230.27	2.43	2.37	266.62	274.60	-1.47	-1.82
Dec 17	282.41	0.33	313.97	-0.78	270.29	-0.69	277.08	28 Dec	75.7		2.6		236.23		2.27		282.69		-2.05	
1724								1725												1725
								Jan 2		147.8		5.0		242.22		2.13		290.89		-2.13
Dec 27	283.59	0.32	316.17	-0.78	277.87	-0.77	287.27	7 Jan 12	210.4	269.8	1.3	-3.6	248.24	254.29	1.96	1.77	299.15	307.30	-2.02	-1.66
Jan 6	284.77	0.31	318.46	-0.79	285.52	-0.85	297.46	17 Jan 22	333.6	42.0	-4.7	-0.1	260.37	266.45	1.55	1.32	314.90	321.04	-0.98	0.06
Jan 16	285.91	0.30	320.81	-0.79	293.23	-0.93	307.63	27 J-F 1	113.8	181.7	4.7	3.3	272.56	278.68	1.07	0.81	324.24	323.15	1.42	2.80
Jan 26	287.01	0.29	323.20	-0.80	300.98	-0.99	317.76	6 Feb 11	241.9	303.1	-1.7	-5.0	284.80	290.94	0.55	0.29	318.27	312.73	3.64	3.55
Feb 5	288.04	0.28	325.60	-0.81	308.77	-1.05	327.86	16 Feb 21	10.8	81.1	-2.5	3.3	297.08	303.23	0.04	-0.20	309.49	309.29	2.74	1.67
Feb 15	288.99	0.27	328.00	-0.83	316.59	-1.09	337.92	26 Feb	150.5		4.7		309.38		-0.44		311.56		0.62	
								Mar 3		214.1		0.4		315.54		-0.65		315.59		-0.31
Feb 25	289.85	0.26	330.37	-0.84	324.42	-1.13	347.92	8 Mar 13	273.6	338.0	-4.2	-4.5	321.69	327.85	-0.85	-1.02	320.84	327.00	-1.07	-1.67
Mar 7	290.58	0.25	332.69	-0.86	332.26	-1.16	357.87	18 Mar 23	49.3	119.7	1.0	5.1	334.01	340.17	-1.17	-1.30	333.88	341.38	-2.09	-2.32
Mar 17	291.18	0.24	334.95	-0.88	340.08	-1.17	7.76	28 M-A 2	185.4	245.9	2.8	-2.5	346.33	352.48	-1.40	-1.47	349.47	358.16	-2.36	-2.19
Mar 27	291.64	0.23	337.10	-0.91	347.88	-1.17	17.59	7 Apr 12	306.2	14.5	-5.3	-2.1	358.64	4.79	-1.51	-1.52	7.44	17.34	-1.82	-1.24
Apr 6	291.95	0.22	339.15	-0.94	355.65	-1.16	27.38	17 Apr 22	88.3	156.7	4.1	4.6	10.94	17.09	-1.51	-1.46	27.76	38.43	-0.47	0.41
Apr 16	292.10	0.21	341.05	-0.97	3.37	-1.13	37.10	27 Apr	218.7		-0.1		23.24		-1.39		48.84		1.27	
								May 2		278.0		-4.6		29.38		-1.30		58.39		1.97
Apr 26	292.08	0.20	342.79	-1.00	11.03	-1.09	46.78	7 May 12	340.6	52.5	-4.4	1.4	35.52	41.67	-1.19	-1.05	66.64	73.37	2.38	2.43
May 6	291.90	0.19	344.34	-1.04	18.63	-1.04	56.41	17 May 22	126.5	191.5	5.3	2.3	47.81	53.94	-0.89	-0.72	78.45	81.74	2.11	1.39
May 16	291.58	0.18	345.67	-1.08	26.14	-0.97	66.01	27 M-J 1	251.2	311.2	-2.9	-5.2	60.08	66.22	-0.54	-0.35	83.11	82.59	0.31	-1.04
May 26	291.11	0.17	346.76	-1.13	33.57	-0.89	75.58	6 Jun 11	16.8	91.3	-1.8	4.2	72.36	78.50	-0.16	0.04	80.53	77.77	-2.46	-3.65
Jun 5	290.53	0.15	347.58	-1.18	40.90	-0.80	85.13	16 Jun 21	162.8	224.6	4.1	-0.7	84.64	90.78	0.24	0.43	75.44	74.51	-4.35	-4.47
Jun 15	289.87	0.14	348.10	-1.23	48.12	-0.70	94.66	26 Jun	283.9		-4.7		96.92		0.62		75.49		-4.08	
								Jul 1		346.1		-3.8		103.07		0.79		78.47		-3.32
Jun 25	289.15	0.13	348.31	-1.28	55.22	-0.59	104.19	6 Jul 11	54.9	129.5	1.7	5.0	109.22	115.37	0.95	1.09	83.39	90.13	-2.33	-1.22
Jul 5	288.42	0.11	348.19	-1.33	62.20	-0.47	113.73	16 Jul 21	197.0	256.8	1.5	-3.4	121.52	127.68	1.21	1.31	98.50	108.17	-0.14	0.78
Jul 15	287.70	0.10	347.75	-1.38	69.05	-0.34	123.28	26 Jul 31	317.7	22.9	-4.8	-0.8	133.83	139.99	1.39	1.44	118.54	128.98	1.42	1.73
Jul 25	287.04	0.08	347.01	-1.42	75.75	-0.20	132.85	5 Aug 10	94.0	166.0	4.4	3.6	146.15	152.31	1.46	1.46	139.05	148.55	1.74	1.50
Aug 4	286.46	0.07	346.01	-1.46	82.30	-0.05	142.46	15 Aug 20	229.5	289.1	-1.6	-5.0	158.47	164.63	1.43	1.37	157.44	165.74	1.08	0.53
Aug 14	286.01	0.05	344.82	-1.49	88.67	0.11	152.11	25 Aug 30	352.9	61.4	-3.1	2.7	170.79	176.94	1.28	1.16	173.47	180.64	-0.11	-0.80
Aug 24	285.69	0.04	343.52	-1.50	94.86	0.28	161.79	4 Sep 9	132.7	200.5	5.0	0.8	183.09	189.24	1.02	0.86	187.19	193.03	-1.52	-2.22
Sep 3	285.52	0.02	342.20	-1.51	100.84	0.47	171.54	14 Sep 19	261.2	322.2	-4.1	-4.8	195.39	201.53	0.68	0.47	197.94	201.56	-2.87	-3.40
Sep 13	285.52	0.01	340.97	-1.50	106.58	0.66	181.33	24 Sep 29	29.8	100.8	0.1	5.0	207.67	213.80	0.25	0.02	203.29	202.35	-3.69	-3.56
Sep 23	285.68	0.00	339.91	-1.49	112.04	0.88	191.18	4 Oct 9	169.8	233.3	3.3	-2.2	219.93	226.05	-0.22	-0.46	198.29	192.47	-2.75	-1.24
Oct 3	286.00	-0.02	339.10	-1.46	117.18	1.11	201.08	14 Oct 19	292.9	356.4	-5.3	-2.8	232.16	238.27	-0.70	-0.95	188.39	188.59	0.41	1.57
Oct 13	286.48	-0.03	338.60	-1.43	121.93	1.36	211.04	24 Oct 29	68.1	139.6	3.4	4.9	244.37	250.46	-1.18	-1.40	192.73	199.23	2.09	2.10
Oct 23	287.10	-0.04	338.42	-1.40	126.22	1.64	221.06	3 Nov 8	204.8	265.4	0.3	-4.4	256.54	262.60	-1.61	-1.79	206.80	214.76	1.81	1.35
Nov 2	287.86	-0.05	338.59	-1.36	129.92	1.95	231.13	13 Nov 18	325.3	32.3	-4.7	0.4	268.66	274.69	-1.95	-2.08	222.78	230.77	0.81	0.24
Nov 12	288.73	-0.07	339.10	-1.32	132.93	2.29	241.23	23 Nov 28	107.1	176.5	5.1	2.7	280.71	286.70	-2.18	-2.24	238.69	246.56	-0.32	-0.85
Nov 22	289.69	-0.08	339.92	-1.28	135.06	2.66	251.38	3 Dec 8	238.3	297.7	-2.6	-5.1	292.67	298.61	-2.27	-2.25	254.41	262.27	-1.32	-1.72
Dec 2	290.74	-0.09	341.03	-1.25	136.13	3.07	261.55	13 Dec 18	359.2	69.8	-2.4	3.5	304.50	310.36	-2.18	-2.07	270.15	278.06	-2.02	-2.18
Dec 12	291.86	-0.10	342.41	-1.22	135.96	3.50	271.73	23 Dec 28	145.2	211.1	4.4	-0.5	316.16	321.90	-1.91	-1.70	285.94	293.59	-2.17	-1.93
1725								1725												1725

JULIAN 12 UT	SATURN LONG.	SATURN LAT.	JUPITER LONG.	JUPITER LAT.	MARS LONG.	MARS LAT.	SUN LONG.	GREGORIAN 12 UT	MOON LONG.		MOON LAT.		VENUS LONG.		VENUS LAT.		MERCURY LONG.		MERCURY LAT.	
1725								1726												1726
Dec 22	293.01	-0.11	344.01	-1.19	134.44	3.91	281.93	2 Jan 7	270.9	330.9	-4.6	-4.0	327.56	333.15	-1.43	-1.12	300.57	305.95	-1.37	-0.41
Jan 1	294.19	-0.12	345.81	-1.16	131.64	4.26	292.12	12 Jan 17	35.0	108.4	1.0	5.0	338.63	344.00	-0.75	-0.34	308.21	305.92	0.96	2.47
Jan 11	295.38	-0.14	347.78	-1.14	127.93	4.47	302.30	22 Jan 27	181.5	244.1	1.8	-3.3	349.23	354.30	0.13	0.65	300.12	294.67	3.46	3.49
Jan 21	296.55	-0.15	349.89	-1.13	123.96	4.50	312.45	1 Feb 6	303.6	5.5	-4.9	-1.4	359.18	3.82	1.21	1.82	292.36	293.26	2.81	1.86
Jan 31	297.68	-0.16	352.11	-1.11	120.48	4.37	322.57	11 Feb 16	72.9	146.7	4.0	4.1	8.20	12.24	2.47	3.16	296.47	301.20	0.89	0.02
Feb 10	298.77	-0.18	354.42	-1.10	118.04	4.11	332.64	21 Feb 26	215.6	276.2	-1.4	-5.0	15.88	19.03	3.87	4.61	306.95	313.45	-0.73	-1.34
Feb 20	299.78	-0.19	356.78	-1.09	116.89	3.78	342.68	3 Mar 8	336.8	41.9	-3.5	2.0	21.61	23.47	5.36	6.08	320.55	328.17	-1.80	-2.11
Mar 2	300.70	-0.20	359.19	-1.09	117.02	3.43	352.66	13 Mar 18	112.1	183.6	5.2	1.4	24.50	24.57	6.76	7.34	336.28	344.91	-2.24	-2.19
Mar 12	301.52	-0.22	1.61	-1.09	118.27	3.09	2.57	23 Mar 28	248.2	308.0	-4.0	-5.0	23.62	21.67	7.75	7.90	354.07	3.77	-1.93	-1.46
Mar 22	302.22	-0.23	4.02	-1.09	120.45	2.78	12.44	2 Apr 7	11.1	80.3	-0.7	4.6	18.96	15.85	7.72	7.18	13.95	24.37	-0.78	0.07
Apr 1	302.78	-0.25	6.41	-1.10	123.39	2.49	22.25	12 Apr 17	151.1	218.5	3.9	-1.8	12.84	10.40	6.33	5.25	34.55	43.81	0.99	1.84
Apr 11	303.20	-0.27	8.76	-1.11	126.95	2.22	32.00	22 Apr 27	280.1	340.4	-5.2	-3.3	8.85	8.30	4.08	2.92	51.57	57.49	2.44	2.70
Apr 21	303.46	-0.29	11.04	-1.13	130.99	1.98	41.71	2 May 7	47.3	119.7	2.5	5.1	8.74	10.06	1.84	0.87	61.38	63.13	2.53	1.90
May 1	303.56	-0.31	13.23	-1.14	135.43	1.76	51.36	12 May 17	188.6	252.1	0.9	-4.2	12.12	14.82	0.03	-0.69	62.78	60.76	0.83	-0.55
May 11	303.50	-0.32	15.32	-1.16	140.20	1.56	60.98	22 May 27	312.0	14.1	-4.7	-0.4	18.03	21.68	-1.29	-1.78	57.97	55.58	-1.98	-3.14
May 21	303.28	-0.34	17.28	-1.19	145.24	1.37	70.56	1 Jun 6	85.4	158.5	4.7	3.2	25.67	29.95	-2.17	-2.48	54.55	55.31	-3.85	-4.06
May 31	302.91	-0.36	19.08	-1.22	150.51	1.19	80.12	11 Jun 16	224.0	284.9	-2.3	-5.0	34.47	39.18	-2.70	-2.85	57.92	62.24	-3.84	-3.29
Jun 10	302.41	-0.38	20.70	-1.25	155.99	1.03	89.66	21 Jun 26	344.5	49.7	-2.6	2.8	44.07	49.10	-2.93	-2.95	68.14	75.51	-2.49	-1.53
Jun 20	301.80	-0.40	22.12	-1.28	161.65	0.87	99.20	1 Jul 6	124.4	195.5	4.7	-0.1	54.26	59.52	-2.91	-2.83	84.26	94.14	-0.51	0.44
Jun 30	301.12	-0.41	23.29	-1.32	167.47	0.73	108.73	11 Jul 16	257.9	317.5	-4.5	-4.1	64.87	70.30	-2.70	-2.54	104.73	115.41	1.20	1.65
Jul 10	300.39	-0.43	24.21	-1.36	173.44	0.58	118.27	21 Jul 26	18.4	87.4	0.5	4.9	75.81	81.39	-2.34	-2.11	125.71	135.40	1.80	1.67
Jul 20	299.65	-0.44	24.83	-1.41	179.54	0.45	127.84	31 J-A 5	162.9	230.4	2.5	-3.2	87.02	92.72	-1.87	-1.60	144.41	152.74	1.32	0.79
Jul 30	298.94	-0.46	25.14	-1.45	185.79	0.32	137.42	10 Aug 15	290.7	350.6	-5.0	-1.8	98.46	104.25	-1.32	-1.04	160.41	167.41	0.15	-0.58
Aug 9	298.30	-0.47	25.13	-1.49	192.16	0.19	147.05	20 Aug 25	54.3	126.3	3.6	4.7	110.08	115.96	-0.75	-0.46	173.66	179.05	-1.37	-2.17
Aug 19	297.75	-0.48	24.78	-1.53	198.65	0.07	156.71	30 Aug	199.6		-0.8		121.88		-0.18		183.33		-2.94	
								Sep 4		263.6		-5.0		127.83		0.09		186.10		-3.61
Aug 29	297.33	-0.49	24.11	-1.56	205.27	-0.05	166.43	9 Sep 14	323.2	25.1	-3.8	1.4	133.82	139.84	0.34	0.58	186.79	184.81	-4.04	-4.01
Sep 8	297.06	-0.49	23.17	-1.59	212.00	-0.16	176.19	19 Sep 24	92.3	164.9	5.2	2.2	145.89	151.97	0.80	1.00	180.27	175.00	-3.27	-1.82
Sep 18	296.94	-0.50	22.00	-1.61	218.85	-0.27	186.01	29 S-O 4	234.3	295.7	-3.7	-5.1	158.09	164.22	1.17	1.31	172.10	173.36	-0.17	1.12
Sep 28	296.99	-0.50	20.70	-1.61	225.82	-0.37	195.89	9 Oct 14	356.0	61.5	-1.3	4.1	170.38	176.55	1.42	1.50	178.28	185.40	1.82	2.01
Oct 8	297.20	-0.51	19.35	-1.61	232.90	-0.47	205.82	19 Oct 24	131.6	202.1	4.5	-1.1	182.75	188.96	1.55	1.58	193.50	201.88	1.86	1.49
Oct 18	297.58	-0.51	18.06	-1.59	240.08	-0.56	215.81	29 Oct	267.4		-5.1		195.19		1.57		210.23		1.00	
								Nov 3		327.3		-3.5		201.44		1.53		218.44		0.45
Oct 28	298.11	-0.52	16.93	-1.55	247.38	-0.65	225.85	8 Nov 13	29.9	99.8	1.8	5.2	207.69	213.95	1.46	1.37	226.49	234.41	-0.11	-0.65
Nov 7	298.78	-0.52	16.05	-1.51	254.77	-0.74	235.93	18 Nov 23	170.5	237.6	1.6	-3.8	220.22	226.49	1.25	1.11	242.23	249.98	-1.16	-1.61
Nov 17	299.58	-0.53	15.47	-1.47	262.25	-0.81	246.06	28 N-D 3	299.5	359.2	-4.7	-0.8	232.78	239.06	0.95	0.78	257.68	265.30	-1.97	-2.22
Nov 27	300.50	-0.53	15.22	-1.42	269.83	-0.88	256.22	8 Dec 13	65.5	138.9	4.3	3.7	245.35	251.64	0.59	0.40	272.77	279.91	-2.30	-2.17
Dec 7	301.50	-0.54	15.32	-1.36	277.48	-0.94	266.39	18 Dec 23	208.0	271.6	-1.8	-5.0	257.93	264.22	0.20	0.00	286.23	290.82	-1.74	-0.89
Dec 17	302.58	-0.55	15.76	-1.31	285.20	-0.99	276.59	28 Dec	331.4		-2.8		270.51		-0.20		292.11		0.44	
1726								1727												1727
								Jan 2		32.3		2.3		276.81		-0.40		288.77		2.03
Dec 27	303.72	-0.56	16.52	-1.26	292.99	-1.03	286.78	7 Jan 12	103.0	177.5	5.0	0.5	283.09	289.38	-0.59	-0.76	282.37	277.35	3.18	3.36
Jan 6	304.89	-0.57	17.59	-1.21	300.82	-1.06	296.96	17 Jan 22	243.7	304.5	-4.4	-4.3	295.67	301.95	-0.92	-1.07	276.05	277.98	2.82	1.99
Jan 16	306.09	-0.58	18.91	-1.17	308.68	-1.07	307.13	27 J-F 1	3.8	67.5	0.2	4.7	308.23	314.51	-1.19	-1.29	282.00	287.34	1.12	0.30
Jan 26	307.27	-0.59	20.47	-1.13	316.57	-1.08	317.27	6 Feb 11	141.5	214.3	3.3	-2.8	320.77	327.04	-1.37	-1.42	293.51	300.29	-0.43	-1.04
Feb 5	308.44	-0.60	22.23	-1.09	324.47	-1.08	327.36	16 Feb 21	277.6	337.0	-5.2	-2.1	333.30	339.55	-1.45	-1.45	307.54	315.23	-1.54	-1.90
Feb 15	309.57	-0.62	24.15	-1.06	332.37	-1.06	337.42	26 Feb	37.6		3.2		345.80		-1.42		323.33		-2.11	
								Mar 3		104.8		5.1		352.03		-1.36		331.87		-2.15
Feb 25	310.64	-0.64	26.22	-1.04	340.24	-1.03	347.43	8 Mar 13	179.7	249.3	0.1	-4.9	358.26	4.47	-1.28	-1.18	340.86	350.33	-2.01	-1.65
Mar 7	311.63	-0.66	28.39	-1.01	348.10	-0.99	357.38	18 Mar 23	310.2	9.9	-4.1	0.8	10.68	16.88	-1.05	-0.90	0.22	10.34	-1.07	-0.27
Mar 17	312.53	-0.68	30.66	-0.99	355.91	-0.94	7.27	28 M-A 2	73.4	143.5	5.1	3.2	23.06	29.23	-0.73	-0.54	20.20	29.09	0.68	1.64
Mar 27	313.32	-0.70	32.98	-0.97	3.67	-0.88	17.11	7 Apr 12	216.7	282.6	-3.1	-5.1	35.39	41.54	-0.34	-0.13	36.25	41.17	2.44	2.91
Apr 6	313.98	-0.72	35.34	-0.96	11.36	-0.81	26.89	17 Apr 22	342.2	43.9	-1.7	3.6	47.67	53.79	0.09	0.31	43.61	43.58	2.93	2.41
Apr 16	314.51	-0.74	37.73	-0.95	18.99	-0.73	36.62	27 Apr	111.3		4.8		59.89		0.53		41.52		1.37	
								May 2		182.4		-0.2		65.98		0.75		38.45		-0.01
Apr 26	314.89	-0.77	40.11	-0.94	26.54	-0.64	46.30	7 May 12	252.2	314.7	-4.9	-3.6	72.05	78.10	0.96	1.16	35.71	34.36	-1.41	-2.54
May 6	315.10	-0.80	42.47	-0.93	34.01	-0.55	55.94	17 May 22	14.4	80.0	1.3	5.0	84.14	90.16	1.34	1.51	34.85	37.18	-3.27	-3.59
May 16	315.16	-0.82	44.79	-0.93	41.39	-0.45	65.54	27 M-J 1	150.5	220.4	2.4	-3.4	96.16	102.14	1.65	1.76	41.15	46.53	-3.54	-3.18
May 26	315.05	-0.85	47.06	-0.93	48.67	-0.34	75.11	6 Jun 11	286.1	346.2	-4.8	-0.9	108.10	114.03	1.85	1.91	53.20	61.07	-2.57	-1.77
Jun 5	314.79	-0.87	49.25	-0.93	55.86	-0.23	84.66	16 Jun 21	47.9	118.0	3.9	4.2	119.94	125.82	1.93	1.91	70.09	80.13	-0.85	0.10
Jun 15	314.38	-0.90	51.34	-0.94	62.95	-0.12	94.20	26 Jun	189.6		-1.2		131.66		1.86		90.84		0.93	
								Jul 1		256.7		-5.0		137.48		1.76		101.70		1.52
Jun 25	313.85	-0.92	53.30	-0.95	69.94	0.00	103.73	6 Jul 11	318.8	18.2	-2.9	2.1	143.25	148.99	1.63	1.45	112.19	122.03	1.82	1.81
Jul 5	313.22	-0.94	55.12	-0.96	76.83	0.12	113.27	16 Jul 21	83.3	156.9	5.1	1.4	154.68	160.32	1.24	0.98	131.11	139.42	1.55	1.07
Jul 15	312.51	-0.96	56.77	-0.97	83.61	0.25	122.82	26 Jul 31	227.4	291.2	-4.2	-4.5	165.91	171.43	0.68	0.34	146.97	153.73	0.44	-0.33
Jul 25	311.77	-0.98	58.22	-0.99	90.28	0.37	132.39	5 Aug 10	351.0	51.4	-0.1	4.5	176.87	182.24	-0.03	-0.44	159.60	164.43	-1.18	-2.09
Aug 4	311.03	-0.99	59.44	-1.00	96.85	0.50	142.00	15 Aug 20	120.6	195.5	4.1	-2.1	187.52	192.68	-0.88	-1.35	167.95	169.74	-2.99	-3.80
Aug 14	310.33	-1.00	60.41	-1.02	103.29	0.63	151.64	25 Aug 30	263.4	324.3	-5.3	-2.4	197.73	202.62	-1.84	-2.36	169.31	166.42	-4.36	-4.41
Aug 24	309.71	-1.00	61.08	-1.04	109.62	0.77	161.33	4 Sep 9	23.5	86.6	2.8	5.3	207.35	211.87	-2.88	-3.42	161.75	157.42	-3.72	-2.33
Sep 3	309.19	-1.01	61.45	-1.06	115.81	0.91	171.06	14 Sep 19	158.9	232.9	1.2	-4.6	216.14	220.13	-3.96	-4.50	155.87	158.19	-0.72	0.65
Sep 13	308.81	-1.01	61.49	-1.07	121.86	1.05	180.86	24 Sep 29	297.4	356.7	-4.3	0.5	223.75	226.93	-5.01	-5.50	163.84	171.55	1.51	1.88
Sep 23	308.58	-1.00	61.19	-1.09	127.76	1.20	190.70	4 Oct 9	57.4	123.8	4.8	3.9	229.58	231.58	-5.93	-6.28	180.15	188.94	1.87	1.60
Oct 3	308.51	-1.00	60.57	-1.10	133.48	1.36	200.60	14 Oct 19	197.4	268.3	-2.3	-5.2	232.81	233.15	-6.53	-6.61	197.62	206.06	1.17	0.66
Oct 13	308.61	-1.00	59.66	-1.11	139.00	1.52	210.56	24 Oct 29	329.8	29.4	-1.9	3.2	232.49	230.85	-6.48	-6.08	214.27	222.25	0.10	-0.46
Oct 23	308.88	-0.99	58.51	-1.10	144.29	1.70	220.57	3 Nov 8	93.3	162.3	5.0	0.7	228.37	225.41	-5.37	-4.38	230.06	237.70	-1.00	-1.49
Nov 2	309.31	-0.99	57.21	-1.09	149.30	1.88	230.64	13 Nov 18	234.9	301.7	-4.6	-3.7	222.46	220.00	-3.17	-1.89	245.21	252.54	-1.92	-2.24
Nov 12	309.90	-0.98	55.85	-1.07	153.98	2.08	240.75	23 Nov 28	1.5	63.3	1.1	4.8	218.37	217.74	-0.64	0.48	259.62	266.22	-2.42	-2.41
Nov 22	310.62	-0.98	54.53	-1.05	158.26	2.30	250.88	3 Dec 8	131.0	201.2	3.0	-2.8	218.13	219.45	1.44	2.21	271.86	275.60	-2.10	-1.37
Dec 2	311.47	-0.98	53.37	-1.01	162.05	2.54	261.06	13 Dec 18	270.9	333.8	-4.9	-1.1	221.58	224.38	2.81	3.26	275.91	271.66	-0.12	1.51
Dec 12	312.42	-0.98	52.44	-0.97	165.23	2.80	271.24	23 Dec 28	33.3	98.8	3.7	4.6	227.73	231.54	3.57	3.76	265.01	260.64	2.79	3.15
1727								1727												1727

JULIAN 12 UT	SATURN LONG.	SATURN LAT.	JUPITER LONG.	JUPITER LAT.	MARS LONG.	MARS LAT.	SUN LONG.	GREGORIAN 12 UT	MOON LONG.	MOON LONG.	MOON LAT.	MOON LAT.	VENUS LONG.	VENUS LONG.	VENUS LAT.	VENUS LAT.	MERCURY LONG.	MERCURY LONG.	MERCURY LAT.	MERCURY LAT.
1727								1728												1728
Dec 22	313.46	-0.98	51.80	-0.93	167.66	3.08	281.44	2 Jan 7	169.8	239.5	-0.5	-4.9	235.72	240.20	3.86	3.87	260.36	263.23	2.78	2.09
Jan 1	314.57	-0.99	51.50	-0.88	169.16	3.37	291.63	12 Jan 17	305.1	5.3	-3.2	1.9	244.94	249.89	3.80	3.66	267.99	273.85	1.31	0.55
Jan 11	315.73	-0.99	51.54	-0.84	169.56	3.67	301.81	22 Jan 27	66.3	135.9	5.1	2.4	255.00	260.26	3.48	3.24	280.40	287.41	-0.15	-0.77
Jan 21	316.92	-1.00	51.91	-0.79	168.69	3.95	311.96	1 Feb 6	208.7	276.1	-3.8	-4.8	265.64	271.12	2.97	2.67	294.80	302.52	-1.29	-1.70
Jan 31	318.13	-1.01	52.61	-0.75	166.53	4.16	322.08	11 Feb 16	338.0	37.2	-0.4	4.3	276.69	282.33	2.35	2.01	310.59	319.01	-1.98	-2.10
Feb 10	319.32	-1.03	53.60	-0.71	163.26	4.26	332.16	21 Feb 26	100.8	174.0	4.7	-1.1	288.02	293.77	1.66	1.30	327.82	337.03	-2.05	-1.80
Feb 20	320.48	-1.04	54.85	-0.67	159.39	4.20	342.19	2 Mar 7	246.6	310.7	-5.3	-2.8	299.57	305.40	0.95	0.61	346.60	356.35	-1.33	-0.61
Mar 1	321.61	-1.06	56.33	-0.64	155.60	3.98	352.17	12 Mar 17	10.3	70.4	2.5	5.3	311.26	317.16	0.27	-0.04	5.83	14.27	0.32	1.37
Mar 11	322.66	-1.08	58.01	-0.61	152.58	3.63	2.10	22 Mar 27	137.3	212.5	2.2	-4.0	323.07	329.01	-0.34	-0.61	20.73	24.51	2.35	3.05
Mar 21	323.64	-1.10	59.85	-0.58	150.74	3.21	11.96	1 Apr 6	282.6	343.7	-4.5	0.1	334.96	340.93	-0.86	-1.08	25.30	23.44	3.27	2.87
Mar 31	324.51	-1.13	61.83	-0.55	150.19	2.78	21.78	11 Apr 16	43.0	105.4	4.5	4.3	346.91	352.91	-1.27	-1.42	20.00	16.62	1.87	0.51
Apr 10	325.28	-1.16	63.93	-0.53	150.86	2.37	31.53	21 Apr 26	175.4	250.2	-1.2	-5.1	358.91	4.92	-1.54	-1.63	14.63	14.62	-0.83	-1.93
Apr 20	325.91	-1.18	66.11	-0.51	152.60	2.00	41.24	1 May 6	316.5	16.0	-2.1	2.9	10.94	16.97	-1.69	-1.71	16.56	20.18	-2.70	-3.11
Apr 30	326.40	-1.22	68.35	-0.49	155.20	1.66	50.90	11 May 16	76.8	142.4	5.0	1.5	23.01	29.05	-1.71	-1.67	25.18	31.36	-3.21	-3.03
May 10	326.75	-1.25	70.64	-0.47	158.53	1.35	60.52	21 May 26	214.3	286.3	-4.1	-4.0	35.09	41.15	-1.60	-1.50	38.59	46.82	-2.60	-1.96
May 20	326.93	-1.28	72.95	-0.45	162.43	1.08	70.10	31 M-J 5	348.9	48.5	1.0	4.7	47.20	53.27	-1.38	-1.24	56.03	66.15	-1.14	-0.24
May 30	326.94	-1.32	75.26	-0.44	166.82	0.83	79.66	10 Jun 15	112.4	181.0	3.5	-2.1	59.34	65.42	-1.08	-0.90	76.92	87.88	0.63	1.34
Jun 9	326.80	-1.35	77.56	-0.42	171.62	0.61	89.20	20 Jun 25	252.8	320.3	-5.0	-1.3	71.50	77.59	-0.71	-0.51	98.50	108.42	1.79	1.92
Jun 19	326.50	-1.38	79.83	-0.41	176.75	0.41	98.73	30 Jun	20.6		3.6		83.69		-0.31		117.51		1.77	
								Jul 5		82.0		4.9		89.80		-0.10		125.73		1.37
Jun 29	326.05	-1.41	82.04	-0.40	182.17	0.22	108.27	10 Jul 15	149.7	220.4	0.4	-4.7	95.92	102.04	0.10	0.31	133.06	139.47	0.75	-0.03
Jul 9	325.49	-1.44	84.18	-0.39	187.85	0.05	117.81	20 Jul 25	289.6	352.8	-3.6	1.7	108.18	114.32	0.50	0.68	144.83	148.95	-0.96	-1.97
Jul 19	324.83	-1.46	86.23	-0.38	193.76	-0.11	127.37	30 J-A 4	52.4	117.0	5.1	3.1	120.47	126.64	0.85	0.99	151.52	152.18	-3.01	-3.96
Jul 29	324.10	-1.48	88.16	-0.37	199.88	-0.25	136.96	9 Aug 14	188.4	259.0	-3.0	-5.1	132.81	138.99	1.12	1.23	150.65	147.13	-4.62	-4.71
Aug 8	323.35	-1.49	89.94	-0.36	206.18	-0.39	146.58	19 Aug 24	324.3	24.7	-0.8	4.1	145.18	151.37	1.32	1.37	142.84	139.81	-4.05	-2.75
Aug 18	322.61	-1.50	91.56	-0.35	212.67	-0.51	156.24	29 Aug	85.3		4.9		157.58		1.41		139.72		-1.21	
								Sep 3		153.6		0.0		163.79		1.41		143.09		0.17
Aug 28	321.92	-1.50	92.99	-0.34	219.32	-0.63	165.95	8 Sep 13	227.6	295.8	-5.1	-3.2	170.00	176.22	1.39	1.34	149.42	157.65	1.17	1.71
Sep 7	321.32	-1.50	94.18	-0.33	226.13	-0.73	175.71	18 Sep 23	357.5	56.9	2.1	5.2	182.44	188.67	1.27	1.17	166.72	175.93	1.84	1.68
Sep 17	320.83	-1.50	95.12	-0.32	233.08	-0.82	185.53	28 S-O 3	119.5	191.7	2.9	-3.2	194.90	201.13	1.05	0.90	184.93	193.62	1.33	0.86
Sep 27	320.49	-1.49	95.78	-0.31	240.17	-0.90	195.40	8 Oct 13	265.8	330.4	-4.7	-0.2	207.36	213.60	0.74	0.56	201.98	210.04	0.32	-0.25
Oct 7	320.30	-1.48	96.12	-0.30	247.40	-0.98	205.33	18 Oct 23	30.1	90.2	4.3	4.5	219.83	226.06	0.37	0.17	217.84	225.39	-0.82	-1.36
Oct 17	320.28	-1.46	96.14	-0.29	254.74	-1.04	215.32	28 Oct	155.6		-0.4		232.29		-0.04		232.71		-1.85	
								Nov 2		230.4		-5.0		238.53		-0.25		239.76		-2.25
Oct 27	320.44	-1.45	95.82	-0.28	262.20	-1.09	225.36	7 Nov 12	301.8	3.4	-2.4	2.8	244.76	250.99	-0.47	-0.68	246.43	252.47	-2.53	-2.63
Nov 6	320.76	-1.44	95.18	-0.26	269.76	-1.13	235.44	17 Nov 22	62.9	125.0	5.0	2.0	257.22	263.44	-0.88	-1.06	257.39	260.24	-2.45	-1.86
Nov 16	321.25	-1.42	94.25	-0.24	277.41	-1.15	245.56	27 N-D 2	193.6	268.3	-3.6	-4.3	269.66	275.88	-1.24	-1.39	259.55	254.57	-0.71	0.92
Nov 26	321.88	-1.41	93.10	-0.22	285.14	-1.17	255.72	7 Dec 12	335.7	35.6	0.8	4.7	282.10	288.31	-1.53	-1.64	247.96	244.39	2.33	2.88
Dec 6	322.66	-1.40	91.80	-0.20	292.93	-1.17	265.90	17 Dec 22	96.5	161.8	3.8	-1.4	294.50	300.70	-1.72	-1.77	245.11	248.85	2.70	2.16
Dec 16	323.55	-1.40	90.44	-0.18	300.77	-1.16	276.09	27 Dec	232.6		-5.1		306.88		-1.79		254.28		1.48	
1728								1729												1729
								Jan 1		304.4		-1.8		313.05		-1.78		260.63		0.77
Dec 26	324.54	-1.39	89.14	-0.15	308.65	-1.13	286.29	6 Jan 11	8.0	67.7	3.6	5.1	319.20	325.33	-1.73	-1.65	267.52	274.75	0.10	-0.52
Jan 5	325.62	-1.39	87.99	-0.13	316.55	-1.10	296.47	16 Jan 21	131.5	200.3	1.2	-4.4	331.45	337.54	-1.53	-1.38	282.25	290.00	-1.06	-1.50
Jan 15	326.76	-1.39	87.07	-0.10	324.45	-1.05	306.64	26 Jan 31	271.2	338.6	-4.2	1.3	343.60	349.63	-1.19	-0.97	298.01	306.31	-1.83	-2.03
Jan 25	327.94	-1.39	86.44	-0.08	332.35	-1.00	316.78	5 Feb 10	39.6	100.4	5.1	3.7	355.63	1.58	-0.72	-0.44	314.92	323.85	-2.07	-1.93
Feb 4	329.15	-1.40	86.13	-0.05	340.22	-0.93	326.88	15 Feb 20	168.1	239.7	-2.1	-5.3	7.49	13.35	-0.14	0.19	333.08	342.42	-1.56	-0.94
Feb 14	330.36	-1.41	86.16	-0.03	348.06	-0.86	336.94	25 Feb	308.2		-1.4		19.15		0.53		351.46		-0.05	
								Mar 2		11.6		3.9		24.89		0.89		359.37		1.04
Feb 24	331.56	-1.42	86.51	-0.01	355.86	-0.78	346.95	7 Mar 12	71.3	134.4	5.0	0.9	30.56	36.14	1.26	1.63	5.05	7.57	2.18	3.10
Mar 6	332.73	-1.44	87.16	0.01	3.59	-0.69	356.90	17 Mar 22	206.3	278.3	-4.6	-3.7	41.63	47.02	2.00	2.37	6.66	3.16	3.53	3.23
Mar 16	333.84	-1.46	88.10	0.03	11.27	-0.59	6.80	27 M-A 1	343.3	43.9	1.7	5.0	52.28	57.40	2.72	3.06	358.96	355.97	2.27	0.97
Mar 26	334.89	-1.49	89.29	0.05	18.87	-0.49	16.64	6 Apr 11	103.8	170.2	3.2	-2.4	62.36	67.13	3.37	3.64	355.13	356.45	-0.31	-1.37
Apr 5	335.85	-1.51	90.71	0.07	26.39	-0.39	26.42	16 Apr 21	245.3	315.1	-4.9	-0.6	71.68	75.96	3.87	4.05	359.60	4.19	-2.16	-2.65
Apr 15	336.71	-1.54	92.31	0.08	33.82	-0.29	36.15	26 Apr	16.9		4.1		79.92		4.17		9.95		-2.88	
								May 1		76.4		4.5		83.51		4.21		16.69		-2.85
Apr 25	337.45	-1.58	94.08	0.10	41.18	-0.18	45.84	6 May 11	137.8	208.1	0.2	-4.7	86.63	89.19	4.16	4.00	24.32	32.79	-2.58	-2.09
May 5	338.06	-1.61	95.98	0.11	48.44	-0.07	55.47	16 May 21	283.6	349.8	-2.9	2.6	91.08	92.18	3.71	3.26	42.10	52.22	-1.40	-0.57
May 15	338.53	-1.65	97.99	0.13	55.62	0.03	65.08	26 May 31	49.7	109.7	5.0	2.3	92.37	91.58	2.64	1.83	62.97	73.95	0.31	1.11
May 25	338.84	-1.69	100.09	0.14	62.71	0.14	74.65	5 Jun 10	173.9	247.2	-3.1	-4.8	89.81	87.22	0.83	-0.30	84.63	94.57	1.70	1.99
Jun 4	338.99	-1.73	102.25	0.15	69.71	0.25	84.20	15 Jun 20	319.9	22.9	0.3	4.7	84.16	81.11	-1.48	-2.59	103.58	111.60	1.98	1.67
Jun 14	338.97	-1.77	104.46	0.17	76.62	0.35	93.73	25 Jun 30	82.4	144.3	4.1	-0.8	78.53	76.76	-3.53	-4.25	118.60	124.49	1.11	0.31
Jun 24	338.79	-1.80	106.70	0.18	83.45	0.45	103.27	5 Jul 10	212.1	285.7	-5.1	-2.6	75.95	76.11	-4.74	-5.03	129.15	132.33	-0.67	-1.80
Jul 4	338.45	-1.84	108.94	0.20	90.20	0.55	112.80	15 Jul 20	354.3	55.0	3.3	5.2	77.17	79.02	-5.15	-5.12	133.75	133.19	-2.97	-4.05
Jul 14	337.98	-1.87	111.17	0.22	96.87	0.65	122.35	25 Jul 30	115.6	181.0	1.7	-3.8	81.53	84.59	-4.99	-4.78	130.70	127.05	-4.77	-4.88
Jul 24	337.38	-1.90	113.36	0.23	103.46	0.75	131.93	4 Aug 9	251.5	322.5	-4.7	0.7	88.10	91.99	-4.50	-4.18	123.71	122.27	-4.26	-3.07
Aug 3	336.69	-1.93	115.51	0.25	109.97	0.85	141.53	14 Aug 19	27.1	86.9	5.0	3.9	96.19	100.66	-3.81	-3.42	123.66	128.05	-1.64	-0.29
Aug 13	335.95	-1.94	117.58	0.27	116.40	0.94	151.17	24 Aug 29	150.1	219.6	-1.4	-5.2	105.35	110.23	-3.01	-2.58	135.01	143.71	0.79	1.48
Aug 23	335.19	-1.95	119.56	0.29	122.76	1.04	160.86	3 Sep 8	290.4	357.5	-2.2	3.5	115.28	120.47	-2.15	-1.73	153.21	162.81	1.78	1.74
Sep 2	334.45	-1.96	121.42	0.31	129.03	1.13	170.59	13 Sep 18	59.0	119.2	5.0	1.3	125.78	131.21	-1.31	-0.90	172.14	181.07	1.48	1.05
Sep 12	333.77	-1.95	123.14	0.34	135.22	1.22	180.38	23 Sep 28	186.5	259.1	-4.1	-4.2	136.74	142.35	-0.50	-0.13	189.60	197.74	0.53	-0.04
Sep 22	333.19	-1.94	124.69	0.36	141.32	1.32	190.22	3 Oct 8	327.6	31.1	1.2	4.9	148.04	153.80	0.22	0.54	205.54	213.00	-0.63	-1.22
Oct 2	332.74	-1.93	126.03	0.39	147.33	1.41	200.12	13 Oct 18	90.8	152.9	3.3	-1.8	159.63	165.51	0.83	1.08	220.14	226.89	-1.76	-2.25
Oct 12	332.43	-1.91	127.15	0.42	153.23	1.50	210.08	23 Oct 28	224.7	297.7	-5.0	-1.1	171.44	177.43	1.30	1.49	233.14	238.60	-2.63	-2.84
Oct 22	332.29	-1.89	128.00	0.45	159.02	1.60	220.09	2 Nov 7	3.0	63.8	4.0	4.6	183.45	189.51	1.63	1.74	242.75	244.65	-2.80	-2.35
Nov 1	332.32	-1.87	128.56	0.48	164.66	1.69	230.15	12 Nov 17	123.2	188.4	0.5	-4.3	195.60	201.72	1.81	1.85	242.99	237.45	-1.31	0.29
Nov 11	332.52	-1.85	128.81	0.52	170.15	1.79	240.25	22 Nov 27	263.4	334.4	-3.6	2.3	207.86	214.03	1.84	1.81	231.17	228.50	1.81	2.56
Nov 21	332.90	-1.82	128.73	0.55	175.46	1.90	250.40	2 Dec 7	36.8	96.3	5.1	2.6	220.21	226.41	1.74	1.64	230.18	234.72	2.58	2.19
Dec 1	333.44	-1.80	128.32	0.59	180.55	2.01	260.56	12 Dec 17	157.1	225.9	-2.6	-5.1	232.62	238.85	1.51	1.36	240.77	247.59	1.61	0.97
Dec 11	334.12	-1.78	127.61	0.62	185.36	2.12	270.75	22 Dec 27	301.4	9.2	-0.4	4.7	245.08	251.32	1.19	1.00	254.80	262.24	0.33	-0.28
1729								1729												1729

JULIAN 12 UT	SATURN LONG.	SATURN LAT.	JUPITER LONG.	JUPITER LAT.	MARS LONG.	MARS LAT.	SUN LONG.	GREGORIAN 12 UT	MOON LONG.		MOON LAT.		VENUS LONG.		VENUS LAT.		MERCURY LONG.		MERCURY LAT.	
1729								1730												1730
Dec 21	334.94	-1.77	126.62	0.65	189.86	2.23	280.94	1 Jan 6	69.5	129.3	4.4	-0.3	257.57	263.82	0.80	0.58	269.85	277.63	-0.83	-1.31
Dec 31	335.88	-1.76	125.44	0.68	193.95	2.35	291.13	11 Jan 16	193.0	264.5	-4.9	-3.5	270.07	276.32	0.37	0.15	285.58	293.75	-1.69	-1.95
Jan 10	336.91	-1.75	124.13	0.71	197.55	2.48	301.31	21 Jan 26	337.8	42.2	2.8	5.3	282.58	288.83	-0.07	-0.28	302.15	310.81	-2.07	-2.03
Jan 20	338.02	-1.74	122.80	0.73	200.53	2.61	311.47	31 J-F 5	101.6	163.7	2.2	-3.3	295.09	301.34	-0.48	-0.67	319.66	328.57	-1.76	-1.24
Jan 30	339.18	-1.74	121.55	0.74	202.74	2.73	321.59	10 Feb 15	231.1	303.0	-5.1	-0.2	307.59	313.84	-0.84	-1.00	337.10	344.42	-0.43	0.66
Feb 9	340.39	-1.74	120.46	0.75	204.01	2.83	331.67	20 Feb 25	12.4	74.1	4.8	4.1	320.09	326.33	-1.13	-1.25	349.26	350.48	1.91	3.05
Feb 19	341.61	-1.74	119.61	0.75	204.16	2.91	341.71	2 Mar 7	134.2	200.1	-0.8	-5.0	332.56	338.80	-1.33	-1.39	347.92	343.20	3.66	3.46
Mar 1	342.84	-1.75	119.04	0.75	203.05	2.93	351.69	12 Mar 17	270.3	340.3	-2.9	3.0	345.02	351.25	-1.43	-1.44	338.90	336.79	2.55	1.33
Mar 11	344.04	-1.77	118.78	0.75	200.69	2.86	1.62	22 Mar 27	45.7	105.6	5.0	1.6	357.47	3.68	-1.42	-1.38	337.18	339.67	0.13	-0.89
Mar 21	345.21	-1.79	118.85	0.74	197.34	2.68	11.49	1 Apr 6	168.0	238.3	-3.6	-4.5	9.88	16.08	-1.31	-1.22	343.76	349.08	-1.68	-2.23
Mar 31	346.32	-1.81	119.22	0.74	193.55	2.37	21.30	11 Apr 16	309.3	16.1	0.6	4.8	22.27	28.45	-1.10	-0.96	355.38	2.50	-2.55	-2.64
Apr 10	347.36	-1.83	119.89	0.73	190.02	1.97	31.06	21 Apr 26	78.0	137.6	3.4	-1.5	34.63	40.80	-0.81	-0.64	10.37	18.98	-2.52	-2.17
Apr 20	348.32	-1.86	120.82	0.72	187.39	1.53	40.77	1 May 6	203.8	277.4	-5.0	-1.9	46.97	53.12	-0.46	-0.26	28.32	38.38	-1.61	-0.87
Apr 30	349.16	-1.90	122.00	0.72	186.01	1.09	50.42	11 May 16	346.9	50.3	3.7	4.7	59.27	65.42	-0.06	0.14	49.04	59.96	-0.02	0.84
May 10	349.89	-1.93	123.38	0.71	185.95	0.68	60.05	21 May 26	110.0	171.0	0.7	-4.1	71.56	77.69	0.34	0.54	70.61	80.50	1.56	2.02
May 20	350.48	-1.97	124.95	0.71	187.12	0.31	69.63	31 M-J 5	241.5	315.9	-4.3	1.7	83.82	89.94	0.73	0.91	89.34	97.02	2.16	1.98
May 30	350.92	-2.02	126.68	0.71	189.34	0.00	79.19	10 Jun 15	22.6	83.3	5.2	2.9	96.06	102.17	1.08	1.23	103.49	108.65	1.49	0.71
Jun 9	351.20	-2.06	128.53	0.71	192.44	-0.27	88.73	20 Jun 25	142.6	206.5	-2.3	-5.2	108.27	114.37	1.36	1.46	112.33	114.31	-0.32	-1.55
Jun 19	351.32	-2.10	130.49	0.71	196.27	-0.51	98.27	30 Jun	280.2		-1.5		120.46		1.54		114.40		-2.85	
								Jul 5		353.0		4.4		126.54		1.60		112.66		-4.01
Jun 29	351.27	-2.15	132.54	0.72	200.69	-0.71	107.80	10 Jul 15	56.5	115.8	4.6	0.1	132.61	138.68	1.62	1.61	109.65	106.54	-4.76	-4.87
Jul 9	351.06	-2.19	134.65	0.72	205.62	-0.88	117.34	20 Jul 25	176.6	244.0	-4.6	-4.2	144.74	150.78	1.57	1.50	104.64	104.92	-4.31	-3.27
Jul 19	350.69	-2.23	136.80	0.73	210.96	-1.03	126.90	30 J-A 4	318.6	28.2	2.0	5.3	156.81	162.83	1.40	1.27	107.75	113.09	-2.00	-0.71
Jul 29	350.19	-2.26	138.99	0.74	216.67	-1.16	136.49	9 Aug 14	89.1	148.7	2.4	-2.8	168.84	174.83	1.10	0.91	120.61	129.72	0.40	1.22
Aug 8	349.56	-2.29	141.17	0.75	222.69	-1.26	146.11	19 Aug 24	212.6	282.8	-5.2	-1.2	180.81	186.76	0.69	0.44	139.60	149.56	1.66	1.77
Aug 18	348.85	-2.31	143.34	0.76	228.98	-1.35	155.77	29 Aug	355.8		4.5		192.70		0.18		159.21		1.60	
								Sep 3		61.5		4.2		198.62		-0.11		168.39		1.24
Aug 28	348.10	-2.33	145.48	0.78	235.52	-1.42	165.48	8 Sep 13	121.0	183.1	-0.5	-4.7	204.51	210.38	-0.41	-0.72	177.08	185.30	0.75	0.19
Sep 7	347.33	-2.33	147.57	0.80	242.27	-1.47	175.24	18 Sep 23	250.6	321.6	-3.5	2.3	216.22	222.03	-1.04	-1.36	193.10	200.48	-0.43	-1.06
Sep 17	346.59	-2.33	149.58	0.82	249.21	-1.51	185.05	28 S-O 3	31.3	93.5	5.0	1.7	227.81	233.54	-1.68	-1.99	207.42	213.86	-1.67	-2.23
Sep 27	345.92	-2.32	151.48	0.85	256.33	-1.53	194.92	8 Oct 13	153.3	219.3	-3.3	-4.8	239.23	244.87	-2.28	-2.55	219.65	224.49	-2.71	-3.05
Oct 7	345.36	-2.31	153.27	0.88	263.60	-1.53	204.85	18 Oct 23	289.7	359.4	-0.2	4.7	250.46	255.97	-2.80	-3.01	227.82	228.72	-3.15	-2.84
Oct 17	344.93	-2.28	154.90	0.91	271.00	-1.52	214.83	28 Oct	65.0		3.6		261.40		-3.19		226.12		-1.91	
								Nov 2		124.9		-1.3		266.73		-3.32		220.25		-0.36
Oct 27	344.66	-2.26	156.34	0.94	278.52	-1.49	224.87	7 Nov 12	186.8	257.1	-4.9	-2.7	271.95	277.03	-3.40	-3.41	214.55	212.85	1.25	2.19
Nov 6	344.56	-2.23	157.58	0.98	286.14	-1.45	234.95	17 Nov 22	328.7	35.6	3.3	4.9	281.95	286.67	-3.36	-3.23	215.45	220.75	2.42	2.19
Nov 16	344.64	-2.19	158.57	1.02	293.83	-1.39	245.07	27 N-D 2	97.5	156.8	0.9	-3.9	291.15	295.33	-3.01	-2.70	227.39	234.65	1.72	1.14
Nov 26	344.89	-2.16	159.28	1.06	301.58	-1.33	255.23	7 Dec 12	221.9	295.7	-4.8	0.7	299.15	302.54	-2.28	-1.73	242.17	249.82	0.54	-0.06
Dec 6	345.32	-2.13	159.70	1.11	309.37	-1.25	265.41	17 Dec 22	6.4	70.0	5.2	3.2	305.37	307.55	-1.05	-0.23	257.55	265.36	-0.62	-1.12
Dec 16	345.91	-2.10	159.81	1.15	317.18	-1.16	275.60	27 Dec	129.5		-2.0		308.92		0.75		273.26		-1.54	
1730								1731												1731
								Jan 1		190.0		-5.3		309.37		1.87		281.30		-1.86
Dec 26	346.64	-2.08	159.59	1.20	325.00	-1.06	285.79	6 Jan 11	258.7	334.0	-2.5	3.8	308.81	307.22	3.10	4.38	289.50	297.87	-2.06	-2.10
Jan 5	347.50	-2.06	159.06	1.24	332.82	-0.96	295.98	16 Jan 21	42.2	102.8	4.8	0.4	304.76	301.77	5.59	6.60	306.35	314.79	-1.94	-1.53
Jan 15	348.47	-2.04	158.24	1.28	340.61	-0.85	306.14	26 Jan 31	162.2	225.1	-4.3	-4.6	298.75	296.20	7.30	7.65	322.77	329.43	-0.82	0.24
Jan 25	349.54	-2.02	157.18	1.32	348.36	-0.74	316.29	5 Feb 10	296.7	11.2	0.9	5.2	294.51	293.82	7.65	7.39	333.39	333.31	1.56	2.88
Feb 4	350.68	-2.01	155.96	1.34	356.07	-0.62	326.39	15 Feb 20	75.9	135.2	2.7	-2.5	294.14	295.38	6.94	6.37	329.31	323.87	3.67	3.57
Feb 14	351.87	-2.01	154.66	1.36	3.71	-0.50	336.45	25 Feb	196.2		-5.1		297.42		5.73		320.02		2.74	
								Mar 2		262.2		-2.1		300.12		5.06		319.00		1.61
Feb 24	353.09	-2.01	153.37	1.36	11.29	-0.39	346.46	7 Mar 12	335.3	46.5	3.9	4.3	303.37	307.06	4.38	3.71	320.54	323.99	0.50	-0.47
Mar 6	354.33	-2.01	152.19	1.36	18.80	-0.27	356.42	17 Mar 22	108.2	167.9	-0.3	-4.5	311.11	315.46	3.06	2.43	328.83	334.69	-1.25	-1.84
Mar 16	355.57	-2.02	151.19	1.35	26.23	-0.16	6.32	27 M-A 1	232.0	300.9	-3.9	1.5	320.06	324.86	1.83	1.26	341.36	348.73	-2.24	-2.43
Mar 26	356.78	-2.04	150.44	1.33	33.58	-0.04	16.16	6 Apr 11	13.2	80.0	5.0	1.9	329.83	334.94	0.74	0.25	356.74	5.40	-2.42	-2.21
Apr 5	357.95	-2.05	149.98	1.31	40.85	0.07	25.95	16 Apr 21	139.8	201.7	-3.1	-4.9	340.17	345.49	-0.19	-0.58	14.70	24.64	-1.78	-1.14
Apr 15	359.07	-2.08	149.82	1.28	48.03	0.17	35.68	26 Apr	269.6		-0.9		350.91		-0.93		35.15		-0.34	
								May 1		340.1		4.4		356.39		-1.24		45.93		0.54
Apr 25	0.10	-2.10	149.96	1.25	55.13	0.27	45.37	6 May 11	49.4	112.2	3.9	-1.1	1.95	7.55	-1.49	-1.70	56.48	66.22	1.36	1.98
May 5	1.05	-2.14	150.41	1.22	62.16	0.37	55.01	16 May 21	171.8	237.2	-4.9	-3.4	13.21	18.90	-1.87	-1.99	74.78	81.97	2.30	2.28
May 15	1.88	-2.17	151.13	1.19	69.10	0.47	64.61	26 May 31	308.5	18.6	2.6	5.1	24.63	30.40	-2.06	-2.10	87.69	91.83	1.90	1.16
May 25	2.60	-2.21	152.10	1.17	75.97	0.55	74.19	5 Jun 10	83.8	143.9	1.3	-3.8	36.20	42.02	-2.09	-2.05	94.22	94.73	0.11	-1.20
Jun 4	3.17	-2.25	153.31	1.14	82.77	0.64	83.74	15 Jun 20	205.0	274.4	-5.0	-0.4	47.87	53.74	-1.97	-1.86	93.44	90.87	-2.59	-3.80
Jun 14	3.59	-2.30	154.71	1.12	89.50	0.72	93.27	25 Jun 30	347.6	55.1	4.9	3.6	59.64	65.55	-1.72	-1.56	88.03	86.11	-4.54	-4.67
Jun 24	3.86	-2.34	156.29	1.10	96.16	0.79	102.81	5 Jul 10	116.7	176.1	-1.7	-5.2	71.49	77.45	-1.37	-1.17	85.94	87.91	-4.22	-3.37
Jul 4	3.95	-2.39	158.01	1.09	102.77	0.86	112.35	15 Jul 20	239.8	312.9	-3.3	3.0	83.43	89.43	-0.96	-0.73	92.05	98.23	-2.27	-1.10
Jul 14	3.88	-2.44	159.87	1.08	109.31	0.93	121.89	25 Jul 30	25.7	89.6	5.0	0.8	95.44	101.48	-0.50	-0.27	106.26	115.70	0.01	0.92
Jul 24	3.64	-2.48	161.82	1.07	115.81	0.99	131.47	4 Aug 9	149.1	209.4	-4.0	-4.8	107.54	113.61	-0.04	0.18	125.90	136.19	1.50	1.75
Aug 3	3.24	-2.52	163.86	1.06	122.26	1.05	141.07	14 Aug 19	276.4	351.7	-0.1	4.9	119.71	125.82	0.40	0.60	146.13	155.54	1.70	1.43
Aug 13	2.71	-2.55	165.96	1.06	128.66	1.11	150.70	24 Aug 29	61.6	122.6	3.0	-2.2	131.95	138.09	0.78	0.95	164.39	172.69	0.98	0.42
Aug 23	2.06	-2.58	168.10	1.06	135.02	1.16	160.39	3 Sep 8	182.0	244.5	-5.0	-2.6	144.26	150.44	1.09	1.22	180.48	187.75	-0.21	-0.87
Sep 2	1.33	-2.60	170.26	1.06	141.33	1.21	170.12	13 Sep 18	314.6	29.5	3.2	4.6	156.63	162.83	1.31	1.39	194.48	200.58	-1.55	-2.21
Sep 12	0.56	-2.61	172.42	1.07	147.60	1.25	179.90	23 Sep 28	95.4	154.9	-0.1	-4.4	169.05	175.28	1.43	1.44	205.88	210.02	-2.79	-3.26
Sep 22	359.78	-2.61	174.56	1.08	153.84	1.30	189.75	3 Oct 8	216.0	281.5	-4.2	0.8	181.52	187.77	1.43	1.39	212.45	212.31	-3.49	-3.31
Oct 2	359.04	-2.60	176.67	1.09	160.02	1.34	199.64	13 Oct 18	353.5	65.2	5.0	2.2	194.02	200.28	1.33	1.24	208.86	202.94	-2.50	-0.99
Oct 12	358.39	-2.59	178.71	1.11	166.16	1.37	209.59	23 Oct 28	127.7	187.5	-3.0	-5.0	206.55	212.82	1.13	0.99	198.06	197.38	0.66	1.78
Oct 22	357.84	-2.56	180.66	1.13	172.26	1.40	219.60	2 Nov 7	251.5	320.1	-1.6	4.0	219.10	225.38	0.84	0.67	200.85	206.85	2.21	2.15
Nov 1	357.44	-2.53	182.50	1.15	178.29	1.43	229.66	12 Nov 17	31.8	98.9	4.4	-0.9	231.66	237.94	0.49	0.30	214.07	221.76	1.80	1.30
Nov 11	357.21	-2.50	184.20	1.18	184.26	1.46	239.76	22 Nov 27	159.2	220.8	-4.9	-3.9	244.22	250.50	0.10	-0.10	229.60	237.46	0.74	0.16
Nov 21	357.15	-2.46	185.74	1.21	190.16	1.48	249.90	2 Dec 7	288.7	359.5	1.8	5.3	256.79	263.07	-0.30	-0.50	245.32	253.16	-0.40	-0.93
Dec 1	357.27	-2.42	187.07	1.25	195.98	1.50	260.07	12 Dec 17	68.2	131.3	1.9	-3.6	269.35	275.63	-0.68	-0.86	261.03	268.95	-1.38	-1.76
Dec 11	357.57	-2.38	188.18	1.29	201.69	1.52	270.25	22 Dec 27	191.0	255.3	-5.2	-1.3	281.91	288.19	-1.02	-1.17	276.94	285.02	-2.03	-2.15
1731								1731												1731

JULIAN 12 UT	SATURN LONG.	SATURN LAT.	JUPITER LONG.	JUPITER LAT.	MARS LONG.	MARS LAT.	SUN LONG.	GREGORIAN 12 UT	MOON LONG.		MOON LAT.		VENUS LONG.		VENUS LAT.		MERCURY LONG.		MERCURY LAT.	
1731								1732												1732
Dec 21	358.04	-2.34	189.03	1.33	207.29	1.52	280.45	1 Jan 6	327.3	38.0	4.5	4.1	294.46	300.73	-1.29	-1.40	293.12	301.08	-2.10	-1.80
Dec 31	358.68	-2.31	189.60	1.37	212.73	1.52	290.64	11 Jan 16	102.8	163.3	-1.2	-5.0	306.99	313.25	-1.48	-1.53	308.47	314.43	-1.20	-0.21
Jan 10	359.45	-2.28	189.87	1.42	218.01	1.51	300.82	21 Jan 26	223.6	291.6	-3.7	2.0	319.50	325.75	-1.55	-1.55	317.46	316.14	1.14	2.59
Jan 20	0.35	-2.25	189.82	1.46	223.06	1.49	310.97	31 J-F 5	6.3	74.5	5.1	1.2	331.98	338.21	-1.52	-1.45	310.98	305.33	3.55	3.56
Jan 30	1.36	-2.23	189.45	1.51	227.86	1.46	321.10	10 Feb 15	136.0	195.6	-3.8	-4.8	344.42	350.61	-1.36	-1.24	302.27	302.36	2.84	1.82
Feb 9	2.46	-2.21	188.79	1.54	232.33	1.41	331.18	20 Feb 25	257.7	329.6	-0.8	4.5	356.79	2.96	-1.10	-0.92	304.92	309.18	0.80	-0.11
Feb 19	3.62	-2.20	187.86	1.58	236.39	1.33	341.22	1 Mar 6	44.2	109.0	3.3	-2.0	9.10	15.23	-0.73	-0.51	314.60	320.86	-0.88	-1.49
Feb 29	4.84	-2.19	186.74	1.60	239.94	1.23	351.21	11 Mar 16	168.7	228.9	-5.0	-2.9	21.33	27.41	-0.28	-0.03	327.80	335.33	-1.94	-2.22
Mar 10	6.08	-2.19	185.49	1.61	242.86	1.08	1.13	21 Mar 26	293.9	8.3	2.5	4.8	33.46	39.49	0.23	0.49	343.40	352.04	-2.31	-2.21
Mar 20	7.33	-2.19	184.20	1.61	244.97	0.88	11.01	31 M-A 5	80.1	141.9	0.2	-4.4	45.48	51.44	0.76	1.02	1.24	11.02	-1.90	-1.38
Mar 30	8.58	-2.20	182.97	1.60	246.11	0.62	20.83	10 Apr 15	201.5	263.7	-4.4	0.2	57.37	63.25	1.29	1.53	21.32	31.89	-0.66	0.21
Apr 9	9.80	-2.21	181.87	1.58	246.10	0.28	30.59	20 Apr 25	332.2	46.4	4.8	2.9	69.10	74.90	1.77	1.98	42.24	51.75	1.11	1.88
Apr 19	10.98	-2.23	180.99	1.55	244.82	-0.14	40.30	30 Apr	114.0		-2.9		80.65		2.17		59.91		2.39	
								May 5		174.2		-5.2		86.34		2.33		66.42		2.55
Apr 29	12.09	-2.25	180.35	1.52	242.38	-0.63	49.96	10 May 15	235.1	300.6	-2.2	3.4	91.97	97.53	2.46	2.54	71.14	73.93	2.32	1.66
May 9	13.13	-2.28	180.01	1.48	239.12	-1.16	59.58	20 May 25	11.6	82.8	4.9	-0.3	103.02	108.42	2.59	2.58	74.72	73.64	0.60	-0.76
May 19	14.07	-2.31	179.97	1.43	235.69	-1.66	69.18	30 M-J 4	146.5	206.3	-4.8	-4.2	113.72	118.90	2.52	2.41	71.24	68.52	-2.19	-3.39
May 29	14.91	-2.34	180.22	1.39	232.82	-2.10	78.73	9 Jun 14	270.0	339.4	0.9	5.2	123.96	128.86	2.23	1.99	66.60	66.27	-4.12	-4.31
Jun 8	15.61	-2.38	180.76	1.35	231.06	-2.43	88.28	19 Jun 24	50.3	117.3	2.6	-3.2	133.58	138.10	1.68	1.29	67.85	71.33	-4.02	-3.37
Jun 18	16.18	-2.43	181.57	1.31	230.66	-2.65	97.81	29 Jun	178.3		-5.2		142.36		0.82		76.58		-2.47	
								Jul 4		239.1		-1.9		146.32		0.27		83.51		-1.43
Jun 28	16.59	-2.47	182.62	1.27	231.62	-2.80	107.34	9 Jul 14	306.9	18.9	3.8	4.5	149.93	153.10	-0.37	-1.11	91.96	101.66	-0.37	0.59
Jul 8	16.83	-2.52	183.89	1.24	233.79	-2.87	116.88	19 Jul 24	87.2	150.4	-0.7	-4.9	155.75	157.77	-1.94	-2.87	112.11	122.67	1.30	1.69
Jul 18	16.91	-2.57	185.35	1.21	236.99	-2.90	126.45	29 J-A 3	210.1	273.3	-3.9	1.3	159.04	159.43	-3.88	-4.95	132.88	142.50	1.77	1.59
Jul 28	16.82	-2.61	186.98	1.18	241.04	-2.88	136.03	8 Aug 13	345.6	57.5	5.1	1.8	158.86	157.32	-6.02	-7.02	151.49	159.84	1.21	0.68
Aug 7	16.55	-2.65	188.76	1.16	245.78	-2.84	145.64	18 Aug 23	122.3	182.9	-3.5	-4.8	154.92	151.98	-7.84	-8.37	167.59	174.73	0.04	-0.67
Aug 17	16.13	-2.69	190.65	1.14	251.09	-2.77	155.31	28 Aug	242.7		-1.2		148.92		-8.53		181.21		-1.41	
								Sep 2		309.4		4.0		146.22		-8.33		186.92		-2.16
Aug 27	15.57	-2.72	192.64	1.12	256.88	-2.68	165.01	7 Sep 12	24.7	94.1	3.9	-1.7	144.28	143.29	-7.81	-7.08	191.65	195.02	-2.86	-3.46
Sep 6	14.90	-2.75	194.71	1.11	263.04	-2.57	174.76	17 Sep 22	156.0	215.4	-4.9	-3.1	143.30	144.25	-6.23	-5.33	196.46	195.26	-3.83	-3.78
Sep 16	14.15	-2.76	196.84	1.10	269.53	-2.44	184.58	27 S-O 2	276.8	347.4	2.0	5.0	146.03	148.50	-4.43	-3.56	191.12	185.48	-3.05	-1.60
Sep 26	13.37	-2.77	199.00	1.09	276.27	-2.30	194.44	7 Oct 12	62.7	128.7	0.8	-4.3	151.57	155.12	-2.74	-1.97	181.66	182.03	0.07	1.33
Oct 6	12.58	-2.77	201.17	1.09	283.23	-2.15	204.37	17 Oct 22	188.7	248.6	-4.6	-0.2	159.09	163.39	-1.26	-0.62	186.31	192.98	1.96	2.08
Oct 16	11.85	-2.75	203.34	1.09	290.36	-1.99	214.35	27 Oct	312.9		4.6		167.97		-0.04		200.75		1.86	
								Nov 1		26.2		3.7		172.79		0.48		208.87		1.44
Oct 26	11.20	-2.73	205.48	1.09	297.62	-1.82	224.38	6 Nov 11	98.8	161.7	-2.5	-5.2	177.81	183.01	0.93	1.32	217.04	225.12	0.93	0.37
Nov 5	10.68	-2.70	207.57	1.10	304.99	-1.64	234.46	16 Nov 21	221.2	283.3	-2.6	2.9	188.35	193.81	1.64	1.91	233.11	241.00	-0.19	-0.73
Nov 15	10.30	-2.66	209.58	1.11	312.44	-1.47	244.58	26 N-D 1	351.1	64.2	5.2	0.6	199.38	205.05	2.12	2.27	248.84	256.65	-1.23	-1.65
Nov 25	10.10	-2.62	211.50	1.12	319.93	-1.29	254.73	6 Dec 11	132.9	193.7	-4.7	-4.4	210.79	216.60	2.36	2.41	264.45	272.24	-1.98	-2.19
Dec 5	10.08	-2.57	213.29	1.14	327.45	-1.11	264.91	16 Dec 21	254.2	320.1	0.3	4.9	222.47	228.38	2.41	2.36	279.96	287.42	-2.23	-2.05
Dec 15	10.24	-2.53	214.92	1.16	334.98	-0.94	275.11	26 Dec 31	30.3	100.8	3.4	-2.6	234.35	240.34	2.27	2.14	294.19	299.39	-1.57	-0.68
1732								1733												1733
Dec 25	10.58	-2.48	216.37	1.19	342.50	-0.76	285.30	5 Jan 10	165.6	225.3	-5.2	-2.2	246.37	252.42	1.98	1.80	301.48	299.02	0.65	2.20
Jan 4	11.10	-2.44	217.61	1.21	349.99	-0.60	295.49	15 Jan 20	288.5	358.6	3.2	4.8	258.50	264.59	1.59	1.36	293.03	287.57	3.32	3.46
Jan 14	11.77	-2.40	218.60	1.24	357.45	-0.44	305.66	25 Jan 30	69.0	135.9	0.0	-4.7	270.70	276.82	1.11	0.86	285.45	286.63	2.87	1.98
Jan 24	12.58	-2.37	219.33	1.28	4.85	-0.29	315.80	4 Feb 9	197.5	257.4	-3.9	0.8	282.95	289.09	0.61	0.35	290.08	295.00	1.05	0.19
Feb 3	13.51	-2.34	219.76	1.31	12.21	-0.15	325.91	14 Feb 19	324.7	37.8	4.9	2.4	295.24	301.39	0.09	-0.15	300.90	307.49	-0.55	-1.18
Feb 13	14.55	-2.31	219.89	1.34	19.50	-0.01	335.97	24 Feb	106.2		-3.2		307.55		-0.39		314.62		-1.67	
								Mar 1		169.6		-4.8		313.70		-0.60		322.24		-2.00
Feb 23	15.67	-2.29	219.70	1.37	26.72	0.12	345.98	6 Mar 11	229.3	291.0	-1.3	3.7	319.86	326.03	-0.80	-0.98	330.33	338.89	-2.18	-2.18
Mar 5	16.86	-2.27	219.21	1.40	33.88	0.23	355.94	16 Mar 21	2.7	76.2	4.4	-1.1	332.19	338.35	-1.14	-1.27	347.95	357.53	-1.99	-1.58
Mar 15	18.10	-2.26	218.43	1.42	40.97	0.34	5.85	26 Mar 31	141.7	202.4	-5.0	-3.3	344.51	350.67	-1.37	-1.45	7.57	17.86	-0.95	-0.13
Mar 25	19.36	-2.26	217.43	1.44	47.98	0.44	15.69	5 Apr 10	261.8	326.7	1.7	5.1	356.83	2.99	-1.50	-1.52	27.94	37.13	0.81	1.71
Apr 4	20.63	-2.26	216.25	1.44	54.93	0.54	25.48	15 Apr 20	41.4	112.9	1.8	-4.1	9.14	15.30	-1.51	-1.47	44.77	50.43	2.42	2.79
Apr 14	21.88	-2.26	214.99	1.44	61.81	0.62	35.22	25 Apr 30	175.5	234.8	-4.8	-0.6	21.44	27.59	-1.41	-1.32	53.87	55.01	2.73	2.17
Apr 24	23.11	-2.27	213.73	1.42	68.62	0.70	44.90	5 May 10	295.9	4.5	4.3	4.4	33.74	39.88	-1.21	-1.08	54.03	51.55	1.14	-0.23
May 4	24.30	-2.29	212.55	1.40	75.37	0.77	54.55	15 May 20	79.5	147.7	-1.6	-5.3	46.02	52.16	-0.93	-0.76	48.67	46.59	-1.66	-2.84
May 14	25.42	-2.31	211.54	1.37	82.06	0.84	64.15	25 May 30	208.1	267.9	-2.9	2.4	58.30	64.44	-0.58	-0.39	46.08	47.42	-3.58	-3.86
May 24	26.46	-2.33	210.75	1.33	88.70	0.90	73.72	4 Jun 9	332.1	43.3	5.3	1.6	70.58	76.72	-0.20	0.00	50.53	55.25	-3.74	-3.29
Jun 3	27.40	-2.36	210.24	1.29	95.29	0.95	83.28	14 Jun 19	116.2	180.8	-4.3	-4.6	82.86	89.00	0.20	0.39	61.42	68.94	-2.58	-1.70
Jun 13	28.24	-2.40	210.01	1.24	101.84	0.99	92.81	24 Jun 29	240.3	302.6	-0.1	4.6	95.15	101.29	0.58	0.75	77.73	87.63	-0.72	0.24
Jun 23	28.94	-2.44	210.08	1.20	108.34	1.03	102.34	4 Jul 9	10.4	82.0	4.0	-1.9	107.44	113.59	0.92	1.06	98.25	109.02	1.05	1.58
Jul 3	29.50	-2.48	210.45	1.15	114.81	1.07	111.88	14 Jul 19	151.3	212.8	-5.1	-2.3	119.74	125.89	1.19	1.29	119.45	129.25	1.81	1.74
Jul 13	29.90	-2.52	211.11	1.11	121.25	1.10	121.43	24 Jul 29	273.0	339.3	2.8	5.0	132.05	138.20	1.38	1.43	138.34	146.71	1.44	0.95
Jul 23	30.14	-2.56	212.02	1.07	127.66	1.12	131.00	3 Aug 8	49.6	119.5	0.8	-4.4	144.36	150.52	1.46	1.47	154.38	161.33	0.31	-0.43
Aug 2	30.20	-2.61	213.17	1.03	134.06	1.14	140.60	13 Aug 18	184.7	244.3	-4.0	0.6	156.67	162.83	1.44	1.39	167.50	172.73	-1.24	-2.09
Aug 12	30.09	-2.65	214.53	0.99	140.43	1.16	150.23	23 Aug 28	307.1	17.6	4.7	3.1	168.98	175.13	1.30	1.19	176.78	179.26	-2.92	-3.65
Aug 22	29.80	-2.69	216.09	0.96	146.79	1.17	159.91	2 Sep 7	88.5	155.3	-2.7	-4.9	181.28	187.43	1.06	0.90	179.63	177.41	-4.16	-4.21
Sep 1	29.36	-2.73	217.80	0.93	153.14	1.18	169.64	12 Sep 17	216.9	276.4	-1.5	3.4	193.58	199.72	0.72	0.52	172.87	167.88	-3.54	-2.15
Sep 11	28.78	-2.76	219.66	0.91	159.48	1.18	179.42	22 Sep 27	343.0	56.6	4.8	-0.3	205.85	211.98	0.30	0.08	165.31	166.77	-0.50	0.85
Sep 21	28.09	-2.78	221.63	0.88	165.82	1.18	189.26	2 Oct 7	126.0	189.3	-4.9	-3.6	218.10	224.22	-0.16	-0.41	171.82	179.11	1.66	1.96
Oct 1	27.33	-2.79	223.69	0.86	172.15	1.17	199.16	12 Oct 17	248.8	309.9	1.5	5.1	230.33	236.43	-0.65	-0.90	187.39	195.95	1.88	1.56
Oct 11	26.53	-2.79	225.82	0.85	178.47	1.16	209.11	22 Oct 27	20.5	95.1	2.8	-3.6	242.53	248.61	-1.14	-1.36	204.44	212.76	1.10	0.57
Oct 21	25.74	-2.78	228.01	0.83	184.80	1.14	219.11	1 Nov 6	161.7	222.1	-5.0	-0.9	254.68	260.75	-1.57	-1.77	220.89	228.85	0.01	-0.54
Oct 31	25.00	-2.76	230.22	0.82	191.12	1.12	229.17	11 Nov 16	281.4	345.3	4.0	4.9	266.80	272.83	-1.93	-2.07	236.67	244.38	-1.06	-1.53
Nov 10	24.37	-2.73	232.44	0.81	197.44	1.09	239.27	21 Nov 26	58.8	132.0	-0.6	-5.2	278.84	284.83	-2.18	-2.25	252.00	259.51	-1.93	-2.22
Nov 20	23.86	-2.69	234.64	0.80	203.75	1.06	249.41	1 Dec 6	195.4	254.5	-3.2	2.0	290.79	296.72	-2.29	-2.28	266.83	273.77	-2.35	-2.29
Nov 30	23.51	-2.65	236.80	0.80	210.05	1.02	259.58	11 Dec 16	315.5	22.6	5.2	2.6	302.62	308.47	-2.22	-2.12	279.89	284.28	-1.93	-1.16
Dec 10	23.34	-2.60	238.90	0.80	216.34	0.97	269.76	21 Dec 26	96.9	166.9	-3.7	-4.7	314.26	319.99	-1.97	-1.77	285.41	281.94	0.11	1.72
Dec 20	23.35	-2.55	240.91	0.80	222.61	0.91	279.95	31 Dec	227.6		-0.3		325.66		-1.52		275.44		2.98	
1733								1733												1733

JULIAN 12 UT	SATURN LONG.	SATURN LAT.	JUPITER LONG.	JUPITER LAT.	MARS LONG.	MARS LAT.	SUN LONG.	GREGORIAN 12 UT	MOON LONG.	MOON LONG.	MOON LAT.	MOON LAT.	VENUS LONG.	VENUS LONG.	VENUS LAT.	VENUS LAT.	MERCURY LONG.	MERCURY LONG.	MERCURY LAT.	MERCURY LAT.
1733								1734												1734
								Jan 5		287.4		4.2		331.24		-1.21		270.46		3.29
Dec 30	23.54	-2.50	242.81	0.80	228.85	0.84	290.15	10 Jan 15	351.6	61.1	4.4	-1.0	336.72	342.08	-0.85	-0.44	269.37	271.54	2.85	2.09
Jan 9	23.92	-2.46	244.57	0.81	235.07	0.75	300.33	20 Jan 25	134.1	200.0	-5.0	-2.5	347.30	352.36	0.03	0.54	275.78	281.29	1.26	0.46
Jan 19	24.47	-2.41	246.16	0.81	241.25	0.66	310.48	30 J-F 4	259.4	321.7	2.6	5.0	357.22	1.86	1.11	1.72	287.59	294.44	-0.26	-0.89
Jan 29	25.17	-2.37	247.56	0.82	247.38	0.54	320.61	9 Feb 14	29.4	100.0	1.5	-4.0	6.21	10.23	2.37	3.07	301.73	309.41	-1.41	-1.80
Feb 8	26.01	-2.33	248.73	0.83	253.44	0.41	330.70	19 Feb 24	169.6	231.7	-4.2	0.5	13.85	16.97	3.80	4.56	317.47	325.93	-2.04	-2.13
Feb 18	26.98	-2.30	249.65	0.84	259.44	0.26	340.74	1 Mar 6	291.6	357.7	4.6	3.7	19.50	21.32	5.32	6.08	334.83	344.16	-2.04	-1.75
Feb 28	28.04	-2.27	250.29	0.85	265.33	0.08	350.73	11 Mar 16	68.4	138.1	-2.1	-5.1	22.29	22.29	6.79	7.39	353.90	3.86	-1.22	-0.47
Mar 10	29.19	-2.24	250.63	0.87	271.11	-0.13	0.66	21 Mar 26	203.3	263.2	-1.8	3.3	21.26	19.25	7.83	8.01	13.61	22.39	0.47	1.47
Mar 20	30.39	-2.23	250.66	0.88	276.73	-0.38	10.53	31 M-A 5	325.0	35.2	5.1	0.7	16.49	13.37	7.85	7.33	29.41	34.03	2.36	2.96
Mar 30	31.64	-2.21	250.38	0.88	282.16	-0.66	20.36	10 Apr 15	107.4	174.4	-4.7	-4.0	10.40	8.02	6.49	5.42	35.96	35.28	3.10	2.66
Apr 9	32.92	-2.21	249.81	0.89	287.34	-0.99	30.12	20 Apr 25	235.9	295.3	1.2	5.0	6.53	6.06	4.26	3.10	32.65	29.34	1.67	0.31
Apr 19	34.20	-2.20	248.96	0.89	292.21	-1.37	39.83	30 Apr	0.2		3.6		6.57		2.02		26.77		-1.08	
								May 5		73.8		-2.7		7.95		1.05		25.87		-2.22
Apr 29	35.47	-2.20	247.91	0.88	296.67	-1.82	49.50	10 May 15	145.2	208.6	-5.2	-1.3	10.07	12.81	0.20	-0.53	26.88	29.69	-2.99	-3.38
May 9	36.71	-2.21	246.70	0.87	300.62	-2.33	59.12	20 May 25	268.2	328.8	3.7	5.1	16.07	19.74	-1.14	-1.65	34.04	39.72	-3.41	-3.15
May 19	37.90	-2.22	245.43	0.86	303.89	-2.92	68.71	30 M-J 4	37.1	112.4	0.5	-4.9	23.76	28.06	-2.06	-2.38	46.58	54.55	-2.63	-1.91
May 29	39.03	-2.24	244.19	0.83	306.32	-3.59	78.27	9 Jun 14	181.0	241.5	-3.5	1.7	32.60	37.33	-2.61	-2.78	63.60	73.62	-1.03	-0.10
Jun 8	40.08	-2.26	243.05	0.80	307.72	-4.31	87.82	19 Jun 24	301.1	4.1	5.0	3.1	42.23	47.27	-2.87	-2.91	84.34	95.25	0.76	1.42
Jun 18	41.02	-2.29	242.10	0.77	307.92	-5.06	97.35	29 Jun	75.6		-2.8		52.43		-2.89		105.84		1.80	
								Jul 4		150.0		-4.8		57.70		-2.82		115.77		1.87
Jun 28	41.86	-2.32	241.39	0.74	306.86	-5.77	106.89	9 Jul 14	214.6	274.0	-0.6	4.0	63.06	68.50	-2.70	-2.55	124.91	133.25	1.66	1.23
Jul 8	42.56	-2.35	240.96	0.70	304.77	-6.32	116.42	19 Jul 24	335.2	41.4	4.5	-0.2	74.01	79.59	-2.36	-2.14	140.77	147.45	0.61	-0.16
Jul 18	43.12	-2.39	240.83	0.66	302.12	-6.61	125.98	29 J-A 3	114.5	185.4	-4.9	-2.7	85.23	90.92	-1.90	-1.64	153.19	157.82	-1.04	-1.99
Jul 28	43.52	-2.43	241.01	0.62	299.63	-6.60	135.57	8 Aug 13	246.9	306.9	2.4	5.0	96.67	102.45	-1.37	-1.09	161.06	162.51	-2.95	-3.83
Aug 7	43.74	-2.47	241.49	0.59	298.00	-6.30	145.18	18 Aug 23	11.1	80.2	2.2	-3.5	108.29	114.16	-0.81	-0.52	161.74	158.67	-4.45	-4.57
Aug 17	43.80	-2.51	242.25	0.55	297.60	-5.81	154.84	28 Aug	152.5		-4.5		120.08		-0.24		154.14		-3.94	
								Sep 2		218.8		0.3		126.03		0.03		150.20		-2.62
Aug 27	43.67	-2.55	243.27	0.52	298.56	-5.22	164.54	7 Sep 12	278.7	340.9	4.6	4.2	132.01	138.03	0.29	0.53	149.04	151.58	-1.03	0.37
Sep 6	43.37	-2.58	244.53	0.49	300.76	-4.60	174.29	17 Sep 22	48.8	119.5	-1.3	-5.2	144.08	150.16	0.75	0.95	157.36	165.21	1.33	1.80
Sep 16	42.91	-2.61	246.01	0.46	304.01	-4.00	184.10	27 S-O 2	188.5	250.9	-2.3	3.1	156.27	162.40	1.13	1.27	173.99	182.96	1.87	1.66
Sep 26	42.31	-2.63	247.67	0.44	308.11	-3.42	193.97	7 Oct 12	310.8	16.3	5.2	1.6	168.55	174.73	1.39	1.48	191.79	200.35	1.27	0.77
Oct 6	41.61	-2.65	249.48	0.41	312.89	-2.90	203.88	17 Oct 22	87.8	157.8	-4.3	-4.4	180.92	187.13	1.54	1.57	208.64	216.67	0.23	-0.34
Oct 16	40.83	-2.65	251.44	0.39	318.18	-2.41	213.86	27 Oct	222.5		0.8		193.36		1.56		224.48		-0.89	
								Nov 1		282.7		4.9		199.60		1.53		232.10		-1.41
Oct 26	40.02	-2.65	253.51	0.37	323.87	-1.98	223.89	6 Nov 11	343.7	53.3	4.1	-1.7	205.85	212.11	1.47	1.38	239.54	246.77	-1.86	-2.23
Nov 5	39.22	-2.63	255.66	0.35	329.87	-1.58	233.97	16 Nov 21	126.9	194.0	-5.2	-1.8	218.38	224.65	1.27	1.14	253.70	260.11	-2.46	-2.51
Nov 15	38.49	-2.61	257.88	0.33	336.10	-1.23	244.09	26 N-D 1	255.4	314.8	3.5	5.1	230.94	237.22	0.99	0.82	265.54	269.07	-2.28	-1.64
Nov 25	37.86	-2.57	260.14	0.32	342.50	-0.91	254.24	6 Dec 11	18.2	91.5	1.4	-4.4	243.51	249.80	0.63	0.44	269.22	264.89	-0.46	1.16
Dec 5	37.37	-2.53	262.43	0.30	349.02	-0.63	264.42	16 Dec 21	164.7	228.3	-3.8	1.4	256.09	262.38	0.24	0.04	258.20	253.90	2.55	3.04
Dec 15	37.04	-2.49	264.70	0.29	355.64	-0.38	274.61	26 Dec 31	287.9	348.1	4.9	3.5	268.68	274.97	-0.16	-0.36	253.82	256.91	2.78	2.17
1734								1735												1735
Dec 25	36.89	-2.44	266.96	0.27	2.30	-0.16	284.81	5 Jan 10	54.6	130.1	-2.0	-4.9	281.26	287.55	-0.55	-0.72	261.88	267.90	1.43	0.69
Jan 4	36.93	-2.39	269.16	0.26	9.01	0.04	294.99	15 Jan 20	200.3	261.2	-0.8	3.9	293.83	300.12	-0.89	-1.04	274.56	281.65	0.00	-0.63
Jan 14	37.16	-2.34	271.30	0.25	15.73	0.21	305.16	25 Jan 30	320.8	23.1	4.6	0.4	306.40	312.68	-1.16	-1.27	289.07	296.79	-1.16	-1.59
Jan 24	37.56	-2.29	273.33	0.24	22.45	0.37	315.31	4 Feb 9	92.8	167.6	-4.6	-3.2	318.95	325.22	-1.35	-1.41	304.81	313.16	-1.90	-2.07
Feb 3	38.14	-2.25	275.25	0.22	29.17	0.50	325.41	14 Feb 19	233.8	293.5	2.3	5.1	331.48	337.73	-1.44	-1.45	321.86	330.93	-2.07	-1.88
Feb 13	38.87	-2.20	277.02	0.21	35.86	0.62	335.48	24 Feb	354.6		2.6		343.98		-1.43		340.34		-1.46	
								Mar 1		60.1		-2.9		350.22		-1.38		349.92		-0.80
Feb 23	39.73	-2.16	278.61	0.20	42.53	0.72	345.50	6 Mar 11	131.8	203.2	-4.9	-0.2	356.45	2.66	-1.31	-1.21	359.26	7.57	0.10	1.17
Mar 5	40.72	-2.13	280.00	0.19	49.17	0.81	355.46	16 Mar 21	266.0	325.8	4.6	4.5	8.87	15.07	-1.08	-0.94	13.87	17.32	2.23	3.05
Mar 15	41.80	-2.10	281.15	0.17	55.78	0.89	5.37	26 Mar 31	29.9	99.0	-0.5	-5.1	21.26	27.43	-0.77	-0.59	17.59	15.12	3.39	3.08
Mar 25	42.96	-2.07	282.05	0.16	62.36	0.95	15.22	5 Apr 10	170.2	237.1	-3.0	2.8	33.60	39.74	-0.39	-0.18	11.30	7.92	2.12	0.80
Apr 4	44.18	-2.05	282.67	0.14	68.90	1.01	25.01	15 Apr 20	297.7	359.0	5.3	2.3	45.88	52.00	0.04	0.26	6.26	6.71	-0.53	-1.63
Apr 14	45.45	-2.04	282.99	0.13	75.41	1.05	34.75	25 Apr 30	67.1	138.5	-3.6	-4.8	58.10	64.19	0.48	0.70	9.09	13.07	-2.42	-2.90
Apr 24	46.73	-2.03	283.00	0.11	81.89	1.09	44.44	5 May 10	206.7	269.8	0.2	4.7	70.26	76.32	0.92	1.12	18.35	24.73	-3.07	-2.97
May 4	48.03	-2.02	282.70	0.09	88.34	1.12	54.08	15 May 20	329.5	33.6	4.3	-0.8	82.35	88.37	1.31	1.48	32.08	40.38	-2.63	-2.06
May 14	49.30	-2.02	282.09	0.07	94.76	1.14	63.69	25 May 30	105.9	176.9	-5.1	-2.4	94.37	100.35	1.63	1.75	49.59	59.66	-1.31	-0.44
May 24	50.55	-2.03	281.22	0.05	101.16	1.16	73.26	4 Jun 9	241.5	302.0	3.1	5.0	106.31	112.24	1.85	1.91	70.40	81.38	0.45	1.21
Jun 3	51.75	-2.03	280.14	0.03	107.54	1.17	82.82	14 Jun 19	2.3	70.2	1.8	-3.7	118.14	124.02	1.94	1.93	92.07	102.06	1.73	1.95
Jun 13	52.89	-2.05	278.92	0.00	113.90	1.18	92.36	24 Jun 29	145.1	213.3	-4.2	1.0	129.87	135.68	1.89	1.80	111.19	119.39	1.87	1.53
Jun 23	53.94	-2.06	277.64	-0.02	120.25	1.18	101.89	4 Jul 9	275.0	334.5	4.8	3.6	141.45	147.18	1.68	1.51	126.66	132.94	0.94	0.16
Jul 3	54.90	-2.08	276.39	-0.04	126.60	1.17	111.42	14 Jul 19	36.7	108.7	-1.4	-5.0	152.87	158.50	1.30	1.05	138.10	141.94	-0.78	-1.84
Jul 13	55.74	-2.11	275.27	-0.07	132.94	1.16	120.97	24 Jul 29	183.1	247.9	-1.4	3.8	164.08	169.59	0.76	0.42	144.17	144.45	-2.94	-3.95
Jul 23	56.44	-2.13	274.34	-0.09	139.28	1.15	130.54	3 Aug 8	307.8	7.8	4.7	0.8	175.03	180.39	0.05	-0.35	142.60	139.02	-4.67	-4.82
Aug 2	57.00	-2.16	273.66	-0.11	145.63	1.13	140.14	13 Aug 18	73.2	147.6	-4.2	-3.9	185.65	190.81	-0.79	-1.26	135.04	132.51	-4.22	-2.99
Aug 12	57.39	-2.20	273.28	-0.13	151.99	1.11	149.77	23 Aug 28	219.0	280.9	2.0	5.1	195.83	200.71	-1.76	-2.28	132.85	136.46	-1.50	-0.10
Aug 22	57.62	-2.23	273.22	-0.14	158.37	1.08	159.45	2 Sep 7	340.5	42.8	3.0	-2.4	205.41	209.90	-2.82	-3.37	142.91	151.26	0.96	1.59
Sep 1	57.66	-2.26	273.47	-0.16	164.76	1.05	169.17	12 Sep 17	111.6	185.5	-5.2	-0.9	214.15	218.09	-3.92	-4.47	160.50	169.88	1.82	1.73
Sep 11	57.53	-2.29	274.02	-0.18	171.17	1.02	178.95	22 Sep 27	252.9	313.1	4.4	4.7	221.67	224.80	-5.01	-5.52	179.04	187.86	1.42	0.97
Sep 21	57.21	-2.32	274.87	-0.19	177.61	0.98	188.79	2 Oct 7	14.0	79.9	0.1	-4.8	227.39	229.31	-5.98	-6.36	196.31	204.42	0.44	-0.13
Oct 1	56.74	-2.34	275.99	-0.20	184.08	0.93	198.68	12 Oct 17	150.7	221.6	-3.7	2.3	230.46	230.70	-6.64	-6.76	212.24	219.77	-0.71	-1.27
Oct 11	56.12	-2.35	277.35	-0.22	190.58	0.88	208.63	22 Oct 27	285.3	345.0	5.3	2.7	229.95	228.22	-6.66	-6.28	227.03	233.98	-1.78	-2.23
Oct 21	55.40	-2.36	278.92	-0.23	197.10	0.82	218.63	1 Nov 6	49.0	118.7	-2.9	-5.1	225.67	222.69	-5.59	-4.61	240.50	246.35	-2.56	-2.72
Oct 31	54.61	-2.36	280.68	-0.24	203.67	0.76	228.68	11 Nov 16	189.1	256.0	-0.6	4.5	219.76	217.36	-3.42	-2.13	251.05	253.67	-2.62	-2.12
Nov 10	53.79	-2.35	282.60	-0.25	210.27	0.70	238.78	21 Nov 26	317.0	17.6	4.4	-0.3	215.82	215.29	-0.88	0.25	252.84	247.81	-1.05	0.55
Nov 20	53.00	-2.33	284.66	-0.26	216.90	0.62	248.91	1 Dec 6	85.9	158.0	-4.8	-3.0	215.76	217.16	1.21	2.00	241.23	237.75	2.05	2.74
Nov 30	52.27	-2.30	286.82	-0.28	223.57	0.54	259.08	11 Dec 16	225.9	289.1	2.7	5.0	219.35	222.21	2.62	3.09	238.65	242.60	2.68	2.21
Dec 10	51.65	-2.27	289.07	-0.29	230.28	0.45	269.26	21 Dec 26	348.7	51.6	2.0	-3.2	225.61	229.45	3.43	3.65	248.23	254.74	1.58	0.90
Dec 20	51.17	-2.23	291.37	-0.30	237.03	0.36	279.46	31 Dec	124.3		-4.6		233.66		3.77		261.74		0.24	
1735								1735												1735

JULIAN 12 UT	SATURN LONG.	SATURN LAT.	JUPITER LONG.	JUPITER LAT.	MARS LONG.	MARS LAT.	SUN LONG.
1735							
Dec 30	50.86	-2.18	293.71	-0.32	243.81	0.25	289.65
Jan 9	50.73	-2.14	296.07	-0.33	250.63	0.14	299.83
Jan 19	50.79	-2.09	298.41	-0.35	257.48	0.01	309.99
Jan 29	51.04	-2.04	300.73	-0.37	264.37	-0.13	320.12
Feb 8	51.46	-1.99	302.99	-0.39	271.28	-0.27	330.21
Feb 18	52.06	-1.95	305.18	-0.41	278.23	-0.43	340.25
Feb 28	52.81	-1.91	307.26	-0.43	285.19	-0.60	350.25
Mar 9	53.69	-1.87	309.23	-0.46	292.18	-0.79	0.19
Mar 19	54.69	-1.84	311.04	-0.49	299.17	-0.98	10.06
Mar 29	55.79	-1.81	312.68	-0.51	306.17	-1.19	19.89
Apr 8	56.96	-1.78	314.11	-0.55	313.16	-1.41	29.66
Apr 18	58.19	-1.76	315.32	-0.58	320.12	-1.63	39.37
Apr 28	59.46	-1.74	316.27	-0.62	327.03	-1.87	49.04
May 8	60.76	-1.73	316.94	-0.66	333.88	-2.11	58.67
May 18	62.06	-1.72	317.31	-0.70	340.62	-2.36	68.25
May 28	63.34	-1.72	317.36	-0.74	347.22	-2.61	77.82
Jun 7	64.60	-1.72	317.10	-0.78	353.63	-2.85	87.36
Jun 17	65.80	-1.72	316.53	-0.82	359.79	-3.10	96.89
Jun 27	66.95	-1.73	315.68	-0.86	5.63	-3.34	106.43
Jul 7	68.01	-1.74	314.60	-0.89	11.04	-3.56	115.97
Jul 17	68.97	-1.75	313.37	-0.92	15.92	-3.77	125.52
Jul 27	69.81	-1.77	312.08	-0.94	20.09	-3.95	135.10
Aug 6	70.52	-1.78	310.80	-0.96	23.39	-4.09	144.72
Aug 16	71.07	-1.81	309.65	-0.97	25.61	-4.18	154.37
Aug 26	71.47	-1.83	308.69	-0.97	26.53	-4.18	164.07
Sep 5	71.69	-1.85	308.00	-0.96	26.00	-4.05	173.82
Sep 15	71.73	-1.87	307.61	-0.95	24.10	-3.75	183.62
Sep 25	71.58	-1.89	307.55	-0.94	21.18	-3.28	193.49
Oct 5	71.26	-1.91	307.82	-0.93	17.97	-2.66	203.40
Oct 15	70.77	-1.92	308.41	-0.91	15.28	-1.97	213.37
Oct 25	70.14	-1.93	309.31	-0.90	13.66	-1.29	223.40
Nov 4	69.41	-1.93	310.48	-0.89	13.37	-0.70	233.48
Nov 14	68.61	-1.92	311.90	-0.88	14.38	-0.20	243.59
Nov 24	67.78	-1.91	313.54	-0.87	16.50	0.20	253.74
Dec 4	66.99	-1.89	315.37	-0.86	19.52	0.51	263.92
Dec 14	66.26	-1.86	317.36	-0.85	23.25	0.76	274.11
1736							
Dec 24	65.65	-1.83	319.48	-0.85	27.54	0.95	284.30
Jan 3	65.18	-1.79	321.71	-0.85	32.25	1.10	294.49
Jan 13	64.88	-1.75	324.02	-0.85	37.29	1.22	304.66
Jan 23	64.77	-1.70	326.38	-0.86	42.59	1.30	314.81
Feb 2	64.85	-1.66	328.78	-0.86	48.07	1.36	324.92
Feb 12	65.11	-1.62	331.19	-0.88	53.71	1.41	334.99
Feb 22	65.56	-1.57	333.59	-0.89	59.46	1.44	345.01
Mar 4	66.16	-1.53	335.96	-0.90	65.30	1.46	354.98
Mar 14	66.92	-1.50	338.27	-0.92	71.21	1.47	4.89
Mar 24	67.82	-1.46	340.51	-0.95	77.18	1.47	14.74
Apr 3	68.82	-1.43	342.65	-0.97	83.18	1.46	24.54
Apr 13	69.93	-1.40	344.67	-1.00	89.22	1.45	34.28
Apr 23	71.11	-1.38	346.55	-1.03	95.29	1.43	43.98
May 3	72.34	-1.36	348.26	-1.07	101.37	1.40	53.62
May 13	73.62	-1.34	349.78	-1.11	107.48	1.37	63.23
May 23	74.92	-1.33	351.08	-1.15	113.62	1.34	72.81
Jun 2	76.22	-1.32	352.12	-1.20	119.77	1.31	82.36
Jun 12	77.51	-1.31	352.89	-1.24	125.94	1.27	91.89
Jun 22	78.77	-1.30	353.36	-1.29	132.14	1.22	101.43
Jul 2	79.98	-1.30	353.51	-1.35	138.36	1.18	110.96
Jul 12	81.13	-1.30	353.34	-1.40	144.62	1.13	120.51
Jul 22	82.20	-1.30	352.85	-1.44	150.90	1.08	130.08
Aug 1	83.16	-1.31	352.06	-1.49	157.22	1.02	139.67
Aug 11	84.00	-1.31	351.02	-1.52	163.58	0.96	149.30
Aug 21	84.71	-1.32	349.80	-1.55	169.99	0.90	158.98
Aug 31	85.27	-1.33	348.48	-1.56	176.44	0.84	168.70
Sep 10	85.66	-1.34	347.17	-1.56	182.94	0.77	178.48
Sep 20	85.88	-1.35	345.95	-1.55	189.49	0.70	188.31
Sep 30	85.91	-1.36	344.93	-1.53	196.10	0.62	198.20
Oct 10	85.75	-1.37	344.18	-1.50	202.77	0.54	208.14
Oct 20	85.42	-1.38	343.73	-1.46	209.50	0.46	218.14
Oct 30	84.92	-1.38	343.62	-1.43	216.29	0.37	228.19
Nov 9	84.28	-1.38	343.86	-1.38	223.14	0.28	238.29
Nov 19	83.54	-1.38	344.42	-1.34	230.06	0.18	248.42
Nov 29	82.73	-1.36	345.31	-1.30	237.04	0.08	258.58
Dec 9	81.91	-1.35	346.47	-1.27	244.10	-0.02	268.77
Dec 19	81.11	-1.33	347.90	-1.23	251.21	-0.13	278.96
1737							

GREGORIAN 12 UT	MOON LONG.		MOON LAT.		VENUS LONG.		VENUS LAT.		MERCURY LONG.		MERCURY LAT.	
1736												1736
Jan 5		196.3		0.5		238.17		3.79		269.04		-0.38
10 Jan 15	261.0	321.6	4.7	3.7	242.93	247.90	3.75	3.63	276.58	284.33	-0.93	-1.39
20 Jan 25	21.1	87.6	-1.1	-5.0	253.03	258.31	3.46	3.24	292.31	300.53	-1.75	-1.99
30 J-F 4	163.0	232.8	-2.2	3.6	263.70	269.19	2.99	2.70	309.03	317.83	-2.08	-1.98
9 Feb 14	294.7	354.0	4.9	1.2	274.77	280.42	2.39	2.05	326.89	336.06	-1.68	-1.11
19 Feb 24	55.2	125.4	-3.9	-4.6	286.13	291.89	1.71	1.36	344.93	352.69	-0.27	0.81
29 Feb	200.7		1.3		297.69		1.02		358.21		2.00	
Mar 5		267.3		5.2		303.53		0.67		0.40		3.04
10 Mar 15	327.3	27.1	3.3	-1.9	309.40	315.30	0.34	0.02	358.98	355.00	3.59	3.38
20 Mar 25	91.6	164.1	-5.3	-2.0	321.23	327.17	-0.27	-0.55	350.63	347.84	2.47	1.21
30 M-A 4	236.8	300.3	4.1	4.8	333.13	339.11	-0.80	-1.03	347.37	349.09	-0.04	-1.10
9 Apr 14	359.6	62.0	0.6	-4.3	345.10	351.10	-1.22	-1.38	352.58	357.44	-1.90	-2.44
19 Apr 24	130.0	202.4	-4.4	1.5	357.11	3.12	-1.51	-1.61	3.38	10.25	-2.73	-2.77
29 Apr	271.2		5.2		9.15		-1.67		17.94		-2.58	
May 4		332.1		2.9		15.18		-1.70		26.43		-2.16
9 May 14	32.6	98.9	-2.3	-5.1	21.22	27.26	-1.70	-1.67	35.72	45.78	-1.54	-0.75
19 May 24	169.2	239.4	-1.4	4.2	33.31	39.37	-1.61	-1.52	56.47	67.43	0.12	0.96
29 M-J 3	304.2	3.8	4.4	0.0	45.42	51.49	-1.40	-1.27	78.14	88.12	1.61	2.00
8 Jun 13	67.1	137.5	-4.5	-3.6	57.56	63.64	-1.11	-0.94	97.14	105.11	2.06	1.83
18 Jun 23	208.0	274.8	2.2	5.0	69.73	75.82	-0.75	-0.55	111.98	117.67	1.31	0.53
28 Jun	336.2		2.1		81.92		-0.35		122.03		-0.46	
Jul 3		36.1		-2.9		88.02		-0.15		124.84		-1.63
8 Jul 13	103.7	176.7	-4.9	-0.3	94.14	100.27	0.06	0.26	125.84	124.88	-2.86	-3.99
18 Jul 23	245.3	308.6	4.6	3.9	106.40	112.54	0.46	0.64	122.19	118.67	-4.76	-4.91
28 J-A 2	8.1	70.1	-0.9	-4.9	118.69	124.85	0.81	0.96	115.81	114.93	-4.35	-3.25
7 Aug 12	141.8	215.2	-3.1	3.2	131.02	137.20	1.10	1.21	116.77	121.41	-1.89	-0.55
17 Aug 22	280.8	341.3	5.1	1.5	143.39	149.58	1.30	1.36	128.48	137.28	0.57	1.34
27 Aug	40.7		-3.6		155.78		1.40		146.92		1.72	
Sep 1		106.0		-5.0		161.99		1.41		156.69		1.76
6 Sep 11	180.5	252.0	0.3	5.1	168.20	174.42	1.40	1.36	166.18	175.24	1.55	1.16
16 Sep 21	314.5	13.7	3.6	-1.5	180.64	186.86	1.29	1.19	183.87	192.07	0.66	0.09
26 S-O 1	75.0	143.6	-5.2	-2.9	193.08	199.31	1.08	0.94	199.89	207.34	-0.51	-1.11
6 Oct 11	218.5	286.8	3.5	5.0	205.54	211.78	0.78	0.60	214.42	221.08	-1.69	-2.21
16 Oct 21	346.9	46.9	0.9	-4.0	218.01	224.24	0.41	0.21	227.17	232.44	-2.64	-2.92
26 Oct 31	111.3	182.2	-4.7	0.5	230.47	236.70	0.00	-0.21	236.35	238.01	-2.96	-2.60
5 Nov 10	255.2	319.7	5.0	3.0	242.93	249.15	-0.42	-0.63	236.20	230.66	-1.65	-0.09
15 Nov 20	19.0	81.7	-2.0	-5.0	255.38	261.60	-0.84	-1.03	224.47	221.91	1.50	2.38
25 Nov 30	149.4	220.7	-2.2	3.6	267.82	274.04	-1.21	-1.37	223.76	228.50	2.53	2.22
5 Dec 10	290.0	351.3	4.5	0.2	280.25	286.46	-1.51	-1.62	234.75	241.72	1.69	1.08
15 Dec 20	51.6	118.2	-4.3	-4.0	292.65	298.84	-1.71	-1.77	249.05	256.57	0.46	-0.15
25 Dec 30	188.4	258.2	1.5	5.0	305.02	311.19	-1.80	-1.79	264.22	272.00	-0.71	-1.20
1737												1737
4 Jan 9	323.2	22.7	2.3	-2.8	317.34	323.48	-1.75	-1.68	279.93	288.03	-1.60	-1.90
14 Jan 19	85.5	156.2	-5.1	-1.1	329.59	335.68	-1.57	-1.42	296.33	304.85	-2.06	-2.07
24 Jan 29	227.3	293.9	4.4	4.2	341.74	347.78	-1.24	-1.03	313.54	322.26	-1.86	-1.41
3 Feb 8	355.3	54.8	-0.6	-4.8	353.78	359.73	-0.78	-0.51	330.61	337.77	-0.65	0.41
13 Feb 18	121.1	194.9	-3.8	2.4	5.64	11.50	-0.21	0.12	342.44	343.35	1.69	2.92
23 Feb 28	264.9	327.8	5.2	1.8	17.31	23.05	0.46	0.82	340.35	335.30	3.66	3.56
5 Mar 10	27.3	88.3	-3.4	-5.2	28.71	34.30	1.19	1.57	331.02	329.19	2.71	1.53
15 Mar 20	158.2	233.3	-0.9	4.8	39.78	45.17	1.95	2.32	329.93	332.75	0.36	-0.65
25 Mar 30	300.4	0.6	3.9	-1.1	50.43	55.55	2.69	3.03	337.10	342.61	-1.45	-2.03
4 Apr 9	60.1	123.8	-5.0	-3.6	60.50	65.26	3.35	3.64	349.03	356.23	-2.40	-2.55
14 Apr 19	196.6	270.3	2.5	5.0	69.80	74.06	3.89	4.09	4.13	12.72	-2.49	-2.21
24 Apr 29	333.9	33.1	1.2	-3.7	78.01	81.57	4.23	4.30	22.00	31.99	-1.72	-1.04
4 May 9	94.3	161.2	-4.8	-0.4	84.66	87.19	4.27	4.14	42.56	53.43	-0.21	0.66
14 May 19	235.2	305.3	4.7	3.2	89.04	90.09	3.87	3.45	64.07	73.97	1.44	1.98
24 May 29	6.2	66.2	-1.8	-5.0	90.22	89.37	2.85	2.06	82.78	90.36	2.22	2.12
3 Jun 8	130.5	200.1	-2.7	3.0	87.54	84.92	1.07	-0.05	96.63	101.48	1.70	0.95
13 Jun 18	272.8	338.5	4.7	0.3	81.84	78.80	-1.23	-2.34	104.75	106.25	-0.08	-1.33
23 Jun 28	38.1	100.6	-4.2	-4.4	76.26	74.54	-3.30	-4.04	105.86	103.80	-2.68	-3.88
3 Jul 8	168.4	239.3	0.7	5.0	73.80	74.03	-4.56	-4.88	100.79	98.05	-4.67	-4.82
13 Jul 18	308.5	10.4	2.7	-2.6	75.14	77.03	-5.02	-5.03	96.75	97.61	-4.34	-3.39
23 Jul 28	70.5	136.6	-5.2	-2.0	79.58	82.66	-4.92	-4.73	100.86	106.44	-2.20	-0.96
2 Aug 7	207.6	277.4	3.9	4.6	86.20	90.11	-4.48	-4.17	114.07	123.25	0.17	1.05
12 Aug 17	342.2	42.1	-0.3	-4.7	94.32	98.80	-3.82	-3.45	133.25	143.37	1.58	1.76
22 Aug 27	104.1	174.3	-4.3	1.4	103.50	108.39	-3.05	-2.63	153.17	162.48	1.66	1.35
1 Sep 6	246.7	313.4	5.3	2.3	113.44	118.64	-2.21	-1.79	171.27	179.56	0.88	0.32
11 Sep 16	14.8	74.3	-3.1	-5.2	123.95	129.38	-1.38	-0.97	187.38	194.74	-0.30	-0.94
21 Sep 26	139.2	213.0	-1.8	4.2	134.90	140.52	-0.58	-0.21	201.63	207.98	-1.58	-2.19
1 Oct 6	284.5	347.5	4.2	-0.8	146.21	151.97	0.14	0.46	213.62	218.25	-2.72	-3.12
11 Oct 16	47.1	107.8	-4.8	-4.0	157.79	163.67	0.76	1.02	221.32	221.96	-3.29	-3.07
21 Oct 26	176.0	251.7	1.6	5.1	169.60	175.58	1.25	1.44	219.22	213.41	-2.24	-0.74
31 Oct	320.0		1.5		181.60		1.59		207.86		0.91	
Nov 5		20.4		-3.5		187.66		1.71		206.29		1.98
10 Nov 15	80.0	142.9	-4.9	-1.1	193.75	199.86	1.79	1.83	209.04	214.53	2.33	2.20
20 Nov 25	214.3	288.9	4.3	3.6	206.01	212.17	1.83	1.81	221.36	228.79	1.79	1.25
30 N-D 5	353.4	52.8	-1.7	-4.9	218.36	224.56	1.74	1.65	236.44	244.17	0.66	0.07
10 Dec 15	114.2	180.1	-3.1	2.4	230.77	237.00	1.53	1.39	251.95	259.77	-0.49	-1.01
20 Dec 25	253.0	323.9	5.0	0.7	243.23	249.47	1.22	1.03	267.65	275.64	-1.45	-1.80
30 Dec	25.5		-4.2		255.72		0.84		283.74		-2.04	
1737												1737

JULIAN 12 UT	SATURN LONG.	SATURN LAT.	JUPITER LONG.	JUPITER LAT.	MARS LONG.	MARS LAT.	SUN LONG.	GREGORIAN 12 UT	MOON LONG.	MOON LONG.	MOON LAT.	MOON LAT.	VENUS LONG.	VENUS LONG.	VENUS LAT.	VENUS LAT.	MERCURY LONG.	MERCURY LONG.	MERCURY LAT.	MERCURY LAT.
1737								1738												1738
								Jan 4		85.6		-4.6		261.97		0.63		291.97		-2.12
Dec 29	80.39	-1.30	349.54	-1.20	258.39	-0.25	289.15	9 Jan 14	149.9	219.0	0.0	4.8	268.22	274.48	0.41	0.19	300.29	308.54	-2.02	-1.68
Jan 8	79.79	-1.27	351.38	-1.18	265.64	-0.36	299.34	19 Jan 24	290.8	357.1	3.3	-2.4	280.74	286.99	-0.02	-0.23	316.33	322.82	-1.03	-0.02
Jan 18	79.33	-1.23	353.38	-1.16	272.94	-0.48	309.50	29 J-F 3	57.2	119.2	-5.2	-2.7	293.25	299.50	-0.44	-0.63	326.62	326.27	1.30	2.69
Jan 28	79.05	-1.20	355.52	-1.14	280.30	-0.61	319.62	8 Feb 13	187.4	258.3	3.2	5.0	305.76	312.01	-0.81	-0.96	321.91	316.29	3.61	3.61
Feb 7	78.95	-1.16	357.75	-1.12	287.71	-0.73	329.72	18 Feb 23	326.7	29.3	0.3	-4.6	318.26	324.50	-1.10	-1.22	312.57	311.85	2.85	1.78
Feb 17	79.04	-1.13	0.07	-1.11	295.17	-0.85	339.76	28 Feb	89.1		-4.6		330.74		-1.31		313.70		0.70	
								Mar 5		154.2		0.4		336.98		-1.38		317.42		-0.26
Feb 27	79.32	-1.09	2.44	-1.10	302.67	-0.98	349.76	10 Mar 15	226.4	296.5	5.1	2.9	343.21	349.43	-1.42	-1.44	322.45	328.44	-1.05	-1.66
Mar 9	79.77	-1.06	4.85	-1.10	310.21	-1.10	359.70	20 Mar 25	0.9	61.2	-2.7	-5.2	355.66	1.87	-1.43	-1.39	335.20	342.61	-2.09	-2.33
Mar 19	80.39	-1.02	7.27	-1.10	317.76	-1.22	9.58	30 M-A 4	121.9	191.1	-2.4	3.5	8.08	14.28	-1.33	-1.24	350.62	359.23	-2.38	-2.23
Mar 29	81.15	-0.99	9.68	-1.10	325.34	-1.33	19.40	9 Apr 14	265.5	332.7	4.6	-0.4	20.47	26.66	-1.13	-1.00	8.45	18.30	-1.86	-1.29
Apr 8	82.05	-0.96	12.06	-1.11	332.91	-1.44	29.18	19 Apr 24	34.0	93.5	-4.6	-4.1	32.84	39.01	-0.84	-0.68	28.70	39.39	-0.53	0.34
Apr 18	83.06	-0.94	14.39	-1.11	340.48	-1.53	38.90	29 Apr	156.5		0.8		45.18		-0.50		49.90		1.21	
								May 4		229.7		5.0		51.34		-0.31		59.62		1.91
Apr 28	84.17	-0.91	16.66	-1.13	348.02	-1.62	48.56	9 May 14	303.2	7.0	1.9	-3.3	57.49	63.64	-0.11	0.10	68.10	75.12	2.33	2.40
May 8	85.35	-0.89	18.83	-1.14	355.51	-1.69	58.19	19 May 24	66.7	126.9	-4.9	-1.5	69.78	75.92	0.30	0.50	80.56	84.28	2.11	1.42
May 18	86.58	-0.87	20.90	-1.16	2.95	-1.74	67.79	29 M-J 3	193.3	268.6	3.8	4.1	82.04	88.16	0.69	0.88	86.14	86.08	0.38	-0.94
May 28	87.86	-0.86	22.83	-1.19	10.30	-1.78	77.35	8 Jun 13	338.8	40.0	-1.4	-4.9	94.28	100.39	1.05	1.20	84.35	81.65	-2.36	-3.59
Jun 7	89.16	-0.84	24.60	-1.21	17.55	-1.80	86.90	18 Jun 23	99.6	162.0	-3.4	1.8	106.49	112.59	1.33	1.45	79.08	77.70	-4.36	-4.54
Jun 17	90.47	-0.83	26.19	-1.24	24.67	-1.80	96.43	28 Jun	232.0		5.1		118.68		1.53		78.17		-4.19	
								Jul 3		306.3		1.3		124.76		1.59		80.67		-3.43
Jun 27	91.76	-0.82	27.56	-1.28	31.64	-1.78	105.96	8 Jul 13	12.5	72.3	-4.1	-4.8	130.83	136.89	1.62	1.62	85.19	91.59	-2.43	-1.31
Jul 7	93.02	-0.81	28.69	-1.31	38.43	-1.74	115.50	18 Jul 23	133.4	199.3	-0.7	4.5	142.95	148.99	1.59	1.53	99.71	109.21	-0.21	0.73
Jul 17	94.23	-0.80	29.56	-1.35	45.00	-1.68	125.06	28 J-A 2	271.2	342.0	4.0	-1.9	155.02	161.03	1.43	1.31	119.49	129.91	1.39	1.72
Jul 27	95.38	-0.79	30.13	-1.39	51.31	-1.59	134.63	7 Aug 12	44.9	104.7	-5.3	-3.1	167.04	173.03	1.15	0.96	140.00	149.54	1.75	1.52
Aug 6	96.45	-0.79	30.38	-1.43	57.31	-1.48	144.25	17 Aug 22	168.9	238.4	2.5	5.2	179.00	184.95	0.74	0.50	158.49	166.85	1.11	0.57
Aug 16	97.41	-0.78	30.31	-1.47	62.93	-1.34	153.90	27 Aug	309.3		1.0		190.88		0.24		174.66		-0.06	
								Sep 1		16.0		-4.3		196.79		-0.05		181.91		-0.75
Aug 26	98.26	-0.78	29.90	-1.51	68.11	-1.17	163.60	6 Sep 11	76.6	137.9	-4.7	-0.2	202.68	208.54	-0.35	-0.66	188.58	194.58	-1.45	-2.15
Sep 5	98.97	-0.78	29.18	-1.54	72.73	-0.96	173.35	16 Sep 21	206.4	277.8	4.7	3.5	214.38	220.18	-0.98	-1.31	199.70	203.63	-2.79	-3.32
Sep 15	99.52	-0.78	28.18	-1.56	76.68	-0.71	183.15	26 S-O 1	345.7	48.7	-2.2	-5.1	225.95	231.67	-1.63	-1.94	205.78	205.37	-3.63	-3.54
Sep 25	99.91	-0.77	26.98	-1.57	79.80	-0.42	193.00	6 Oct 11	108.3	172.5	-2.6	2.8	237.35	242.98	-2.24	-2.53	201.79	195.99	-2.81	-1.36
Oct 5	100.12	-0.77	25.66	-1.57	81.88	-0.07	202.92	16 Oct 21	245.3	316.0	4.8	0.2	248.55	254.05	-2.79	-3.01	191.35	190.82	0.31	1.54
Oct 15	100.14	-0.77	24.31	-1.56	82.72	0.34	212.89	26 Oct 31	20.4	80.9	-4.5	-4.1	259.47	264.78	-3.20	-3.35	194.43	200.61	2.10	2.13
Oct 25	99.98	-0.77	23.05	-1.54	82.16	0.81	222.92	5 Nov 10	140.9	209.1	0.4	4.8	269.98	275.05	-3.44	-3.48	208.01	215.88	1.85	1.40
Nov 4	99.63	-0.76	21.96	-1.50	80.15	1.31	232.99	15 Nov 20	284.3	352.3	2.6	-3.1	279.95	284.64	-3.44	-3.33	223.86	231.82	0.85	0.28
Nov 14	99.12	-0.76	21.12	-1.46	76.94	1.80	243.11	25 Nov 30	53.9	113.2	-4.9	-1.7	289.09	293.24	-3.13	-2.84	239.73	247.59	-0.28	-0.81
Nov 24	98.48	-0.75	20.60	-1.41	73.13	2.22	253.25	5 Dec 10	175.2	247.3	3.4	4.6	297.02	300.35	-2.43	-1.91	255.45	263.32	-1.29	-1.69
Dec 4	97.73	-0.73	20.42	-1.36	69.49	2.53	263.43	15 Dec 20	321.8	26.7	-0.8	-5.0	303.13	305.24	-1.24	-0.43	271.23	279.18	-2.00	-2.17
Dec 14	96.92	-0.72	20.58	-1.31	66.77	2.71	273.62	25 Dec 30	86.5	146.5	-3.7	1.4	306.54	306.90	0.54	1.65	287.12	294.89	-2.17	-1.93
1738								1739												1739
Dec 24	96.09	-0.70	21.08	-1.25	65.32	2.78	283.81	4 Jan 9	211.7	285.8	5.2	2.3	306.24	304.56	2.89	4.17	302.08	307.85	-1.40	-0.48
Jan 3	95.30	-0.68	21.91	-1.20	65.22	2.77	294.00	14 Jan 19	357.4	59.6	-3.8	-5.0	302.04	299.03	5.39	6.42	310.73	309.20	0.84	2.35
Jan 13	94.59	-0.66	23.02	-1.16	66.33	2.71	304.17	24 Jan 29	119.0	181.5	-1.1	4.1	296.03	293.56	7.13	7.50	303.78	298.05	3.43	3.55
Jan 23	94.00	-0.63	24.39	-1.12	68.45	2.62	314.32	3 Feb 8	250.1	323.4	4.6	-1.1	291.95	291.36	7.53	7.30	295.17	295.55	2.91	1.96
Feb 2	93.55	-0.60	25.98	-1.08	71.38	2.52	324.43	13 Feb 18	31.2	91.5	-5.2	-3.4	291.78	293.11	6.88	6.34	298.39	302.86	0.97	0.08
Feb 12	93.28	-0.58	27.77	-1.04	74.94	2.41	334.50	23 Feb 28	152.3	218.7	1.9	5.2	295.21	297.97	5.74	5.09	308.43	314.80	-0.70	-1.32
Feb 22	93.19	-0.55	29.72	-1.01	79.00	2.31	344.52	5 Mar 10	289.2	359.7	1.9	-3.9	301.27	305.01	4.43	3.78	321.79	329.33	-1.80	-2.11
Mar 4	93.29	-0.52	31.80	-0.98	83.46	2.20	354.49	15 Mar 20	63.7	123.3	-4.7	-0.6	309.10	313.48	3.13	2.51	337.39	345.96	-2.25	-2.21
Mar 14	93.58	-0.50	33.99	-0.96	88.22	2.10	4.40	25 Mar 30	187.3	257.5	4.3	4.0	318.10	322.92	1.92	1.36	355.07	4.74	-1.96	-1.50
Mar 24	94.03	-0.47	36.25	-0.94	93.22	1.99	14.26	4 Apr 9	327.7	34.4	-1.6	-5.0	327.91	333.04	0.83	0.35	14.90	25.36	-0.83	0.01
Apr 3	94.65	-0.45	38.58	-0.92	98.43	1.89	24.06	14 Apr 19	95.4	155.8	-2.7	2.4	338.28	343.62	-0.09	-0.50	35.64	45.10	0.93	1.77
Apr 13	95.42	-0.43	40.94	-0.90	103.79	1.80	33.80	24 Apr 29	224.2	296.7	5.0	0.9	349.05	354.55	-0.85	-1.16	53.15	59.44	2.38	2.65
Apr 23	96.32	-0.41	43.32	-0.89	109.29	1.70	43.50	4 May 9	4.7	67.8	-4.3	-4.2	0.11	5.72	-1.43	-1.64	63.77	66.03	2.52	1.93
May 3	97.33	-0.39	45.69	-0.88	114.91	1.61	53.15	14 May 19	127.2	190.1	0.2	4.6	11.38	17.09	-1.82	-1.95	66.20	64.57	0.90	-0.45
May 13	98.43	-0.37	48.04	-0.88	120.62	1.52	62.76	24 May 29	262.7	335.0	3.4	-2.7	22.83	28.60	-2.03	-2.07	61.89	59.31	-1.89	-3.11
May 23	99.61	-0.35	50.35	-0.87	126.43	1.43	72.34	3 Jun 8	40.0	100.3	-5.1	-2.0	34.40	40.23	-2.08	-2.04	57.86	58.14	-3.87	-4.13
Jun 2	100.84	-0.33	52.59	-0.87	132.31	1.34	81.90	13 Jun 18	159.9	226.4	3.1	5.0	46.08	51.96	-1.97	-1.87	60.30	64.23	-3.93	-3.39
Jun 12	102.11	-0.32	54.76	-0.87	138.26	1.25	91.44	23 Jun 28	301.5	11.6	0.1	-4.9	57.85	63.77	-1.74	-1.58	69.80	76.90	-2.58	-1.61
Jun 22	103.41	-0.30	56.82	-0.88	144.29	1.16	100.97	3 Jul 8	73.6	132.7	-3.9	1.0	69.71	75.67	-1.41	-1.21	85.42	95.16	-0.58	0.39
Jul 2	104.71	-0.29	58.75	-0.88	150.38	1.08	110.51	13 Jul 18	194.3	264.5	5.0	3.2	81.65	87.65	-1.00	-0.78	105.66	116.33	1.16	1.63
Jul 12	106.00	-0.27	60.54	-0.89	156.54	0.99	120.05	23 Jul 28	339.5	46.2	-3.2	-5.1	93.67	99.70	-0.55	-0.32	126.67	136.41	1.79	1.68
Jul 22	107.26	-0.26	62.15	-0.90	162.78	0.90	129.62	2 Aug 7	106.1	166.1	-1.5	3.7	105.76	111.83	-0.09	0.13	145.49	153.90	1.33	0.83
Aug 1	108.47	-0.25	63.55	-0.91	169.08	0.81	139.21	12 Aug 17	230.9	303.2	5.0	-0.1	117.92	124.03	0.35	0.55	161.66	168.77	0.19	-0.53
Aug 11	109.62	-0.23	64.72	-0.93	175.45	0.72	148.84	22 Aug 27	15.8	79.1	-5.0	-3.6	130.16	136.30	0.74	0.91	175.17	180.75	-1.30	-2.08
Aug 21	110.68	-0.22	65.63	-0.94	181.90	0.62	158.51	1 Sep 6	138.4	201.2	1.5	5.1	142.46	148.64	1.06	1.19	185.28	188.40	-2.84	-3.51
Aug 31	111.64	-0.21	66.24	-0.96	188.42	0.53	168.24	11 Sep 16	269.2	341.5	2.7	-3.3	154.83	161.03	1.29	1.37	189.54	188.07	-3.96	-3.98
Sep 10	112.48	-0.20	66.54	-0.97	195.02	0.44	178.00	21 Sep 26	50.3	110.9	-4.8	-0.8	167.24	173.47	1.42	1.44	183.85	178.42	-3.33	-1.95
Sep 20	113.18	-0.18	66.51	-0.98	201.70	0.34	187.83	1 Oct 6	171.6	238.2	4.0	4.4	179.71	185.95	1.44	1.40	174.90	175.46	-0.28	1.06
Sep 30	113.73	-0.17	66.15	-0.99	208.46	0.24	197.72	11 Oct 16	308.3	18.5	-0.8	-4.9	192.20	198.46	1.34	1.26	179.88	186.71	1.81	2.03
Oct 10	114.11	-0.16	65.46	-1.00	215.30	0.15	207.66	21 Oct 26	83.0	142.5	-2.8	2.2	204.73	211.00	1.15	1.02	194.66	202.96	1.89	1.52
Oct 20	114.31	-0.14	64.49	-1.00	222.23	0.05	217.66	31 Oct	206.2		5.0		217.27		0.87		211.27		1.03	
								Nov 5		276.8		1.7		223.55		0.71		219.46		0.49
Oct 30	114.32	-0.13	63.31	-0.99	229.24	-0.05	227.71	10 Nov 15	347.0	53.7	-4.0	-4.4	229.83	236.10	0.53	0.34	227.52	235.44	-0.07	-0.62
Nov 9	114.14	-0.12	61.98	-0.98	236.33	-0.15	237.79	20 Nov 25	114.8	174.8	0.0	4.5	242.38	248.67	0.14	-0.06	243.27	251.04	-1.13	-1.58
Nov 19	113.78	-0.10	60.62	-0.96	243.52	-0.25	247.93	30 N-D 5	242.5	315.7	4.0	-1.9	254.95	261.23	-0.26	-0.45	258.77	266.44	-1.94	-2.19
Nov 29	113.27	-0.08	59.33	-0.93	250.78	-0.36	258.09	10 Dec 15	24.3	87.3	-5.2	-2.3	267.51	273.79	-0.64	-0.82	274.00	281.27	-2.29	-2.17
Dec 9	112.61	-0.07	58.21	-0.89	258.12	-0.46	268.27	20 Dec 25	146.6	208.6	2.9	5.2	280.07	286.35	-0.99	-1.14	287.84	292.83	-1.76	-0.95
Dec 19	111.86	-0.05	57.34	-0.85	265.54	-0.55	278.46	30 Dec	280.1		1.2		292.62		-1.27		294.77		0.33	
1739								1739												1739

JULIAN 12 UT	SATURN LONG.	SATURN LAT.	JUPITER LONG.	JUPITER LAT.	MARS LONG.	MARS LAT.	SUN LONG.
1739							
Dec 29	111.05	-0.03	56.77	-0.81	273.03	-0.65	288.66
Jan 8	110.23	-0.01	56.53	-0.77	280.59	-0.75	298.84
Jan 18	109.44	0.01	56.64	-0.72	288.21	-0.83	309.00
Jan 28	108.74	0.02	57.08	-0.68	295.88	-0.92	319.13
Feb 7	108.16	0.04	57.83	-0.64	303.59	-1.00	329.22
Feb 17	107.73	0.06	58.87	-0.61	311.34	-1.07	339.27
Feb 27	107.47	0.08	60.16	-0.57	319.12	-1.13	349.27
Mar 8	107.39	0.09	61.68	-0.54	326.90	-1.18	359.21
Mar 18	107.49	0.11	63.38	-0.51	334.69	-1.22	9.10
Mar 28	107.78	0.12	65.25	-0.48	342.47	-1.25	18.93
Apr 7	108.24	0.13	67.24	-0.46	350.22	-1.26	28.70
Apr 17	108.86	0.15	69.35	-0.44	357.94	-1.26	38.42
Apr 27	109.63	0.16	71.53	-0.41	5.60	-1.25	48.10
May 7	110.52	0.17	73.78	-0.40	13.21	-1.22	57.72
May 17	111.52	0.19	76.06	-0.38	20.74	-1.17	67.32
May 27	112.62	0.20	78.36	-0.36	28.18	-1.12	76.89
Jun 6	113.79	0.21	80.67	-0.35	35.52	-1.04	86.43
Jun 16	115.01	0.23	82.95	-0.33	42.76	-0.96	95.97
Jun 26	116.28	0.24	85.20	-0.32	49.87	-0.86	105.50
Jul 6	117.57	0.25	87.39	-0.30	56.84	-0.74	115.04
Jul 16	118.86	0.27	89.51	-0.29	63.67	-0.62	124.60
Jul 26	120.14	0.28	91.52	-0.28	70.33	-0.48	134.18
Aug 5	121.40	0.30	93.42	-0.27	76.81	-0.33	143.79
Aug 15	122.60	0.32	95.17	-0.26	83.09	-0.16	153.44
Aug 25	123.74	0.33	96.74	-0.25	89.13	0.02	163.13
Sep 4	124.79	0.35	98.11	-0.23	94.91	0.22	172.87
Sep 14	125.75	0.37	99.25	-0.22	100.39	0.44	182.68
Sep 24	126.57	0.39	100.12	-0.21	105.50	0.68	192.53
Oct 4	127.26	0.42	100.71	-0.20	110.17	0.95	202.44
Oct 14	127.80	0.44	100.97	-0.18	114.32	1.24	212.41
Oct 24	128.16	0.46	100.91	-0.16	117.82	1.58	222.43
Nov 3	128.34	0.49	100.52	-0.15	120.52	1.94	232.50
Nov 13	128.33	0.51	99.80	-0.13	122.23	2.36	242.61
Nov 23	128.14	0.54	98.82	-0.11	122.78	2.80	252.76
Dec 3	127.77	0.56	97.62	-0.08	121.99	3.26	262.93
Dec 13	127.24	0.59	96.30	-0.06	119.81	3.70	273.12
1740							
Dec 23	126.58	0.61	94.95	-0.04	116.49	4.05	283.32
Jan 2	125.82	0.63	93.68	-0.01	112.56	4.25	293.51
Jan 12	125.01	0.65	92.58	0.01	108.75	4.28	303.68
Jan 22	124.19	0.67	91.72	0.03	105.78	4.15	313.83
Feb 1	123.42	0.68	91.16	0.05	104.03	3.92	323.94
Feb 11	122.73	0.69	90.93	0.07	103.60	3.63	334.02
Feb 21	122.16	0.70	91.02	0.09	104.37	3.34	344.04
Mar 3	121.75	0.71	91.44	0.11	106.17	3.05	354.01
Mar 13	121.50	0.71	92.16	0.12	108.81	2.78	3.93
Mar 23	121.43	0.72	93.15	0.14	112.11	2.53	13.78
Apr 2	121.54	0.72	94.38	0.15	115.94	2.31	23.59
Apr 12	121.83	0.72	95.84	0.17	120.20	2.10	33.34
Apr 22	122.29	0.73	97.47	0.18	124.81	1.91	43.03
May 2	122.91	0.73	99.26	0.19	129.69	1.73	52.68
May 12	123.67	0.73	101.18	0.21	134.82	1.56	62.30
May 22	124.56	0.74	103.20	0.22	140.15	1.40	71.87
Jun 1	125.56	0.74	105.31	0.23	145.65	1.25	81.43
Jun 11	126.64	0.75	107.47	0.25	151.30	1.11	90.97
Jun 21	127.80	0.75	109.68	0.26	157.10	0.98	100.50
Jul 1	129.02	0.76	111.91	0.28	163.02	0.84	110.04
Jul 11	130.28	0.77	114.14	0.29	169.07	0.72	119.58
Jul 21	131.56	0.79	116.36	0.31	175.23	0.59	129.15
Jul 31	132.84	0.80	118.54	0.32	181.50	0.47	138.74
Aug 10	134.11	0.81	120.66	0.34	187.88	0.36	148.37
Aug 20	135.36	0.83	122.71	0.36	194.37	0.24	158.04
Aug 30	136.55	0.85	124.66	0.39	200.97	0.13	167.76
Sep 9	137.67	0.87	126.49	0.41	207.68	0.02	177.53
Sep 19	138.71	0.90	128.16	0.43	214.49	-0.09	187.35
Sep 29	139.65	0.93	129.65	0.46	221.40	-0.19	197.24
Oct 9	140.46	0.95	130.94	0.49	228.42	-0.30	207.18
Oct 19	141.13	0.98	131.99	0.52	235.55	-0.40	217.17
Oct 29	141.64	1.02	132.77	0.56	242.77	-0.49	227.22
Nov 8	141.98	1.05	133.25	0.59	250.09	-0.58	237.31
Nov 18	142.14	1.08	133.41	0.63	257.51	-0.67	247.43
Nov 28	142.11	1.12	133.25	0.67	265.01	-0.75	257.60
Dec 8	141.90	1.15	132.76	0.70	272.60	-0.82	267.78
Dec 18	141.51	1.19	131.97	0.74	280.25	-0.89	277.97
1741							

GREGORIAN 12 UT	MOON LONG.		MOON LAT.		VENUS LONG.		VENUS LAT.		MERCURY LONG.		MERCURY LAT.	
1740												1740
Jan 4		354.0		-4.6		298.89		-1.38		292.16		1.91
9 Jan 14	59.7	119.8	-4.3	0.6	305.16	311.41	-1.47	-1.52	286.01	280.54	3.14	3.41
19 Jan 24	179.3	244.2	4.8	4.0	317.66	323.91	-1.56	-1.56	278.62	280.05	2.91	2.08
29 J-F 3	318.4	30.6	-2.3	-5.2	330.15	336.37	-1.53	-1.48	283.75	288.86	1.20	0.36
8 Feb 13	93.1	152.2	-1.8	3.3	342.58	348.78	-1.39	-1.28	294.89	301.56	-0.39	-1.02
18 Feb 23	213.7	281.7	5.1	1.0	354.96	1.13	-1.14	-0.97	308.73	316.36	-1.53	-1.90
28 Feb	356.5		-4.6		7.28		-0.78		324.41		-2.12	
Mar 4		65.2		-3.7		13.41		-0.57		332.91		-2.17
9 Mar 14	125.3	185.5	1.2	4.9	19.51	25.59	-0.33	-0.09	341.87	351.31	-2.03	-1.68
19 Mar 24	250.0	320.4	3.3	-2.5	31.64	37.67	0.17	0.44	1.20	11.36	-1.11	-0.33
29 M-A 3	33.3	98.0	-4.9	-1.0	43.67	49.63	0.71	0.97	21.34	30.45	0.61	1.56
8 Apr 13	157.4	220.3	3.8	4.6	55.56	61.44	1.24	1.49	37.95	43.33	2.36	2.85
18 Apr 23	288.1	359.3	-0.1	-4.8	67.29	73.09	1.73	1.96	46.30	46.83	2.90	2.43
28 Apr	68.3		-3.1		78.84		2.15		45.23		1.44	
May 3		129.7		2.0		84.53		2.32		42.35		0.09
8 May 13	190.1	256.7	5.1	2.4	90.16	95.72	2.46	2.56	39.47	37.73	-1.34	-2.52
18 May 23	327.3	37.1	-3.5	-4.8	101.20	106.60	2.61	2.62	37.77	39.66	-3.30	-3.65
28 M-J 2	101.6	161.3	-0.2	4.4	111.89	117.07	2.57	2.47	43.24	48.32	-3.62	-3.27
7 Jun 12	224.1	294.8	4.5	-1.0	122.12	127.01	2.31	2.08	54.72	62.37	-2.66	-1.85
17 Jun 22	6.4	72.8	-5.2	-2.7	131.72	136.22	1.78	1.41	71.21	81.12	-0.92	0.04
27 Jun	133.9		2.6		140.47		0.95		91.76		0.88	
Jul 2		193.7		5.3		144.41		0.41		102.61		1.49
7 Jul 12	259.8	333.8	2.2	-4.0	147.99	151.12	-0.23	-0.96	113.15	123.05	1.80	1.81
17 Jul 22	44.0	106.7	-4.6	0.3	153.73	155.70	-1.79	-2.72	132.22	140.63	1.56	1.10
27 J-A 1	166.1	227.4	4.6	4.4	156.91	157.24	-3.73	-4.81	148.30	155.20	0.48	-0.27
6 Aug 11	297.3	12.5	-1.2	-5.2	156.60	154.98	-5.90	-6.91	161.25	166.31	-1.11	-1.99
16 Aug 21	79.3	139.5	-2.1	3.0	152.53	149.55	-7.75	-8.30	170.13	172.31	-2.88	-3.69
26 Aug 31	199.1	262.9	5.1	1.7	146.51	143.86	-8.48	-8.29	172.36	169.90	-4.27	-4.38
5 Sep 10	335.9	49.5	-4.1	-4.0	141.97	141.05	-7.80	-7.10	165.38	160.70	-3.78	-2.46
15 Sep 20	112.8	172.1	1.0	4.7	141.12	142.13	-6.27	-5.40	158.50	160.17	-0.83	0.58
25 Sep 30	233.5	300.3	3.6	-1.8	143.96	146.47	-4.51	-3.66	165.36	172.80	1.49	1.89
5 Oct 10	14.5	84.3	-5.0	-1.2	149.57	153.14	-2.85	-2.08	181.26	189.99	1.89	1.63
15 Oct 20	145.0	205.2	3.7	4.8	157.12	161.44	-1.38	-0.74	198.64	207.07	1.20	0.69
25 Oct 30	269.6	339.2	0.6	-4.5	166.04	170.87	-0.16	0.36	215.28	223.28	0.14	-0.42
4 Nov 9	51.8	117.2	-3.6	1.9	175.89	181.09	0.82	1.22	231.11	238.78	-0.96	-1.45
14 Nov 19	176.9	239.5	5.1	3.0	186.44	191.91	1.56	1.83	246.33	253.73	-1.88	-2.21
24 Nov 29	307.5	18.3	-2.8	-5.1	197.48	203.15	2.05	2.21	260.91	267.67	-2.40	-2.39
4 Dec 9	87.0	149.0	-0.7	4.3	208.89	214.70	2.32	2.38	273.57	277.75	-2.11	-1.42
14 Dec 19	209.1	275.2	4.8	0.0	220.57	226.49	2.38	2.35	278.72	275.15	-0.22	1.39
24 Dec 29	346.7	56.1	-5.0	-3.3	232.46	238.46	2.27	2.15	268.58	263.63	2.75	3.19
1741												1741
3 Jan 8	120.6	180.6	2.3	5.3	244.49	250.54	2.00	1.82	262.72	265.13	2.86	2.17
13 Jan 18	242.3	312.6	2.8	-3.2	256.62	262.72	1.62	1.40	269.60	275.28	1.39	0.61
23 Jan 28	25.7	92.0	-4.9	-0.2	268.83	274.96	1.16	0.91	281.70	288.63	-0.11	-0.74
2 Feb 7	153.2	212.8	4.4	4.5	281.09	287.24	0.66	0.40	295.95	303.62	-1.27	-1.69
12 Feb 17	277.1	351.2	-0.3	-5.0	293.39	299.55	0.15	-0.10	311.65	320.04	-1.97	-2.11
22 Feb 27	63.1	126.1	-2.5	2.8	305.71	311.87	-0.33	-0.55	328.83	338.03	-2.07	-1.83
4 Mar 9	185.7	246.2	5.0	2.1	318.03	324.20	-0.76	-0.94	347.61	357.41	-1.36	-0.66
14 Mar 19	314.0	29.8	-3.4	-4.4	330.37	336.53	-1.10	-1.24	7.02	15.70	0.25	1.29
24 Mar 29	98.4	159.0	0.8	4.7	342.70	348.86	-1.35	-1.43	22.57	26.87	2.26	2.98
3 Apr 8	218.7	281.5	3.8	-1.2	355.02	1.18	-1.49	-1.51	28.29	26.98	3.23	2.89
13 Apr 18	352.6	66.9	-5.0	-1.8	7.34	13.50	-1.51	-1.48	23.84	20.38	1.95	0.61
23 Apr 28	131.8	191.4	3.6	4.9	19.65	25.80	-1.42	-1.34	18.04	17.58	-0.76	-1.91
3 May 8	252.8	319.0	1.2	-4.1	31.95	38.09	-1.23	-1.11	19.09	22.33	-2.71	-3.15
13 May 18	31.6	102.1	-4.2	1.5	44.24	50.38	-0.96	-0.80	27.03	32.97	-3.27	-3.10
23 May 28	164.1	224.1	5.1	3.5	56.52	62.66	-0.62	-0.44	40.00	48.06	-2.67	-2.02
2 Jun 7	288.6	358.2	-2.1	-5.3	68.80	74.94	-0.24	-0.04	57.12	67.12	-1.21	-0.31
12 Jun 17	69.5	135.6	-1.4	4.1	81.08	87.22	0.15	0.35	77.83	88.78	0.58	1.30
22 Jun 27	195.9	257.8	4.9	0.7	93.37	99.51	0.54	0.72	99.46	109.47	1.76	1.91
2 Jul 7	326.4	37.5	-4.6	-3.9	105.66	111.81	0.88	1.03	118.66	127.01	1.77	1.39
12 Jul 17	105.5	168.1	1.8	5.2	117.96	124.11	1.16	1.28	134.49	141.08	0.79	0.02
22 Jul 27	227.9	293.1	3.1	-2.4	130.26	136.42	1.36	1.43	146.67	151.07	-0.88	-1.87
1 Aug 6	5.6	75.6	-5.0	-0.8	142.57	148.72	1.46	1.47	154.00	155.09	-2.89	-3.84
11 Aug 16	139.9	200.1	4.2	4.5	154.88	161.03	1.45	1.40	154.00	150.77	-4.53	-4.70
21 Aug 26	260.8	330.4	0.2	-4.7	167.19	173.34	1.33	1.22	146.41	142.91	-4.13	-2.88
31 Aug	44.8		-3.1		179.48		1.10		142.17		-1.33	
Sep 5		111.7		2.5		185.63		0.94		144.95		0.09
10 Sep 15	173.0	232.4	5.0	2.4	191.77	197.90	0.76	0.57	150.86	158.84	1.13	1.70
20 Sep 25	295.5	9.1	-2.9	-4.7	204.03	210.16	0.35	0.13	167.79	176.94	1.85	1.71
30 S-O 5	82.3	146.0	0.3	4.6	216.28	222.39	-0.11	-0.36	185.93	194.62	1.36	0.89
10 Oct 15	205.7	265.8	4.0	-0.7	228.50	234.60	-0.60	-0.85	202.99	211.08	0.35	-0.22
20 Oct 25	332.4	47.7	-4.9	-2.5	240.69	246.77	-1.09	-1.32	218.90	226.50	-0.78	-1.32
30 Oct	117.7		3.4		252.83		-1.54		233.87		-1.80	
Nov 4		178.9		5.1		258.89		-1.74		241.00		-2.21
9 Nov 14	238.5	301.0	1.6	-3.7	264.94	270.96	-1.92	-2.06	247.79	254.01	-2.49	-2.60
19 Nov 24	10.8	84.9	-4.8	0.8	276.97	282.95	-2.18	-2.26	259.22	262.53	-2.44	-1.89
29 N-D 4	151.3	211.1	5.1	3.8	288.91	294.83	-2.31	-2.31	262.51	258.12	-0.80	0.80
9 Dec 14	272.2	338.4	-1.5	-5.2	300.72	306.57	-2.26	-2.17	251.43	247.20	2.28	2.91
19 Dec 24	49.7	120.4	-2.4	3.7	312.36	318.08	-2.03	-1.84	247.30	250.61	2.77	2.23
29 Dec	183.5		5.0		323.74		-1.60		255.78		1.54	
1741												1741

JULIAN	SATURN		JUPITER		MARS		SUN	GREGORIAN	MOON				VENUS				MERCURY			
12 UT	LONG.	LAT.	LONG.	LAT.	LONG.	LAT.	LONG.	12 UT	LONG.		LAT.		LONG.		LAT.		LONG.		LAT.	
1741								1742												1742
								Jan 3		243.1		1.2		329.31		-1.30		261.97		0.83
Dec 28	140.97	1.22	130.93	0.77	287.98	-0.95	288.17	8 Jan 13	307.6	17.4	-4.1	-4.4	334.79	340.14	-0.95	-0.54	268.75	275.91	0.14	-0.49
Jan 7	140.30	1.24	129.71	0.80	295.75	-1.00	298.35	18 Jan 23	87.7	154.4	1.1	5.0	345.36	350.40	-0.08	0.43	283.36	291.08	-1.03	-1.48
Jan 17	139.54	1.27	128.39	0.82	303.58	-1.04	308.51	28 J-F 2	215.0	276.0	3.3	-1.9	355.26	359.88	1.00	1.61	299.06	307.34	-1.82	-2.03
Jan 27	138.73	1.29	127.07	0.84	311.43	-1.07	318.65	7 Feb 12	345.0	56.6	-5.0	-1.5	4.22	8.22	2.27	2.98	315.94	324.87	-2.08	-1.94
Feb 6	137.93	1.30	125.86	0.85	319.31	-1.09	328.74	17 Feb 22	124.2	187.2	3.9	4.5	11.80	14.89	3.72	4.50	334.12	343.52	-1.59	-0.98
Feb 16	137.17	1.31	124.82	0.85	327.19	-1.09	338.79	27 Feb	246.6		0.5		17.37		5.29		352.70		-0.12	
								Mar 4		310.6		-4.3		19.14		6.07		0.89		0.95
Feb 26	136.49	1.31	124.04	0.85	335.07	-1.09	348.79	9 Mar 14	23.7	94.7	-3.7	2.1	20.05	19.98	6.80	7.44	7.02	10.15	2.08	3.01
Mar 8	135.94	1.31	123.54	0.85	342.93	-1.07	358.73	19 Mar 24	159.1	219.4	5.0	2.5	18.88	16.81	7.91	8.11	9.90	6.86	3.48	3.26
Mar 18	135.54	1.31	123.37	0.84	350.76	-1.04	8.62	29 M-A 3	279.3	347.2	-2.6	-5.0	14.00	10.87	7.98	7.47	2.73	359.43	2.36	1.07
Mar 28	135.31	1.30	123.50	0.83	358.55	-1.00	18.45	8 Apr 13	62.4	131.0	-0.5	4.6	7.93	5.62	6.64	5.58	358.13	359.01	-0.24	-1.34
Apr 7	135.25	1.29	123.95	0.82	6.28	-0.95	28.23	18 Apr 23	192.5	251.8	4.2	-0.4	4.21	3.81	4.43	3.28	1.79	6.08	-2.16	-2.68
Apr 17	135.37	1.28	124.68	0.81	13.96	-0.89	37.95	28 Apr	313.9		-4.8		4.39		2.20		11.61		-2.92	
								May 3		25.5		-3.5		5.83		1.22		18.16		-2.90
Apr 27	135.67	1.27	125.67	0.80	21.56	-0.81	47.63	8 May 13	100.0	165.4	2.9	5.2	8.01	10.80	0.37	-0.37	25.63	33.96	-2.63	-2.14
May 7	136.13	1.26	126.89	0.79	29.08	-0.73	57.26	18 May 23	225.1	285.2	2.0	-3.3	14.09	17.79	-1.00	-1.52	43.16	53.19	-1.46	-0.63
May 17	136.75	1.26	128.32	0.79	36.51	-0.64	66.85	28 M-J 2	350.7	64.2	-5.1	-0.3	21.84	26.16	-1.94	-2.27	63.88	74.87	0.26	1.06
May 27	137.51	1.25	129.92	0.78	43.85	-0.54	76.42	7 Jun 12	135.9	198.3	4.9	4.0	30.72	35.47	-2.53	-2.70	85.61	95.66	1.66	1.97
Jun 6	138.39	1.25	131.67	0.78	51.10	-0.43	85.97	17 Jun 22	257.7	320.5	-1.0	-5.0	40.38	45.43	-2.82	-2.86	104.81	112.99	1.97	1.68
Jun 16	139.38	1.24	133.54	0.78	58.24	-0.32	95.50	27 Jun	29.3		-3.2		50.61		-2.86		120.17		1.13	
								Jul 2		102.2		3.0		55.88		-2.80		126.29		0.37
Jun 26	140.46	1.24	135.52	0.78	65.27	-0.20	105.04	7 Jul 12	170.0	230.2	5.0	1.4	61.25	66.70	-2.70	-2.55	131.23	134.76	-0.59	-1.69
Jul 6	141.61	1.24	137.58	0.78	72.19	-0.07	114.58	17 Jul 22	291.2	357.9	-3.6	-4.7	72.21	77.79	-2.38	-2.17	136.61	136.48	-2.84	-3.93
Jul 16	142.82	1.25	139.70	0.79	78.99	0.06	124.13	27 J-A 1	68.4	138.7	0.3	4.9	83.43	89.13	-1.94	-1.69	134.34	130.77	-4.70	-4.89
Jul 26	144.06	1.25	141.86	0.79	85.68	0.19	133.71	6 Aug 11	202.5	262.0	3.4	-1.6	94.87	100.66	-1.42	-1.14	127.14	125.15	-4.35	-3.20
Aug 5	145.33	1.26	144.04	0.80	92.24	0.33	143.32	16 Aug 21	326.4	36.8	-5.0	-2.2	106.49	112.37	-0.86	-0.58	125.93	129.80	-1.76	-0.38
Aug 15	146.61	1.28	146.22	0.81	98.66	0.48	152.97	26 Aug 31	106.9	173.5	3.5	4.7	118.28	124.23	-0.30	-0.03	136.38	144.85	0.74	1.46
Aug 25	147.86	1.29	148.38	0.83	104.94	0.63	162.66	5 Sep 10	234.2	294.8	0.6	-4.1	130.21	136.23	0.23	0.48	154.24	163.80	1.78	1.76
Sep 4	149.09	1.31	150.50	0.84	111.06	0.79	172.40	15 Sep 20	3.6	75.9	-4.2	1.4	142.28	148.35	0.70	0.91	173.12	182.06	1.50	1.08
Sep 14	150.27	1.33	152.57	0.86	117.00	0.95	182.20	25 Sep 30	143.9	206.7	5.1	2.7	154.46	160.58	1.08	1.24	190.61	198.79	0.57	0.00
Sep 24	151.38	1.36	154.55	0.89	122.75	1.13	192.06	5 Oct 10	266.0	329.1	-2.4	-5.2	166.74	172.91	1.36	1.46	206.63	214.15	-0.59	-1.17
Oct 4	152.40	1.39	156.42	0.91	128.28	1.31	201.96	15 Oct 20	41.9	114.2	-1.5	4.3	179.10	185.31	1.52	1.56	221.35	228.20	-1.72	-2.20
Oct 14	153.31	1.42	158.17	0.94	133.54	1.51	211.93	25 Oct 30	179.0	239.1	4.5	-0.1	191.53	197.77	1.56	1.54	234.58	240.25	-2.58	-2.80
Oct 24	154.10	1.45	159.75	0.98	138.50	1.72	221.95	4 Nov 9	298.7	5.3	-4.6	-4.2	204.02	210.28	1.48	1.40	244.72	247.11	-2.77	-2.36
Nov 3	154.74	1.49	161.14	1.01	143.09	1.95	232.01	14 Nov 19	80.4	150.7	2.0	5.3	216.54	222.82	1.30	1.17	246.10	241.04	-1.39	0.18
Nov 13	155.23	1.53	162.31	1.05	147.23	2.20	242.12	24 Nov 29	212.4	271.5	2.3	-2.9	229.10	235.38	1.02	0.85	234.52	231.14	1.74	2.57
Nov 23	155.54	1.57	163.22	1.09	150.81	2.48	252.27	4 Dec 9	333.2	42.9	-5.3	-1.4	241.67	247.96	0.67	0.48	232.21	236.36	2.63	2.25
Dec 3	155.67	1.61	163.86	1.14	153.72	2.79	262.44	14 Dec 19	118.1	185.1	4.6	4.2	254.25	260.54	0.28	0.08	242.18	248.86	1.67	1.02
Dec 13	155.61	1.65	164.20	1.18	155.78	3.12	272.63	24 Dec 29	244.7	304.9	-0.6	-4.8	266.84	273.13	-0.12	-0.32	255.98	263.36	0.37	-0.25
1742								1743												1743
Dec 23	155.37	1.69	164.22	1.23	156.82	3.47	282.82	3 Jan 8	9.7	81.4	-3.8	2.1	279.42	285.71	-0.51	-0.69	270.93	278.68	-0.80	-1.28
Jan 2	154.96	1.73	163.91	1.28	156.65	3.82	293.01	13 Jan 18	154.2	217.7	5.0	1.6	292.00	298.28	-0.85	-1.01	286.62	294.78	-1.67	-1.94
Jan 12	154.41	1.76	163.30	1.32	155.17	4.14	303.19	23 Jan 28	276.9	340.0	-3.3	-4.8	304.56	310.84	-1.14	-1.25	303.18	311.84	-2.07	-2.03
Jan 22	153.73	1.79	162.41	1.36	152.43	4.39	313.34	2 Feb 7	47.9	119.7	-0.6	4.6	317.12	323.39	-1.34	-1.40	320.74	329.72	-1.78	-1.28
Feb 1	152.97	1.82	161.31	1.39	148.78	4.49	323.45	12 Feb 17	188.5	249.2	3.5	-1.4	329.65	335.90	-1.44	-1.45	338.41	346.03	-0.49	0.57
Feb 11	152.17	1.83	160.05	1.41	144.83	4.42	333.53	22 Feb 27	309.9	16.9	-4.9	-2.9	342.15	348.39	-1.44	-1.40	351.37	353.26	1.80	2.94
Feb 21	151.37	1.84	158.75	1.42	141.31	4.19	343.55	4 Mar 9	87.0	156.9	3.0	4.9	354.63	0.85	-1.33	-1.23	351.36	346.94	3.62	3.51
Mar 3	150.63	1.84	157.48	1.42	138.80	3.85	353.52	14 Mar 19	221.3	280.8	0.8	-4.0	7.06	13.26	-1.11	-0.97	342.46	339.89	2.66	1.44
Mar 13	149.98	1.84	156.35	1.41	137.55	3.45	3.44	24 Mar 29	344.4	55.2	-4.7	0.5	19.45	25.63	-0.81	-0.63	339.80	341.90	0.21	-0.84
Mar 23	149.44	1.83	155.41	1.40	137.58	3.04	13.31	3 Apr 8	125.9	192.2	5.0	3.1	31.79	37.94	-0.43	-0.23	345.69	350.77	-1.66	-2.24
Apr 2	149.06	1.82	154.73	1.37	138.74	2.66	23.11	13 Apr 18	253.2	313.2	-2.2	-5.2	44.08	50.20	-0.01	0.21	356.88	3.86	-2.57	-2.68
Apr 12	148.85	1.80	154.35	1.35	140.87	2.31	32.86	23 Apr 28	20.5	94.1	-2.5	3.8	56.30	62.40	0.44	0.66	11.61	20.11	-2.56	-2.22
Apr 22	148.80	1.78	154.27	1.32	143.78	1.99	42.56	3 May 8	163.2	225.8	4.8	0.3	68.47	74.52	0.88	1.08	29.36	39.34	-1.66	-0.93
May 2	148.94	1.76	154.49	1.29	147.33	1.70	52.21	13 May 18	285.2	347.2	-4.4	-4.7	80.56	86.58	1.28	1.45	49.96	60.89	-0.08	0.79
May 12	149.24	1.74	155.00	1.26	151.40	1.44	61.83	23 May 28	58.3	132.6	0.9	5.2	92.58	98.56	1.61	1.74	71.61	81.63	1.51	1.98
May 22	149.71	1.72	155.79	1.23	155.89	1.20	71.41	2 Jun 7	198.4	258.5	2.7	-2.6	104.52	110.44	1.84	1.91	90.64	98.53	2.14	1.97
Jun 1	150.33	1.70	156.82	1.20	160.73	0.99	80.97	12 Jun 17	318.2	23.0	-5.2	-2.2	116.35	122.22	1.95	1.95	105.24	110.69	1.51	0.76
Jun 11	151.09	1.69	158.07	1.17	165.88	0.79	90.51	22 Jun 27	97.0	169.4	3.9	4.4	128.07	133.88	1.92	1.84	114.73	117.12	-0.24	-1.43
Jun 21	151.97	1.68	159.52	1.15	171.29	0.60	100.04	2 Jul 7	231.8	291.1	-0.4	-4.6	139.65	145.37	1.72	1.56	117.67	116.31	-2.72	-3.90
Jul 1	152.95	1.67	161.13	1.13	176.92	0.43	109.57	12 Jul 17	352.8	60.8	-4.1	1.3	151.05	156.68	1.36	1.12	113.46	110.17	-4.72	-4.91
Jul 11	154.02	1.66	162.89	1.12	182.76	0.27	119.12	22 Jul 27	135.5	204.0	5.0	1.8	162.25	167.76	0.83	0.51	107.84	107.55	-4.41	-3.40
Jul 21	155.17	1.65	164.77	1.10	188.78	0.12	128.69	1 Aug 6	264.1	324.6	-3.2	-4.9	173.19	178.54	0.14	-0.26	109.84	114.73	-2.12	-0.81
Jul 31	156.36	1.65	166.74	1.09	194.98	-0.02	138.27	11 Aug 16	29.2	99.7	-1.3	4.2	183.79	188.93	-0.70	-1.18	121.93	130.82	0.34	1.18
Aug 10	157.60	1.66	168.80	1.09	201.33	-0.15	147.90	21 Aug 26	172.5	236.8	3.9	-1.3	193.94	198.80	-1.68	-2.20	140.60	150.53	1.65	1.77
Aug 20	158.86	1.67	170.91	1.08	207.85	-0.28	157.57	31 Aug	296.3		-4.9		203.48		-2.75		160.18		1.62	
								Sep 5		359.4		-3.4		207.95		-3.31		169.38		1.27
Aug 30	160.12	1.68	173.05	1.08	214.50	-0.39	167.29	10 Sep 15	67.4	138.7	2.3	5.1	212.16	216.07	-3.88	-4.45	178.10	186.37	0.79	0.22
Sep 9	161.36	1.69	175.21	1.08	221.30	-0.50	177.06	20 Sep 25	207.3	268.5	1.2	-3.9	219.60	222.69	-5.00	-5.54	194.22	201.66	-0.38	-1.01
Sep 19	162.57	1.71	177.37	1.09	228.23	-0.60	186.88	30 S-O 5	329.1	35.9	-4.9	-0.4	225.21	227.06	-6.02	-6.44	208.69	215.24	-1.61	-2.17
Sep 29	163.73	1.73	179.51	1.10	235.29	-0.70	196.76	10 Oct 15	106.6	176.2	4.8	3.6	228.12	228.26	-6.75	-6.90	221.19	226.27	-2.65	-2.99
Oct 9	164.82	1.76	181.59	1.11	242.47	-0.78	206.70	20 Oct 25	240.4	300.1	-1.9	-5.2	227.42	225.60	-6.83	-6.48	229.95	231.36	-3.10	-2.83
Oct 19	165.81	1.79	183.61	1.13	249.77	-0.86	216.69	30 Oct	2.9		-3.2		223.01		-5.81		229.38		-1.98	
								Nov 4		73.9		3.0		220.00		-4.84		223.85		-0.47
Oct 29	166.70	1.83	185.54	1.15	257.18	-0.93	226.73	9 Nov 14	145.7	211.6	5.1	0.7	217.10	214.76	-3.65	-2.37	217.78	215.34	1.16	2.18
Nov 8	167.46	1.86	187.34	1.18	264.69	-0.99	236.82	19 Nov 24	272.6	332.4	-4.2	-4.8	213.31	212.87	-1.12	0.02	217.35	222.28	2.45	2.24
Nov 18	168.07	1.91	189.00	1.20	272.29	-1.03	246.95	29 N-D 4	38.3	112.8	-0.1	5.0	213.43	214.90	0.99	1.80	228.72	235.86	1.77	1.19
Nov 28	168.52	1.95	190.48	1.24	279.98	-1.07	257.10	9 Dec 14	183.0	245.4	3.1	-2.3	217.15	220.05	2.44	2.93	243.31	250.91	0.58	-0.02
Dec 8	168.79	1.99	191.76	1.27	287.73	-1.09	267.29	19 Dec 24	304.9	5.9	-5.1	-2.7	223.49	227.38	3.29	3.53	258.61	266.40	-0.58	-1.09
Dec 18	168.89	2.04	192.80	1.31	295.55	-1.11	277.48	29 Dec	75.5		3.1		231.62		3.67		274.30		-1.51	
1743								1743												1743

JULIAN 12 UT	SATURN LONG.	SATURN LAT.	JUPITER LONG.	JUPITER LAT.	MARS LONG.	MARS LAT.	SUN LONG.
1743							
Dec 28	168.80	2.09	193.57	1.35	303.41	-1.11	287.67
Jan 7	168.53	2.13	194.06	1.40	311.30	-1.10	297.86
Jan 17	168.10	2.17	194.24	1.44	319.20	-1.08	308.02
Jan 27	167.52	2.20	194.10	1.48	327.11	-1.05	318.15
Feb 6	166.84	2.23	193.65	1.52	335.02	-1.00	328.26
Feb 16	166.08	2.25	192.91	1.56	342.89	-0.95	338.31
Feb 26	165.28	2.27	191.93	1.59	350.73	-0.89	348.31
Mar 7	164.50	2.27	190.76	1.61	358.53	-0.81	358.26
Mar 17	163.78	2.27	189.50	1.61	6.26	-0.73	8.15
Mar 27	163.15	2.25	188.22	1.61	13.93	-0.65	17.98
Apr 6	162.64	2.24	187.02	1.59	21.53	-0.55	27.76
Apr 16	162.28	2.21	185.97	1.57	29.04	-0.45	37.49
Apr 26	162.08	2.19	185.15	1.54	36.48	-0.35	47.16
May 6	162.06	2.16	184.59	1.50	43.82	-0.25	56.80
May 16	162.21	2.13	184.33	1.46	51.07	-0.14	66.40
May 26	162.53	2.10	184.36	1.41	58.24	-0.03	75.96
Jun 5	163.00	2.07	184.69	1.37	65.31	0.08	85.51
Jun 15	163.63	2.05	185.30	1.33	72.29	0.19	95.05
Jun 25	164.39	2.03	186.18	1.29	79.17	0.30	104.58
Jul 5	165.26	2.01	187.29	1.25	85.97	0.41	114.12
Jul 15	166.25	1.99	188.61	1.22	92.68	0.52	123.67
Jul 25	167.31	1.98	190.12	1.19	99.30	0.63	133.24
Aug 4	168.45	1.97	191.79	1.16	105.82	0.74	142.85
Aug 14	169.64	1.96	193.59	1.14	112.26	0.85	152.50
Aug 24	170.87	1.97	195.52	1.11	118.60	0.96	162.18
Sep 3	172.12	1.97	197.53	1.10	124.84	1.08	171.93
Sep 13	173.36	1.98	199.62	1.08	130.98	1.19	181.72
Sep 23	174.59	1.99	201.76	1.07	137.00	1.30	191.57
Oct 3	175.78	2.01	203.93	1.07	142.90	1.42	201.48
Oct 13	176.91	2.04	206.10	1.07	148.67	1.54	211.44
Oct 23	177.97	2.06	208.27	1.07	154.28	1.66	221.46
Nov 2	178.94	2.10	210.40	1.07	159.70	1.79	231.52
Nov 12	179.79	2.13	212.47	1.08	164.91	1.93	241.63
Nov 22	180.52	2.17	214.46	1.09	169.87	2.07	251.77
Dec 2	181.09	2.22	216.35	1.10	174.51	2.22	261.95
Dec 12	181.50	2.26	218.10	1.12	178.78	2.39	272.13
1744							
Dec 22	181.74	2.31	219.68	1.14	182.58	2.56	282.33
Jan 1	181.79	2.36	221.08	1.16	185.80	2.75	292.52
Jan 11	181.66	2.41	222.25	1.19	188.31	2.94	302.69
Jan 21	181.37	2.45	223.18	1.21	189.94	3.14	312.85
Jan 31	180.91	2.49	223.83	1.24	190.50	3.32	322.97
Feb 10	180.32	2.52	224.18	1.27	189.84	3.47	333.04
Feb 20	179.62	2.55	224.22	1.30	187.88	3.55	343.07
Mar 2	178.86	2.56	223.95	1.33	184.80	3.53	353.05
Mar 12	178.08	2.57	223.38	1.36	181.04	3.36	2.97
Mar 22	177.31	2.57	222.54	1.37	177.26	3.05	12.83
Apr 1	176.61	2.56	221.49	1.38	174.16	2.65	22.64
Apr 11	176.00	2.54	220.28	1.39	172.21	2.21	32.39
Apr 21	175.52	2.52	219.01	1.38	171.56	1.77	42.10
May 1	175.19	2.49	217.77	1.36	172.17	1.36	51.75
May 11	175.02	2.45	216.63	1.33	173.88	0.99	61.37
May 21	175.01	2.42	215.67	1.30	176.51	0.67	70.95
May 31	175.18	2.38	214.95	1.26	179.91	0.38	80.51
Jun 10	175.51	2.34	214.51	1.22	183.94	0.13	90.05
Jun 20	176.00	2.31	214.36	1.17	188.49	-0.09	99.58
Jun 30	176.64	2.28	214.51	1.13	193.47	-0.29	109.11
Jul 10	177.40	2.25	214.96	1.08	198.82	-0.47	118.66
Jul 20	178.28	2.23	215.68	1.04	204.49	-0.62	128.22
Jul 30	179.27	2.20	216.66	1.00	210.44	-0.76	137.81
Aug 9	180.33	2.19	217.87	0.96	216.64	-0.89	147.43
Aug 19	181.47	2.18	219.29	0.93	223.07	-0.99	157.10
Aug 29	182.65	2.17	220.89	0.90	229.70	-1.08	166.81
Sep 8	183.87	2.17	222.65	0.87	236.51	-1.16	176.58
Sep 18	185.10	2.17	224.54	0.84	243.50	-1.23	186.40
Sep 28	186.33	2.18	226.54	0.82	250.64	-1.28	196.28
Oct 8	187.54	2.19	228.63	0.80	257.92	-1.31	206.21
Oct 18	188.71	2.21	230.79	0.78	265.33	-1.33	216.20
Oct 28	189.82	2.23	232.98	0.77	272.86	-1.34	226.24
Nov 7	190.85	2.26	235.20	0.76	280.47	-1.33	236.33
Nov 17	191.79	2.29	237.42	0.75	288.18	-1.31	246.46
Nov 27	192.60	2.33	239.62	0.74	295.94	-1.28	256.61
Dec 7	193.28	2.37	241.77	0.74	303.76	-1.23	266.79
Dec 17	193.81	2.41	243.86	0.73	311.61	-1.17	276.99
1745							

GREGORIAN 12 UT	MOON LONG.		MOON LAT.		VENUS LONG.		VENUS LAT.		MERCURY LONG.		MERCURY LAT.	
1744												
Jan 3		151.2		4.6		236.15		3.72		282.34		-1.84
8 Jan 13	218.1	278.2	-0.1	-4.4	240.93	245.92	3.69	3.60	290.55	298.93	-2.05	-2.10
18 Jan 23	337.9	41.3	-4.2	0.5	251.07	256.36	3.45	3.24	307.46	315.99	-1.95	-1.56
28 J-F 2	114.0	187.9	4.9	2.2	261.76	267.27	3.00	2.73	324.15	331.14	-0.87	0.15
7 Feb 12	251.3	310.7	-3.1	-5.0	272.86	278.52	2.42	2.10	335.65	336.29	1.45	2.77
17 Feb 22	12.2	78.8	-1.7	3.7	284.23	290.00	1.77	1.42	332.89	327.55	3.63	3.62
27 Feb	152.5		4.4		295.81		1.08		323.32		2.84	
Mar 3		222.4		-1.0		301.66		0.74		321.77		1.72
8 Mar 13	283.5	343.7	-4.9	-3.8	307.54	313.45	0.41	0.09	322.85	325.97	0.58	-0.42
18 Mar 23	48.2	117.9	1.6	5.2	319.38	325.33	-0.21	-0.49	330.55	336.23	-1.23	-1.84
28 M-A 2	189.8	255.4	1.8	-3.7	331.29	337.27	-0.74	-0.97	342.76	350.01	-2.25	-2.45
7 Apr 12	315.3	17.7	-5.1	-1.2	343.27	349.27	-1.17	-1.34	357.93	6.49	-2.45	-2.24
17 Apr 22	86.2	157.1	4.4	4.2	355.29	1.31	-1.48	-1.58	15.72	25.61	-1.82	-1.[illegible]9
27 Apr	225.3		-1.4		7.34		-1.65		36.08		-0.40	
May 2		287.4		-5.1		13.38		-1.69		46.88		0.48
7 May 12	347.5	53.4	-3.6	2.1	19.42	25.47	-1.70	-1.67	57.51	67.40	1.30	1.93
17 May 22	125.5	195.0	5.2	1.4	31.52	37.58	-1.62	-1.53	76.17	83.62	2.26	2.26
27 M-J 1	259.2	319.3	-3.9	-4.9	43.64	49.71	-1.43	-1.29	89.65	94.16	1.90	1.20
6 Jun 11	20.8	91.2	-0.8	4.5	55.78	61.86	-1.14	-0.97	96.98	97.96	0.18	-1.09
16 Jun 21	164.6	230.9	3.6	-1.9	67.95	74.04	-0.79	-0.60	97.09	94.74	-2.47	-3.72
26 Jun	292.1		-5.0		80.14		-0.40		91.82		-4.53	
Jul 1		351.7		-2.9		86.25		-0.19		89.52		-4.73
6 Jul 11	55.9	130.2	2.4	4.9	92.36	98.49	0.02	0.22	88.84	90.29	-4.33	-3.49
16 Jul 21	202.0	265.0	0.4	-4.3	104.62	110.76	0.41	0.60	93.96	99.78	-2.38	-1.19
26 Jul 31	324.7	25.2	-4.3	0.1	116.91	123.07	0.77	0.93	107.51	116.76	-0.06	0.87
5 Aug 10	93.2	168.9	4.7	2.9	129.23	135.41	1.07	1.19	126.86	137.13	1.48	1.75
15 Aug 20	237.3	298.0	-2.8	-5.1	141.59	147.78	1.28	1.35	147.08	156.53	1.71	1.44
25 Aug 30	357.7	60.7	-2.1	3.2	153.98	160.18	1.40	1.41	165.42	173.78	1.01	0.46
4 Sep 9	132.0	206.0	4.8	-0.4	166.39	172.61	1.41	1.37	181.63	188.98	-0.16	-0.82
14 Sep 19	270.8	330.4	-4.9	-4.1	178.83	185.05	1.31	1.22	195.81	202.05	-1.49	-2.14
24 Sep 29	31.8	98.3	1.0	5.1	191.27	197.50	1.11	0.97	207.53	211.95	-2.72	-3.18
4 Oct 9	170.8	241.0	2.7	-3.4	203.73	209.95	0.82	0.64	214.77	215.16	-3.43	-3.30
14 Oct 19	303.0	3.0	-5.2	-1.7	216.18	222.41	0.46	0.26	212.26	206.51	-2.55	-1.11
24 Oct 29	67.7	137.4	3.8	4.7	228.64	234.87	0.05	-0.16	201.15	199.72	0.56	1.75
3 Nov 8	208.3	274.4	-0.7	-5.0	241.09	247.32	-0.38	-0.59	202.62	208.29	2.23	2.19
13 Nov 18	334.6	36.5	-3.8	1.4	253.54	259.76	-0.80	-0.99	215.32	222.91	1.85	1.35
23 Nov 28	105.6	176.5	5.1	2.1	265.98	272.19	-1.18	-1.34	230.70	238.53	0.78	0.20
3 Dec 8	244.2	306.8	-3.6	-4.9	278.40	284.61	-1.49	-1.61	246.36	254.20	-0.37	-0.89
13 Dec 18	6.3	71.6	-1.1	4.0	290.81	296.99	-1.70	-1.77	262.06	269.99	-1.36	-1.74
23 Dec 28	144.7	214.4	4.0	-1.4	303.17	309.34	-1.81	-1.81	278.00	286.11	-2.01	-2.14
1745												
2 Jan 7	278.5	338.7	-4.9	-3.1	315.49	321.62	-1.78	-1.71	294.27	302.34	-2.10	-1.82
12 Jan 17	39.1	108.7	2.0	5.0	327.74	333.83	-1.61	-1.47	309.93	316.24	-1.24	-0.29
22 Jan 27	183.4	250.5	1.0	-4.2	339.89	345.92	-1.29	-1.09	319.86	319.29	1.02	2.48
1 Feb 6	311.7	10.9	-4.5	-0.2	351.92	357.88	-0.85	-0.58	314.63	308.86	3.51	3.62
11 Feb 16	73.8	147.0	4.5	3.7	3.79	9.65	-0.28	0.05	305.28	304.84	2.94	1.92
21 Feb 26	220.6	284.7	-2.4	-5.2	15.45	21.19	0.39	0.75	306.99	310.95	0.88	-0.06
3 Mar 8	344.2	44.4	-2.4	2.8	26.86	32.44	1.13	1.51	316.16	322.28	-0.85	-1.48
13 Mar 18	110.7	185.4	5.2	0.6	37.93	43.31	1.89	2.27	329.11	336.54	-1.94	-2.23
23 Mar 28	256.0	317.5	-4.8	-4.3	48.56	53.68	2.65	3.01	344.54	353.10	-2.33	-2.24
2 Apr 7	16.9	79.8	0.5	4.9	58.62	63.38	3.34	3.65	2.25	11.99	-1.94	-1.42
12 Apr 17	149.3	222.9	3.6	-2.7	67.90	72.15	3.92	4.14	22.27	32.86	-0.71	0.15
22 Apr 27	289.7	349.5	-5.2	-2.0	76.08	79.61	4.30	4.38	43.31	52.99	1.04	1.82
2 May 7	50.7	117.4	3.3	5.0	82.68	85.17	4.38	4.27	61.40	68.24	2.34	2.52
12 May 17	188.3	258.8	0.3	-4.8	86.98	87.97	4.03	3.64	73.35	76.60	2.31	1.69
22 May 27	322.0	21.6	-3.9	1.0	88.04	87.12	3.06	2.29	77.88	77.26	0.67	-0.65
1 Jun 6	86.3	156.4	4.9	2.8	85.24	82.57	1.32	0.20	75.13	72.38	-2.09	-3.34
11 Jun 16	226.6	293.1	-3.1	-4.9	79.49	76.46	-0.97	-2.09	70.15	69.35	-4.14	-4.38
21 Jun 26	353.6	54.7	-1.3	3.7	73.96	72.31	-3.07	-3.83	70.43	73.46	-4.12	-3.47
1 Jul 6	123.9	195.6	4.5	-0.8	71.63	71.92	-4.37	-4.72	78.34	84.95	-2.57	-1.52
11 Jul 16	263.3	326.1	-4.9	-3.2	73.09	75.03	-4.89	-4.92	93.16	102.69	-0.44	0.53
21 Jul 26	25.4	89.6	1.8	5.1	77.61	80.73	-4.85	-4.68	113.05	123.59	1.26	1.68
31 J-A 5	162.7	233.7	1.9	-3.9	84.29	88.22	-4.45	-4.16	133.82	143.49	1.77	1.61
10 Aug 15	298.2	358.3	-4.7	-0.5	92.45	96.94	-3.83	-3.47	152.54	160.96	1.23	0.71
20 Aug 25	58.3	126.5	4.3	4.4	101.65	106.54	-3.08	-2.68	168.79	176.03	0.08	-0.61
30 Aug	201.5		-1.7		111.60		-2.27		182.64		-1.35	
Sep 4		270.1		-5.2		116.80		-1.86		188.51		-2.08
9 Sep 14	331.5	30.7	-2.7	2.5	122.12	127.54	-1.45	-1.05	193.46	197.15	-2.77	-3.37
19 Sep 24	93.1	164.6	5.3	1.7	133.07	138.68	-0.66	-0.28	199.02	198.34	-3.75	-3.75
29 S-O 4	239.1	304.4	-4.4	-4.5	144.37	150.13	0.07	0.39	194.65	188.99	-3.10	-1.72
9 Oct 14	4.0	64.3	0.1	4.6	155.95	161.83	0.69	0.96	184.59	184.24	-0.04	1.28
19 Oct 24	129.8	203.1	4.2	-1.8	167.76	173.73	1.19	1.39	187.99	194.35	1.96	2.10
29 Oct	274.9		-5.2		179.75		1.55		201.94		1.89	
Nov 3		337.1		-2.2		185.81		1.67		209.98		1.48
8 Nov 13	36.5	99.7	2.9	5.1	191.90	198.01	1.76	1.81	218.10	226.16	0.97	0.41
18 Nov 23	168.1	240.9	1.2	-4.4	204.16	210.32	1.82	1.80	234.14	242.03	-0.16	-0.70
28 N-D 3	308.8	8.8	-4.0	0.7	216.51	222.71	1.75	1.66	249.88	257.71	-1.19	-1.62
8 Dec 13	70.0	137.0	4.7	3.4	228.92	235.15	1.55	1.41	265.54	273.37	-1.96	-2.17
18 Dec 23	207.1	277.3	-2.4	-4.9	241.38	247.62	1.25	1.07	281.16	288.74	-2.22	-2.06
28 Dec	341.0		-1.4		253.87		0.88		295.73		-1.59	
1745												

JULIAN 12 UT	SATURN LONG.	SATURN LAT.	JUPITER LONG.	JUPITER LAT.	MARS LONG.	MARS LAT.	SUN LONG.	GREGORIAN 12 UT	MOON LONG.	MOON LONG.	MOON LAT.	MOON LAT.	VENUS LONG.	VENUS LONG.	VENUS LAT.	VENUS LAT.	MERCURY LONG.	MERCURY LONG.	MERCURY LAT.	MERCURY LAT.
1745								1746												1746
								Jan 2		40.5		3.5		260.13		0.67		301.30		-0.75
Dec 27	194.18	2.46	245.84	0.73	319.47	-1.11	287.18	7 Jan 12	105.1	175.6	4.8	0.0	266.38	272.64	0.46	0.24	304.02	302.32	0.54	2.08
Jan 6	194.37	2.51	247.71	0.74	327.34	-1.03	297.36	17 Jan 22	245.6	312.0	-4.8	-3.5	278.90	285.16	0.02	-0.19	296.69	290.91	3.27	3.52
Jan 16	194.38	2.55	249.43	0.74	335.19	-0.94	307.53	27 J-F 1	12.6	73.1	1.6	5.1	291.42	297.67	-0.39	-0.59	288.20	288.85	2.97	2.07
Jan 26	194.21	2.60	250.98	0.74	343.01	-0.85	317.67	6 Feb 11	141.7	214.6	2.8	-3.4	303.93	310.18	-0.77	-0.93	291.95	296.62	1.13	0.25
Feb 5	193.88	2.64	252.32	0.75	350.80	-0.75	327.77	16 Feb 21	282.7	345.1	-5.0	-0.8	316.43	322.68	-1.07	-1.20	302.34	308.81	-0.52	-1.16
Feb 15	193.39	2.68	253.43	0.76	358.53	-0.65	337.82	26 Feb	44.4		4.1		328.92		-1.29		315.85		-1.66	
								Mar 3		107.3		4.8		335.16		-1.37		323.40		-2.00
Feb 25	192.78	2.71	254.28	0.76	6.21	-0.54	347.83	8 Mar 13	179.7	252.8	-0.6	-5.2	341.39	347.62	-1.42	-1.44	331.43	339.94	-2.19	-2.20
Mar 7	192.08	2.73	254.85	0.77	13.81	-0.43	357.78	18 Mar 23	317.7	17.5	-3.1	2.1	353.85	0.06	-1.43	-1.40	348.96	358.50	-2.01	-1.62
Mar 17	191.32	2.74	255.11	0.78	21.34	-0.32	7.67	28 M-A 2	77.3	143.3	5.2	2.7	6.27	12.48	-1.34	-1.26	8.53	18.85	-1.00	-0.19
Mar 27	190.55	2.74	255.07	0.79	28.79	-0.21	17.51	7 Apr 12	218.2	289.2	-3.7	-4.7	18.67	24.86	-1.15	-1.03	29.04	38.44	0.74	1.64
Apr 6	189.80	2.74	254.71	0.79	36.16	-0.10	27.29	17 Apr 22	350.9	50.1	-0.2	4.3	31.05	37.22	-0.88	-0.72	46.37	52.42	2.35	2.73
Apr 16	189.12	2.72	254.07	0.79	43.44	0.00	37.02	27 Apr	112.0		4.5		43.39		-0.54		56.35		2.70	
								May 2		181.2		-0.8		49.55		-0.35		58.03		2.19
Apr 26	188.54	2.69	253.17	0.79	50.64	0.11	46.69	7 May 12	256.3	323.6	-5.1	-2.4	55.71	61.85	-0.15	0.05	57.56	55.41	1.21	-0.14
May 6	188.09	2.66	252.07	0.78	57.76	0.21	56.33	17 May 22	23.3	83.7	2.7	5.0	68.00	74.13	0.26	0.46	52.55	50.19	-1.58	-2.80
May 16	187.79	2.63	250.84	0.76	64.79	0.31	65.93	27 M-J 1	148.6	220.1	1.9	-3.8	80.26	86.38	0.65	0.84	49.25	50.12	-3.60	-3.93
May 26	187.64	2.59	249.57	0.74	71.74	0.41	75.50	6 Jun 11	292.8	356.2	-4.2	0.6	92.50	98.60	1.01	1.17	52.80	57.16	-3.82	-3.38
Jun 5	187.67	2.55	248.34	0.72	78.61	0.51	85.05	16 Jun 21	55.6	118.9	4.6	3.8	104.71	110.80	1.31	1.43	63.02	70.29	-2.67	-1.78
Jun 15	187.86	2.51	247.24	0.69	85.41	0.60	94.58	26 Jun	186.9		-1.7		116.89		1.52		78.88		-0.79	
								Jul 1		258.9		-5.0		122.97		1.59		88.63		0.18
Jun 25	188.21	2.47	246.35	0.66	92.13	0.68	104.11	6 Jul 11	327.2	27.9	-1.7	3.4	129.04	135.10	1.63	1.63	99.17	109.93	1.00	1.56
Jul 5	188.72	2.43	245.71	0.62	98.79	0.77	113.65	16 Jul 21	88.9	155.8	5.0	0.9	141.15	147.19	1.61	1.55	120.39	130.26	1.80	1.75
Jul 15	189.36	2.39	245.35	0.59	105.38	0.85	123.21	26 Jul 31	226.3	296.1	-4.5	-3.9	153.22	159.23	1.47	1.34	139.42	147.88	1.45	0.97
Jul 25	190.14	2.36	245.30	0.55	111.90	0.92	132.78	5 Aug 10	0.0	59.7	1.4	5.0	165.23	171.22	1.19	1.01	155.65	162.73	0.35	-0.38
Aug 4	191.03	2.33	245.56	0.52	118.37	1.00	142.38	15 Aug 20	123.4	194.3	3.5	-2.6	177.18	183.13	0.80	0.56	169.04	174.48	-1.17	-2.00
Aug 14	192.01	2.31	246.11	0.48	124.78	1.07	152.03	25 Aug 30	265.2	331.2	-5.2	-1.2	189.06	194.97	0.30	0.02	178.80	181.65	-2.82	-3.55
Aug 24	193.08	2.29	246.94	0.45	131.13	1.13	161.71	4 Sep 9	32.0	92.2	3.9	5.0	200.85	206.71	-0.28	-0.60	182.47	180.74	-4.07	-4.17
Sep 3	194.22	2.27	248.03	0.42	137.43	1.20	171.45	14 Sep 19	159.6	233.4	0.5	-4.9	212.53	218.33	-0.92	-1.25	176.47	171.28	-3.60	-2.27
Sep 13	195.40	2.26	249.35	0.39	143.67	1.26	181.24	24 Sep 29	302.4	4.7	-3.6	1.8	224.09	229.80	-1.57	-1.90	168.09	168.85	-0.62	0.79
Sep 23	196.61	2.26	250.88	0.37	149.85	1.32	191.09	4 Oct 9	64.2	126.1	5.1	3.2	235.47	241.09	-2.21	-2.50	173.41	180.40	1.64	1.97
Oct 3	197.83	2.26	252.59	0.34	155.97	1.38	200.99	14 Oct 19	197.3	271.9	-2.8	-4.9	246.65	252.13	-2.77	-3.01	188.54	197.02	1.91	1.59
Oct 13	199.04	2.27	254.46	0.32	162.02	1.44	210.95	24 Oct 29	337.4	37.3	-0.6	4.1	257.54	262.84	-3.21	-3.37	205.48	213.78	1.14	0.61
Oct 23	200.23	2.28	256.45	0.30	168.00	1.49	220.96	3 Nov 8	97.2	161.8	4.6	0.1	268.02	273.06	-3.48	-3.54	221.91	229.88	0.05	-0.50
Nov 2	201.37	2.29	258.55	0.28	173.89	1.55	231.03	13 Nov 18	236.0	308.4	-4.8	-2.8	277.94	282.61	-3.52	-3.43	237.71	245.45	-1.03	-1.50
Nov 12	202.45	2.32	260.73	0.26	179.69	1.60	241.14	23 Nov 28	10.6	70.0	2.5	5.0	287.03	291.14	-3.25	-2.98	253.10	260.67	-1.90	-2.19
Nov 22	203.45	2.34	262.97	0.25	185.38	1.65	251.28	3 Dec 8	131.6	199.3	2.4	-3.2	294.88	298.16	-2.59	-2.08	268.08	275.16	-2.33	-2.28
Dec 2	204.34	2.37	265.25	0.23	190.93	1.70	261.45	13 Dec 18	274.2	342.6	-4.5	0.4	300.89	302.93	-1.44	-0.64	281.52	286.31	-1.94	-1.21
Dec 12	205.12	2.41	267.54	0.21	196.34	1.75	271.64	23 Dec 28	42.8	103.4	4.6	4.1	304.15	304.42	0.32	1.43	288.09	285.35	0.01	1.60
1746								1747												1747
Dec 22	205.75	2.45	269.82	0.20	201.55	1.79	281.83	2 Jan 7	168.0	238.3	-1.0	-5.1	303.66	301.90	2.67	3.96	279.06	273.61	2.93	3.33
Jan 1	206.23	2.49	272.07	0.19	206.54	1.83	292.02	12 Jan 17	310.7	15.2	-2.2	3.3	299.31	296.29	5.19	6.22	271.87	273.56	2.94	2.18
Jan 11	206.55	2.54	274.26	0.17	211.25	1.86	302.20	22 Jan 27	74.8	137.9	5.1	1.6	293.34	290.93	6.96	7.34	277.49	282.78	1.33	0.52
Jan 21	206.69	2.58	276.38	0.16	215.62	1.89	312.35	1 Feb 6	206.2	277.2	-4.1	-4.5	289.41	288.92	7.40	7.20	288.94	295.69	-0.22	-0.86
Jan 31	206.65	2.63	278.40	0.14	219.56	1.91	322.48	11 Feb 16	345.4	46.9	0.9	5.0	289.43	290.85	6.82	6.31	302.91	310.53	-1.39	-1.79
Feb 10	206.45	2.67	280.28	0.13	222.98	1.91	332.56	21 Feb 26	107.3	174.1	3.9	-1.7	293.03	295.84	5.73	5.11	318.55	326.97	-2.05	-2.14
Feb 20	206.08	2.71	282.01	0.11	225.73	1.88	342.59	3 Mar 8	245.5	314.6	-5.3	-1.9	299.19	302.96	4.47	3.84	335.83	345.15	-2.06	-1.77
Mar 2	205.57	2.74	283.56	0.10	227.66	1.83	352.57	13 Mar 18	18.7	78.6	3.6	5.1	307.09	311.50	3.21	2.60	354.89	4.89	-1.26	-0.53
Mar 12	204.94	2.76	284.90	0.08	228.58	1.73	2.49	23 Mar 28	140.9	212.0	1.3	-4.4	316.15	320.99	2.01	1.45	14.76	23.77	0.40	1.39
Mar 22	204.23	2.78	286.00	0.07	228.34	1.57	12.36	2 Apr 7	284.5	350.2	-4.0	1.4	326.00	331.14	0.93	0.44	31.14	36.23	2.28	2.89
Apr 1	203.47	2.78	286.84	0.05	226.82	1.32	22.17	12 Apr 17	51.2	110.9	5.0	3.5	336.40	341.76	0.00	-0.41	38.73	38.64	3.06	2.68
Apr 11	202.71	2.78	287.40	0.03	224.14	0.99	31.93	22 Apr 27	176.4	251.0	-2.0	-5.0	347.19	352.70	-0.77	-1.09	36.43	33.22	1.74	0.40
Apr 21	201.98	2.77	287.65	0.01	220.68	0.58	41.63	2 May 7	321.6	24.1	-1.0	3.9	358.27	3.89	-1:36	-1.58	30.42	29.09	-1.01	-2.20
May 1	201.33	2.75	287.58	-0.01	217.07	0.12	51.29	12 May 17	83.6	144.5	4.6	0.6	9.56	15.27	-1.77	-1.90	29.65	32.05	-3.01	-3.43
May 11	200.77	2.72	287.21	-0.03	214.05	-0.34	60.91	22 May 27	213.9	289.6	-4.5	-3.3	21.02	26.80	-2.00	-2.05	36.05	41.44	-3.48	-3.22
May 21	200.35	2.68	286.54	-0.06	212.16	-0.75	70.49	1 Jun 6	356.7	57.0	2.3	5.0	32.60	38.43	-2.06	-2.04	48.06	55.83	-2.71	-1.98
May 31	200.08	2.64	285.62	-0.08	211.60	-1.09	80.05	11 Jun 16	116.7	180.2	2.7	-2.7	44.29	50.17	-1.98	-1.88	64.72	74.61	-1.10	-0.17
Jun 10	199.97	2.60	284.50	-0.11	212.37	-1.37	89.59	21 Jun 26	252.9	326.4	-4.9	-0.1	56.07	61.99	-1.76	-1.61	85.26	96.16	0.71	1.38
Jun 20	200.03	2.55	283.25	-0.13	214.33	-1.59	99.12	1 Jul 6	30.1	89.6	4.6	4.3	67.93	73.89	-1.44	-1.25	106.80	116.80	1.77	1.86
Jun 30	200.24	2.51	281.97	-0.15	217.29	-1.76	108.66	11 Jul 16	151.0	218.1	-0.4	-4.9	79.87	85.87	-1.04	-0.82	126.04	134.48	1.67	1.25
Jul 10	200.62	2.47	280.74	-0.18	221.09	-1.89	118.20	21 Jul 26	291.6	1.2	-3.0	3.0	91.89	97.92	-0.60	-0.37	142.13	148.96	0.65	-0.10
Jul 20	201.15	2.43	279.66	-0.20	225.58	-1.99	127.76	31 J-A 5	62.3	122.6	5.2	2.1	103.97	110.05	-0.14	0.08	154.90	159.78	-0.96	-1.89
Jul 30	201.81	2.39	278.78	-0.22	230.63	-2.05	137.35	10 Aug 15	187.3	257.3	-3.5	-4.9	116.14	122.24	0.30	0.51	163.34	165.19	-2.83	-3.71
Aug 9	202.60	2.36	278.17	-0.23	236.16	-2.09	146.97	20 Aug 25	328.9	34.3	0.3	4.9	128.37	134.51	0.70	0.87	164.90	162.21	-4.36	-4.54
Aug 19	203.50	2.33	277.86	-0.25	242.08	-2.10	156.63	30 Aug	94.2		4.2		140.66		1.03		157.75		-4.01	
								Sep 4		156.7		-1.0		146.84		1.16		153.44		-2.75
Aug 29	204.50	2.30	277.87	-0.26	248.34	-2.10	166.35	9 Sep 14	225.6	296.4	-5.2	-2.6	153.02	159.22	1.27	1.35	151.64	153.55	-1.15	0.29
Sep 8	205.57	2.28	278.20	-0.27	254.89	-2.07	176.11	19 Sep 24	4.3	66.4	3.2	5.1	165.43	171.66	1.41	1.44	158.87	166.44	1.29	1.79
Sep 18	206.70	2.26	278.83	-0.28	261.68	-2.03	185.92	29 S-O 4	126.3	192.6	1.7	-3.8	177.89	184.13	1.44	1.41	175.08	183.99	1.88	1.68
Sep 28	207.88	2.25	279.75	-0.29	268.68	-1.96	195.80	9 Oct 14	264.9	334.0	-4.4	0.8	190.38	196.64	1.36	1.28	192.80	201.36	1.30	0.81
Oct 8	209.08	2.25	280.94	-0.30	275.86	-1.89	205.72	19 Oct 24	38.1	98.1	4.9	3.6	202.90	209.17	1.18	1.05	209.65	217.70	0.26	-0.30
Oct 18	210.29	2.25	282.36	-0.31	283.18	-1.79	215.71	29 Oct	159.5		-1.4		215.44		0.91		225.54		-0.85	
								Nov 3		230.4		-5.0		221.72		0.75		233.19		-1.37
Oct 28	211.49	2.25	283.99	-0.32	290.62	-1.69	225.75	8 Nov 13	303.8	9.8	-1.6	3.7	227.99	234.27	0.57	0.38	240.68	247.97	-1.82	-2.19
Nov 7	212.66	2.26	285.80	-0.33	298.15	-1.57	235.83	18 Nov 23	71.0	130.3	4.7	0.9	240.55	246.83	0.19	-0.01	255.00	261.57	-2.43	-2.48
Nov 17	213.78	2.28	287.76	-0.34	305.75	-1.45	245.95	28 N-D 3	194.6	269.0	-4.1	-3.9	253.11	259.39	-0.21	-0.41	267.26	271.23	-2.27	-1.67
Nov 27	214.82	2.30	289.85	-0.35	313.40	-1.32	256.11	8 Dec 13	340.9	43.9	1.9	5.1	265.68	271.96	-0.60	-0.79	272.04	268.37	-0.55	1.04
Dec 7	215.78	2.32	292.04	-0.37	321.07	-1.18	266.29	18 Dec 23	103.5	163.8	2.9	-2.3	278.23	284.51	-0.96	-1.11	261.74	256.85	2.49	3.07
Dec 17	216.63	2.35	294.31	-0.38	328.76	-1.04	276.48	28 Dec	231.7		-5.1		290.78		-1.25		256.12		2.86	
1747								1747												1747

JULIAN 12 UT	SATURN LONG.	SATURN LAT.	JUPITER LONG.	JUPITER LAT.	MARS LONG.	MARS LAT.	SUN LONG.
1747							
Dec 27	217.36	2.38	296.64	-0.39	336.43	-0.90	286.68
Jan 6	217.94	2.42	298.99	-0.40	344.08	-0.75	296.86
Jan 16	218.37	2.46	301.36	-0.42	351.70	-0.61	307.03
Jan 26	218.63	2.50	303.70	-0.44	359.27	-0.47	317.17
Feb 5	218.72	2.54	306.02	-0.46	6.77	-0.34	327.27
Feb 15	218.64	2.58	308.27	-0.48	14.22	-0.21	337.33
Feb 25	218.39	2.61	310.44	-0.50	21.60	-0.08	347.34
Mar 6	217.98	2.65	312.51	-0.52	28.90	0.04	357.29
Mar 16	217.44	2.67	314.44	-0.55	36.12	0.16	7.19
Mar 26	216.80	2.69	316.23	-0.58	43.27	0.26	17.04
Apr 5	216.08	2.70	317.83	-0.61	50.35	0.37	26.82
Apr 15	215.33	2.70	319.22	-0.65	57.34	0.46	36.55
Apr 25	214.58	2.69	320.38	-0.68	64.27	0.55	46.23
May 5	213.87	2.67	321.28	-0.72	71.12	0.63	55.87
May 15	213.24	2.65	321.89	-0.76	77.91	0.71	65.47
May 25	212.72	2.62	322.20	-0.81	84.63	0.78	75.04
Jun 4	212.34	2.58	322.19	-0.85	91.30	0.84	84.59
Jun 14	212.10	2.54	321.86	-0.89	97.91	0.90	94.13
Jun 24	212.03	2.49	321.23	-0.94	104.48	0.96	103.66
Jul 4	212.11	2.45	320.33	-0.97	110.99	1.01	113.20
Jul 14	212.36	2.40	319.21	-1.01	117.47	1.05	122.75
Jul 24	212.77	2.36	317.96	-1.03	123.91	1.09	132.32
Aug 3	213.32	2.32	316.66	-1.05	130.32	1.12	141.92
Aug 13	214.01	2.28	315.40	-1.07	136.70	1.16	151.56
Aug 23	214.82	2.25	314.28	-1.07	143.06	1.18	161.25
Sep 2	215.74	2.22	313.37	-1.07	149.39	1.20	170.98
Sep 12	216.74	2.19	312.74	-1.06	155.71	1.22	180.77
Sep 22	217.82	2.17	312.42	-1.04	162.00	1.24	190.61
Oct 2	218.96	2.15	312.43	-1.03	168.28	1.25	200.51
Oct 12	220.13	2.14	312.78	-1.01	174.53	1.25	210.47
Oct 22	221.33	2.13	313.44	-0.99	180.77	1.25	220.48
Nov 1	222.53	2.13	314.41	-0.97	186.98	1.25	230.54
Nov 11	223.71	2.14	315.64	-0.96	193.18	1.24	240.64
Nov 21	224.85	2.14	317.12	-0.94	199.34	1.22	250.79
Dec 1	225.94	2.16	318.81	-0.93	205.46	1.20	260.95
Dec 11	226.95	2.17	320.69	-0.92	211.54	1.17	271.14
1748							
Dec 21	227.87	2.20	322.72	-0.91	217.58	1.13	281.34
Dec 31	228.68	2.22	324.87	-0.91	223.54	1.08	291.52
Jan 10	229.36	2.25	327.12	-0.91	229.44	1.02	301.71
Jan 20	229.89	2.28	329.45	-0.91	235.24	0.95	311.86
Jan 30	230.26	2.32	331.83	-0.91	240.93	0.85	321.98
Feb 9	230.47	2.35	334.24	-0.92	246.48	0.74	332.07
Feb 19	230.50	2.39	336.65	-0.93	251.87	0.60	342.10
Mar 1	230.37	2.42	339.05	-0.94	257.04	0.44	352.08
Mar 11	230.07	2.45	341.41	-0.96	261.95	0.23	2.01
Mar 21	229.63	2.47	343.72	-0.98	266.53	-0.01	11.88
Mar 31	229.07	2.49	345.94	-1.00	270.69	-0.31	21.70
Apr 10	228.40	2.50	348.07	-1.02	274.32	-0.67	31.46
Apr 20	227.68	2.51	350.07	-1.05	277.28	-1.11	41.16
Apr 30	226.93	2.50	351.92	-1.09	279.40	-1.62	50.82
May 10	226.19	2.49	353.60	-1.12	280.49	-2.23	60.45
May 20	225.51	2.47	355.08	-1.16	280.40	-2.91	70.03
May 30	224.91	2.44	356.33	-1.21	279.08	-3.62	79.59
Jun 9	224.43	2.41	357.33	-1.25	276.68	-4.30	89.13
Jun 19	224.08	2.37	358.05	-1.30	273.73	-4.84	98.66
Jun 29	223.88	2.33	358.46	-1.35	270.92	-5.17	108.19
Jul 9	223.85	2.28	358.55	-1.40	268.93	-5.28	117.74
Jul 19	223.97	2.24	358.32	-1.45	268.20	-5.19	127.30
Jul 29	224.26	2.19	357.77	-1.50	268.87	-4.97	136.88
Aug 8	224.69	2.15	356.92	-1.54	270.82	-4.67	146.50
Aug 18	225.28	2.11	355.84	-1.57	273.91	-4.33	156.16
Aug 28	225.99	2.08	354.59	-1.59	277.91	-3.97	165.87
Sep 7	226.82	2.04	353.26	-1.60	282.65	-3.60	175.63
Sep 17	227.76	2.02	351.96	-1.60	287.98	-3.25	185.44
Sep 27	228.78	1.99	350.78	-1.58	293.78	-2.89	195.31
Oct 7	229.87	1.97	349.80	-1.56	299.93	-2.56	205.24
Oct 17	231.01	1.95	349.10	-1.52	306.37	-2.23	215.22
Oct 27	232.19	1.94	348.72	-1.48	313.03	-1.92	225.26
Nov 6	233.38	1.93	348.68	-1.44	319.84	-1.63	235.34
Nov 16	234.56	1.93	348.99	-1.40	326.77	-1.35	245.46
Nov 26	235.73	1.93	349.62	-1.35	333.79	-1.09	255.62
Dec 6	236.85	1.94	350.57	-1.31	340.86	-0.85	265.80
Dec 16	237.90	1.95	351.80	-1.27	347.95	-0.62	275.99
1749							

GREGORIAN 12 UT	MOON LONG.		MOON LAT.		VENUS LONG.		VENUS LAT.		MERCURY LONG.		MERCURY LAT.	
1748												1748
Jan 2		307.2		-0.9		297.05		-1.36		258.77		2.24
7 Jan 12	16.0	76.7	4.5	4.5	303.32	309.58	-1.45	-1.52	263.45	269.29	1.50	0.75
17 Jan 22	136.3	199.4	0.1	-4.7	315.83	322.07	-1.56	-1.57	275.83	282.84	0.04	-0.59
27 J-F 1	270.2	344.0	-3.8	2.4	328.31	334.54	-1.55	-1.50	290.20	297.88	-1.14	-1.58
6 Feb 11	49.3	108.9	5.3	2.5	340.75	346.95	-1.42	-1.31	305.86	314.19	-1.89	-2.07
16 Feb 21	170.3	237.1	-2.9	-5.2	353.14	359.30	-1.17	-1.01	322.87	331.93	-2.08	-1.90
26 Feb	308.9		-0.7		5.46		-0.83		341.36		-1.50	
Mar 2		19.1		4.6		11.59		-0.62		350.99		-0.85
7 Mar 12	81.4	141.2	4.3	-0.4	17.69	23.78	-0.39	-0.14	0.46	9.03	0.04	1.08
17 Mar 22	206.3	276.2	-4.8	-3.3	29.83	35.86	0.11	0.38	15.73	19.74	2.13	2.96
27 M-A 1	346.6	52.8	2.6	5.0	41.86	47.82	0.65	0.92	20.64	18.72	3.35	3.10
6 Apr 11	112.9	174.6	1.9	-3.3	53.75	59.64	1.19	1.45	15.13	11.60	2.21	0.90
16 Apr 21	244.2	315.3	-4.7	0.2	65.48	71.28	1.70	1.93	9.54	9.52	-0.45	-1.60
26 Apr	22.8		4.7		77.03		2.13		11.49		-2.43	
May 1		85.3		3.7		82.72		2.31		15.13		-2.93
6 May 11	144.7	210.0	-1.1	-4.9	88.35	93.91	2.46	2.57	20.13	26.29	-3.12	-3.03
16 May 21	283.2	353.2	-2.3	3.4	99.39	104.78	2.63	2.65	33.46	41.60	-2.69	-2.12
26 May 31	57.4	117.3	4.8	1.1	110.07	115.24	2.62	2.53	50.67	60.64	-1.37	-0.50
5 Jun 10	177.8	247.3	-3.8	-4.5	120.28	125.16	2.38	2.17	71.32	82.30	0.39	1.17
15 Jun 20	321.9	29.4	1.2	5.1	129.85	134.34	1.88	1.52	93.03	103.12	1.70	1.93
25 Jun 30	90.6	149.7	3.2	-1.9	138.56	142.48	1.08	0.54	112.36	120.71	1.87	1.54
5 Jul 10	212.9	285.9	-5.2	-2.0	146.03	149.13	-0.08	-0.81	128.14	134.61	0.97	0.21
15 Jul 20	359.4	63.7	4.1	4.8	151.69	153.61	-1.64	-2.57	140.02	144.16	-0.70	-1.73
25 Jul 30	123.0	183.4	0.4	-4.4	154.76	155.02	-3.59	-4.67	146.77	147.49	-2.81	-3.83
4 Aug 9	250.0	324.4	-4.5	1.5	154.31	152.62	-5.77	-6.80	146.06	142.71	-4.59	-4.81
14 Aug 19	35.0	96.4	5.3	2.8	150.11	147.11	-7.65	-8.21	138.57	135.55	-4.30	-3.12
24 Aug 29	155.8	219.0	-2.5	-5.2	144.08	141.48	-8.41	-8.25	135.26	138.30	-1.62	-0.19
3 Sep 8	288.6	2.0	-1.7	4.2	139.66	138.80	-7.78	-7.11	144.34	152.45	0.91	1.57
13 Sep 18	68.6	128.3	4.4	-0.1	138.94	140.00	-6.31	-5.45	161.55	170.88	1.82	1.74
23 Sep 28	189.8	256.7	-4.6	-3.9	141.88	144.43	-4.59	-3.75	180.03	188.85	1.44	1.00
3 Oct 8	327.4	37.9	1.9	5.0	147.56	151.16	-2.95	-2.20	197.32	205.46	0.48	-0.09
13 Oct 18	100.8	160.4	2.0	-3.0	155.16	159.49	-1.50	-0.85	213.31	220.89	-0.67	-1.22
23 Oct 28	225.5	295.6	-4.9	-0.6	164.10	168.94	-0.27	0.25	228.21	235.24	-1.74	-2.18
2 Nov 7	5.6	72.0	4.5	3.8	173.98	179.18	0.72	1.12	241.88	247.92	-2.51	-2.68
12 Nov 17	132.2	193.5	-1.0	-4.8	184.53	190.00	1.47	1.75	252.91	255.99	-2.60	-2.14
22 Nov 27	263.0	334.7	-3.1	2.9	195.58	201.25	1.98	2.15	255.81	251.36	-1.13	0.43
2 Dec 7	42.2	104.7	5.0	1.3	206.99	212.81	2.27	2.34	244.68	240.53	1.98	2.76
12 Dec 17	163.9	228.2	-3.7	-4.9	218.68	224.60	2.36	2.33	240.80	244.32	2.73	2.28
22 Dec 27	301.3	12.7	0.2	5.1	230.57	236.58	2.27	2.16	249.70	256.05	1.64	0.95
1749												1749
1 Jan 6	76.9	136.8	3.5	-1.7	242.61	248.67	2.02	1.85	262.95	270.19	0.28	-0.35
11 Jan 16	196.9	264.6	-5.2	-3.0	254.75	260.85	1.65	1.44	277.68	285.40	-0.90	-1.37
21 Jan 26	339.8	48.9	3.5	5.0	266.97	273.10	1.20	0.96	293.35	301.56	-1.74	-1.98
31 J-F 5	110.0	169.3	0.8	-4.1	279.24	285.39	0.71	0.45	310.05	318.85	-2.08	-2.00
10 Feb 15	231.6	302.4	-4.8	0.4	291.54	297.70	0.20	-0.05	327.94	337.17	-1.70	-1.15
20 Feb 25	17.3	83.0	5.1	3.0	303.87	310.03	-0.28	-0.51	346.18	354.23	-0.34	0.73
2 Mar 7	142.4	203.0	-2.1	-5.1	316.20	322.37	-0.71	-0.90	0.20	3.03	1.90	2.94
12 Mar 17	268.3	341.0	-2.5	3.5	328.54	334.71	-1.07	-1.21	2.28	358.72	3.54	3.41
22 Mar 27	53.1	115.5	4.5	0.1	340.88	347.05	-1.32	-1.41	354.34	351.18	2.57	1.32
1 Apr 6	174.9	238.4	-4.3	-4.2	353.22	359.38	-1.47	-1.50	350.24	351.53	0.04	-1.06
11 Apr 16	306.8	19.2	1.1	5.0	5.54	11.70	-1.51	-1.48	354.66	359.25	-1.90	-2.46
21 Apr 26	87.0	147.2	2.2	-2.8	17.86	24.01	-1.43	-1.36	4.98	11.68	-2.76	-2.81
1 May 6	208.5	275.7	-5.0	-1.4	30.16	36.31	-1.26	-1.13	19.23	27.60	-2.62	-2.21
11 May 16	346.0	56.0	4.1	4.2	42.45	48.60	-0.99	-0.83	36.77	46.75	-1.60	-0.81
21 May 26	119.4	179.0	-0.8	-4.8	54.74	60.88	-0.66	-0.48	57.39	68.35	0.06	0.90
31 M-J 5	243.6	314.3	-3.8	2.2	67.02	73.16	-0.28	-0.09	79.12	89.22	1.57	1.96
10 Jun 15	24.8	90.7	5.2	1.7	79.30	85.45	0.11	0.31	98.38	106.52	2.05	1.83
20 Jun 25	151.2	211.9	-3.5	-5.1	91.59	97.74	0.50	0.68	113.60	119.54	1.33	0.58
30 Jun	280.4		-0.8		103.88		0.85		124.21		-0.39	
Jul 5		353.6		4.8		110.03		1.00		127.39		-1.52
10 Jul 15	61.8	123.9	3.9	-1.3	116.17	122.33	1.14	1.26	128.82	128.29	-2.73	-3.87
20 Jul 25	183.3	246.3	-5.1	-3.6	128.48	134.63	1.35	1.42	125.89	122.38	-4.70	-4.93
30 J-A 4	318.6	31.9	2.6	5.1	140.78	146.93	1.46	1.47	119.16	117.73	-4.45	-3.38
9 Aug 14	96.6	156.4	1.2	-3.8	153.08	159.24	1.46	1.42	118.99	123.13	-2.01	-0.64
19 Aug 24	216.4	282.4	-4.9	-0.6	165.39	171.53	1.35	1.25	129.84	138.41	0.51	1.31
29 Aug	357.5		4.8		177.68		1.13		147.94		1.72	
Sep 3		68.3		3.3		183.82		0.98		157.66		1.77
8 Sep 13	129.8	189.1	-1.9	-5.0	189.95	196.09	0.81	0.62	167.15	176.23	1.57	1.19
18 Sep 23	251.0	320.3	-3.0	2.8	202.22	208.34	0.40	0.18	184.88	193.12	0.69	0.13
28 S-O 3	35.6	102.5	4.7	0.3	214.45	220.56	-0.06	-0.30	200.98	208.49	-0.47	-1.07
8 Oct 13	162.2	222.8	-4.2	-4.4	226.66	232.76	-0.55	-0.80	215.64	222.40	-1.64	-2.16
18 Oct 23	287.6	359.2	0.3	4.9	238.84	244.92	-1.05	-1.28	228.64	234.12	-2.59	-2.87
28 Oct	71.8		2.6		250.98		-1.51		238.36		-2.92	
Nov 2		134.9		-2.7		257.03		-1.71		240.51		-2.60
7 Nov 12	194.6	257.9	-5.1	-2.0	263.07	269.10	-1.90	-2.05	239.33	234.25	-1.71	-0.20
17 Nov 22	326.0	37.8	3.6	4.6	275.10	281.08	-2.18	-2.27	227.81	224.53	1.42	2.38
27 N-D 2	105.8	166.6	-0.5	-4.8	287.03	292.94	-2.32	-2.34	225.76	230.11	2.57	2.28
7 Dec 12	227.5	294.7	-4.2	1.3	298.83	304.67	-2.30	-2.22	236.13	242.97	1.75	1.14
17 Dec 22	5.4	74.7	5.2	2.3	310.45	316.18	-2.09	-1.91	250.21	257.68	0.50	-0.11
27 Dec	138.5		-3.3		321.83		-1.68		265.30		-0.68	
1749												1749

JULIAN 12 UT	SATURN LONG.	SATURN LAT.	JUPITER LONG.	JUPITER LAT.	MARS LONG.	MARS LAT.	SUN LONG.	GREGORIAN 12 UT	MOON LONG.	MOON LONG.	MOON LAT.	MOON LAT.	VENUS LONG.	VENUS LONG.	VENUS LAT.	VENUS LAT.	MERCURY LONG.	MERCURY LONG.	MERCURY LAT.	MERCURY LAT.
1749								1750												1750
								Jan 1		198.2		-5.2		327.39		-1.39		273.05		-1.17
Dec 26	238.88	1.96	353.27	-1.24	355.06	-0.42	286.18	6 Jan 11	261.7	333.0	-1.7	4.2	332.86	338.21	-1.04	-0.64	280.97	289.06	-1.58	-1.89
Jan 5	239.76	1.98	354.96	-1.21	2.16	-0.23	296.37	16 Jan 21	44.1	109.6	4.4	-0.9	343.41	348.45	-0.19	0.32	297.37	305.89	-2.06	-2.07
Jan 15	240.52	2.00	356.84	-1.18	9.23	-0.06	306.54	26 Jan 31	170.5	230.5	-4.9	-4.0	353.29	357.89	0.89	1.50	314.62	323.42	-1.88	-1.44
Jan 25	241.14	2.03	358.87	-1.16	16.28	0.10	316.68	5 Feb 10	297.5	12.0	1.6	5.1	2.22	6.19	2.17	2.89	331.93	339.39	-0.71	0.33
Feb 4	241.62	2.05	1.03	-1.14	23.28	0.24	326.78	15 Feb 20	81.1	143.2	1.6	-3.5	9.76	12.81	3.64	4.43	344.58	346.18	1.58	2.82
Feb 14	241.94	2.08	3.29	-1.12	30.24	0.37	336.84	25 Feb	202.8		-4.9		15.25		5.24		343.83		3.61	
								Mar 2		264.3		-1.2		16.96		6.05		339.02		3.60
Feb 24	242.09	2.11	5.62	-1.11	37.16	0.49	346.85	7 Mar 12	335.2	50.3	4.2	3.7	17.80	17.67	6.81	7.48	334.49	332.18	2.81	1.64
Mar 6	242.07	2.14	8.00	-1.10	44.02	0.59	356.81	17 Mar 22	115.9	176.0	-1.6	-4.9	16.50	14.37	7.98	8.21	332.45	334.88	0.45	-0.60
Mar 16	241.89	2.16	10.41	-1.09	50.83	0.68	6.71	27 M-A 1	235.8	300.1	-3.2	2.1	11.52	8.38	8.09	7.60	338.94	344.23	-1.43	-2.03
Mar 26	241.55	2.18	12.83	-1.09	57.59	0.76	16.55	6 Apr 11	14.0	86.6	4.9	0.6	5.47	3.22	6.79	5.74	350.49	357.55	-2.42	-2.58
Apr 5	241.07	2.20	15.23	-1.09	64.30	0.83	26.34	16 Apr 21	149.2	208.7	-4.2	-4.5	1.90	1.58	4.60	3.45	5.34	13.83	-2.53	-2.26
Apr 15	240.48	2.21	17.61	-1.10	70.96	0.90	36.08	26 Apr	270.3		-0.2		2.22		2.37		23.03		-1.77	
								May 1		338.1		4.7		3.73		1.39		32.95		-1.09
Apr 25	239.80	2.22	19.93	-1.11	77.57	0.95	45.76	6 May 11	52.4	120.9	3.3	-2.5	5.96	8.80	0.53	-0.22	43.48	54.36	-0.27	0.60
May 5	239.07	2.21	22.18	-1.12	84.14	1.00	55.40	16 May 21	181.5	242.0	-5.2	-2.6	12.13	15.86	-0.85	-1.39	65.08	75.11	1.39	1.94
May 15	238.32	2.20	24.34	-1.14	90.66	1.04	65.00	26 May 31	306.8	17.4	3.1	5.0	19.93	24.27	-1.82	-2.17	84.09	91.90	2.19	2.11
May 25	237.60	2.19	26.38	-1.15	97.15	1.07	74.58	5 Jun 10	89.2	153.7	0.2	-4.7	28.85	33.62	-2.44	-2.63	98.43	103.61	1.71	0.99
Jun 4	236.94	2.16	28.28	-1.18	103.61	1.10	84.13	15 Jun 20	213.5	276.5	-4.5	0.5	38.54	43.61	-2.76	-2.82	107.27	109.21	0.00	-1.22
Jun 14	236.38	2.13	30.02	-1.20	110.04	1.12	93.67	25 Jun 30	345.3	56.3	5.1	3.0	48.79	54.07	-2.83	-2.78	109.27	107.54	-2.55	-3.78
Jun 24	235.93	2.10	31.57	-1.23	116.45	1.13	103.20	5 Jul 10	124.1	185.6	-2.9	-5.2	59.45	64.90	-2.69	-2.56	104.62	101.63	-4.63	-4.86
Jul 4	235.62	2.06	32.90	-1.26	122.85	1.14	112.74	15 Jul 20	246.0	313.0	-2.2	3.5	70.42	76.00	-2.39	-2.19	99.85	100.16	-4.44	-3.52
Jul 14	235.47	2.02	33.99	-1.30	129.22	1.15	122.28	25 Jul 30	24.8	93.7	4.8	-0.3	81.64	87.34	-1.97	-1.73	102.91	108.06	-2.32	-1.05
Jul 24	235.47	1.98	34.80	-1.33	135.59	1.15	131.86	4 Aug 9	157.6	217.4	-4.7	-4.1	93.08	98.87	-1.47	-1.19	115.38	124.35	0.11	1.01
Aug 3	235.64	1.94	35.32	-1.37	141.95	1.15	141.46	14 Aug 19	279.8	351.3	0.9	5.0	104.70	110.58	-0.92	-0.64	134.24	144.32	1.56	1.76
Aug 13	235.96	1.90	35.51	-1.41	148.32	1.14	151.10	24 Aug 29	63.7	129.2	2.2	-3.3	116.49	122.43	-0.36	-0.09	154.13	163.47	1.67	1.37
Aug 23	236.44	1.86	35.37	-1.44	154.69	1.13	160.78	3 Sep 8	190.2	249.8	-4.9	-1.5	128.41	134.43	0.18	0.42	172.30	180.63	0.91	0.36
Sep 2	237.06	1.82	34.90	-1.48	161.06	1.11	170.51	13 Sep 18	315.5	30.5	3.7	4.2	140.47	146.54	0.65	0.86	188.51	195.94	-0.25	-0.89
Sep 12	237.80	1.79	34.12	-1.50	167.45	1.09	180.29	23 Sep 28	100.7	163.1	-1.3	-4.9	152.64	158.77	1.04	1.20	202.92	209.38	-1.52	-2.12
Sep 22	238.66	1.76	33.08	-1.52	173.85	1.06	190.14	3 Oct 8	222.6	283.4	-3.4	1.6	164.92	171.09	1.33	1.43	215.19	220.07	-2.65	-3.05
Oct 2	239.62	1.73	31.84	-1.53	180.26	1.03	200.04	13 Oct 18	353.1	68.7	5.1	1.3	177.27	183.48	1.51	1.55	223.50	224.66	-3.24	-3.06
Oct 12	240.66	1.71	30.50	-1.52	186.69	0.99	209.99	23 Oct 28	135.7	196.0	-4.0	-4.7	189.70	195.94	1.56	1.54	222.50	217.00	-2.30	-0.85
Oct 22	241.76	1.69	29.16	-1.51	193.15	0.95	220.00	2 Nov 7	255.6	319.1	-0.6	4.3	202.18	208.44	1.49	1.42	211.08	208.77	0.82	1.96
Nov 1	242.91	1.68	27.92	-1.48	199.62	0.90	230.06	12 Nov 17	31.8	105.2	4.0	-2.1	214.71	220.98	1.32	1.19	210.92	216.04	2.36	2.24
Nov 11	244.09	1.66	26.87	-1.44	206.12	0.85	240.16	22 Nov 27	168.9	228.4	-5.2	-2.9	227.26	233.54	1.05	0.89	222.67	229.98	1.84	1.30
Nov 21	245.27	1.66	26.10	-1.39	212.63	0.79	250.30	2 Dec 7	289.9	357.0	2.5	5.3	239.83	246.12	0.71	0.52	237.56	245.26	0.70	0.11
Dec 1	246.45	1.65	25.64	-1.34	219.17	0.72	260.46	12 Dec 17	70.1	139.7	1.1	-4.5	252.41	258.70	0.33	0.13	253.01	260.81	-0.46	-0.98
Dec 11	247.59	1.65	25.52	-1.29	225.73	0.64	270.65	22 Dec 27	201.0	261.1	-4.6	-0.1	265.00	271.29	-0.07	-0.27	268.69	276.67	-1.42	-1.78
1750								1751												1751
Dec 21	248.68	1.66	25.76	-1.24	232.31	0.55	280.84	1 Jan 6	326.2	36.1	4.7	3.8	277.58	283.87	-0.47	-0.65	284.79	293.05	-2.02	-2.12
Dec 31	249.71	1.66	26.33	-1.19	238.90	0.46	291.03	11 Jan 16	107.1	172.7	-2.2	-5.2	290.16	296.45	-0.82	-0.97	301.41	309.75	-2.03	-1.70
Jan 10	250.65	1.67	27.21	-1.14	245.51	0.35	301.21	21 Jan 26	232.5	295.0	-2.5	2.9	302.73	309.01	-1.11	-1.23	317.72	324.54	-1.08	-0.10
Jan 20	251.48	1.69	28.37	-1.09	252.14	0.23	311.37	31 J-F 5	4.5	75.0	5.0	0.5	315.29	321.56	-1.32	-1.39	328.89	329.28	1.19	2.58
Jan 30	252.19	1.70	29.79	-1.05	258.77	0.09	321.49	10 Feb 15	142.6	204.8	-4.5	-4.1	327.82	334.08	-1.44	-1.45	325.51	319.92	3.56	3.66
Feb 9	252.76	1.72	31.42	-1.01	265.41	-0.05	331.57	20 Feb 25	264.5	330.8	0.5	4.8	340.33	346.57	-1.45	-1.41	315.77	314.51	2.95	1.88
Feb 19	253.19	1.74	33.24	-0.98	272.06	-0.22	341.61	2 Mar 7	43.6	112.6	2.8	-2.9	352.81	359.03	-1.35	-1.26	315.92	319.32	0.78	-0.20
Mar 1	253.45	1.76	35.21	-0.95	278.69	-0.40	351.59	12 Mar 17	176.6	236.6	-4.9	-1.7	5.25	11.45	-1.14	-1.01	324.11	329.93	-1.02	-1.65
Mar 11	253.54	1.78	37.31	-0.92	285.31	-0.60	1.52	22 Mar 27	297.7	8.4	3.4	4.6	17.64	23.82	-0.85	-0.67	336.55	343.85	-2.09	-2.35
Mar 21	253.47	1.80	39.51	-0.90	291.91	-0.83	11.40	1 Apr 6	82.2	148.5	-0.7	-4.9	29.99	36.14	-0.48	-0.27	351.78	0.31	-2.41	-2.26
Mar 31	253.24	1.81	41.79	-0.88	298.47	-1.07	21.22	11 Apr 16	209.6	268.9	-3.6	1.4	42.28	48.40	-0.06	0.16	9.47	19.26	-1.90	-1.34
Apr 10	252.86	1.83	44.11	-0.86	304.98	-1.34	30.98	21 Apr 26	333.0	47.0	5.1	2.3	54.51	60.60	0.39	0.61	29.63	40.34	-0.59	0.28
Apr 20	252.35	1.84	46.48	-0.84	311.40	-1.63	40.69	1 May 6	119.3	182.6	-3.8	-4.9	66.68	72.74	0.83	1.04	50.93	60.81	1.15	1.85
Apr 30	251.73	1.84	48.85	-0.83	317.71	-1.94	50.35	11 May 16	242.1	302.6	-1.0	4.0	78.78	84.79	1.24	1.42	69.51	76.82	2.29	2.38
May 10	251.03	1.84	51.21	-0.82	323.88	-2.28	59.97	21 May 26	10.4	85.4	4.7	-1.2	90.79	96.77	1.59	1.72	82.59	86.72	2.10	1.46
May 20	250.30	1.83	53.55	-0.81	329.83	-2.65	69.57	31 M-J 5	154.6	215.4	-5.2	-3.2	102.73	108.66	1.83	1.91	89.04	89.47	0.45	-0.83
May 30	249.56	1.82	55.84	-0.81	335.53	-3.04	79.12	10 Jun 15	275.0	338.5	2.0	5.3	114.56	120.43	1.96	1.97	88.11	85.54	-2.24	-3.51
Jun 9	248.86	1.80	58.07	-0.81	340.86	-3.45	88.67	20 Jun 25	49.1	122.6	2.1	-4.0	126.27	132.08	1.94	1.88	82.80	81.01	-4.35	-4.60
Jun 19	248.23	1.78	60.21	-0.81	345.73	-3.89	98.21	30 Jun	188.0		-4.7		137.85		1.77		80.96		-4.29	
								Jul 5		247.5		-0.5		143.57		1.62		82.97		-3.55
Jun 29	247.70	1.75	62.25	-0.81	349.99	-4.34	107.74	10 Jul 15	309.2	16.4	4.3	4.3	149.24	154.87	1.42	1.19	87.06	93.11	-2.54	-1.41
Jul 9	247.29	1.72	64.15	-0.81	353.46	-4.80	117.28	20 Jul 25	87.9	158.0	-1.4	-5.1	160.43	165.93	0.91	0.59	100.96	110.26	-0.29	0.68
Jul 19	247.02	1.68	65.90	-0.82	355.95	-5.24	126.84	30 J-A 4	220.1	280.0	-2.6	2.5	171.36	176.69	0.23	-0.18	120.45	130.85	1.36	1.71
Jul 29	246.91	1.64	67.47	-0.82	357.24	-5.62	136.42	9 Aug 14	345.5	55.5	5.0	1.3	181.93	187.05	-0.61	-1.09	140.95	150.53	1.75	1.53
Aug 8	246.96	1.61	68.83	-0.83	357.18	-5.89	146.04	19 Aug 24	125.8	191.8	-4.2	-4.2	192.05	196.89	-1.59	-2.12	159.53	167.95	1.13	0.60
Aug 18	247.17	1.57	69.95	-0.84	355.78	-5.98	155.70	29 Aug	251.7		0.3		201.55		-2.67		175.83		-0.02	
								Sep 3		313.7		4.6		205.99		-3.24		183.17		-0.69
Aug 28	247.54	1.53	70.80	-0.85	353.37	-5.80	165.40	8 Sep 13	23.5	94.5	3.5	-2.3	210.17	214.04	-3.83	-4.41	189.95	196.08	-1.39	-2.07
Sep 7	248.06	1.50	71.35	-0.86	350.56	-5.34	175.16	18 Sep 23	162.0	224.2	-5.0	-1.9	217.53	220.56	-4.99	-5.55	201.41	205.62	-2.71	-3.23
Sep 17	248.72	1.47	71.58	-0.87	348.12	-4.67	184.97	28 S-O 3	283.6	349.2	3.1	4.9	223.02	224.80	-6.06	-6.51	208.17	208.29	-3.56	-3.51
Sep 27	249.50	1.44	71.48	-0.88	346.70	-3.88	194.84	8 Oct 13	62.4	132.4	0.2	-4.8	225.77	225.81	-6.85	-7.03	205.23	199.56	-2.86	-1.48
Oct 7	250.39	1.41	71.04	-0.89	346.55	-3.09	204.76	18 Oct 23	196.4	256.1	-3.8	1.1	224.87	222.97	-6.99	-6.66	194.42	193.16	0.21	1.49
Oct 17	251.37	1.39	70.29	-0.89	347.70	-2.36	214.74	28 Oct	316.7		5.0		220.32		-6.02		196.19		2.10	
								Nov 2		26.3		3.2		217.30		-5.06		202.04		2.16
Oct 27	252.43	1.37	69.27	-0.89	349.98	-1.71	224.77	7 Nov 12	101.0	168.5	-3.2	-5.1	214.42	212.15	-3.88	-2.61	209.25	217.02	1.89	1.44
Nov 6	253.54	1.35	68.05	-0.87	353.19	-1.17	234.85	17 Nov 22	229.3	288.5	-1.3	3.7	210.78	210.44	-1.35	-0.21	224.94	232.87	0.89	0.32
Nov 16	254.70	1.33	66.71	-0.86	357.13	-0.71	244.98	27 N-D 2	351.6	64.4	5.0	-0.1	211.09	212.63	0.77	1.60	240.77	248.63	-0.25	-0.78
Nov 26	255.88	1.32	65.35	-0.83	1.65	-0.33	255.13	7 Dec 12	138.3	202.5	-5.1	-3.5	214.94	217.89	2.26	2.77	256.48	264.36	-1.26	-1.67
Dec 6	257.06	1.32	64.09	-0.80	6.60	-0.01	265.30	17 Dec 22	261.7	322.3	1.6	5.1	221.38	225.29	3.15	3.41	272.29	280.28	-1.97	-2.15
Dec 16	258.22	1.31	63.02	-0.77	11.89	0.25	275.50	27 Dec	28.5		3.0		229.56		3.57		288.29		-2.16	
1751								1751												1751

JULIAN 12 UT	SATURN LONG.	SATURN LAT.	JUPITER LONG.	JUPITER LAT.	MARS LONG.	MARS LAT.	SUN LONG.	GREGORIAN 12 UT	MOON LONG.	MOON LONG.	MOON LAT.	MOON LAT.	VENUS LONG.	VENUS LONG.	VENUS LAT.	VENUS LAT.	MERCURY LONG.	MERCURY LONG.	MERCURY LAT.	MERCURY LAT.
1751								1752												1752
								Jan 1		102.6		-3.3		234.13		3.64		296.16		-1.94
Dec 26	259.34	1.31	62.20	-0.73	17.44	0.46	285.69	6 Jan 11	173.6	234.9	-4.8	-0.7	238.93	243.93	3.64	3.56	303.55	309.67	-1.43	-0.55
Jan 5	260.41	1.31	61.70	-0.69	23.18	0.64	295.88	16 Jan 21	294.4	357.9	4.0	4.6	249.10	254.40	3.43	3.24	313.14	312.37	0.73	2.23
Jan 15	261.40	1.31	61.53	-0.65	29.07	0.79	306.05	26 Jan 31	66.9	140.1	-0.5	-5.0	259.82	265.34	3.01	2.75	307.44	301.54	3.38	3.61
Jan 25	262.30	1.32	61.71	-0.61	35.06	0.91	316.19	5 Feb 10	207.0	266.7	-2.8	2.3	270.94	276.61	2.46	2.15	298.10	297.95	3.01	2.06
Feb 4	263.09	1.32	62.21	-0.57	41.14	1.01	326.30	15 Feb 20	328.4	35.4	5.0	2.0	282.34	288.11	1.82	1.48	300.38	304.58	1.05	0.13
Feb 14	263.75	1.33	63.02	-0.53	47.28	1.09	336.36	25 Feb	105.9		-3.7		293.93		1.14		309.96		-0.66	
								Mar 1		176.1		-4.4		299.79		0.80		316.18		-1.30
Feb 24	264.27	1.34	64.11	-0.50	53.46	1.15	346.37	6 Mar 11	239.1	298.7	0.2	4.5	305.67	311.59	0.48	0.16	323.07	330.53	-1.79	-2.12
Mar 5	264.64	1.36	65.45	-0.47	59.66	1.20	356.33	16 Mar 21	3.9	74.3	4.0	-1.6	317.53	323.48	-0.14	-0.43	338.51	347.03	-2.27	-2.23
Mar 15	264.85	1.37	67.00	-0.44	65.88	1.24	6.24	26 Mar 31	144.3	210.3	-5.1	-2.2	329.46	335.44	-0.69	-0.92	356.09	5.71	-1.99	-1.54
Mar 25	264.89	1.38	68.73	-0.41	72.12	1.27	16.08	5 Apr 10	270.6	331.8	3.0	5.2	341.44	347.45	-1.12	-1.30	15.86	26.34	-0.89	-0.05
Apr 4	264.76	1.39	70.62	-0.39	78.35	1.29	25.87	15 Apr 20	41.1	113.4	1.2	-4.5	353.47	359.50	-1.44	-1.55	36.71	46.35	0.86	1.70
Apr 14	264.48	1.40	72.63	-0.36	84.59	1.30	35.61	25 Apr 30	181.0	243.1	-4.2	0.8	5.54	11.58	-1.63	-1.68	54.67	61.31	2.32	2.61
Apr 24	264.06	1.40	74.74	-0.34	90.83	1.30	45.29	5 May 10	302.5	6.5	4.9	3.9	17.62	23.67	-1.69	-1.67	66.06	68.82	2.50	1.95
May 4	263.52	1.40	76.93	-0.32	97.07	1.30	54.94	15 May 20	79.5	151.5	-2.2	-5.2	29.73	35.79	-1.62	-1.55	69.51	68.30	0.97	-0.35
May 14	262.87	1.40	79.18	-0.30	103.31	1.29	64.55	25 May 30	215.7	275.4	-1.7	3.5	41.86	47.93	-1.45	-1.32	65.82	63.11	-1.80	-3.06
May 24	262.17	1.39	81.46	-0.29	109.55	1.28	74.12	4 Jun 9	335.7	43.1	5.2	1.0	54.00	60.08	-1.17	-1.01	61.29	61.09	-3.89	-4.20
Jun 3	261.43	1.38	83.76	-0.27	115.80	1.26	83.67	14 Jun 19	118.3	187.7	-4.7	-3.8	66.17	72.27	-0.83	-0.64	62.77	66.29	-4.03	-3.49
Jun 13	260.70	1.37	86.05	-0.25	122.05	1.24	93.21	24 Jun 29	248.8	308.2	1.3	4.9	78.37	84.47	-0.44	-0.23	71.51	78.32	-2.68	-1.70
Jun 23	260.02	1.35	88.32	-0.24	128.31	1.21	102.74	4 Jul 9	10.6	81.3	3.5	-2.4	90.59	96.71	-0.03	0.17	86.61	96.18	-0.66	0.33
Jul 3	259.42	1.32	90.55	-0.22	134.58	1.18	112.28	14 Jul 19	156.2	221.8	-4.9	-0.9	102.84	108.98	0.37	0.56	106.59	117.25	1.12	1.61
Jul 13	258.92	1.30	92.72	-0.21	140.87	1.15	121.83	24 Jul 29	281.3	342.1	3.8	4.7	115.13	121.28	0.74	0.90	127.61	137.40	1.79	1.68
Jul 23	258.55	1.27	94.81	-0.20	147.18	1.11	131.39	3 Aug 8	47.5	120.3	0.3	-4.7	127.45	133.62	1.04	1.16	146.54	155.03	1.35	0.86
Aug 2	258.33	1.24	96.80	-0.18	153.51	1.07	141.00	13 Aug 18	192.1	254.3	-3.1	2.1	139.80	145.99	1.26	1.34	162.88	170.10	0.23	-0.47
Aug 12	258.27	1.21	98.66	-0.17	159.87	1.03	150.63	23 Aug 28	314.0	17.6	5.0	2.6	152.19	158.39	1.39	1.41	176.63	182.39	-1.23	-2.00
Aug 22	258.37	1.17	100.37	-0.16	166.26	0.98	160.31	2 Sep 7	86.0	158.5	-3.1	-4.7	164.59	170.81	1.41	1.38	187.16	190.60	-2.75	-3.41
Sep 1	258.63	1.14	101.89	-0.14	172.69	0.92	170.04	12 Sep 17	225.9	286.0	-0.1	4.4	177.02	183.24	1.32	1.24	192.18	191.22	-3.87	-3.95
Sep 11	259.04	1.11	103.21	-0.13	179.16	0.87	179.82	22 Sep 27	347.6	54.8	4.4	-0.8	189.46	195.68	1.13	1.00	187.41	181.91	-3.38	-2.06
Sep 21	259.61	1.08	104.29	-0.11	185.67	0.81	189.66	2 Oct 7	125.4	195.0	-5.1	-2.6	201.91	208.13	0.85	0.69	177.81	177.66	-0.39	1.00
Oct 1	260.30	1.06	105.10	-0.09	192.22	0.74	199.55	12 Oct 17	258.2	317.9	2.8	5.2	214.36	220.59	0.50	0.30	181.54	188.06	1.80	2.05
Oct 11	261.12	1.03	105.61	-0.08	198.83	0.67	209.50	22 Oct 27	22.6	93.6	2.0	-3.9	226.81	233.04	0.10	-0.12	195.84	204.05	1.92	1.56
Oct 21	262.04	1.01	105.79	-0.06	205.48	0.60	219.51	1 Nov 6	164.0	229.5	-4.6	0.4	239.26	245.48	-0.33	-0.55	212.32	220.49	1.07	0.53
Oct 31	263.05	0.99	105.65	-0.04	212.19	0.52	229.56	11 Nov 16	290.0	350.6	4.8	4.3	251.70	257.92	-0.75	-0.95	228.54	236.47	-0.04	-0.58
Nov 10	264.13	0.97	105.18	-0.02	218.95	0.43	239.66	21 Nov 26	59.1	132.8	-1.3	-5.2	264.14	270.35	-1.14	-1.31	244.31	252.10	-1.09	-1.55
Nov 20	265.26	0.96	104.40	0.01	225.77	0.34	249.80	1 Dec 6	200.7	262.5	-2.2	3.2	276.56	282.76	-1.46	-1.59	259.86	267.58	-1.91	-2.17
Nov 30	266.43	0.94	103.36	0.03	232.64	0.25	259.97	11 Dec 16	322.0	24.6	5.1	1.8	288.95	295.14	-1.70	-1.77	275.22	282.61	-2.27	-2.17
Dec 10	267.60	0.93	102.13	0.05	239.57	0.15	270.15	21 Dec 26	97.1	170.9	-4.1	-4.1	301.32	307.48	-1.81	-1.82	289.39	294.77	-1.78	-1.00
Dec 20	268.78	0.92	100.79	0.08	246.56	0.04	280.34	31 Dec	235.3		1.0		313.63		-1.80		297.32		0.23	
1752								1753												1753
								Jan 5		295.1		4.8		319.77		-1.74		295.46		1.79
Dec 30	269.93	0.92	99.45	0.10	253.60	-0.08	290.54	10 Jan 15	355.0	60.6	3.7	-1.6	325.88	331.97	-1.64	-1.51	289.66	283.84	3.08	3.46
Jan 9	271.03	0.91	98.22	0.12	260.70	-0.19	300.72	20 Jan 25	135.7	206.9	-5.0	-1.2	338.03	344.06	-1.35	-1.15	281.30	282.22	3.00	2.18
Jan 19	272.07	0.91	97.17	0.14	267.85	-0.32	310.87	30 J-F 4	268.4	327.9	3.7	4.7	350.06	356.02	-0.91	-0.64	285.56	290.44	1.28	0.42
Jan 29	273.03	0.91	96.38	0.16	275.05	-0.45	321.00	9 Feb 14	29.7	98.5	0.8	-4.3	1.93	7.80	-0.35	-0.03	296.30	302.86	-0.34	-0.99
Feb 8	273.89	0.91	95.89	0.18	282.30	-0.59	331.09	19 Feb 24	173.6	240.8	-3.6	2.0	13.60	19.34	0.32	0.68	309.95	317.50	-1.51	-1.89
Feb 18	274.64	0.91	95.73	0.20	289.60	-0.73	341.13	1 Mar 6	300.7	1.4	5.0	3.0	25.00	30.58	1.06	1.44	325.50	333.95	-2.12	-2.18
Feb 28	275.25	0.91	95.89	0.21	296.94	-0.88	351.12	11 Mar 16	66.3	137.6	-2.5	-5.1	36.07	41.45	1.83	2.22	342.88	352.30	-2.05	-1.72
Mar 10	275.72	0.91	96.37	0.22	304.31	-1.03	1.05	21 Mar 26	209.7	273.3	-0.6	4.4	46.70	51.81	2.61	2.98	2.18	12.37	-1.16	-0.39
Mar 20	276.03	0.91	97.15	0.23	311.70	-1.18	10.92	31 M-A 5	333.0	36.3	4.7	0.0	56.75	61.50	3.33	3.65	22.45	31.77	0.54	1.49
Mar 30	276.19	0.92	98.19	0.25	319.12	-1.32	20.75	10 Apr 15	104.9	176.3	-4.9	-3.4	66.01	70.25	3.93	4.17	39.59	45.39	2.29	2.79
Apr 9	276.17	0.92	99.47	0.26	326.54	-1.47	30.51	20 Apr 25	244.0	305.0	2.5	5.3	74.15	77.66	4.35	4.47	48.88	49.97	2.87	2.45
Apr 19	276.00	0.92	100.96	0.27	333.95	-1.61	40.22	30 Apr	5.8		2.7		80.70		4.49		48.87		1.51	
								May 5		73.1		-3.2		83.16		4.41		46.25		0.18
Apr 29	275.68	0.92	102.62	0.28	341.33	-1.75	49.89	10 May 15	144.4	213.3	-5.0	-0.2	84.92	85.86	4.19	3.83	43.30	41.23	-1.26	-2.49
May 9	275.22	0.92	104.43	0.29	348.67	-1.87	59.51	20 May 25	277.0	336.8	4.6	4.5	85.87	84.89	3.27	2.52	40.80	42.23	-3.32	-3.71
May 19	274.64	0.91	106.37	0.30	355.95	-1.99	69.10	30 M-J 4	40.1	111.7	-0.4	-5.0	82.94	80.24	1.57	0.46	45.42	50.16	-3.69	-3.35
May 29	273.97	0.90	108.40	0.31	3.14	-2.09	78.66	9 Jun 14	183.1	248.4	-2.8	2.8	77.15	74.15	-0.72	-1.85	56.29	63.71	-2.74	-1.93
Jun 8	273.26	0.89	110.52	0.32	10.21	-2.17	88.20	19 Jun 24	309.3	9.3	5.1	2.1	71.69	70.09	-2.83	-3.62	72.36	82.12	-0.99	-0.03
Jun 18	272.52	0.88	112.68	0.34	17.13	-2.23	97.74	29 Jun	76.2		-3.4		69.47		-4.19		92.68		0.83	
								Jul 4		150.9		-4.5		69.83		-4.56		103.52		1.46
Jun 28	271.80	0.86	114.89	0.35	23.86	-2.28	107.27	9 Jul 14	220.0	282.1	0.6	4.7	71.05	73.04	-4.76	-4.82	114.09	124.06	1.78	1.81
Jul 8	271.14	0.84	117.11	0.36	30.35	-2.30	116.81	19 Jul 24	341.7	43.3	3.8	-1.0	75.65	78.80	-4.77	-4.63	133.31	141.82	1.57	1.12
Jul 18	270.57	0.82	119.34	0.38	36.56	-2.30	126.37	29 J-A 3	114.4	189.2	-5.0	-1.8	82.39	86.34	-4.42	-4.15	149.60	156.63	0.52	-0.22
Jul 28	270.10	0.80	121.54	0.40	42.41	-2.27	135.96	8 Aug 13	254.9	315.0	3.5	4.8	90.59	95.09	-3.84	-3.49	162.85	168.13	-1.03	-1.90
Aug 7	269.78	0.78	123.70	0.41	47.83	-2.21	145.57	18 Aug 23	14.8	79.3	1.2	-3.9	99.81	104.71	-3.12	-2.73	172.23	174.78	-2.77	-3.57
Aug 17	269.61	0.75	125.80	0.43	52.71	-2.11	155.23	28 Aug	153.3		-4.2		109.77		-2.33		175.29		-4.17	
								Sep 2		225.6		1.6		114.97		-1.92		173.29		-4.35
Aug 27	269.59	0.73	127.82	0.45	56.91	-1.97	164.93	7 Sep 12	288.1	347.7	5.1	3.3	120.29	125.72	-1.52	-1.12	169.00	164.08	-3.84	-2.58
Sep 6	269.74	0.70	129.74	0.48	60.29	-1.79	174.69	17 Sep 22	49.4	117.4	-2.0	-5.2	131.24	136.85	-0.73	-0.36	161.24	162.24	-0.95	0.50
Sep 16	270.06	0.68	131.53	0.50	62.63	-1.55	184.50	27 S-O 2	191.5	259.8	-1.4	4.2	142.54	148.30	-0.01	0.32	166.94	174.08	1.46	1.89
Sep 26	270.52	0.65	133.15	0.53	63.73	-1.24	194.36	7 Oct 12	320.4	20.9	4.8	0.6	154.12	159.99	0.62	0.89	182.39	191.05	1.91	1.66
Oct 6	271.13	0.63	134.59	0.56	63.41	-0.85	204.28	17 Oct 22	86.0	156.5	-4.5	-4.1	165.92	171.89	1.13	1.34	199.66	208.09	1.24	0.73
Oct 16	271.87	0.61	135.82	0.59	61.62	-0.38	214.26	27 Oct	228.0		1.9		177.91		1.51		216.30		0.18	
								Nov 1		292.6		5.3		183.97		1.64		224.31		-0.38
Oct 26	272.72	0.59	136.80	0.63	58.60	0.14	224.29	6 Nov 11	352.2	55.4	3.0	-2.5	190.05	196.17	1.73	1.79	232.16	239.86	-0.92	-1.42
Nov 5	273.68	0.57	137.50	0.66	54.95	0.66	234.36	16 Nov 21	124.6	195.2	-5.2	-1.0	202.31	208.47	1.81	1.80	247.45	254.91	-1.84	-2.17
Nov 15	274.72	0.56	137.90	0.70	51.46	1.12	244.48	26 N-D 1	262.8	324.3	4.3	4.5	214.65	220.86	1.75	1.68	262.18	269.08	-2.37	-2.37
Nov 25	275.82	0.54	137.98	0.74	48.87	1.49	254.64	6 Dec 11	24.5	91.8	0.1	-4.7	227.07	233.29	1.57	1.44	275.22	279.81	-2.11	-1.46
Dec 5	276.96	0.53	137.73	0.78	47.57	1.75	264.81	16 Dec 21	163.9	232.4	-3.3	2.3	239.53	245.77	1.28	1.11	281.41	278.55	-0.31	1.27
Dec 15	278.14	0.52	137.16	0.82	47.62	1.91	275.00	26 Dec 31	296.2	355.9	5.0	2.3	252.02	258.28	0.92	0.71	272.18	266.74	2.69	3.23
1753								1753												1753

JULIAN 12 UT	SATURN LONG.	SATURN LAT.	JUPITER LONG.	JUPITER LAT.	MARS LONG.	MARS LAT.	SUN LONG.	GREGORIAN 12 UT	MOON LONG.	MOON LONG.	MOON LAT.	MOON LAT.	VENUS LONG.	VENUS LONG.	VENUS LAT.	VENUS LAT.	MERCURY LONG.	MERCURY LONG.	MERCURY LAT.	MERCURY LAT.
1753								1754												1754
Dec 25	279.32	0.50	136.30	0.85	48.89	2.01	285.20	5 Jan 10	58.1	129.9	-2.9	-4.8	264.53	270.79	0.50	0.29	265.18	267.11	2.94	2.26
Jan 4	280.48	0.49	135.21	0.89	51.16	2.06	295.38	15 Jan 20	202.5	267.9	0.1	4.6	277.05	283.31	0.07	-0.14	271.26	276.73	1.46	0.67
Jan 14	281.62	0.48	133.96	0.91	54.24	2.07	305.56	25 Jan 30	328.8	28.2	3.9	-0.7	289.57	295.83	-0.35	-0.54	283.02	289.86	-0.07	-0.71
Jan 24	282.70	0.48	132.64	0.93	57.95	2.06	315.70	4 Feb 9	93.6	168.6	-4.9	-2.6	302.09	308.35	-0.73	-0.89	297.11	304.73	-1.25	-1.68
Feb 3	283.72	0.47	131.34	0.95	62.15	2.03	325.80	14 Feb 19	239.3	301.8	3.3	5.0	314.60	320.85	-1.04	-1.17	312.72	321.08	-1.97	-2.11
Feb 13	284.65	0.46	130.16	0.95	66.74	1.99	335.87	24 Feb	1.2		1.5		327.10		-1.27		329.84		-2.08	
								Mar 1		61.9		-3.6		333.34		-1.35		339.03		-1.85
Feb 23	285.47	0.46	129.19	0.95	71.62	1.94	345.89	6 Mar 11	131.1	206.6	-4.8	0.8	339.58	345.81	-1.41	-1.43	348.61	358.45	-1.40	-0.71
Mar 5	286.17	0.45	128.47	0.95	76.74	1.89	355.85	16 Mar 21	274.2	334.5	5.1	3.6	352.03	358.25	-1.43	-1.41	8.18	17.09	0.19	1.20
Mar 15	286.74	0.44	128.06	0.94	82.04	1.84	5.75	26 Mar 31	34.1	97.9	-1.5	-5.2	4.47	10.67	-1.36	-1.28	24.32	29.14	2.18	2.90
Mar 25	287.16	0.44	127.96	0.93	87.49	1.78	15.60	5 Apr 10	169.8	243.2	-2.5	3.8	16.87	23.06	-1.18	-1.06	31.15	30.42	3.19	2.91
Apr 4	287.42	0.43	128.17	0.91	93.06	1.72	25.39	15 Apr 20	307.5	6.8	5.0	1.0	29.25	35.43	-0.91	-0.75	27.65	24.20	2.02	0.71
Apr 14	287.52	0.43	128.68	0.90	98.73	1.66	35.14	25 Apr 30	68.6	135.9	-4.1	-4.6	41.60	47.76	-0.58	-0.39	21.56	20.66	-0.69	-1.88
Apr 24	287.46	0.42	129.47	0.89	104.49	1.60	44.82	5 May 10	208.3	278.0	1.0	5.1	53.92	60.07	-0.19	0.01	21.71	24.57	-2.72	-3.19
May 4	287.24	0.41	130.51	0.87	110.31	1.53	54.47	15 May 20	339.5	39.5	3.2	-2.0	66.21	72.35	0.21	0.41	28.95	34.63	-3.33	-3.17
May 14	286.88	0.40	131.78	0.86	116.20	1.47	64.08	25 May 30	105.1	175.1	-5.1	-1.9	78.48	84.60	0.61	0.80	41.44	49.32	-2.74	-2.09
May 24	286.38	0.39	133.24	0.86	122.15	1.40	73.65	4 Jun 9	245.8	311.3	3.9	4.6	90.71	96.82	0.98	1.14	58.23	68.11	-1.28	-0.37
Jun 3	285.77	0.38	134.88	0.85	128.15	1.33	83.20	14 Jun 19	11.1	73.7	0.4	-4.3	102.92	109.02	1.29	1.41	78.74	89.69	0.52	1.26
Jun 13	285.09	0.37	136.65	0.84	134.20	1.26	92.74	24 Jun 29	143.4	214.1	-3.9	1.7	115.11	121.18	1.51	1.58	100.41	110.50	1.73	1.90
Jun 23	284.36	0.36	138.55	0.84	140.30	1.19	102.27	4 Jul 9	281.6	343.5	5.0	2.4	127.25	133.31	1.63	1.64	119.80	128.26	1.77	1.40
Jul 3	283.63	0.34	140.54	0.84	146.45	1.12	111.81	14 Jul 19	43.2	109.8	-2.6	-5.0	139.36	145.40	1.63	1.58	135.89	142.64	0.82	0.08
Jul 13	282.92	0.33	142.61	0.84	152.66	1.05	121.36	24 Jul 29	182.6	251.8	-0.8	4.4	151.43	157.44	1.50	1.38	148.44	153.11	-0.80	-1.77
Jul 23	282.28	0.31	144.73	0.85	158.92	0.97	130.93	3 Aug 8	315.7	15.4	4.1	-0.5	163.43	169.41	1.24	1.06	156.38	157.90	-2.77	-3.71
Aug 2	281.74	0.29	146.89	0.85	165.23	0.90	140.52	13 Aug 18	76.8	147.6	-4.7	-3.4	175.38	181.32	0.85	0.62	157.26	154.37	-4.43	-4.68
Aug 12	281.31	0.27	149.07	0.86	171.60	0.82	150.16	23 Aug 28	221.3	287.7	2.8	5.1	187.25	193.15	0.36	0.08	150.02	146.11	-4.20	-3.01
Aug 22	281.03	0.26	151.24	0.87	178.04	0.73	159.84	2 Sep 7	348.5	47.8	1.8	-3.3	199.02	204.87	-0.22	-0.53	144.72	146.90	-1.45	0.01
Sep 1	280.91	0.24	153.39	0.89	184.53	0.65	169.56	12 Sep 17	112.2	186.2	-5.1	-0.2	210.69	216.48	-0.86	-1.19	152.35	160.07	1.08	1.68
Sep 11	280.95	0.22	155.50	0.90	191.10	0.57	179.35	22 Sep 27	258.5	321.6	5.0	3.9	222.23	227.94	-1.52	-1.85	168.87	177.96	1.86	1.73
Sep 21	281.15	0.20	157.54	0.92	197.73	0.48	189.18	2 Oct 7	20.9	81.7	-1.1	-5.1	233.60	239.20	-2.17	-2.47	186.93	195.62	1.39	0.92
Oct 1	281.52	0.19	159.49	0.95	204.44	0.39	199.07	12 Oct 17	149.4	224.4	-3.3	3.1	244.75	250.22	-2.75	-3.00	204.00	212.11	0.39	-0.18
Oct 11	282.03	0.17	161.33	0.97	211.22	0.30	209.02	22 Oct 27	293.6	354.2	5.1	1.3	255.61	260.89	-3.22	-3.40	219.96	227.60	-0.74	-1.28
Oct 21	282.69	0.16	163.03	1.00	218.07	0.20	219.02	1 Nov 6	53.9	117.6	-3.7	-4.9	266.06	271.08	-3.52	-3.59	235.03	242.22	-1.76	-2.17
Oct 31	283.47	0.14	164.56	1.04	225.00	0.10	229.08	11 Nov 16	187.8	261.4	0.0	4.9	275.93	280.57	-3.60	-3.53	249.12	255.51	-2.45	-2.57
Nov 10	284.37	0.13	165.89	1.08	232.01	0.00	239.18	21 Nov 26	326.8	26.2	3.3	-1.7	284.96	289.04	-3.37	-3.11	260.99	264.73	-2.43	-1.92
Nov 20	285.36	0.12	166.99	1.12	239.10	-0.10	249.31	1 Dec 6	88.3	155.3	-5.0	-2.6	292.74	295.97	-2.75	-2.26	265.36	261.61	-0.88	0.68
Nov 30	286.42	0.10	167.83	1.16	246.26	-0.20	259.47	11 Dec 16	226.5	296.7	3.3	4.7	298.63	300.60	-1.63	-0.85	254.96	250.13	2.21	2.93
Dec 10	287.55	0.09	168.39	1.20	253.50	-0.30	269.66	21 Dec 26	358.7	58.6	0.5	-4.1	301.74	301.92	0.10	1.21	249.57	252.43	2.83	2.30
Dec 20	288.71	0.08	168.64	1.25	260.81	-0.41	279.85	31 Dec	124.4		-4.3		301.06		2.44		257.32		1.61	
1754								1755												1755
								Jan 5		194.3		1.0		299.21		3.73		263.34		0.88
Dec 30	289.89	0.07	168.57	1.30	268.20	-0.52	290.04	10 Jan 15	264.4	330.2	5.0	2.6	296.57	293.54	4.97	6.02	270.00	277.09	0.19	-0.45
Jan 9	291.07	0.05	168.18	1.34	275.65	-0.62	300.23	20 Jan 25	30.0	92.1	-2.5	-5.1	290.62	288.29	6.77	7.18	284.49	292.16	-1.01	-1.47
Jan 19	292.23	0.04	167.49	1.38	283.16	-0.73	310.38	30 J-F 4	162.0	233.3	-1.6	4.1	286.87	286.47	7.27	7.10	300.11	308.37	-1.81	-2.03
Jan 29	293.35	0.03	166.54	1.42	290.73	-0.83	320.51	9 Feb 14	300.6	2.5	4.4	-0.3	287.08	288.57	6.75	6.28	316.95	325.89	-2.08	-1.96
Feb 8	294.42	0.02	165.38	1.45	298.35	-0.92	330.60	19 Feb 24	61.9	127.2	-4.6	-4.1	290.83	293.70	5.73	5.13	335.15	344.61	-1.61	-1.03
Feb 18	295.41	0.01	164.11	1.47	306.01	-1.02	340.64	1 Mar 6	200.6	271.2	1.9	5.3	297.10	300.91	4.51	3.89	353.92	2.35	-0.18	0.87
Feb 28	296.31	-0.01	162.82	1.47	313.70	-1.10	350.63	11 Mar 16	334.9	34.5	2.2	-3.1	305.07	309.52	3.28	2.68	8.91	12.62	1.98	2.92
Mar 10	297.09	-0.02	161.58	1.47	321.42	-1.18	0.57	21 Mar 26	95.1	164.1	-5.3	-1.4	314.19	319.05	2.10	1.55	13.03	10.52	3.43	3.28
Mar 20	297.76	-0.03	160.49	1.46	329.15	-1.25	10.45	31 M-A 5	239.2	307.2	4.6	4.2	324.08	329.24	1.02	0.54	6.53	3.00	2.44	1.17
Mar 30	298.28	-0.05	159.62	1.44	336.87	-1.31	20.27	10 Apr 15	7.8	67.1	-0.8	-4.8	334.51	339.88	0.09	-0.32	1.26	1.69	-0.16	-1.30
Apr 9	298.65	-0.06	159.02	1.41	344.59	-1.35	30.04	20 Apr 25	130.2	202.3	-3.9	2.1	345.33	350.85	-0.68	-1.01	4.07	8.05	-2.15	-2.70
Apr 19	298.87	-0.07	158.71	1.38	352.27	-1.38	39.75	30 Apr	276.6		5.1		356.43		-1.29		13.33		-2.96	
								May 5		341.1		1.6		2.06		-1.52		19.68		-2.95
Apr 29	298.92	-0.09	158.70	1.35	359.92	-1.40	49.42	10 May 15	40.3	101.1	-3.4	-4.9	7.74	13.46	-1.71	-1.86	26.97	35.17	-2.69	-2.20
May 9	298.81	-0.11	159.00	1.31	7.51	-1.40	59.05	20 May 25	167.3	241.0	-0.9	4.5	19.21	24.99	-1.96	-2.02	44.24	54.17	-1.52	-0.69
May 19	298.55	-0.12	159.59	1.28	15.03	-1.39	68.64	30 M-J 4	312.1	13.5	3.5	-1.5	30.80	36.64	-2.05	-2.03	64.80	75.78	0.20	1.01
May 29	298.14	-0.14	160.43	1.25	22.47	-1.36	78.20	9 Jun 14	73.2	136.9	-4.9	-3.1	42.50	48.38	-1.98	-1.89	86.57	96.73	1.62	1.94
Jun 8	297.61	-0.15	161.52	1.22	29.80	-1.31	87.75	19 Jun 24	205.9	279.0	2.6	4.8	54.29	60.21	-1.78	-1.63	106.01	114.34	1.96	1.69
Jun 18	296.98	-0.17	162.82	1.19	37.02	-1.25	97.28	29 Jun	345.6		0.7		66.15		-1.47		121.70		1.16	
								Jul 4		45.4		-4.0		72.11		-1.28		128.03		0.42
Jun 28	296.28	-0.18	164.31	1.17	44.11	-1.17	106.81	9 Jul 14	107.3	174.4	-4.6	0.2	78.09	84.09	-1.08	-0.87	133.23	137.10	-0.51	-1.58
Jul 8	295.54	-0.20	165.96	1.15	51.05	-1.07	116.36	19 Jul 24	245.2	315.1	4.9	3.0	90.11	96.14	-0.64	-0.42	139.35	139.67	-2.72	-3.80
Jul 18	294.81	-0.21	167.75	1.13	57.81	-0.95	125.91	29 J-A 3	17.8	77.6	-2.3	-5.2	102.19	108.26	-0.19	0.04	137.91	134.49	-4.62	-4.89
Jul 28	294.11	-0.23	169.65	1.12	64.39	-0.82	135.49	8 Aug 13	142.9	213.4	-2.4	3.6	114.35	120.46	0.25	0.46	130.65	128.13	-4.44	-3.33
Aug 7	293.50	-0.24	171.65	1.11	70.74	-0.67	145.11	18 Aug 23	283.6	349.2	4.8	0.1	126.58	132.72	0.66	0.84	128.31	131.63	-1.89	-0.47
Aug 17	292.98	-0.25	173.71	1.10	76.85	-0.50	154.76	28 Aug	49.4		-4.5		138.87		0.99		137.81		0.68	
								Sep 2		110.8		-4.5		145.04		1.13		146.02		1.43
Aug 27	292.60	-0.26	175.83	1.10	82.66	-0.31	164.46	7 Sep 12	180.2	252.7	0.9	5.2	151.22	157.42	1.25	1.33	155.28	164.79	1.77	1.77
Sep 6	292.37	-0.27	177.99	1.10	88.12	-0.10	174.22	17 Sep 22	320.1	22.0	2.7	-2.8	163.63	169.85	1.40	1.43	174.10	183.05	1.52	1.11
Sep 16	292.29	-0.28	180.15	1.10	93.18	0.14	184.02	27 S-O 2	81.5	145.5	-5.3	-2.2	176.08	182.32	1.44	1.42	191.62	199.83	0.60	0.04
Sep 26	292.39	-0.29	182.30	1.11	97.74	0.41	193.88	7 Oct 12	218.7	290.8	3.9	4.5	188.57	194.82	1.37	1.30	207.70	215.27	-0.55	-1.12
Oct 6	292.64	-0.30	184.42	1.12	101.70	0.71	203.80	17 Oct 22	354.5	54.3	-0.4	-4.7	201.08	207.35	1.20	1.08	222.54	229.47	-1.67	-2.15
Oct 16	293.06	-0.31	186.49	1.13	104.91	1.05	213.77	27 Oct	114.6		-4.2		213.62		0.94		235.98		-2.53	
								Nov 1		181.9		1.1		219.89		0.78		241.84		-2.75
Oct 26	293.63	-0.31	188.49	1.15	107.21	1.44	223.80	6 Nov 11	257.5	326.7	5.1	1.9	226.17	232.44	0.61	0.42	246.61	249.46	-2.75	-2.37
Nov 5	294.34	-0.32	190.38	1.17	108.40	1.88	233.88	16 Nov 21	27.6	87.1	-3.2	-4.9	238.72	245.00	0.23	0.03	249.10	244.59	-1.46	0.06
Nov 15	295.16	-0.33	192.15	1.19	108.28	2.35	243.99	26 N-D 1	149.4	219.9	-1.5	4.0	251.28	257.56	-0.17	-0.37	237.96	233.90	1.67	2.58
Nov 25	296.10	-0.34	193.76	1.22	106.75	2.83	254.14	6 Dec 11	295.0	0.5	3.9	-1.3	263.84	270.12	-0.56	-0.75	234.32	238.05	2.68	2.31
Dec 5	297.12	-0.35	195.19	1.25	103.91	3.28	264.32	16 Dec 21	60.0	120.9	-4.8	-3.4	276.40	282.67	-0.92	-1.08	243.62	250.14	1.73	1.07
Dec 15	298.22	-0.36	196.40	1.29	100.18	3.63	274.50	26 Dec 31	186.2	258.7	1.9	5.0	288.94	295.21	-1.22	-1.34	257.17	264.49	0.42	-0.21
1755								1755												1755

JULIAN 12 UT	SATURN LONG.	SATURN LAT.	JUPITER LONG.	JUPITER LAT.	MARS LONG.	MARS LAT.	SUN LONG.
1755							
Dec 25	299.36	-0.37	197.37	1.33	96.24	3.82	284.70
Jan 4	300.54	-0.38	198.07	1.37	92.89	3.85	294.89
Jan 14	301.73	-0.39	198.47	1.41	90.66	3.75	305.06
Jan 24	302.91	-0.40	198.56	1.45	89.74	3.57	315.21
Feb 3	304.07	-0.42	198.34	1.49	90.11	3.35	325.31
Feb 13	305.18	-0.43	197.81	1.53	91.59	3.12	335.38
Feb 23	306.23	-0.45	197.00	1.56	93.97	2.90	345.40
Mar 4	307.19	-0.46	195.96	1.59	97.07	2.69	355.37
Mar 14	308.06	-0.48	194.77	1.60	100.75	2.49	5.27
Mar 24	308.81	-0.50	193.49	1.60	104.89	2.31	15.13
Apr 3	309.44	-0.52	192.23	1.60	109.39	2.13	24.92
Apr 13	309.92	-0.54	191.06	1.58	114.18	1.97	34.66
Apr 23	310.25	-0.57	190.07	1.55	119.22	1.82	44.36
May 3	310.42	-0.59	189.31	1.51	124.45	1.68	54.01
May 13	310.43	-0.61	188.83	1.47	129.85	1.55	63.61
May 23	310.28	-0.64	188.65	1.43	135.41	1.42	73.19
Jun 2	309.97	-0.66	188.76	1.39	141.08	1.29	82.74
Jun 12	309.53	-0.68	189.17	1.34	146.88	1.17	92.28
Jun 22	308.96	-0.70	189.85	1.30	152.79	1.05	101.82
Jul 2	308.31	-0.72	190.78	1.26	158.80	0.94	111.35
Jul 12	307.59	-0.74	191.95	1.22	164.90	0.83	120.90
Jul 22	306.85	-0.75	193.33	1.19	171.10	0.72	130.47
Aug 1	306.12	-0.76	194.88	1.16	177.40	0.61	140.06
Aug 11	305.44	-0.77	196.59	1.13	183.79	0.50	149.69
Aug 21	304.84	-0.78	198.43	1.11	190.27	0.39	159.37
Aug 31	304.36	-0.79	200.38	1.09	196.84	0.29	169.09
Sep 10	304.02	-0.79	202.42	1.07	203.52	0.18	178.87
Sep 20	303.84	-0.79	204.53	1.06	210.28	0.08	188.70
Sep 30	303.81	-0.79	206.68	1.05	217.15	-0.03	198.59
Oct 10	303.96	-0.79	208.86	1.04	224.11	-0.13	208.54
Oct 20	304.27	-0.79	211.04	1.04	231.16	-0.23	218.54
Oct 30	304.74	-0.79	213.20	1.04	238.31	-0.33	228.59
Nov 9	305.36	-0.79	215.32	1.04	245.56	-0.43	238.68
Nov 19	306.12	-0.79	217.38	1.05	252.89	-0.52	248.82
Nov 29	306.99	-0.79	219.35	1.06	260.31	-0.61	258.98
Dec 9	307.97	-0.79	221.20	1.07	267.81	-0.70	269.17
Dec 19	309.02	-0.80	222.91	1.09	275.39	-0.78	279.36
1756							
Dec 29	310.14	-0.81	224.45	1.10	283.03	-0.86	289.55
Jan 8	311.31	-0.81	225.79	1.13	290.74	-0.92	299.73
Jan 18	312.50	-0.82	226.90	1.15	298.50	-0.98	309.90
Jan 28	313.70	-0.83	227.76	1.18	306.30	-1.04	320.02
Feb 7	314.88	-0.85	228.33	1.20	314.13	-1.08	330.11
Feb 17	316.03	-0.86	228.60	1.23	321.98	-1.11	340.16
Feb 27	317.13	-0.88	228.56	1.26	329.83	-1.13	350.15
Mar 9	318.17	-0.90	228.21	1.28	337.68	-1.13	0.09
Mar 19	319.11	-0.92	227.57	1.30	345.51	-1.13	9.97
Mar 29	319.95	-0.95	226.67	1.32	353.30	-1.11	19.80
Apr 8	320.68	-0.97	225.57	1.32	1.06	-1.08	29.57
Apr 18	321.27	-1.00	224.35	1.32	8.76	-1.04	39.29
Apr 28	321.72	-1.03	223.08	1.31	16.39	-0.98	48.95
May 8	322.01	-1.06	221.85	1.28	23.96	-0.91	58.58
May 18	322.14	-1.09	220.75	1.26	31.44	-0.83	68.18
May 28	322.11	-1.12	219.85	1.22	38.82	-0.75	77.74
Jun 7	321.91	-1.15	219.20	1.18	46.11	-0.65	87.28
Jun 17	321.57	-1.18	218.83	1.14	53.29	-0.54	96.82
Jun 27	321.09	-1.21	218.76	1.09	60.35	-0.42	106.35
Jul 7	320.49	-1.23	218.99	1.05	67.30	-0.29	115.89
Jul 17	319.81	-1.25	219.51	1.00	74.12	-0.16	125.45
Jul 27	319.08	-1.27	220.31	0.96	80.80	-0.02	135.02
Aug 6	318.33	-1.28	221.35	0.93	87.34	0.13	144.64
Aug 16	317.61	-1.29	222.62	0.89	93.71	0.29	154.29
Aug 26	316.94	-1.29	224.10	0.86	99.92	0.45	163.98
Sep 5	316.37	-1.29	225.75	0.83	105.93	0.63	173.74
Sep 15	315.93	-1.29	227.55	0.80	111.72	0.82	183.54
Sep 25	315.63	-1.29	229.48	0.77	117.27	1.02	193.39
Oct 5	315.49	-1.28	231.51	0.75	122.53	1.24	203.31
Oct 15	315.52	-1.27	233.63	0.73	127.44	1.47	213.29
Oct 25	315.72	-1.26	235.80	0.72	131.94	1.73	223.31
Nov 4	316.08	-1.25	238.02	0.70	135.94	2.01	233.39
Nov 14	316.61	-1.24	240.25	0.69	139.31	2.32	243.50
Nov 24	317.28	-1.23	242.47	0.68	141.93	2.67	253.65
Dec 4	318.08	-1.23	244.67	0.67	143.60	3.04	263.83
Dec 14	319.00	-1.22	246.81	0.67	144.15	3.44	274.01
1757							

GREGORIAN 12 UT	MOON LONG.	MOON LONG.	MOON LAT.	MOON LAT.	VENUS LONG.	VENUS LONG.	VENUS LAT.	VENUS LAT.	MERCURY LONG.	MERCURY LONG.	MERCURY LAT.	MERCURY LAT.
1756												1756
5 Jan 10	330.5	32.8	1.1	-3.9	301.48	307.74	-1.44	-1.51	272.02	279.75	-0.77	-1.26
15 Jan 20	92.6	156.3	-4.8	-0.5	313.99	320.23	-1.56	-1.57	287.67	295.81	-1.65	-1.93
25 Jan 30	224.8	296.8	4.6	3.6	326.47	332.70	-1.56	-1.52	304.21	312.88	-2.07	-2.04
4 Feb 9	4.1	64.5	-2.0	-5.2	338.92	345.12	-1.44	-1.34	321.80	330.85	-1.80	-1.32
14 Feb 19	125.9	193.4	-3.0	2.8	351.30	357.47	-1.21	-1.05	339.68	347.58	-0.55	0.49
24 Feb 29	264.2	333.2	5.1	0.7	3.63	9.76	-0.87	-0.67	353.39	355.94	1.70	2.84
5 Mar 10	36.5	96.2	-4.4	-4.8	15.87	21.95	-0.44	-0.20	354.71	350.68	3.56	3.54
15 Mar 20	160.5	232.2	0.0	5.0	28.01	34.04	0.06	0.32	346.11	343.12	2.75	1.54
25 Mar 30	302.7	7.8	3.2	-2.4	40.04	46.01	0.60	0.87	342.55	344.22	0.29	-0.80
4 Apr 9	68.5	128.8	-5.2	-2.7	51.94	57.83	1.14	1.41	347.69	352.52	-1.65	-2.24
14 Apr 19	197.0	271.4	3.1	4.8	63.67	69.47	1.66	1.90	358.44	5.25	-2.60	-2.72
24 Apr 29	339.4	41.2	0.0	-4.5	75.22	80.91	2.11	2.30	12.87	21.26	-2.60	-2.26
4 May 9	100.7	163.0	-4.3	0.4	86.54	92.09	2.46	2.58	30.41	40.32	-1.72	-0.99
14 May 19	235.4	309.4	4.9	2.3	97.57	102.95	2.65	2.69	50.89	61.81	-0.13	0.73
24 May 29	14.0	73.9	-3.0	-4.9	108.24	113.40	2.67	2.59	72.60	82.74	1.46	1.94
3 Jun 8	133.9	199.4	-1.8	3.5	118.43	123.31	2.46	2.26	91.91	99.99	2.11	1.97
13 Jun 18	274.4	345.5	4.4	-1.0	127.99	132.46	1.98	1.63	106.93	112.66	1.52	0.80
23 Jun 28	47.2	106.7	-4.8	-3.7	136.66	140.56	1.20	0.68	117.02	119.82	-0.16	-1.33
3 Jul 8	168.6	237.7	1.4	5.1	144.08	147.15	0.06	-0.66	120.81	119.87	-2.59	-3.79
13 Jul 18	312.4	19.5	1.8	-3.8	149.67	151.53	-1.48	-2.41	117.25	113.87	-4.66	-4.93
23 Jul 28	79.6	140.3	-5.0	-1.1	152.61	152.81	-3.43	-4.52	111.13	110.29	-4.51	-3.52
2 Aug 7	205.5	277.0	4.2	4.3	152.03	150.28	-5.63	-6.68	112.03	116.45	-2.24	-0.90
12 Aug 17	348.6	52.1	-1.5	-5.2	147.72	144.69	-7.54	-8.12	123.28	131.94	0.27	1.14
22 Aug 27	111.9	175.3	-3.4	2.1	141.67	139.11	-8.34	-8.20	141.60	151.49	1.64	1.78
1 Sep 6	244.3	315.4	5.3	1.5	137.36	136.57	-7.76	-7.11	161.14	170.36	1.63	1.29
11 Sep 16	22.9	83.9	-4.1	-4.9	136.77	137.89	-6.34	-5.51	179.11	187.42	0.82	0.26
21 Sep 26	144.7	212.4	-0.6	4.5	139.81	142.41	-4.66	-3.84	195.32	202.82	-0.34	-0.96
1 Oct 6	283.7	352.2	3.8	-1.9	145.57	149.19	-3.05	-2.31	209.93	216.59	-1.56	-2.11
11 Oct 16	55.9	115.6	-5.1	-2.9	153.21	157.55	-1.61	-0.97	222.69	227.98	-2.59	-2.93
21 Oct 26	178.9	251.1	2.4	5.0	162.17	167.02	-0.39	0.14	231.99	233.90	-3.06	-2.83
31 Oct	322.2		0.6		172.07		0.61		232.54		-2.03	
Nov 5		27.3		-4.3		177.27		1.02		227.44		-0.58
10 Nov 15	88.2	147.9	-4.3	0.0	182.63	188.10	1.37	1.67	221.11	217.94	1.08	2.17
20 Nov 25	215.0	290.1	4.6	3.0	193.68	199.35	1.91	2.09	219.32	223.86	2.48	2.29
30 N-D 5	358.9	61.0	-2.7	-5.0	205.10	210.92	2.22	2.30	230.08	237.08	1.82	1.24
10 Dec 15	120.4	181.8	-2.1	3.1	216.80	222.72	2.33	2.32	244.46	252.01	0.63	0.02
20 Dec 25	252.9	327.9	4.8	-0.4	228.69	234.70	2.26	2.17	259.68	267.45	-0.55	-1.06
30 Dec	33.7		-4.8		240.74		2.03		275.34		-1.49	
1757												1757
Jan 4		93.7		-3.9		246.80		1.87		283.38		-1.82
9 Jan 14	153.4	217.8	1.0	5.1	252.88	258.99	1.68	1.47	291.59	299.99	-2.04	-2.09
19 Jan 24	291.4	3.9	2.7	-3.5	265.11	271.24	1.25	1.01	308.56	317.17	-1.96	-1.58
29 J-F 3	66.8	126.2	-5.1	-1.5	277.38	283.54	0.76	0.51	325.49	332.78	-0.92	0.07
8 Feb 13	188.1	255.9	3.8	4.9	289.69	295.86	0.25	0.01	337.81	339.16	1.34	2.65
18 Feb 23	329.4	38.1	-0.6	-5.1	302.03	308.20	-0.23	-0.46	336.41	331.26	3.57	3.66
28 Feb	98.8		-3.6		314.37		-0.67		326.71		2.95	
Mar 5		159.2		1.5		320.54		-0.86		324.65		1.83
10 Mar 15	224.8	295.0	5.2	2.4	326.72	332.89	-1.03	-1.17	325.27	328.02	0.66	-0.36
20 Mar 25	6.0	70.9	-3.6	-4.9	339.07	345.24	-1.30	-1.39	332.33	337.81	-1.20	-1.83
30 M-A 4	130.5	193.7	-1.0	4.0	351.41	357.57	-1.46	-1.50	344.18	351.31	-2.26	-2.48
9 Apr 14	263.4	333.8	4.3	-1.2	3.74	9.90	-1.51	-1.49	359.12	7.60	-2.48	-2.28
19 Apr 24	41.2	102.8	-5.0	-3.0	16.06	22.22	-1.44	-1.37	16.75	26.58	-1.87	-1.25
29 Apr	162.8		2.0		28.37		-1.28		37.01		-0.46	
May 4		230.2		5.0		34.52		-1.16		47.82		0.42
9 May 14	302.6	11.2	1.3	-4.0	40.67	46.81	-1.02	-0.87	58.52	68.55	1.24	1.88
19 May 24	74.9	134.5	-4.4	-0.2	52.96	59.10	-0.70	-0.52	77.51	85.20	2.23	2.24
29 M-J 3	196.7	268.4	4.4	3.7	65.24	71.39	-0.33	-0.13	91.53	96.39	1.90	1.24
8 Jun 13	341.2	46.9	-2.2	-5.1	77.53	83.67	0.07	0.26	99.63	101.07	0.25	-0.99
18 Jun 23	107.6	166.9	-2.3	2.8	89.81	95.96	0.46	0.64	100.63	98.57	-2.35	-3.62
28 Jun	232.6		5.1		102.10		0.81		95.64		-4.50	
Jul 3		307.3		0.6		108.25		0.97		93.02		-4.78
8 Jul 13	18.2	80.7	-4.7	-4.2	114.39	120.54	1.11	1.23	91.85	92.76	-4.43	-3.61
18 Jul 23	140.0	201.0	0.6	4.9	126.69	132.84	1.33	1.41	95.95	101.37	-2.50	-1.29
28 J-A 2	270.3	345.5	3.6	-2.8	138.99	145.14	1.46	1.48	108.81	117.85	-0.13	0.82
7 Aug 12	53.2	113.4	-5.2	-1.8	151.29	157.44	1.47	1.44	127.84	138.06	1.45	1.74
17 Aug 22	173.0	237.1	3.4	5.1	163.59	169.73	1.37	1.28	148.03	157.51	1.72	1.46
27 Aug	309.0		0.4		175.87		1.16		166.44		1.04	
Sep 1		22.3		-4.9		182.01		1.02		174.85		0.50
6 Sep 11	86.4	145.6	-3.8	1.1	188.14	194.27	0.85	0.66	182.76	190.19	-0.12	-0.77
16 Sep 21	207.7	275.2	5.0	3.1	200.40	206.52	0.45	0.23	197.12	203.49	-1.43	-2.07
26 S-O 1	347.4	57.1	-2.9	-4.9	212.63	218.73	-0.01	-0.25	209.15	213.82	-2.64	-3.11
6 Oct 11	118.3	178.5	-1.2	3.7	224.83	230.92	-0.50	-0.75	217.01	217.91	-3.37	-3.28
16 Oct 21	244.4	314.2	4.6	-0.4	237.00	243.08	-1.00	-1.24	215.58	210.10	-2.60	-1.22
26 Oct 31	24.8	90.2	-4.8	-3.1	249.14	255.18	-1.47	-1.68	204.35	202.18	0.46	1.71
5 Nov 10	149.8	212.7	1.8	5.0	261.21	267.23	-1.87	-2.04	204.47	209.78	2.24	2.23
15 Nov 20	282.6	353.1	2.1	-3.6	273.23	279.20	-2.17	-2.27	216.61	224.09	1.89	1.39
25 Nov 30	60.4	122.1	-4.6	-0.3	285.15	291.06	-2.34	-2.36	231.80	239.60	0.82	0.24
5 Dec 10	181.8	248.5	4.3	4.3	296.94	302.77	-2.34	-2.27	247.41	255.23	-0.33	-0.86
15 Dec 20	321.5	30.8	-1.5	-5.2	308.55	314.27	-2.15	-1.98	263.10	271.03	-1.33	-1.71
25 Dec 30	94.3	153.9	-2.6	2.6	319.92	325.47	-1.76	-1.48	279.06	287.20	-1.99	-2.13
1757												1757

JULIAN 12 UT	SATURN LONG.	SATURN LAT.	JUPITER LONG.	JUPITER LAT.	MARS LONG.	MARS LAT.	SUN LONG.	GREGORIAN 12 UT	MOON LONG.		MOON LAT.		VENUS LONG.		VENUS LAT.		MERCURY LONG.		MERCURY LAT.	
1757								1758												1758
Dec 24	320.01	-1.22	248.88	0.66	143.41	3.84	284.21	4 Jan 9	215.2	285.8	5.3	1.7	330.93	336.27	-1.14	-0.75	295.41	303.57	-2.09	-1.83
Jan 3	321.10	-1.22	250.84	0.66	141.33	4.20	294.40	14 Jan 19	360.0	66.5	-4.4	-4.5	341.47	346.50	-0.30	0.21	311.33	317.97	-1.28	-0.36
Jan 13	322.25	-1.22	252.68	0.66	138.10	4.46	304.57	24 Jan 29	127.0	186.4	0.3	4.7	351.33	355.92	0.77	1.39	322.16	322.33	0.92	2.36
Jan 23	323.43	-1.23	254.37	0.66	134.20	4.56	314.72	3 Feb 8	250.4	324.0	4.3	-1.8	0.22	4.17	2.07	2.79	318.25	312.45	3.45	3.67
Feb 2	324.64	-1.24	255.87	0.67	130.34	4.48	324.83	13 Feb 18	37.0	100.2	-5.3	-2.1	7.71	10.72	3.56	4.36	308.39	307.42	3.04	2.03
Feb 12	325.84	-1.25	257.16	0.67	127.24	4.24	334.90	23 Feb 28	159.4	220.4	3.0	5.2	13.12	14.77	5.19	6.02	309.13	312.78	0.97	0.00
Feb 22	327.03	-1.27	258.21	0.67	125.31	3.91	344.92	5 Mar 10	287.6	2.3	1.4	-4.4	15.55	15.35	6.81	7.51	317.77	323.73	-0.81	-1.46
Mar 4	328.18	-1.28	259.00	0.68	124.67	3.55	354.89	15 Mar 20	72.0	132.6	-4.0	0.9	14.11	11.92	8.04	8.29	330.43	337.77	-1.94	-2.24
Mar 14	329.27	-1.30	259.50	0.68	125.26	3.18	4.80	25 Mar 30	192.5	256.4	4.8	3.6	9.03	5.89	8.20	7.73	345.69	354.18	-2.35	-2.26
Mar 24	330.29	-1.33	259.69	0.69	126.90	2.83	14.65	4 Apr 9	326.2	39.6	-2.1	-5.0	3.01	0.83	6.92	5.89	3.27	12.96	-1.97	-1.47
Apr 3	331.22	-1.35	259.57	0.69	129.40	2.51	24.45	14 Apr 19	105.2	164.6	-1.3	3.5	359.58	359.34	4.76	3.62	23.21	33.82	-0.77	0.09
Apr 13	332.04	-1.38	259.15	0.69	132.60	2.21	34.20	24 Apr 29	226.9	294.1	4.8	0.4	0.06	1.62	2.54	1.56	44.35	54.19	0.98	1.76
Apr 23	332.73	-1.41	258.44	0.68	136.37	1.95	43.89	4 May 9	5.2	75.1	-4.6	-3.4	3.90	6.79	0.69	-0.06	62.84	69.98	2.29	2.48
May 3	333.30	-1.45	257.48	0.68	140.59	1.71	53.54	14 May 19	137.1	197.1	1.7	5.0	10.15	13.92	-0.71	-1.25	75.46	79.15	2.30	1.71
May 13	333.71	-1.48	256.35	0.66	145.19	1.49	63.15	24 May 29	262.9	333.2	2.8	-3.1	18.01	22.38	-1.70	-2.07	80.93	80.77	0.74	-0.55
May 23	333.97	-1.52	255.10	0.65	150.11	1.29	72.73	3 Jun 8	43.4	108.7	-4.9	-0.6	26.97	31.76	-2.35	-2.55	78.97	76.27	-1.99	-3.28
Jun 2	334.07	-1.56	253.83	0.62	155.28	1.10	82.28	13 Jun 18	168.7	230.8	4.2	4.7	36.70	41.77	-2.69	-2.77	73.78	72.54	-4.14	-4.45
Jun 12	334.00	-1.59	252.63	0.60	160.69	0.92	91.82	23 Jun 28	300.7	12.4	-0.5	-5.1	46.96	52.25	-2.79	-2.76	73.12	75.68	-4.22	-3.58
Jun 22	333.77	-1.63	251.57	0.57	166.29	0.76	101.35	3 Jul 8	79.6	141.2	-3.1	2.3	57.63	63.09	-2.68	-2.56	80.16	86.45	-2.67	-1.61
Jul 2	333.39	-1.66	250.73	0.54	172.08	0.60	110.89	13 Jul 18	200.8	266.1	5.3	2.6	68.62	74.21	-2.41	-2.22	94.40	103.75	-0.51	0.47
Jul 12	332.87	-1.69	250.16	0.50	178.03	0.46	120.44	23 Jul 28	339.5	50.4	-3.7	-4.8	79.85	85.55	-2.00	-1.77	114.00	124.52	1.23	1.66
Jul 22	332.25	-1.72	249.87	0.47	184.14	0.32	130.00	2 Aug 7	113.8	173.4	-0.1	4.5	91.29	97.08	-1.51	-1.24	134.77	144.48	1.77	1.62
Aug 1	331.55	-1.74	249.90	0.44	190.39	0.18	139.60	12 Aug 17	234.2	303.1	4.6	-0.7	102.91	108.78	-0.97	-0.69	153.59	162.08	1.25	0.74
Aug 11	330.80	-1.76	250.23	0.41	196.77	0.05	149.23	22 Aug 27	18.4	86.2	-5.2	-2.5	114.69	120.64	-0.41	-0.14	169.98	177.31	0.13	-0.56
Aug 21	330.05	-1.77	250.86	0.38	203.29	-0.07	158.90	1 Sep 6	146.7	206.2	2.7	5.1	126.61	132.63	0.12	0.37	184.03	190.07	-1.28	-2.00
Aug 31	329.32	-1.77	251.76	0.35	209.95	-0.19	168.62	11 Sep 16	269.3	341.5	2.1	-3.8	138.67	144.74	0.60	0.81	195.23	199.21	-2.69	-3.28
Sep 10	328.68	-1.77	252.92	0.32	216.72	-0.30	178.39	21 Sep 26	55.9	119.9	-4.2	0.6	150.83	156.95	1.00	1.16	201.49	201.33	-3.68	-3.72
Sep 20	328.13	-1.76	254.30	0.29	223.62	-0.40	188.22	1 Oct 6	179.3	240.2	4.6	3.9	163.10	169.27	1.30	1.41	198.11	192.55	-3.15	-1.83
Sep 30	327.72	-1.75	255.88	0.27	230.64	-0.50	198.11	11 Oct 16	306.3	20.3	-1.3	-5.0	175.45	181.65	1.49	1.53	187.65	186.57	-0.15	1.22
Oct 10	327.46	-1.73	257.64	0.25	237.77	-0.60	208.05	21 Oct 26	91.1	152.3	-1.6	3.4	187.87	194.11	1.55	1.54	189.74	195.75	1.95	2.12
Oct 20	327.37	-1.72	259.55	0.23	245.02	-0.69	218.05	31 Oct	212.2		4.9		200.35		1.50		203.16		1.93	
								Nov 5		275.9		1.1		206.61		1.43		211.11		1.52
Oct 30	327.45	-1.70	261.58	0.21	252.37	-0.77	228.10	10 Nov 15	344.9	57.9	-4.3	-3.9	212.87	219.14	1.34	1.22	219.17	227.21	1.01	0.44
Nov 9	327.70	-1.68	263.71	0.19	259.82	-0.84	238.19	20 Nov 25	124.3	184.1	1.5	5.0	225.42	231.70	1.08	0.92	235.17	243.06	-0.12	-0.67
Nov 19	328.12	-1.66	265.92	0.17	267.37	-0.91	248.33	30 N-D 5	246.1	313.5	3.3	-2.4	237.99	244.28	0.75	0.56	250.91	258.76	-1.16	-1.59
Nov 29	328.70	-1.64	268.19	0.16	275.00	-0.96	258.49	10 Dec 15	24.1	93.6	-5.2	-1.1	250.57	256.87	0.37	0.17	266.61	274.48	-1.93	-2.15
Dec 9	329.42	-1.63	270.48	0.14	282.70	-1.01	268.67	20 Dec 25	156.3	216.1	4.0	5.0	263.16	269.45	-0.03	-0.23	282.33	290.02	-2.21	-2.06
Dec 19	330.27	-1.61	272.78	0.12	290.47	-1.04	278.86	30 Dec	281.4		0.4		275.75		-0.42		297.21		-1.62	
1758								1759												1759
								Jan 4		352.4		-4.9		282.04		-0.61		303.14		-0.81
Dec 29	331.22	-1.61	275.07	0.11	298.29	-1.07	289.06	9 Jan 14	62.4	127.5	-3.7	1.9	288.33	294.61	-0.78	-0.94	306.45	305.50	0.43	1.96
Jan 8	332.27	-1.60	277.31	0.09	306.15	-1.08	299.24	19 Jan 24	187.9	249.0	5.2	3.2	300.90	307.18	-1.08	-1.21	300.33	294.33	3.21	3.56
Jan 18	333.39	-1.60	279.50	0.08	314.04	-1.08	309.40	29 J-F 3	318.4	31.6	-2.8	-5.0	313.46	319.73	-1.31	-1.38	291.04	291.17	3.06	2.17
Jan 28	334.57	-1.60	281.61	0.06	321.95	-1.08	319.53	8 Feb 13	98.7	160.4	-0.6	4.2	326.00	332.26	-1.43	-1.46	293.87	298.28	1.21	0.31
Feb 7	335.77	-1.60	283.60	0.05	329.85	-1.05	329.62	18 Feb 23	220.0	283.4	4.7	0.1	338.51	344.75	-1.45	-1.42	303.82	310.15	-0.48	-1.13
Feb 17	336.99	-1.61	285.46	0.03	337.75	-1.02	339.67	28 Feb	356.8		-4.8		350.99		-1.37		317.10		-1.64	
								Mar 5		69.5		-2.9		357.22		-1.28		324.57		-2.01
Feb 27	338.20	-1.62	287.16	0.01	345.61	-0.98	349.67	10 Mar 15	133.1	193.0	2.5	5.0	3.44	9.64	-1.17	-1.04	332.54	340.99	-2.20	-2.22
Mar 9	339.39	-1.64	288.67	0.00	353.44	-0.92	359.61	20 Mar 25	253.1	319.9	2.5	-3.1	15.83	22.02	-0.89	-0.71	349.97	359.48	-2.04	-1.65
Mar 19	340.54	-1.65	289.97	-0.02	1.22	-0.86	9.49	30 M-A 4	35.5	105.2	-4.6	0.4	28.19	34.34	-0.52	-0.32	9.49	19.84	-1.05	-0.25
Mar 29	341.62	-1.68	291.02	-0.04	8.94	-0.79	19.32	9 Apr 14	166.2	225.8	4.6	4.0	40.48	46.61	-0.11	0.12	30.12	39.70	0.67	1.57
Apr 8	342.63	-1.70	291.81	-0.07	16.60	-0.71	29.09	19 Apr 24	288.0	358.3	-0.8	-5.0	52.72	58.81	0.34	0.57	47.92	54.34	2.28	2.68
Apr 18	343.55	-1.73	292.31	-0.09	24.18	-0.62	38.82	29 Apr	73.1		-2.2		64.89		0.79		58.73		2.67	
								May 4		138.9		3.3		70.95		1.00		60.94		2.21
Apr 28	344.35	-1.77	292.49	-0.11	31.68	-0.52	48.49	9 May 14	198.7	259.6	5.0	1.6	76.99	83.00	1.21	1.39	60.99	59.23	1.27	-0.04
May 8	345.03	-1.80	292.37	-0.14	39.09	-0.42	58.11	19 May 24	325.1	37.5	-3.8	-4.5	89.00	94.98	1.56	1.71	56.47	53.90	-1.49	-2.76
May 18	345.57	-1.84	291.93	-0.16	46.41	-0.32	67.71	29 M-J 3	108.8	171.4	1.1	5.0	100.94	106.86	1.82	1.91	52.54	52.94	-3.62	-3.99
May 28	345.95	-1.88	291.21	-0.19	53.64	-0.21	77.28	8 Jun 13	231.2	295.0	3.7	-1.7	112.76	118.63	1.97	1.99	55.18	59.14	-3.91	-3.47
Jun 7	346.18	-1.92	290.24	-0.22	60.77	-0.09	86.82	18 Jun 23	4.0	75.7	-5.3	-1.9	124.47	130.28	1.97	1.91	64.69	71.69	-2.76	-1.86
Jun 17	346.25	-1.97	289.08	-0.24	67.81	0.02	96.36	28 Jun	142.7		3.8		136.04		1.81		80.07		-0.87	
								Jul 3		203.2		5.1		141.76		1.67		89.66		0.12
Jun 27	346.14	-2.01	287.82	-0.27	74.75	0.14	105.89	8 Jul 13	264.6	332.4	1.1	-4.3	147.43	153.05	1.48	1.26	100.11	110.85	0.96	1.53
Jul 7	345.88	-2.05	286.54	-0.29	81.59	0.26	115.43	18 Jul 23	43.5	112.1	-4.2	1.4	158.61	164.10	0.98	0.67	121.34	131.26	1.79	1.75
Jul 17	345.47	-2.09	285.33	-0.31	88.33	0.38	124.99	28 J-A 2	175.3	235.1	5.1	3.4	169.51	174.84	0.31	-0.09	140.49	149.04	1.47	1.00
Jul 27	344.92	-2.12	284.28	-0.33	94.97	0.50	134.57	7 Aug 12	299.4	11.5	-2.0	-5.1	180.07	185.18	-0.52	-1.00	156.91	164.10	0.39	-0.32
Aug 6	344.28	-2.14	283.46	-0.35	101.51	0.62	144.17	17 Aug 22	81.9	146.8	-1.2	4.0	190.15	194.98	-1.50	-2.04	170.56	176.18	-1.10	-1.91
Aug 16	343.55	-2.16	282.91	-0.36	107.94	0.74	153.83	27 Aug	207.4		4.7		199.61		-2.60		180.76		-2.72	
								Sep 1		267.7		0.6		204.03		-3.18		183.94		-3.44
Aug 26	342.79	-2.18	282.67	-0.37	114.25	0.87	163.52	6 Sep 11	336.3	50.7	-4.5	-3.4	208.18	212.01	-3.77	-4.38	185.21	183.97	-3.98	-4.13
Sep 5	342.03	-2.18	282.75	-0.38	120.45	1.00	173.26	16 Sep 21	118.4	180.2	2.1	5.0	215.45	218.43	-4.97	-5.55	180.06	174.76	-3.64	-2.39
Sep 15	341.32	-2.18	283.15	-0.39	126.52	1.13	183.07	26 S-O 1	239.6	302.0	2.7	-2.5	220.83	222.53	-6.10	-6.57	170.98	171.05	-0.73	0.72
Sep 25	340.69	-2.17	283.85	-0.39	132.44	1.27	192.92	6 Oct 11	14.7	88.6	-4.9	-0.2	223.42	223.36	-6.95	-7.16	175.07	181.74	1.62	1.98
Oct 5	340.17	-2.15	284.84	-0.40	138.21	1.41	202.83	16 Oct 21	153.0	212.9	4.5	4.2	222.32	220.34	-7.15	-6.85	189.70	198.10	1.93	1.63
Oct 15	339.79	-2.13	286.09	-0.41	143.80	1.56	212.80	26 Oct 31	272.7	338.3	-0.3	-4.8	217.64	214.61	-6.22	-5.28	206.52	214.81	1.18	0.65
Oct 25	339.57	-2.11	287.57	-0.41	149.19	1.72	222.82	5 Nov 10	53.4	124.4	-3.0	3.1	211.77	209.56	-4.11	-2.84	222.93	230.91	0.09	-0.47
Nov 4	339.52	-2.08	289.25	-0.42	154.33	1.88	232.89	15 Nov 20	186.2	245.6	5.1	2.0	208.28	208.03	-1.58	-0.44	238.75	246.51	-0.99	-1.47
Nov 14	339.65	-2.05	291.11	-0.43	159.17	2.06	243.01	25 Nov 30	307.5	16.5	-3.4	-5.0	208.76	210.37	0.56	1.40	254.20	261.82	-1.86	-2.16
Nov 24	339.95	-2.03	293.11	-0.44	163.67	2.25	253.15	5 Dec 10	90.9	158.3	0.3	5.0	212.74	215.74	2.07	2.60	269.31	276.52	-2.31	-2.27
Dec 4	340.43	-2.00	295.24	-0.45	167.74	2.46	263.32	15 Dec 20	218.3	279.0	4.0	-1.1	219.27	223.22	3.00	3.29	283.09	288.26	-1.95	-1.26
Dec 14	341.05	-1.98	297.46	-0.46	171.27	2.68	273.52	25 Dec 30	344.5	55.5	-5.2	-2.8	227.52	232.11	3.47	3.56	290.65	288.65	-0.09	1.48
1759								1759												1759

JULIAN 12 UT	SATURN		JUPITER		MARS		SUN	GREGORIAN 12 UT	MOON				VENUS				MERCURY			
	LONG.	LAT.	LONG.	LAT.	LONG.	LAT.	LONG.		LONG.		LAT.		LONG.		LAT.		LONG.		LAT.	
1759								1760												1760
Dec 24	341.82	-1.95	299.76	-0.47	174.15	2.93	283.71	4 Jan 9	126.9	190.7	3.3	5.1	236.93	241.95	3.58	3.52	282.70	276.86	2.87	3.37
Jan 3	342.71	-1.94	302.10	-0.48	176.22	3.18	293.90	14 Jan 19	250.3	314.0	1.5	-3.8	247.13	252.45	3.40	3.23	274.49	275.66	3.02	2.27
Jan 13	343.71	-1.92	304.47	-0.49	177.29	3.45	304.08	24 Jan 29	23.2	93.8	-4.7	0.6	257.89	263.41	3.02	2.77	279.24	284.31	1.41	0.58
Jan 23	344.79	-1.91	306.84	-0.51	177.19	3.70	314.22	3 Feb 8	161.2	222.3	4.9	3.5	269.02	274.70	2.49	2.19	290.31	296.96	-0.18	-0.83
Feb 2	345.94	-1.90	309.19	-0.53	175.81	3.91	324.34	13 Feb 18	282.8	350.9	-1.5	-5.0	280.44	286.23	1.87	1.54	304.10	311.66	-1.37	-1.78
Feb 12	347.13	-1.90	311.50	-0.54	173.18	4.03	334.41	23 Feb 28	62.5	130.7	-1.9	3.6	292.06	297.92	1.20	0.87	319.63	328.02	-2.05	-2.15
Feb 22	348.36	-1.90	313.75	-0.57	169.61	4.02	344.44	4 Mar 9	194.3	253.9	4.7	0.8	303.81	309.73	0.54	0.22	336.85	346.14	-2.08	-1.80
Mar 3	349.59	-1.91	315.91	-0.59	165.71	3.85	354.41	14 Mar 19	317.0	29.4	-4.1	-4.0	315.68	321.64	-0.08	-0.36	355.88	5.92	-1.30	-0.58
Mar 13	350.81	-1.92	317.96	-0.62	162.17	3.54	4.33	24 Mar 29	100.9	166.0	1.7	5.0	327.62	333.62	-0.63	-0.86	15.88	25.11	0.33	1.31
Mar 23	352.00	-1.94	319.88	-0.64	159.61	3.14	14.18	3 Apr 8	226.7	286.3	2.8	-2.2	339.62	345.64	-1.07	-1.25	32.81	38.35	2.20	2.82
Apr 2	353.15	-1.96	321.63	-0.67	158.30	2.70	23.98	13 Apr 18	353.2	68.2	-5.1	-1.0	351.66	357.70	-1.40	-1.52	41.40	41.88	3.02	2.69
Apr 12	354.23	-1.98	323.20	-0.71	158.28	2.27	33.73	23 Apr 28	137.6	199.7	4.4	4.4	3.73	9.78	-1.61	-1.66	40.15	37.13	1.80	0.50
Apr 22	355.23	-2.01	324.56	-0.75	159.42	1.87	43.43	3 May 8	259.0	320.5	0.0	-4.6	15.83	21.88	-1.68	-1.67	34.16	32.44	-0.93	-2.16
May 2	356.14	-2.04	325.68	-0.78	161.56	1.51	53.08	13 May 18	31.2	106.1	-3.8	2.4	27.94	34.01	-1.63	-1.56	32.54	34.50	-3.02	-3.47
May 12	356.93	-2.08	326.54	-0.83	164.51	1.19	62.69	23 May 28	172.4	232.4	5.3	2.3	40.08	46.15	-1.46	-1.34	38.14	43.22	-3.55	-3.30
May 22	357.59	-2.12	327.10	-0.87	168.13	0.90	72.27	2 Jun 7	292.2	356.9	-3.0	-5.2	52.22	58.31	-1.20	-1.04	49.59	57.15	-2.79	-2.06
Jun 1	358.11	-2.16	327.36	-0.92	172.30	0.65	81.83	12 Jun 17	70.0	142.4	-0.8	4.8	64.39	70.49	-0.87	-0.68	65.86	75.62	-1.18	-0.23
Jun 11	358.47	-2.21	327.29	-0.96	176.93	0.42	91.37	22 Jun 27	205.5	264.8	4.2	-0.6	76.59	82.70	-0.48	-0.28	86.18	97.07	0.66	1.35
Jun 21	358.68	-2.25	326.91	-1.01	181.94	0.21	100.90	2 Jul 7	327.1	35.2	-4.9	-3.6	88.81	94.93	-0.07	0.13	107.74	117.82	1.75	1.85
Jul 1	358.71	-2.30	326.23	-1.05	187.28	0.02	110.43	12 Jul 17	108.2	176.9	2.6	5.1	101.06	107.20	0.33	0.52	127.14	135.69	1.68	1.27
Jul 11	358.58	-2.34	325.28	-1.09	192.90	-0.15	119.98	22 Jul 27	237.5	298.1	1.8	-3.3	113.35	119.50	0.70	0.86	143.47	150.44	0.68	-0.05
Jul 21	358.29	-2.38	324.13	-1.12	198.78	-0.31	129.54	1 Aug 6	4.1	74.3	-4.9	-0.2	125.66	131.83	1.01	1.14	156.55	161.66	-0.89	-1.79
Jul 31	357.85	-2.42	322.86	-1.15	204.88	-0.45	139.13	11 Aug 16	145.2	209.8	4.7	3.6	138.01	144.20	1.24	1.32	165.53	167.77	-2.72	-3.59
Aug 10	357.28	-2.45	321.55	-1.16	211.19	-0.58	148.76	21 Aug 26	269.3	332.9	-1.2	-4.9	150.39	156.59	1.38	1.41	167.94	165.68	-4.26	-4.51
Aug 20	356.60	-2.48	320.31	-1.17	217.69	-0.69	158.43	31 Aug	42.7		-2.6		162.79		1.41		161.38		-4.06	
								Sep 5		113.0		3.2		169.00		1.39		156.77		-2.87
Aug 30	355.86	-2.50	319.22	-1.17	224.36	-0.80	168.15	10 Sep 15	180.4	241.6	4.8	1.0	175.21	181.43	1.34	1.26	154.34	155.60	-1.27	0.21
Sep 9	355.09	-2.50	318.36	-1.16	231.20	-0.89	177.92	20 Sep 25	301.7	9.5	-3.8	-4.5	187.65	193.87	1.16	1.04	160.43	167.71	1.25	1.79
Sep 19	354.33	-2.50	317.78	-1.15	238.19	-0.98	187.74	30 S-O 5	81.9	150.4	0.9	5.0	200.09	206.31	0.89	0.72	176.20	185.04	1.90	1.71
Sep 29	353.62	-2.49	317.53	-1.13	245.33	-1.05	197.63	10 Oct 15	213.9	273.2	3.0	-2.0	212.54	218.76	0.54	0.35	193.81	202.37	1.33	0.84
Oct 9	353.00	-2.48	317.61	-1.11	252.60	-1.11	207.57	20 Oct 25	335.6	47.6	-5.2	-2.0	224.99	231.21	0.14	-0.07	210.67	218.73	0.30	-0.26
Oct 19	352.50	-2.45	318.02	-1.09	259.98	-1.15	217.56	30 Oct	120.4		4.0		237.43		-0.29		226.59		-0.81	
								Nov 4		185.9		4.7		243.65		-0.50		234.27		-1.33
Oct 29	352.16	-2.42	318.75	-1.06	267.49	-1.19	227.61	9 Nov 14	246.3	305.7	0.3	-4.4	249.87	256.08	-0.71	-0.92	241.80	249.16	-1.78	-2.15
Nov 8	351.98	-2.39	319.77	-1.04	275.09	-1.21	237.70	19 Nov 24	11.3	86.0	-4.5	1.5	262.30	268.50	-1.11	-1.28	256.29	263.01	-2.40	-2.46
Nov 18	351.97	-2.36	321.06	-1.02	282.77	-1.22	247.82	29 N-D 4	157.2	219.5	5.3	2.6	274.71	280.91	-1.44	-1.57	268.94	273.31	-2.27	-1.71
Nov 28	352.15	-2.32	322.59	-1.01	290.53	-1.21	257.99	9 Dec 14	278.7	339.8	-2.6	-5.3	287.10	293.29	-1.68	-1.77	274.75	271.78	-0.64	0.92
Dec 8	352.50	-2.29	324.33	-0.99	298.34	-1.20	268.17	19 Dec 24	48.6	124.0	-1.9	4.3	299.46	305.63	-1.82	-1.84	265.34	259.92	2.42	3.10
Dec 18	353.02	-2.25	326.25	-0.98	306.20	-1.17	278.36	29 Dec	192.0		4.4		311.77		-1.82		258.53		2.93	
1760								1761												1761
								Jan 3		252.0		-0.3		317.91		-1.77		260.69		2.33
Dec 28	353.69	-2.22	328.31	-0.97	314.08	-1.13	288.56	8 Jan 13	311.8	15.9	-4.6	-4.1	324.02	330.11	-1.68	-1.56	265.07	270.72	1.57	0.81
Jan 7	354.50	-2.19	330.49	-0.96	321.98	-1.07	298.74	18 Jan 23	87.0	160.6	1.7	5.1	336.17	342.21	-1.40	-1.20	277.13	284.05	0.09	-0.56
Jan 17	355.43	-2.17	332.77	-0.96	329.87	-1.01	308.90	28 J-F 2	224.9	284.1	1.9	-3.0	348.20	354.16	-0.97	-0.71	291.34	298.97	-1.11	-1.56
Jan 27	356.46	-2.15	335.11	-0.96	337.75	-0.94	319.04	7 Feb 12	346.5	53.8	-4.9	-1.0	0.08	5.94	-0.42	-0.10	306.93	315.22	-1.89	-2.07
Feb 6	357.57	-2.14	337.51	-0.96	345.60	-0.86	329.13	17 Feb 22	125.6	195.3	4.4	3.8	11.74	17.48	0.24	0.61	323.88	332.94	-2.09	-1.92
Feb 16	358.75	-2.13	339.92	-0.97	353.40	-0.77	339.18	27 Feb	256.6		-1.1		23.14		0.98		342.37		-1.53	
								Mar 4		316.8		-4.8		28.73		1.38		352.05		-0.90
Feb 26	359.96	-2.12	342.34	-0.98	1.15	-0.68	349.19	9 Mar 14	23.0	92.9	-3.2	2.6	34.21	39.59	1.77	2.17	1.63	10.43	-0.03	1.00
Mar 8	1.20	-2.12	344.74	-0.99	8.83	-0.58	359.13	19 Mar 24	163.2	228.4	5.0	1.2	44.84	49.94	2.56	2.94	17.52	22.05	2.04	2.88
Mar 18	2.45	-2.13	347.09	-1.01	16.45	-0.48	9.02	29 M-A 3	288.1	350.9	-3.7	-4.8	54.87	59.61	3.31	3.64	23.58	22.24	3.30	3.11
Mar 28	3.68	-2.14	349.39	-1.02	23.99	-0.38	18.85	8 Apr 13	61.0	131.9	0.0	4.9	64.12	68.34	3.95	4.20	18.96	15.35	2.28	1.00
Apr 7	4.88	-2.15	351.60	-1.05	31.45	-0.27	28.63	18 Apr 23	199.0	260.5	3.4	-1.8	72.23	75.72	4.41	4.54	12.93	12.45	-0.37	-1.56
Apr 17	6.03	-2.17	353.71	-1.07	38.82	-0.16	38.35	28 Apr	320.3		-5.2		78.72		4.59		13.99		-2.43	
								May 3		26.7		-2.9		81.14		4.54		17.26		-2.96
Apr 27	7.11	-2.20	355.69	-1.10	46.11	-0.06	48.02	8 May 13	99.9	169.7	3.4	5.0	82.86	83.75	4.35	4.01	21.98	27.90	-3.17	-3.09
May 7	8.11	-2.23	357.52	-1.13	53.31	0.05	57.66	18 May 23	232.9	292.5	0.7	-4.2	83.70	82.65	3.48	2.75	34.88	42.84	-2.75	-2.19
May 17	9.01	-2.26	359.18	-1.17	60.42	0.16	67.25	28 M-J 2	353.9	64.1	-4.8	0.4	80.65	77.91	1.81	0.71	51.77	61.63	-1.44	-0.57
May 27	9.79	-2.30	0.63	-1.21	67.45	0.26	76.82	7 Jun 12	138.6	205.3	5.2	3.0	74.81	71.83	-0.46	-1.60	72.24	83.20	0.33	1.12
Jun 6	10.44	-2.34	1.84	-1.25	74.39	0.36	86.36	17 Jun 22	265.7	325.3	-2.3	-5.2	69.42	67.88	-2.60	-3.40	93.99	104.16	1.67	1.92
Jun 16	10.94	-2.38	2.80	-1.30	81.25	0.46	95.90	27 Jun	29.4		-2.6		67.32		-4.00		113.51		1.87	
								Jul 2		102.7		3.5		67.74		-4.39		121.99		1.55
Jun 26	11.29	-2.43	3.47	-1.35	88.03	0.56	105.43	7 Jul 12	175.8	239.0	4.6	0.0	69.02	71.05	-4.62	-4.71	129.58	136.23	1.00	0.27
Jul 6	11.48	-2.48	3.84	-1.40	94.74	0.65	114.97	17 Jul 22	298.3	359.5	-4.4	-4.3	73.70	76.87	-4.68	-4.57	141.86	146.30	-0.63	-1.63
Jul 16	11.49	-2.52	3.88	-1.45	101.37	0.75	124.52	27 J-A 1	66.7	141.4	0.8	5.0	80.48	84.45	-4.38	-4.13	149.27	150.42	-2.69	-3.70
Jul 26	11.34	-2.57	3.59	-1.50	107.92	0.84	134.10	6 Aug 11	210.9	271.5	2.2	-2.9	88.72	93.23	-3.84	-3.51	149.43	146.37	-4.49	-4.79
Aug 5	11.02	-2.61	2.99	-1.54	114.41	0.92	143.71	16 Aug 21	331.5	35.5	-5.0	-1.7	97.96	102.87	-3.15	-2.77	142.17	138.69	-4.37	-3.24
Aug 15	10.55	-2.65	2.10	-1.58	120.82	1.01	153.35	26 Aug 31	105.5	178.8	3.9	4.1	107.94	113.14	-2.38	-1.98	137.77	140.23	-1.74	-0.28
Aug 25	9.95	-2.68	0.98	-1.61	127.16	1.09	163.05	5 Sep 10	244.0	303.6	-0.9	-4.8	118.46	123.89	-1.58	-1.19	145.83	153.66	0.86	1.55
Sep 4	9.26	-2.70	359.71	-1.63	133.44	1.18	172.79	15 Sep 20	6.0	73.3	-3.7	1.8	129.42	135.02	-0.80	-0.43	162.62	171.89	1.83	1.76
Sep 14	8.50	-2.71	358.37	-1.63	139.63	1.26	182.59	25 Sep 30	144.6	214.1	5.1	1.6	140.71	146.47	-0.08	0.25	181.02	189.85	1.47	1.03
Sep 24	7.71	-2.72	357.08	-1.62	145.75	1.34	192.44	5 Oct 10	275.9	336.1	-3.6	-5.0	152.28	158.16	0.55	0.83	198.33	206.50	0.51	-0.05
Oct 4	6.95	-2.71	355.92	-1.60	151.79	1.42	202.35	15 Oct 20	42.0	112.5	-0.8	4.6	164.08	170.05	1.07	1.28	214.37	221.99	-0.63	-1.18
Oct 14	6.25	-2.69	354.99	-1.57	157.74	1.50	212.32	25 Oct 30	182.4	247.5	3.9	-1.5	176.07	182.12	1.46	1.60	229.37	236.48	-1.69	-2.13
Oct 24	5.64	-2.67	354.34	-1.53	163.58	1.58	222.34	4 Nov 9	307.4	9.5	-5.1	-3.5	188.20	194.32	1.70	1.77	243.24	249.45	-2.47	-2.64
Nov 3	5.17	-2.64	354.02	-1.49	169.30	1.66	232.40	14 Nov 19	79.7	151.7	2.5	5.2	200.46	206.62	1.80	1.79	254.71	258.22	-2.58	-2.16
Nov 13	4.85	-2.60	354.04	-1.45	174.89	1.74	242.51	24 Nov 29	218.3	279.9	1.2	-4.0	212.80	219.00	1.76	1.69	258.68	254.85	-1.20	0.32
Nov 23	4.71	-2.56	354.41	-1.40	180.32	1.82	252.66	4 Dec 9	339.5	44.5	-5.0	-0.6	225.22	231.44	1.59	1.46	248.21	243.44	1.91	2.76
Dec 3	4.75	-2.52	355.10	-1.35	185.56	1.91	262.83	14 Dec 19	118.4	189.4	4.8	3.4	237.68	243.92	1.31	1.14	243.04	246.11	2.79	2.35
Dec 13	4.97	-2.48	356.10	-1.31	190.57	1.99	273.02	24 Dec 29	252.4	312.1	-1.9	-5.1	250.17	256.43	0.95	0.76	251.21	257.39	1.71	1.01
1761								1761												1761

JULIAN 12 UT	SATURN LONG.	SATURN LAT.	JUPITER LONG.	JUPITER LAT.	MARS LONG.	MARS LAT.	SUN LONG.	GREGORIAN 12 UT	MOON LONG.		MOON LAT.		VENUS LONG.		VENUS LAT.		MERCURY LONG.		MERCURY LAT.	
1761								1762												1762
Dec 23	5.36	-2.44	357.38	-1.27	195.31	2.08	283.22	3 Jan 8	12.7	81.2	-3.1	2.7	262.69	268.95	0.55	0.33	264.19	271.35	0.33	-0.31
Jan 2	5.92	-2.40	358.89	-1.23	199.70	2.17	293.40	13 Jan 18	157.0	224.9	4.8	0.3	275.21	281.47	0.12	-0.10	278.79	286.47	-0.87	-1.35
Jan 12	6.63	-2.36	0.62	-1.20	203.68	2.26	303.58	23 Jan 28	285.4	344.9	-4.2	-4.4	287.73	294.00	-0.30	-0.50	294.40	302.59	-1.73	-1.98
Jan 22	7.48	-2.33	2.53	-1.17	207.13	2.34	313.73	2 Feb 7	47.6	119.5	0.1	4.8	300.26	306.51	-0.69	-0.86	311.07	319.87	-2.08	-2.01
Feb 1	8.45	-2.30	4.58	-1.15	209.94	2.42	323.84	12 Feb 17	194.1	258.5	2.6	-2.8	312.77	319.02	-1.01	-1.14	328.98	338.26	-1.72	-1.19
Feb 11	9.51	-2.28	6.76	-1.13	211.94	2.49	333.92	22 Feb 27	317.9	18.9	-5.0	-2.1	325.27	331.52	-1.25	-1.34	347.40	355.70	-0.40	0.64
Feb 21	10.65	-2.26	9.03	-1.11	212.95	2.52	343.95	4 Mar 9	84.8	158.3	3.4	4.6	337.76	343.99	-1.40	-1.43	2.11	5.54	1.80	2.85
Mar 3	11.84	-2.25	11.37	-1.10	212.80	2.51	353.92	14 Mar 19	229.2	290.8	-0.6	-4.8	350.22	356.44	-1.44	-1.42	5.46	2.40	3.48	3.43
Mar 13	13.08	-2.24	13.76	-1.09	211.38	2.44	3.84	24 Mar 29	350.7	54.5	-4.0	1.1	2.66	8.87	-1.37	-1.30	358.09	354.62	2.65	1.42
Mar 23	14.34	-2.24	16.17	-1.08	208.76	2.26	13.70	3 Apr 8	123.7	196.0	5.2	2.3	15.07	21.27	-1.20	-1.09	353.22	354.06	0.12	-1.02
Apr 2	15.59	-2.24	18.58	-1.08	205.27	1.98	23.50	13 Apr 18	262.5	322.6	-3.4	-5.2	27.45	33.63	-0.95	-0.79	356.83	1.12	-1.89	-2.47
Apr 12	16.83	-2.25	20.98	-1.08	201.52	1.60	33.25	23 Apr 28	24.4	92.2	-1.6	4.1	39.81	45.98	-0.62	-0.43	6.63	13.14	-2.79	-2.86
Apr 22	18.04	-2.27	23.34	-1.08	198.23	1.17	42.95	3 May 8	163.1	232.0	4.5	-0.9	52.13	58.28	-0.24	-0.04	20.54	28.78	-2.68	-2.27
May 2	19.19	-2.28	25.65	-1.09	195.99	0.72	52.61	13 May 18	294.7	354.6	-5.0	-3.9	64.43	70.56	0.17	0.37	37.84	47.72	-1.65	-0.87
May 12	20.28	-2.31	27.88	-1.10	195.04	0.31	62.22	23 May 28	59.7	131.3	1.6	5.2	76.70	82.82	0.57	0.76	58.30	69.26	0.00	0.85
May 22	21.28	-2.34	30.02	-1.12	195.42	-0.06	71.80	2 Jun 7	201.3	266.2	1.8	-3.7	88.93	95.04	0.95	1.11	80.09	90.29	1.53	1.93
Jun 1	22.18	-2.37	32.04	-1.13	196.99	-0.38	81.36	12 Jun 17	326.6	27.6	-5.0	-1.2	101.14	107.24	1.26	1.39	99.59	107.90	2.03	1.83
Jun 11	22.95	-2.40	33.92	-1.15	199.57	-0.65	90.90	22 Jun 27	97.0	170.6	4.3	3.9	113.32	119.40	1.50	1.58	115.17	121.35	1.35	0.62
Jun 21	23.60	-2.44	35.62	-1.18	203.00	-0.87	100.43	2 Jul 7	237.6	299.3	-1.5	-5.0	125.47	131.52	1.63	1.65	126.31	129.84	-0.31	-1.41
Jul 1	24.09	-2.49	37.14	-1.20	207.12	-1.06	109.96	12 Jul 17	358.8	62.3	-3.2	2.0	137.57	143.61	1.64	1.60	131.70	131.60	-2.60	-3.74
Jul 11	24.43	-2.53	38.43	-1.23	211.82	-1.22	119.51	22 Jul 27	135.9	208.4	5.0	0.8	149.63	155.64	1.52	1.42	129.55	126.12	-4.62	-4.94
Jul 21	24.60	-2.58	39.48	-1.27	217.00	-1.35	129.08	1 Aug 6	272.1	332.0	-4.1	-4.5	161.63	167.61	1.28	1.11	122.62	120.65	-4.54	-3.50
Jul 31	24.60	-2.62	40.24	-1.30	222.59	-1.46	138.66	11 Aug 16	32.1	99.2	-0.3	4.5	173.57	179.51	0.91	0.68	121.31	124.93	-2.14	-0.74
Aug 10	24.42	-2.67	40.70	-1.33	228.53	-1.54	148.29	21 Aug 26	174.7	244.1	3.3	-2.5	185.43	191.33	0.42	0.14	131.25	139.57	0.45	1.28
Aug 20	24.08	-2.71	40.83	-1.37	234.77	-1.61	157.96	31 Aug	305.2		-5.1		197.20		-0.15		148.97		1.71	
								Sep 5		4.8		-2.5		203.04		-0.47		158.64		1.78
Aug 30	23.59	-2.74	40.63	-1.40	241.28	-1.65	167.67	10 Sep 15	67.1	137.7	2.9	5.0	208.85	214.63	-0.79	-1.13	168.12	177.21	1.59	1.22
Sep 9	22.97	-2.77	40.11	-1.43	248.02	-1.68	177.45	20 Sep 25	212.2	277.9	0.1	-4.7	220.37	226.07	-1.46	-1.80	185.89	194.16	0.73	0.17
Sep 19	22.25	-2.79	39.28	-1.45	254.96	-1.68	187.27	30 S-O 5	337.7	38.6	-4.3	0.6	231.72	237.32	-2.12	-2.43	202.07	209.63	-0.43	-1.02
Sep 29	21.48	-2.80	38.19	-1.46	262.07	-1.68	197.15	10 Oct 15	104.3	176.6	5.0	3.1	242.85	248.31	-2.73	-2.99	216.85	223.70	-1.59	-2.11
Oct 9	20.68	-2.79	36.93	-1.46	269.35	-1.65	207.09	20 Oct 25	247.7	310.3	-3.1	-5.2	253.68	258.95	-3.22	-3.42	230.07	235.75	-2.53	-2.82
Oct 19	19.92	-2.78	35.58	-1.45	276.76	-1.61	217.08	30 Oct	10.0		-2.0		264.09		-3.56		240.30		-2.89	
								Nov 4		74.0		3.5		269.09		-3.65		242.91		-2.60
Oct 29	19.22	-2.76	34.25	-1.43	284.27	-1.56	227.12	9 Nov 14	143.2	214.4	4.9	-0.2	273.92	278.53	-3.67	-3.62	242.36	237.80	-1.77	-0.32
Nov 8	18.63	-2.73	33.03	-1.40	291.88	-1.49	237.21	19 Nov 24	281.4	341.9	-4.9	-4.0	282.89	286.94	-3.48	-3.25	231.24	227.28	1.33	2.37
Nov 18	18.18	-2.69	32.03	-1.36	299.56	-1.41	247.34	29 N-D 4	43.2	111.5	1.0	5.1	290.59	293.76	-2.91	-2.44	227.85	231.77	2.61	2.33
Nov 28	17.89	-2.65	31.31	-1.32	307.30	-1.32	257.49	9 Dec 14	182.5	250.7	2.5	-3.2	296.36	298.26	-1.83	-1.06	237.55	244.24	1.81	1.19
Dec 8	17.78	-2.60	30.91	-1.27	315.06	-1.22	267.68	19 Dec 24	314.0	13.5	-4.9	-1.5	299.31	299.40	-0.12	0.98	251.39	258.80	0.55	-0.07
Dec 18	17.85	-2.56	30.85	-1.21	322.84	-1.11	277.87	29 Dec	77.9		3.7		298.45		2.21		266.38		-0.64	
1762								1763												1763
								Jan 3		150.4		4.3		296.51		3.51		274.11		-1.15
Dec 28	18.11	-2.51	31.15	-1.16	330.62	-0.99	288.06	8 Jan 13	220.7	285.4	-1.0	-4.9	293.82	290.78	4.75	5.81	282.01	290.09	-1.56	-1.87
Jan 7	18.54	-2.46	31.77	-1.11	338.38	-0.87	298.25	18 Jan 23	346.0	46.0	-3.3	1.6	287.90	285.66	6.58	7.01	298.40	306.94	-2.05	-2.07
Jan 17	19.14	-2.42	32.70	-1.07	346.11	-0.75	308.41	28 J-F 2	114.5	189.2	5.0	1.5	284.33	284.03	7.13	6.99	315.70	324.56	-1.89	-1.47
Jan 27	19.89	-2.38	33.91	-1.02	353.80	-0.63	318.54	7 Feb 12	257.1	318.9	-3.9	-4.6	284.73	286.31	6.67	6.23	333.21	340.95	-0.76	0.24
Feb 6	20.77	-2.35	35.36	-0.98	1.44	-0.50	328.64	17 Feb 22	18.1	80.3	-0.6	4.2	288.63	291.57	5.71	5.14	346.61	348.89	1.48	2.71
Feb 16	21.76	-2.32	37.02	-0.94	9.01	-0.38	338.69	27 Feb	152.6		4.0		295.01		4.55		347.22		3.55	
								Mar 4		226.8		-2.0		298.86		3.94		342.75		3.63
Feb 26	22.85	-2.29	38.86	-0.91	16.51	-0.26	348.69	9 Mar 14	291.7	351.5	-5.2	-2.8	303.06	307.53	3.34	2.75	338.05	335.28	2.91	1.75
Mar 8	24.01	-2.27	40.85	-0.88	23.94	-0.14	358.64	19 Mar 24	51.3	116.8	2.5	5.2	312.23	317.12	2.18	1.64	335.07	337.10	0.53	-0.54
Mar 18	25.23	-2.26	42.96	-0.85	31.29	-0.02	8.53	29 M-A 3	191.1	262.6	1.1	-4.6	322.16	327.33	1.12	0.63	340.85	345.91	-1.40	-2.03
Mar 28	26.48	-2.25	45.17	-0.83	38.56	0.09	18.37	8 Apr 13	324.8	24.0	-4.5	0.1	332.62	338.01	0.18	-0.23	351.99	358.91	-2.43	-2.61
Apr 7	27.75	-2.24	47.45	-0.81	45.75	0.19	28.15	18 Apr 23	86.3	155.1	4.7	3.9	343.47	349.00	-0.60	-0.93	6.58	14.97	-2.57	-2.30
Apr 17	29.02	-2.24	49.77	-0.79	52.86	0.30	37.87	28 Apr	228.9		-2.3		354.59		-1.22		24.08		-1.82	
								May 3		296.7		-5.3		0.23		-1.46		33.92		-1.15
Apr 27	30.27	-2.25	52.12	-0.78	59.89	0.39	47.55	8 May 13	356.8	57.5	-2.3	3.0	5.91	11.64	-1.66	-1.81	44.40	55.28	-0.33	0.55
May 7	31.48	-2.26	54.49	-0.76	66.84	0.48	57.19	18 May 23	123.5	194.2	5.1	0.8	17.40	23.19	-1.93	-2.00	66.06	76.22	1.33	1.90
May 17	32.65	-2.27	56.84	-0.75	73.72	0.57	66.79	28 M-J 2	265.3	329.3	-4.6	-4.1	29.00	34.84	-2.03	-2.02	85.38	93.39	2.16	2.10
May 27	33.74	-2.30	59.16	-0.74	80.53	0.65	76.35	7 Jun 12	28.8	92.7	0.6	4.8	40.71	46.60	-1.98	-1.90	100.18	105.66	1.72	1.03
Jun 6	34.74	-2.32	61.43	-0.74	87.27	0.73	85.91	17 Jun 22	162.3	232.8	3.2	-2.7	52.50	58.43	-1.79	-1.66	109.69	112.05	0.07	-1.11
Jun 16	35.64	-2.35	63.64	-0.73	93.95	0.80	95.44	27 Jun	300.0		-5.0		64.37		-1.50		112.56		-2.42	
								Jul 2		1.0		-1.6		70.34		-1.32		111.21		-3.67
Jun 26	36.41	-2.38	65.76	-0.73	100.57	0.87	104.97	7 Jul 12	61.6	129.9	3.4	4.7	76.32	82.32	-1.12	-0.91	108.44	105.29	-4.58	-4.90
Jul 6	37.06	-2.42	67.76	-0.73	107.14	0.93	114.51	17 Jul 22	201.5	269.9	-0.3	-4.8	88.33	94.36	-0.69	-0.47	103.07	102.83	-4.54	-3.64
Jul 16	37.55	-2.46	69.63	-0.73	113.65	0.99	124.06	27 J-A 1	333.3	32.7	-3.5	1.4	100.41	106.48	-0.24	-0.01	105.04	109.75	-2.44	-1.15
Jul 26	37.88	-2.50	71.34	-0.74	120.12	1.04	133.64	6 Aug 11	96.1	168.5	5.0	2.3	112.57	118.67	0.21	0.42	116.72	125.47	0.03	0.96
Aug 5	38.03	-2.54	72.87	-0.74	126.55	1.09	143.25	16 Aug 21	240.0	305.1	-3.6	-4.8	124.79	130.92	0.61	0.80	135.24	145.27	1.54	1.76
Aug 15	38.02	-2.59	74.18	-0.74	132.94	1.14	152.89	26 Aug 31	5.6	65.3	-0.8	4.0	137.07	143.24	0.96	1.10	155.08	164.45	1.69	1.39
Aug 25	37.82	-2.62	75.24	-0.75	139.29	1.18	162.58	5 Sep 10	132.5	207.3	4.6	-1.2	149.42	155.61	1.22	1.31	173.31	181.69	0.94	0.39
Sep 4	37.46	-2.66	76.03	-0.76	145.61	1.22	172.32	15 Sep 20	276.7	338.7	-5.2	-3.1	161.82	168.04	1.38	1.42	189.62	197.12	-0.21	-0.84
Sep 14	36.95	-2.69	76.51	-0.76	151.89	1.25	182.11	25 Sep 30	37.9	99.6	2.1	5.3	174.26	180.50	1.44	1.42	204.18	210.75	-1.47	-2.06
Sep 24	36.31	-2.71	76.68	-0.77	158.13	1.28	191.96	5 Oct 10	170.3	245.2	2.1	-4.1	186.75	193.00	1.38	1.31	216.72	221.82	-2.59	-2.99
Oct 4	35.58	-2.72	76.50	-0.77	164.35	1.31	201.87	15 Oct 20	311.5	11.2	-4.7	-0.3	199.26	205.52	1.22	1.11	225.60	227.25	-3.19	-3.04
Oct 14	34.79	-2.73	76.00	-0.77	170.53	1.34	211.83	25 Oct 30	71.2	135.9	4.4	4.5	211.79	218.06	0.97	0.82	225.70	220.58	-2.34	-0.96
Oct 24	33.99	-2.72	75.19	-0.77	176.66	1.36	221.85	4 Nov 9	208.7	281.4	-1.3	-5.2	224.33	230.61	0.65	0.46	214.38	211.35	0.72	1.93
Nov 3	33.22	-2.70	74.12	-0.76	182.76	1.37	231.92	14 Nov 19	344.3	43.6	-2.6	2.6	236.89	243.17	0.27	0.07	212.87	217.60	2.38	2.28
Nov 13	32.53	-2.68	72.87	-0.75	188.81	1.38	242.02	24 Nov 29	106.2	174.0	5.1	1.7	249.44	255.72	-0.13	-0.33	224.00	231.18	1.89	1.34
Nov 23	31.95	-2.64	71.51	-0.73	194.79	1.39	252.16	4 Dec 9	246.8	315.7	-4.1	-4.2	262.00	268.28	-0.52	-0.71	238.69	246.34	0.75	0.15
Dec 3	31.51	-2.60	70.17	-0.70	200.72	1.39	262.34	14 Dec 19	16.1	76.8	0.4	4.6	274.56	280.85	-0.89	-1.05	254.07	261.85	-0.42	-0.95
Dec 13	31.25	-2.55	68.94	-0.67	206.56	1.39	272.52	24 Dec 29	143.1	212.9	3.7	-2.0	287.10	293.37	-1.20	-1.32	269.72	277.71	-1.40	-1.76
1763								1763												1763

JULIAN 12 UT	SATURN LONG.	SATURN LAT.	JUPITER LONG.	JUPITER LAT.	MARS LONG.	MARS LAT.	SUN LONG.	GREGORIAN 12 UT	MOON LONG.		MOON LAT.		VENUS LONG.		VENUS LAT.		MERCURY LONG.		MERCURY LAT.	
1763								1764												1764
Dec 23	31.17	-2.51	67.92	-0.64	212.30	1.37	282.72	3 Jan 8	283.7	348.3	-5.0	-1.8	299.64	305.90	-1.43	-1.50	285.83	294.11	-2.01	-2.11
Jan 2	31.27	-2.46	67.16	-0.60	217.94	1.35	292.91	13 Jan 18	47.7	111.4	3.2	4.9	312.15	318.40	-1.55	-1.58	302.52	310.95	-2.03	-1.72
Jan 12	31.56	-2.41	66.72	-0.56	223.43	1.32	303.08	23 Jan 28	181.5	251.7	0.5	-4.6	324.63	330.86	-1.57	-1.53	319.07	326.19	-1.12	-0.18
Jan 22	32.03	-2.36	66.62	-0.52	228.76	1.27	313.23	2 Feb 7	318.8	19.9	-3.8	1.3	337.08	343.28	-1.47	-1.37	331.05	332.17	1.08	2.46
Feb 1	32.65	-2.32	66.86	-0.49	233.88	1.21	323.35	12 Feb 17	80.0	147.7	5.0	3.2	349.47	355.64	-1.25	-1.10	329.05	323.60	3.50	3.70
Feb 11	33.43	-2.27	67.42	-0.45	238.74	1.13	333.43	22 Feb 27	220.4	289.2	-3.0	-5.1	1.80	7.93	-0.92	-0.72	319.07	317.28	3.06	1.99
Feb 21	34.34	-2.24	68.29	-0.42	243.29	1.02	343.46	3 Mar 8	352.3	51.7	-1.2	3.8	14.04	20.13	-0.50	-0.25	318.24	321.28	0.87	-0.14
Mar 2	35.35	-2.21	69.42	-0.39	247.45	0.88	353.44	13 Mar 18	113.8	185.4	5.0	0.0	26.19	32.22	0.00	0.27	325.82	331.46	-0.98	-1.64
Mar 12	36.46	-2.18	70.79	-0.36	251.11	0.70	3.35	23 Mar 28	258.9	324.6	-5.1	-3.4	38.22	44.19	0.54	0.82	337.94	345.13	-2.10	-2.36
Mar 22	37.64	-2.16	72.37	-0.33	254.14	0.47	13.22	2 Apr 7	24.8	84.3	1.8	5.2	50.12	56.01	1.09	1.36	352.96	1.41	-2.43	-2.29
Apr 1	38.87	-2.14	74.12	-0.31	256.40	0.18	23.03	12 Apr 17	149.5	223.9	3.1	-3.3	61.86	67.66	1.62	1.87	10.50	20.23	-1.94	-1.39
Apr 11	40.14	-2.13	76.02	-0.29	257.69	-0.19	32.78	22 Apr 27	295.8	358.1	-4.9	-0.6	73.41	79.10	2.09	2.29	30.56	41.28	-0.64	0.22
Apr 21	41.42	-2.12	78.05	-0.27	257.85	-0.64	42.48	2 May 7	57.3	118.6	4.1	4.7	84.72	90.28	2.45	2.58	51.94	61.97	1.09	1.80
May 1	42.70	-2.12	80.16	-0.25	256.75	-1.17	52.14	12 May 17	187.0	262.2	-0.3	-5.0	95.75	101.13	2.67	2.72	70.89	78.45	2.24	2.35
May 11	43.96	-2.12	82.35	-0.23	254.49	-1.75	61.76	22 May 27	330.5	30.6	-2.8	2.3	106.41	111.57	2.71	2.65	84.54	89.05	2.10	1.48
May 21	45.19	-2.13	84.59	-0.21	251.38	-2.34	71.34	1 Jun 6	90.6	154.9	5.0	2.3	116.59	121.45	2.53	2.34	91.83	92.73	0.52	-0.73
May 31	46.36	-2.14	86.87	-0.19	248.09	-2.85	80.90	11 Jun 16	225.9	299.2	-3.5	-4.4	126.13	130.58	2.08	1.75	91.79	89.43	-2.13	-3.43
Jun 10	47.45	-2.16	89.15	-0.18	245.34	-3.24	90.44	21 Jun 26	3.4	62.8	0.3	4.5	134.76	138.64	1.33	0.82	86.58	84.42	-4.34	-4.66
Jun 20	48.46	-2.18	91.43	-0.16	243.68	-3.49	99.97	1 Jul 6	125.4	192.9	4.0	-1.2	142.13	145.17	0.21	-0.50	83.87	85.37	-4.39	-3.66
Jun 30	49.36	-2.20	93.68	-0.14	243.40	-3.61	109.51	11 Jul 16	264.8	334.1	-5.0	-2.1	147.64	149.45	-1.32	-2.25	89.00	94.68	-2.65	-1.50
Jul 10	50.14	-2.23	95.89	-0.13	244.48	-3.63	119.05	21 Jul 26	35.3	95.8	3.1	5.0	150.47	150.59	-3.27	-4.37	102.24	111.34	-0.36	0.62
Jul 20	50.79	-2.26	98.04	-0.11	246.79	-3.59	128.62	31 J-A 5	162.0	232.1	1.3	-4.2	149.74	147.93	-5.49	-6.55	121.42	131.78	1.33	1.70
Jul 30	51.28	-2.29	100.10	-0.10	250.13	-3.49	138.21	10 Aug 15	302.5	7.2	-4.2	1.0	145.33	142.28	-7.43	-8.02	141.90	151.51	1.75	1.55
Aug 9	51.60	-2.33	102.05	-0.08	254.32	-3.37	147.83	20 Aug 25	67.0	130.0	4.9	3.8	139.27	136.76	-8.26	-8.14	160.55	169.04	1.16	0.63
Aug 19	51.75	-2.37	103.87	-0.07	259.20	-3.22	157.50	30 Aug	200.1		-2.2		135.07		-7.73		176.98		0.02	
								Sep 4		271.3		-5.2		134.35		-7.11		184.40		-0.64
Aug 29	51.72	-2.40	105.53	-0.05	264.64	-3.05	167.21	9 Sep 14	338.0	39.3	-1.6	3.6	134.62	135.79	-6.36	-5.55	191.28	197.55	-1.33	-2.00
Sep 8	51.51	-2.43	107.00	-0.03	270.54	-2.87	176.97	19 Sep 24	99.2	165.7	5.1	0.9	137.75	140.39	-4.73	-3.92	203.07	207.54	-2.63	-3.15
Sep 18	51.14	-2.46	108.26	-0.02	276.81	-2.67	186.79	29 S-O 4	239.2	308.9	-4.7	-3.9	143.58	147.23	-3.15	-2.41	210.47	211.10	-3.49	-3.48
Sep 28	50.61	-2.48	109.27	0.00	283.38	-2.47	196.67	9 Oct 14	11.8	71.4	1.4	5.1	151.26	155.62	-1.72	-1.08	208.58	203.14	-2.90	-1.59
Oct 8	49.95	-2.50	110.01	0.02	290.18	-2.26	206.60	19 Oct 24	132.7	203.0	3.5	-2.4	160.25	165.10	-0.50	0.03	197.59	195.59	0.10	1.45
Oct 18	49.21	-2.51	110.44	0.04	297.17	-2.04	216.59	29 Oct	277.9		-5.0		170.15		0.51		198.03		2.11	
								Nov 3		344.3		-1.0		175.37		0.92		203.51		2.20
Oct 28	48.40	-2.50	110.55	0.07	304.30	-1.83	226.63	8 Nov 13	44.6	104.2	3.9	4.8	180.72	186.20	1.28	1.59	210.52	218.17	1.93	1.48
Nov 7	47.59	-2.49	110.32	0.09	311.54	-1.62	236.72	18 Nov 23	168.0	241.5	0.5	-4.7	191.78	197.46	1.84	2.03	226.04	233.93	0.94	0.36
Nov 17	46.82	-2.47	109.77	0.12	318.87	-1.41	246.85	28 N-D 3	314.8	17.8	-3.1	2.1	203.21	209.03	2.17	2.26	241.81	249.66	-0.21	-0.75
Nov 27	46.14	-2.44	108.93	0.14	326.24	-1.20	257.00	8 Dec 13	77.2	138.3	5.0	2.7	214.91	220.83	2.31	2.30	257.52	265.40	-1.23	-1.64
Dec 7	45.57	-2.40	107.83	0.17	333.63	-1.00	267.18	18 Dec 23	205.2	279.9	-2.8	-4.7	226.80	232.81	2.26	2.17	273.35	281.37	-1.95	-2.13
Dec 17	45.15	-2.36	106.57	0.19	341.04	-0.81	277.38	28 Dec	349.5		0.0		238.85		2.05		289.43		-2.15	
1764								1765												1765
								Jan 2		50.1		4.4		244.92		1.89		297.41		-1.95
Dec 27	44.91	-2.31	105.23	0.21	348.43	-0.62	287.57	7 Jan 12	110.3	174.2	4.3	-0.6	251.01	257.12	1.71	1.51	304.97	311.42	-1.46	-0.61
Jan 6	44.85	-2.27	103.91	0.24	355.80	-0.45	297.75	17 Jan 22	244.0	317.0	-5.0	-2.6	263.24	269.38	1.29	1.05	315.44	315.42	0.63	2.11
Jan 16	44.98	-2.22	102.71	0.26	3.13	-0.28	307.92	27 J-F 1	22.4	82.1	3.0	5.2	275.52	281.68	0.81	0.56	311.06	305.09	3.32	3.65
Jan 26	45.30	-2.17	101.72	0.27	10.42	-0.13	318.06	6 Feb 11	144.5	212.1	2.0	-3.7	287.84	294.01	0.31	0.06	301.13	300.43	3.11	2.16
Feb 5	45.79	-2.12	101.00	0.29	17.65	0.01	328.16	16 Feb 21	283.1	352.1	-4.7	0.5	300.18	306.36	-0.18	-0.41	302.45	306.34	1.14	0.19
Feb 15	46.44	-2.08	100.58	0.30	24.82	0.14	338.21	26 Feb	54.2		4.9		312.54		-0.62		311.52		-0.62	
								Mar 3		114.3		4.2		318.71		-0.81		317.59		-1.28
Feb 25	47.24	-2.04	100.49	0.31	31.93	0.27	348.22	8 Mar 13	180.2	251.3	-1.2	-5.3	324.89	331.07	-0.99	-1.14	324.37	331.74	-1.79	-2.12
Mar 7	48.17	-2.00	100.73	0.32	38.98	0.38	358.17	18 Mar 23	321.0	25.7	-2.3	3.3	337.25	343.42	-1.27	-1.37	339.65	348.10	-2.28	-2.26
Mar 17	49.20	-1.97	101.27	0.33	45.97	0.48	8.06	28 M-A 2	85.9	147.6	5.2	1.7	349.60	355.77	-1.44	-1.49	357.11	6.69	-2.03	-1.59
Mar 27	50.32	-1.94	102.11	0.34	52.88	0.58	17.90	7 Apr 12	217.8	290.5	-4.1	-4.3	1.93	8.10	-1.50	-1.49	16.81	27.30	-0.94	-0.11
Apr 6	51.52	-1.92	103.20	0.35	59.74	0.66	27.68	17 Apr 22	356.9	58.4	1.0	4.9	14.26	20.42	-1.45	-1.39	37.76	47.57	0.80	1.64
Apr 16	52.76	-1.90	104.52	0.36	66.53	0.74	37.41	27 Apr	118.0		3.8		26.58		-1.30		56.14		2.26	
								May 2		182.6		-1.5		32.73		-1.19		63.10		2.57
Apr 26	54.04	-1.88	106.03	0.36	73.26	0.81	47.08	7 May 12	256.7	328.1	-5.1	-1.4	38.88	45.03	-1.05	-0.90	68.26	71.48	2.48	1.97
May 6	55.33	-1.87	107.72	0.37	79.94	0.87	56.72	17 May 22	31.1	90.9	3.7	4.8	51.17	57.32	-0.74	-0.56	72.68	71.93	1.03	-0.26
May 16	56.62	-1.87	109.55	0.38	86.57	0.92	66.32	27 M-J 1	151.4	219.8	1.0	-4.2	63.46	69.61	-0.37	-0.17	69.71	66.96	-1.70	-3.01
May 26	57.89	-1.87	111.50	0.39	93.15	0.97	75.89	6 Jun 11	295.5	3.6	-3.6	1.9	75.75	81.89	0.02	0.22	64.81	64.14	-3.90	-4.26
Jun 5	59.12	-1.87	113.54	0.40	99.69	1.01	85.44	16 Jun 21	64.2	123.8	5.0	3.0	88.03	94.18	0.42	0.60	65.34	68.42	-4.12	-3.59
Jun 15	60.30	-1.88	115.66	0.41	106.19	1.05	94.98	26 Jun	186.6		-2.3		100.32		0.78		73.28		-2.78	
								Jul 1		258.6		-5.0		106.47		0.94		79.79		-1.79
Jun 25	61.41	-1.89	117.83	0.42	112.65	1.08	104.51	6 Jul 11	332.8	37.2	-0.5	4.4	112.61	118.76	1.08	1.21	87.84	97.23	-0.73	0.27
Jul 5	62.42	-1.90	120.03	0.43	119.09	1.10	114.04	16 Jul 21	96.8	157.8	4.5	0.0	124.91	131.06	1.32	1.40	107.54	118.16	1.07	1.59
Jul 15	63.33	-1.92	122.24	0.45	125.50	1.13	123.60	26 Jul 31	224.1	297.5	-4.8	-3.4	137.20	143.35	1.45	1.48	128.54	138.39	1.78	1.69
Jul 25	64.11	-1.94	124.45	0.46	131.89	1.14	133.17	5 Aug 10	8.0	69.7	2.7	5.2	149.50	155.64	1.48	1.45	147.59	156.15	1.37	0.89
Aug 4	64.76	-1.96	126.64	0.48	138.27	1.15	142.78	15 Aug 20	129.6	193.6	2.5	-3.1	161.79	167.93	1.39	1.31	164.08	171.40	0.28	-0.42
Aug 14	65.25	-1.99	128.78	0.50	144.64	1.16	152.42	25 Aug 30	263.1	335.1	-5.1	-0.2	174.07	180.21	1.19	1.06	178.06	183.98	-1.16	-1.92
Aug 24	65.57	-2.02	130.85	0.52	151.00	1.16	162.11	4 Sep 9	41.4	101.5	4.7	4.4	186.34	192.46	0.89	0.71	188.97	192.72	-2.66	-3.31
Sep 3	65.71	-2.04	132.84	0.54	157.36	1.16	171.84	14 Sep 19	163.4	231.5	-0.5	-5.0	198.58	204.70	0.50	0.28	194.72	194.27	-3.79	-3.91
Sep 13	65.67	-2.07	134.72	0.56	163.71	1.15	181.64	24 Sep 29	302.3	10.9	-3.0	2.9	210.81	216.91	0.05	-0.20	190.91	185.45	-3.42	-2.18
Sep 23	65.46	-2.09	136.46	0.59	170.07	1.14	191.48	4 Oct 9	73.6	133.4	5.2	2.1	223.00	229.09	-0.45	-0.70	180.83	179.95	-0.50	0.94
Oct 3	65.07	-2.11	138.03	0.62	176.43	1.12	201.39	14 Oct 19	198.8	270.7	-3.4	-4.7	235.16	241.23	-0.96	-1.20	183.27	189.45	1.79	2.07
Oct 13	64.53	-2.13	139.42	0.65	182.80	1.10	211.35	24 Oct 29	340.3	45.1	0.4	4.7	247.29	253.33	-1.43	-1.65	197.04	205.16	1.95	1.60
Oct 23	63.86	-2.14	140.57	0.69	189.17	1.07	221.36	3 Nov 8	105.4	166.2	3.8	-1.0	259.35	265.36	-1.85	-2.02	213.38	221.53	1.11	0.57
Nov 2	63.09	-2.14	141.48	0.72	195.54	1.04	231.42	13 Nov 18	236.1	309.8	-5.0	-2.0	271.36	277.32	-2.17	-2.28	229.57	237.50	0.00	-0.55
Nov 12	62.28	-2.13	142.10	0.76	201.93	1.00	241.53	23 Nov 28	16.5	78.2	3.4	4.8	283.27	289.17	-2.35	-2.39	245.35	253.16	-1.06	-1.51
Nov 22	61.47	-2.11	142.41	0.80	208.31	0.95	251.67	3 Dec 8	137.5	200.8	1.2	-3.8	295.04	300.87	-2.38	-2.32	260.94	268.70	-1.88	-2.14
Dec 2	60.70	-2.09	142.40	0.84	214.70	0.90	261.84	13 Dec 18	274.6	347.2	-4.2	1.5	306.64	312.36	-2.21	-2.05	276.41	283.92	-2.25	-2.16
Dec 12	60.02	-2.06	142.06	0.88	221.09	0.83	272.03	23 Dec 28	50.9	110.7	5.0	3.2	318.00	323.55	-1.84	-1.57	290.90	296.62	-1.80	-1.06
1765								1765												1765

JULIAN 12 UT	SATURN		JUPITER		MARS		SUN	GREGORIAN 12 UT	MOON				VENUS				MERCURY			
	LONG.	LAT.	LONG.	LAT.	LONG.	LAT.	LONG.		LONG.		LAT.		LONG.		LAT.		LONG.		LAT.	
1765								1766												1766
Dec 22	59.46	-2.02	141.41	0.92	227.47	0.76	282.22	2 Jan 7	170.7	237.5	-1.9	-5.2	329.00	334.33	-1.24	-0.85	299.76	298.65	0.13	1.67
Jan 1	59.06	-1.98	140.49	0.96	233.85	0.68	292.42	12 Jan 17	313.0	22.7	-1.4	4.2	339.52	344.54	-0.41	0.10	293.30	287.23	3.02	3.49
Jan 11	58.84	-1.94	139.35	0.99	240.21	0.58	302.59	22 Jan 27	83.9	143.4	4.7	0.4	349.35	353.92	0.66	1.28	284.08	284.47	3.09	2.27
Jan 21	58.80	-1.89	138.07	1.02	246.56	0.47	312.74	1 Feb 6	205.8	275.9	-4.5	-4.2	358.20	2.14	1.96	2.69	287.43	292.05	1.36	0.48
Jan 31	58.95	-1.84	136.75	1.03	252.88	0.35	322.87	11 Feb 16	350.1	56.4	2.0	5.3	5.64	8.62	3.47	4.29	297.74	304.18	-0.30	-0.96
Feb 10	59.28	-1.80	135.48	1.04	259.17	0.20	332.95	21 Feb 26	116.1	177.1	2.8	-2.6	10.98	12.57	5.14	5.99	311.18	318.66	-1.50	-1.89
Feb 20	59.79	-1.75	134.35	1.04	265.41	0.04	342.97	3 Mar 8	243.1	314.7	-5.3	-1.2	13.28	13.00	6.81	7.54	326.60	335.01	-2.13	-2.19
Mar 2	60.46	-1.71	133.44	1.04	271.60	-0.15	352.96	13 Mar 18	25.7	88.7	4.4	4.5	11.70	9.45	8.09	8.37	343.89	353.28	-2.07	-1.75
Mar 12	61.27	-1.68	132.80	1.03	277.71	-0.36	2.88	23 Mar 28	148.3	212.6	0.0	-4.7	6.53	3.39	8.30	7.84	3.15	13.36	-1.20	-0.44
Mar 22	62.21	-1.64	132.47	1.02	283.73	-0.61	12.74	2 Apr 7	282.1	352.7	-3.7	2.2	0.55	358.42	7.05	6.04	23.54	33.05	0.47	1.41
Apr 1	63.26	-1.61	132.44	1.00	289.62	-0.89	22.56	12 Apr 17	59.7	120.2	5.1	2.2	357.26	357.10	4.91	3.78	41.16	47.36	2.22	2.73
Apr 11	64.39	-1.58	132.73	0.99	295.35	-1.20	32.31	22 Apr 27	181.4	250.1	-2.9	-4.9	357.89	359.51	2.70	1.72	51.34	52.99	2.84	2.45
Apr 21	65.59	-1.56	133.31	0.97	300.88	-1.56	42.01	2 May 7	321.3	29.4	-0.3	4.5	1.85	4.77	0.85	0.09	52.40	50.11	1.57	0.27
May 1	66.85	-1.54	134.16	0.95	306.14	-1.97	51.67	12 May 17	92.5	152.0	3.9	-0.8	8.18	11.98	-0.57	-1.12	47.17	44.82	-1.17	-2.45
May 11	68.13	-1.53	135.26	0.94	311.04	-2.43	61.29	22 May 27	216.3	289.0	-4.8	-2.8	16.10	20.49	-1.58	-1.96	43.95	44.91	-3.33	-3.76
May 21	69.43	-1.51	136.57	0.93	315.49	-2.95	70.87	1 Jun 6	359.6	64.3	3.1	4.9	25.10	29.90	-2.25	-2.48	47.68	52.07	-3.77	-3.44
May 31	70.73	-1.51	138.07	0.92	319.35	-3.53	80.44	11 Jun 16	124.6	184.7	1.4	-3.6	34.85	39.94	-2.63	-2.72	57.91	65.09	-2.83	-2.01
Jun 10	72.00	-1.50	139.73	0.91	322.45	-4.17	89.98	21 Jun 26	253.2	327.9	-4.7	0.7	45.14	50.44	-2.76	-2.74	73.54	83.15	-1.06	-0.09
Jun 20	73.24	-1.50	141.54	0.90	324.61	-4.84	99.51	1 Jul 6	36.1	97.8	5.0	3.4	55.82	61.29	-2.67	-2.56	93.61	104.43	0.78	1.43
Jun 30	74.42	-1.50	143.45	0.90	325.61	-5.53	109.05	11 Jul 16	156.9	219.4	-1.6	-5.2	66.82	72.41	-2.42	-2.24	115.03	125.06	1.77	1.80
Jul 10	75.53	-1.50	145.46	0.89	325.35	-6.16	118.59	21 Jul 26	291.6	5.7	-2.4	3.8	78.06	83.75	-2.03	-1.80	134.39	142.99	1.58	1.15
Jul 20	76.55	-1.51	147.54	0.89	323.87	-6.64	128.15	31 J-A 5	70.7	130.3	4.9	0.8	89.50	95.29	-1.55	-1.29	150.87	158.02	0.55	-0.16
Jul 30	77.47	-1.52	149.67	0.90	321.50	-6.87	137.74	10 Aug 15	190.3	256.1	-4.1	-4.7	101.12	106.99	-1.02	-0.75	164.41	169.89	-0.96	-1.82
Aug 9	78.25	-1.53	151.83	0.90	318.90	-6.76	147.36	20 Aug 25	330.2	41.7	1.0	5.2	112.90	118.84	-0.47	-0.20	174.26	177.17	-2.67	-3.46
Aug 19	78.90	-1.55	154.00	0.91	316.81	-6.35	157.03	30 Aug	103.6		3.1		124.82		0.06		178.12		-4.08	
								Sep 4		162.9		-2.1		130.83		0.31		176.60		-4.30
Aug 29	79.39	-1.56	156.16	0.92	315.77	-5.73	166.74	9 Sep 14	225.5	294.4	-5.2	-2.2	136.86	142.93	0.55	0.76	172.60	167.52	-3.88	-2.70
Sep 8	79.70	-1.58	158.30	0.94	316.05	-5.01	176.50	19 Sep 24	8.1	75.7	3.9	4.6	149.02	155.14	0.96	1.12	164.10	164.41	-1.07	0.42
Sep 18	79.84	-1.59	160.38	0.96	317.61	-4.27	186.32	29 S-O 4	135.6	196.6	0.2	-4.4	161.28	167.45	1.27	1.38	168.58	175.40	1.43	1.89
Sep 28	79.80	-1.61	162.40	0.98	320.29	-3.57	196.19	9 Oct 14	262.8	333.3	-4.1	1.4	173.63	179.83	1.47	1.52	183.54	192.12	1.93	1.69
Oct 8	79.57	-1.62	164.32	1.00	323.91	-2.93	206.12	19 Oct 24	44.4	108.1	5.0	2.4	186.05	192.28	1.55	1.54	200.69	209.11	1.27	0.76
Oct 18	79.17	-1.63	166.12	1.03	328.25	-2.36	216.11	29 Oct	167.5		-2.7		198.52		1.51		217.32		0.21	
								Nov 3		231.9		-5.0		204.78		1.44		225.34		-0.35
Oct 28	78.61	-1.64	167.77	1.06	333.16	-1.85	226.15	8 Nov 13	301.5	11.7	-1.1	4.2	211.04	217.31	1.36	1.24	233.20	240.93	-0.89	-1.38
Nov 7	77.93	-1.64	169.24	1.09	338.51	-1.41	236.23	18 Nov 23	78.9	139.6	4.1	-0.6	223.59	229.87	1.11	0.96	248.56	256.07	-1.81	-2.14
Nov 17	77.16	-1.63	170.50	1.13	344.19	-1.02	246.35	28 N-D 3	200.3	268.8	-4.7	-3.5	236.15	242.44	0.79	0.60	263.42	270.46	-2.34	-2.36
Nov 27	76.34	-1.62	171.53	1.17	350.12	-0.69	256.51	8 Dec 13	340.6	48.7	2.5	5.1	248.73	255.03	0.41	0.21	276.82	281.78	-2.11	-1.50
Dec 7	75.53	-1.60	172.29	1.22	356.24	-0.39	266.69	18 Dec 23	111.9	171.2	1.7	-3.4	261.32	267.61	0.01	-0.19	283.99	281.87	-0.40	1.15
Dec 17	74.76	-1.58	172.76	1.26	2.50	-0.14	276.88	28 Dec	234.5		-5.0		273.90		-0.38		275.81		2.62	
1766								1767												1767
								Jan 2		307.0		-0.3		280.20		-0.57		269.96		3.26
Dec 27	74.08	-1.55	172.92	1.31	8.86	0.08	287.08	7 Jan 12	18.9	83.8	4.9	3.8	286.49	292.78	-0.75	-0.91	267.75	269.16	3.02	2.34
Jan 6	73.53	-1.51	172.76	1.36	15.29	0.27	297.26	17 Jan 22	144.0	203.8	-1.3	-5.1	299.06	305.34	-1.06	-1.18	272.97	278.23	1.53	0.73
Jan 16	73.14	-1.48	172.28	1.40	21.77	0.44	307.42	27 J-F 1	270.5	345.5	-3.4	3.1	311.62	317.90	-1.29	-1.37	284.37	291.10	-0.02	-0.68
Jan 26	72.93	-1.44	171.51	1.44	28.27	0.58	317.56	6 Feb 11	55.5	117.2	5.1	1.2	324.16	330.43	-1.42	-1.45	298.28	305.85	-1.23	-1.66
Feb 5	72.91	-1.40	170.50	1.47	34.78	0.70	327.66	16 Feb 21	176.4	238.1	-3.8	-5.0	336.68	342.93	-1.46	-1.43	313.79	322.11	-1.97	-2.12
Feb 15	73.07	-1.36	169.31	1.50	41.30	0.80	337.72	26 Feb	308.1		-0.2		349.17		-1.38		330.85		-2.09	
								Mar 3		23.3		5.0		355.40		-1.31		340.02		-1.88
Feb 25	73.42	-1.32	168.03	1.51	47.81	0.89	347.73	8 Mar 13	90.0	149.6	3.3	-1.8	1.62	7.83	-1.20	-1.08	349.60	359.48	-1.44	-0.76
Mar 7	73.94	-1.28	166.74	1.52	54.31	0.96	357.68	18 Mar 23	209.9	274.5	-5.1	-2.9	14.02	20.21	-0.93	-0.76	9.31	18.44	0.12	1.13
Mar 17	74.62	-1.25	165.54	1.51	60.78	1.03	7.58	28 M-A 2	346.7	59.6	3.2	4.7	26.38	32.54	-0.57	-0.37	26.01	31.30	2.09	2.82
Mar 27	75.44	-1.21	164.51	1.49	67.24	1.08	17.42	7 Apr 12	122.7	182.1	0.4	-4.1	38.68	44.81	-0.15	0.07	33.89	33.75	3.14	2.91
Apr 6	76.39	-1.18	163.71	1.47	73.68	1.12	27.20	17 Apr 22	244.9	312.7	-4.4	0.6	50.92	57.01	0.29	0.52	31.41	28.06	2.08	0.80
Apr 16	77.44	-1.16	163.19	1.44	80.09	1.15	36.93	27 Apr	25.2		4.9		63.09		0.75		25.18		-0.61	
								May 2		94.0		2.6		69.15		0.96		23.85		-1.84
Apr 26	78.57	-1.13	162.96	1.40	86.48	1.17	46.62	7 May 12	154.5	215.3	-2.5	-5.0	75.19	81.21	1.17	1.36	24.46	26.91	-2.73	-3.23
May 6	79.78	-1.11	163.04	1.37	92.85	1.19	56.25	17 May 22	281.8	351.9	-1.8	3.8	87.21	93.19	1.54	1.69	30.95	36.34	-3.39	-3.23
May 16	81.03	-1.09	163.41	1.33	99.21	1.20	65.86	27 M-J 1	62.4	126.7	4.4	-0.4	99.14	105.07	1.81	1.91	42.93	50.62	-2.81	-2.17
May 26	82.32	-1.07	164.07	1.30	105.55	1.21	75.43	6 Jun 11	186.3	250.0	-4.6	-4.0	110.97	116.84	1.97	2.00	59.37	69.12	-1.35	-0.43
Jun 5	83.62	-1.06	164.98	1.27	111.88	1.21	84.98	16 Jun 21	320.2	30.9	1.7	5.2	122.67	128.47	1.99	1.94	79.67	90.60	0.47	1.22
Jun 15	84.92	-1.05	166.12	1.24	118.20	1.20	94.51	26 Jun	97.6		2.1		134.24		1.85		101.35		1.70	
								Jul 1		158.6		-3.2		139.95		1.72		111.52		1.88
Jun 25	86.20	-1.04	167.47	1.21	124.52	1.19	104.05	6 Jul 11	218.9	286.4	-5.2	-1.3	145.62	151.23	1.54	1.32	120.92	129.50	1.77	1.42
Jul 5	87.44	-1.03	169.01	1.18	130.84	1.18	113.58	16 Jul 21	359.4	68.3	4.5	4.2	156.78	162.27	1.06	0.75	137.26	144.17	0.86	0.13
Jul 15	88.63	-1.03	170.69	1.16	137.16	1.16	123.14	26 Jul 31	131.1	190.5	-0.9	-5.0	167.67	172.99	0.40	0.00	150.16	155.08	-0.73	-1.67
Jul 25	89.74	-1.03	172.51	1.15	143.49	1.14	132.71	5 Aug 10	252.8	324.4	-3.9	2.1	178.21	183.30	-0.43	-0.90	158.68	160.60	-2.65	-3.59
Aug 4	90.76	-1.02	174.44	1.13	149.83	1.11	142.31	15 Aug 20	38.1	103.6	5.2	1.6	188.27	193.07	-1.41	-1.95	160.42	157.92	-4.33	-4.64
Aug 14	91.67	-1.03	176.46	1.12	156.20	1.08	151.96	25 Aug 30	163.7	223.4	-3.6	-5.0	197.68	202.07	-2.52	-3.11	153.66	149.41	-4.26	-3.13
Aug 24	92.46	-1.03	178.54	1.11	162.58	1.04	161.64	4 Sep 9	288.6	3.2	-1.0	4.6	206.19	209.99	-3.72	-4.33	147.40	148.93	-1.58	-0.08
Sep 3	93.11	-1.03	180.67	1.11	168.99	1.00	171.37	14 Sep 19	74.9	137.0	3.6	-1.5	213.38	216.30	-4.95	-5.56	153.91	161.32	1.03	1.67
Sep 13	93.59	-1.03	182.83	1.11	175.42	0.96	181.17	24 Sep 29	196.3	257.7	-4.9	-3.3	218.64	220.27	-6.13	-6.63	169.97	179.00	1.87	1.75
Sep 23	93.90	-1.04	184.99	1.11	181.89	0.91	191.01	4 Oct 9	326.1	41.5	2.4	4.9	221.07	220.93	-7.03	-7.28	187.94	196.62	1.42	0.96
Oct 3	94.03	-1.04	187.14	1.12	188.39	0.85	200.91	14 Oct 19	109.5	169.4	0.6	-4.0	219.79	217.72	-7.30	-7.02	205.02	213.14	0.42	-0.14
Oct 13	93.98	-1.05	189.25	1.13	194.94	0.80	210.87	24 Oct 29	229.7	293.8	-4.6	-0.1	214.97	211.95	-6.41	-5.48	221.02	228.69	-0.70	-1.24
Oct 23	93.74	-1.05	191.30	1.14	201.51	0.73	220.88	3 Nov 8	4.9	78.1	4.7	3.0	209.15	207.01	-4.33	-3.06	236.16	243.43	-1.72	-2.12
Nov 2	93.33	-1.05	193.27	1.16	208.13	0.66	230.94	13 Nov 18	142.2	201.7	-2.4	-5.1	205.81	205.64	-1.81	-0.66	250.42	256.97	-2.42	-2.54
Nov 12	92.76	-1.04	195.13	1.18	214.80	0.59	241.04	23 Nov 28	264.4	331.9	-2.4	3.3	206.45	208.14	0.35	1.20	262.69	266.84	-2.42	-1.94
Nov 22	92.07	-1.03	196.86	1.20	221.51	0.51	251.18	3 Dec 8	43.6	112.7	4.8	-0.1	210.57	213.61	1.89	2.44	268.10	265.03	-0.96	0.57
Dec 2	91.29	-1.02	198.42	1.23	228.26	0.42	261.35	13 Dec 18	173.9	234.4	-4.6	-4.4	217.17	221.16	2.86	3.17	258.55	253.18	2.13	2.95
Dec 12	90.47	-1.01	199.79	1.26	235.05	0.32	271.54	23 Dec 28	300.8	11.2	0.8	5.2	225.48	230.09	3.37	3.48	251.94	254.32	2.90	2.38
1767								1767												1767

JULIAN 12 UT	SATURN LONG	SATURN LAT	JUPITER LONG	JUPITER LAT	MARS LONG	MARS LAT	SUN LONG	GREGORIAN 12 UT	MOON LONG · LAT	VENUS LONG · LAT	MERCURY LONG · LAT
1767								**1768**			**1768**
Dec 22	89.65	-0.99	200.93	1.30	241.89	0.22	281.73	2 Jan 7	81.0 145.6 2.7 -3.0	234.94 239.97 3.51 3.48	258.91 264.73 1.68 0.94
Jan 1	88.88	-0.97	201.83	1.34	248.78	0.11	291.92	12 Jan 17	205.4 268.2 -5.3 -2.1	245.17 250.50 3.38 3.22	271.28 278.28 0.23 -0.41
Jan 11	88.21	-0.94	202.44	1.38	255.71	-0.01	302.10	22 Jan 27	338.8 50.2 3.9 4.6	255.95 261.49 3.03 2.79	285.62 293.25 -0.98 -1.45
Jan 21	87.68	-0.91	202.76	1.42	262.68	-0.14	312.26	1 Feb 6	116.4 177.8 -0.4 -4.8	267.11 272.80 2.52 2.23	301.17 309.40 -1.80 -2.02
Jan 31	87.30	-0.88	202.76	1.46	269.69	-0.28	322.38	11 Feb 16	237.6 303.6 -4.2 1.1	278.54 284.34 1.92 1.59	317.97 326.90 -2.09 -1.97
Feb 10	87.10	-0.85	202.45	1.50	276.74	-0.42	332.46	21 Feb 26	17.7 87.6 5.1 2.0	290.17 296.05 1.26 0.93	336.17 345.68 -1.64 -1.07
Feb 20	87.09	-0.82	201.84	1.53	283.83	-0.58	342.49	2 Mar 7	150.2 210.0 -3.3 -5.0	301.95 307.87 0.61 0.29	355.10 3.77 -0.24 0.79
Mar 1	87.26	-0.79	200.96	1.56	290.94	-0.74	352.47	12 Mar 17	271.0 340.9 -1.6 4.0	313.82 319.80 -0.02 -0.30	10.72 14.99 1.88 2.83
Mar 11	87.62	-0.76	199.88	1.58	298.09	-0.92	2.40	22 Mar 27	56.2 122.8 4.0 -1.3	325.78 331.78 -0.57 -0.81	16.05 14.10 3.37 3.29
Mar 21	88.14	-0.73	198.65	1.59	305.25	-1.10	12.27	1 Apr 6	183.2 242.9 -4.8 -3.5	337.79 343.81 -1.02 -1.21	10.36 6.66 2.52 1.28
Mar 31	88.83	-0.70	197.37	1.59	312.42	-1.29	22.09	11 Apr 16	306.4 19.6 1.7 5.0	349.84 355.88 -1.37 -1.49	4.51 4.48 -0.08 -1.26
Apr 10	89.65	-0.68	196.13	1.57	319.59	-1.48	31.85	21 Apr 26	93.1 156.3 1.1 -3.9	1.93 7.98 -1.58 -1.64	6.45 10.10 -2.15 -2.72
Apr 20	90.60	-0.65	195.01	1.55	326.75	-1.67	41.55	1 May 6	215.9 277.1 -4.7 -0.6	14.03 20.09 -1.67 -1.67	15.11 21.24 -3.00 -3.00
Apr 30	91.65	-0.63	194.08	1.52	333.87	-1.87	51.21	11 May 16	344.0 58.3 4.4 3.6	26.15 32.22 -1.63 -1.57	28.36 36.40 -2.75 -2.27
May 10	92.79	-0.61	193.39	1.48	340.93	-2.07	60.84	21 May 26	127.8 188.8 -2.2 -5.2	38.29 44.37 -1.48 -1.37	45.34 55.16 -1.58 -0.75
May 20	93.99	-0.59	192.99	1.44	347.91	-2.26	70.42	31 M-J 5	248.9 313.1 -2.9 2.7	50.44 56.53 -1.23 -1.08	65.73 76.69 0.14 0.96
May 30	95.24	-0.57	192.89	1.40	354.77	-2.44	79.98	10 Jun 15	23.2 95.6 5.1 0.6	62.62 68.71 -0.90 -0.72	87.53 97.77 1.58 1.92
Jun 9	96.53	-0.56	193.08	1.35	1.47	-2.62	89.52	20 Jun 25	160.9 220.8 -4.5 -4.6	74.81 80.92 -0.52 -0.32	107.17 115.65 1.95 1.69
Jun 19	97.83	-0.54	193.57	1.31	7.97	-2.78	99.05	30 Jun	283.1 0.1	87.03 -0.12	123.18 1.18
								Jul 5	351.2 4.9	93.15 0.09	129.72 0.46
Jun 29	99.13	-0.53	194.32	1.26	14.20	-2.93	108.59	10 Jul 15	62.4 130.9 3.4 -2.5	99.28 105.42 0.29 0.48	135.16 139.34 -0.44 -1.48
Jul 9	100.40	-0.52	195.33	1.22	20.10	-3.06	118.13	20 Jul 25	193.0 253.1 -5.3 -2.6	111.57 117.72 0.66 0.83	141.97 142.73 -2.59 -3.67
Jul 19	101.64	-0.50	196.55	1.19	25.56	-3.17	127.69	30 J-A 4	319.2 30.7 3.1 4.9	123.88 130.05 0.98 1.11	141.40 138.19 -4.52 -4.87
Jul 29	102.83	-0.49	197.98	1.15	30.48	-3.24	137.27	9 Aug 14	100.1 164.7 0.2 -4.6	136.22 142.41 1.22 1.31	134.21 131.22 -4.51 -3.45
Aug 8	103.94	-0.48	199.58	1.12	34.71	-3.29	146.90	19 Aug 24	224.7 286.5 -4.3 0.5	148.60 154.79 1.37 1.41	130.78 133.54 -2.01 -0.56
Aug 18	104.96	-0.48	201.34	1.10	38.08	-3.28	156.56	29 Aug	357.1 4.9	160.99 1.42	139.28 0.62
								Sep 3	69.7 2.6	167.20 1.40	147.22 1.41
Aug 28	105.87	-0.47	203.22	1.07	40.38	-3.22	166.26	8 Sep 13	135.9 197.4 -2.9 -5.0	173.41 179.62 1.35 1.28	156.54 165.78 1.77 1.78
Sep 7	106.66	-0.46	205.20	1.05	41.38	-3.07	176.03	18 Sep 23	256.9 321.7 -1.9 3.4	185.84 192.06 1.19 1.07	175.08 184.04 1.55 1.14
Sep 17	107.30	-0.45	207.26	1.04	40.95	-2.82	185.84	28 S-O 3	36.2 107.2 4.4 -0.8	198.28 204.50 0.93 0.76	192.63 200.86 0.64 0.07
Sep 27	107.77	-0.44	209.39	1.02	39.08	-2.43	195.71	8 Oct 13	170.2 229.8 -4.8 -3.6	210.72 216.94 0.59 0.39	208.77 216.38 -0.51 -1.08
Oct 7	108.08	-0.44	211.56	1.01	36.10	-1.92	205.64	18 Oct 23	290.2 358.9 1.3 5.0	223.16 229.38 0.19 -0.02	223.71 230.73 -1.62 -2.10
Oct 17	108.20	-0.43	213.74	1.01	32.68	-1.33	215.62	28 Oct	74.7 1.8	235.60 -0.24	237.35 -2.48
								Nov 2	142.5 -3.8	241.82 -0.46	243.39 -2.71
Oct 27	108.13	-0.42	215.93	1.00	29.64	-0.72	225.66	7 Nov 12	203.2 262.6 -4.8 -1.0	248.03 254.25 -0.67 -0.88	248.44 251.72 -2.72 -2.38
Nov 6	107.88	-0.41	218.09	1.00	27.63	-0.16	235.74	17 Nov 22	325.5 37.4 4.1 4.3	260.46 266.66 -1.07 -1.25	251.99 248.09 -1.52 -0.06
Nov 16	107.45	-0.40	220.20	1.01	26.94	0.31	245.86	27 N-D 2	111.5 176.1 -1.6 -5.2	272.87 279.06 -1.41 -1.56	241.47 236.77 1.58 2.57
Nov 26	106.88	-0.38	222.24	1.01	27.57	0.68	256.02	7 Dec 12	235.6 296.6 -3.2 2.1	285.26 291.44 -1.67 -1.76	236.52 239.21 2.73 2.38
Dec 6	106.17	-0.37	224.18	1.02	29.35	0.97	266.20	17 Dec 22	2.9 75.8 5.3 1.6	297.61 303.77 -1.82 -1.85	245.11 251.46 1.79 1.13
Dec 16	105.39	-0.35	226.00	1.03	32.09	1.18	276.39	27 Dec	146.4 -4.2	309.92 -1.84	258.38 0.46
1768								**1769**			**1769**
								Jan 1	208.3 -4.8	316.05 -1.80	265.64 -0.17
Dec 26	104.57	-0.33	227.67	1.05	35.57	1.33	286.58	6 Jan 11	268.1 332.5 -0.5 4.5	322.16 328.25 -1.72 -1.60	273.12 280.82 -0.74 -1.24
Jan 5	103.75	-0.32	229.17	1.07	39.63	1.44	296.77	16 Jan 21	41.9 113.2 4.1 -1.8	334.32 340.35 -1.45 -1.26	288.71 296.84 -1.64 -1.92
Jan 15	103.00	-0.30	230.45	1.09	44.14	1.52	306.93	26 Jan 31	179.8 239.8 -5.2 -2.8	346.35 352.30 -1.03 -0.78	305.23 313.91 -2.06 -2.04
Jan 25	102.34	-0.28	231.50	1.11	49.00	1.57	317.08	5 Feb 10	301.6 10.3 2.5 5.1	358.22 4.08 -0.49 -0.17	322.85 331.96 -1.82 -1.35
Feb 4	101.81	-0.25	232.28	1.13	54.13	1.60	327.18	15 Feb 20	81.0 149.2 1.0 -4.3	9.89 15.62 0.17 0.53	340.92 349.07 -0.61 0.41
Feb 14	101.45	-0.23	232.78	1.16	59.46	1.62	337.24	25 Feb	212.0 -4.3	21.29 0.91	355.33 1.60
								Mar 2	271.6 0.1	26.87 1.31	358.50 2.74
Feb 24	101.26	-0.21	232.97	1.18	64.96	1.62	347.25	7 Mar 12	336.9 49.4 4.6 3.2	32.35 37.72 1.71 2.11	357.96 354.39 3.50 3.55
Mar 6	101.26	-0.19	232.85	1.20	70.59	1.61	357.21	17 Mar 22	119.0 183.6 -2.5 -5.0	42.97 48.07 2.52 2.91	349.82 346.46 2.84 1.65
Mar 16	101.44	-0.17	232.42	1.22	76.32	1.60	7.10	27 M-A 1	243.9 304.5 -2.0 3.1	52.99 57.72 3.28 3.64	345.41 346.64 0.38 -0.74
Mar 26	101.80	-0.15	231.71	1.24	82.13	1.58	16.95	6 Apr 11	14.2 88.2 4.8 -0.2	62.21 66.43 3.96 4.23	349.76 354.33 -1.63 -2.25
Apr 5	102.33	-0.13	230.75	1.25	88.00	1.55	26.74	16 Apr 21	155.3 216.8 -4.8 -3.8	70.30 73.76 4.46 4.62	0.04 6.69 -2.62 -2.75
Apr 15	103.02	-0.12	229.61	1.25	93.92	1.52	36.47	26 Apr	276.0 1.0	76.73 4.69	14.16 -2.65
								May 1	339.3 5.0	79.11 4.67	22.43 -2.31
Apr 25	103.84	-0.10	228.36	1.24	99.89	1.49	46.15	6 May 11	52.7 125.7 2.7 -3.4	80.79 81.63 4.51 4.19	31.48 41.30 -1.77 -1.04
May 5	104.78	-0.08	227.10	1.23	105.89	1.45	55.79	16 May 21	189.7 249.3 -5.0 -1.3	81.52 80.41 3.69 2.98	51.81 62.73 -0.19 0.67
May 15	105.83	-0.07	225.90	1.20	111.93	1.41	65.39	26 May 31	309.5 16.4 3.8 4.9	78.34 75.56 2.06 0.97	73.57 83.81 1.41 1.91
May 25	106.96	-0.05	224.84	1.17	118.00	1.36	74.97	5 Jun 10	91.3 161.3 -0.7 -5.2	72.46 69.51 -0.20 -1.35	93.13 101.40 2.09 1.96
Jun 4	108.16	-0.04	224.00	1.14	124.11	1.31	84.52	15 Jun 20	222.7 282.1 -3.5 1.6	67.15 65.67 -2.36 -3.19	108.56 114.54 1.54 0.84
Jun 14	109.41	-0.02	223.42	1.09	130.24	1.26	94.05	25 Jun 30	344.9 54.9 5.2 2.6	65.17 65.64 -3.80 -4.23	119.23 122.41 -0.09 -1.22
Jun 24	110.69	-0.01	223.13	1.05	136.41	1.21	103.59	5 Jul 10	128.8 195.1 -3.7 -4.9	66.98 69.05 -4.48 -4.60	123.83 123.32 -2.46 -3.66
Jul 4	111.99	0.01	223.15	1.01	142.61	1.15	113.12	15 Jul 20	254.8 316.0 -0.8 4.1	71.75 74.95 -4.60 -4.51	120.99 117.61 -4.58 -4.95
Jul 14	113.28	0.02	223.46	0.96	148.85	1.09	122.67	25 Jul 30	22.4 93.8 4.5 -0.9	78.58 82.57 -4.34 -4.11	114.52 113.14 -4.60 -3.65
Jul 24	114.56	0.04	224.05	0.92	155.14	1.03	132.25	4 Aug 9	164.7 227.5 -5.1 -2.9	86.85 91.38 -3.84 -3.52	114.30 118.23 -2.36 -1.00
Aug 3	115.79	0.05	224.92	0.88	161.46	0.97	141.85	14 Aug 19	287.1 351.8 2.1 5.1	96.12 101.03 -3.18 -2.81	124.69 133.10 0.20 1.10
Aug 13	116.97	0.07	226.03	0.85	167.83	0.90	151.48	24 Aug 29	61.3 132.0 1.7 -3.9	106.10 111.31 -2.43 -2.04	142.62 152.46 1.62 1.78
Aug 23	118.08	0.08	227.37	0.81	174.25	0.83	161.17	3 Sep 8	198.8 259.0 -4.4 -0.1	116.64 122.07 -1.65 -1.26	162.10 171.34 1.65 1.32
Sep 2	119.10	0.10	228.90	0.78	180.72	0.76	170.90	13 Sep 18	320.4 29.5 4.4 3.8	127.59 133.20 -0.88 -0.51	180.12 188.47 0.85 0.30
Sep 12	120.00	0.12	230.60	0.75	187.25	0.68	180.68	23 Sep 28	100.5 168.6 -1.9 -5.0	138.88 144.64 -0.15 0.18	196.41 203.97 -0.30 -0.91
Sep 22	120.78	0.13	232.44	0.73	193.84	0.61	190.53	3 Oct 8	231.4 290.7 -2.2 2.8	150.45 156.33 0.49 0.76	211.15 217.91 -1.50 -2.06
Oct 2	121.40	0.15	234.41	0.70	200.50	0.52	200.43	13 Oct 18	355.4 68.2 5.0 0.7	162.25 168.22 1.01 1.23	224.15 229.65 -2.53 -2.88
Oct 12	121.87	0.17	236.48	0.68	207.21	0.44	210.38	23 Oct 28	138.7 203.3 -4.6 -4.1	174.23 180.28 1.41 1.56	233.97 236.34 -3.02 -2.82
Oct 22	122.16	0.19	238.62	0.66	213.99	0.35	220.39	2 Nov 7	263.4 323.5 0.7 4.9	186.36 192.47 1.67 1.75	235.60 231.00 -2.08 -0.70
Nov 1	122.27	0.21	240.82	0.65	220.84	0.26	230.45	12 Nov 17	32.1 106.9 3.6 -2.8	198.61 204.77 1.78 1.79	224.52 220.66 0.98 2.14
Nov 11	122.18	0.23	243.05	0.63	227.76	0.16	240.55	22 Nov 27	175.2 236.5 -5.2 -1.7	210.95 217.15 1.76 1.69	221.38 225.50 2.52 2.34
Nov 21	121.92	0.25	245.30	0.62	234.76	0.06	250.69	2 Dec 7	295.6 358.0 3.5 5.2	223.37 229.59 1.60 1.48	231.47 238.33 1.88 1.29
Dec 1	121.48	0.27	247.53	0.61	241.82	-0.04	260.86	12 Dec 17	70.0 144.5 0.5 -5.0	235.83 242.07 1.34 1.17	245.62 253.12 0.67 0.06
Dec 11	120.89	0.30	249.72	0.60	248.95	-0.15	271.04	22 Dec 27	209.5 268.9 -3.8 1.3	248.32 254.58 0.99 0.80	260.75 268.50 -0.52 -1.03
1769								**1769**			

JULIAN 12 UT	SATURN LONG.	SATURN LAT.	JUPITER LONG.	JUPITER LAT.	MARS LONG.	MARS LAT.	SUN LONG.	GREGORIAN 12 UT	MOON LONG.		MOON LAT.		VENUS LONG.		VENUS LAT.		MERCURY LONG.		MERCURY LAT.	
1769								1770												1770
Dec 21	120.18	0.32	251.86	0.59	256.15	-0.26	281.24	1 Jan 6	329.1	34.5	5.0	3.4	260.84	267.10	0.59	0.38	276.38	284.41	-1.47	-1.81
Dec 31	119.40	0.34	253.91	0.59	263.41	-0.37	291.43	11 Jan 16	108.3	180.2	-2.9	-5.0	273.37	279.63	0.16	-0.05	292.62	301.04	-2.02	-2.09
Jan 10	118.58	0.35	255.86	0.59	270.74	-0.48	301.61	21 Jan 26	242.2	301.5	-1.1	3.8	285.90	292.16	-0.26	-0.46	309.64	318.32	-1.97	-1.61
Jan 20	117.77	0.37	257.67	0.58	278.14	-0.60	311.77	31 J-F 5	4.4	72.7	4.7	0.0	298.43	304.69	-0.65	-0.82	326.79	334.36	-0.97	0.00
Jan 30	117.02	0.39	259.32	0.58	285.59	-0.71	321.89	10 Feb 15	146.1	214.0	-4.8	-3.1	310.94	317.20	-0.98	-1.11	339.87	341.90	1.23	2.54
Feb 9	116.38	0.40	260.77	0.58	293.09	-0.83	331.97	20 Feb 25	274.0	335.1	1.9	5.0	323.45	329.70	-1.23	-1.32	339.83	334.98	3.50	3.69
Feb 19	115.86	0.41	262.02	0.58	300.63	-0.94	342.01	2 Mar 7	41.5	111.7	2.4	-3.4	335.94	342.18	-1.38	-1.42	330.20	327.65	3.04	1.94
Mar 1	115.51	0.42	263.01	0.58	308.22	-1.05	351.99	12 Mar 17	182.6	246.3	-4.6	-0.2	348.41	354.63	-1.44	-1.42	327.78	330.14	0.75	-0.31
Mar 11	115.33	0.43	263.74	0.58	315.83	-1.16	1.92	22 Mar 27	305.8	10.2	4.3	4.3	0.85	7.07	-1.38	-1.32	334.17	339.43	-1.17	-1.83
Mar 21	115.34	0.44	264.17	0.59	323.46	-1.26	11.80	1 Apr 6	80.1	150.4	-1.2	-5.1	13.27	19.47	-1.23	-1.11	345.64	352.64	-2.27	-2.50
Mar 31	115.53	0.45	264.29	0.58	331.10	-1.35	21.61	11 Apr 16	217.2	277.9	-2.5	2.7	25.66	31.84	-0.98	-0.83	0.34	8.73	-2.52	-2.32
Apr 10	115.90	0.46	264.10	0.58	338.74	-1.43	31.37	21 Apr 26	338.6	47.0	5.2	1.7	38.02	44.19	-0.66	-0.47	17.79	27.55	-1.91	-1.30
Apr 20	116.43	0.46	263.61	0.58	346.35	-1.50	41.08	1 May 6	119.3	187.6	-4.2	-4.5	50.35	56.50	-0.28	-0.08	37.94	48.75	-0.52	0.36
Apr 30	117.11	0.47	262.84	0.57	353.94	-1.55	50.74	11 May 16	250.3	309.8	0.4	4.8	62.64	68.78	0.12	0.33	59.51	69.67	1.19	1.83
May 10	117.93	0.48	261.84	0.56	1.47	-1.59	60.36	21 May 26	13.0	85.2	4.2	-1.8	74.91	81.04	0.53	0.73	78.81	86.73	2.19	2.22
May 20	118.87	0.49	260.67	0.54	8.94	-1.62	69.95	31 M-J 5	157.8	222.6	-5.3	-2.1	87.15	93.26	0.91	1.08	93.34	98.53	1.91	1.27
May 30	119.92	0.50	259.41	0.52	16.33	-1.63	79.51	10 Jun 15	282.7	342.7	3.2	5.2	99.36	105.45	1.24	1.37	102.17	104.05	0.32	-0.88
Jun 9	121.04	0.51	258.14	0.50	23.61	-1.61	89.05	20 Jun 25	49.2	124.1	1.5	-4.5	111.54	117.62	1.48	1.57	104.07	102.33	-2.23	-3.52
Jun 19	122.23	0.52	256.97	0.47	30.76	-1.58	98.59	30 Jun	194.4		-4.1		123.68		1.63		99.48		-4.46	
								Jul 5		256.0		1.0		129.74		1.66		96.62		-4.82
Jun 29	123.48	0.53	255.95	0.44	37.77	-1.53	108.12	10 Jul 15	315.3	17.2	4.8	3.8	135.78	141.81	1.65	1.62	94.97	95.34	-4.53	-3.73
Jul 9	124.75	0.54	255.17	0.41	44.61	-1.46	117.66	20 Jul 25	87.1	162.3	-2.0	-5.0	147.84	153.84	1.55	1.45	98.03	103.03	-2.61	-1.38
Jul 19	126.04	0.55	254.66	0.38	51.24	-1.37	127.23	30 J-A 4	228.8	288.5	-1.3	3.6	159.83	165.81	1.32	1.15	110.15	118.96	-0.20	0.77
Jul 29	127.32	0.57	254.46	0.35	57.64	-1.26	136.81	9 Aug 14	348.9	53.7	4.8	0.7	171.76	177.70	0.96	0.73	128.83	139.01	1.43	1.73
Aug 8	128.59	0.58	254.56	0.32	63.76	-1.12	146.43	19 Aug 24	126.0	198.7	-4.5	-3.4	183.61	189.50	0.48	0.21	148.98	158.49	1.73	1.48
Aug 18	129.82	0.60	254.97	0.29	69.56	-0.96	156.09	29 Aug	261.5		1.8		195.37		-0.09		167.47		1.06	
								Sep 3		321.1		4.9		201.21		-0.40		175.92		0.53
Aug 28	130.99	0.62	255.67	0.27	74.97	-0.77	165.79	8 Sep 13	24.1	91.9	2.9	-2.7	207.01	212.78	-0.73	-1.07	183.89	191.38	-0.08	-0.72
Sep 7	132.09	0.64	256.64	0.24	79.90	-0.55	175.55	18 Sep 23	164.5	232.9	-4.9	-0.5	218.52	224.21	-1.41	-1.74	198.40	204.89	-1.37	-2.00
Sep 17	133.09	0.67	257.86	0.22	84.26	-0.30	185.36	28 S-O 3	293.4	354.4	4.2	4.6	229.85	235.43	-2.08	-2.40	210.72	215.63	-2.57	-3.04
Sep 27	133.98	0.69	259.31	0.19	87.92	0.00	195.23	8 Oct 13	60.9	131.2	-0.3	-5.0	240.95	246.39	-2.70	-2.98	219.16	220.55	-3.31	-3.25
Oct 7	134.74	0.72	260.95	0.17	90.70	0.34	205.15	18 Oct 23	201.4	265.4	-3.0	2.5	251.75	257.00	-3.22	-3.43	218.80	213.69	-2.64	-1.33
Oct 17	135.35	0.74	262.75	0.15	92.41	0.74	215.14	28 Oct	325.1		5.2		262.13		-3.59		207.64		0.36	
								Nov 2		29.0		2.4		267.11		-3.70		204.75		1.67
Oct 27	135.80	0.77	264.71	0.13	92.86	1.18	225.17	7 Nov 12	99.4	170.1	-3.6	-4.8	271.91	276.49	-3.74	-3.71	206.41	211.32	2.25	2.26
Nov 6	136.07	0.80	266.78	0.11	91.88	1.67	235.25	17 Nov 22	236.3	297.3	0.0	4.6	280.82	284.83	-3.60	-3.39	217.92	225.28	1.94	1.44
Nov 16	136.16	0.83	268.94	0.10	89.50	2.17	245.37	27 N-D 2	357.5	65.0	4.5	-0.8	288.44	291.56	-3.06	-2.61	232.93	240.67	0.87	0.28
Nov 26	136.05	0.86	271.18	0.08	86.04	2.62	255.52	7 Dec 12	138.6	207.2	-5.2	-2.6	294.09	295.91	-2.02	-1.27	248.46	256.27	-0.30	-0.83
Dec 6	135.77	0.89	273.46	0.06	82.13	2.96	265.70	17 Dec 22	269.7	329.2	2.9	5.2	296.88	296.88	-0.35	0.74	264.13	272.07	-1.30	-1.69
Dec 16	135.31	0.92	275.77	0.05	78.59	3.16	275.89	27 Dec	31.2		2.2		295.84		1.97		280.11		-1.97	
1770								1771												1771
								Jan 1		102.7		-3.8		293.82		3.27		288.27		-2.12
Dec 26	134.72	0.95	278.09	0.03	76.07	3.22	286.08	6 Jan 11	177.0	242.2	-4.4	0.6	291.07	288.02	4.52	5.59	296.53	304.77	-2.09	-1.85
Jan 5	134.00	0.97	280.38	0.01	74.85	3.18	296.27	16 Jan 21	302.3	2.0	4.7	4.0	285.20	283.04	6.38	6.83	312.70	319.64	-1.31	-0.43
Jan 15	133.21	0.99	282.63	0.00	74.96	3.08	306.44	26 Jan 31	66.7	141.3	-1.1	-5.0	281.81	281.61	6.98	6.88	324.34	325.24	0.81	2.24
Jan 25	132.40	1.01	284.81	-0.02	76.23	2.94	316.58	5 Feb 10	213.5	275.6	-1.6	3.4	282.40	284.05	6.59	6.18	321.79	316.09	3.39	3.70
Feb 4	131.60	1.03	286.90	-0.04	78.47	2.79	326.69	15 Feb 20	335.0	36.3	4.8	1.2	286.44	289.44	5.69	5.15	311.61	310.09	3.14	2.14
Feb 14	130.87	1.03	288.88	-0.05	81.48	2.64	336.75	25 Feb	104.3		-4.1		292.93		4.58		311.36		1.05	
								Mar 2		179.5		-3.9		296.82		3.99		314.68		0.06
Feb 24	130.23	1.04	290.72	-0.07	85.10	2.49	346.76	7 Mar 12	247.8	308.0	1.6	5.0	301.05	305.55	3.40	2.83	319.43	325.21	-0.78	-1.45
Mar 6	129.74	1.04	292.39	-0.09	89.19	2.35	356.72	17 Mar 22	8.3	72.5	3.3	-2.1	310.27	315.18	2.26	1.72	331.78	339.02	-1.94	-2.25
Mar 16	129.40	1.04	293.87	-0.11	93.66	2.22	6.63	27 M-A 1	143.3	216.1	-5.1	-1.0	320.25	325.44	1.21	0.72	346.85	355.28	-2.37	-2.29
Mar 26	129.24	1.04	295.12	-0.13	98.43	2.09	16.47	6 Apr 11	280.5	340.2	4.2	4.8	330.74	336.13	0.27	-0.14	4.30	13.94	-2.01	-1.51
Apr 5	129.25	1.04	296.13	-0.16	103.45	1.96	26.26	16 Apr 21	42.9	110.8	0.4	-4.7	341.61	347.15	-0.52	-0.85	24.15	34.77	-0.82	0.03
Apr 15	129.45	1.03	296.87	-0.18	108.66	1.85	36.00	26 Apr	182.3		-3.7		352.75		-1.15		45.37		0.91	
								May 1		250.8		2.1		358.40		-1.40		55.37		1.70
Apr 25	129.82	1.03	297.31	-0.21	114.03	1.73	45.68	6 May 11	312.4	12.8	5.3	3.0	4.09	9.82	-1.60	-1.77	64.23	71.66	2.24	2.45
May 5	130.35	1.03	297.44	-0.24	119.54	1.62	55.33	16 May 21	79.2	150.2	-2.8	-5.1	15.59	21.38	-1.89	-1.97	77.49	81.61	2.28	1.73
May 15	131.04	1.02	297.25	-0.27	125.17	1.52	64.94	26 May 31	219.7	284.1	-0.7	4.4	27.20	33.05	-2.01	-2.01	83.86	84.19	0.80	-0.45
May 25	131.85	1.02	296.76	-0.30	130.90	1.41	74.51	5 Jun 10	344.1	46.6	4.6	0.0	38.92	44.81	-1.98	-1.91	82.75	80.18	-1.88	-3.20
Jun 4	132.79	1.02	295.98	-0.33	136.73	1.31	84.06	15 Jun 20	117.5	189.3	-4.9	-3.1	50.72	56.65	-1.81	-1.68	77.51	75.85	-4.13	-4.50
Jun 14	133.82	1.02	294.97	-0.36	142.64	1.21	93.60	25 Jun 30	255.3	316.6	2.5	5.1	62.59	68.56	-1.53	-1.35	75.92	77.99	-4.31	-3.69
Jun 24	134.94	1.02	293.79	-0.38	148.64	1.11	103.13	5 Jul 10	16.4	82.4	2.5	-3.1	74.54	80.54	-1.16	-0.95	82.06	88.00	-2.78	-1.70
Jul 4	136.12	1.03	292.51	-0.41	154.71	1.02	112.67	15 Jul 20	156.7	226.5	-4.7	0.2	86.55	92.59	-0.74	-0.51	95.68	104.83	-0.59	0.42
Jul 14	137.36	1.04	291.23	-0.43	160.87	0.92	122.22	25 Jul 30	289.2	349.0	4.5	4.1	98.63	104.70	-0.29	-0.06	114.97	125.45	1.19	1.64
Jul 24	138.62	1.05	290.05	-0.45	167.10	0.82	131.78	4 Aug 9	50.0	120.1	-0.6	-4.9	110.78	116.88	0.16	0.37	135.71	145.47	1.77	1.63
Aug 3	139.90	1.06	289.04	-0.46	173.41	0.72	141.39	14 Aug 19	195.3	261.8	-2.2	3.3	123.00	129.13	0.57	0.76	154.62	163.18	1.28	0.77
Aug 13	141.17	1.07	288.26	-0.48	179.80	0.63	151.02	24 Aug 29	322.2	21.8	4.9	1.6	135.28	141.44	0.92	1.07	171.15	178.57	0.17	-0.51
Aug 23	142.43	1.09	287.78	-0.48	186.27	0.53	160.70	3 Sep 8	85.6	159.0	-3.6	-4.4	147.62	153.81	1.19	1.29	185.40	191.58	-1.22	-1.93
Sep 2	143.64	1.11	287.60	-0.49	192.82	0.43	170.43	13 Sep 18	232.1	295.3	1.1	5.0	160.02	166.23	1.37	1.41	196.94	201.19	-2.60	-3.19
Sep 12	144.80	1.13	287.76	-0.50	199.46	0.33	180.21	23 Sep 28	354.9	56.1	3.6	-1.6	172.46	178.69	1.43	1.43	203.86	204.20	-3.60	-3.68
Sep 22	145.88	1.16	288.23	-0.50	206.18	0.23	190.05	3 Oct 8	123.3	197.4	-5.2	-1.9	184.93	191.18	1.39	1.33	201.51	196.12	-3.18	-1.94
Oct 2	146.86	1.19	289.00	-0.50	212.99	0.13	199.95	13 Oct 18	266.7	327.7	3.9	5.0	197.44	203.70	1.24	1.13	190.80	188.99	-0.26	1.17
Oct 12	147.74	1.22	290.05	-0.51	219.88	0.03	209.90	23 Oct 28	27.8	92.2	0.9	-4.3	209.97	216.23	1.00	0.85	191.56	197.20	1.95	2.15
Oct 22	148.47	1.25	291.36	-0.51	226.87	-0.07	219.90	2 Nov 7	162.3	234.3	-4.4	1.5	222.51	228.78	0.69	0.51	204.41	212.24	1.96	1.56
Nov 1	149.06	1.28	292.89	-0.51	233.94	-0.17	229.96	12 Nov 17	299.7	359.5	5.2	3.3	235.06	241.33	0.32	0.12	220.26	228.26	1.05	0.48
Nov 11	149.48	1.32	294.62	-0.52	241.10	-0.27	240.05	22 Nov 27	62.0	130.5	-2.1	-5.2	247.61	253.89	-0.08	-0.28	236.20	244.09	-0.09	-0.63
Nov 21	149.73	1.36	296.53	-0.52	248.35	-0.37	250.19	2 Dec 7	201.1	269.5	-1.5	4.1	260.16	266.44	-0.48	-0.67	251.95	259.80	-1.13	-1.56
Dec 1	149.79	1.40	298.57	-0.53	255.68	-0.47	260.36	12 Dec 17	331.6	31.5	4.7	0.5	272.72	278.99	-0.85	-1.02	267.67	275.58	-1.91	-2.13
Dec 11	149.66	1.44	300.73	-0.53	263.09	-0.56	270.54	22 Dec 27	97.9	169.7	-4.5	-3.7	285.26	291.53	-1.17	-1.30	283.49	291.28	-2.20	-2.06
1771								1771												1771

JULIAN 12 UT	SATURN LONG.	LAT.	JUPITER LONG.	LAT.	MARS LONG.	LAT.	SUN LONG.	GREGORIAN 12 UT	MOON LONG.		LAT.		VENUS LONG.		LAT.		MERCURY LONG.		LAT.	
1771								1772												1772
Dec 21	149.35	1.47	302.98	-0.54	270.58	-0.66	280.74	1 Jan 6	238.7	303.2	2.0	5.0	297.79	304.06	-1.41	-1.49	298.65	304.90	-1.64	-0.86
Dec 31	148.88	1.51	305.30	-0.55	278.14	-0.75	290.93	11 Jan 16	3.2	64.7	2.6	-2.5	310.31	316.56	-1.55	-1.58	308.76	308.56	0.33	1.84
Jan 10	148.27	1.54	307.66	-0.56	285.77	-0.83	301.11	21 Jan 26	135.6	208.5	-4.9	-0.4	322.79	329.02	-1.58	-1.55	303.95	297.84	3.15	3.60
Jan 20	147.55	1.56	310.04	-0.58	293.45	-0.91	311.27	31 J-F 5	274.6	336.0	4.4	4.2	335.24	341.45	-1.49	-1.40	294.00	293.57	3.16	2.27
Jan 30	146.77	1.58	312.42	-0.59	301.18	-0.98	321.40	10 Feb 15	35.3	99.8	-0.3	-4.7	347.64	353.81	-1.28	-1.14	295.87	299.99	1.29	0.37
Feb 9	145.96	1.60	314.78	-0.61	308.95	-1.05	331.48	20 Feb 25	174.2	245.7	-3.0	3.0	359.97	6.10	-0.97	-0.77	305.33	311.52	-0.43	-1.11
Feb 19	145.17	1.61	317.09	-0.63	316.74	-1.10	341.52	1 Mar 6	308.9	8.4	5.1	1.9	12.22	18.31	-0.55	-0.31	318.37	325.76	-1.63	-2.01
Feb 29	144.45	1.61	319.33	-0.65	324.55	-1.14	351.51	11 Mar 16	68.7	136.9	-3.3	-5.0	24.37	30.41	-0.06	0.21	333.66	342.06	-2.21	-2.24
Mar 10	143.84	1.61	321.48	-0.68	332.37	-1.18	1.44	21 Mar 26	212.5	281.1	0.3	5.0	36.41	42.38	0.48	0.76	350.99	0.46	-2.07	-1.69
Mar 20	143.36	1.60	323.51	-0.70	340.18	-1.20	11.32	31 M-A 5	341.8	41.2	3.9	-1.1	48.31	54.20	1.04	1.32	10.45	20.82	-1.10	-0.31
Mar 30	143.04	1.59	325.41	-0.73	347.97	-1.21	21.14	10 Apr 15	104.2	175.5	-5.1	-2.9	60.05	65.85	1.58	1.83	31.19	40.93	0.60	1.50
Apr 9	142.89	1.58	327.14	-0.77	355.73	-1.20	30.90	20 Apr 25	249.5	314.6	3.4	5.1	71.60	77.29	2.06	2.27	49.42	56.19	2.22	2.63
Apr 19	142.92	1.56	328.69	-0.80	3.44	-1.18	40.61	30 Apr	14.0		1.4		82.91		2.44		61.01		2.65	
								May 5		75.3		-3.8		88.46		2.59		63.73		2.22
Apr 29	143.13	1.55	330.01	-0.84	11.09	-1.15	50.28	10 May 15	141.9	214.2	-4.8	0.5	93.94	99.31	2.69	2.75	64.31	62.98	1.33	0.05
May 9	143.50	1.53	331.10	-0.88	18.68	-1.10	59.90	20 May 25	284.8	346.8	5.1	3.5	104.58	109.74	2.76	2.71	60.40	57.69	-1.40	-2.71
May 19	144.04	1.52	331.91	-0.93	26.18	-1.04	69.49	30 M-J 4	46.6	111.3	-1.6	-5.1	114.75	119.60	2.60	2.43	55.95	55.87	-3.63	-4.05
May 29	144.72	1.50	332.43	-0.97	33.60	-0.97	79.06	9 Jun 14	181.0	252.0	-2.3	3.6	124.26	128.70	2.18	1.86	57.65	61.22	-4.00	-3.57
Jun 8	145.54	1.49	332.63	-1.02	40.91	-0.88	88.60	19 Jun 24	318.4	18.4	4.7	0.7	132.87	136.71	1.45	0.95	66.42	73.14	-2.86	-1.95
Jun 18	146.47	1.49	332.52	-1.07	48.11	-0.78	98.13	29 Jun	80.3		-4.1		140.18		0.36		81.29		-0.94	
								Jul 4		149.4		-4.2		143.18		-0.35		90.72		0.06
Jun 28	147.49	1.48	332.08	-1.12	55.20	-0.67	107.67	9 Jul 14	220.1	288.2	1.3	5.0	145.62	147.37	-1.16	-2.09	101.06	111.77	0.91	1.50
Jul 8	148.60	1.48	331.36	-1.16	62.15	-0.55	117.21	19 Jul 24	350.8	50.4	2.7	-2.3	148.33	148.38	-3.11	-4.22	122.28	132.24	1.77	1.75
Jul 18	149.78	1.48	330.37	-1.20	68.96	-0.42	126.77	29 J-A 3	116.0	188.4	-5.0	-1.2	147.46	145.59	-5.34	-6.41	141.55	150.17	1.48	1.03
Jul 28	151.00	1.48	329.19	-1.23	75.61	-0.27	136.35	8 Aug 13	258.1	322.7	4.2	4.3	142.94	139.88	-7.31	-7.92	158.13	165.43	0.43	-0.27
Aug 7	152.25	1.49	327.90	-1.25	82.09	-0.12	145.96	18 Aug 23	22.7	83.5	-0.1	-4.6	136.89	134.42	-8.17	-8.08	172.03	177.83	-1.03	-1.83
Aug 17	153.52	1.50	326.59	-1.26	88.39	0.05	155.62	28 Aug	153.4		-3.8		132.79		-7.69		182.64		-2.62	
								Sep 2		227.3		2.3		132.15		-7.10		186.15		-3.34
Aug 27	154.78	1.51	325.37	-1.27	94.47	0.23	165.33	7 Sep 12	294.4	355.8	5.2	2.1	132.48	133.70	-6.38	-5.60	187.84	187.11	-3.89	-4.09
Sep 6	156.02	1.53	324.31	-1.26	100.32	0.43	175.08	17 Sep 22	54.9	118.5	-3.0	-5.2	135.71	138.38	-4.80	-4.00	183.59	178.28	-3.68	-2.50
Sep 16	157.22	1.55	323.49	-1.25	105.89	0.64	184.89	27 S-O 2	191.8	264.8	-0.7	4.8	141.59	145.27	-3.24	-2.51	173.97	173.33	-0.85	0.64
Sep 26	158.36	1.57	322.97	-1.23	111.13	0.87	194.75	7 Oct 12	328.7	28.2	4.1	-0.8	149.32	153.69	-1.83	-1.19	176.79	183.12	1.60	1.99
Oct 6	159.42	1.60	322.78	-1.21	115.99	1.12	204.67	17 Oct 22	88.5	155.3	-4.9	-3.7	158.33	163.19	-0.61	-0.08	190.89	199.19	1.96	1.66
Oct 16	160.38	1.63	322.92	-1.18	120.39	1.40	214.65	27 Oct	230.2		2.7		168.25		0.40		207.57		1.21	
								Nov 1		300.4		5.2		173.47		0.82		215.84		0.68
Oct 26	161.23	1.67	323.39	-1.16	124.23	1.70	224.68	6 Nov 11	1.5	60.9	1.7	-3.4	178.83	184.31	1.19	1.51	223.96	231.93	0.13	-0.43
Nov 5	161.94	1.70	324.18	-1.13	127.36	2.05	234.75	16 Nov 21	123.9	193.6	-5.0	-0.5	189.89	195.56	1.76	1.97	239.79	247.57	-0.96	-1.43
Nov 15	162.49	1.74	325.26	-1.10	129.64	2.43	244.88	26 N-D 1	267.5	334.0	5.0	3.6	201.32	207.14	2.12	2.22	255.29	262.95	-1.83	-2.13
Nov 25	162.88	1.79	326.61	-1.08	130.87	2.84	255.03	6 Dec 11	33.4	94.9	-1.3	-4.9	213.02	218.95	2.28	2.28	270.52	277.84	-2.29	-2.26
Dec 5	163.09	1.83	328.18	-1.06	130.88	3.28	265.20	16 Dec 21	161.4	232.3	-3.0	2.9	224.92	230.93	2.25	2.17	284.62	290.14	-1.96	-1.30
Dec 15	163.12	1.88	329.96	-1.05	129.52	3.72	275.39	26 Dec 31	303.2	6.0	4.8	0.9	236.98	243.05	2.06	1.91	293.11	291.86	-0.18	1.36
1772								1773												1773
Dec 25	162.97	1.92	331.91	-1.03	126.85	4.10	285.59	5 Jan 10	65.6	130.6	-3.9	-4.5	249.14	255.25	1.74	1.54	286.34	280.21	2.79	3.40
Jan 4	162.63	1.96	334.00	-1.02	123.21	4.35	295.77	15 Jan 20	200.1	270.5	0.5	4.9	261.38	267.52	1.33	1.10	277.21	277.85	3.11	2.36
Jan 14	162.14	2.00	336.21	-1.01	119.23	4.44	305.95	25 Jan 30	337.2	37.4	3.0	-2.1	273.67	279.83	0.86	0.61	281.06	285.88	1.49	0.64
Jan 24	161.52	2.03	338.51	-1.01	115.67	4.36	316.09	4 Feb 9	98.7	167.9	-5.1	-2.1	285.99	292.16	0.36	0.11	291.72	298.25	-0.13	-0.80
Feb 3	160.79	2.06	340.87	-1.01	113.12	4.14	326.20	14 Feb 19	239.3	307.2	3.8	4.6	298.34	304.52	-0.13	-0.36	305.30	312.80	-1.35	-1.77
Feb 13	160.01	2.07	343.27	-1.01	111.84	3.84	336.26	24 Feb	9.7		0.1		310.70		-0.57		320.72		-2.05	
								Mar 1		69.1		-4.4		316.88		-0.77		329.07		-2.16
Feb 23	159.21	2.08	345.69	-1.01	111.86	3.51	346.28	6 Mar 11	133.5	206.3	-4.4	1.4	323.07	329.25	-0.95	-1.11	337.86	347.13	-2.10	-1.83
Mar 5	158.44	2.09	348.11	-1.02	113.02	3.19	356.24	16 Mar 21	277.5	341.9	5.3	2.5	335.43	341.61	-1.24	-1.34	356.85	6.92	-1.34	-0.63
Mar 15	157.74	2.08	350.50	-1.03	115.13	2.88	6.15	26 Mar 31	41.8	102.0	-2.7	-5.3	347.78	353.96	-1.42	-1.47	16.98	26.40	0.26	1.24
Mar 25	157.15	2.07	352.85	-1.05	118.01	2.60	16.00	5 Apr 10	170.0	245.0	-1.9	4.3	0.13	6.30	-1.50	-1.49	34.40	40.37	2.12	2.75
Apr 4	156.69	2.06	355.14	-1.06	121.51	2.34	25.79	15 Apr 20	313.9	15.1	4.4	-0.4	12.46	18.63	-1.46	-1.40	43.94	45.00	2.98	2.69
Apr 14	156.39	2.04	357.34	-1.09	125.51	2.11	35.53	25 Apr 30	74.3	136.7	-4.7	-4.2	24.79	30.94	-1.32	-1.21	43.77	41.01	1.86	0.59
Apr 24	156.26	2.01	359.43	-1.11	129.90	1.89	45.22	5 May 10	208.0	282.8	1.6	5.2	37.10	43.25	-1.08	-0.94	37.97	35.90	-0.85	-2.12
May 4	156.31	1.99	1.39	-1.14	134.63	1.70	54.86	15 May 20	348.1	47.5	1.9	-3.1	49.39	55.54	-0.77	-0.60	35.54	37.05	-3.03	-3.52
May 14	156.53	1.96	3.20	-1.17	139.62	1.51	64.47	25 May 30	108.0	173.4	-5.0	-1.4	61.69	67.83	-0.41	-0.22	40.30	45.07	-3.62	-3.38
May 24	156.91	1.94	4.82	-1.21	144.85	1.34	74.05	4 Jun 9	246.8	318.8	4.3	3.8	73.97	80.11	-0.02	0.18	51.17	58.51	-2.87	-2.14
Jun 3	157.45	1.91	6.24	-1.25	150.28	1.18	83.60	14 Jun 19	20.8	80.3	-1.2	-4.8	86.26	92.40	0.37	0.56	67.02	76.64	-1.25	-0.30
Jun 13	158.14	1.89	7.42	-1.29	155.88	1.03	93.14	24 Jun 29	143.3	211.8	-3.4	2.2	98.54	104.69	0.74	0.91	87.11	97.98	0.60	1.31
Jun 23	158.95	1.88	8.34	-1.34	161.63	0.88	102.67	4 Jul 9	285.1	352.7	4.9	1.1	110.83	116.98	1.06	1.19	108.68	118.82	1.73	1.84
Jul 3	159.88	1.86	8.96	-1.39	167.53	0.74	112.20	14 Jul 19	52.7	114.0	-3.8	-4.7	123.12	129.27	1.30	1.38	128.23	136.88	1.68	1.29
Jul 13	160.90	1.85	9.28	-1.44	173.57	0.61	121.75	24 Jul 29	180.5	251.1	-0.3	4.7	135.42	141.56	1.44	1.48	144.76	151.88	0.72	0.00
Jul 23	162.00	1.84	9.27	-1.48	179.73	0.48	131.32	3 Aug 8	321.7	25.1	3.4	-2.0	147.71	153.85	1.48	1.46	158.16	163.49	-0.82	-1.70
Aug 2	163.17	1.83	8.93	-1.53	186.01	0.35	140.91	13 Aug 18	84.8	149.3	-5.1	-2.8	159.99	166.13	1.41	1.33	167.63	170.25	-2.61	-3.47
Aug 12	164.38	1.83	8.28	-1.57	192.42	0.23	150.55	23 Aug 28	219.3	289.8	3.2	5.0	172.27	178.40	1.23	1.09	170.87	169.07	-4.16	-4.46
Aug 22	165.63	1.84	7.35	-1.61	198.94	0.11	160.23	2 Sep 7	356.2	56.8	0.5	-4.3	184.53	190.65	0.93	0.75	165.01	160.19	-4.11	-2.98
Sep 1	166.88	1.85	6.20	-1.63	205.57	0.00	169.95	12 Sep 17	117.6	186.1	-4.7	0.4	196.77	202.88	0.55	0.33	157.16	157.75	-1.40	0.12
Sep 11	168.13	1.86	4.90	-1.64	212.32	-0.11	179.74	22 Sep 27	258.6	326.7	5.2	3.0	208.99	215.09	0.10	-0.15	162.06	169.02	1.21	1.78
Sep 21	169.35	1.87	3.57	-1.64	219.18	-0.22	189.57	2 Oct 7	29.2	88.7	-2.4	-5.3	221.17	227.25	-0.40	-0.65	177.33	186.09	1.91	1.74
Oct 1	170.53	1.89	2.28	-1.63	226.15	-0.32	199.46	12 Oct 17	151.9	224.4	-2.6	3.6	233.33	239.39	-0.91	-1.16	194.83	203.38	1.36	0.88
Oct 11	171.64	1.92	1.15	-1.61	233.22	-0.42	209.41	22 Oct 27	297.1	1.5	4.7	0.0	245.44	251.48	-1.39	-1.62	211.68	219.75	0.34	-0.23
Oct 21	172.68	1.95	0.26	-1.57	240.41	-0.52	219.41	1 Nov 6	61.6	121.5	-4.5	-4.4	257.50	263.50	-1.82	-2.00	227.63	235.35	-0.78	-1.29
Oct 31	173.61	1.98	359.67	-1.53	247.69	-0.61	229.47	11 Nov 16	187.8	263.2	0.7	5.1	269.49	275.45	-2.16	-2.28	242.92	250.34	-1.75	-2.12
Nov 10	174.42	2.02	359.40	-1.49	255.07	-0.69	239.57	21 Nov 26	333.4	34.8	2.3	-2.9	281.38	287.29	-2.36	-2.41	257.55	264.41	-2.36	-2.44
Nov 20	175.10	2.06	359.48	-1.44	262.55	-0.77	249.70	1 Dec 6	94.2	155.9	-5.0	-1.9	293.15	298.97	-2.41	-2.36	270.57	275.31	-2.27	-1.74
Nov 30	175.62	2.11	359.91	-1.39	270.11	-0.84	259.86	11 Dec 16	225.5	301.0	3.7	4.2	304.74	310.44	-2.27	-2.12	277.35	275.11	-0.72	0.80
Dec 10	175.97	2.15	0.65	-1.34	277.75	-0.91	270.05	21 Dec 26	7.5	67.3	-1.0	-4.7	316.08	321.62	-1.91	-1.65	268.97	263.11	2.35	3.12
Dec 20	176.14	2.20	1.70	-1.30	285.46	-0.96	280.24	31 Dec	127.8		-3.7		327.06		-1.34		261.05		3.00	
1773								1773												1773

JULIAN	SATURN		JUPITER		MARS		SUN	GREGORIAN	MOON				VENUS				MERCURY			
12 UT	LONG.	LAT.	LONG.	LAT.	LONG.	LAT.	LONG.	12 UT	LONG.		LAT.		LONG.		LAT.		LONG.		LAT.	
1773								1774												1774
								Jan 5		192.3		1.5		332.38		-0.96		262.70		2.41
Dec 30	176.13	2.25	3.02	-1.26	293.23	-1.01	290.43	10 Jan 15	264.3	337.0	5.0	1.5	337.56	342.57	-0.52	-0.02	266.74	272.18	1.65	0.87
Jan 9	175.94	2.29	4.58	-1.22	301.05	-1.04	300.62	20 Jan 25	40.1	99.7	-3.7	-4.9	347.37	351.93	0.54	1.16	278.46	285.28	0.13	-0.53
Jan 19	175.58	2.34	6.33	-1.19	308.90	-1.07	310.77	30 J-F 4	162.7	230.6	-0.9	4.4	356.19	0.09	1.84	2.58	292.50	300.08	-1.09	-1.55
Jan 29	175.07	2.37	8.27	-1.16	316.78	-1.08	320.90	9 Feb 14	302.8	11.0	4.0	-1.7	3.56	6.51	3.37	4.21	307.99	316.26	-1.88	-2.07
Feb 8	174.43	2.40	10.34	-1.13	324.67	-1.08	330.99	19 Feb 24	71.8	132.7	-5.2	-3.4	8.82	10.36	5.08	5.96	324.90	333.94	-2.10	-1.94
Feb 18	173.70	2.43	12.53	-1.11	332.56	-1.07	341.03	1 Mar 6	199.5	270.0	2.4	5.2	11.00	10.64	6.80	7.56	343.37	353.08	-1.56	-0.94
Feb 28	172.92	2.44	14.81	-1.09	340.44	-1.05	351.02	11 Mar 16	339.7	43.7	1.1	-4.1	9.26	6.95	8.14	8.44	2.77	11.78	-0.09	0.92
Mar 10	172.13	2.45	17.15	-1.08	348.29	-1.02	0.96	21 Mar 26	103.5	166.9	-4.9	-0.5	4.00	0.87	8.39	7.95	19.23	24.26	1.95	2.80
Mar 20	171.38	2.44	19.54	-1.07	356.10	-0.97	10.83	31 M-A 5	237.9	308.9	4.8	3.6	358.07	356.01	7.18	6.17	26.40	25.65	3.24	3.11
Mar 30	170.70	2.43	21.95	-1.06	3.87	-0.92	20.66	10 Apr 15	14.7	75.8	-2.0	-5.2	354.92	354.85	5.06	3.93	22.74	19.15	2.35	1.10
Apr 9	170.14	2.41	24.35	-1.06	11.57	-0.85	30.43	20 Apr 25	135.8	203.0	-3.0	2.7	355.71	357.40	2.86	1.88	16.42	15.49	-0.29	-1.52
Apr 19	169.71	2.39	26.74	-1.06	19.20	-0.78	40.14	30 Apr	277.2		4.9		359.79		1.01		16.58		-2.43	
								May 5		346.0		0.4		2.76		0.24		19.48		-2.99
Apr 29	169.43	2.36	29.09	-1.06	26.76	-0.69	49.81	10 May 15	48.3	107.9	-4.3	-4.5	6.20	10.03	-0.42	-0.99	23.88	29.55	-3.22	-3.15
May 9	169.32	2.33	31.38	-1.07	34.23	-0.60	59.44	20 May 25	169.7	241.1	0.0	4.8	14.18	18.59	-1.46	-1.85	36.32	44.11	-2.82	-2.26
May 19	169.39	2.30	33.59	-1.08	41.62	-0.50	69.02	30 M-J 4	315.6	21.0	2.7	-2.7	23.22	28.04	-2.16	-2.40	52.89	62.63	-1.51	-0.63
May 29	169.62	2.26	35.71	-1.09	48.91	-0.40	78.59	9 Jun 14	81.2	140.9	-5.0	-2.2	33.01	38.11	-2.57	-2.67	73.16	84.10	0.27	1.07
Jun 8	170.02	2.23	37.70	-1.10	56.10	-0.29	88.14	19 Jun 24	205.6	280.1	3.2	4.6	43.32	48.62	-2.72	-2.71	94.93	105.18	1.63	1.90
Jun 18	170.57	2.20	39.55	-1.12	63.19	-0.17	97.67	29 Jun	352.1		-0.6		54.02		-2.66		114.64		1.86	
								Jul 4		54.4		-4.7		59.49		-2.56		123.24		1.56
Jun 28	171.26	2.17	41.22	-1.14	70.18	-0.05	107.20	9 Jul 14	113.9	175.3	-3.9	1.0	65.02	70.62	-2.43	-2.26	130.98	137.81	1.03	0.31
Jul 8	172.08	2.15	42.70	-1.17	77.06	0.08	116.75	19 Jul 24	243.6	318.5	5.0	2.2	76.27	81.97	-2.06	-1.84	143.66	148.36	-0.55	-1.53
Jul 18	173.00	2.13	43.95	-1.19	83.83	0.20	126.30	29 J-A 3	26.5	86.9	-3.5	-5.0	87.71	93.50	-1.59	-1.34	151.67	153.24	-2.57	-3.57
Jul 28	174.03	2.11	44.94	-1.22	90.48	0.34	135.89	8 Aug 13	147.2	211.7	-1.4	3.9	99.33	105.20	-1.07	-0.80	152.71	149.99	-4.39	-4.76
Aug 7	175.13	2.10	45.65	-1.25	97.02	0.47	145.50	18 Aug 23	282.7	355.1	4.6	-1.1	111.11	117.05	-0.53	-0.26	145.82	141.95	-4.43	-3.37
Aug 17	176.29	2.09	46.06	-1.28	103.42	0.61	155.15	28 Aug	59.4		-5.1		123.02		0.01		140.40		-1.87	
								Sep 2		119.1		-3.7		129.03		0.26		142.24		-0.38
Aug 27	177.49	2.09	46.13	-1.31	109.70	0.75	164.86	7 Sep 12	181.9	250.2	1.7	5.3	135.06	141.13	0.49	0.71	147.37	154.91	0.80	1.53
Sep 6	178.72	2.09	45.87	-1.34	115.83	0.90	174.61	17 Sep 22	321.5	29.8	1.9	-3.8	147.22	153.33	0.91	1.08	163.71	172.91	1.83	1.78
Sep 16	179.96	2.09	45.29	-1.36	121.80	1.06	184.41	27 S-O 2	91.3	151.6	-5.0	-1.0	159.47	165.63	1.23	1.35	182.02	190.84	1.50	1.07
Sep 26	181.20	2.10	44.41	-1.38	127.60	1.22	194.28	7 Oct 12	218.4	289.6	4.3	4.1	171.81	178.01	1.44	1.50	199.34	207.53	0.55	-0.01
Oct 6	182.40	2.12	43.28	-1.39	133.19	1.39	204.19	17 Oct 22	358.7	63.0	-1.4	-5.1	184.22	190.45	1.54	1.54	215.43	223.09	-0.59	-1.14
Oct 16	183.55	2.14	42.00	-1.39	138.55	1.57	214.17	27 Oct	122.8		-3.2		196.69		1.51		230.52		-1.65	
								Nov 1		185.4		2.1		202.94		1.46		237.70		-2.09
Oct 26	184.64	2.17	40.64	-1.37	143.64	1.76	224.20	6 Nov 11	256.8	328.4	5.0	1.1	209.20	215.47	1.37	1.27	244.56	250.93	-2.43	-2.61
Nov 5	185.65	2.20	39.32	-1.35	148.40	1.97	234.27	16 Nov 21	34.1	95.4	-4.1	-4.5	221.75	228.03	1.14	0.99	256.44	260.37	-2.56	-2.17
Nov 15	186.54	2.23	38.14	-1.32	152.76	2.20	244.39	26 N-D 1	154.9	221.0	-0.3	4.4	234.31	240.60	0.82	0.64	261.43	258.28	-1.27	0.20
Nov 25	187.32	2.27	37.18	-1.28	156.64	2.44	254.54	6 Dec 11	295.8	5.5	3.4	-2.4	246.90	253.19	0.45	0.26	251.79	246.46	1.82	2.77
Dec 5	187.95	2.31	36.51	-1.23	159.92	2.71	264.71	16 Dec 21	68.1	127.6	-5.0	-2.4	259.48	265.77	0.06	-0.14	245.37	247.96	2.85	2.42
Dec 15	188.43	2.36	36.17	-1.18	162.47	3.01	274.90	26 Dec 31	188.5	258.5	2.8	5.0	272.07	278.36	-0.34	-0.53	252.76	258.76	1.77	1.07
1774								1775												1775
Dec 25	188.74	2.40	36.18	-1.13	164.10	3.32	285.10	5 Jan 10	333.9	40.6	0.1	-4.7	284.65	290.94	-0.71	-0.88	265.44	272.53	0.37	-0.27
Jan 4	188.87	2.45	36.53	-1.08	164.64	3.65	295.28	15 Jan 20	100.9	160.5	-4.1	0.6	297.23	303.51	-1.03	-1.16	279.91	287.55	-0.85	-1.33
Jan 14	188.82	2.50	37.21	-1.03	163.93	3.96	305.45	25 Jan 30	224.1	297.0	4.9	3.2	309.79	316.06	-1.27	-1.35	295.45	303.62	-1.71	-1.97
Jan 24	188.59	2.54	38.18	-0.98	161.90	4.21	315.60	4 Feb 9	10.3	74.0	-3.1	-5.2	322.33	328.60	-1.42	-1.45	312.09	320.89	-2.08	-2.02
Feb 3	188.20	2.59	39.43	-0.94	158.73	4.36	325.70	14 Feb 19	133.3	194.7	-1.8	3.5	334.85	341.10	-1.46	-1.44	330.01	339.34	-1.74	-1.23
Feb 13	187.67	2.62	40.92	-0.90	154.88	4.35	335.77	24 Feb	261.8		5.0		347.34		-1.40		348.59		-0.45	
								Mar 1		335.2		-0.1		353.58		-1.33		357.13		0.56
Feb 23	187.01	2.65	42.60	-0.87	151.04	4.17	345.79	6 Mar 11	44.9	106.1	-5.0	-3.9	359.80	6.01	-1.23	-1.11	3.94	7.94	1.70	2.75
Mar 5	186.28	2.67	44.46	-0.83	147.89	3.85	355.75	16 Mar 21	166.1	231.0	1.1	5.1	12.21	18.40	-0.96	-0.80	8.53	6.01	3.42	3.44
Mar 15	185.51	2.68	46.47	-0.80	145.90	3.46	5.66	26 Mar 31	300.9	12.3	2.8	-3.3	24.57	30.73	-0.61	-0.41	1.87	358.17	2.73	1.53
Mar 25	184.73	2.68	48.59	-0.78	145.19	3.04	15.52	5 Apr 10	78.0	137.8	-5.0	-1.3	36.88	43.01	-0.20	0.02	356.32	356.71	0.20	-0.97
Apr 4	184.00	2.67	50.79	-0.75	145.72	2.63	25.31	15 Apr 20	200.2	269.3	3.8	4.5	49.12	55.22	0.25	0.47	359.08	3.06	-1.87	-2.49
Apr 14	183.35	2.65	53.07	-0.73	147.32	2.25	35.05	25 Apr 30	339.8	47.9	-0.7	-4.9	61.30	67.36	0.70	0.92	8.33	14.65	-2.83	-2.90
Apr 24	182.81	2.63	55.39	-0.72	149.81	1.91	44.75	5 May 10	110.1	169.9	-3.3	1.7	73.40	79.42	1.13	1.33	21.89	29.98	-2.73	-2.32
May 4	182.41	2.60	57.73	-0.70	153.03	1.60	54.39	15 May 20	236.3	308.5	5.0	1.8	85.42	91.40	1.51	1.67	38.93	48.71	-1.71	-0.93
May 14	182.16	2.56	60.08	-0.69	156.84	1.32	64.00	25 May 30	17.6	82.0	-3.8	-4.6	97.35	103.28	1.80	1.90	59.22	70.17	-0.06	0.79
May 24	182.08	2.52	62.42	-0.68	161.14	1.07	73.58	4 Jun 9	141.8	203.3	-0.5	4.2	109.18	115.05	1.98	2.01	81.04	91.34	1.48	1.90
Jun 3	182.17	2.48	64.72	-0.67	165.83	0.84	83.13	14 Jun 19	274.2	347.3	4.0	-1.8	120.88	126.68	2.01	1.98	100.78	109.24	2.02	1.83
Jun 13	182.42	2.44	66.98	-0.66	170.87	0.63	92.67	24 Jun 29	53.7	114.8	-5.1	-2.6	132.43	138.15	1.89	1.77	116.69	123.10	1.37	0.67
Jun 23	182.83	2.41	69.16	-0.65	176.20	0.44	102.21	4 Jul 9	174.1	238.8	2.5	5.2	143.81	149.42	1.60	1.39	128.33	132.21	-0.24	-1.31
Jul 3	183.40	2.37	71.24	-0.65	181.78	0.27	111.74	14 Jul 19	313.0	24.6	1.1	-4.5	154.96	160.44	1.13	0.83	134.46	134.81	-2.47	-3.62
Jul 13	184.10	2.34	73.22	-0.65	187.59	0.11	121.29	24 Jul 29	87.9	147.2	-4.4	0.2	165.84	171.15	0.48	0.09	133.15	129.86	-4.52	-4.93
Jul 23	184.92	2.31	75.05	-0.65	193.60	-0.05	130.86	3 Aug 8	207.8	276.2	4.7	4.0	176.35	181.43	-0.34	-0.81	126.16	123.69	-4.61	-3.63
Aug 2	185.86	2.28	76.72	-0.65	199.79	-0.19	140.45	13 Aug 18	351.5	60.1	-2.4	-5.3	186.38	191.16	-1.32	-1.86	123.75	126.82	-2.26	-0.84
Aug 12	186.88	2.26	78.20	-0.65	206.17	-0.32	150.09	23 Aug 28	120.7	180.1	-2.2	3.1	195.75	200.12	-2.43	-3.03	132.72	140.77	0.38	1.24
Aug 22	187.98	2.25	79.46	-0.65	212.70	-0.44	159.77	2 Sep 7	243.5	314.7	5.2	0.9	204.21	207.96	-3.66	-4.29	150.02	159.63	1.70	1.79
Sep 1	189.14	2.24	80.46	-0.65	219.39	-0.55	169.49	12 Sep 17	28.6	93.6	-4.7	-4.1	211.31	214.17	-4.93	-5.55	169.09	178.20	1.61	1.24
Sep 11	190.33	2.23	81.18	-0.65	226.23	-0.66	179.26	22 Sep 27	152.9	214.4	0.8	4.9	216.45	218.00	-6.15	-6.68	186.90	195.20	0.76	0.20
Sep 21	191.56	2.23	81.60	-0.66	233.21	-0.75	189.10	2 Oct 7	281.1	353.3	3.5	-2.5	218.72	218.48	-7.12	-7.39	203.14	210.75	-0.38	-0.97
Oct 1	192.78	2.23	81.69	-0.66	240.32	-0.84	198.98	12 Oct 17	63.8	125.6	-5.0	-1.5	217.24	215.09	-7.44	-7.19	218.04	224.97	-1.54	-2.05
Oct 11	193.99	2.24	81.45	-0.66	247.55	-0.91	208.93	22 Oct 27	185.5	250.6	3.4	4.8	212.30	209.27	-6.60	-5.68	231.47	237.33	-2.48	-2.77
Oct 21	195.18	2.26	80.88	-0.65	254.91	-0.98	218.93	1 Nov 6	320.0	31.0	0.1	-4.7	206.52	204.45	-4.54	-3.28	242.16	245.21	-2.85	-2.60
Oct 31	196.31	2.28	80.01	-0.64	262.37	-1.03	228.98	11 Nov 16	97.3	157.1	-3.4	1.5	203.34	203.25	-2.03	-0.88	245.28	241.31	-1.83	-0.43
Nov 10	197.36	2.30	78.89	-0.63	269.93	-1.08	239.08	21 Nov 26	219.2	288.5	4.9	2.5	204.14	205.90	0.14	1.00	234.74	230.13	1.24	2.36
Nov 20	198.33	2.33	77.61	-0.62	277.58	-1.11	249.21	1 Dec 6	359.1	67.0	-3.3	-4.8	208.39	211.48	1.71	2.28	230.03	233.50	2.65	2.39
Nov 30	199.19	2.36	76.25	-0.59	285.31	-1.13	259.37	11 Dec 16	129.4	188.9	-0.7	4.0	215.08	219.09	2.72	3.04	239.00	245.53	1.87	1.24
Dec 10	199.93	2.40	74.92	-0.57	293.10	-1.14	269.56	21 Dec 26	254.6	327.3	4.5	-1.0	223.44	228.08	3.27	3.40	252.59	259.93	0.59	-0.03
Dec 20	200.51	2.44	73.73	-0.54	300.94	-1.14	279.75	31 Dec	37.1		-5.1		232.94		3.45		267.47		-0.61	
1775								1775												1775

JULIAN 12 UT	SATURN LONG.	SATURN LAT.	JUPITER LONG.	JUPITER LAT.	MARS LONG.	MARS LAT.	SUN LONG.
1775							
Dec 30	200.94	2.49	72.76	-0.50	308.82	-1.12	289.94
Jan 9	201.20	2.54	72.07	-0.47	316.72	-1.09	300.12
Jan 19	201.29	2.58	71.70	-0.43	324.63	-1.06	310.29
Jan 29	201.19	2.63	71.66	-0.40	332.53	-1.01	320.41
Feb 8	200.93	2.67	71.97	-0.37	340.41	-0.95	330.50
Feb 18	200.50	2.71	72.59	-0.34	348.26	-0.88	340.55
Feb 28	199.95	2.74	73.50	-0.31	356.06	-0.80	350.54
Mar 9	199.28	2.76	74.68	-0.28	3.82	-0.72	0.48
Mar 19	198.54	2.78	76.08	-0.25	11.50	-0.63	10.36
Mar 29	197.78	2.78	77.69	-0.23	19.12	-0.53	20.18
Apr 8	197.02	2.77	79.46	-0.21	26.65	-0.43	29.96
Apr 18	196.31	2.76	81.38	-0.19	34.11	-0.33	39.68
Apr 28	195.68	2.74	83.41	-0.17	41.48	-0.23	49.34
May 8	195.17	2.71	85.53	-0.15	48.76	-0.12	58.97
May 18	194.80	2.67	87.72	-0.13	55.95	-0.01	68.57
May 28	194.59	2.63	89.95	-0.11	63.05	0.10	78.13
Jun 7	194.54	2.59	92.22	-0.10	70.06	0.21	87.68
Jun 17	194.65	2.55	94.49	-0.08	76.99	0.31	97.21
Jun 27	194.93	2.50	96.75	-0.07	83.82	0.42	106.74
Jul 7	195.36	2.46	98.99	-0.05	90.58	0.53	116.28
Jul 17	195.95	2.42	101.17	-0.03	97.24	0.63	125.84
Jul 27	196.66	2.39	103.29	-0.02	103.82	0.73	135.42
Aug 6	197.50	2.36	105.32	0.00	110.32	0.83	145.03
Aug 16	198.44	2.33	107.24	0.02	116.74	0.94	154.68
Aug 26	199.47	2.31	109.01	0.03	123.07	1.04	164.38
Sep 5	200.57	2.29	110.63	0.05	129.31	1.14	174.13
Sep 15	201.72	2.27	112.05	0.07	135.46	1.24	183.93
Sep 25	202.91	2.27	113.24	0.09	141.50	1.34	193.79
Oct 5	204.13	2.26	114.19	0.12	147.44	1.44	203.71
Oct 15	205.34	2.27	114.85	0.14	153.26	1.55	213.68
Oct 25	206.53	2.27	115.20	0.16	158.93	1.65	223.70
Nov 4	207.69	2.29	115.22	0.19	164.45	1.77	233.78
Nov 14	208.79	2.31	114.92	0.22	169.78	1.88	243.89
Nov 24	209.82	2.33	114.29	0.25	174.88	2.00	254.04
Dec 4	210.75	2.36	113.38	0.27	179.72	2.13	264.22
Dec 14	211.56	2.39	112.24	0.30	184.23	2.26	274.40
1776							
Dec 24	212.25	2.42	110.95	0.32	188.34	2.40	284.60
Jan 3	212.79	2.46	109.61	0.35	191.95	2.55	294.79
Jan 13	213.17	2.51	108.31	0.37	194.95	2.71	304.96
Jan 23	213.37	2.55	107.16	0.38	197.18	2.86	315.11
Feb 2	213.41	2.59	106.23	0.40	198.46	3.01	325.22
Feb 12	213.27	2.63	105.58	0.41	198.64	3.13	335.29
Feb 22	212.96	2.67	105.24	0.42	197.55	3.20	345.31
Mar 4	212.51	2.70	105.22	0.43	195.20	3.18	355.28
Mar 14	211.92	2.73	105.53	0.43	191.85	3.05	5.19
Mar 24	211.25	2.74	106.14	0.44	188.03	2.78	15.04
Apr 3	210.51	2.75	107.02	0.44	184.45	2.41	24.84
Apr 13	209.75	2.75	108.16	0.44	181.75	1.98	34.58
Apr 23	209.00	2.74	109.52	0.45	180.28	1.53	44.28
May 3	208.32	2.72	111.07	0.45	180.12	1.11	53.93
May 13	207.72	2.69	112.78	0.46	181.19	0.73	63.54
May 23	207.25	2.66	114.63	0.47	183.31	0.40	73.11
Jun 2	206.91	2.62	116.60	0.47	186.30	0.10	82.67
Jun 12	206.73	2.58	118.65	0.48	190.02	-0.15	92.21
Jun 22	206.71	2.53	120.77	0.49	194.34	-0.37	101.74
Jul 2	206.86	2.49	122.94	0.50	199.15	-0.57	111.28
Jul 12	207.16	2.45	125.13	0.51	204.37	-0.73	120.82
Jul 22	207.62	2.40	127.34	0.53	209.96	-0.88	130.39
Aug 1	208.23	2.36	129.54	0.54	215.86	-1.01	139.99
Aug 11	208.96	2.33	131.71	0.56	222.03	-1.12	149.62
Aug 21	209.82	2.29	133.82	0.58	228.45	-1.21	159.29
Aug 31	210.77	2.27	135.87	0.60	235.08	-1.29	169.01
Sep 10	211.80	2.24	137.83	0.62	241.91	-1.35	178.79
Sep 20	212.91	2.22	139.67	0.65	248.92	-1.39	188.62
Sep 30	214.07	2.21	141.37	0.68	256.09	-1.42	198.51
Oct 10	215.25	2.20	142.89	0.71	263.41	-1.43	208.45
Oct 20	216.46	2.20	144.21	0.74	270.84	-1.43	218.45
Oct 30	217.66	2.20	145.30	0.78	278.39	-1.42	228.50
Nov 9	218.83	2.20	146.12	0.82	286.03	-1.39	238.59
Nov 19	219.96	2.22	146.66	0.86	293.75	-1.35	248.72
Nov 29	221.03	2.23	146.88	0.90	301.52	-1.29	258.88
Dec 9	222.02	2.25	146.79	0.94	309.34	-1.22	269.06
Dec 19	222.91	2.28	146.36	0.99	317.18	-1.15	279.26
1777							

GREGORIAN 12 UT	MOON LONG.		MOON LAT.		VENUS LONG.		VENUS LAT.		MERCURY LONG.		MERCURY LAT.	
1776												
Jan 5		101.3		-2.9		237.99		3.43		275.18		-1.12
10 Jan 15	161.2	221.9	2.3	5.2	243.20	248.55	3.35	3.21	283.05	291.12	-1.54	-1.86
20 Jan 25	291.5	5.9	2.2	-4.1	254.01	259.56	3.03	2.80	299.43	307.97	-2.04	-2.07
30 J-F 4	73.2	134.2	-4.7	-0.1	265.19	270.89	2.55	2.26	316.76	325.69	-1.90	-1.50
9 Feb 14	193.5	256.7	4.5	4.5	276.64	282.45	1.96	1.64	334.47	342.46	-0.81	0.17
19 Feb 24	329.6	43.3	-1.3	-5.3	288.29	294.17	1.32	0.99	348.57	351.49	1.37	2.60
29 Feb	107.3		-2.5		300.08		0.67		350.52		3.48	
Mar 5		166.6		2.7		306.02		0.35		346.47		3.65
10 Mar 15	227.2	293.5	5.2	1.9	311.97	317.95	0.05	-0.24	341.69	338.50	3.00	1.86
20 Mar 25	8.0	78.8	-4.1	-4.2	323.94	329.95	-0.51	-0.75	337.80	339.41	0.62	-0.49
30 M-A 4	139.9	199.5	0.5	4.6	335.96	341.99	-0.97	-1.16	342.84	347.65	-1.38	-2.03
9 Apr 14	262.8	332.0	3.9	-1.7	348.02	354.07	-1.33	-1.46	353.53	0.30	-2.45	-2.64
19 Apr 24	45.7	112.3	-5.0	-1.7	0.12	6.17	-1.56	-1.62	7.85	16.13	-2.60	-2.35
29 Apr	171.9		3.3		12.23		-1.66		25.15		-1.87	
May 4		233.5		4.9		18.29		-1.66		34.90		-1.20
9 May 14	300.2	11.1	0.9	-4.4	24.36	30.43	-1.64	-1.58	45.33	56.20	-0.39	0.49
19 May 24	81.7	144.4	-3.7	1.4	36.50	42.58	-1.50	-1.39	67.04	77.31	1.28	1.86
29 M-J 3	204.2	269.2	4.9	3.2	48.66	54.75	-1.26	-1.11	86.63	94.84	2.13	2.09
8 Jun 13	339.1	49.6	-2.7	-5.0	60.84	66.93	-0.94	-0.76	101.86	107.63	1.72	1.07
18 Jun 23	115.8	176.0	-1.0	3.9	73.04	79.14	-0.57	-0.37	112.00	114.78	0.14	-1.01
28 Jun	237.5		4.9		85.26		-0.16		115.74		-2.30	
Jul 3		306.6		0.0		91.38		0.04		114.79		-3.55
8 Jul 13	18.4	86.2	-5.0	-3.4	97.50	103.64	0.24	0.44	112.24	108.99	-4.52	-4.92
18 Jul 23	148.5	208.0	2.0	5.2	109.78	115.94	0.62	0.79	106.38	105.60	-4.63	-3.76
28 J-A 2	272.4	345.3	3.0	-3.3	122.09	128.26	0.95	1.08	107.26	111.50	-2.56	-1.25
7 Aug 12	56.7	120.8	-5.0	-0.5	134.43	140.61	1.20	1.29	118.11	126.61	-0.04	0.92
17 Aug 22	180.7	241.0	4.3	4.8	146.80	153.00	1.36	1.40	136.25	146.23	1.52	1.76
27 Aug	309.0		-0.3		159.20		1.42		156.04		1.70	
Sep 1		24.3		-5.1		165.40		1.41		165.42		1.41
6 Sep 11	93.0	153.9	-2.8	2.4	171.60	177.82	1.37	1.30	174.32	182.74	0.97	0.43
16 Sep 21	213.3	275.7	5.1	2.5	184.03	190.25	1.21	1.10	190.72	198.28	-0.17	-0.79
26 S-O 1	347.2	62.1	-3.4	-4.5	196.46	202.68	0.96	0.80	205.42	212.10	-1.41	-2.00
6 Oct 11	127.0	186.5	0.3	4.5	208.90	215.12	0.63	0.44	218.21	223.53	-2.52	-2.93
16 Oct 21	247.0	312.3	4.1	-0.9	221.34	227.55	0.23	0.02	227.62	229.74	-3.14	-3.02
26 Oct 31	25.9	97.8	-5.0	-2.0	233.77	239.98	-0.19	-0.41	228.79	224.14	-2.38	-1.07
5 Nov 10	159.6	219.3	3.2	5.0	246.20	252.41	-0.63	-0.84	217.78	214.05	0.62	1.89
15 Nov 20	282.4	350.7	1.5	-4.0	258.62	264.82	-1.03	-1.22	214.91	219.21	2.40	2.33
25 Nov 30	63.9	131.3	-4.2	1.2	271.02	277.21	-1.39	-1.51	225.37	232.41	1.94	1.39
5 Dec 10	191.4	252.8	4.9	3.6	283.40	289.58	-1.66	-1.75	239.84	247.44	0.79	0.19
15 Dec 20	319.5	30.0	-2.0	-5.2	295.76	301.92	-1.82	-1.86	255.13	262.90	-0.39	-0.92
25 Dec 30	100.1	163.5	-1.5	3.8	308.06	314.19	-1.86	-1.82	270.76	278.74	-1.37	-1.74
1777												
4 Jan 9	223.3	287.7	5.1	0.9	320.30	326.39	-1.75	-1.64	286.87	295.17	-1.99	-2.10
14 Jan 19	358.2	68.6	-4.6	-4.0	332.45	338.49	-1.50	-1.31	303.62	312.12	-2.03	-1.74
24 Jan 29	134.4	195.2	1.6	5.2	344.48	350.44	-1.10	-0.84	320.39	327.78	-1.16	-0.25
3 Feb 8	255.8	324.2	3.5	-2.4	356.35	2.22	-0.56	-0.25	333.13	334.94	0.98	2.34
13 Feb 18	37.5	105.3	-5.1	-1.1	8.02	13.76	0.09	0.46	332.51	327.31	3.42	3.72
23 Feb 28	167.6	227.2	4.0	4.8	19.42	25.00	0.84	1.24	322.48	320.17	3.15	2.10
5 Mar 10	289.9	2.4	0.6	-4.7	30.48	35.85	1.64	2.05	320.65	323.32	0.95	-0.09
15 Mar 20	75.7	140.1	-3.2	2.1	41.10	46.19	2.47	2.87	327.59	333.03	-0.95	-1.63
25 Mar 30	200.2	260.1	5.0	2.8	51.11	55.83	3.26	3.63	339.36	346.43	-2.10	-2.38
4 Apr 9	325.9	41.2	-2.7	-4.8	60.31	64.50	3.96	4.26	354.16	2.53	-2.46	-2.33
14 Apr 19	111.8	173.4	0.0	4.4	68.36	71.80	4.51	4.69	11.55	21.22	-1.99	-1.43
24 Apr 29	232.9	294.6	4.3	-0.4	74.74	77.08	4.79	4.79	31.50	42.22	-0.70	0.16
4 May 9	4.0	79.2	-4.8	-2.6	78.70	79.49	4.66	4.37	52.94	63.10	1.03	1.75
14 May 19	146.0	206.0	3.0	5.1	79.32	78.15	3.90	3.21	72.21	80.03	2.20	2.32
24 May 29	266.5	331.3	1.9	-3.5	76.03	73.21	2.30	1.22	86.42	91.29	2.09	1.51
3 Jun 8	43.2	115.4	-4.7	0.6	70.09	67.17	0.05	-1.10	94.49	95.86	0.58	-0.63
13 Jun 18	178.7	238.4	4.9	4.0	64.87	63.45	-2.13	-2.97	95.36	93.27	-2.01	-3.34
23 Jun 28	301.5	9.9	-1.2	-5.2	63.02	63.55	-3.61	-4.06	90.40	87.92	-4.31	-4.70
3 Jul 8	81.8	149.6	-2.3	3.5	64.94	67.06	-4.34	-4.48	86.88	87.86	-4.49	-3.78
13 Jul 18	210.5	271.4	5.2	1.5	69.79	73.02	-4.51	-4.44	91.03	96.32	-2.76	-1.60
23 Jul 28	338.5	49.4	-4.1	-4.5	76.68	80.69	-4.30	-4.09	103.56	112.45	-0.44	0.57
2 Aug 7	118.7	182.5	1.0	5.0	84.98	89.53	-3.83	-3.53	122.40	132.72	1.30	1.69
12 Aug 17	242.3	305.9	3.7	-1.6	94.27	99.20	-3.20	-2.85	142.84	152.49	1.76	1.56
22 Aug 27	17.3	88.1	-5.1	-1.7	104.27	109.48	-2.48	-2.10	161.57	170.11	1.18	0.67
1 Sep 6	153.7	214.7	3.7	4.8	114.81	120.25	-1.71	-1.33	178.12	185.62	0.06	-0.59
11 Sep 16	274.7	342.2	1.0	-4.2	125.77	131.38	-0.95	-0.58	192.59	198.99	-1.27	-1.93
21 Sep 26	56.6	125.0	-3.8	1.7	137.06	142.81	-0.23	0.11	204.68	209.40	-2.55	-3.08
1 Oct 6	187.3	246.8	5.0	3.0	148.62	154.49	0.42	0.70	212.68	213.80	-3.42	-3.45
11 Oct 16	308.5	20.3	-2.2	-5.0	160.41	166.38	0.95	1.18	211.86	206.74	-2.93	-1.69
21 Oct 26	94.8	160.0	-0.6	4.3	172.39	178.43	1.37	1.52	200.87	198.15	-0.01	1.39
31 Oct	220.1		4.4		184.52		1.64		199.94		2.11	
Nov 5		279.7		0.0		190.63		1.72		205.03		2.23
10 Nov 15	344.4	59.0	-4.6	-3.4	196.76	202.92	1.77	1.78	211.82	219.35	1.98	1.53
20 Nov 25	131.0	193.3	2.7	5.2	209.10	215.30	1.76	1.70	227.14	235.00	0.98	0.40
30 N-D 5	252.7	314.1	2.3	-3.0	221.52	227.74	1.62	1.50	242.86	250.70	-0.17	-0.71
10 Dec 15	22.3	96.9	-5.1	-0.2	233.98	240.22	1.37	1.21	258.55	266.45	-1.20	-1.61
20 Dec 25	165.2	225.6	4.8	4.2	246.47	252.73	1.03	0.84	274.41	282.46	-1.93	-2.12
30 Dec	285.9		-0.7		258.99		0.63		290.57		-2.15	
1777												

JULIAN 12 UT	SATURN LONG.	SATURN LAT.	JUPITER LONG.	JUPITER LAT.	MARS LONG.	MARS LAT.	SUN LONG.
1777							
Dec 29	223.68	2.31	145.64	1.02	325.02	-1.06	289.45
Jan 8	224.32	2.34	144.65	1.06	332.86	-0.97	299.63
Jan 18	224.80	2.37	143.47	1.09	340.68	-0.87	309.79
Jan 28	225.13	2.41	142.17	1.11	348.46	-0.76	319.92
Feb 7	225.28	2.45	140.86	1.12	356.19	-0.65	330.01
Feb 17	225.26	2.49	139.62	1.13	3.86	-0.54	340.06
Feb 27	225.08	2.52	138.55	1.13	11.47	-0.42	350.06
Mar 9	224.73	2.55	137.71	1.12	19.01	-0.31	360.00
Mar 19	224.24	2.58	137.14	1.11	26.47	-0.20	9.88
Mar 29	223.64	2.60	136.89	1.09	33.84	-0.09	19.71
Apr 8	222.95	2.61	136.94	1.07	41.14	0.02	29.48
Apr 18	222.21	2.61	137.30	1.05	48.35	0.13	39.20
Apr 28	221.46	2.60	137.95	1.03	55.48	0.23	48.87
May 8	220.73	2.59	138.86	1.01	62.52	0.33	58.50
May 18	220.07	2.57	140.01	1.00	69.49	0.43	68.10
May 28	219.51	2.53	141.36	0.98	76.37	0.52	77.66
Jun 7	219.07	2.50	142.90	0.97	83.19	0.61	87.21
Jun 17	218.77	2.46	144.60	0.96	89.93	0.69	96.74
Jun 27	218.63	2.42	146.42	0.95	96.61	0.77	106.28
Jul 7	218.65	2.37	148.36	0.95	103.22	0.84	115.81
Jul 17	218.83	2.33	150.38	0.94	109.78	0.92	125.37
Jul 27	219.17	2.28	152.47	0.94	116.27	0.98	134.95
Aug 6	219.66	2.24	154.61	0.95	122.72	1.05	144.56
Aug 16	220.29	2.20	156.77	0.95	129.11	1.11	154.21
Aug 26	221.05	2.17	158.94	0.96	135.46	1.17	163.91
Sep 5	221.92	2.13	161.09	0.97	141.75	1.22	173.65
Sep 15	222.89	2.11	163.21	0.99	148.00	1.27	183.46
Sep 25	223.94	2.08	165.28	1.00	154.20	1.32	193.31
Oct 5	225.05	2.06	167.27	1.02	160.34	1.37	203.23
Oct 15	226.21	2.05	169.16	1.05	166.43	1.41	213.20
Oct 25	227.40	2.04	170.91	1.08	172.46	1.45	223.22
Nov 4	228.59	2.04	172.52	1.11	178.42	1.49	233.29
Nov 14	229.78	2.04	173.93	1.14	184.30	1.53	243.40
Nov 24	230.93	2.04	175.13	1.18	190.08	1.56	253.55
Dec 4	232.04	2.05	176.09	1.23	195.76	1.59	263.72
Dec 14	233.08	2.06	176.77	1.27	201.31	1.62	273.91
1778							
Dec 24	234.03	2.08	177.15	1.32	206.71	1.64	284.10
Jan 3	234.87	2.10	177.22	1.36	211.92	1.65	294.29
Jan 13	235.60	2.13	176.97	1.41	216.90	1.66	304.47
Jan 23	236.18	2.15	176.41	1.45	221.61	1.65	314.61
Feb 2	236.61	2.18	175.57	1.49	225.97	1.63	324.73
Feb 12	236.88	2.22	174.51	1.52	229.90	1.59	334.80
Feb 22	236.98	2.25	173.28	1.54	233.29	1.53	344.82
Mar 4	236.91	2.28	171.99	1.55	236.01	1.44	354.79
Mar 14	236.68	2.30	170.72	1.55	237.89	1.30	4.71
Mar 24	236.29	2.33	169.56	1.54	238.76	1.10	14.56
Apr 3	235.77	2.35	168.59	1.52	238.44	0.83	24.37
Apr 13	235.15	2.36	167.86	1.49	236.86	0.47	34.12
Apr 23	234.45	2.36	167.42	1.45	234.15	0.04	43.81
May 3	233.71	2.36	167.27	1.42	230.71	-0.44	53.46
May 13	232.97	2.35	167.43	1.38	227.23	-0.93	63.08
May 23	232.26	2.33	167.88	1.34	224.44	-1.36	72.65
Jun 2	231.63	2.30	168.60	1.31	222.80	-1.72	82.21
Jun 12	231.10	2.27	169.58	1.27	222.53	-1.99	91.75
Jun 22	230.70	2.23	170.78	1.24	223.60	-2.19	101.28
Jul 2	230.44	2.20	172.18	1.21	225.83	-2.33	110.82
Jul 12	230.34	2.15	173.75	1.19	229.06	-2.42	120.37
Jul 22	230.40	2.11	175.48	1.17	233.11	-2.47	129.93
Aug 1	230.62	2.07	177.33	1.15	237.83	-2.49	139.52
Aug 11	231.00	2.03	179.28	1.14	243.10	-2.48	149.15
Aug 21	231.52	1.99	181.32	1.12	248.84	-2.44	158.82
Aug 31	232.18	1.95	183.42	1.12	254.96	-2.39	168.54
Sep 10	232.97	1.92	185.56	1.11	261.39	-2.31	178.31
Sep 20	233.86	1.89	187.72	1.11	268.10	-2.22	188.14
Sep 30	234.85	1.86	189.88	1.11	275.03	-2.12	198.02
Oct 10	235.91	1.84	192.02	1.12	282.14	-2.00	207.96
Oct 20	237.04	1.82	194.12	1.13	289.41	-1.87	217.96
Oct 30	238.20	1.81	196.16	1.14	296.79	-1.72	228.00
Nov 9	239.38	1.80	198.10	1.16	304.27	-1.57	238.10
Nov 19	240.57	1.79	199.93	1.18	311.80	-1.42	248.22
Nov 29	241.73	1.79	201.61	1.21	319.39	-1.26	258.39
Dec 9	242.87	1.79	203.13	1.23	326.99	-1.10	268.57
Dec 19	243.95	1.80	204.44	1.27	334.60	-0.94	278.76
1779							

GREGORIAN 12 UT	MOON LONG.		MOON LAT.		VENUS LONG.		VENUS LAT.		MERCURY LONG.		MERCURY LAT.	
1778												
Jan 4		350.6		-5.0		265.26		0.42		298.63		-1.96
9 Jan 14	61.2	133.2	-3.2	3.0	271.52	277.79	0.21	0.00	306.36	313.10	-1.49	-0.68
19 Jan 24	197.9	257.4	5.2	1.9	284.05	290.32	-0.21	-0.41	317.64	318.36	0.53	1.99
29 J-F 3	320.4	29.1	-3.5	-4.8	296.59	302.85	-0.61	-0.78	314.62	308.70	3.24	3.68
8 Feb 13	99.8	168.0	0.1	4.8	309.11	315.37	-0.94	-1.08	304.28	303.03	3.21	2.27
18 Feb 23	229.6	289.8	3.8	-1.1	321.62	327.87	-1.20	-1.30	304.59	308.17	1.22	0.25
28 Feb	356.9		-5.0		334.12		-1.37		313.12		-0.58	
Mar 5		68.4		-2.4		340.36		-1.42		319.04		-1.26
10 Mar 15	137.1	201.4	3.3	4.8	346.59	352.82	-1.44	-1.43	325.69	332.96	-1.78	-2.13
20 Mar 25	261.2	323.5	1.2	-3.8	359.04	5.26	-1.39	-1.33	340.80	349.18	-2.30	-2.28
30 M-A 4	35.1	107.0	-4.3	1.2	11.47	17.67	-1.25	-1.14	358.14	7.67	-2.06	-1.63
9 Apr 14	172.8	234.0	4.9	3.1	23.86	30.04	-1.01	-0.86	17.76	28.26	-0.99	-0.17
19 Apr 24	293.4	359.3	-1.9	-5.1	36.22	42.39	-0.70	-0.52	38.79	48.75	0.73	1.57
29 Apr	73.9		-1.5		48.56		-0.32		57.55		2.21	
May 4		144.1		4.1		54.71		-0.12		64.82		2.52
9 May 14	206.8	266.3	4.6	0.3	60.86	66.99	0.08	0.28	70.36	74.03	2.46	1.98
19 May 24	327.2	37.0	-4.4	-4.2	73.13	79.25	0.49	0.69	75.73	75.46	1.09	-0.16
29 M-J 3	112.1	179.3	2.0	5.3	85.37	91.48	0.87	1.05	73.56	70.85	-1.60	-2.94
8 Jun 13	239.7	299.2	2.6	-2.6	97.58	103.67	1.21	1.35	68.44	67.32	-3.89	-4.31
18 Jun 23	3.2	75.7	-5.3	-1.3	109.75	115.83	1.47	1.56	68.02	70.64	-4.21	-3.69
28 Jun	148.9		4.5		121.89		1.62		75.11		-2.88	
Jul 3		212.7		4.4		127.95		1.66		81.30		-1.88
8 Jul 13	272.0	333.7	-0.2	-4.7	133.99	140.02	1.66	1.64	89.10	98.30	-0.81	0.21
18 Jul 23	41.1	114.2	-3.9	2.2	146.04	152.04	1.58	1.49	108.49	119.08	1.03	1.56
28 J-A 2	183.8	244.9	5.2	2.1	158.03	164.00	1.36	1.20	129.48	139.37	1.77	1.70
7 Aug 12	305.0	10.3	-3.0	-5.0	169.95	175.89	1.01	0.79	148.63	157.25	1.39	0.91
17 Aug 22	80.1	151.5	-0.7	4.5	181.80	187.68	0.54	0.27	165.27	172.68	0.32	-0.37
27 Aug	217.0		3.9		193.54		-0.02		179.46		-1.10	
Sep 1		276.6		-0.9		199.37		-0.34		185.53		-1.85
6 Sep 11	339.4	48.6	-4.8	-3.0	205.17	210.94	-0.66	-1.00	190.73	194.77	-2.57	-3.22
16 Sep 21	119.0	187.1	2.8	4.9	216.66	222.34	-1.35	-1.69	197.16	197.22	-3.70	-3.86
26 S-O 1	248.9	308.7	1.3	-3.6	227.97	233.55	-2.03	-2.36	194.35	189.03	-3.45	-2.28
6 Oct 11	15.6	87.7	-4.7	0.5	239.05	244.48	-2.67	-2.96	183.96	182.36	-0.62	0.87
16 Oct 21	156.8	220.9	4.9	3.3	249.82	255.06	-3.22	-3.44	185.08	190.88	1.77	2.08
26 Oct 31	280.5	342.2	-1.7	-5.1	260.17	265.12	-3.62	-3.75	198.28	206.29	1.98	1.63
5 Nov 10	53.3	126.4	-2.4	3.7	269.90	274.45	-3.81	-3.80	214.45	222.57	1.15	0.60
15 Nov 20	192.7	253.6	4.8	0.6	278.75	282.71	-3.71	-3.52	230.60	238.53	0.04	-0.51
25 Nov 30	312.8	17.4	-4.2	-4.7	286.28	289.35	-3.22	-2.79	246.39	254.21	-1.03	-1.48
5 Dec 10	91.7	163.7	1.0	5.3	291.82	293.56	-2.22	-1.49	262.02	269.82	-1.86	-2.12
15 Dec 20	226.6	285.9	3.0	-2.3	294.44	294.34	-0.58	0.50	277.59	285.20	-2.24	-2.16
25 Dec 30	346.5	54.4	-5.3	-2.3	293.21	291.11	1.73	3.03	292.36	298.41	-1.81	-1.11
1779												
4 Jan 9	129.8	198.8	4.0	4.6	288.31	285.26	4.29	5.37	302.10	301.74	0.04	1.55
14 Jan 19	259.2	318.8	0.1	-4.4	282.48	280.40	6.17	6.65	296.93	290.70	2.94	3.52
24 Jan 29	22.2	92.7	-4.3	1.2	279.28	279.18	6.82	6.76	286.98	286.81	3.18	2.37
3 Feb 8	166.8	232.0	5.1	2.3	280.06	281.79	6.51	6.13	289.37	293.71	1.44	0.54
13 Feb 18	291.3	353.1	-2.8	-5.0	284.25	287.30	5.67	5.15	299.22	305.52	-0.26	-0.94
23 Feb 28	59.8	131.4	-1.5	4.1	290.84	294.77	4.60	4.03	312.42	319.83	-1.48	-1.89
5 Mar 10	202.0	263.9	4.1	-0.7	299.03	303.56	3.46	2.90	327.71	336.07	-2.13	-2.21
15 Mar 20	323.8	29.2	-4.7	-3.6	308.31	313.24	2.34	1.81	344.91	354.27	-2.10	-1.78
25 Mar 30	98.7	169.4	2.1	5.0	318.32	323.53	1.30	0.81	4.12	14.35	-1.24	-0.50
4 Apr 9	235.5	295.4	1.5	-3.5	328.85	334.25	0.36	-0.05	24.61	34.29	0.40	1.34
14 Apr 19	357.5	66.9	-4.9	-0.5	339.74	345.29	-0.43	-0.77	42.68	49.25	2.15	2.67
24 Apr 29	137.9	205.7	4.7	3.7	350.90	356.56	-1.07	-1.33	53.71	55.89	2.80	2.46
4 May 9	267.8	327.4	-1.5	-5.1	2.26	8.00	-1.55	-1.72	55.83	53.93	1.62	0.37
14 May 19	32.9	105.7	-3.3	3.0	13.77	19.57	-1.85	-1.94	51.08	48.51	-1.08	-2.40
24 May 29	176.1	240.0	5.1	1.0	25.40	31.25	-1.99	-2.00	47.22	47.72	-3.34	-3.81
3 Jun 8	299.8	0.7	-4.0	-5.0	37.12	43.02	-1.97	-1.91	50.04	54.06	-3.85	-3.53
13 Jun 18	70.0	144.6	-0.1	5.0	48.93	54.86	-1.82	-1.70	59.59	66.51	-2.92	-2.09
23 Jun 28	212.1	273.0	3.3	-1.9	60.81	66.78	-1.55	-1.38	74.75	84.19	-1.14	-0.15
3 Jul 8	332.4	35.7	-5.1	-3.0	72.76	78.76	-1.20	-0.99	94.56	105.34	0.73	1.39
13 Jul 18	108.4	182.2	3.2	4.8	84.77	90.81	-0.78	-0.56	115.96	126.05	1.75	1.80
23 Jul 28	246.1	305.5	0.4	-4.2	96.85	102.92	-0.33	-0.11	135.45	144.13	1.59	1.17
2 Aug 7	6.3	72.7	-4.5	0.3	109.00	115.10	0.11	0.32	152.11	159.39	0.59	-0.11
12 Aug 17	147.2	217.7	4.9	2.6	121.21	127.34	0.53	0.71	165.92	171.59	-0.90	-1.73
22 Aug 27	278.8	338.6	-2.6	-5.0	133.49	139.65	0.89	1.04	176.21	179.46	-2.57	-3.35
1 Sep 6	41.9	111.2	-2.1	3.6	145.82	152.01	1.17	1.27	180.84	179.81	-3.98	-4.25
11 Sep 16	185.0	251.1	4.4	-0.5	158.21	164.42	1.35	1.40	176.17	171.03	-3.92	-2.81
21 Sep 26	310.8	12.7	-4.6	-4.0	170.65	176.88	1.43	1.43	167.06	166.68	-1.19	0.34
1 Oct 6	79.4	150.4	1.4	5.1	183.12	189.36	1.40	1.34	170.28	176.75	1.39	1.89
11 Oct 16	220.7	283.2	2.0	-3.4	195.62	201.88	1.26	1.16	184.71	193.20	1.95	1.72
21 Oct 26	343.1	48.3	-5.1	-1.3	208.14	214.41	1.03	0.89	201.73	210.13	1.31	0.80
31 Oct	118.3		4.3		220.68		0.72		218.33		0.25	
Nov 5		188.6		4.2		226.95		0.55		226.36		-0.31
10 Nov 15	254.5	314.7	-1.1	-5.0	233.23	239.50	0.36	0.16	234.24	241.99	-0.85	-1.34
20 Nov 25	16.2	85.5	-3.8	2.1	245.78	252.05	-0.04	-0.24	249.65	257.22	-1.77	-2.11
30 N-D 5	157.6	225.0	5.3	1.6	258.33	264.61	-0.44	-0.63	264.64	271.80	-2.32	-2.34
10 Dec 15	287.1	346.6	-3.7	-5.1	270.88	277.15	-0.82	-0.99	278.37	283.68	-2.12	-1.54
20 Dec 25	50.7	124.0	-1.0	4.6	283.42	289.69	-1.14	-1.28	286.47	285.08	-0.49	1.03
30 Dec	195.7		3.7		295.95		-1.39		279.45		2.54	
1779												

JULIAN 12 UT	SATURN LONG.	SATURN LAT.	JUPITER LONG.	JUPITER LAT.	MARS LONG.	MARS LAT.	SUN LONG.
1779							
Dec 29	244.95	1.81	205.52	1.30	342.19	-0.78	288.95
Jan 8	245.86	1.82	206.34	1.34	349.75	-0.63	299.14
Jan 18	246.66	1.84	206.88	1.38	357.28	-0.48	309.30
Jan 28	247.34	1.86	207.11	1.42	4.75	-0.33	319.43
Feb 7	247.87	1.88	207.03	1.45	12.16	-0.19	329.53
Feb 17	248.24	1.90	206.63	1.49	19.51	-0.06	339.57
Feb 27	248.45	1.93	205.94	1.52	26.80	0.06	349.57
Mar 8	248.50	1.95	205.01	1.55	34.00	0.18	359.52
Mar 18	248.38	1.97	203.88	1.56	41.14	0.29	9.40
Mar 28	248.10	1.99	202.63	1.57	48.20	0.39	19.24
Apr 7	247.67	2.01	201.35	1.56	55.19	0.49	29.01
Apr 17	247.12	2.02	200.14	1.54	62.11	0.58	38.73
Apr 27	246.48	2.02	199.06	1.52	68.96	0.66	48.41
May 7	245.77	2.02	198.19	1.48	75.74	0.73	58.04
May 17	245.03	2.01	197.58	1.44	82.46	0.80	67.64
May 27	244.30	2.00	197.26	1.40	89.13	0.86	77.21
Jun 6	243.61	1.98	197.23	1.35	95.74	0.92	86.75
Jun 16	243.01	1.95	197.51	1.31	102.31	0.97	96.29
Jun 26	242.52	1.92	198.07	1.26	108.83	1.01	105.82
Jul 6	242.15	1.89	198.89	1.22	115.31	1.05	115.36
Jul 16	241.94	1.85	199.96	1.18	121.76	1.09	124.91
Jul 26	241.88	1.81	201.24	1.14	128.18	1.12	134.49
Aug 5	241.98	1.77	202.72	1.11	134.57	1.14	144.10
Aug 15	242.25	1.73	204.37	1.08	140.94	1.16	153.74
Aug 25	242.66	1.69	206.17	1.06	147.30	1.18	163.44
Sep 4	243.23	1.66	208.08	1.03	153.64	1.19	173.18
Sep 14	243.92	1.63	210.09	1.01	159.96	1.20	182.98
Sep 24	244.74	1.60	212.18	1.00	166.28	1.20	192.83
Oct 4	245.66	1.57	214.33	0.98	172.58	1.20	202.74
Oct 14	246.67	1.54	216.51	0.97	178.88	1.20	212.71
Oct 24	247.75	1.52	218.71	0.97	185.16	1.19	222.73
Nov 3	248.88	1.51	220.89	0.96	191.43	1.17	232.80
Nov 13	250.04	1.49	223.05	0.96	197.69	1.15	242.91
Nov 23	251.22	1.48	225.15	0.96	203.93	1.12	253.06
Dec 3	252.40	1.48	227.17	0.97	210.15	1.08	263.23
Dec 13	253.55	1.47	229.09	0.98	216.34	1.04	273.42
1780							
Dec 23	254.66	1.47	230.88	0.99	222.50	0.98	283.62
Jan 2	255.71	1.48	232.51	1.00	228.62	0.92	293.80
Jan 12	256.68	1.48	233.96	1.02	234.69	0.84	303.98
Jan 22	257.55	1.49	235.18	1.04	240.69	0.75	314.13
Feb 1	258.30	1.50	236.17	1.06	246.62	0.64	324.24
Feb 11	258.93	1.52	236.88	1.08	252.45	0.51	334.32
Feb 21	259.40	1.53	237.31	1.10	258.17	0.35	344.35
Mar 3	259.72	1.55	237.42	1.12	263.73	0.17	354.32
Mar 13	259.88	1.56	237.22	1.14	269.11	-0.05	4.23
Mar 23	259.87	1.58	236.72	1.16	274.26	-0.31	14.10
Apr 2	259.70	1.59	235.94	1.17	279.12	-0.61	23.90
Apr 12	259.37	1.60	234.93	1.17	283.61	-0.98	33.65
Apr 22	258.91	1.61	233.76	1.17	287.61	-1.40	43.35
May 2	258.33	1.61	232.50	1.16	291.02	-1.90	53.00
May 12	257.66	1.61	231.24	1.14	293.65	-2.49	62.61
May 22	256.95	1.60	230.07	1.11	295.32	-3.15	72.19
Jun 1	256.21	1.59	229.06	1.08	295.87	-3.88	81.75
Jun 11	255.49	1.58	228.28	1.04	295.18	-4.63	91.29
Jun 21	254.83	1.55	227.78	1.00	293.33	-5.33	100.82
Jul 1	254.26	1.53	227.56	0.96	290.70	-5.85	110.35
Jul 11	253.80	1.50	227.65	0.92	287.93	-6.12	119.90
Jul 21	253.48	1.47	228.04	0.88	285.76	-6.12	129.46
Jul 31	253.31	1.43	228.71	0.84	284.74	-5.88	139.05
Aug 10	253.30	1.40	229.64	0.80	285.06	-5.48	148.68
Aug 20	253.45	1.36	230.82	0.76	286.71	-5.01	158.35
Aug 30	253.76	1.33	232.21	0.73	289.52	-4.51	168.06
Sep 9	254.22	1.30	233.80	0.70	293.29	-4.00	177.83
Sep 19	254.82	1.27	235.55	0.67	297.84	-3.52	187.66
Sep 29	255.56	1.24	237.43	0.65	302.99	-3.06	197.54
Oct 9	256.41	1.21	239.44	0.62	308.61	-2.63	207.48
Oct 19	257.35	1.19	241.54	0.60	314.59	-2.23	217.47
Oct 29	258.39	1.17	243.71	0.59	320.85	-1.86	227.51
Nov 8	259.48	1.15	245.93	0.57	327.33	-1.52	237.60
Nov 18	260.62	1.14	248.18	0.55	333.95	-1.21	247.73
Nov 28	261.79	1.12	250.43	0.54	340.69	-0.92	257.89
Dec 8	262.97	1.11	252.66	0.53	347.51	-0.67	268.07
Dec 18	264.14	1.11	254.85	0.52	354.37	-0.43	278.27
1781							

GREGORIAN 12 UT	MOON LONG.		MOON LAT.		VENUS LONG.		VENUS LAT.		MERCURY LONG.		MERCURY LAT.	
1780												1780
Jan 4		259.4		-1.6		302.22		-1.48		273.27		3.28
9 Jan 14	319.3	19.5	-5.0	-3.4	308.47	314.72	-1.55	-1.58	270.42	271.29	3.10	2.43
19 Jan 24	87.0	162.8	2.3	4.9	320.96	327.19	-1.59	-1.57	274.74	279.76	1.61	0.79
29 J-F 3	231.7	292.5	0.7	-4.0	333.40	339.61	-1.51	-1.43	285.74	292.37	0.03	-0.64
8 Feb 13	352.0	54.1	-4.6	-0.3	345.80	351.98	-1.32	-1.18	299.47	306.98	-1.21	-1.65
18 Feb 23	125.1	200.3	4.6	3.0	358.14	4.27	-1.01	-0.82	314.88	323.16	-1.96	-2.12
28 Feb	265.5		-2.5		10.39		-0.60		331.87		-2.11	
Mar 4		325.1		-5.0		16.48		-0.37		341.01		-1.90
9 Mar 14	25.7	90.8	-2.5	3.0	22.55	28.59	-0.11	0.15	350.59	0.50	-1.47	-0.81
19 Mar 24	164.0	235.8	4.8	-0.2	34.59	40.56	0.43	0.71	10.42	19.74	0.05	1.05
29 M-A 3	298.1	357.8	-4.6	-4.3	46.49	52.39	0.99	1.27	27.63	33.37	2.01	2.75
8 Apr 13	61.0	129.5	0.7	5.1	58.24	64.04	1.54	1.80	36.52	36.98	3.09	2.91
18 Apr 23	202.1	269.5	2.7	-3.1	69.78	75.47	2.03	2.25	35.12	31.95	2.14	0.90
28 Apr	329.9		-5.2		81.09		2.44		28.90		-0.52	
May 3		31.2		-2.0		86.64		2.59		27.17		-1.80
8 May 13	98.2	169.0	3.8	4.7	92.11	97.49	2.70	2.77	27.32	29.34	-2.73	-3.27
18 May 23	238.5	302.0	-0.5	-4.9	102.75	107.90	2.80	2.76	33.03	38.13	-3.45	-3.31
28 M-J 2	1.8	66.1	-4.1	1.2	112.90	117.75	2.67	2.51	44.47	51.95	-2.89	-2.24
7 Jun 12	137.1	207.6	5.2	2.2	122.40	126.82	2.28	1.97	60.53	70.14	-1.42	-0.50
17 Jun 22	273.1	334.0	-3.4	-5.1	130.96	134.79	1.58	1.09	80.60	91.50	0.41	1.17
27 Jun	34.5		-1.6		138.22		0.51		102.29		1.67	
Jul 2		103.0		4.0		141.18		-0.19		112.53		1.87
7 Jul 12	176.5	244.3	4.2	-1.1	143.58	145.29	-1.00	-1.92	122.02	130.71	1.78	1.43
17 Jul 22	306.5	6.1	-4.9	-3.5	146.18	146.16	-2.95	-4.05	138.59	145.64	0.89	0.18
27 J-A 1	68.8	141.6	1.6	5.0	145.17	143.23	-5.19	-6.27	151.82	156.98	-0.66	-1.58
6 Aug 11	214.7	279.1	1.2	-3.9	140.54	137.47	-7.18	-7.80	160.88	163.19	-2.54	-3.47
16 Aug 21	339.2	39.0	-4.6	-0.7	134.50	132.07	-8.08	-8.01	163.46	161.38	-4.23	-4.59
26 Aug 31	105.2	180.5	4.3	3.6	130.51	129.93	-7.65	-7.08	157.30	152.79	-4.30	-3.25
5 Sep 10	250.8	312.4	-2.1	-5.1	130.33	131.61	-6.39	-5.63	150.18	151.06	-1.71	-0.18
15 Sep 20	11.9	73.7	-2.8	2.5	133.66	136.36	-4.85	-4.08	155.53	162.61	0.99	1.66
25 Sep 30	143.4	218.3	5.1	0.5	139.61	143.31	-3.33	-2.61	171.09	180.03	1.88	1.77
5 Oct 10	284.9	344.9	-4.6	-4.5	147.38	151.76	-1.93	-1.30	188.94	197.62	1.45	0.99
15 Oct 20	45.4	110.4	0.2	4.8	156.41	161.28	-0.72	-0.19	206.02	214.16	0.46	-0.10
25 Oct 30	182.3	254.2	3.5	-2.7	166.34	171.57	0.30	0.73	222.06	229.77	-0.66	-1.20
4 Nov 9	317.5	17.2	-5.3	-2.4	176.93	182.41	1.10	1.42	237.29	244.62	-1.68	-2.09
14 Nov 19	80.5	149.1	3.2	5.1	188.00	193.68	1.69	1.91	251.70	258.39	-2.38	-2.51
24 Nov 29	220.4	288.3	0.3	-4.7	199.43	205.26	2.07	2.18	264.34	268.87	-2.41	-1.96
4 Dec 9	349.2	50.0	-4.2	0.6	211.14	217.07	2.25	2.26	270.72	268.37	-1.03	0.45
14 Dec 19	117.5	188.4	5.0	2.9	223.05	229.06	2.24	2.17	262.17	256.33	2.05	2.95
24 Dec 29	257.2	321.1	-2.9	-5.0	235.10	241.18	2.07	1.93	254.42	256.28	2.96	2.46
1781												1781
3 Jan 8	20.7	84.2	-1.8	3.4	247.27	253.39	1.77	1.58	260.54	266.17	1.75	1.00
13 Jan 18	156.1	226.9	4.5	-0.6	259.51	265.66	1.37	1.14	272.58	279.49	0.28	-0.38
23 Jan 28	292.3	353.2	-4.8	-3.6	271.81	277.98	0.90	0.66	286.77	294.35	-0.95	-1.43
2 Feb 7	53.0	120.4	1.3	5.0	284.15	290.32	0.41	0.16	302.24	310.44	-1.79	-2.02
12 Feb 17	195.0	263.7	1.9	-3.7	296.50	302.68	-0.08	-0.31	318.99	327.90	-2.09	-1.99
22 Feb 27	326.0	25.3	-4.8	-0.9	308.87	315.05	-0.52	-0.73	337.18	346.73	-1.67	-1.11
4 Mar 9	86.8	158.2	4.0	4.3	321.24	327.42	-0.91	-1.07	356.26	5.14	-0.30	0.71
14 Mar 19	232.9	298.7	-1.6	-5.2	333.61	339.79	-1.21	-1.32	12.46	17.25	1.79	2.74
24 Mar 29	358.7	58.3	-3.1	2.1	345.97	352.15	-1.40	-1.46	18.94	17.59	3.31	3.29
3 Apr 8	122.9	196.8	5.3	1.6	358.33	4.50	-1.49	-1.49	14.17	10.39	2.59	1.38
13 Apr 18	269.2	332.0	-4.3	-4.7	10.67	16.83	-1.47	-1.41	7.87	7.37	0.01	-1.21
23 Apr 28	31.2	92.9	-0.3	4.5	22.99	29.15	-1.34	-1.23	8.92	12.22	-2.14	-2.75
3 May 8	161.0	234.9	4.3	-1.8	35.31	41.46	-1.11	-0.97	16.94	22.84	-3.04	-3.06
13 May 18	303.6	4.1	-5.3	-2.7	47.61	53.76	-0.81	-0.64	29.77	37.65	-2.81	-2.33
23 May 28	64.4	129.6	2.6	5.2	59.91	66.05	-0.45	-0.26	46.46	56.16	-1.65	-0.82
2 Jun 7	200.1	271.7	1.3	-4.4	72.19	78.34	-0.06	0.14	66.65	77.59	0.08	0.91
12 Jun 17	336.5	36.0	-4.3	0.2	84.48	90.62	0.33	0.52	88.47	98.80	1.55	1.90
22 Jun 27	99.2	168.2	4.6	3.5	96.76	102.91	0.70	0.87	108.31	116.93	1.94	1.70
2 Jul 7	238.8	306.9	-2.3	-5.0	109.05	115.20	1.03	1.16	124.61	131.34	1.21	0.51
12 Jul 17	8.3	68.6	-1.9	3.1	121.34	127.48	1.28	1.37	137.02	141.49	-0.37	-1.38
22 Jul 27	135.9	207.5	4.8	0.2	133.63	139.77	1.44	1.48	144.48	145.68	-2.47	-3.54
1 Aug 6	276.3	340.4	-4.6	-3.7	145.92	152.06	1.49	1.47	144.78	141.86	-4.42	-4.84
11 Aug 16	40.0	102.6	1.1	4.9	158.20	164.33	1.43	1.36	137.83	134.41	-4.57	-3.57
21 Aug 26	174.3	246.2	2.8	-3.2	170.47	176.60	1.26	1.13	133.36	135.52	-2.14	-0.66
31 Aug	312.0		-4.9		182.72		0.97		140.81		0.56	
Sep 5		12.9		-1.2		188.84		0.80		148.46		1.38
10 Sep 15	72.3	138.5	3.8	4.8	194.95	201.06	0.60	0.38	157.41	166.78	1.77	1.80
20 Sep 25	213.1	283.3	-0.7	-5.1	207.16	213.26	0.15	-0.09	176.06	185.03	1.57	1.17
30 S-O 5	345.8	45.1	-3.4	1.8	219.34	225.42	-0.35	-0.60	193.63	201.89	0.67	0.11
10 Oct 15	106.2	176.0	5.2	2.6	231.49	237.54	-0.86	-1.11	209.83	217.48	-0.47	-1.04
20 Oct 25	251.2	318.4	-3.8	-4.9	243.59	249.62	-1.35	-1.58	224.87	231.96	-1.57	-2.05
30 Oct	18.5		-0.7		255.64		-1.79		238.70		-2.43	
Nov 4		78.1		4.2		261.64		-1.98		244.90		-2.67
9 Nov 14	142.1	214.4	4.7	-0.8	267.62	273.57	-2.15	-2.28	250.20	253.89	-2.69	-2.38
19 Nov 24	287.8	351.6	-5.2	-2.9	279.50	285.40	-2.37	-2.43	254.77	251.52	-1.58	-0.17
29 N-D 4	50.8	112.7	2.3	5.1	291.26	297.07	-2.44	-2.41	245.03	239.76	1.49	2.56
9 Dec 14	179.9	252.6	2.1	-3.8	302.83	308.53	-2.32	-2.18	238.82	241.62	2.78	2.44
19 Dec 24	322.5	23.4	-4.4	0.0	314.16	319.70	-1.99	-1.74	246.63	252.81	1.86	1.19
29 Dec	83.7		4.4		325.13		-1.43		259.62		0.51	
1781												1781

JULIAN 12 UT	SATURN LONG.	SATURN LAT.	JUPITER LONG.	JUPITER LAT.	MARS LONG.	MARS LAT.	SUN LONG.	GREGORIAN 12 UT	MOON LONG.		MOON LAT.		VENUS LONG.		VENUS LAT.		MERCURY LONG.		MERCURY LAT.	
1781								1782												1782
								Jan 3		149.2		4.0		330.44		-1.06		266.80		-0.13
Dec 28	265.28	1.10	256.98	0.51	1.26	-0.22	288.46	8 Jan 13	218.8	289.9	-1.5	-5.0	335.61	340.61	-0.63	-0.13	274.23	281.89	-0.71	-1.21
Jan 7	266.37	1.10	259.02	0.51	8.16	-0.03	298.65	18 Jan 23	355.4	54.9	-2.1	2.9	345.40	349.94	0.42	1.05	289.76	297.88	-1.62	-1.91
Jan 17	267.39	1.10	260.94	0.50	15.05	0.14	308.81	28 J-F 2	117.9	187.3	5.0	1.0	354.18	358.06	1.73	2.47	306.26	314.93	-2.06	-2.05
Jan 27	268.32	1.10	262.72	0.50	21.93	0.29	318.94	7 Feb 12	257.7	325.5	-4.4	-4.0	1.49	4.40	3.27	4.12	323.89	333.05	-1.84	-1.38
Feb 6	269.15	1.10	264.34	0.49	28.78	0.42	329.04	17 Feb 22	27.2	87.0	0.9	4.9	6.66	8.15	5.01	5.91	342.13	350.52	-0.66	0.33
Feb 16	269.85	1.11	265.75	0.49	35.59	0.54	339.09	27 Feb	153.7		3.5		8.72		6.78		357.18		1.50	
								Mar 4		226.2		-2.6		8.29		7.57		0.94		2.63
Feb 26	270.43	1.11	266.95	0.49	42.37	0.65	349.09	9 Mar 14	295.6	359.3	-5.2	-1.5	6.83	4.46	8.18	8.51	1.09	358.04	3.43	3.56
Mar 8	270.85	1.12	267.89	0.49	49.11	0.74	359.04	19 Mar 24	58.9	120.5	3.6	5.1	1.49	358.37	8.47	8.05	353.57	349.91	2.92	1.76
Mar 18	271.12	1.13	268.55	0.48	55.80	0.82	8.93	29 M-A 3	191.1	264.9	0.5	-4.9	355.62	353.62	7.29	6.30	348.38	349.16	0.47	-0.69
Mar 28	271.22	1.13	268.92	0.48	62.45	0.89	18.76	8 Apr 13	331.4	32.0	-3.7	1.4	352.61	352.61	5.20	4.08	351.91	356.19	-1.61	-2.26
Apr 7	271.16	1.14	268.97	0.48	69.06	0.95	28.54	18 Apr 23	91.4	155.8	5.1	3.4	353.55	355.30	3.02	2.04	1.68	8.15	-2.65	-2.79
Apr 17	270.94	1.14	268.72	0.47	75.63	1.00	38.27	28 Apr	229.6		-2.9		357.74		1.16		15.48		-2.69	
								May 3		302.2		-5.0		0.75		0.39		23.62		-2.37
Apr 27	270.57	1.15	268.16	0.46	82.16	1.04	47.94	8 May 13	5.3	64.5	-1.0	3.9	4.23	8.09	-0.28	-0.86	32.56	42.29	-1.83	-1.10
May 7	270.07	1.15	267.33	0.45	88.66	1.08	57.58	18 May 23	125.4	192.9	4.8	0.2	12.27	16.70	-1.34	-1.74	52.74	63.64	-0.25	0.62
May 17	269.46	1.14	266.29	0.43	95.12	1.11	67.17	28 M-J 2	268.1	337.4	-4.9	-3.1	21.35	26.18	-2.07	-2.32	74.53	84.87	1.36	1.87
May 27	268.78	1.14	265.09	0.41	101.55	1.13	76.74	7 Jun 12	37.9	97.6	2.0	5.0	31.16	36.27	-2.50	-2.62	94.34	102.77	2.07	1.96
Jun 6	268.05	1.13	263.82	0.39	107.97	1.14	86.29	17 Jun 22	161.2	231.6	2.7	-3.1	41.49	46.81	-2.68	-2.68	110.14	116.37	1.55	0.88
Jun 16	267.32	1.11	262.56	0.37	114.36	1.16	95.83	27 Jun	305.5		-4.6		52.21		-2.64		121.35		-0.02	
								Jul 2		10.7		-0.1		57.68		-2.56		124.89		-1.12
Jun 26	266.62	1.09	261.42	0.34	120.73	1.16	105.36	7 Jul 12	70.1	132.1	4.3	4.3	63.22	68.82	-2.43	-2.27	126.74	126.67	-2.34	-3.54
Jul 6	265.98	1.07	260.45	0.31	127.09	1.16	114.90	17 Jul 22	198.9	270.7	-0.8	-5.0	74.48	80.18	-2.08	-1.87	124.68	121.37	-4.50	-4.94
Jul 16	265.44	1.05	259.73	0.29	133.45	1.16	124.45	27 J-A 1	340.9	42.6	-2.4	2.8	85.92	91.71	-1.63	-1.38	118.01	116.10	-4.68	-3.77
Jul 26	265.02	1.02	259.28	0.26	139.80	1.15	134.03	6 Aug 11	102.8	168.2	5.1	1.7	97.54	103.41	-1.12	-0.85	116.68	120.09	-2.49	-1.10
Aug 5	264.75	0.99	259.15	0.23	146.16	1.14	143.64	16 Aug 21	238.0	308.7	-4.0	-4.4	109.32	115.26	-0.58	-0.31	126.15	134.29	0.13	1.06
Aug 15	264.62	0.97	259.32	0.20	152.52	1.12	153.28	26 Aug 31	14.3	74.3	0.6	4.8	121.23	127.23	-0.05	0.20	143.66	153.44	1.61	1.78
Aug 25	264.66	0.94	259.81	0.18	158.90	1.10	162.97	5 Sep 10	136.6	206.0	4.1	-1.7	133.26	139.32	0.44	0.66	163.07	172.32	1.67	1.34
Sep 4	264.86	0.91	260.58	0.15	165.29	1.07	172.71	15 Sep 20	277.3	344.7	-5.3	-2.0	145.41	151.52	0.87	1.04	181.13	189.51	0.88	0.33
Sep 14	265.22	0.88	261.62	0.13	171.69	1.04	182.51	25 Sep 30	46.6	106.3	3.4	5.2	157.66	163.81	1.20	1.32	197.49	205.10	-0.26	-0.86
Sep 24	265.73	0.86	262.90	0.11	178.12	1.00	192.36	5 Oct 10	171.9	245.0	1.4	-4.5	169.99	176.19	1.42	1.49	212.35	219.21	-1.45	-2.00
Oct 4	266.38	0.83	264.40	0.09	184.57	0.96	202.27	15 Oct 20	315.3	18.8	-4.2	1.1	182.40	188.62	1.53	1.54	225.58	231.27	-2.47	-2.82
Oct 14	267.15	0.81	266.09	0.07	191.04	0.91	212.23	25 Oct 30	78.7	139.5	5.0	3.8	194.86	201.11	1.51	1.47	235.88	238.68	-2.97	-2.80
Oct 24	268.03	0.79	267.95	0.05	197.54	0.86	222.24	4 Nov 9	208.8	283.9	-1.9	-5.1	207.37	213.64	1.39	1.29	238.55	234.51	-2.13	-0.80
Nov 3	269.01	0.77	269.94	0.03	204.07	0.81	232.31	14 Nov 19	351.1	51.7	-1.4	3.6	219.91	226.19	1.16	1.02	228.01	223.50	0.88	2.12
Nov 13	270.07	0.75	272.05	0.02	210.63	0.74	242.42	24 Nov 29	111.3	174.3	4.9	1.0	232.48	238.76	0.86	0.68	223.53	227.19	2.54	2.39
Nov 23	271.19	0.74	274.25	0.00	217.22	0.67	252.56	4 Dec 9	247.1	321.2	-4.5	-3.4	245.06	251.35	0.50	0.30	232.90	239.61	1.93	1.35
Dec 3	272.34	0.72	276.51	-0.02	223.84	0.59	262.73	14 Dec 19	24.9	84.3	1.8	4.9	257.64	263.94	0.10	-0.10	246.80	254.24	0.72	0.10
Dec 13	273.52	0.71	278.81	-0.03	230.49	0.51	272.92	24 Dec 29	145.1	211.1	3.0	-2.4	270.23	276.52	-0.30	-0.49	261.83	269.56	-0.48	-1.00
1782								1783												1783
Dec 23	274.69	0.70	281.13	-0.05	237.17	0.41	283.12	3 Jan 8	285.6	356.3	-4.9	-0.4	282.81	289.11	-0.67	-0.84	277.42	285.44	-1.45	-1.79
Jan 2	275.85	0.69	283.46	-0.06	243.88	0.31	293.31	13 Jan 18	57.4	117.3	4.2	4.5	295.39	301.68	-0.99	-1.13	293.65	302.08	-2.01	-2.09
Jan 12	276.97	0.68	285.75	-0.08	250.61	0.19	303.48	23 Jan 28	180.6	249.7	-0.1	-4.8	307.96	314.23	-1.25	-1.44	310.71	319.46	-1.98	-1.63
Jan 22	278.04	0.68	288.00	-0.10	257.37	0.06	313.63	2 Feb 7	323.1	29.4	-3.0	2.7	320.51	326.77	-1.41	-1.45	328.05	335.88	-1.01	-0.08
Feb 1	279.03	0.67	290.17	-0.12	264.16	-0.07	323.75	12 Feb 17	89.3	151.1	5.2	2.4	333.03	339.28	-1.46	-1.45	341.83	344.52	1.13	2.43
Feb 11	279.93	0.67	292.25	-0.13	270.96	-0.23	333.83	22 Feb 27	218.1	289.0	-3.4	-4.9	345.53	351.76	-1.41	-1.35	343.16	338.68	3.43	3.71
Feb 21	280.71	0.67	294.21	-0.15	277.78	-0.39	343.86	4 Mar 9	358.7	61.5	0.1	4.7	357.99	4.20	-1.26	-1.14	333.76	330.75	3.14	2.05
Mar 3	281.38	0.66	296.02	-0.18	284.62	-0.57	353.83	14 Mar 19	121.4	186.4	4.4	-0.8	10.40	16.59	-1.00	-0.84	330.39	332.35	0.84	-0.25
Mar 13	281.90	0.66	297.66	-0.20	291.45	-0.77	3.75	24 Mar 29	257.2	327.3	-5.2	-2.7	22.77	28.93	-0.66	-0.46	336.07	341.09	-1.14	-1.82
Mar 23	282.27	0.66	299.10	-0.22	298.29	-0.98	13.62	3 Apr 8	32.7	93.2	3.0	5.2	35.08	41.21	-0.25	-0.03	347.13	353.99	-2.28	-2.52
Apr 2	282.49	0.66	300.32	-0.25	305.11	-1.21	23.42	13 Apr 18	154.4	223.6	2.1	-3.7	47.33	53.42	0.20	0.43	1.59	9.87	-2.55	-2.36
Apr 12	282.54	0.66	301.28	-0.27	311.90	-1.45	33.17	23 Apr 28	296.5	3.6	-4.5	0.6	59.50	65.57	0.66	0.88	18.85	28.54	-1.96	-1.35
Apr 22	282.43	0.65	301.96	-0.30	318.64	-1.71	42.88	3 May 8	65.6	125.2	4.8	4.0	71.61	77.63	1:09	1.30	38.88	49.67	-0.57	0.30
May 2	282.16	0.65	302.35	-0.33	325.30	-1.98	52.53	13 May 18	189.0	262.4	-1.1	-5.1	83.63	89.61	1.48	1.65	60.49	70.77	1.13	1.79
May 12	281.75	0.64	302.42	-0.37	331.87	-2.26	62.15	23 May 28	334.4	38.2	-1.8	3.4	95.56	101.49	1.79	1.90	80.09	88.22	2.16	2.20
May 22	281.22	0.64	302.18	-0.40	338.28	-2.56	71.73	2 Jun 7	98.1	158.3	4.9	1.4	107.39	113.25	1.98	2.02	95.09	100.60	1.91	1.30
Jun 1	280.58	0.63	301.63	-0.43	344.51	-2.87	81.29	12 Jun 17	225.7	301.4	-4.0	-3.9	119.09	124.88	2.03	2.00	104.61	106.93	0.38	-0.78
Jun 11	279.88	0.61	300.80	-0.46	350.47	-3.18	90.83	22 Jun 27	10.4	71.4	1.5	4.9	130.63	136.34	1.93	1.82	107.40	106.04	-2.11	-3.41
Jun 21	279.15	0.60	299.75	-0.49	356.09	-3.50	100.37	2 Jul 7	130.9	193.1	3.3	-1.9	142.00	147.61	1.66	1.46	103.32	100.29	-4.41	-4.85
Jul 1	278.42	0.59	298.54	-0.52	1.26	-3.82	109.90	12 Jul 17	264.3	339.0	-5.1	-1.0	153.15	158.61	1.21	0.91	98.21	98.03	-4.62	-3.85
Jul 11	277.74	0.57	297.26	-0.55	5.85	-4.13	119.44	22 Jul 27	44.3	104.1	4.2	4.7	164.00	169.30	0.57	0.18	100.21	104.76	-2.73	-1.48
Jul 21	277.13	0.55	295.99	-0.57	9.69	-4.43	129.01	1 Aug 6	164.6	230.1	0.4	-4.5	174.49	179.56	-0.24	-0.71	111.53	120.11	-0.28	0.72
Jul 31	276.62	0.53	294.82	-0.58	12.59	-4.69	138.60	11 Aug 16	303.3	14.7	-3.7	2.3	184.49	189.25	-1.22	-1.77	129.84	139.97	1.40	1.73
Aug 10	276.24	0.51	293.85	-0.59	14.34	-4.89	148.22	21 Aug 26	77.0	136.7	5.2	2.8	193.82	198.16	-2.35	-2.96	149.93	159.47	1.73	1.50
Aug 20	276.01	0.49	293.13	-0.60	14.72	-4.99	157.89	31 Aug	200.0		-2.7		202.22		-3.59		168.48		1.09	
								Sep 5		268.9		-5.2		205.94		-4.24		176.98		0.56
Aug 30	275.94	0.47	292.71	-0.61	13.70	-4.93	167.60	10 Sep 15	341.3	48.4	-0.7	4.6	209.24	212.05	-4.89	-5.54	185.00	192.56	-0.03	-0.67
Sep 9	276.02	0.45	292.60	-0.61	11.44	-4.66	177.36	20 Sep 25	108.8	170.1	4.6	-0.1	214.25	215.73	-6.16	-6.73	199.67	206.27	-1.32	-1.94
Sep 19	276.28	0.43	292.82	-0.61	8.47	-4.16	187.19	30 S-O 5	237.5	308.3	-4.9	-3.4	216.36	216.03	-7.19	-7.50	212.25	217.38	-2.50	-2.97
Sep 29	276.69	0.41	293.36	-0.61	5.59	-3.49	197.06	10 Oct 15	17.4	80.9	2.5	5.2	214.71	212.48	-7.57	-7.35	221.24	223.09	-3.25	-3.22
Oct 9	277.24	0.39	294.20	-0.61	3.50	-2.74	207.00	20 Oct 25	140.6	205.0	2.4	-3.1	209.63	206.61	-6.77	-5.88	221.94	217.25	-2.67	-1.43
Oct 19	277.94	0.37	295.31	-0.61	2.63	-2.01	216.99	30 Oct	276.5		-4.8		203.90		-4.74		211.03		0.25	
								Nov 4		346.6		-0.1		201.92		-3.49		207.44		1.62
Oct 29	278.75	0.35	296.68	-0.61	3.09	-1.35	227.03	9 Nov 14	52.0	112.7	4.6	4.1	200.89	200.88	-2.24	-1.09	208.43	212.91	2.26	2.30
Nov 8	279.68	0.34	298.26	-0.61	4.77	-0.79	237.11	19 Nov 24	173.1	241.9	-0.6	-4.9	201.85	203.67	-0.07	0.80	219.27	226.49	1.98	1.49
Nov 18	280.69	0.32	300.04	-0.61	7.45	-0.33	247.24	29 N-D 4	315.7	23.1	-2.4	3.1	206.21	209.35	1.53	2.11	234.06	241.76	0.91	0.32
Nov 28	281.77	0.31	301.98	-0.61	10.95	0.04	257.40	9 Dec 14	85.3	144.7	4.9	1.6	212.99	217.03	2.57	2.92	249.52	257.31	-0.26	-0.80
Dec 8	282.90	0.29	304.06	-0.61	15.08	0.34	267.57	19 Dec 24	207.2	280.2	-3.5	-4.4	221.41	226.06	3.16	3.31	265.16	273.10	-1.27	-1.67
Dec 18	284.07	0.28	306.25	-0.62	19.70	0.59	277.77	29 Dec	353.5		1.0		230.95		3.38		281.15		-1.95	
1783								1783												1783

JULIAN 12 UT	SATURN LONG.	SATURN LAT.	JUPITER LONG.	JUPITER LAT.	MARS LONG.	MARS LAT.	SUN LONG.
1783							
Dec 28	285.25	0.27	308.53	-0.63	24.70	0.78	287.96
Jan 7	286.43	0.26	310.87	-0.64	29.98	0.93	298.15
Jan 17	287.58	0.25	313.24	-0.65	35.48	1.06	308.32
Jan 27	288.68	0.24	315.63	-0.66	41.15	1.15	318.45
Feb 6	289.72	0.23	318.02	-0.67	46.95	1.23	328.55
Feb 16	290.69	0.22	320.37	-0.69	52.85	1.28	338.61
Feb 26	291.55	0.21	322.67	-0.71	58.82	1.33	348.61
Mar 7	292.30	0.20	324.90	-0.73	64.85	1.36	358.56
Mar 17	292.92	0.19	327.04	-0.76	70.91	1.37	8.46
Mar 27	293.40	0.18	329.06	-0.79	77.01	1.38	18.29
Apr 6	293.72	0.17	330.93	-0.82	83.12	1.39	28.07
Apr 16	293.89	0.16	332.64	-0.85	89.26	1.38	37.81
Apr 26	293.89	0.14	334.15	-0.89	95.41	1.37	47.48
May 6	293.73	0.13	335.44	-0.93	101.57	1.36	57.12
May 16	293.42	0.12	336.48	-0.98	107.74	1.34	66.72
May 26	292.97	0.11	337.25	-1.02	113.93	1.31	76.29
Jun 5	292.40	0.09	337.72	-1.07	120.12	1.28	85.84
Jun 15	291.75	0.08	337.87	-1.12	126.34	1.25	95.37
Jun 25	291.03	0.07	337.70	-1.17	132.57	1.21	104.90
Jul 5	290.30	0.05	337.22	-1.22	138.82	1.17	114.44
Jul 15	289.58	0.04	336.44	-1.26	145.09	1.13	123.99
Jul 25	288.91	0.02	335.42	-1.30	151.40	1.08	133.57
Aug 4	288.32	0.01	334.21	-1.32	157.73	1.03	143.17
Aug 14	287.85	-0.01	332.91	-1.34	164.10	0.98	152.82
Aug 24	287.51	-0.02	331.60	-1.35	170.51	0.92	162.50
Sep 3	287.33	-0.03	330.40	-1.35	176.96	0.86	172.24
Sep 13	287.31	-0.05	329.37	-1.34	183.45	0.79	182.03
Sep 23	287.45	-0.06	328.60	-1.32	190.00	0.72	191.88
Oct 3	287.75	-0.07	328.14	-1.30	196.59	0.65	201.78
Oct 13	288.22	-0.08	328.01	-1.27	203.25	0.57	211.74
Oct 23	288.82	-0.10	328.21	-1.24	209.95	0.49	221.75
Nov 2	289.56	-0.11	328.75	-1.21	216.72	0.41	231.82
Nov 12	290.42	-0.12	329.60	-1.18	223.55	0.32	241.92
Nov 22	291.38	-0.13	330.73	-1.16	230.43	0.22	252.06
Dec 2	292.43	-0.14	332.13	-1.13	237.38	0.12	262.24
Dec 12	293.54	-0.15	333.75	-1.11	244.40	0.02	272.42
1784							
Dec 22	294.69	-0.16	335.56	-1.09	251.47	-0.09	282.61
Jan 1	295.87	-0.17	337.54	-1.07	258.60	-0.21	292.81
Jan 11	297.05	-0.19	339.66	-1.06	265.79	-0.33	302.99
Jan 21	298.23	-0.20	341.89	-1.05	273.04	-0.45	313.14
Jan 31	299.37	-0.21	344.20	-1.04	280.35	-0.58	323.26
Feb 10	300.46	-0.23	346.57	-1.04	287.70	-0.71	333.34
Feb 20	301.48	-0.24	348.98	-1.04	295.10	-0.84	343.37
Mar 2	302.42	-0.25	351.40	-1.05	302.54	-0.98	353.35
Mar 12	303.25	-0.27	353.81	-1.05	310.02	-1.11	3.27
Mar 22	303.97	-0.29	356.20	-1.06	317.52	-1.24	13.14
Apr 1	304.55	-0.30	358.54	-1.08	325.03	-1.37	22.95
Apr 11	304.98	-0.32	0.81	-1.10	332.54	-1.49	32.71
Apr 21	305.26	-0.34	3.00	-1.12	340.04	-1.60	42.41
May 1	305.38	-0.36	5.07	-1.14	347.51	-1.70	52.07
May 11	305.33	-0.38	7.00	-1.17	354.94	-1.79	61.69
May 21	305.13	-0.40	8.78	-1.20	2.30	-1.87	71.27
May 31	304.78	-0.42	10.37	-1.24	9.57	-1.92	80.83
Jun 10	304.29	-0.44	11.75	-1.28	16.73	-1.96	90.37
Jun 20	303.70	-0.46	12.89	-1.32	23.74	-1.98	99.90
Jun 30	303.02	-0.47	13.76	-1.36	30.59	-1.98	109.44
Jul 10	302.29	-0.49	14.34	-1.41	37.22	-1.96	118.98
Jul 20	301.55	-0.50	14.60	-1.46	43.59	-1.91	128.54
Jul 30	300.84	-0.52	14.53	-1.50	49.66	-1.83	138.13
Aug 9	300.19	-0.53	14.14	-1.55	55.36	-1.73	147.75
Aug 19	299.63	-0.54	13.44	-1.59	60.60	-1.60	157.42
Aug 29	299.19	-0.54	12.46	-1.62	65.28	-1.42	167.13
Sep 8	298.90	-0.55	11.27	-1.64	69.27	-1.21	176.89
Sep 18	298.76	-0.56	9.96	-1.65	72.41	-0.95	186.71
Sep 28	298.79	-0.56	8.62	-1.64	74.50	-0.63	196.59
Oct 8	298.99	-0.56	7.36	-1.63	75.33	-0.25	206.52
Oct 18	299.35	-0.57	6.26	-1.60	74.74	0.20	216.50
Oct 28	299.87	-0.57	5.42	-1.56	72.69	0.70	226.54
Nov 7	300.52	-0.57	4.88	-1.51	69.47	1.21	236.62
Nov 17	301.31	-0.58	4.67	-1.47	65.70	1.67	246.75
Nov 27	302.22	-0.58	4.82	-1.42	62.17	2.03	256.91
Dec 7	303.21	-0.59	5.30	-1.37	59.61	2.27	267.08
Dec 17	304.29	-0.60	6.11	-1.32	58.35	2.41	277.27
1785							

GREGORIAN 12 UT	MOON LONG.	MOON LONG.	MOON LAT.	MOON LAT.	VENUS LONG.	VENUS LONG.	VENUS LAT.	VENUS LAT.	MERCURY LONG.	MERCURY LONG.	MERCURY LAT.	MERCURY LAT.
1784												1784
Jan 3		57.9		4.9		236.02		3.38		289.34		-2.11
8 Jan 13	117.9	177.6	3.5	-1.5	241.24	246.60	3.32	3.20	297.64	305.96	-2.09	-1.86
18 Jan 23	243.5	318.6	-5.1	-1.9	252.07	257.63	3.03	2.82	314.03	321.24	-1.35	-0.49
28 J-F 2	29.4	91.1	4.0	4.9	263.28	268.98	2.57	2.30	326.43	328.04	0.71	2.12
7 Feb 12	150.5	212.3	0.8	-4.2	274.75	280.56	2.00	1.70	325.27	319.77	3.31	3.72
17 Feb 22	281.6	356.1	-4.5	1.5	286.41	292.30	1.38	1.05	314.94	312.88	3.24	2.25
27 Feb	63.4		5.3		298.22		0.73		313.68		1.14	
Mar 3		123.4		3.1		304.16		0.42		316.64		0.12
8 Mar 13	183.9	249.2	-2.2	-5.3	310.12	316.10	0.11	-0.18	321.13	326.73	-0.74	-1.43
18 Mar 23	320.5	32.2	-1.7	4.2	322.10	328.11	-0.45	-0.70	333.16	340.29	-1.94	-2.26
28 M-A 2	95.9	155.4	4.7	0.3	334.14	340.17	-0.92	-1.12	348.04	356.39	-2.39	-2.32
7 Apr 12	219.0	288.0	-4.5	-4.0	346.21	352.26	-1.29	-1.42	5.34	14.92	-2.05	-1.56
17 Apr 22	358.8	66.6	1.8	5.1	358.31	4.37	-1.53	-1.60	25.10	35.72	-0.87	-0.03
27 Apr	127.6		2.5		10.43		-1.65		46.38		0.85	
May 2		188.2		-2.6		16.50		-1.66		56.52		1.64
7 May 12	256.1	327.2	-5.0	-0.8	22.57	28.64	-1.64	-1.59	65.59	73.28	2.19	2.41
17 May 22	35.9	99.7	4.3	4.1	34.72	40.80	-1.51	-1.41	79.44	83.95	2.27	1.75
27 M-J 1	159.2	222.7	-0.4	-4.7	46.88	52.97	-1.28	-1.14	86.67	87.48	0.86	-0.36
6 Jun 11	294.8	5.8	-3.2	2.7	59.06	65.16	-0.97	-0.80	86.46	84.08	-1.76	-3.11
16 Jun 21	71.2	132.0	5.0	1.8	71.26	77.37	-0.61	-0.41	81.30	79.26	-4.11	-4.55
26 Jun	191.7		-3.3		83.48		-0.21		78.84		-4.41	
Jul 1		259.2		-4.9		89.60		0.00		80.41		-3.80
6 Jul 11	333.7	42.7	0.2	4.9	95.73	101.86	0.20	0.39	84.03	89.60	-2.89	-1.80
16 Jul 21	104.9	164.1	3.7	-1.2	108.01	114.16	0.58	0.76	96.99	105.93	-0.67	0.36
26 Jul 31	225.9	297.3	-5.1	-2.9	120.31	126.48	0.91	1.05	115.95	126.38	1.15	1.62
5 Aug 10	11.9	77.7	3.5	5.0	132.65	138.83	1.18	1.27	136.65	146.44	1.77	1.64
15 Aug 20	137.6	197.2	1.2	-3.9	145.01	151.20	1.35	1.40	155.64	164.25	1.30	0.80
25 Aug 30	262.2	336.0	-4.9	0.5	157.40	163.60	1.42	1.41	172.30	179.80	0.21	-0.46
4 Sep 9	48.3	110.9	5.2	3.4	169.80	176.01	1.38	1.32	186.73	193.04	-1.16	-1.86
14 Sep 19	170.1	232.1	-1.8	-5.2	182.22	188.44	1.24	1.13	198.59	203.11	-2.52	-3.11
24 Sep 29	300.2	14.0	-2.6	3.6	194.65	200.87	0.99	0.84	206.14	206.98	-3.52	-3.63
4 Oct 9	82.7	142.9	4.8	0.6	207.08	213.30	0.67	0.48	204.82	199.71	-3.20	-2.04
14 Oct 19	203.4	268.9	-4.1	-4.4	219.51	225.73	0.28	0.07	194.05	191.53	-0.38	1.10
24 Oct 29	339.1	50.7	0.9	5.0	231.94	238.15	-0.15	-0.37	193.46	198.70	1.94	2.17
3 Nov 8	115.3	174.7	2.7	-2.3	244.36	250.57	-0.58	-0.79	205.69	213.40	2.00	1.60
13 Nov 18	238.3	307.4	-5.0	-1.6	256.78	262.98	-1.00	-1.19	221.35	229.32	1.09	0.52
23 Nov 28	17.7	85.7	4.0	4.3	269.17	275.37	-1.36	-1.51	237.24	245.12	-0.05	-0.60
3 Dec 8	146.9	207.1	-0.3	-4.5	281.55	287.73	-1.64	-1.75	252.98	260.84	-1.10	-1.54
13 Dec 18	274.8	346.5	-3.8	2.1	293.90	300.06	-1.82	-1.86	268.73	276.68	-1.88	-2.11
23 Dec 28	55.1	119.0	5.1	2.0	306.20	312.33	-1.87	-1.85	284.64	292.52	-2.19	-2.06
1785												1785
2 Jan 7	178.4	241.0	-3.1	-5.1	318.44	324.53	-1.78	-1.68	300.05	306.60	-1.67	-0.92
12 Jan 17	312.7	25.0	-0.8	4.7	330.59	336.62	-1.54	-1.37	310.97	311.51	0.24	1.72
22 Jan 27	90.7	151.3	4.1	-0.9	342.62	348.58	-1.16	-0.91	307.52	301.42	3.07	3.62
1 Feb 6	210.8	276.5	-5.0	-3.7	354.49	0.36	-0.63	-0.32	297.07	296.08	3.25	2.38
11 Feb 16	351.1	62.1	2.6	5.2	6.16	11.90	0.02	0.38	297.94	301.76	1.38	0.44
21 Feb 26	124.4	183.6	1.5	-3.6	17.56	23.14	0.76	1.16	306.89	312.93	-0.39	-1.08
3 Mar 8	244.8	313.8	-5.1	-0.7	28.62	33.99	1.58	1.99	319.66	326.96	-1.62	-2.01
13 Mar 18	29.2	97.0	4.8	3.6	39.23	44.32	2.41	2.83	334.80	343.14	-2.22	-2.26
23 Mar 28	156.9	216.8	-1.5	-5.0	49.23	53.94	3.23	3.61	352.02	1.44	-2.10	-1.73
2 Apr 7	280.8	352.4	-3.2	2.8	58.41	62.59	3.97	4.28	11.41	21.78	-1.14	-0.36
12 Apr 17	66.0	130.0	4.8	0.8	66.42	69.84	4.55	4.76	32.22	42.13	0.54	1.43
22 Apr 27	189.2	251.5	-3.9	-4.6	72.75	75.06	4.89	4.91	50.86	57.96	2.15	2.58
2 May 7	318.7	31.1	0.1	4.8	76.63	77.36	4.81	4.55	63.19	66.40	2.62	2.22
12 May 17	100.8	161.9	2.9	-2.2	77.14	75.91	4.10	3.44	67.51	66.64	1.38	0.14
22 May 27	222.3	288.0	-5.0	-2.2	73.74	70.88	2.55	1.48	64.31	61.54	-1.30	-2.65
1 Jun 6	357.8	68.8	3.5	4.6	67.76	64.86	0.31	-0.85	59.47	58.92	-3.63	-4.10
11 Jun 16	133.9	193.6	-0.1	-4.5	62.61	61.26	-1.89	-2.75	60.23	63.36	-4.08	-3.67
21 Jun 26	256.6	326.1	-4.3	1.2	60.89	61.48	-3.42	-3.89	68.21	74.64	-2.96	-2.04
1 Jul 6	37.0	104.4	5.2	2.4	62.91	65.08	-4.20	-4.37	82.54	91.78	-1.02	-0.01
11 Jul 16	165.9	225.9	-2.9	-5.2	67.84	71.11	-4.42	-4.37	102.02	112.68	0.86	1.47
21 Jul 26	292.6	5.3	-1.8	4.2	74.79	78.81	-4.25	-4.06	123.21	133.22	1.76	1.75
31 J-A 5	74.8	138.2	4.5	-0.6	83.13	87.68	-3.82	-3.54	142.59	151.29	1.50	1.05
10 Aug 15	197.8	259.5	-4.8	-4.2	92.44	97.37	-3.23	-2.89	159.34	166.74	0.47	-0.22
20 Aug 25	330.2	44.2	1.6	5.2	102.45	107.66	-2.53	-2.15	173.46	179.43	-0.97	-1.75
30 Aug	110.4		1.9		112.99		-1.77		184.46		-2.53	
Sep 4		170.9		-3.3		118.43		-1.39		188.27		-3.24
9 Sep 14	230.4	294.8	-5.1	-1.5	123.95	129.56	-1.02	-0.65	190.38	190.14	-3.79	-4.03
19 Sep 24	8.8	81.4	4.3	3.9	135.24	140.99	-0.30	0.04	187.08	181.85	-3.70	-2.61
29 S-O 4	144.2	203.5	-1.2	-4.8	146.80	152.67	0.35	0.64	177.08	175.72	-0.98	0.56
9 Oct 14	264.4	331.9	-3.6	1.9	158.58	164.55	0.89	1.12	178.58	184.54	1.57	2.00
19 Oct 24	47.4	116.4	5.0	1.0	170.55	176.60	1.32	1.48	192.11	200.31	1.99	1.70
29 Oct	176.7		-3.8		182.67		1.60		208.63		1.25	
Nov 3		236.7		-4.7		188.78		1.69		216.88		0.72
8 Nov 13	300.2	10.5	-0.5	4.5	194.92	201.08	1.75	1.77	224.99	232.96	0.16	-0.39
18 Nov 23	84.4	149.4	3.3	-2.1	207.26	213.45	1.75	1.71	240.83	248.62	-0.92	-1.40
28 N-D 3	208.9	271.0	-5.0	-2.8	219.67	225.89	1.63	1.52	256.37	264.08	-1.80	-2.10
8 Dec 13	337.9	49.5	2.9	5.0	232.13	238.37	1.39	1.24	271.71	279.14	-2.27	-2.25
18 Dec 23	119.4	181.2	0.4	-4.4	244.63	250.88	1.06	0.88	286.10	291.94	-1.97	-1.34
28 Dec	241.4		-4.6		257.15		0.68		295.45		-0.27	
1785												1785

JULIAN 12 UT	SATURN LONG.	SATURN LAT.	JUPITER LONG.	JUPITER LAT.	MARS LONG.	MARS LAT.	SUN LONG.	GREGORIAN 12 UT	MOON LONG. (1)	MOON LONG. (2)	MOON LAT. (1)	MOON LAT. (2)	VENUS LONG. (1)	VENUS LONG. (2)	VENUS LAT. (1)	VENUS LAT. (2)	MERCURY LONG. (1)	MERCURY LONG. (2)	MERCURY LAT. (1)	MERCURY LAT. (2)
1785								1786												1786
								Jan 2		306.9		0.4		263.41		0.47		294.95		1.23
Dec 27	305.42	-0.61	7.20	-1.27	58.42	2.46	287.47	7 Jan 12	17.0	87.2	5.1	3.1	269.68	275.95	0.26	0.04	289.97	283.65	2.71	3.42
Jan 6	306.59	-0.62	8.57	-1.23	59.70	2.46	297.65	17 Jan 22	152.6	212.7	-2.6	-5.3	282.21	288.48	-0.17	-0.37	280.05	280.12	3.19	2.46
Jan 16	307.78	-0.63	10.16	-1.19	61.96	2.42	307.82	27 J-F 1	274.8	344.5	-2.5	3.5	294.75	301.02	-0.56	-0.74	282.94	287.50	1.57	0.70
Jan 26	308.98	-0.64	11.95	-1.16	65.02	2.36	317.96	6 Feb 11	56.2	123.1	4.8	0.0	307.28	313.54	-0.91	-1.05	293.16	299.57	-0.09	-0.77
Feb 5	310.15	-0.65	13.90	-1.13	68.69	2.29	328.06	16 Feb 21	185.0	244.7	-4.6	-4.4	319.79	326.05	-1.18	-1.28	306.53	313.95	-1.33	-1.76
Feb 15	311.28	-0.67	15.99	-1.11	72.85	2.21	338.11	26 Feb	309.7		0.7		332.29		-1.36		321.82		-2.05	
								Mar 3		23.4		5.1		338.54		-1.41		330.12		-2.17
Feb 25	312.36	-0.69	18.19	-1.08	77.39	2.13	348.12	8 Mar 13	94.0	157.3	2.4	-3.0	344.78	351.01	-1.43	-1.43	338.88	348.12	-2.11	-1.86
Mar 7	313.36	-0.70	20.48	-1.07	82.22	2.05	358.07	18 Mar 23	217.2	277.8	-5.0	-2.0	357.23	3.45	-1.40	-1.35	357.83	7.91	-1.38	-0.68
Mar 17	314.28	-0.72	22.82	-1.05	87.29	1.96	7.97	28 M-A 2	346.7	62.1	3.6	4.3	9.66	15.87	-1.27	-1.17	18.05	27.64	0.20	1.16
Mar 27	315.08	-0.75	25.21	-1.04	92.55	1.88	17.81	7 Apr 12	129.7	190.4	-0.9	-4.7	22.06	28.25	-1.04	-0.90	35.94	42.31	2.05	2.68
Apr 6	315.76	-0.77	27.61	-1.03	97.96	1.80	27.59	17 Apr 22	250.0	312.8	-3.7	1.3	34.43	40.60	-0.73	-0.56	46.38	48.00	2.93	2.69
Apr 16	316.30	-0.79	30.01	-1.03	103.49	1.72	37.32	27 Apr	25.2		5.0		46.77		-0.37		47.30		1.91	
								May 2		99.4		1.5		52.92		-0.17		44.87		0.68
Apr 26	316.70	-0.82	32.38	-1.03	109.14	1.64	47.01	7 May 12	163.5	223.1	-3.7	-4.8	59.07	65.21	0.03	0.24	41.83	39.46	-0.76	-2.07
May 6	316.94	-0.85	34.71	-1.03	114.87	1.56	56.64	17 May 22	283.8	350.0	-1.0	4.2	71.34	77.47	0.45	0.65	38.65	39.71	-3.04	-3.57
May 16	317.01	-0.87	36.99	-1.03	120.68	1.48	66.25	27 M-J 1	64.1	134.6	4.0	-1.8	83.59	89.70	0.84	1.02	42.55	46.98	-3.69	-3.46
May 26	316.93	-0.90	39.18	-1.04	126.57	1.40	75.82	6 Jun 11	196.1	256.0	-5.1	-3.2	95.80	101.89	1.18	1.32	52.80	59.90	-2.95	-2.22
Jun 5	316.68	-0.93	41.26	-1.05	132.52	1.32	85.37	16 Jun 21	319.5	29.0	2.3	5.2	107.97	114.04	1.45	1.55	68.22	77.68	-1.32	-0.36
Jun 15	316.29	-0.95	43.22	-1.06	138.53	1.24	94.91	26 Jun	101.8		1.1		120.11		1.62		88.05		0.55	
								Jul 1		168.0		-4.3		126.16		1.66		98.88		1.27
Jun 25	315.77	-0.98	45.04	-1.08	144.61	1.16	104.44	6 Jul 11	228.1	289.8	-4.8	-0.3	132.21	138.23	1.67	1.66	109.61	119.81	1.71	1.84
Jul 5	315.15	-1.00	46.67	-1.10	150.74	1.07	113.97	16 Jul 21	357.2	68.3	4.7	3.8	144.25	150.25	1.60	1.52	129.30	138.04	1.69	1.31
Jul 15	314.45	-1.02	48.10	-1.12	156.94	0.99	123.53	26 Jul 31	137.6	200.3	-2.1	-5.3	156.24	162.20	1.40	1.25	146.04	153.28	0.75	0.05
Jul 25	313.71	-1.04	49.31	-1.15	163.20	0.91	133.10	5 Aug 10	260.2	325.4	-2.9	2.8	168.15	174.08	1.06	0.85	159.72	165.25	-0.75	-1.62
Aug 4	312.97	-1.05	50.25	-1.17	169.52	0.82	142.70	15 Aug 20	36.5	106.4	5.1	0.6	179.98	185.86	0.60	0.34	169.67	172.64	-2.51	-3.36
Aug 14	312.26	-1.06	50.90	-1.20	175.91	0.74	152.35	25 Aug 30	171.7	232.1	-4.4	-4.5	191.72	197.55	0.04	-0.27	173.70	172.38	-4.05	-4.40
Aug 24	311.63	-1.06	51.24	-1.22	182.36	0.65	162.03	4 Sep 9	293.3	3.0	0.1	4.8	203.34	209.09	-0.60	-0.94	168.63	163.69	-4.13	-3.09
Sep 3	311.10	-1.06	51.25	-1.25	188.89	0.56	171.76	14 Sep 19	75.8	142.6	3.0	-2.6	214.81	220.48	-1.28	-1.63	160.10	160.00	-1.52	0.03
Sep 13	310.70	-1.06	50.92	-1.27	195.49	0.47	181.56	24 Sep 29	204.6	264.0	-5.0	-2.2	226.10	231.66	-1.98	-2.32	163.76	170.37	1.17	1.78
Sep 23	310.45	-1.06	50.27	-1.29	202.17	0.37	191.40	4 Oct 9	328.0	41.8	3.1	4.6	237.15	242.57	-2.64	-2.94	178.49	187.16	1.93	1.76
Oct 3	310.36	-1.06	49.34	-1.30	208.92	0.28	201.30	14 Oct 19	113.6	177.2	-0.4	-4.7	247.89	253.11	-3.22	-3.45	195.86	204.39	1.40	0.91
Oct 13	310.44	-1.05	48.18	-1.31	215.75	0.18	211.26	24 Oct 29	237.1	297.1	-3.9	0.9	258.20	263.13	-3.65	-3.79	212.69	220.78	0.37	-0.19
Oct 23	310.69	-1.05	46.87	-1.30	222.67	0.09	221.27	3 Nov 8	4.7	80.5	4.9	2.2	267.89	272.41	-3.88	-3.89	228.68	236.41	-0.74	-1.25
Nov 2	311.11	-1.04	45.51	-1.28	229.66	-0.01	231.33	13 Nov 18	149.3	210.4	-3.5	-4.9	276.67	280.59	-3.82	-3.65	244.02	251.50	-1.71	-2.08
Nov 12	311.67	-1.04	44.21	-1.26	236.74	-0.11	241.44	23 Nov 28	269.7	331.9	-1.4	3.8	284.11	287.13	-3.37	-2.97	258.79	265.78	-2.33	-2.42
Nov 22	312.38	-1.03	43.06	-1.22	243.90	-0.22	251.58	3 Dec 8	42.9	117.6	4.6	-1.2	289.53	291.19	-2.42	-1.70	272.14	277.23	-2.26	-1.76
Dec 2	313.22	-1.03	42.16	-1.18	251.13	-0.32	261.74	13 Dec 18	183.2	242.8	-5.1	-3.5	291.98	291.78	-0.81	0.26	279.84	278.33	-0.79	0.68
Dec 12	314.16	-1.03	41.55	-1.13	258.45	-0.42	271.93	23 Dec 28	303.4	8.9	1.7	5.3	290.55	288.38	1.48	2.78	272.60	266.40	2.26	3.13
1786								1787												1787
Dec 22	315.20	-1.03	41.28	-1.08	265.84	-0.52	282.12	2 Jan 7	81.5	153.0	2.1	-4.0	285.54	282.50	4.04	5.13	263.68	264.78	3.07	2.49
Jan 1	316.30	-1.03	41.35	-1.03	273.30	-0.62	292.31	12 Jan 17	215.6	275.1	-4.9	-0.9	279.76	277.77	5.95	6.45	268.47	273.68	1.72	0.93
Jan 11	317.46	-1.04	41.76	-0.99	280.83	-0.72	302.49	22 Jan 27	338.8	47.7	4.3	4.4	276.76	276.76	6.66	6.63	279.80	286.52	0.18	-0.49
Jan 21	318.65	-1.05	42.50	-0.94	288.41	-0.82	312.64	1 Feb 6	119.3	186.7	-1.4	-5.2	277.73	279.54	6.41	6.07	293.68	301.20	-1.07	-1.53
Jan 31	319.85	-1.06	43.52	-0.90	296.05	-0.91	322.76	11 Feb 16	247.2	308.3	-3.1	2.1	282.06	285.16	5.63	5.14	309.07	317.30	-1.87	-2.07
Feb 10	321.05	-1.07	44.81	-0.85	303.73	-0.99	332.85	21 Feb 26	16.3	86.9	5.1	1.4	288.75	292.73	4.62	4.07	325.92	334.94	-2.11	-1.96
Feb 20	322.22	-1.09	46.33	-0.82	311.45	-1.07	342.88	3 Mar 8	155.7	219.3	-4.1	-4.5	297.02	301.57	3.52	2.96	344.37	354.11	-1.59	-0.99
Mar 2	323.35	-1.10	48.05	-0.78	319.20	-1.14	352.86	13 Mar 18	278.8	343.2	-0.2	4.4	306.35	311.30	2.42	1.89	3.89	13.09	-0.16	0.84
Mar 12	324.41	-1.12	49.93	-0.75	326.96	-1.20	2.79	23 Mar 28	55.1	125.2	3.6	-2.1	316.40	321.63	1.38	0.90	20.87	26.38	1.86	2.71
Mar 22	325.40	-1.15	51.94	-0.72	334.72	-1.25	12.66	2 Apr 7	190.5	251.2	-5.0	-2.3	326.96	332.38	0.45	0.03	29.10	28.95	3.18	3.11
Apr 1	326.29	-1.17	54.07	-0.70	342.48	-1.28	22.47	12 Apr 17	311.4	20.0	2.7	4.9	337.87	343.44	-0.35	-0.70	26.49	22.99	2.40	1.19
Apr 11	327.07	-1.20	56.28	-0.68	350.21	-1.31	32.23	22 Apr 27	94.1	161.9	0.3	-4.6	349.06	354.72	-1.00	-1.27	20.01	18.66	-0.21	-1.47
Apr 21	327.72	-1.23	58.56	-0.66	357.90	-1.32	41.94	2 May 7	224.0	283.3	-4.0	0.6	0.43	6.18	-1.49	-1.67	19.30	21.79	-2.43	-3.02
May 1	328.23	-1.26	60.87	-0.64	5.55	-1.31	51.60	12 May 17	345.8	58.4	4.8	3.1	11.96	17.76	-1.81	-1.91	25.87	31.27	-3.27	-3.22
May 11	328.59	-1.30	63.21	-0.62	13.14	-1.29	61.22	22 May 27	131.9	196.7	-3.1	-5.1	23.60	29.46	-1.96	-1.98	37.82	45.43	-2.89	-2.33
May 21	328.80	-1.33	65.54	-0.61	20.65	-1.26	70.81	1 Jun 6	256.6	316.4	-1.7	3.5	35.33	41.23	-1.97	-1.91	54.05	63.65	-1.57	-0.69
May 31	328.83	-1.36	67.86	-0.60	28.07	-1.20	80.37	11 Jun 16	22.4	97.1	5.0	-0.2	47.15	53.08	-1.83	-1.72	74.09	85.01	0.22	1.02
Jun 10	328.71	-1.40	70.15	-0.59	35.39	-1.14	89.91	21 Jun 26	168.0	229.9	-5.1	-3.8	59.03	65.00	-1.58	-1.41	95.87	106.20	1.60	1.88
Jun 20	328.43	-1.43	72.38	-0.58	42.59	-1.06	99.45	1 Jul 6	289.2	351.4	1.3	5.1	70.98	76.98	-1.23	-1.03	115.76	124.48	1.86	1.57
Jun 30	328.00	-1.46	74.53	-0.57	49.67	-0.96	108.98	11 Jul 16	60.7	134.9	3.0	-3.4	83.00	89.03	-0.82	-0.60	132.34	139.34	1.06	0.36
Jul 10	327.45	-1.49	76.59	-0.57	56.60	-0.85	118.52	21 Jul 26	202.2	262.1	-5.0	-1.2	95.08	101.14	-0.38	-0.16	145.39	150.35	-0.49	-1.44
Jul 20	326.79	-1.51	78.53	-0.56	63.36	-0.72	128.08	31 J-A 5	322.8	28.5	3.9	4.7	107.22	113.32	0.06	0.28	153.99	155.96	-2.46	-3.45
Jul 30	326.07	-1.53	80.32	-0.56	69.95	-0.58	137.67	10 Aug 15	99.6	171.2	-0.5	-5.0	119.43	125.56	0.48	0.67	155.87	153.54	-4.28	-4.71
Aug 9	325.32	-1.55	81.95	-0.56	76.34	-0.42	147.29	20 Aug 25	234.8	294.3	-3.2	1.8	131.70	137.86	0.85	1.00	149.48	145.30	-4.47	-3.48
Aug 19	324.58	-1.55	83.37	-0.55	82.49	-0.25	156.95	30 Aug	358.2		5.0		144.03		1.14		143.15		-2.00	
								Sep 4		67.2		2.2		150.21		1.25		144.35		-0.48
Aug 29	323.88	-1.56	84.57	-0.55	88.39	-0.05	166.66	9 Sep 14	138.0	205.7	-3.6	-4.6	156.41	162.62	1.33	1.39	148.98	156.19	0.74	1.51
Sep 8	323.27	-1.56	85.51	-0.55	93.97	0.16	176.42	19 Sep 24	266.4	327.3	-0.4	4.2	168.84	175.07	1.42	1.43	164.83	173.94	1.83	1.80
Sep 18	322.76	-1.55	86.16	-0.55	99.20	0.39	186.23	29 S-O 4	35.4	106.4	4.1	-1.5	181.30	187.55	1.41	1.36	183.02	191.84	1.52	1.10
Sep 28	322.40	-1.54	86.51	-0.55	103.99	0.65	196.10	9 Oct 14	175.0	238.6	-5.0	-2.5	193.80	200.06	1.28	1.18	200.34	208.55	0.58	0.02
Oct 8	322.19	-1.53	86.52	-0.54	108.26	0.94	206.03	19 Oct 24	298.0	1.8	2.5	5.1	206.32	212.58	1.06	0.92	216.48	224.18	-0.55	-1.10
Oct 18	322.16	-1.52	86.20	-0.54	111.88	1.27	216.01	29 Oct	74.0		1.2		218.85		0.76		231.65		-1.61	
								Nov 3		144.9		-4.4		225.12		0.59		238.90		-2.05
Oct 28	322.29	-1.50	85.55	-0.53	114.71	1.63	226.05	8 Nov 13	210.2	270.7	-4.3	0.4	231.39	237.67	0.40	0.20	245.86	252.37	-2.39	-2.58
Nov 7	322.60	-1.49	84.62	-0.52	116.57	2.04	236.13	18 Nov 23	330.4	38.0	4.7	3.9	243.94	250.22	0.00	-0.20	258.12	262.42	-2.54	-2.18
Nov 17	323.06	-1.47	83.46	-0.50	117.26	2.49	246.25	28 N-D 3	112.7	181.8	-2.3	-5.3	256.49	262.77	-0.40	-0.59	264.07	261.61	-1.33	0.09
Nov 27	323.68	-1.46	82.15	-0.48	116.62	2.96	256.41	8 Dec 13	243.7	302.8	-2.0	3.2	269.04	275.31	-0.78	-0.95	255.39	249.58	1.73	2.76
Dec 7	324.44	-1.45	80.79	-0.46	114.57	3.42	266.59	18 Dec 23	4.6	75.6	5.2	1.0	281.58	287.85	-1.11	-1.25	247.81	249.88	2.90	2.49
Dec 17	325.32	-1.44	79.49	-0.43	111.34	3.81	276.78	28 Dec	150.5		-4.8		294.11		-1.37		254.36		1.84	
1787								1787												1787

JULIAN 12 UT	SATURN LONG.	SATURN LAT.	JUPITER LONG.	JUPITER LAT.	MARS LONG.	MARS LAT.	SUN LONG.	GREGORIAN 12 UT	MOON LONG.	MOON LAT.	VENUS LONG.	VENUS LAT.	MERCURY LONG.	MERCURY LAT.
1787								1788						1788
								Jan 2	216.5	-4.0	300.37	-1.47	260.16	1.13
Dec 27	326.31	-1.43	78.34	-0.40	107.43	4.06	286.97	7 Jan 12	276.2 336.0	0.9 4.9	306.63 312.88	-1.54 -1.58	266.71 273.72	0.42 -0.23
Jan 6	327.37	-1.43	77.43	-0.37	103.60	4.13	297.16	17 Jan 22	40.7 113.9	3.7 -2.5	319.12 325.35	-1.60 -1.58	281.05 288.64	-0.82 -1.31
Jan 16	328.51	-1.43	76.81	-0.34	100.56	4.06	307.32	27 J-F 1	186.7 249.4	-5.1 -1.4	331.56 337.77	-1.53 -1.46	296.51 304.65	-1.70 -1.96
Jan 26	329.69	-1.43	76.51	-0.31	98.73	3.87	317.46	6 Feb 11	308.6 10.9	3.5 4.9	343.97 350.14	-1.35 -1.22	313.11 321.91	-2.08 -2.03
Feb 5	330.90	-1.44	76.55	-0.28	98.23	3.62	327.57	16 Feb 21	78.6 152.0	0.5 -4.7	356.30 2.44	-1.06 -0.87	331.03 340.41	-1.77 -1.27
Feb 15	332.11	-1.45	76.92	-0.25	98.95	3.35	337.62	26 Feb	221.0	-3.4	8.56	-0.65	349.76	-0.51
								Mar 2	281.3	1.6	14.65	-0.42	358.51	0.48
Feb 25	333.31	-1.46	77.60	-0.22	100.70	3.08	347.64	7 Mar 12	341.9 47.7	4.9 2.8	20.72 26.76	-0.17 0.09	5.69 10.24	1.60 2.65
Mar 6	334.49	-1.48	78.57	-0.20	103.31	2.83	357.59	17 Mar 22	117.6 188.9	-3.0 -4.8	32.77 38.74	0.37 0.65	11.49 9.56	3.35 3.44
Mar 16	335.61	-1.50	79.79	-0.17	106.59	2.60	7.49	27 M-A 1	253.6 313.0	-0.5 4.1	44.67 50.57	0.94 1.22	5.67 1.80	2.80 1.63
Mar 26	336.67	-1.52	81.23	-0.15	110.40	2.38	17.33	6 Apr 11	16.6 86.0	4.5 -0.7	56.42 62.22	1.49 1.76	359.54 359.47	0.29 -0.92
Apr 5	337.64	-1.55	82.87	-0.13	114.64	2.18	27.12	16 Apr 21	156.5 224.1	-5.0 -2.9	67.97 73.66	2.00 2.23	1.44 5.09	-1.86 -2.50
Apr 15	338.52	-1.58	84.66	-0.11	119.23	2.00	36.85	26 Apr	285.3	2.4	79.28	2.42	10.09	-2.86
								May 1	345.6	5.2	84.82	2.59	16.20	-2.95
Apr 25	339.27	-1.61	86.60	-0.09	124.10	1.83	46.54	6 May 11	53.0 125.2	2.1 -3.9	90.29 95.66	2.72 2.80	23.27 31.22	-2.78 -2.38
May 5	339.90	-1.65	88.64	-0.07	129.20	1.67	56.18	16 May 21	194.1 257.5	-4.7 0.1	100.93 106.06	2.83 2.82	40.04 49.71	-1.77 -1.00
May 15	340.39	-1.69	90.76	-0.05	134.50	1.52	65.78	26 May 31	317.1 19.6	4.6 4.5	111.06 115.89	2.74 2.59	60.15 71.08	-0.12 0.74
May 25	340.72	-1.73	92.95	-0.04	139.97	1.37	75.36	5 Jun 10	91.0 163.9	-1.3 -5.3	120.53 124.94	2.38 2.08	81.99 92.38	1.44 1.87
Jun 4	340.88	-1.77	95.19	-0.02	145.59	1.24	84.91	15 Jun 20	229.6 290.0	-2.4 2.9	129.07 132.87	1.70 1.23	101.94 110.55	2.00 1.83
Jun 14	340.89	-1.81	97.44	-0.01	151.34	1.10	94.44	25 Jun 30	349.7 55.3	5.3 1.9	136.27 139.20	0.66 -0.03	118.18 124.79	1.39 0.71
Jun 24	340.73	-1.84	99.71	0.01	157.22	0.98	103.98	5 Jul 10	129.8 201.0	-4.2 -4.3	141.55 143.21	-0.83 -1.75	130.28 134.47	-0.17 -1.21
Jul 4	340.41	-1.88	101.95	0.03	163.20	0.85	113.52	15 Jul 20	263.2 322.5	0.6 4.7	144.05 143.97	-2.77 -3.89	137.12 137.91	-2.35 -3.49
Jul 14	339.95	-1.92	104.17	0.04	169.30	0.73	123.06	25 Jul 30	23.9 92.9	4.1 -1.5	142.90 140.90	-5.03 -6.12	136.65 133.58	-4.42 -4.90
Jul 24	339.37	-1.95	106.33	0.06	175.51	0.62	132.64	4 Aug 9	168.3 235.8	-5.1 -1.7	138.17 135.09	-7.04 -7.68	129.76 126.83	-4.68 -3.75
Aug 3	338.69	-1.97	108.42	0.08	181.81	0.50	142.24	14 Aug 19	295.8 355.9	3.3 4.9	132.14 129.77	-7.98 -7.93	126.28 128.79	-2.39 -0.94
Aug 13	337.95	-1.99	110.42	0.10	188.22	0.39	151.88	24 Aug 29	59.9 131.7	1.2 -4.3	128.26 127.75	-7.60 -7.06	134.24 142.00	0.31 1.21
Aug 23	337.19	-2.00	112.30	0.11	194.73	0.28	161.56	3 Sep 8	205.1 268.8	-3.7 1.4	128.21 129.54	-6.39 -5.66	151.09 160.63	1.69 1.80
Sep 2	336.45	-2.00	114.03	0.13	201.34	0.17	171.30	13 Sep 18	328.3 30.7	4.9 3.2	131.64 134.37	-4.91 -4.15	170.07 179.18	1.63 1.27
Sep 12	335.76	-2.00	115.60	0.15	208.06	0.06	181.08	23 Sep 28	97.8 170.4	-2.3 -5.0	137.64 141.36	-3.41 -2.71	187.90 196.23	0.79 0.24
Sep 22	335.16	-1.99	116.96	0.18	214.87	-0.05	190.92	3 Oct 8	239.8 300.7	-0.9 4.0	145.45 149.85	-2.04 -1.41	204.21 211.87	-0.34 -0.93
Oct 2	334.69	-1.98	118.09	0.20	221.79	-0.15	200.82	13 Oct 18	1.3 67.1	4.7 0.1	154.50 159.38	-0.83 -0.29	219.21 226.23	-1.49 -2.00
Oct 12	334.36	-1.96	118.96	0.23	228.80	-0.25	210.77	23 Oct 28	137.0 207.7	-4.8 -3.4	164.44 169.67	0.19 0.63	232.84 238.87	-2.43 -2.72
Oct 22	334.20	-1.94	119.54	0.25	235.92	-0.35	220.78	2 Nov 7	272.6 332.3	2.2 5.2	175.04 180.52	1.01 1.34	243.97 247.42	-2.82 -2.59
Nov 1	334.21	-1.91	119.80	0.28	243.14	-0.45	230.84	12 Nov 17	35.4 105.2	2.8 -3.2	186.11 191.79	1.62 1.84	248.09 244.75	-1.88 -0.54
Nov 11	334.40	-1.89	119.74	0.31	250.44	-0.54	240.94	22 Nov 27	176.1 243.1	-5.0 -0.4	197.55 203.37	2.01 2.14	238.29 233.10	1.14 2.33
Nov 21	334.75	-1.87	119.35	0.34	257.84	-0.63	251.08	2 Dec 7	304.5 4.5	4.4 4.7	209.26 215.19	2.21 2.24	232.30 235.29	2.69 2.45
Dec 1	335.27	-1.84	118.65	0.37	265.33	-0.71	261.25	12 Dec 17	71.0 144.4	-0.3 -5.1	221.17 227.18	2.22 2.17	240.50 246.85	1.93 1.30
Dec 11	335.94	-1.82	117.67	0.40	272.90	-0.79	271.43	22 Dec 27	213.7 276.7	-2.9 2.5	233.23 239.30	2.08 1.95	253.80 261.08	0.64 0.01
1788								1789						1789
Dec 21	336.75	-1.81	116.49	0.43	280.54	-0.86	281.63	1 Jan 6	336.5 37.8	5.2 2.5	245.40 251.52	1.79 1.61	268.57 276.25	-0.58 -1.10
Dec 31	337.67	-1.79	115.18	0.45	288.24	-0.93	291.82	11 Jan 16	108.3 183.0	-3.4 -4.6	257.65 263.79	1.40 1.18	284.10 292.16	-1.52 -1.84
Jan 10	338.69	-1.78	113.84	0.47	296.00	-0.98	302.00	21 Jan 26	249.1 309.5	0.2 4.5	269.95 276.12	0.95 0.71	300.45 309.01	-2.04 -2.07
Jan 20	339.79	-1.77	112.58	0.49	303.81	-1.03	312.16	31 J-F 5	9.1 72.9	4.2 -0.7	282.29 288.47	0.46 0.22	317.82 326.80	-1.91 -1.53
Jan 30	340.96	-1.77	111.47	0.50	311.65	-1.07	322.28	10 Feb 15	146.8 219.9	-5.0 -2.0	294.65 300.84	-0.02 -0.26	335.69 343.92	-0.86 0.09
Feb 9	342.16	-1.77	110.61	0.51	319.51	-1.09	332.36	20 Feb 25	282.7 342.2	3.2 4.9	307.03 313.22	-0.48 -0.68	350.43 353.97	1.27 2.49
Feb 19	343.38	-1.78	110.03	0.52	327.38	-1.11	342.40	2 Mar 7	43.0 110.1	1.6 -3.8	319.41 325.60	-0.87 -1.03	353.70 350.15	3.41 3.65
Mar 1	344.61	-1.79	109.77	0.53	335.25	-1.11	352.38	12 Mar 17	185.3 254.7	-4.2 1.2	331.78 337.97	-1.17 -1.29	345.39 341.83	3.08 1.97
Mar 11	345.82	-1.80	109.84	0.53	343.10	-1.10	2.31	22 Mar 27	315.3 15.3	4.9 3.6	344.16 350.34	-1.38 -1.45	340.64 341.81	0.71 -0.43
Mar 21	346.99	-1.82	110.22	0.53	350.93	-1.08	12.19	1 Apr 6	78.9 149.0	-1.7 -5.1	356.52 2.69	-1.48 -1.49	344.90 349.44	-1.35 -2.03
Mar 31	348.12	-1.84	110.89	0.53	358.71	-1.04	22.00	11 Apr 16	222.4 287.7	-1.5 3.9	8.86 15.03	-1.47 -1.43	355.12 1.72	-2.47 -2.67
Apr 10	349.17	-1.86	111.84	0.53	6.45	-0.99	31.76	21 Apr 26	347.4 49.5	4.9 0.8	21.20 27.36	-1.35 -1.26	9.14 17.30	-2.64 -2.39
Apr 20	350.13	-1.89	113.02	0.53	14.12	-0.93	41.47	1 May 6	116.7 188.2	-4.5 -4.1	33.52 39.67	-1.14 -1.00	26.22 35.89	-1.92 -1.26
Apr 30	350.99	-1.93	114.42	0.54	21.72	-0.87	51.13	11 May 16	257.6 319.7	1.7 5.2	45.82 51.97	-0.84 -0.67	46.26 57.11	-0.45 0.43
May 10	351.74	-1.96	116.01	0.54	29.24	-0.79	60.75	21 May 26	19.8 85.4	3.3 -2.4	58.12 64.27	-0.49 -0.30	68.00 78.37	1.23 1.81
May 20	352.34	-2.00	117.75	0.54	36.68	-0.70	70.34	31 M-J 5	156.1 226.1	-5.2 -1.1	70.41 76.56	-0.10 0.09	87.85 96.24	2.10 2.07
May 30	352.81	-2.04	119.62	0.55	44.02	-0.60	79.90	10 Jun 15	291.2 351.4	4.2 4.8	82.70 88.84	0.29 0.48	103.49 109.52	1.73 1.10
Jun 9	353.11	-2.09	121.60	0.55	51.26	-0.49	89.44	20 Jun 25	53.3 123.3	0.5 -4.7	94.98 101.13	0.67 0.84	114.21 117.38	0.20 -0.91
Jun 19	353.25	-2.13	123.66	0.56	58.40	-0.38	98.98	30 Jun	195.3	-3.5	107.27	1.00	118.79	-2.17
								Jul 5	262.0	2.1	113.41	1.14	118.27	-3.43
Jun 29	353.23	-2.18	125.79	0.57	65.42	-0.26	108.51	10 Jul 15	323.9 23.5	5.1 2.8	119.56 125.70	1.26 1.35	116.00 112.74	-4.44 -4.92
Jul 9	353.03	-2.22	127.96	0.58	72.33	-0.13	118.05	20 Jul 25	88.6 162.5	-2.7 -4.8	131.84 137.99	1.43 1.47	109.79 108.48	-4.71 -3.89
Jul 19	352.69	-2.26	130.15	0.59	79.11	0.01	127.62	30 J-A 4	233.0 296.3	-0.2 4.4	144.13 150.27	1.49 1.48	109.58 113.33	-2.68 -1.36
Jul 29	352.20	-2.29	132.35	0.60	85.76	0.15	137.20	9 Aug 14	356.2 56.8	4.3 -0.2	156.40 162.54	1.45 1.38	119.55 127.79	-0.12 0.87
Aug 8	351.59	-2.32	134.54	0.62	92.28	0.29	146.82	19 Aug 24	125.9 201.3	-4.8 -2.6	168.67 174.80	1.29 1.16	137.27 147.20	1.50 1.76
Aug 18	350.89	-2.35	136.69	0.64	98.65	0.44	156.48	29 Aug	268.6	2.9	180.92	1.01	157.00	1.71
								Sep 3	329.4	5.0	187.03	0.84	166.40	1.43
Aug 28	350.13	-2.36	138.79	0.66	104.86	0.60	166.18	8 Sep 13	28.9 91.9	1.9 -3.3	193.14 199.25	0.65 0.43	175.32 183.78	1.00 0.47
Sep 7	349.36	-2.37	140.81	0.68	110.90	0.77	175.94	18 Sep 23	164.6 238.5	-4.7 0.7	205.35 211.44	0.20 -0.04	191.81 199.43	-0.13 -0.75
Sep 17	348.62	-2.37	142.74	0.70	116.74	0.95	185.76	28 S-O 3	302.5 2.1	4.9 3.8	217.52 223.59	-0.29 -0.55	206.65 213.42	-1.36 -1.94
Sep 27	347.94	-2.36	144.54	0.73	122.35	1.14	195.62	8 Oct 13	62.8 129.2	-1.2 -5.1	229.65 235.70	-0.81 -1.06	219.67 225.18	-2.46 -2.86
Oct 7	347.36	-2.34	146.19	0.76	127.71	1.34	205.55	18 Oct 23	203.2 273.5	-2.3 3.7	241.74 247.77	-1.31 -1.55	229.57 232.13	-3.09 -3.00
Oct 17	346.91	-2.32	147.65	0.79	132.76	1.56	215.53	28 Oct	335.0	5.1	253.78	-1.76	231.78	-2.42
								Nov 2	34.8	1.3	259.78	-1.96	227.67	-1.17
Oct 27	346.62	-2.29	148.91	0.83	137.44	1.79	225.56	7 Nov 12	98.6 168.0	-4.0 -4.6	265.75 271.70	-2.13 -2.27	221.26 216.87	0.51 1.85
Nov 6	346.50	-2.26	149.93	0.87	141.68	2.05	235.64	17 Nov 22	240.4 306.8	1.0 5.2	277.62 283.51	-2.38 -2.45	217.03 220.88	2.42 2.37
Nov 16	346.56	-2.23	150.68	0.91	145.38	2.33	245.76	27 N-D 2	6.8 68.6	3.6 -1.7	289.36 295.17	-2.47 -2.45	226.78 233.67	1.99 1.45
Nov 26	346.79	-2.20	151.13	0.95	148.41	2.65	255.92	7 Dec 12	136.4 207.0	-5.2 -2.0	300.92 306.61	-2.37 -2.25	241.00 248.55	0.84 0.23
Dec 6	347.20	-2.17	151.26	0.99	150.61	2.99	266.09	17 Dec 22	276.1 338.9	3.8 4.8	312.23 317.76	-2.07 -1.83	256.21 263.95	-0.35 -0.89
Dec 16	347.77	-2.14	151.07	1.04	151.80	3.35	276.29	27 Dec	38.6	0.9	323.19	-1.53	271.80	-1.35
1789								1789						1789

JULIAN 12 UT	SATURN LONG.	SATURN LAT.	JUPITER LONG.	JUPITER LAT.	MARS LONG.	MARS LAT.	SUN LONG.	GREGORIAN 12 UT	MOON LONG.		MOON LAT.		VENUS LONG.		VENUS LAT.		MERCURY LONG.		MERCURY LAT.	
1789								1790												1790
								Jan 1		104.0		-4.2		328.49		-1.17		279.78		-1.72
Dec 26	348.48	-2.11	150.56	1.08	151.81	3.73	286.48	6 Jan 11	175.5	245.0	-4.0	1.5	333.65	338.63	-0.74	-0.25	287.91	296.22	-1.98	-2.10
Jan 5	349.33	-2.09	149.76	1.12	150.49	4.09	296.67	16 Jan 21	310.2	10.5	5.0	2.9	343.41	347.93	0.30	0.92	304.70	313.27	-2.04	-1.76
Jan 15	350.29	-2.07	148.72	1.15	147.89	4.37	306.83	26 Jan 31	71.4	141.3	-2.2	-5.0	352.15	356.00	1.61	2.36	321.67	329.32	-1.20	-0.32
Jan 25	351.35	-2.05	147.50	1.18	144.31	4.53	316.97	5 Feb 10	214.5	281.3	-0.9	4.2	359.40	2.27	3.17	4.04	335.11	337.59	0.88	2.23
Feb 4	352.48	-2.04	146.19	1.20	140.35	4.51	327.08	15 Feb 20	343.2	42.5	4.3	0.0	4.48	5.91	4.94	5.86	335.87	331.01	3.34	3.73
Feb 14	353.67	-2.03	144.89	1.21	136.74	4.32	337.14	25 Feb	106.0		-4.5		6.42		6.76		325.98		3.24	
								Mar 2		179.8		-3.4		5.91		7.57		323.18		2.21
Feb 24	354.89	-2.03	143.69	1.21	134.08	4.01	347.15	7 Mar 12	252.1	315.9	2.6	5.1	4.37	1.94	8.21	8.56	323.15	325.44	1.05	-0.02
Mar 6	356.13	-2.04	142.67	1.20	132.68	3.63	357.11	17 Mar 22	15.6	75.6	2.2	-3.0	358.94	355.84	8.55	8.14	329.42	334.64	-0.92	-1.61
Mar 16	357.37	-2.04	141.91	1.19	132.56	3.23	7.01	27 M-A 1	142.8	218.3	-5.1	-0.2	353.14	351.22	7.40	6.42	340.81	347.75	-2.11	-2.40
Mar 26	358.58	-2.06	141.42	1.17	133.60	2.85	16.86	6 Apr 11	287.8	349.0	4.8	4.1	350.28	350.36	5.33	4.23	355.38	3.66	-2.49	-2.36
Apr 5	359.76	-2.07	141.25	1.15	135.61	2.50	26.65	16 Apr 21	48.3	110.7	-0.8	-5.0	351.37	353.19	3.17	2.20	12.60	22.20	-2.03	-1.48
Apr 15	0.89	-2.10	141.38	1.13	138.43	2.19	36.38	26 Apr	181.2		-3.3		355.68		1.32		32.44		-0.75	
								May 1		255.6		3.1		358.74		0.54		43.15		0.10
Apr 25	1.93	-2.12	141.82	1.11	141.89	1.90	46.07	6 May 11	321.7	21.3	5.2	1.7	2.25	6.14	-0.14	-0.73	53.92	64.20	0.97	1.69
May 5	2.89	-2.15	142.53	1.08	145.88	1.64	55.71	16 May 21	82.0	148.0	-3.5	-5.0	10.35	14.80	-1.22	-1.64	73.50	81.55	2.16	2.30
May 15	3.75	-2.19	143.50	1.06	150.29	1.40	65.32	26 May 31	220.1	291.4	0.0	5.0	19.47	24.31	-1.97	-2.23	88.23	93.44	2.09	1.53
May 25	4.48	-2.23	144.70	1.05	155.06	1.18	74.89	5 Jun 10	354.1	53.7	3.7	-1.3	29.31	34.43	-2.43	-2.56	97.05	98.88	0.64	-0.53
Jun 4	5.07	-2.27	146.11	1.03	160.13	0.98	84.44	15 Jun 20	117.7	186.9	-5.0	-2.7	39.66	44.99	-2.63	-2.65	98.83	97.07	-1.89	-3.23
Jun 14	5.51	-2.31	147.69	1.01	165.46	0.80	93.98	25 Jun 30	258.2	325.5	3.2	4.9	50.40	55.88	-2.62	-2.55	94.26	91.52	-4.26	-4.73
Jun 24	5.80	-2.36	149.41	1.00	171.01	0.63	103.51	5 Jul 10	25.8	87.1	1.1	-3.8	61.42	67.02	-2.44	-2.29	90.01	90.46	-4.58	-3.90
Jul 4	5.92	-2.41	151.27	0.99	176.76	0.47	113.05	15 Jul 20	155.4	226.1	-4.4	0.8	72.68	78.39	-2.11	-1.90	93.14	98.02	-2.88	-1.70
Jul 14	5.87	-2.45	153.23	0.99	182.69	0.31	122.60	25 Jul 30	294.8	358.1	4.9	3.0	84.13	89.92	-1.67	-1.43	104.94	113.59	-0.52	0.51
Jul 24	5.65	-2.50	155.27	0.98	188.79	0.17	132.17	4 Aug 9	57.6	122.3	-1.9	-5.0	95.75	101.62	-1.17	-0.90	123.40	133.66	1.26	1.67
Aug 3	5.27	-2.54	157.37	0.98	195.05	0.03	141.78	14 Aug 19	194.3	264.4	-1.7	3.9	107.53	113.46	-0.64	-0.37	143.78	153.46	1.76	1.58
Aug 13	4.75	-2.57	159.51	0.99	201.46	-0.09	151.41	24 Aug 29	329.7	30.0	4.5	0.2	119.43	125.43	-0.10	0.15	162.59	171.17	1.21	0.70
Aug 23	4.12	-2.60	161.67	0.99	208.00	-0.22	161.09	3 Sep 8	90.4	159.3	-4.4	-4.1	131.46	137.52	0.39	0.61	179.24	186.81	0.10	-0.54
Sep 2	3.40	-2.62	163.84	1.00	214.69	-0.33	170.83	13 Sep 18	233.2	301.1	1.9	5.2	143.61	149.71	0.82	1.00	193.88	200.40	-1.21	-1.87
Sep 12	2.62	-2.63	165.99	1.01	221.51	-0.44	180.61	23 Sep 28	2.9	62.1	2.5	-2.7	155.85	162.00	1.16	1.29	206.25	211.20	-2.48	-3.00
Sep 22	1.85	-2.63	168.09	1.03	228.45	-0.54	190.45	3 Oct 8	124.9	197.5	-5.2	-1.2	168.17	174.37	1.40	1.47	214.82	216.41	-3.35	-3.41
Oct 2	1.10	-2.63	170.14	1.04	235.52	-0.64	200.34	13 Oct 18	271.1	335.8	4.6	4.4	180.58	186.80	1.52	1.53	215.04	210.32	-2.95	-1.79
Oct 12	0.43	-2.61	172.10	1.07	242.71	-0.73	210.30	23 Oct 28	35.4	95.3	-0.4	-4.8	193.03	199.28	1.52	1.47	204.24	200.82	-0.12	1.34
Oct 22	359.87	-2.59	173.96	1.09	250.01	-0.81	220.30	2 Nov 7	161.3	235.9	-4.0	2.2	205.54	211.81	1.40	1.31	201.95	206.60	2.11	2.26
Nov 1	359.45	-2.55	175.68	1.12	257.42	-0.88	230.36	12 Nov 17	307.1	8.7	5.2	2.0	218.08	224.36	1.19	1.05	213.15	220.55	2.02	1.57
Nov 11	359.20	-2.52	177.23	1.15	264.93	-0.94	240.46	22 Nov 27	68.0	130.4	-3.1	-5.1	230.64	236.93	0.89	0.72	228.27	236.08	1.02	0.44
Nov 21	359.11	-2.48	178.59	1.19	272.52	-0.99	250.59	2 Dec 7	199.3	273.5	-1.0	4.6	243.22	249.51	0.54	0.34	243.91	251.74	-0.14	-0.68
Dec 1	359.21	-2.44	179.72	1.23	280.20	-1.03	260.76	12 Dec 17	341.0	40.7	3.9	-1.0	255.81	262.10	0.14	-0.06	259.59	267.49	-1.17	-1.59
Dec 11	359.49	-2.40	180.60	1.27	287.95	-1.07	270.94	22 Dec 27	101.7	167.5	-4.8	-3.3	268.39	274.68	-0.25	-0.45	275.46	283.53	-1.91	-2.10
1790								1791												1791
Dec 21	359.94	-2.36	181.20	1.32	295.76	-1.09	281.13	1 Jan 6	238.1	309.7	2.5	4.9	280.98	287.27	-0.63	-0.81	291.69	299.83	-2.14	-1.97
Dec 31	0.56	-2.33	181.50	1.36	303.62	-1.10	291.33	11 Jan 16	13.3	72.7	1.2	-3.6	293.56	299.84	-0.96	-1.10	307.71	314.73	-1.52	-0.74
Jan 10	1.31	-2.30	181.47	1.41	311.50	-1.09	301.50	21 Jan 26	136.9	206.0	-4.7	0.1	306.12	312.40	-1.22	-1.32	319.75	321.17	0.43	1.87
Jan 20	2.20	-2.27	181.14	1.45	319.41	-1.08	311.66	31 J-F 5	276.5	344.0	4.8	3.3	318.67	324.94	-1.40	-1.44	318.11	312.36	3.16	3.69
Jan 30	3.20	-2.24	180.49	1.50	327.32	-1.05	321.79	10 Feb 15	44.7	105.5	-1.8	-5.0	331.20	337.46	-1.47	-1.46	307.54	305.72	3.30	2.37
Feb 9	4.28	-2.23	179.59	1.53	335.22	-1.02	331.87	20 Feb 25	173.8	245.2	-2.5	3.5	343.70	349.94	-1.43	-1.37	306.82	310.06	1.31	0.32
Feb 19	5.44	-2.21	178.47	1.56	343.10	-0.97	341.91	2 Mar 7	313.7	16.9	4.8	0.5	356.16	2.38	-1.28	-1.17	314.76	320.51	-0.54	-1.24
Mar 1	6.65	-2.20	177.23	1.57	350.95	-0.91	351.90	12 Mar 17	76.3	139.9	-4.2	-4.6	8.59	14.78	-1.03	-0.88	327.04	334.21	-1.77	-2.14
Mar 11	7.89	-2.20	175.93	1.58	358.75	-0.84	1.83	22 Mar 27	212.0	283.7	0.9	5.2	20.96	27.12	-0.70	-0.51	341.96	350.28	-2.32	-2.30
Mar 21	9.15	-2.20	174.69	1.57	6.49	-0.77	11.70	1 Apr 6	348.8	49.1	2.9	-2.4	33.27	39.41	-0.30	-0.08	359.17	8.65	-2.09	-1.67
Mar 31	10.40	-2.21	173.57	1.55	14.18	-0.68	21.52	11 Apr 16	109.0	176.0	-5.3	-2.3	45.52	51.62	0.15	0.38	18.71	29.21	-1.04	-0.23
Apr 10	11.62	-2.22	172.66	1.53	21.78	-0.59	31.29	21 Apr 26	250.8	320.5	4.0	4.6	57.71	63.77	0.61	0.84	39.80	49.90	0.67	1.51
Apr 20	12.81	-2.23	172.01	1.50	29.31	-0.50	41.00	1 May 6	22.2	81.5	0.0	-4.5	69.81	75.84	1.05	1.26	58.92	66.48	2.15	2.48
Apr 30	13.93	-2.26	171.65	1.46	36.76	-0.40	50.67	11 May 16	143.3	213.7	-4.4	1.1	81.84	87.82	1.45	1.62	72.38	76.49	2.44	1.99
May 10	14.98	-2.28	171.58	1.42	44.11	-0.29	60.29	21 May 26	288.9	355.2	5.2	2.3	93.77	99.69	1.77	1.89	78.68	78.90	1.14	-0.07
May 20	15.94	-2.31	171.82	1.38	51.38	-0.18	69.88	31 M-J 5	54.8	114.9	-2.9	-5.1	105.59	111.46	1.98	2.03	77.37	74.77	-1.49	-2.86
May 30	16.79	-2.35	172.34	1.35	58.55	-0.07	79.44	10 Jun 15	179.6	252.5	-1.8	4.0	117.29	123.08	2.05	2.03	72.16	70.62	-3.88	-4.36
Jun 9	17.51	-2.39	173.13	1.31	65.63	0.04	88.98	20 Jun 25	325.4	28.1	4.1	-0.8	128.83	134.53	1.97	1.87	70.82	72.97	-4.30	-3.80
Jun 19	18.10	-2.43	174.17	1.27	72.61	0.15	98.52	30 Jun	87.4		-4.7		140.19		1.72		77.03		-2.98	
								Jul 5		149.9		-3.7		145.79		1.52		82.88		-1.98
Jun 29	18.53	-2.47	175.43	1.24	79.51	0.27	108.05	10 Jul 15	217.7	291.1	1.8	5.0	151.32	156.78	1.28	0.99	90.41	99.40	-0.89	0.14
Jul 9	18.80	-2.52	176.88	1.21	86.30	0.38	117.59	20 Jul 25	359.7	60.0	1.4	-3.5	162.16	167.45	0.66	0.28	109.47	120.00	0.99	1.54
Jul 19	18.90	-2.57	178.50	1.19	93.00	0.49	127.15	30 J-A 4	120.9	186.7	-4.9	-0.7	172.63	177.69	-0.15	-0.62	130.41	140.34	1.77	1.70
Jul 29	18.82	-2.61	180.26	1.17	99.61	0.61	136.74	9 Aug 14	256.9	328.1	4.5	3.7	182.61	187.35	-1.13	-1.68	149.66	158.35	1.41	0.94
Aug 8	18.58	-2.66	182.14	1.15	106.12	0.73	146.35	19 Aug 24	32.3	92.0	-1.6	-5.0	191.89	196.20	-2.26	-2.88	166.43	173.93	0.35	-0.32
Aug 18	18.18	-2.70	184.12	1.13	112.53	0.84	156.01	29 Aug	155.7		-3.2		200.23		-3.52		180.83		-1.04	
								Sep 3		225.2		2.8		203.91		-4.18		187.05		-1.77
Aug 28	17.64	-2.73	186.18	1.12	118.84	0.96	165.72	8 Sep 13	295.9	3.0	5.1	0.9	207.17	209.93	-4.86	-5.53	192.45	196.75	-2.49	-3.13
Sep 7	16.98	-2.75	188.29	1.11	125.03	1.08	175.47	18 Sep 23	64.1	124.6	-4.1	-4.9	212.06	213.47	-6.18	-6.77	199.52	200.07	-3.62	-3.81
Sep 17	16.24	-2.77	190.45	1.11	131.12	1.20	185.28	28 S-O 3	192.1	264.5	-0.1	5.0	214.01	213.59	-7.26	-7.60	197.71	192.63	-3.46	-2.38
Sep 27	15.45	-2.78	192.61	1.11	137.07	1.32	195.15	8 Oct 13	333.2	36.3	3.4	-2.1	212.18	209.88	-7.70	-7.50	187.20	184.89	-0.74	0.80
Oct 7	14.67	-2.77	194.78	1.11	142.88	1.45	205.07	18 Oct 23	95.9	158.4	-5.2	-3.0	206.99	203.97	-6.94	-6.06	186.97	192.37	1.75	2.10
Oct 17	13.92	-2.76	196.91	1.12	148.54	1.59	215.05	28 Oct	230.1		3.2		201.31		-4.93		199.54		2.01	
								Nov 2		303.2		4.9		199.41		-3.69		207.43		1.67
Oct 27	13.26	-2.74	199.00	1.13	154.01	1.72	225.08	7 Nov 12	8.4	68.8	0.4	-4.3	198.47	198.54	-2.45	-1.29	215.54	223.63	1.19	0.64
Nov 6	12.72	-2.70	201.01	1.14	159.26	1.87	235.15	17 Nov 22	128.5	193.9	-4.6	0.2	199.58	201.46	-0.27	0.61	231.63	239.56	0.08	-0.48
Nov 16	12.33	-2.67	202.93	1.16	164.26	2.02	245.28	27 N-D 2	268.8	340.0	5.0	2.6	204.06	207.24	1.35	1.95	247.42	255.26	-0.99	-1.45
Nov 26	12.10	-2.63	204.73	1.18	168.95	2.19	255.43	7 Dec 12	41.9	101.4	-2.6	-5.0	210.92	214.99	2.43	2.79	263.08	270.92	-1.83	-2.10
Dec 6	12.05	-2.58	206.37	1.20	173.27	2.37	265.60	17 Dec 22	162.6	231.2	-2.3	3.4	219.39	224.06	3.05	3.22	278.75	286.46	-2.22	-2.15
Dec 16	12.19	-2.54	207.83	1.23	177.12	2.56	275.79	27 Dec	306.9		4.4		228.96		3.31		293.79		-1.83	
1791								1791												1791

JULIAN 12 UT	SATURN LONG	LAT	JUPITER LONG	LAT	MARS LONG	LAT	SUN LONG	GREGORIAN 12 UT	MOON LONG	LONG	LAT	LAT	VENUS LONG	LONG	LAT	LAT	MERCURY LONG	LONG	LAT	LAT
1791								**1792**												**1792**
								Jan 1	14.5		-0.6		234.05		3.33		300.13		-1.16	
Dec 26	12.51	-2.49	209.08	1.26	180.40	2.76	285.99	6 Jan 11	74.5	134.7	-4.6	-3.9	239.29	244.66	3.28	3.18	304.33	304.71	-0.06	1.42
Jan 5	13.00	-2.45	210.10	1.29	182.98	2.98	296.17	16 Jan 21	198.5	270.0	1.1	5.0	250.14	255.71	3.02	2.83	300.51	294.25	2.86	3.54
Jan 15	13.66	-2.41	210.84	1.33	184.67	3.21	306.34	26 Jan 31	343.4	47.3	1.9	-3.4	261.36	267.08	2.59	2.33	290.00	289.25	3.27	2.47
Jan 25	14.45	-2.37	211.30	1.37	185.32	3.43	316.49	5 Feb 10	106.8	169.3	-5.0	-1.3	272.85	278.67	2.05	1.74	291.38	295.43	1.53	0.61
Feb 4	15.37	-2.34	211.45	1.41	184.75	3.62	326.59	15 Feb 20	236.5	308.7	4.1	4.3	284.53	290.42	1.43	1.11	300.73	306.89	-0.21	-0.91
Feb 14	16.40	-2.31	211.28	1.44	182.88	3.75	336.66	25 Feb	17.8		-1.3		296.35		0.79		313.69		-1.47	
								Mar 1	79.1		-5.1		302.30		0.48		321.03		-1.88	
Feb 24	17.51	-2.29	210.80	1.47	179.85	3.77	346.67	6 Mar 11	139.6	205.6	-3.6	2.0	308.27	314.25	0.18	-0.11	328.84	337.14	-2.14	-2.22
Mar 5	18.69	-2.27	210.05	1.50	176.10	3.65	356.63	16 Mar 21	275.9	346.0	5.3	1.6	320.26	326.27	-0.39	-0.64	345.94	355.26	-2.12	-1.81
Mar 15	19.92	-2.26	209.05	1.52	172.27	3.38	6.54	26 Mar 31	50.8	110.7	-3.9	-5.0	332.30	338.34	-0.87	-1.07	5.09	15.33	-1.29	-0.55
Mar 25	21.18	-2.26	207.88	1.53	169.08	3.01	16.39	5 Apr 10	173.5	243.7	-0.9	4.6	344.39	350.44	-1.24	-1.39	25.66	35.49	0.34	1.27
Apr 4	22.45	-2.25	206.62	1.53	167.00	2.57	26.18	15 Apr 20	315.0	21.5	3.9	-1.6	356.50	2.56	-1.50	-1.58	44.14	51.07	2.08	2.61
Apr 14	23.71	-2.26	205.35	1.52	166.22	2.13	35.92	25 Apr 30	83.0	142.9	-5.1	-3.3	8.63	14.70	-1.63	-1.65	55.97	58.67	2.76	2.46
Apr 24	24.95	-2.27	204.17	1.50	166.70	1.71	45.61	5 May 10	209.1	282.9	2.3	5.0	20.77	26.85	-1.64	-1.60	59.16	57.68	1.67	0.45
May 4	26.14	-2.28	203.14	1.47	168.28	1.34	55.25	15 May 20	352.5	55.4	0.9	-4.1	32.93	39.01	-1.53	-1.43	55.01	52.28	-0.99	-2.34
May 14	27.27	-2.30	202.34	1.43	170.80	1.00	64.86	25 May 30	115.2	176.4	-4.6	-0.4	45.10	51.19	-1.31	-1.17	50.61	50.64	-3.34	-3.86
May 24	28.32	-2.33	201.80	1.39	174.09	0.70	74.43	4 Jun 9	246.8	321.7	4.6	3.1	57.28	63.38	-1.01	-0.83	52.51	56.13	-3.93	-3.62
Jun 3	29.28	-2.36	201.56	1.35	178.00	0.44	83.98	14 Jun 19	27.9	88.4	-2.4	-5.0	69.48	75.59	-0.65	-0.45	61.32	67.97	-3.01	-2.18
Jun 13	30.13	-2.39	201.61	1.30	182.44	0.20	93.53	24 Jun 29	147.9	211.8	-2.5	2.8	81.71	87.83	-0.25	-0.05	75.99	85.26	-1.22	-0.22
Jun 23	30.85	-2.43	201.96	1.26	187.31	-0.01	103.06	4 Jul 9	285.8	358.6	4.8	-0.2	93.95	100.08	0.16	0.35	95.51	106.25	0.68	1.36
Jul 3	31.43	-2.47	202.60	1.21	192.54	-0.20	112.59	14 Jul 19	61.6	121.1	-4.6	-4.1	106.22	112.37	0.54	0.72	116.89	127.03	1.73	1.80
Jul 13	31.85	-2.51	203.49	1.17	198.10	-0.37	122.14	24 Jul 29	182.0	249.4	0.6	4.9	118.53	124.69	0.88	1.03	136.50	145.27	1.60	1.19
Jul 23	32.11	-2.55	204.62	1.13	203.93	-0.52	131.71	3 Aug 8	324.4	33.5	2.6	-3.2	130.86	137.03	1.15	1.25	153.34	160.72	0.63	-0.06
Aug 2	32.20	-2.60	205.96	1.10	210.01	-0.66	141.31	13 Aug 18	94.2	154.2	-5.1	-1.8	143.22	149.41	1.33	1.39	167.39	173.24	-0.83	-1.65
Aug 12	32.11	-2.64	207.49	1.06	216.32	-0.78	150.94	23 Aug 28	218.0	288.5	3.6	4.8	155.60	161.80	1.42	1.42	178.10	181.66	-2.47	-3.25
Aug 22	31.85	-2.68	209.19	1.03	222.82	-0.89	160.62	2 Sep 7	1.4	66.6	-0.7	-5.0	168.00	174.20	1.39	1.34	103.46	102.92	-3.00	-4.19
Sep 1	31.42	-2.71	211.02	1.01	229.52	-0.98	170.35	12 Sep 17	126.3	188.5	-3.9	1.3	180.41	186.63	1.26	1.15	179.69	174.58	-3.93	-2.91
Sep 11	30.86	-2.74	212.97	0.99	236.39	-1.06	180.13	22 Sep 27	256.1	327.4	5.2	2.4	192.84	199.05	1.03	0.88	170.14	169.05	-1.32	0.25
Sep 21	30.18	-2.76	215.01	0.97	243.41	-1.13	189.96	2 Oct 7	36.5	98.6	-3.5	-5.1	205.26	211.48	0.71	0.52	172.06	178.15	1.36	1.89
Oct 1	29.42	-2.78	217.13	0.95	250.59	-1.19	199.85	12 Oct 17	158.6	224.6	-1.4	4.0	217.69	223.90	0.32	0.12	185.91	194.29	1.97	1.75
Oct 11	28.63	-2.78	219.29	0.94	257.90	-1.23	209.81	22 Oct 27	295.4	5.0	4.4	-1.0	230.11	236.32	-0.10	-0.32	202.78	211.15	1.34	0.84
Oct 21	27.83	-2.77	221.49	0.93	265.33	-1.26	219.81	1 Nov 6	70.1	130.1	-5.0	-3.5	242.53	248.73	-0.54	-0.75	219.35	227.38	0.29	-0.27
Oct 31	27.09	-2.75	223.69	0.92	272.87	-1.28	229.86	11 Nov 16	192.0	262.5	1.7	5.1	254.94	261.14	-0.96	-1.15	235.27	243.05	-0.81	-1.31
Nov 10	26.44	-2.72	225.87	0.92	280.50	-1.28	239.96	21 Nov 26	334.4	40.8	1.5	-3.8	267.33	273.52	-1.33	-1.49	250.74	258.35	-1.74	-2.08
Nov 20	25.91	-2.68	228.02	0.92	288.21	-1.27	250.10	1 Dec 6	102.6	162.0	-4.6	-0.7	279.71	285.88	-1.63	-1.74	265.85	273.12	-2.29	-2.33
Nov 30	25.54	-2.64	230.11	0.92	295.99	-1.24	260.26	11 Dec 16	227.1	301.4	4.2	3.7	292.05	298.21	-1.82	-1.87	279.87	285.51	-2.12	-1.57
Dec 10	25.34	-2.59	232.12	0.92	303.82	-1.21	270.45	21 Dec 26	11.9	75.1	-2.0	-5.0	304.35	310.48	-1.89	-1.87	288.84	288.19	-0.57	0.91
Dec 20	25.33	-2.55	234.01	0.93	311.68	-1.16	280.64	31 Dec	134.9		-2.7		316.58		-1.81		283.07		2.45	
1792								**1793**												**1793**
								Jan 5	195.2		2.4		322.67		-1.72		276.68		3.29	
Dec 30	25.50	-2.50	235.77	0.94	319.56	-1.10	290.83	10 Jan 15	264.2	339.8	5.0	0.6	328.73	334.76	-1.59	-1.42	273.20	273.51	3.18	2.52
Jan 9	25.86	-2.45	237.36	0.95	327.44	-1.03	301.02	20 Jan 25	47.4	108.1	-4.5	-4.3	340.76	346.72	-1.22	-0.98	276.58	281.34	1.69	0.85
Jan 19	26.38	-2.40	238.75	0.96	335.31	-0.95	311.17	30 J-F 4	167.5	230.4	0.2	4.7	352.63	358.49	-0.70	-0.39	287.15	293.66	0.08	-0.61
Jan 29	27.07	-2.36	239.92	0.98	343.15	-0.87	321.30	9 Feb 14	302.6	16.6	3.5	-2.8	4.30	10.04	-0.06	0.30	300.68	308.13	-1.19	-1.64
Feb 8	27.89	-2.32	240.84	1.00	350.96	-0.78	331.39	19 Feb 24	81.1	140.5	-5.2	-2.2	15.70	21.27	0.69	1.09	315.97	324.22	-1.96	-2.13
Feb 18	28.84	-2.29	241.49	1.02	358.71	-0.68	341.43	1 Mar 6	201.4	267.8	3.1	5.2	26.75	32.12	1.51	1.93	332.89	342.01	-2.12	-1.92
Feb 28	29.89	-2.26	241.83	1.03	6.41	-0.58	351.42	11 Mar 16	341.0	51.7	0.4	-4.8	37.35	42.44	2.36	2.78	351.58	1.51	-1.51	-0.86
Mar 10	31.03	-2.23	241.87	1.05	14.04	-0.47	1.36	21 Mar 26	113.4	173.1	-4.1	0.7	47.35	52.04	3.20	3.59	11.51	21.00	-0.01	0.97
Mar 20	32.23	-2.21	241.60	1.07	21.59	-0.36	11.23	31 M-A 5	237.3	306.7	5.0	3.2	56.50	60.66	3.97	4.30	29.19	35.36	1.93	2.67
Mar 30	33.48	-2.20	241.02	1.08	29.06	-0.25	21.05	10 Apr 15	18.5	85.1	-2.9	-5.1	64.48	67.87	4.59	4.82	39.04	40.09	3.04	2.90
Apr 9	34.75	-2.19	240.18	1.09	36.45	-0.15	30.82	20 Apr 25	145.1	206.9	-1.7	3.5	70.75	73.01	4.98	5.03	38.74	35.83	2.19	0.99
Apr 19	36.03	-2.19	239.13	1.09	43.76	-0.04	40.53	30 Apr	275.2		4.7		74.54		4.96		32.68		-0.43	
								May 5	345.8		-0.2		75.21		4.73		30.60		-1.75	
Apr 29	37.30	-2.19	237.92	1.08	50.98	0.07	50.20	10 May 15	54.6	117.4	-4.8	-3.5	74.93	73.63	4.31	3.66	30.29	31.88	-2.73	-3.31
May 9	38.55	-2.20	236.66	1.07	58.12	0.17	59.83	20 May 25	177.0	242.5	1.3	5.0	71.41	68.53	2.79	1.73	35.18	39.97	-3.51	-3.38
May 19	39.74	-2.21	235.41	1.04	65.16	0.28	69.41	30 M-J 4	314.3	24.0	2.2	-3.5	65.40	62.53	0.56	-0.60	46.05	53.31	-2.96	-2.32
May 29	40.88	-2.22	234.28	1.02	72.13	0.37	78.98	9 Jun 14	89.0	149.2	-4.7	-0.9	60.33	59.04	-1.65	-2.53	61.71	71.18	-1.49	-0.57
Jun 8	41.94	-2.24	233.32	0.98	79.02	0.47	88.52	19 Jun 24	210.1	280.1	4.0	4.3	58.75	59.39	-3.22	-3.72	81.54	92.41	0.35	1.13
Jun 18	42.91	-2.27	232.61	0.94	85.83	0.57	98.06	29 Jun	353.3		-1.4		60.88		-4.05		103.21		1.65	
								Jul 4	60.4		-5.1		63.08		-4.25		113.52		1.86	
Jun 28	43.76	-2.30	232.17	0.90	92.56	0.66	107.59	9 Jul 14	122.1	181.3	-2.9	2.1	65.89	69.19	-4.32	-4.30	123.10	131.89	1.78	1.45
Jul 8	44.48	-2.33	232.03	0.86	99.22	0.74	117.13	19 Jul 24	245.1	318.8	5.2	1.6	72.89	76.93	-4.20	-4.03	139.89	147.09	0.92	0.23
Jul 18	45.05	-2.36	232.20	0.82	105.82	0.83	126.69	29 J-A 3	31.0	94.9	-4.3	-4.6	81.26	85.83	-3.81	-3.55	153.43	158.81	-0.59	-1.49
Jul 28	45.47	-2.40	232.66	0.78	112.34	0.91	136.27	8 Aug 13	154.5	214.6	-0.1	4.5	90.59	95.53	-3.25	-2.92	163.00	165.68	-2.43	-3.35
Aug 7	45.73	-2.44	233.40	0.74	118.81	0.99	145.89	18 Aug 23	282.1	357.3	4.3	-1.9	100.62	105.84	-2.57	-2.20	166.40	164.77	-4.12	-4.53
Aug 17	45.80	-2.48	234.41	0.71	125.20	1.06	155.54	28 Aug	66.9		-5.3		111.17		-1.83		160.94		-4.33	
								Sep 2	127.9		-2.5		116.60		-1.46		156.24		-3.35	
Aug 27	45.70	-2.52	235.65	0.67	131.54	1.14	165.24	7 Sep 12	187.2	249.9	2.7	5.3	122.13	127.74	-1.09	-0.72	153.07	153.29	-1.84	-0.28
Sep 6	45.42	-2.55	237.10	0.64	137.81	1.21	174.99	17 Sep 22	320.4	34.8	1.4	-4.5	133.42	139.16	-0.37	-0.03	157.21	163.95	0.93	1.64
Sep 16	44.98	-2.58	238.73	0.61	144.02	1.28	184.80	27 S-O 2	100.7	160.1	-4.3	0.4	144.97	150.83	0.28	0.57	172.24	181.09	1.89	1.80
Sep 26	44.40	-2.60	240.53	0.59	150.16	1.35	194.66	7 Oct 12	221.1	287.2	4.7	3.8	156.75	162.71	0.83	1.07	189.96	198.63	1.48	1.02
Oct 6	43.71	-2.62	242.46	0.56	156.22	1.42	204.58	17 Oct 22	359.1	70.5	-2.1	-5.1	168.72	174.76	1.27	1.44	207.03	215.18	0.49	-0.06
Oct 16	42.93	-2.63	244.50	0.54	162.21	1.48	214.56	27 Oct	132.9		-1.9		180.83		1.57		223.11		-0.62	
								Nov 1	192.5		3.2		186.94		1.67		230.84		-1.16	
Oct 26	42.12	-2.62	246.63	0.52	168.11	1.55	224.59	6 Nov 11	256.9	325.9	4.9	0.6	193.07	199.23	1.73	1.76	238.40	245.79	-1.64	-2.05
Nov 5	41.33	-2.61	248.82	0.51	173.91	1.61	234.66	16 Nov 21	37.0	104.3	-4.5	-3.7	205.41	211.60	1.75	1.71	252.96	259.78	-2.34	-2.48
Nov 15	40.58	-2.58	251.06	0.49	179.59	1.67	244.78	26 N-D 1	164.4	225.9	1.2	4.9	217.82	224.04	1.64	1.54	265.94	270.35	-1.74	-1.98
Nov 25	39.94	-2.55	253.32	0.48	185.13	1.74	254.93	6 Dec 11	294.4	5.0	2.9	-2.9	230.28	236.53	1.41	1.27	273.23	271.60	-1.09	0.33
Dec 5	39.43	-2.51	255.58	0.46	190.52	1.80	265.11	16 Dec 21	73.6	136.6	-4.9	-1.0	242.78	249.04	1.10	0.91	265.80	259.59	1.95	2.95
Dec 15	39.08	-2.47	257.81	0.45	195.71	1.86	275.30	26 Dec 31	196.0	260.8	3.8	4.7	255.30	261.57	0.72	0.51	256.99	258.32	3.03	2.54
1793								**1793**												**1793**

JULIAN 12 UT	SATURN LONG.	SATURN LAT.	JUPITER LONG.	JUPITER LAT.	MARS LONG.	MARS LAT.	SUN LONG.	GREGORIAN 12 UT	MOON LONG.		MOON LAT.		VENUS LONG.		VENUS LAT.		MERCURY LONG.		MERCURY LAT.	
1793								1794												1794
Dec 25	38.90	-2.42	260.00	0.44	200.67	1.92	285.49	5 Jan 10	333.0	43.3	-0.5	-5.0	267.84	274.11	0.30	0.09	262.23	267.63	1.82	1.06
Jan 4	38.92	-2.37	262.11	0.43	205.34	1.98	295.68	15 Jan 20	108.2	168.4	-3.3	1.9	280.38	286.65	-0.12	-0.33	273.90	280.72	0.33	-0.34
Jan 14	39.12	-2.32	264.13	0.42	209.66	2.03	305.85	25 Jan 30	228.7	297.3	5.2	2.6	292.91	299.18	-0.52	-0.71	287.92	295.46	-0.93	-1.41
Jan 24	39.50	-2.27	266.03	0.42	213.54	2.08	316.00	4 Feb 9	11.7	79.9	-3.7	-4.9	305.45	311.71	-0.87	-1.02	303.31	311.49	-1.78	-2.02
Feb 3	40.06	-2.22	267.78	0.41	216.88	2.12	326.11	14 Feb 19	141.4	200.6	-0.5	4.3	317.97	324.22	-1.15	-1.26	320.01	328.91	-2.10	-2.00
Feb 13	40.76	-2.18	269.35	0.40	219.55	2.14	336.17	24 Feb	263.2		4.7		330.47		-1.34		338.19		-1.69	
								Mar 1		335.2		-0.8		336.72		-1.40		347.77		-1.15
Feb 23	41.61	-2.14	270.72	0.40	221.37	2.14	346.19	6 Mar 11	49.5	114.4	-5.2	-2.8	342.96	349.20	-1.43	-1.43	357.39	6.47	-0.36	0.63
Mar 5	42.58	-2.11	271.86	0.39	222.17	2.09	356.15	16 Mar 21	173.9	234.1	2.4	5.2	355.42	1.64	-1.41	-1.36	14.12	19.40	1.70	2.65
Mar 15	43.65	-2.08	272.75	0.39	221.79	1.99	6.06	26 Mar 31	299.6	13.7	2.3	-3.8	7.86	14.06	-1.29	-1.19	21.70	20.97	3.24	3.28
Mar 25	44.81	-2.05	273.35	0.38	220.13	1.81	15.91	5 Apr 10	85.4	147.2	-4.5	0.2	20.26	26.45	-1.07	-0.93	17.94	14.18	2.65	1.47
Apr 4	46.02	-2.03	273.65	0.37	217.32	1.53	25.71	15 Apr 20	206.6	269.3	4.5	4.1	32.64	38.81	-0.77	-0.60	11.34	10.39	0.10	-1.16
Apr 14	47.28	-2.01	273.64	0.36	213.76	1.16	35.44	25 Apr 30	337.8	51.8	-1.2	-5.0	44.98	51.14	-0.41	-0.21	11.49	14.41	-2.13	-2.77
Apr 24	48.57	-2.00	273.31	0.35	210.11	0.73	45.14	5 May 10	119.3	179.2	-2.0	3.0	57.29	63.43	-0.01	0.20	18.83	24.49	-3.09	-3.11
May 4	49.86	-2.00	272.70	0.34	207.10	0.28	54.78	15 May 20	240.3	306.3	4.9	1.3	69.56	75.68	0.40	0.60	31.22	38.93	-2.87	-2.39
May 14	51.14	-1.99	271.82	0.32	205.22	-0.15	64.39	25 May 30	17.0	88.3	-4.2	-4.0	81.80	87.91	0.80	0.98	47.59	57.18	-1.71	-0.88
May 24	52.39	-2.00	270.73	0.30	204.67	-0.52	73.97	4 Jun 9	151.7	211.3	1.0	4.8	94.01	100.11	1.15	1.30	67.59	78.50	0.02	0.86
Jun 3	53.60	-2.00	269.50	0.28	205.44	-0.83	83.52	14 Jun 19	275.6	345.0	3.5	-2.3	106.19	112.26	1.43	1.53	89.40	99.81	1.51	1.87
Jun 13	54.74	-2.01	268.23	0.26	207.37	-1.09	93.06	24 Jun 29	55.7	122.8	-5.1	-1.4	118.32	124.37	1.61	1.66	109.43	118.18	1.93	1.70
Jun 23	55.81	-2.03	266.99	0.24	210.28	-1.30	102.59	4 Jul 9	183.4	244.4	3.7	5.0	130.42	136.44	1.68	1.67	126.01	132.92	1.23	0.55
Jul 3	56.78	-2.05	265.88	0.21	214.02	-1.47	112.13	14 Jul 19	312.6	24.3	0.5	-4.8	142.45	148.45	1.63	1.55	138.82	143.56	-0.30	-1.29
Jul 13	57.63	-2.07	264.96	0.18	218.42	-1.61	121.67	24 Jul 29	92.8	155.7	-3.7	1.6	154.43	160.40	1.44	1.29	146.90	148.52	-2.35	-3.41
Jul 23	58.36	-2.10	264.29	0.16	223.39	-1.71	131.25	3 Aug 8	215.2	278.9	5.2	3.3	166.35	172.27	1.11	0.90	148.06	145.49	-4.31	-4.80
Aug 2	58.94	-2.13	263.92	0.13	228.82	-1.79	140.84	13 Aug 18	351.1	63.0	-2.9	-5.1	178.17	184.04	0.66	0.40	141.49	137.71	-4.62	-3.69
Aug 12	59.35	-2.16	263.86	0.11	234.64	-1.85	150.47	23 Aug 28	127.8	188.0	-0.9	4.0	189.89	195.71	0.11	-0.20	136.05	137.60	-2.27	-0.77
Aug 22	59.60	-2.19	264.11	0.08	240.81	-1.88	160.15	2 Sep 7	248.0	315.0	4.9	0.2	201.50	207.25	-0.53	-0.87	142.41	149.73	0.50	1.35
Sep 1	59.67	-2.22	264.66	0.06	247.26	-1.89	169.87	12 Sep 17	30.1	99.6	-5.0	-3.2	212.95	218.61	-1.22	-1.58	158.51	167.80	1.76	1.81
Sep 11	59.56	-2.25	265.50	0.04	253.96	-1.89	179.65	22 Sep 27	161.1	220.5	2.1	5.1	224.22	229.77	-1.93	-2.28	177.05	186.02	1.59	1.20
Sep 21	59.27	-2.27	266.61	0.02	260.88	-1.86	189.49	2 Oct 7	282.3	352.8	2.9	-3.0	235.25	240.65	-2.61	-2.92	194.64	202.91	0.70	0.15
Oct 1	58.81	-2.30	267.95	0.00	267.99	-1.82	199.37	12 Oct 17	68.2	134.1	-4.7	-0.1	245.96	251.16	-3.21	-3.46	210.88	218.58	-0.43	-0.99
Oct 11	58.21	-2.31	269.51	-0.01	275.25	-1.77	209.32	22 Oct 27	193.7	253.9	4.3	4.3	256.23	261.14	-3.67	-3.84	226.01	233.18	-1.53	-2.00
Oct 21	57.50	-2.32	271.25	-0.03	282.65	-1.69	219.32	1 Nov 6	318.4	31.5	-0.4	-4.9	265.87	270.37	-3.94	-3.97	240.02	246.37	-2.39	-2.63
Oct 31	56.72	-2.32	273.15	-0.05	290.15	-1.61	229.37	11 Nov 16	104.3	166.8	-2.4	2.9	274.59	278.47	-3.92	-3.78	251.91	255.98	-2.67	-2.39
Nov 10	55.90	-2.31	275.18	-0.06	297.75	-1.51	239.47	21 Nov 26	226.4	288.9	5.0	1.9	281.94	284.90	-3.53	-3.15	257.43	254.87	-1.63	-0.28
Nov 20	55.10	-2.29	277.32	-0.08	305.40	-1.40	249.60	1 Dec 6	356.5	69.8	-3.7	-4.4	287.24	288.82	-2.62	-1.92	248.63	242.87	1.39	2.55
Nov 30	54.36	-2.27	279.55	-0.09	313.11	-1.29	259.77	11 Dec 16	138.3	198.7	0.8	4.8	289.52	289.22	-1.05	0.01	241.22	243.50	2.82	2.51
Dec 10	53.72	-2.23	281.83	-0.11	320.83	-1.17	269.95	21 Dec 26	259.6	325.6	3.9	-1.5	287.90	285.65	1.23	2.53	248.19	254.18	1.92	1.24
Dec 20	53.22	-2.19	284.15	-0.13	328.57	-1.04	280.15	31 Dec	35.8		-5.2		282.78		3.80		260.87		0.56	
1794								1795												1795
								Jan 5		106.5		-2.0		279.75		4.90		267.97		-0.09
Dec 30	52.89	-2.15	286.48	-0.14	336.29	-0.91	290.34	10 Jan 15	170.7	230.4	3.5	5.2	277.07	275.16	5.73	6.26	275.35	282.97	-0.68	-1.19
Jan 9	52.74	-2.10	288.81	-0.16	343.99	-0.77	300.52	20 Jan 25	294.1	4.0	1.3	-4.4	274.24	274.35	6.50	6.50	290.82	298.91	-1.60	-1.90
Jan 19	52.77	-2.05	291.10	-0.18	351.65	-0.64	310.68	30 J-F 4	74.6	141.3	-4.3	1.2	275.41	277.29	6.31	6.00	307.28	315.95	-2.06	-2.05
Jan 29	53.00	-2.00	293.34	-0.19	359.26	-0.51	320.81	9 Feb 14	202.5	262.8	5.1	3.8	279.88	283.04	5.60	5.14	324.93	334.13	-1.85	-1.42
Feb 8	53.40	-1.96	295.50	-0.21	6.82	-0.38	330.90	19 Feb 24	330.2	43.3	-1.9	-5.2	286.67	290.69	4.63	4.11	343.31	351.91	-0.71	0.25
Feb 18	53.97	-1.91	297.57	-0.23	14.31	-0.25	340.94	1 Mar 6	111.8	174.7	-1.5	3.8	295.01	299.60	3.57	3.03	358.95	3.28	1.40	2.53
Feb 28	54.70	-1.87	299.50	-0.26	21.73	-0.12	350.93	11 Mar 16	234.4	296.4	4.9	1.0	304.39	309.36	2.49	1.97	4.10	1.62	3.35	3.56
Mar 10	55.57	-1.84	301.29	-0.28	29.07	-0.01	0.88	21 Mar 26	8.0	81.8	-4.5	-3.6	314.48	319.73	1.47	0.99	357.34	353.44	2.99	1.87
Mar 20	56.55	-1.80	302.89	-0.30	36.34	0.11	10.76	31 M-A 5	147.0	207.4	1.8	5.0	325.07	330.50	0.54	0.12	351.47	351.79	0.56	-0.63
Mar 30	57.64	-1.77	304.30	-0.33	43.52	0.22	20.58	10 Apr 15	267.1	332.1	3.1	-2.3	336.01	341.58	-0.27	-0.62	354.15	358.13	-1.59	-2.27
Apr 9	58.80	-1.75	305.47	-0.36	50.63	0.32	30.35	20 Apr 25	46.9	118.4	-4.9	-0.5	347.21	352.89	-0.93	-1.20	3.38	9.66	-2.68	-2.83
Apr 19	60.03	-1.72	306.38	-0.39	57.66	0.42	40.07	30 Apr	180.6		4.2		358.61		-1.43		16.84		-2.74	
								May 5		240.1		4.4		4.36		-1.62		24.84		-2.42
Apr 29	61.29	-1.71	307.01	-0.42	64.62	0.51	49.74	10 May 15	301.3	9.9	0.1	-4.7	10.14	15.96	-1.77	-1.87	33.66	43.30	-1.88	-1.16
May 9	62.59	-1.69	307.34	-0.46	71.50	0.59	59.37	20 May 25	85.2	153.0	-3.0	2.7	21.80	27.66	-1.94	-1.97	53.67	64.56	-0.31	0.56
May 19	63.88	-1.68	307.35	-0.49	78.31	0.67	68.96	30 M-J 4	213.2	273.4	5.1	2.3	33.54	39.44	-1.96	-1.92	75.48	85.92	1.32	1.83
May 29	65.17	-1.68	307.05	-0.53	85.06	0.75	78.52	9 Jun 14	337.5	49.0	-3.2	-4.9	45.36	51.30	-1.84	-1.73	95.51	104.11	2.05	1.95
Jun 8	66.43	-1.68	306.44	-0.56	91.75	0.81	88.07	19 Jun 24	121.8	185.9	0.2	4.8	57.25	63.22	-1.60	-1.44	111.66	118.13	1.56	0.92
Jun 18	67.65	-1.68	305.57	-0.60	98.38	0.88	97.60	29 Jun	245.6		4.2		69.20		-1.27		123.39		0.04	
								Jul 4		308.0		-0.8		75.20		-1.07		127.27		-1.02
Jun 28	68.80	-1.68	304.48	-0.63	104.96	0.94	107.13	9 Jul 14	15.8	87.8	-5.1	-2.8	81.22	87.25	-0.87	-0.65	129.53	129.90	-2.21	-3.41
Jul 8	69.87	-1.69	303.24	-0.66	111.49	0.99	116.68	19 Jul 24	156.5	217.9	3.2	5.2	93.30	99.36	-0.43	-0.20	128.29	125.14	-4.40	-4.92
Jul 18	70.84	-1.71	301.95	-0.68	117.97	1.04	126.23	29 J-A 3	278.4	344.6	1.9	-3.7	105.44	111.53	0.02	0.23	121.57	119.17	-4.75	-3.89
Jul 28	71.70	-1.72	300.69	-0.70	124.42	1.08	135.81	8 Aug 13	55.3	125.1	-4.7	0.5	117.64	123.77	0.44	0.63	119.16	122.02	-2.61	-1.21
Aug 7	72.43	-1.74	299.56	-0.71	130.83	1.12	145.43	18 Aug 23	189.7	249.6	4.9	4.0	129.91	136.06	0.81	0.97	127.66	135.51	0.06	1.02
Aug 17	73.01	-1.76	298.63	-0.72	137.20	1.16	155.08	28 Aug	312.4		-1.2		142.23		1.11		144.72		1.59	
								Sep 2		23.1		-5.1		148.41		1.22		154.42		1.79
Aug 27	73.42	-1.78	297.97	-0.72	143.55	1.19	164.77	7 Sep 12	94.2	160.5	-2.1	3.4	154.61	160.81	1.31	1.38	164.04	173.30	1.68	1.37
Sep 6	73.67	-1.80	297.61	-0.72	149.87	1.22	174.52	17 Sep 22	222.0	281.7	4.9	1.4	167.03	173.26	1.42	1.43	182.13	190.54	0.91	0.37
Sep 16	73.73	-1.82	297.57	-0.72	156.17	1.24	184.33	27 S-O 2	348.3	62.4	-3.9	-4.1	179.49	185.73	1.41	1.37	198.56	206.22	-0.22	-0.82
Sep 26	73.61	-1.84	297.86	-0.71	162.43	1.26	194.18	7 Oct 12	131.5	194.4	1.3	4.9	191.98	198.24	1.30	1.21	213.54	220.48	-1.40	-1.95
Oct 6	73.30	-1.85	298.47	-0.71	168.68	1.28	204.10	17 Oct 22	254.1	315.2	3.2	-1.8	204.50	210.76	1.09	0.95	226.98	232.84	-2.42	-2.77
Oct 16	72.84	-1.87	299.37	-0.70	174.89	1.29	214.07	27 Oct	26.0		-5.0		217.03		0.80		237.72		-2.93	
								Nov 1		100.8		-1.1		223.29		0.63		240.93		-2.79
Oct 26	72.23	-1.87	300.54	-0.70	181.08	1.30	224.09	6 Nov 11	166.9	227.3	4.1	4.6	229.56	235.84	0.44	0.25	241.38	237.95	-2.17	-0.90
Nov 5	71.50	-1.88	301.97	-0.70	187.24	1.30	234.17	16 Nov 21	286.7	350.6	0.4	-4.4	242.11	248.38	0.05	-0.15	231.55	226.45	0.77	2.08
Nov 15	70.71	-1.87	303.60	-0.69	193.35	1.30	244.28	26 N-D 1	64.6	137.4	-3.7	2.3	254.66	260.93	-0.36	-0.55	225.77	228.94	2.57	2.44
Nov 25	69.89	-1.86	305.43	-0.69	199.43	1.29	254.43	6 Dec 11	200.5	259.9	5.2	2.6	267.20	273.47	-0.74	-0.92	234.36	240.90	1.99	1.40
Dec 5	69.08	-1.84	307.41	-0.69	205.46	1.27	264.61	16 Dec 21	320.8	28.1	-2.7	-5.2	279.74	286.01	-1.08	-1.23	247.99	255.37	0.77	0.14
Dec 15	68.35	-1.81	309.52	-0.69	211.42	1.25	274.79	26 Dec 31	102.7	172.1	-0.7	4.7	292.27	298.53	-1.35	-1.45	262.92	270.61	-0.45	-0.98
1795								1795												1795

JULIAN	SATURN		JUPITER		MARS		SUN	GREGORIAN	MOON				VENUS				MERCURY			
12 UT	LONG.	LAT.	LONG.	LAT.	LONG.	LAT.	LONG.	12 UT	LONG.	LONG.	LAT.	LAT.	LONG.	LONG.	LAT.	LAT.	LONG.	LONG.	LAT.	LAT.
1795								1796												1796
Dec 25	67.72	-1.78	311.74	-0.70	217.31	1.22	284.99	5 Jan 10	232.9	292.8	4.4	-0.3	304.79	311.03	-1.53	-1.58	278.46	286.47	-1.42	-1.77
Jan 4	67.23	-1.74	314.04	-0.70	223.12	1.18	295.18	15 Jan 20	356.9	66.9	-4.9	-3.6	317.28	323.51	-1.60	-1.59	294.68	303.12	-2.00	-2.09
Jan 14	66.91	-1.70	316.39	-0.71	228.82	1.13	305.35	25 Jan 30	139.4	205.0	2.6	5.2	329.73	335.93	-1.55	-1.49	311.77	320.58	-1.98	-1.66
Jan 24	66.77	-1.65	318.78	-0.72	234.39	1.06	315.50	4 Feb 9	264.7	326.9	2.2	-3.1	342.13	348.31	-1.39	-1.26	329.29	337.34	-1.05	-0.15
Feb 3	66.82	-1.61	321.17	-0.73	239.81	0.97	325.61	14 Feb 19	35.0	105.7	-5.0	-0.3	354.47	0.61	-1.10	-0.92	343.71	347.03	1.03	2.31
Feb 13	67.06	-1.57	323.56	-0.75	245.03	0.86	335.68	24 Feb 29	174.7	237.0	4.7	4.0	6.73	12.83	-0.71	-0.48	346.38	342.37	3.35	3.71
Feb 23	67.48	-1.53	325.91	-0.77	250.02	0.73	345.70	5 Mar 10	296.8	3.0	-0.7	-4.9	18.90	24.94	-0.23	0.04	337.39	333.97	3.22	2.16
Mar 4	68.07	-1.49	328.20	-0.79	254.70	0.56	355.67	15 Mar 20	74.2	143.5	-2.8	3.0	30.95	36.92	0.31	0.60	333.11	334.64	0.94	-0.19
Mar 14	68.81	-1.45	330.42	-0.81	259.02	0.36	5.58	25 Mar 30	208.5	268.5	4.9	1.5	42.86	48.76	0.88	1.17	338.03	342.82	-1.11	-1.81
Mar 24	69.69	-1.42	332.54	-0.84	262.86	0.10	15.44	4 Apr 9	330.1	40.8	-3.5	-4.5	54.61	60.41	1.45	1.72	348.66	355.38	-2.29	-2.55
Apr 3	70.68	-1.39	334.54	-0.87	266.11	-0.21	25.24	14 Apr 19	113.1	179.5	0.8	4.9	66.16	71.84	1.97	2.20	2.85	11.04	-2.58	-2.40
Apr 13	71.77	-1.36	336.39	-0.90	268.61	-0.60	34.98	24 Apr 29	241.2	300.6	3.4	-1.6	77.46	83.01	2.41	2.59	19.93	29.53	-2.00	-1.40
Apr 23	72.94	-1.34	338.06	-0.93	270.19	-1.07	44.67	4 May 9	5.5	79.6	-5.0	-2.0	88.47	93.84	2.72	2.82	39.82	50.60	-0.63	0.24
May 3	74.17	-1.31	339.54	-0.97	270.66	-1.63	54.33	14 May 19	150.6	213.9	3.9	4.8	99.10	104.23	2.87	2.87	61.46	71.85	1.07	1.74
May 13	75.44	-1.30	340.79	-1.02	269.90	-2.25	63.93	24 May 29	273.5	334.0	0.7	-4.2	109.22	114.04	2.80	2.67	81.32	89.66	2.12	2.18
May 23	76.74	-1.28	341.79	-1.06	267.94	-2.92	73.51	3 Jun 8	42.9	118.1	-4.4	1.5	118.66	123.06	2.47	2.19	96.77	102.58	1.91	1.32
Jun 2	78.04	-1.27	342.51	-1.11	265.08	-3.54	83.07	13 Jun 18	186.2	246.9	5.2	2.9	127.17	130.95	1.83	1.37	106.95	109.69	0.44	-0.69
Jun 12	79.33	-1.26	342.93	-1.16	261.95	-4.05	92.60	23 Jun 28	306.3	9.5	-2.3	-5.3	134.32	137.21	0.81	0.13	110.61	109.65	-1.99	-3.29
Jun 22	80.60	-1.25	343.03	-1.21	259.25	-4.38	102.14	3 Jul 8	81.4	155.2	-1.8	4.3	139.52	141.12	-0.66	-1.57	107.15	104.02	-4.34	-4.86
Jul 2	81.81	-1.25	342.80	-1.26	257.60	-4.53	111.67	13 Jul 18	219.9	279.3	4.6	0.2	141.91	141.75	-2.60	-3.71	101.55	100.84	-4.70	-3.97
Jul 12	82.97	-1.25	342.26	-1.30	257.30	-4.52	121.22	23 Jul 28	340.4	47.1	-4.5	-4.2	140.61	138.55	-4.86	-5.96	102.47	106.56	-2.85	-1.59
Jul 22	84.04	-1.25	341.44	-1.35	258.38	-4.40	130.78	2 Aug 7	120.1	190.6	1.7	5.2	135.78	132.70	-6.90	-7.55	112.96	121.28	-0.36	0.67
Aug 1	85.02	-1.26	340.37	-1.38	260.70	-4.21	140.38	12 Aug 17	252.2	312.0	2.4	-2.7	129.78	127.45	-7.87	-7.84	130.86	140.92	1.37	1.72
Aug 11	85.88	-1.26	339.14	-1.41	264.08	-3.98	150.01	22 Aug 27	16.6	86.0	-5.1	-1.2	126.01	125.56	-7.54	-7.03	150.88	160.43	1.74	1.52
Aug 21	86.61	-1.27	337.82	-1.42	268.32	-3.72	159.68	1 Sep 6	157.7	224.1	4.3	4.1	126.08	127.47	-6.39	-5.69	169.48	178.03	1.12	0.60
Aug 31	87.19	-1.28	336.53	-1.43	273.25	-3.46	169.41	11 Sep 16	283.9	346.0	-0.5	-4.7	129.61	132.37	-4.95	-4.22	186.10	193.72	0.01	-0.62
Sep 10	87.60	-1.29	335.34	-1.42	278.74	-3.18	179.18	21 Sep 26	54.6	125.0	-3.4	2.4	135.67	139.41	-3.49	-2.80	200.91	207.61	-1.26	-1.88
Sep 20	87.84	-1.30	334.36	-1.41	284.67	-2.90	189.01	1 Oct 6	193.7	256.2	5.0	1.7	143.52	147.93	-2.14	-1.51	213.74	219.08	-2.44	-2.90
Sep 30	87.89	-1.30	333.64	-1.38	290.95	-2.62	198.90	11 Oct 16	315.8	21.8	-3.3	-4.8	152.59	157.48	-0.93	-0.40	223.24	225.54	-3.19	-3.19
Oct 10	87.76	-1.31	333.24	-1.36	297.52	-2.35	208.84	21 Oct 26	93.6	163.1	0.0	4.8	162.55	167.78	0.09	0.53	224.96	220.78	-2.70	-1.53
Oct 20	87.45	-1.32	333.17	-1.32	304.30	-2.08	218.84	31 Oct	227.9		3.6		173.15		0.92		214.48		0.14	
								Nov 5		287.8		-1.3		178.63		1.25		210.24		1.57
Oct 30	86.97	-1.32	333.44	-1.29	311.24	-1.81	228.88	10 Nov 15	348.9	59.1	-5.0	-2.8	184.22	189.90	1.54	1.77	210.54	214.55	2.27	2.34
Nov 9	86.35	-1.32	334.04	-1.26	318.30	-1.56	238.98	20 Nov 25	132.4	199.4	3.4	5.0	195.66	201.49	1.96	2.09	220.65	227.72	2.03	1.54
Nov 19	85.62	-1.31	334.95	-1.23	325.46	-1.31	249.11	30 N-D 5	260.8	319.9	1.0	-3.9	207.37	213.31	2.18	2.21	235.20	242.86	0.96	0.36
Nov 29	84.82	-1.30	336.14	-1.19	332.67	-1.08	259.27	10 Dec 15	23.7	97.2	-4.9	0.5	219.29	225.30	2.21	2.16	250.58	258.36	-0.23	-0.77
Dec 9	83.99	-1.29	337.59	-1.17	339.91	-0.86	269.45	20 Dec 25	170.0	233.7	5.2	3.3	231.35	237.43	2.08	1.96	266.20	274.14	-1.25	-1.64
Dec 19	83.19	-1.26	339.25	-1.14	347.16	-0.65	279.65	30 Dec	293.1		-1.9		243.53		1.81		282.19		-1.94	
1796								1797												1797
								Jan 4		353.3		-5.2		249.65		1.64		290.40		-2.10
Dec 29	82.46	-1.24	341.11	-1.12	354.40	-0.46	289.84	9 Jan 14	60.3	135.5	-2.8	3.6	255.78	261.93	1.44	1.22	298.73	307.12	-2.09	-1.87
Jan 8	81.84	-1.21	343.12	-1.10	1.63	-0.28	300.02	19 Jan 24	205.6	266.5	4.8	0.5	268.09	274.26	0.99	0.76	315.33	322.79	-1.38	-0.56
Jan 18	81.36	-1.18	345.26	-1.09	8.82	-0.12	310.18	29 J-F 3	325.8	28.6	-4.2	-4.6	280.44	286.62	0.51	0.27	328.43	330.71	0.61	2.00
Jan 28	81.05	-1.14	347.51	-1.08	15.97	0.04	320.32	8 Feb 13	98.4	172.9	0.7	5.1	292.81	299.00	0.03	-0.21	328.66	323.46	3.22	3.73
Feb 7	80.93	-1.11	349.84	-1.07	23.08	0.18	330.41	18 Feb 23	239.2	298.5	2.6	-2.4	305.19	311.38	-0.43	-0.63	318.36	315.78	3.33	2.36
Feb 17	80.99	-1.07	352.22	-1.07	30.13	0.31	340.46	28 Feb	359.8		-5.0		317.58		-0.82		316.08		1.24	
								Mar 5		65.9		-1.9		323.77		-0.99		318.67		0.19
Feb 27	81.25	-1.04	354.63	-1.07	37.12	0.42	350.45	10 Mar 15	137.2	208.6	3.8	4.3	329.96	336.15	-1.14	-1.26	322.89	328.29	-0.70	-1.41
Mar 9	81.68	-1.00	357.05	-1.07	44.06	0.53	0.39	20 Mar 25	271.2	330.9	-0.4	-4.5	342.34	348.53	-1.36	-1.43	334.57	341.58	-1.94	-2.27
Mar 19	82.27	-0.97	359.46	-1.08	50.94	0.62	10.28	30 M-A 4	35.5	104.6	-3.9	1.7	354.71	0.89	-1.47	-1.49	349.24	357.50	-2.41	-2.35
Mar 29	83.02	-0.94	1.84	-1.09	57.76	0.70	20.11	9 Apr 14	175.6	242.5	5.1	1.9	7.06	13.24	-1.48	-1.43	6.39	15.91	-2.08	-1.60
Apr 8	83.90	-0.91	4.17	-1.10	64.52	0.78	29.88	19 Apr 24	302.7	4.3	-3.2	-5.0	19.40	25.57	-1.37	-1.28	26.04	36.65	-0.92	-0.09
Apr 18	84.89	-0.89	6.43	-1.12	71.23	0.85	39.60	29 Apr	72.8		-0.9		31.73		-1.17		47.37		0.79	
								May 4		143.9		4.5		37.89		-1.03		57.63		1.58
Apr 28	85.99	-0.86	8.59	-1.14	77.88	0.91	49.27	9 May 14	212.3	275.1	4.0	-1.1	44.04	50.19	-0.88	-0.71	66.89	74.84	2.14	2.38
May 8	87.16	-0.84	10.64	-1.16	84.49	0.96	58.90	19 May 24	334.6	39.2	-5.0	-3.6	56.34	62.49	-0.53	-0.34	81.31	86.20	2.26	1.76
May 18	88.39	-0.82	12.55	-1.19	91.05	1.00	68.49	29 M-J 3	111.5	182.3	2.5	5.2	68.64	74.78	-0.15	0.05	89.36	90.65	0.91	-0.27
May 28	89.66	-0.80	14.29	-1.22	97.57	1.04	78.06	8 Jun 13	247.0	307.1	1.4	-3.7	80.92	87.07	0.25	0.44	90.06	87.94	-1.65	-3.02
Jun 7	90.95	-0.79	15.85	-1.26	104.06	1.07	87.60	18 Jun 23	7.6	75.9	-5.1	-0.6	93.21	99.35	0.63	0.80	85.13	82.77	-4.07	-4.59
Jun 17	92.26	-0.77	17.19	-1.29	110.52	1.10	97.14	28 Jun	150.5		4.9		105.49		0.96		81.86		-4.50	
								Jul 3		218.9		3.6		111.63		1.11		82.92		-3.92
Jun 27	93.55	-0.76	18.28	-1.33	116.94	1.12	106.67	8 Jul 13	280.2	339.6	-1.6	-5.1	117.78	123.92	1.23	1.34	86.08	91.26	-3.00	-1.90
Jul 7	94.82	-0.75	19.11	-1.38	123.35	1.13	116.21	18 Jul 23	42.2	114.1	-3.3	2.8	130.06	136.20	1.42	1.47	98.34	107.06	-0.75	0.30
Jul 17	96.04	-0.74	19.63	-1.42	129.74	1.14	125.77	28 J-A 2	188.5	253.2	4.9	0.8	142.34	148.47	1.50	1.49	116.94	127.32	1.11	1.60
Jul 27	97.19	-0.74	19.84	-1.47	136.12	1.15	135.34	7 Aug 12	312.8	13.1	-4.0	-4.7	154.61	160.74	1.46	1.40	137.58	147.40	1.77	1.65
Aug 6	98.27	-0.73	19.71	-1.51	142.49	1.15	144.95	17 Aug 22	78.8	153.0	-0.2	4.8	166.87	172.99	1.31	1.20	156.65	165.32	1.32	0.84
Aug 16	99.25	-0.73	19.26	-1.55	148.85	1.15	154.61	27 Aug	224.5		2.9		179.11		1.05		173.43		0.25	
								Sep 1		286.0		-2.3		185.23		0.89		181.01		-0.41
Aug 26	100.11	-0.72	18.50	-1.59	155.22	1.14	164.30	6 Sep 11	345.6	48.4	-5.0	-2.5	191.33	197.43	0.69	0.48	188.04	194.48	-1.10	-1.79
Sep 5	100.83	-0.72	17.48	-1.62	161.58	1.13	174.05	16 Sep 21	117.0	191.0	3.2	4.6	203.53	209.61	0.26	0.01	200.20	204.96	-2.45	-3.02
Sep 15	101.41	-0.72	16.26	-1.63	167.96	1.11	183.85	26 S-O 1	258.2	318.1	-0.2	-4.5	215.69	221.76	-0.24	-0.50	208.33	209.65	-3.45	-3.58
Sep 25	101.82	-0.71	14.94	-1.64	174.34	1.09	193.70	6 Oct 11	19.5	85.5	-4.2	0.9	227.82	233.86	-0.76	-1.02	208.04	203.29	-3.22	-2.14
Oct 5	102.05	-0.71	13.60	-1.63	180.73	1.06	203.62	16 Oct 21	156.2	227.2	5.1	2.3	239.90	245.92	-1.27	-1.51	197.40	194.18	-0.49	1.03
Oct 15	102.09	-0.71	12.36	-1.61	187.14	1.03	213.59	26 Oct 31	290.5	350.2	-3.1	-5.2	251.93	257.91	-1.73	-1.94	195.45	200.25	1.93	2.19
Oct 25	101.95	-0.70	11.30	-1.57	193.56	0.99	223.61	5 Nov 10	54.6	124.2	-1.7	4.1	263.88	269.82	-2.12	-2.27	206.99	214.58	2.04	1.65
Nov 4	101.63	-0.70	10.51	-1.53	200.00	0.95	233.68	15 Nov 20	194.7	261.4	4.5	-0.7	275.74	281.62	-2.38	-2.46	222.45	230.38	1.13	0.56
Nov 14	101.14	-0.69	10.03	-1.48	206.45	0.90	243.80	25 Nov 30	322.0	23.0	-4.9	-4.1	287.47	293.27	-2.50	-2.49	238.29	246.16	-0.01	-0.56
Nov 24	100.51	-0.68	9.89	-1.43	212.91	0.84	253.94	5 Dec 10	91.4	163.5	1.6	5.3	299.01	304.70	-2.42	-2.31	254.01	261.88	-1.07	-1.51
Dec 4	99.77	-0.67	10.10	-1.38	219.38	0.77	264.11	15 Dec 20	231.5	294.2	2.0	-3.4	310.31	315.83	-2.14	-1.91	269.79	277.76	-1.86	-2.09
Dec 14	98.96	-0.65	10.64	-1.33	225.87	0.70	274.30	25 Dec 30	353.8	57.1	-5.2	-1.5	321.25	326.54	-1.62	-1.27	285.77	293.74	-2.18	-2.06
1797								1797												1797

JULIAN 12 UT	SATURN LONG.	LAT.	JUPITER LONG.	LAT.	MARS LONG.	LAT.	SUN LONG.	GREGORIAN 12 UT	MOON LONG.		LAT.		VENUS LONG.		LAT.		MERCURY LONG.		LAT.	
1797								1798												1798
Dec 24	98.14	-0.63	11.50	-1.28	232.37	0.62	284.50	4 Jan 9	129.6	201.9	4.3	4.0	331.68	336.65	-0.85	-0.37	301.42	308.24	-1.69	-0.97
Jan 3	97.34	-0.61	12.65	-1.24	238.87	0.52	294.68	14 Jan 19	266.3	326.6	-1.2	-4.9	341.41	345.92	0.18	0.80	313.09	314.34	0.15	1.60
Jan 13	96.62	-0.59	14.06	-1.20	245.38	0.41	304.86	24 Jan 29	26.5	93.0	-3.6	1.8	350.12	353.94	1.49	2.24	311.03	305.05	2.98	3.63
Jan 23	96.01	-0.57	15.69	-1.16	251.88	0.29	315.01	3 Feb 8	168.4	238.3	5.0	1.1	357.31	0.13	3.06	3.94	300.26	298.69	3.34	2.48
Feb 2	95.54	-0.54	17.51	-1.13	258.39	0.16	325.12	13 Feb 18	299.7	359.2	-3.8	-4.7	2.29	3.66	4.86	5.81	300.09	303.59	1.47	0.50
Feb 12	95.25	-0.52	19.48	-1.10	264.88	0.01	335.19	23 Feb 28	60.7	130.7	-0.7	4.4	4.10	3.52	6.73	7.57	308.49	314.37	-0.35	-1.06
Feb 22	95.14	-0.49	21.59	-1.07	271.35	-0.16	345.21	5 Mar 10	206.3	272.6	3.3	-2.1	1.91	359.42	8.23	8.61	320.98	328.19	-1.61	-2.01
Mar 4	95.21	-0.46	23.81	-1.05	277.80	-0.36	355.18	15 Mar 20	332.3	32.6	-5.0	-2.8	356.39	353.31	8.61	8.23	335.94	344.22	-2.23	-2.28
Mar 14	95.47	-0.44	26.10	-1.03	284.20	-0.57	5.10	25 Mar 30	97.0	169.7	2.6	4.9	350.66	348.82	7.50	6.54	353.05	2.43	-2.12	-1.76
Mar 24	95.91	-0.42	28.45	-1.01	290.55	-0.81	14.95	4 Apr 9	242.3	305.3	0.3	-4.5	347.96	348.12	5.46	4.37	12.36	22.74	-1.19	-0.42
Apr 3	96.51	-0.39	30.83	-1.00	296.83	-1.08	24.76	14 Apr 19	4.9	67.5	-4.5	0.3	349.20	351.08	3.32	2.35	33.24	43.29	0.47	1.36
Apr 13	97.26	-0.37	33.23	-0.99	303.01	-1.37	34.51	24 Apr 29	135.4	208.2	4.9	3.1	353.63	356.73	1.47	0.68	52.25	59.66	2.09	2.53
Apr 23	98.14	-0.35	35.61	-0.99	309.06	-1.70	44.20	4 May 9	276.5	337.2	-2.8	-5.2	0.28	4.20	0.00	-0.60	65.28	68.95	2.59	2.23
May 3	99.13	-0.33	37.98	-0.99	314.93	-2.06	53.85	14 May 19	38.0	104.3	-2.3	3.4	8.43	12.91	-1.10	-1.53	70.57	70.19	1.42	0.23
May 13	100.22	-0.31	40.30	-0.99	320.57	-2.46	63.47	24 May 29	174.9	245.0	4.9	-0.1	17.59	22.45	-1.87	-2.15	68.18	65.43	-1.20	-2.58
May 23	101.39	-0.30	42.55	-0.99	325.89	-2.90	73.05	3 Jun 8	309.2	9.1	-4.8	-4.3	27.46	32.60	-2.36	-2.50	63.08	62.09	-3.62	-4.15
Jun 2	102.61	-0.28	44.72	-1.00	330.82	-3.38	82.60	13 Jun 18	72.5	143.0	0.8	5.1	37.84	43.18	-2.59	-2.62	62.90	65.60	-4.17	-3.77
Jun 12	103.88	-0.26	46.78	-1.00	335.21	-3.90	92.15	23 Jun 28	213.7	280.0	2.6	-3.1	48.59	54.08	-2.60	-2.54	70.07	76.18	-3.06	-2.13
Jun 22	105.18	-0.25	48.71	-1.02	338.93	-4.45	101.68	3 Jul 8	341.3	41.5	-5.1	-1.9	59.63	65.23	-2.44	-2.30	83.83	92.88	-1.10	-0.07
Jul 2	106.48	-0.23	50.48	-1.03	341.76	-5.02	111.21	13 Jul 18	109.0	182.4	3.7	4.4	70.89	76.60	-2.13	-1.93	102.99	113.60	0.81	1.44
Jul 12	107.77	-0.22	52.08	-1.05	343.51	-5.57	120.76	23 Jul 28	250.8	313.6	-0.7	-4.8	82.35	88.14	-1.71	-1.47	124.14	134.20	1.75	1.76
Jul 22	109.03	-0.21	53.47	-1.07	344.01	-6.07	130.32	2 Aug 7	13.3	75.4	-3.8	1.2	93.97	99.84	-1.22	-0.96	143.62	152.39	1.51	1.08
Aug 1	110.25	-0.19	54.62	-1.09	343.18	-6.43	139.92	12 Aug 17	147.3	220.9	5.0	1.7	105.74	111.68	-0.69	-0.42	160.52	168.02	0.51	-0.17
Aug 11	111.40	-0.18	55.51	-1.11	341.19	-6.55	149.55	22 Aug 27	286.0	346.5	-3.6	-4.8	117.64	123.64	-0.16	0.09	174.86	180.98	-0.91	-1.68
Aug 21	112.47	-0.17	56.10	-1.13	338.58	-6.36	159.22	1 Sep 6	46.1	111.3	-1.1	4.0	129.66	135.72	0.34	0.56	186.23	190.32	-2.44	-3.14
Aug 31	113.45	-0.15	56.37	-1.16	336.05	-5.88	168.94	11 Sep 16	186.2	257.4	3.9	-1.8	141.80	147.91	0.77	0.96	192.81	193.07	-3.70	-3.97
Sep 10	114.30	-0.14	56.32	-1.18	334.31	-5.19	178.71	21 Sep 26	319.6	19.1	-5.0	-3.1	154.04	160.19	1.12	1.26	190.50	185.45	-3.71	-2.70
Sep 20	115.02	-0.13	55.92	-1.20	333.79	-4.39	188.53	1 Oct 6	80.3	149.1	2.1	5.1	166.36	172.54	1.37	1.45	180.29	178.23	-1.10	0.48
Sep 30	115.59	-0.11	55.22	-1.21	334.57	-3.60	198.42	11 Oct 16	224.3	291.9	1.0	-4.3	178.75	184.97	1.50	1.52	180.46	186.00	1.54	2.00
Oct 10	115.99	-0.10	54.23	-1.22	336.57	-2.87	208.36	21 Oct 26	352.2	52.3	-4.7	-0.2	191.21	197.45	1.52	1.48	193.35	201.43	2.01	1.73
Oct 20	116.21	-0.08	53.03	-1.22	339.58	-2.22	218.35	31 Oct	116.6		4.6		203.71		1.42		209.70		1.29	
								Nov 5		188.0		3.9		209.97		1.33		217.91		0.76
Oct 30	116.24	-0.07	51.70	-1.21	343.40	-1.65	228.40	10 Nov 15	260.6	324.8	-2.3	-5.3	216.24	222.52	1.22	1.08	226.01	233.99	0.20	-0.36
Nov 9	116.09	-0.05	50.34	-1.18	347.86	-1.17	238.49	20 Nov 25	24.3	86.9	-2.7	2.8	228.81	235.09	0.93	0.76	241.86	249.67	-0.89	-1.37
Nov 19	115.75	-0.03	49.07	-1.16	352.80	-0.76	248.62	30 N-D 5	154.9	226.4	5.2	0.8	241.38	247.67	0.58	0.39	257.44	265.19	-1.77	-2.08
Nov 29	115.25	-0.02	47.96	-1.12	358.13	-0.41	258.78	10 Dec 15	295.1	356.5	-4.5	-4.4	253.97	260.26	0.19	-0.01	272.88	280.42	-2.25	-2.24
Dec 9	114.61	0.00	47.11	-1.08	3.74	-0.12	268.96	20 Dec 25	56.9	123.5	0.2	4.8	266.55	272.85	-0.21	-0.41	287.54	293.68	-1.98	-1.38
Dec 19	113.87	0.02	46.57	-1.03	9.56	0.13	279.15	30 Dec	194.2		3.3		279.14		-0.59		297.70		-0.35	
1798								1799												1799
								Jan 4		263.5		-2.5		285.43		-0.77		297.93		1.11
Dec 29	113.06	0.04	46.36	-0.98	15.55	0.34	289.35	9 Jan 14	328.2	28.0	-5.1	-2.2	291.72	298.00	-0.93	-1.08	293.55	287.17	2.62	3.43
Jan 8	112.24	0.05	46.50	-0.93	21.65	0.52	299.53	19 Jan 24	90.6	161.8	3.1	4.7	304.29	310.56	-1.20	-1.30	283.00	282.49	3.27	2.55
Jan 18	111.45	0.07	46.98	-0.89	27.85	0.67	309.69	29 J-F 3	233.0	299.0	-0.1	-4.6	316.84	323.11	-1.38	-1.44	284.89	289.16	1.66	0.77
Jan 28	110.74	0.09	47.77	-0.84	34.10	0.80	319.82	8 Feb 13	0.5	60.0	-3.8	0.9	329.37	335.63	-1.47	-1.47	294.63	300.91	-0.04	-0.74
Feb 7	110.14	0.11	48.85	-0.80	40.40	0.90	329.92	18 Feb 23	126.3	200.7	4.9	2.4	341.88	348.11	-1.44	-1.39	307.77	315.12	-1.31	-1.76
Feb 17	109.69	0.12	50.18	-0.76	46.72	0.99	339.96	28 Feb	270.2		-3.4		354.34		-1.31		322.93		-2.05	
								Mar 5		333.1		-4.9		0.56		-1.20		331.18		-2.18
Feb 27	109.41	0.14	51.73	-0.73	53.06	1.06	349.96	10 Mar 15	32.6	93.5	-1.3	3.7	6.77	12.96	-1.07	-0.92	339.90	349.11	-2.13	-1.89
Mar 9	109.30	0.15	53.47	-0.70	59.41	1.11	359.91	20 Mar 25	163.9	238.9	4.6	-1.1	19.15	25.31	-0.74	-0.55	358.80	8.89	-1.42	-0.74
Mar 19	109.39	0.17	55.37	-0.67	65.75	1.16	9.79	30 M-A 4	305.6	6.0	-5.1	-3.3	31.46	37.60	-0.35	-0.13	19.10	28.85	0.13	1.09
Mar 29	109.65	0.18	57.40	-0.64	72.08	1.20	19.62	9 Apr 14	65.3	129.2	1.8	5.2	43.72	49.83	0.10	0.33	37.42	44.16	1.97	2.62
Apr 8	110.09	0.19	59.54	-0.62	78.41	1.22	29.40	19 Apr 24	202.5	275.6	2.1	-4.0	55.91	61.97	0.56	0.79	48.71	50.88	2.88	2.68
Apr 18	110.69	0.20	61.76	-0.59	84.73	1.24	39.12	29 Apr	339.2		-4.9		68.02		1.01		50.72		1.95	
								May 4		38.4		-0.7		74.04		1.22		48.68		0.77
Apr 28	111.44	0.22	64.03	-0.57	91.04	1.25	48.80	9 May 14	99.6	166.9	4.3	4.5	80.05	86.03	1.42	1.60	45.72	43.13	-0.67	-2.02
May 8	112.31	0.23	66.34	-0.56	97.34	1.26	58.43	19 May 24	240.9	310.5	-1.4	-5.2	91.98	97.90	1.75	1.88	41.90	42.49	-3.04	-3.61
May 18	113.30	0.24	68.67	-0.54	103.63	1.26	68.03	29 M-J 3	11.5	71.3	-3.0	2.3	103.80	109.66	1.98	2.04	44.90	48.96	-3.76	-3.55
May 28	114.39	0.25	71.00	-0.53	109.91	1.25	77.60	8 Jun 13	135.9	205.9	5.2	1.7	115.49	121.28	2.07	2.06	54.48	61.34	-3.04	-2.30
Jun 7	115.55	0.27	73.30	-0.51	116.20	1.24	87.14	18 Jun 23	278.1	343.7	-4.1	-4.5	127.03	132.73	2.01	1.91	69.45	78.74	-1.40	-0.43
Jun 17	116.77	0.28	75.57	-0.50	122.48	1.22	96.68	28 Jun	43.3		-0.1		138.38		1.77		89.01		0.49	
								Jul 3		105.7		4.5		143.98		1.58		99.80		1.23
Jun 27	118.03	0.29	77.78	-0.49	128.77	1.20	106.21	8 Jul 13	174.2	244.9	3.9	-1.8	149.51	154.96	1.35	1.07	110.54	120.79	1.68	1.83
Jul 7	119.31	0.30	79.90	-0.49	135.06	1.18	115.75	18 Jul 23	313.6	15.7	-5.0	-2.3	160.33	165.61	0.74	0.37	130.36	139.19	1.69	1.33
Jul 17	120.61	0.32	81.93	-0.48	141.37	1.15	125.31	28 J-A 2	75.6	142.1	2.8	4.9	170.78	175.82	-0.05	-0.52	147.29	154.65	0.79	0.10
Jul 27	121.89	0.33	83.84	-0.47	147.69	1.11	134.89	7 Aug 12	213.4	282.7	0.7	-4.4	180.72	185.45	-1.03	-1.58	161.25	166.97	-0.68	-1.53
Aug 6	123.15	0.35	85.59	-0.47	154.03	1.08	144.49	17 Aug 22	347.5	47.3	-4.0	0.7	189.97	194.25	-2.17	-2.79	171.64	174.94	-2.41	-3.25
Aug 16	124.36	0.37	87.17	-0.46	160.40	1.04	154.14	27 Aug	109.2		4.8		198.24		-3.45		176.43		-3.95	
								Sep 1		180.1		3.2		201.89		-4.12		175.58		-4.34
Aug 26	125.50	0.39	88.55	-0.45	166.80	0.99	163.84	6 Sep 11	252.3	318.7	-2.9	-5.0	205.11	207.80	-4.81	-5.51	172.20	167.24	-4.15	-3.19
Sep 5	126.57	0.41	89.69	-0.45	173.22	0.94	173.58	16 Sep 21	20.1	79.4	-1.5	3.5	209.87	211.20	-6.18	-6.80	163.14	162.35	-1.65	-0.06
Sep 15	127.53	0.43	90.56	-0.44	179.68	0.89	183.38	26 S-O 1	144.7	218.8	5.0	-0.2	211.65	211.14	-7.33	-7.69	165.53	171.76	1.12	1.77
Sep 25	128.38	0.45	91.15	-0.44	186.18	0.83	193.23	6 Oct 11	289.8	352.9	-5.0	-3.6	209.64	207.28	-7.82	-7.64	179.68	188.25	1.94	1.79
Oct 5	129.08	0.47	91.41	-0.43	192.72	0.77	203.14	16 Oct 21	52.3	113.0	1.4	5.1	204.35	201.34	-7.10	-6.23	196.90	205.41	1.43	0.95
Oct 15	129.64	0.49	91.35	-0.42	199.31	0.71	213.11	26 Oct 31	181.8	257.1	3.0	-3.4	198.73	196.90	-5.12	-3.89	213.71	221.80	0.41	-0.15
Oct 25	130.02	0.52	90.95	-0.41	205.94	0.63	223.13	5 Nov 10	325.3	25.7	-5.0	-1.0	196.04	196.21	-2.65	-1.50	229.71	237.48	-0.70	-1.22
Nov 4	130.22	0.55	90.24	-0.40	212.62	0.56	233.19	15 Nov 20	85.1	148.4	3.9	4.9	197.31	199.26	-0.47	0.42	245.12	252.64	-1.67	-2.05
Nov 14	130.24	0.57	89.25	-0.39	219.35	0.48	243.30	25 Nov 30	220.0	294.1	-0.3	-5.1	201.90	205.13	1.17	1.79	260.01	267.11	-2.30	-2.40
Nov 24	130.06	0.60	88.05	-0.37	226.13	0.39	253.45	5 Dec 10	358.7	57.9	-3.2	1.9	208.84	212.94	2.28	2.66	273.67	279.09	-2.26	-1.79
Dec 4	129.71	0.62	86.72	-0.35	232.96	0.29	263.62	15 Dec 20	119.4	185.8	5.1	2.5	217.36	222.05	2.94	3.13	282.22	281.44	-0.86	0.57
Dec 14	129.20	0.65	85.37	-0.32	239.85	0.19	273.81	25 Dec 30	258.4	329.2	-3.5	-4.6	226.97	232.07	3.24	3.27	276.22	269.78	2.16	3.13
1799								1799												1799

JULIAN 12 UT	SATURN LONG.	LAT.	JUPITER LONG.	LAT.	MARS LONG.	LAT.	SUN LONG.
1799							
Dec 24	128.55	0.67	84.09	-0.30	246.78	0.08	284.01
Jan 3	127.81	0.70	82.99	-0.27	253.77	-0.03	294.19
Jan 13	127.00	0.72	82.14	-0.24	260.80	-0.15	304.37
Jan 23	126.18	0.73	81.59	-0.21	267.89	-0.28	314.52
Feb 2	125.40	0.75	81.37	-0.18	275.02	-0.42	324.63
Feb 12	124.70	0.76	81.48	-0.16	282.20	-0.56	334.71
Feb 22	124.12	0.76	81.91	-0.13	289.42	-0.71	344.73
Mar 3	123.68	0.77	82.66	-0.11	296.68	-0.87	354.70
Mar 13	123.41	0.77	83.67	-0.09	303.96	-1.03	4.62
Mar 23	123.32	0.78	84.94	-0.07	311.27	-1.19	14.48
Apr 2	123.41	0.78	86.42	-0.05	318.59	-1.36	24.29
Apr 12	123.68	0.78	88.08	-0.03	325.91	-1.52	34.04
Apr 22	124.12	0.78	89.90	-0.01	333.22	-1.69	43.74
May 2	124.72	0.78	91.85	0.01	340.49	-1.84	53.39
May 12	125.46	0.78	93.90	0.02	347.72	-2.00	63.00
May 22	126.34	0.79	96.03	0.04	354.87	-2.14	72.59
Jun 1	127.32	0.79	98.22	0.06	1.91	-2.27	82.14
Jun 11	128.39	0.80	100.45	0.07	8.81	-2.38	91.68
Jun 21	129.55	0.80	102.70	0.09	15.53	-2.48	101.22
Jul 1	130.76	0.81	104.95	0.10	22.03	-2.56	110.75
Jul 11	132.01	0.82	107.19	0.12	28.24	-2.61	120.30
Jul 21	133.28	0.83	109.39	0.14	34.10	-2.64	129.86
Jul 31	134.57	0.85	111.53	0.15	39.51	-2.64	139.45
Aug 10	135.84	0.86	113.60	0.17	44.37	-2.60	149.08
Aug 20	137.09	0.88	115.56	0.19	48.53	-2.53	158.75
Aug 30	138.28	0.90	117.40	0.21	51.83	-2.41	168.46
Sep 9	139.42	0.92	119.09	0.23	54.06	-2.22	178.23
Sep 19	140.47	0.95	120.60	0.26	55.01	-1.96	188.06
Sep 29	141.42	0.97	121.91	0.28	54.52	-1.61	197.93
Oct 9	142.24	1.00	122.97	0.31	52.57	-1.16	207.87
Oct 19	142.93	1.03	123.77	0.34	49.48	-0.63	217.87
Oct 29	143.46	1.07	124.27	0.37	45.90	-0.08	227.91
Nov 8	143.82	1.10	124.46	0.40	42.64	0.45	238.00
Nov 18	144.00	1.13	124.31	0.43	40.39	0.88	248.13
Nov 28	143.99	1.17	123.84	0.47	39.46	1.22	258.28
Dec 8	143.80	1.20	123.06	0.50	39.86	1.46	268.47
Dec 18	143.43	1.24	122.03	0.53	41.43	1.63	278.66
1800							
Dec 28	142.90	1.27	120.81	0.55	43.97	1.73	288.85
Jan 7	142.25	1.30	119.48	0.58	47.28	1.80	299.04
Jan 17	141.49	1.32	118.16	0.60	51.18	1.83	309.20
Jan 27	140.69	1.34	116.93	0.61	55.55	1.84	319.33
Feb 6	139.89	1.35	115.88	0.62	60.27	1.84	329.43
Feb 16	139.12	1.36	115.08	0.63	65.28	1.82	339.48
Feb 26	138.43	1.37	114.58	0.63	70.50	1.80	349.48
Mar 8	137.87	1.37	114.40	0.63	75.90	1.76	359.43
Mar 18	137.45	1.36	114.54	0.63	81.43	1.73	9.32
Mar 28	137.19	1.35	114.99	0.63	87.07	1.68	19.15
Apr 7	137.11	1.34	115.72	0.63	92.80	1.64	28.93
Apr 17	137.21	1.33	116.72	0.62	98.61	1.59	38.65
Apr 27	137.49	1.32	117.96	0.62	104.48	1.54	48.33
May 7	137.93	1.31	119.40	0.62	110.40	1.48	57.97
May 17	138.53	1.30	121.01	0.62	116.37	1.43	67.56
May 27	139.27	1.30	122.78	0.62	122.38	1.37	77.13
Jun 6	140.14	1.29	124.67	0.63	128.44	1.31	86.68
Jun 16	141.11	1.29	126.66	0.63	134.54	1.25	96.21
Jun 26	142.18	1.29	128.74	0.64	140.68	1.18	105.75
Jul 6	143.33	1.29	130.87	0.64	146.86	1.12	115.29
Jul 16	144.53	1.29	133.04	0.65	153.10	1.05	124.84
Jul 26	145.77	1.30	135.23	0.66	159.37	0.98	134.42
Aug 5	147.03	1.31	137.43	0.68	165.71	0.91	144.03
Aug 15	148.31	1.32	139.60	0.69	172.09	0.83	153.67
Aug 25	149.57	1.33	141.74	0.71	178.53	0.76	163.36
Sep 4	150.80	1.35	143.82	0.73	185.03	0.68	173.11
Sep 14	151.98	1.37	145.81	0.75	191.59	0.59	182.90
Sep 24	153.10	1.40	147.70	0.78	198.22	0.51	192.75
Oct 4	154.13	1.43	149.46	0.81	204.92	0.42	202.66
Oct 14	155.05	1.46	151.07	0.84	211.68	0.33	212.62
Oct 24	155.85	1.49	152.48	0.87	218.52	0.24	222.64
Nov 3	156.52	1.53	153.67	0.91	225.43	0.14	232.71
Nov 13	157.02	1.57	154.62	0.95	232.42	0.04	242.81
Nov 23	157.35	1.61	155.29	0.99	239.48	-0.06	252.96
Dec 3	157.50	1.65	155.65	1.04	246.62	-0.16	263.13
Dec 13	157.46	1.69	155.70	1.08	253.82	-0.27	273.32
1801							

GREGORIAN 12 UT	MOON LONG. LONG. LAT. LAT.	VENUS LONG. LONG. LAT. LAT.	MERCURY LONG. LONG. LAT. LAT.
1800			
4 Jan 9	30.7 90.6 -0.3 4.2	237.33 242.71 3.24 3.16	266.41 266.95 3.14 2.58
14 Jan 19	155.5 224.6 4.2 -1.1	248.20 253.78 3.02 2.83	270.25 275.22 1.80 0.99
24 Jan 29	296.0 2.5 -5.0 -2.4	259.44 265.17 2.61 2.36	281.18 287.79 0.23 -0.45
3 Feb 8	62.2 124.4 2.6 5.0	270.95 276.78 2.09 1.79	294.87 302.33 -1.04 -1.52
13 Feb 18	193.2 263.7 1.4 -4.1	282.64 288.55 1.48 1.17	310.16 318.35 -1.87 -2.07
23 Feb 28	332.1 34.5 -4.3 0.5	294.48 300.43 0.86 0.54	326.94 335.94 -2.12 -1.98
5 Mar 10	94.1 159.9 4.7 3.9	306.41 312.40 0.24 -0.05	345.37 355.14 -1.62 -1.03
15 Mar 20	232.0 301.9 -2.1 -5.2	318.41 324.44 -0.33 -0.58	4.99 14.37 -0.22 0.77
25 Mar 30	6.4 66.3 -1.9 3.3	330.47 336.51 -0.82 -1.02	22.45 28.41 1.78 2.63
4 Apr 9	127.2 196.9 5.2 1.0	342.57 348.62 -1.20 -1.35	31.70 32.16 3.12 3.09
14 Apr 19	270.9 338.2 -4.7 -4.0	354.68 0.75 -1.47 -1.56	30.18 26.86 2.45 1.29
24 Apr 29	39.3 98.6 1.1 5.0	6.82 12.90 -1.62 -1.64	23.70 21.95 -0.12 -1.42
4 May 9	162.1 235.2 3.8 -2.5	18.98 25.06 -1.64 -1.60	22.13 24.21 -2.42 -3.05
14 May 19	308.6 12.4 -5.1 -1.3	31.14 37.23 -1.54 -1.45	27.93 33.04 -3.33 -3.28
24 May 29	71.7 132.2 3.6 5.0	43.32 49.41 -1.33 -1.20	39.36 46.77 -2.96 -2.40
3 Jun 8	198.9 274.0 0.7 -4.7	55.50 61.60 -1.04 -0.87	55.23 64.69 -1.64 -0.76
13 Jun 18	344.2 45.1 -3.4 1.7	67.71 73.81 -0.69 -0.49	75.04 85.91 0.16 0.98
23 Jun 28	104.7 167.7 5.0 3.0	79.93 86.05 -0.29 -0.09	96.79 107.19 1.57 1.86
3 Jul 8	237.4 311.7 -2.8 -4.8	92.18 98.31 0.11 0.31	116.85 125.68 1.85 1.58
13 Jul 18	17.8 77.4 -0.5 4.1	104.45 110.59 0.50 0.68	133.67 140.82 1.09 0.41
23 Jul 28	138.8 205.0 4.5 -0.3	116.75 122.91 0.85 0.99	147.06 152.26 -0.42 -1.35
2 Aug 7	276.6 347.6 -4.9 -2.8	129.07 135.25 1.12 1.23	156.20 158.56 -2.34 -3.32
12 Aug 17	50.0 109.9 2.5 5.1	141.43 147.61 1.32 1.38	158.92 157.01 -4.16 -4.65
22 Aug 27	174.6 243.9 2.2 -3.6	153.81 160.00 1.41 1.42	153.13 148.73 -4.50 -3.59
1 Sep 6	314.9 21.4 -4.6 0.2	166.20 172.40 1.40 1.35	146.00 146.55 -2.13 -0.58
11 Sep 16	81.6 143.3 4.6 4.3	178.61 184.82 1.28 1.18	150.65 157.51 0.68 1.49
21 Sep 26	212.0 283.5 -1.2 -5.2	191.03 197.24 1.06 0.91	165.96 174.99 1.83 1.82
1 Oct 6	351.3 53.9 -2.4 3.1	203.45 209.66 0.75 0.57	184.03 192.84 1.55 1.13
11 Oct 16	113.5 178.1 5.3 1.8	215.87 222.08 0.37 0.16	201.35 209.57 0.62 0.06
21 Oct 26	250.7 321.6 -4.2 -4.4	228.29 234.49 -0.05 -0.27	217.53 225.26 -0.51 -1.06
31 Oct	25.8 0.7	240.70 -0.49	232.78 -1.57
Nov 5	85.9 4.8	246.90 -0.71	240.08 -2.01
10 Nov 15	146.3 214.6 4.1 -1.5	253.10 259.29 -0.92 -1.12	247.13 253.79 -2.35 -2.54
20 Nov 25	289.7 357.8 -5.2 -1.7	265.49 271.67 -1.30 -1.46	259.75 264.40 -2.52 -2.19
30 N-D 5	58.9 118.4 3.4 5.0	277.85 284.03 -1.61 -1.73	266.61 264.85 -1.39 -0.02
10 Dec 15	180.7 252.6 1.4 -4.2	290.19 296.35 -1.81 -1.87	259.01 252.82 1.63 2.75
20 Dec 25	327.4 32.0 -3.7 1.5	302.49 308.61 -1.90 -1.89	250.35 251.88 2.96 2.56
30 Dec	91.5 4.9	314.72 -1.84	256.01 1.91
1801			
Jan 4	151.9 3.3	320.80 -1.76	261.60 1.19
9 Jan 14	217.1 291.2 -2.0 -5.0	326.86 332.90 -1.64 -1.48	268.01 274.93 0.47 -0.19
19 Jan 24	2.9 64.6 -0.8 4.0	338.89 344.85 -1.28 -1.04	282.19 289.75 -0.79 -1.29
29 J-F 3	124.3 187.1 4.6 0.3	350.76 356.62 -0.77 -0.47	297.58 305.70 -1.68 -1.96
8 Feb 13	255.5 329.1 -4.6 -3.4	2.43 8.17 -0.14 0.23	314.14 322.92 -2.08 -2.04
18 Feb 23	36.5 96.6 2.3 5.2	13.83 19.40 0.61 1.01	332.05 341.46 -1.79 -1.31
28 Feb	157.9 2.7	24.88 1.43	350.91 -0.57
Mar 5	224.1 -3.0	30.24 1.86	359.85 0.41
10 Mar 15	294.8 5.3 -5.1 -0.4	35.47 40.55 2.30 2.73	7.37 12.44 1.51 2.55
20 Mar 25	68.7 128.5 4.6 4.6	45.45 50.14 3.16 3.57	14.33 13.02 3.27 3.42
30 M-A 4	192.8 263.0 -0.3 -5.1	54.59 58.73 3.96 4.32	9.47 5.52 2.86 1.73
9 Apr 14	333.4 39.6 -3.1 2.7	62.53 65.89 4.63 4.89	2.89 2.34 0.39 -0.86
19 Apr 24	100.5 161.2 5.2 2.4	68.75 70.97 5.07 5.15	3.89 7.19 -1.84 -2.52
29 Apr	229.6 -3.4	72.45 5.11	11.92 -2.90
May 4	302.4 -4.8	73.06 4.91	17.80 -2.99
9 May 14	10.2 72.8 0.1 4.6	72.71 71.36 4.51 3.89	24.69 32.49 -2.84 -2.44
19 May 24	132.4 195.4 4.2 -0.7	69.09 66.18 3.04 1.99	41.17 50.73 -1.84 -1.06
29 M-J 3	268.1 340.7 -5.0 -2.2	63.05 60.20 0.82 -0.35	61.09 71.98 -0.18 0.69
8 Jun 13	45.2 105.4 3.1 4.9	58.05 56.84 -1.41 -2.31	82.92 93.39 1.40 1.84
18 Jun 23	165.3 231.7 1.7 -3.7	56.60 57.31 -3.02 -3.55	103.07 111.82 1.98 1.83
28 Jun	307.2 -4.2	58.85 -3.90	119.61 1.40
Jul 3	17.1 1.1	61.09 -4.12	126.42 0.75
8 Jul 13	78.6 138.1 4.8 3.5	63.93 67.26 -4.22 -4.23	132.15 136.64 -0.11 -1.12
18 Jul 23	199.7 270.0 -1.5 -5.1	71.00 75.06 -4.15 -4.00	139.66 140.88 -2.23 -3.35
28 J-A 2	345.2 51.4 -1.4 3.9	79.40 83.98 -3.80 -3.55	140.05 137.27 -4.31 -4.86
7 Aug 12	111.3 171.5 4.8 0.8	88.75 93.70 -3.26 -2.95	133.40 130.07 -4.73 -3.87
17 Aug 22	236.3 309.0 -4.3 -4.1	98.80 104.02 -2.61 -2.25	128.25 130.84 -2.52 -1.05
27 Aug	21.3 1.9	109.35 -1.89	135.82 0.24
Sep 1	84.2 5.2	114.79 -1.52	143.26 1.17
6 Sep 11	143.8 206.5 3.1 -2.4	120.31 125.92 -1.15 -0.79	152.17 161.63 1.68 1.81
16 Sep 21	274.8 347.3 -5.3 -1.2	131.60 137.35 -0.44 -0.11	171.06 180.17 1.65 1.30
26 S-O 1	55.4 116.1 4.4 4.7	143.15 149.01 0.21 0.50	188.50 197.26 0.82 0.28
6 Oct 11	176.9 243.6 0.3 -4.7	154.93 160.88 0.77 1.01	205.27 212.97 -0.30 -0.89
16 Oct 21	314.1 23.9 -3.8 2.1	166.89 172.92 1.22 1.39	220.38 227.47 -1.44 -1.95
26 Oct 31	88.1 147.8 5.2 2.1	179.00 185.10 1.53 1.64	234.19 240.38 -2.38 -2.68
5 Nov 10	211.4 282.3 -2.7 -5.0	191.23 197.39 1.71 1.74	245.71 249.55 -2.78 -2.58
15 Nov 20	352.7 58.8 -0.5 4.4	203.56 209.76 1.74 1.71	250.79 248.11 -1.92 -0.64
25 Nov 30	120.0 180.0 4.3 -0.2	215.97 222.19 1.65 1.56	241.89 236.19 1.03 2.30
5 Dec 10	247.8 321.5 -4.7 -2.8	228.43 234.68 1.44 1.30	234.67 237.14 2.72 2.51
15 Dec 20	29.7 92.4 2.8 4.9	240.93 247.19 1.13 0.95	242.04 248.21 1.99 1.35
25 Dec 30	151.9 213.7 1.9 -3.2	253.45 259.72 0.76 0.56	255.04 262.24 0.69 0.05
1801			

JULIAN	SATURN		JUPITER		MARS		SUN	GREGORIAN	MOON				VENUS				MERCURY			
12 UT	LONG.	LAT.	LONG.	LAT.	LONG.	LAT.	LONG.	12 UT	LONG.		LAT.		LONG.		LAT.		LONG.		LAT.	
1801								1802												1802
Dec 23	157.24	1.73	155.42	1.13	261.10	-0.38	283.51	4 Jan 9	285.7	359.6	-4.7	0.6	265.99	272.26	0.35	0.13	269.68	277.32	-0.55	-1.07
Jan 2	156.85	1.77	154.83	1.17	268.45	-0.48	293.70	14 Jan 19	64.8	125.1	4.8	3.7	278.53	284.81	-0.08	-0.28	285.15	293.19	-1.50	-1.83
Jan 12	156.31	1.81	153.97	1.21	275.86	-0.59	303.87	24 Jan 29	184.6	249.5	-1.2	-5.0	291.07	297.34	-0.48	-0.67	301.48	310.03	-2.03	-2.07
Jan 22	155.64	1.84	152.87	1.24	283.34	-0.70	314.03	3 Feb 8	324.3	35.9	-2.3	3.7	303.61	309.88	-0.84	-0.99	318.86	327.89	-1.93	-1.55
Feb 1	154.89	1.86	151.62	1.26	290.87	-0.81	324.14	13 Feb 18	98.3	157.6	5.0	1.2	316.14	322.40	-1.12	-1.24	336.89	345.32	-0.91	0.02
Feb 11	154.10	1.88	150.31	1.28	298.45	-0.91	334.22	23 Feb 28	218.9	287.4	-3.9	-4.7	328.65	334.90	-1.32	-1.39	352.22	356.33	1.17	2.38
Feb 21	153.30	1.89	149.03	1.28	306.07	-1.01	344.25	5 Mar 10	2.1	70.3	1.1	5.2	341.14	347.38	-1.42	-1.43	356.77	353.77	3.32	3.64
Mar 3	152.55	1.89	147.87	1.28	313.73	-1.11	354.22	15 Mar 20	130.7	190.8	3.4	-1.8	353.61	359.83	-1.42	-1.38	349.14	345.27	3.16	2.08
Mar 13	151.88	1.88	146.92	1.27	321.41	-1.20	4.14	25 Mar 30	255.4	326.2	-5.3	-2.2	6.05	12.26	-1.31	-1.22	343.59	344.31	0.81	-0.37
Mar 23	151.33	1.87	146.22	1.25	329.10	-1.28	14.00	4 Apr 9	38.6	103.2	3.9	4.8	18.46	24.65	-1.10	-0.96	347.03	351.29	-1.33	-2.03
Apr 2	150.93	1.86	145.81	1.23	336.79	-1.35	23.81	14 Apr 19	162.6	225.4	0.7	-4.2	30.84	37.02	-0.81	-0.64	356.75	3.18	-2.49	-2.70
Apr 12	150.70	1.84	145.72	1.21	344.47	-1.40	33.56	24 Apr 29	293.9	4.9	-4.3	1.3	43.18	49.34	-0.45	-0.26	10.45	18.50	-2.68	-2.44
Apr 22	150.64	1.82	145.93	1.18	352.13	-1.45	43.26	4 May 9	73.5	135.0	5.1	2.8	55.50	61.64	-0.05	0.15	27.31	36.90	-1.98	-1.32
May 2	150.75	1.80	146.43	1.16	359.74	-1.47	52.92	14 May 19	195.2	262.2	-2.3	-5.0	67.78	73.90	0.36	0.56	47.20	58.02	-0.50	0.37
May 12	151.03	1.78	147.21	1.13	7.30	-1.49	62.53	24 May 29	333.2	42.3	-1.3	4.1	80.02	86.13	0.76	0.95	68.94	79.42	1.18	1.77
May 22	151.48	1.76	148.24	1.11	14.79	-1.48	72.12	3 Jun 8	106.9	166.5	4.3	0.0	92.23	98.32	1.12	1.27	89.03	97.61	2.08	2.06
Jun 1	152.08	1.74	149.49	1.09	22.19	-1.46	81.67	13 Jun 18	229.3	300.6	-4.5	-3.5	104.40	110.47	1.41	1.52	105.06	111.34	1.74	1.13
Jun 11	152.82	1.73	150.94	1.07	29.48	-1.43	91.21	23 Jun 28	11.9	78.1	2.3	5.0	116.53	122.59	1.61	1.66	116.35	119.88	0.27	-0.82
Jun 21	153.69	1.71	152.55	1.06	36.65	-1.37	100.75	3 Jul 8	139.3	198.8	2.1	-3.0	128.63	134.65	1.69	1.69	121.72	121.65	-2.05	-3.30
Jul 1	154.66	1.70	154.31	1.04	43.68	-1.29	110.28	13 Jul 18	265.3	339.6	-5.0	-0.3	140.66	146.66	1.65	1.58	119.71	116.51	-4.35	-4.91
Jul 11	155.72	1.69	156.19	1.03	50.54	-1.20	119.83	23 Jul 28	49.3	112.1	4.7	4.0	152.64	158.60	1.47	1.33	113.29	111.47	-4.78	-4.01
Jul 21	156.86	1.69	158.17	1.03	57.22	-1.09	129.39	2 Aug 7	171.4	232.6	-0.9	-4.9	164.54	170.46	1.16	0.96	112.00	115.23	-2.81	-1.46
Jul 31	158.05	1.69	160.22	1.02	63.68	-0.96	138.98	12 Aug 17	303.1	17.9	-3.3	3.1	176.36	182.23	0.72	0.46	121.05	129.00	-0.20	0.82
Aug 10	159.28	1.69	162.33	1.02	69.89	-0.81	148.61	22 Aug 27	84.6	144.8	5.1	1.5	188.07	193.89	0.18	-0.13	138.32	148.17	1.48	1.76
Aug 20	160.54	1.70	164.48	1.02	75.80	-0.63	158.28	1 Sep 6	204.2	268.4	-3.6	-5.1	199.67	205.41	-0.46	-0.80	157.96	167.37	1.72	1.45
Aug 30	161.79	1.71	166.64	1.03	81.38	-0.44	167.99	11 Sep 16	341.6	54.7	0.0	5.1	211.11	216.76	-1.16	-1.52	176.33	184.82	1.03	0.50
Sep 9	163.04	1.72	168.80	1.04	86.55	-0.21	177.76	21 Sep 26	118.1	177.3	3.6	-1.4	222.35	227.89	-1.88	-2.23	192.89	200.56	-0.09	-0.70
Sep 19	164.25	1.74	170.94	1.05	91.22	0.04	187.58	1 Oct 6	238.7	306.1	-5.1	-3.0	233.36	238.75	-2.57	-2.90	207.85	214.72	-1.31	-1.89
Sep 29	165.42	1.76	173.04	1.06	95.28	0.34	197.46	11 Oct 16	19.9	89.5	3.2	4.9	244.04	249.22	-3.20	-3.46	221.10	226.80	-2.40	-2.80
Oct 9	166.51	1.79	175.06	1.08	98.61	0.67	207.39	21 Oct 26	150.2	210.3	0.9	-3.9	254.27	259.16	-3.69	-3.87	231.46	234.43	-3.03	-2.97
Oct 19	167.52	1.82	177.00	1.10	101.02	1.05	217.38	31 Oct	275.1		-4.6		263.86		-4.00		234.67		-2.45	
								Nov 5		344.8		0.5		268.33		-4.05		231.14		-1.27
Oct 29	168.42	1.85	178.82	1.13	102.31	1.48	227.42	10 Nov 15	57.0	122.5	4.9	3.0	272.51	276.35	-4.03	-3.91	224.80	219.81	0.40	1.81
Nov 8	169.19	1.89	180.49	1.16	102.30	1.95	237.51	20 Nov 25	181.9	244.7	-2.0	-5.0	279.77	282.67	-3.68	-3.32	219.26	222.61	2.44	2.42
Nov 18	169.82	1.93	181.99	1.19	100.87	2.44	247.64	30 N-D 5	313.3	23.7	-2.0	3.7	284.94	286.45	-2.82	-2.14	228.23	234.95	2.04	1.50
Nov 28	170.28	1.98	183.29	1.23	98.10	2.91	257.79	10 Dec 15	92.3	154.2	4.5	0.1	287.05	286.65	-1.28	-0.24	242.18	249.67	0.89	0.27
Dec 8	170.58	2.02	184.36	1.27	94.40	3.29	267.97	20 Dec 25	214.1	280.8	-4.3	-4.1	285.23	282.91	0.97	2.27	257.28	265.01	-0.32	-0.86
Dec 18	170.69	2.07	185.16	1.31	90.48	3.53	278.17	30 Dec	352.3		1.6		280.01		3.54		272.84		-1.32	
1802								1803												1803
								Jan 4		61.5		5.1		276.99		4.65		280.81		-1.70
Dec 28	170.63	2.12	185.68	1.36	87.12	3.62	288.36	9 Jan 14	126.0	185.7	2.4	-2.8	274.38	272.55	5.51	6.06	288.94	297.27	-1.96	-2.09
Jan 7	170.38	2.16	185.89	1.40	84.88	3.57	298.54	19 Jan 24	247.5	318.4	-5.2	-1.3	271.73	271.94	6.32	6.36	305.78	314.40	-2.04	-1.77
Jan 17	169.96	2.20	185.78	1.45	83.96	3.44	308.71	29 J-F 3	31.1	97.4	4.5	4.3	273.09	275.05	6.21	5.93	322.92	330.80	-1.24	-0.38
Jan 27	169.40	2.23	185.35	1.49	84.34	3.26	318.84	8 Feb 13	158.5	217.9	-0.6	-4.8	277.70	280.91	5.56	5.12	337.01	340.13	0.78	2.11
Feb 6	168.73	2.26	184.64	1.53	85.82	3.06	328.94	18 Feb 23	282.7	356.8	-4.0	2.2	284.59	288.64	4.64	4.13	339.13	334.71	3.26	3.73
Feb 16	167.97	2.29	183.67	1.56	88.22	2.87	339.00	28 Feb	68.5		5.3		293.00		3.61		329.56		3.33	
								Mar 5		131.5		1.9		297.61		3.09		326.29		2.32
Feb 26	167.18	2.30	182.51	1.58	91.34	2.68	349.00	10 Mar 15	190.8	251.6	-3.3	-5.2	302.43	307.42	2.56	2.05	325.76	327.63	1.14	0.04
Mar 8	166.40	2.30	181.25	1.59	95.04	2.50	358.95	20 Mar 25	319.7	35.1	-1.1	4.6	312.56	317.81	1.55	1.08	331.31	336.30	-0.88	-1.60
Mar 18	165.67	2.30	179.96	1.59	99.20	2.33	8.84	30 M-A 4	103.8	164.2	3.9	-1.1	323.18	328.62	0.63	0.20	342.30	349.10	-2.11	-2.42
Mar 28	165.02	2.29	178.75	1.58	103.71	2.17	18.68	9 Apr 14	223.8	287.2	-4.9	-3.6	334.14	339.72	-0.18	-0.54	356.62	4.81	-2.51	-2.40
Apr 7	164.50	2.27	177.68	1.56	108.52	2.02	28.46	19 Apr 24	358.1	72.3	2.3	5.0	345.36	351.05	-0.85	-1.13	13.66	23.20	-2.07	-1.53
Apr 17	164.12	2.25	176.84	1.53	113.56	1.88	38.19	29 Apr	137.2		1.1		356.77		-1.37		33.39		-0.81	
								May 4		196.5		-3.7		2.53		-1.57		44.08		0.04
Apr 27	163.91	2.22	176.26	1.50	118.79	1.75	47.87	9 May 14	258.2	324.7	-4.7	-0.3	8.33	14.14	-1.72	-1.84	54.90	65.29	0.91	1.64
May 7	163.86	2.19	175.97	1.46	124.19	1.62	57.50	19 May 24	37.0	107.6	4.7	3.2	19.99	25.86	-1.91	-1.95	74.76	83.03	2.12	2.27
May 17	163.99	2.16	175.99	1.42	129.73	1.50	67.10	29 M-J 3	169.2	229.3	-1.9	-5.0	31.75	37.65	-1.95	-1.91	89.98	95.51	2.08	1.55
May 27	164.29	2.13	176.30	1.38	135.39	1.39	76.67	8 Jun 13	294.3	3.7	-2.6	3.1	43.57	49.51	-1.85	-1.75	99.51	101.78	0.70	-0.44
Jun 6	164.75	2.10	176.89	1.34	141.16	1.27	86.22	18 Jun 23	75.0	141.0	4.8	0.3	55.47	61.44	-1.62	-1.47	102.19	100.81	-1.77	-3.12
Jun 16	165.36	2.08	177.75	1.30	147.04	1.16	95.76	28 Jun	200.9		-4.3		67.43		-1.30		98.13		-4.20	
								Jul 3		263.2		-4.5		73.43		-1.11		95.21		-4.75
Jun 26	166.10	2.05	178.84	1.27	153.01	1.05	105.29	8 Jul 13	332.1	43.0	0.7	5.1	79.44	85.47	-0.91	-0.69	93.27	93.18	-4.66	-4.01
Jul 6	166.96	2.03	180.15	1.24	159.07	0.94	114.83	18 Jul 23	111.1	173.2	2.8	-2.6	91.52	97.58	-0.47	-0.25	95.34	99.78	-3.00	-1.81
Jul 16	167.93	2.02	181.65	1.21	165.21	0.84	124.38	28 J-A 2	233.0	298.8	-5.3	-2.2	103.66	109.75	-0.03	0.18	106.36	114.75	-0.60	0.45
Jul 26	168.99	2.00	183.30	1.18	171.45	0.73	133.95	7 Aug 12	11.1	81.1	3.9	4.7	115.86	121.98	0.39	0.59	124.41	134.61	1.23	1.66
Aug 5	170.12	1.99	185.10	1.16	177.77	0.63	143.56	17 Aug 22	145.3	205.1	-0.2	-4.7	128.12	134.27	0.77	0.93	144.73	154.43	1.76	1.59
Aug 15	171.31	1.99	187.01	1.14	184.18	0.53	153.21	27 Aug	266.3		-4.4		140.44		1.08		163.60		1.23	
								Sep 1		336.0		1.2		146.62		1.20		172.23		0.73
Aug 25	172.53	1.99	189.01	1.13	190.68	0.42	162.89	6 Sep 11	50.2	117.2	5.2	2.3	152.81	159.01	1.29	1.37	180.35	187.99	0.14	-0.49
Sep 4	173.77	1.99	191.09	1.11	197.26	0.32	172.63	16 Sep 21	178.1	237.6	-3.0	-5.2	165.22	171.45	1.41	1.43	195.14	201.77	-1.15	-1.80
Sep 14	175.02	2.00	193.22	1.11	203.94	0.22	182.43	26 S-O 1	301.1	14.5	-1.9	4.0	177.68	183.92	1.42	1.38	207.78	212.94	-2.41	-2.93
Sep 24	176.25	2.01	195.38	1.10	210.70	0.11	192.27	6 Oct 11	87.8	151.3	4.2	-0.8	190.17	196.42	1.32	1.23	216.87	218.91	-3.28	-3.36
Oct 4	177.44	2.03	197.55	1.10	217.57	0.01	202.18	16 Oct 21	210.7	271.2	-4.7	-3.9	202.68	208.94	1.12	0.98	218.11	213.86	-2.97	-1.88
Oct 14	178.58	2.05	199.71	1.10	224.52	-0.09	212.14	26 Oct 31	337.8	53.1	1.5	5.0	215.20	221.47	0.83	0.66	207.69	203.60	-0.24	1.27
Oct 24	179.65	2.08	201.83	1.11	231.56	-0.19	222.15	5 Nov 10	123.2	184.0	1.4	-3.5	227.74	234.01	0.48	0.29	204.03	208.22	2.10	2.29
Nov 3	180.63	2.11	203.90	1.12	238.70	-0.29	232.22	15 Nov 20	243.7	306.6	-4.8	-1.0	240.28	246.55	0.09	-0.11	214.51	221.76	2.06	1.62
Nov 13	181.49	2.15	205.90	1.13	245.93	-0.39	242.32	25 Nov 30	16.2	90.5	4.3	3.6	252.83	259.10	-0.31	-0.51	229.40	237.16	1.07	0.48
Nov 23	182.23	2.19	207.78	1.15	253.24	-0.48	252.46	5 Dec 10	156.5	216.1	-1.8	-5.0	265.37	271.64	-0.70	-0.88	244.96	252.78	-0.10	-0.65
Dec 3	182.82	2.23	209.54	1.17	260.64	-0.58	262.63	15 Dec 20	277.7	343.9	-3.1	2.5	277.91	284.17	-1.05	-1.20	260.62	268.52	-1.14	-1.56
Dec 13	183.25	2.28	211.14	1.19	268.12	-0.67	272.82	25 Dec 30	55.2	126.0	5.1	0.8	290.43	296.69	-1.33	-1.44	276.51	284.60	-1.89	-2.09
1803								1803												1803

JULIAN 12 UT	SATURN LONG.	LAT.	JUPITER LONG.	LAT.	MARS LONG.	LAT.	SUN LONG.	GREGORIAN 12 UT	MOON LONG.		LAT.		VENUS LONG.		LAT.		MERCURY LONG.		LAT.	
1803								1804												1804
Dec 23	183.50	2.33	212.54	1.22	275.67	-0.75	283.01	4 Jan 9	188.5	248.4	-4.2	-4.8	302.95	309.19	-1.52	-1.58	292.79	301.01	-2.13	-1.97
Jan 2	183.58	2.37	213.73	1.25	283.30	-0.83	293.20	14 Jan 19	313.2	22.8	-0.1	4.9	315.43	321.67	-1.61	-1.60	309.02	316.29	-1.55	-0.79
Jan 12	183.47	2.42	214.68	1.28	290.98	-0.90	303.38	24 Jan 29	93.4	159.6	3.4	-2.3	327.89	334.10	-1.57	-1.51	321.76	323.86	0.34	1.75
Jan 22	183.19	2.47	215.35	1.31	298.72	-0.97	313.53	3 Feb 8	220.0	281.5	-5.3	-2.9	340.29	346.47	-1.42	-1.29	321.52	316.03	3.07	3.70
Feb 1	182.75	2.51	215.72	1.35	306.50	-1.03	323.65	13 Feb 18	350.4	62.2	3.1	5.0	352.64	358.78	-1.14	-0.96	310.89	308.53	3.39	2.48
Feb 11	182.17	2.54	215.78	1.39	314.31	-1.08	333.73	23 Feb 28	129.7	192.2	0.4	-4.4	4.90	11.00	-0.76	-0.53	309.14	312.02	1.41	0.39
Feb 21	181.48	2.57	215.53	1.42	322.14	-1.12	343.76	4 Mar 9	251.9	316.0	-4.6	0.2	17.07	23.11	-0.29	-0.02	316.46	322.02	-0.49	-1.22
Mar 2	180.73	2.58	214.98	1.45	329.97	-1.14	353.74	14 Mar 19	29.0	100.3	5.0	2.8	29.13	35.10	0.25	0.54	328.41	335.48	-1.77	-2.14
Mar 12	179.95	2.59	214.15	1.47	337.81	-1.16	3.66	24 Mar 29	164.3	224.5	-2.7	-5.1	41.04	46.94	0.83	1.12	343.15	351.39	-2.33	-2.33
Mar 22	179.18	2.59	213.11	1.49	345.63	-1.16	13.53	3 Apr 8	284.7	352.5	-2.3	3.3	52.79	58.59	1.40	1.68	0.21	9.64	-2.12	-1.71
Apr 1	178.47	2.58	211.91	1.49	353.42	-1.15	23.34	13 Apr 18	68.0	136.4	4.5	-0.5	64.34	70.03	1.94	2.18	19.67	30.16	-1.09	-0.29
Apr 11	177.85	2.56	210.64	1.49	1.17	-1.13	33.09	23 Apr 28	197.6	257.1	-4.6	-4.0	75.64	81.19	2.40	2.58	40.80	51.03	0.60	1.45
Apr 21	177.35	2.54	209.38	1.47	8.86	-1.09	42.80	3 May 8	319.3	30.9	0.9	5.0	86.65	92.01	2.73	2.84	60.25	68.08	2.10	2.44
May 1	177.00	2.51	208.23	1.45	16.49	-1.04	52.46	13 May 18	105.6	170.6	1.9	-3.4	97.27	102.39	2.90	2.91	74.33	78.84	2.41	2.00
May 11	176.81	2.47	207.26	1.42	24.05	-0.98	62.07	23 May 28	230.3	290.7	-4.9	-1.4	107.37	112.18	2.86	2.75	81.50	82.22	1.19	0.02
May 21	176.79	2.44	206.53	1.38	31.52	-0.90	71.66	2 Jun 7	356.1	69.9	3.9	4.2	116.79	121.17	2.57	2.30	81.10	78.68	-1.38	-2.77
May 31	176.94	2.40	206.07	1.34	38.91	-0.81	81.22	12 Jun 17	141.3	203.3	-1.4	-5.0	125.27	129.02	1.95	1.51	75.95	74.04	-3.85	-4.40
Jun 10	177.25	2.36	205.90	1.29	46.18	-0.72	90.76	22 Jun 27	263.1	326.0	-3.5	1.9	132.37	135.22	0.96	0.29	73.74	75.40	-4.39	-3.91
Jun 20	177.72	2.33	206.03	1.24	53.35	-0.61	100.29	2 Jul 7	34.8	108.0	5.2	1.5	137.48	139.03	-0.49	-1.39	79.02	84.52	-3.09	-2.07
Jun 30	178.34	2.30	206.46	1.20	60.40	-0.49	109.83	12 Jul 17	175.1	235.4	-4.0	-4.9	139.76	139.54	-2.42	-3.53	91.75	100.53	-0.97	0.08
Jul 10	179.09	2.27	207.16	1.16	67.32	-0.36	119.37	22 Jul 27	296.6	3.3	-0.8	4.5	138.33	136.20	-4.69	-5.80	110.46	120.94	0.94	1.52
Jul 20	179.96	2.24	208.12	1.12	74.11	-0.23	128.93	1 Aug 6	74.2	144.2	4.1	-1.7	133.40	130.31	-6.74	-7.42	131.34	141.31	1.76	1.71
Jul 30	180.93	2.22	209.31	1.08	80.75	-0.08	138.52	11 Aug 16	207.6	267.4	-5.2	-3.2	127.42	125.15	-7.75	-7.75	150.67	159.43	1.43	0.97
Aug 9	181.99	2.20	210.71	1.04	87.23	0.07	148.14	21 Aug 26	331.8	42.3	2.4	5.2	123.77	123.38	-7.48	-7.00	167.58	175.17	0.39	-0.27
Aug 19	183.12	2.19	212.29	1.01	93.54	0.24	157.81	31 Aug	112.6		1.1		123.96		-6.39		182.16		-0.98	
								Sep 5		178.6		-4.2		125.40		-5.71		188.52		-1.70
Aug 29	184.30	2.18	214.02	0.98	99.65	0.41	167.52	10 Sep 15	239.4	300.1	-4.7	-0.3	127.58	130.39	-5.00	-4.28	194.10	198.66	-2.40	-3.04
Sep 8	185.51	2.18	215.89	0.96	105.54	0.60	177.28	20 Sep 25	8.9	81.7	4.6	3.4	133.71	137.47	-3.57	-2.89	201.78	202.81	-3.53	-3.75
Sep 18	186.74	2.18	217.87	0.93	111.19	0.80	187.10	30 S-O 5	149.3	211.8	-2.2	-5.0	141.59	146.01	-2.23	-1.61	200.98	196.21	-3.47	-2.47
Sep 28	187.97	2.19	219.94	0.92	116.55	1.02	196.98	10 Oct 15	271.3	334.3	-2.5	2.7	150.69	155.58	-1.04	-0.50	190.53	187.52	-0.86	0.72
Oct 8	189.18	2.20	222.07	0.90	121.57	1.26	206.91	20 Oct 25	47.5	119.8	4.8	0.0	160.65	165.89	-0.01	0.43	188.93	193.90	1.73	2.11
Oct 18	190.36	2.22	224.25	0.89	126.18	1.52	216.89	30 Oct	184.1		-4.5		171.26		0.83		200.82		2.05	
								Nov 4		244.3		-4.1		176.75		1.17		208.59		1.71
Oct 28	191.47	2.24	226.45	0.88	130.29	1.80	226.94	9 Nov 14	304.0	10.6	0.5	4.8	182.34	188.02	1.46	1.71	216.63	224.68	1.23	0.68
Nov 7	192.51	2.27	228.66	0.87	133.79	2.12	237.02	19 Nov 24	86.2	156.0	2.6	-3.2	193.78	199.61	1.90	2.04	232.67	240.59	0.11	-0.44
Nov 17	193.45	2.30	230.84	0.86	136.55	2.47	247.14	29 N-D 4	217.5	276.9	-5.0	-1.7	205.49	211.43	2.14	2.19	248.45	256.30	-0.96	-1.42
Nov 27	194.28	2.33	232.98	0.86	138.38	2.85	257.30	9 Dec 14	338.4	48.6	3.5	4.8	217.41	223.43	2.19	2.16	264.15	272.02	-1.80	-2.08
Dec 7	194.98	2.37	235.05	0.86	139.10	3.26	267.48	19 Dec 24	123.7	190.2	-0.7	-5.1	229.48	235.56	2.08	1.97	279.89	287.69	-2.21	-2.15
Dec 17	195.52	2.42	237.03	0.87	138.53	3.68	277.67	29 Dec	250.0		-3.8		241.66		1.83		295.17		-1.84	
1804								1805												1805
								Jan 3		310.2		1.3		247.78		1.67		301.79		-1.20
Dec 27	195.91	2.46	238.89	0.87	136.61	4.07	287.87	8 Jan 13	15.0	87.2	5.2	2.5	253.92	260.08	1.47	1.26	306.47	307.56	-0.14	1.30
Jan 6	196.11	2.51	240.61	0.88	133.49	4.37	298.05	18 Jan 23	159.5	222.8	-3.7	-5.1	266.24	272.41	1.04	0.80	304.03	297.86	2.76	3.54
Jan 16	196.15	2.56	242.15	0.89	129.63	4.52	308.22	28 J-F 2	282.3	345.2	-1.2	4.0	278.59	284.78	0.56	0.32	293.11	291.79	3.35	2.58
Jan 26	196.00	2.60	243.49	0.90	125.72	4.49	318.36	7 Feb 12	53.6	125.3	4.6	-0.9	290.97	297.16	0.08	-0.15	293.47	297.20	1.62	0.68
Feb 5	195.68	2.65	244.60	0.92	122.51	4.30	328.46	17 Feb 22	193.6	254.5	-5.1	-3.4	303.35	309.55	-0.38	-0.59	302.29	308.30	-0.16	-0.88
Feb 15	195.21	2.68	245.46	0.93	120.44	4.00	338.51	27 Feb	315.1		1.8		315.75		-0.78		314.98		-1.45	
								Mar 4		22.3		5.1		321.95		-0.96		322.23		-1.88
Feb 25	194.62	2.71	246.03	0.95	119.66	3.66	348.52	9 Mar 14	92.8	162.1	1.9	-3.8	328.14	334.33	-1.11	-1.24	329.98	338.22	-2.14	-2.24
Mar 7	193.92	2.74	246.29	0.96	120.14	3.30	358.47	19 Mar 24	226.4	286.0	-4.7	-0.6	340.52	346.71	-1.34	-1.42	346.97	356.25	-2.14	-1.84
Mar 17	193.17	2.75	246.25	0.98	121.67	2.96	8.37	29 M-A 3	349.6	60.9	4.2	3.9	352.90	359.08	-1.46	-1.48	6.06	16.30	-1.33	-0.61
Mar 27	192.39	2.75	245.90	0.99	124.09	2.65	18.21	8 Apr 13	131.4	197.4	-1.7	-5.0	5.26	11.44	-1.48	-1.44	26.69	36.66	0.27	1.20
Apr 6	191.64	2.74	245.26	0.99	127.23	2.36	27.99	18 Apr 23	258.5	318.4	-2.6	2.4	17.61	23.77	-1.38	-1.30	45.55	52.81	2.01	2.55
Apr 16	190.95	2.73	244.36	1.00	130.93	2.10	37.72	28 Apr	26.0		5.0		29.94		-1.19		58.13		2.72	
								May 3		99.9		0.8		36.10		-1.06		61.33		2.45
Apr 26	190.36	2.70	243.26	0.99	135.10	1.86	47.40	8 May 13	168.5	231.1	-4.4	-4.3	42.26	48.41	-0.91	-0.75	62.35	61.34	1.71	0.54
May 6	189.90	2.67	242.03	0.98	139.64	1.65	57.04	18 May 23	290.5	352.4	0.3	4.7	54.56	60.71	-0.57	-0.38	58.91	56.11	-0.89	-2.28
May 16	189.57	2.64	240.76	0.97	144.50	1.45	66.64	28 M-J 2	64.1	138.1	3.5	-2.7	66.86	73.00	-0.19	0.01	54.10	53.66	-3.33	-3.91
May 26	189.41	2.60	239.54	0.94	149.61	1.26	76.21	7 Jun 12	203.7	263.9	-5.2	-2.0	79.14	85.29	0.21	0.40	55.07	58.27	-4.01	-3.71
Jun 5	189.42	2.56	238.44	0.91	154.96	1.09	85.76	17 Jun 22	323.4	28.6	3.2	5.2	91.43	97.57	0.59	0.77	63.12	69.48	-3.10	-2.27
Jun 15	189.59	2.51	237.54	0.88	160.50	0.93	95.29	27 Jun	102.8		0.3		103.71		0.93		77.26		-1.30	
								Jul 2		174.6		-4.9		109.85		1.08		86.35		-0.29
Jun 25	189.93	2.47	236.89	0.84	166.21	0.77	104.83	7 Jul 12	237.1	296.4	-4.0	0.9	115.99	122.13	1.21	1.32	96.48	107.16	0.63	1.32
Jul 5	190.41	2.43	236.53	0.80	172.09	0.63	114.36	17 Jul 22	358.0	66.6	5.0	3.4	128.27	134.41	1.40	1.46	117.81	128.00	1.71	1.79
Jul 15	191.05	2.40	236.47	0.76	178.11	0.49	123.91	27 J-A 1	141.0	209.2	-3.0	-5.1	140.55	146.68	1.50	1.50	137.53	146.38	1.61	1.22
Jul 25	191.81	2.37	236.72	0.72	184.28	0.35	133.49	6 Aug 11	269.4	329.7	-1.5	3.6	152.82	158.94	1.48	1.42	154.54	162.03	0.66	-0.01
Aug 4	192.69	2.34	237.25	0.69	190.57	0.22	143.09	16 Aug 21	34.7	105.4	4.9	0.0	165.07	171.19	1.34	1.23	168.82	174.84	-0.27	-1.57
Aug 14	193.66	2.31	238.07	0.65	197.00	0.10	152.73	26 Aug 31	177.7	242.0	-4.9	-3.5	177.31	183.42	1.09	0.93	179.92	183.78	-2.38	-3.14
Aug 24	194.72	2.29	239.14	0.62	203.55	-0.02	162.42	5 Sep 10	301.5	4.7	1.4	5.0	189.52	195.62	0.74	0.53	185.98	185.93	-3.78	-4.12
Sep 3	195.85	2.28	240.44	0.58	210.22	-0.14	172.15	15 Sep 20	73.2	144.1	2.6	-3.3	201.71	207.79	0.31	0.07	183.15	178.17	-3.94	-3.00
Sep 13	197.02	2.27	241.94	0.56	217.01	-0.25	181.94	25 Sep 30	212.6	273.8	-4.7	-0.8	213.87	219.93	-0.18	-0.44	173.32	171.53	-1.45	0.16
Sep 23	198.23	2.26	243.63	0.53	223.92	-0.35	191.79	5 Oct 10	334.1	41.5	4.0	4.3	225.98	232.02	-0.71	-0.97	173.92	179.61	1.31	1.89
Oct 3	199.45	2.26	245.46	0.50	230.94	-0.45	201.69	15 Oct 20	112.3	181.4	-1.0	-5.0	238.05	244.07	-1.22	-1.47	187.13	195.40	1.99	1.78
Oct 13	200.67	2.27	247.43	0.48	238.07	-0.55	211.65	25 Oct 30	245.8	305.2	-2.8	2.2	250.07	256.05	-1.70	-1.91	203.83	212.18	1.38	0.87
Oct 23	201.86	2.28	249.50	0.46	245.32	-0.64	221.66	4 Nov 9	8.2	79.7	5.1	1.7	262.01	267.95	-2.10	-2.26	220.38	228.41	0.32	-0.24
Nov 2	203.01	2.29	251.66	0.44	252.66	-0.72	231.72	14 Nov 19	151.1	217.0	-4.1	-4.5	273.86	279.74	-2.39	-2.47	236.30	244.10	-0.78	-1.27
Nov 12	204.09	2.31	253.88	0.42	260.11	-0.80	241.83	24 Nov 29	277.9	337.4	0.0	4.5	285.57	291.37	-2.52	-2.52	251.82	259.48	-1.71	-2.05
Nov 22	205.10	2.34	256.13	0.41	267.64	-0.87	251.97	4 Dec 9	44.0	118.4	4.2	-1.9	297.11	302.78	-2.47	-2.37	267.04	274.41	-2.26	-2.31
Dec 2	206.01	2.37	258.39	0.39	275.26	-0.93	262.14	14 Dec 19	188.3	250.8	-5.3	-2.4	308.38	313.90	-2.21	-2.00	281.34	287.27	-2.12	-1.60
Dec 12	206.79	2.41	260.65	0.38	282.95	-0.98	272.33	24 Dec 29	310.0	11.2	2.9	5.3	319.31	324.59	-1.72	-1.38	291.10	291.18	-0.64	0.79
1805								1805												1805

Left Table

NEW MOONS				FULL MOONS			
JD	DATE	TIME	LONG.	JD	DATE	TIME	LONG.
22+				22+			
		1582					
99172	Oct 26	15;13	212.85	99187	Nov 10	6;30	47.56
99202	Nov 25	9;55	242.87	99216	Dec 9	17;30	77.42
99232	Dec 25	4;09	273.16				
		1583					
				99246	Jan 8	6;08	107.51
99261	Jan 23	20;21	303.38	99275	Feb 6	20;10	137.56
99291	Feb 22	9;40	333.25	99305	Mar 8	11;15	167.32
99320	Mar 23	20;16	2.58	99335	Apr 7	3;04	196.63
99350	Apr 22	4;52	31.34	99364	May 6	19;13	225.47
99379	May 21	12;24	59.63	99394	Jun 5	11;01	253.93
99408	Jun 19	19;43	87.64	99424	Jul 5	1;38	282.17
99438	Jul 19	3;39	115.60	99453	Aug 3	14;34	310.39
99467	Aug 17	13;01	143.79	99483	Sep 2	1;53	338.82
99497	Sep 16	0;44	172.42	99512	Oct 1	12;17	7.63
99526	Oct 15	15;28	201.64	99541	Oct 30	22;30	36.92
99556	Nov 14	9;16	231.47	99571	Nov 29	9;03	66.65
99586	Dec 14	4;58	261.73	99600	Dec 28	19;59	96.64
		1584					
99616	Jan 13	0;30	292.12	99630	Jan 27	7;15	126.64
99645	Feb 11	17;55	322.27	99659	Feb 25	18;56	156.40
99675	Mar 12	8;23	351.93	99689	Mar 26	7;23	185.75
99704	Apr 10	20;01	21.00	99718	Apr 24	20;59	214.67
99734	May 10	5;21	49.50	99748	May 24	11;43	243.21
99763	Jun 8	13;04	77.61	99778	Jun 23	3;05	271.52
99792	Jul 7	19;56	105.53	99807	Jul 22	18;17	299.78
99822	Aug 6	2;56	133.54	99837	Aug 21	8;49	328.21
99851	Sep 4	11;11	161.88	99866	Sep 19	22;35	356.99
99880	Oct 3	21;50	190.75	99896	Oct 19	11;40	26.22
99910	Nov 2	11;40	220.24	99926	Nov 18	0;06	55.89
99940	Dec 2	4;40	250.28	99955	Dec 17	11;44	85.85
99969	Dec 31	23;43	280.63				

Right Table

NEW MOONS				FULL MOONS			
JD	DATE	TIME	LONG.	JD	DATE	TIME	LONG.
23+		**1585**	23+				
	Jan 30	19;08	310.95		Jan 15	22;32	115.86
00029	Mar 1	13;19	340.92	00014	Feb 14	8;42	145.67
00059	Mar 31	5;13	10.35	00043	Mar 15	18;43	175.10
00088	Apr 29	18;21	39.17	00073	Apr 14	5;10	204.06
00118	May 29	4;45	67.48	00102	May 13	16;34	232.61
00147	Jun 27	12;57	95.49	00132	Jun 12	5;08	260.88
00176	Jul 26	19;56	123.44	00161	Jul 11	18;57	289.07
00206	Aug 25	2;57	151.61	00191	Aug 10	9;57	317.43
00235	Sep 23	11;08	180.21	00221	Sep 9	1;54	346.13
00264	Oct 22	21;26	209.38	00250	Oct 8	18;16	15.31
00294	Nov 21	10;14	239.11	00280	Nov 7	10;11	44.96
00324	Dec 21	1;29	269.25	00310	Dec 7	0;43	74.95
		1586					
00353	Jan 19	18;40	299.52	00339	Jan 5	13;28	105.05
00383	Feb 18	13;00	329.64	00369	Feb 4	0;32	134.99
00413	Mar 20	7;18	359.34	00398	Mar 5	10;25	164.56
00443	Apr 19	0;11	28.49	00427	Apr 3	19;35	193.65
00472	May 18	14;34	57.10	00457	May 3	4;29	222.25
00502	Jun 17	2;08	85.29	00486	Jun 1	13;40	250.47
00531	Jul 16	11;28	113.31	00515	Jun 30	23;50	278.54
00560	Aug 14	19;38	141.43	00545	Jul 30	11;47	306.71
00590	Sep 13	3;46	169.87	00575	Aug 29	2;10	335.21
00619	Oct 12	12;41	198.81	00604	Sep 27	18;57	4.23
00648	Nov 10	22;49	228.27	00634	Oct 27	13;07	33.79
00678	Dec 10	10;24	258.17	00664	Nov 26	7;06	63.80
		1587		00693	Dec 25	23;32	94.01
00707	Jan 8	23;38	288.29	00723	Jan 24	13;49	124.15
00737	Feb 7	14;41	318.37	00753	Feb 23	1;58	153.95
00767	Mar 9	7;19	348.18	00782	Mar 24	12;13	183.26
00797	Apr 8	0;37	17.53	00811	Apr 22	20;52	212.01
00826	May 7	17;18	46.38	00841	May 22	4;27	240.30
00856	Jun 6	8;18	74.81	00870	Jun 20	11;50	268.31
00885	Jul 5	21;19	102.98	00899	Jul 19	20;12	296.29
00915	Aug 4	8;43	131.15	00929	Aug 18	6;45	324.53
00944	Sep 2	19;05	159.55	00958	Sep 16	20;17	353.25
00974	Oct 2	4;59	188.35	00988	Oct 16	12;47	22.56
01003	Oct 31	14;48	217.64	01018	Nov 15	7;18	52.43
01033	Nov 30	0;52	247.35	01048	Dec 15	2;23	82.67
01062	Dec 29	11;34	277.34				
		1588		01077	Jan 13	20;36	113.00
01091	Jan 27	23;20	307.35	01107	Feb 12	12;50	143.10
01121	Feb 26	12;27	337.16	01137	Mar 13	2;25	172.71
01151	Mar 27	2;47	6.58	01166	Apr 11	13;09	201.72
01180	Apr 25	17;53	35.54	01195	May 10	21;29	230.18
01210	May 25	9;10	64.09	01225	Jun 9	4;21	258.24
01240	Jun 24	0;15	92.38	01254	Jul 8	10;59	286.15
01269	Jul 23	14;53	120.63	01283	Aug 6	18;36	314.19
01299	Aug 22	4;47	149.05	01313	Sep 5	4;16	342.60
01328	Sep 20	17;41	177.80	01342	Oct 4	16;37	11.56
01358	Oct 20	5;30	206.99	01372	Nov 3	7;47	41.12
01387	Nov 18	16;28	236.61	01402	Dec 3	1;30	71.20
01417	Dec 18	3;02	266.53				
		1589		01431	Jan 1	20;52	101.56
01446	Jan 16	13;40	296.54	01461	Jan 31	16;21	131.88
01476	Feb 15	0;34	326.38	01491	Mar 2	10;07	161.83
01505	Mar 16	11;48	355.84	01521	Apr 1	0;49	191.19
01534	Apr 14	23;25	24.84	01550	Apr 30	12;12	219.92
01564	May 14	11;43	53.41	01579	May 29	20;57	248.16
01594	Jun 13	1;07	81.70	01609	Jun 28	4;13	276.12
01623	Jul 12	15;50	109.93	01638	Jul 27	11;11	304.08
01653	Aug 11	7;34	138.32	01667	Aug 25	18;48	332.28
01682	Sep 9	23;28	167.04	01697	Sep 24	3;48	0.92
01712	Oct 9	14;38	196.18	01726	Oct 23	14;49	30.13
01742	Nov 8	4;38	225.77	01756	Nov 22	4;25	59.91
01771	Dec 7	17;28	255.69	01785	Dec 21	20;52	90.10

NEW MOONS / FULL MOONS

Left block

NEW MOONS JD	DATE	TIME	LONG.	FULL MOONS JD	DATE	TIME	LONG.
23+				23+			
			1590				
01801	Jan 6	5;23	285.75	01815	Jan 20	15;37	120.44
01830	Feb 4	16;24	315.69	01845	Feb 19	11;02	150.59
01860	Mar 6	2;30	345.26	01875	Mar 21	5;08	180.28
01889	Apr 4	11;51	14.35	01904	Apr 19	20;34	209.35
01918	May 3	21;01	42.94	01934	May 19	9;08	237.87
01948	Jun 2	6;56	71.20	01963	Jun 17	19;20	266.01
01977	Jul 1	18;34	99.32	01993	Jul 17	3;59	294.00
02007	Jul 31	8;21	127.56	02022	Aug 15	11;53	322.11
02037	Aug 30	0;03	156.13	02051	Sep 13	19;47	350.55
02066	Sep 28	16;53	185.16	02081	Oct 13	4;26	19.49
02096	Oct 28	9;58	214.69	02110	Nov 11	14;40	48.97
02126	Nov 27	2;30	244.65	02140	Dec 11	3;11	78.91
02155	Dec 26	17;48	274.82				
			1591				
				02169	Jan 9	18;11	109.11
02185	Jan 25	7;18	304.92	02199	Feb 8	11;10	139.27
02214	Feb 23	18;38	334.69	02229	Mar 10	4;56	169.11
02244	Mar 25	4;01	3.94	02258	Apr 8	22;12	198.45
02273	Apr 23	12;04	32.66	02288	May 8	14;07	227.25
02302	May 22	19;47	60.94	02318	Jun 7	4;15	255.63
02332	Jun 21	4;12	88.99	02347	Jul 6	16;26	283.77
02361	Jul 20	14;13	117.04	02377	Aug 5	2;51	311.90
02391	Aug 19	2;24	145.35	02406	Sep 3	12;00	340.25
02420	Sep 17	17;01	174.12	02435	Oct 2	20;42	9.02
02450	Oct 17	9;54	203.46	02465	Nov 1	5;51	38.29
02480	Nov 16	4;26	233.34	02494	Nov 30	16;11	68.03
02509	Dec 15	23;17	263.59	02524	Dec 30	4;00	98.06
			1592				
02539	Jan 14	16;46	293.89	02553	Jan 28	17;14	128.14
02569	Feb 13	7;34	323.92	02583	Feb 27	7;35	157.99
02598	Mar 13	19;23	353.45	02612	Mar 27	22;47	187.44
02628	Apr 12	4;45	22.39	02642	Apr 26	14;34	216.41
02657	May 11	12;34	50.82	02672	May 26	6;29	244.98
02686	Jun 9	19;47	78.89	02701	Jun 24	21;46	273.27
02716	Jul 9	3;14	106.84	02731	Jul 24	11;42	301.49
02745	Aug 7	11;46	134.91	02760	Aug 22	24;00	329.85
02774	Sep 5	22;14	163.36	02790	Sep 21	11;02	358.54
02804	Oct 5	11;29	192.36	02819	Oct 20	21;31	27.68
02834	Nov 4	3;54	221.98	02849	Nov 19	8;03	57.29
02863	Dec 3	22;59	252.13	02878	Dec 18	18;54	87.23
			1593				
02893	Jan 2	18;56	282.53	02908	Jan 17	5;59	117.26
02923	Feb 1	13;36	312.81	02937	Feb 15	17;16	147.11
02953	Mar 3	5;32	342.67	02967	Mar 17	5;01	176.58
02982	Apr 1	18;27	11.94	02996	Apr 15	17;42	205.61
03012	May 1	4;45	40.62	03026	May 15	7;37	234.24
03041	May 30	13;07	68.84	03055	Jun 13	22;35	262.59
03070	Jun 28	20;17	96.80	03085	Jul 13	13;56	290.85
03100	Jul 28	3;07	124.75	03115	Aug 12	4;59	319.21
03129	Aug 26	10;40	152.95	03144	Sep 10	19;23	347.88
03158	Sep 24	20;04	181.62	03174	Oct 10	9;06	16.97
03188	Oct 24	8;20	210.89	03203	Nov 8	22;09	46.53
03217	Nov 22	23;52	240.76	03233	Dec 8	10;24	76.43
03247	Dec 22	18;06	271.03				
			1594				
				03262	Jan 6	21;41	106.47
03277	Jan 21	13;30	301.40	03292	Feb 5	8;03	136.38
03307	Feb 20	8;21	331.52	03321	Mar 6	17;54	165.94
03337	Mar 22	1;22	1.14	03351	Apr 5	3;48	195.03
03366	Apr 20	15;50	30.17	03380	May 4	14;24	223.68
03396	May 20	3;31	58.64	03410	Jun 3	2;07	252.00
03425	Jun 18	12;44	86.73	03439	Jul 2	15;10	280.17
03454	Jul 17	20;15	114.68	03469	Aug 1	5;34	308.44
03484	Aug 16	3;12	142.75	03498	Aug 30	21;13	337.01
03513	Sep 14	10;48	171.20	03528	Sep 29	13;43	6.04
03542	Oct 13	20;06	200.17	03558	Oct 29	6;16	35.57
03572	Nov 12	7;43	229.71	03587	Nov 27	21;50	65.49
03601	Dec 11	21;47	259.73	03617	Dec 27	11;39	95.60
			1595				
03631	Jan 10	13;59	289.97	03646	Jan 25	23;33	125.63
03661	Feb 9	7;42	320.16	03676	Feb 24	9;52	155.35
03691	Mar 11	1;58	350.40	03705	Mar 25	19;10	184.60
03720	Apr 9	19;33	19.35	03735	Apr 24	3;56	213.34
03750	May 9	11;12	48.14	03764	May 23	12;42	241.66
03780	Jun 8	0;09	76.46	03793	Jun 21	22;06	269.74
03809	Jul 7	10;35	104.53	03823	Jul 21	8;58	297.83
03838	Aug 5	19;20	132.60	03852	Aug 19	22;05	326.18
03868	Sep 4	3;33	160.93	03882	Sep 18	13;50	355.01
03897	Oct 3	12;19	189.69	03912	Oct 18	7;41	24.40
03926	Nov 1	21;51	218.99	03942	Nov 17	2;11	54.29
03956	Dec 1	8;46	248.76	03971	Dec 16	19;43	84.48
03985	Dec 30	21;07	278.81				

Right block

NEW MOONS JD	DATE	TIME	LONG.	FULL MOONS JD	DATE	TIME	LONG.
23+				23+			
			1596				
				04001	Jan 15	11;12	114.70
04015	Jan 29	11;05	308.92	04031	Feb 14	0;23	144.65
04045	Feb 28	2;45	338.82	04060	Mar 14	11;27	174.13
04074	Mar 28	19;36	8.31	04089	Apr 12	20;43	203.07
04104	Apr 27	12;32	37.33	04119	May 12	4;35	231.49
04134	May 27	4;21	65.88	04148	Jun 10	11;49	259.56
04163	Jun 25	18;23	94.12	04177	Jul 9	19;27	287.52
04193	Jul 25	6;39	122.28	04207	Aug 8	4;44	315.62
04222	Aug 23	17;41	150.60	04236	Sep 6	16;45	344.14
04252	Sep 22	4;01	179.27	04266	Oct 6	7;55	13.24
04281	Oct 21	14;04	208.41	04296	Nov 5	1;43	42.93
04311	Nov 20	0;08	238.00	04325	Dec 4	20;51	73.09
04340	Dec 19	10;31	267.93				
			1597				
				04355	Jan 3	15;46	103.44
04369	Jan 17	21;39	297.95	04385	Feb 2	9;08	133.66
04399	Feb 16	9;54	327.83	04415	Mar 4	0;03	163.46
04428	Mar 17	23;23	357.37	04444	Apr 2	12;05	192.69
04458	Apr 16	13;52	26.46	04473	May 1	21;26	221.33
04488	May 16	4;52	55.12	04503	May 31	4;49	249.50
04517	Jun 14	19;59	83.47	04532	Jun 29	11;21	277.43
04547	Jul 14	10;53	111.71	04561	Jul 28	18;20	305.39
04577	Aug 13	1;21	140.06	04591	Aug 27	2;55	333.64
04606	Sep 11	15;03	168.71	04620	Sep 25	13;56	2.38
04636	Oct 11	3;42	197.78	04650	Oct 25	3;47	31.73
04665	Nov 9	15;18	227.28	04679	Nov 23	20;22	61.65
04695	Dec 9	2;10	257.14	04709	Dec 23	15;05	91.95
			1598				
04724	Jan 7	12;46	287.14	04739	Jan 22	10;42	122.32
04753	Feb 5	23;25	317.05	04769	Feb 21	5;28	152.43
04783	Mar 7	10;17	346.64	04798	Mar 22	21;43	182.01
04812	Apr 5	21;24	15.77	04828	Apr 21	10;39	210.96
04842	May 5	9;02	44.45	04857	May 20	20;32	239.35
04871	Jun 3	21;36	72.80	04887	Jun 19	4;24	267.38
04901	Jul 3	11;32	101.01	04916	Jul 18	11;25	295.32
04931	Aug 2	2;53	129.33	04945	Aug 16	18;42	323.41
04960	Aug 31	19;00	157.93	04975	Sep 15	3;02	351.90
04990	Sep 30	10;54	186.95	05004	Oct 14	13;06	20.91
05020	Oct 30	1;48	216.42	05034	Nov 13	1;25	50.49
05049	Nov 28	15;26	246.27	05063	Dec 12	16;27	80.55
05079	Dec 28	3;55	276.33				
			1599				
				05093	Jan 11	10;06	110.86
05108	Jan 26	15;23	306.34	05123	Feb 10	5;16	141.10
05138	Feb 25	1;50	336.05	05153	Mar 12	0;05	170.95
05167	Mar 26	11;19	5.29	05182	Apr 10	16;49	200.24
05196	Apr 24	20;14	34.02	05212	May 10	6;43	228.95
05226	May 24	5;25	62.35	05241	Jun 8	18;01	257.20
05255	Jun 22	15;55	90.47	05271	Jul 8	3;25	285.22
05285	Jul 22	4;31	118.63	05300	Aug 6	11;44	313.28
05314	Aug 20	19;22	147.06	05329	Sep 4	19;43	341.61
05344	Sep 19	11;54	175.94	05359	Oct 4	4;04	10.38
05374	Oct 19	5;12	205.32	05388	Nov 2	13;33	39.68
05403	Nov 17	22;21	235.17	05418	Dec 2	0;54	69.47
05433	Dec 17	14;33	265.31	05447	Dec 31	14;38	99.59
			1600				
05463	Jan 16	5;06	295.49	05477	Jan 30	6;35	129.77
05492	Feb 14	17;31	325.40	05506	Feb 28	23;53	159.73
05522	Mar 15	3;42	354.84	05536	Mar 29	17;18	189.24
05551	Apr 13	12;09	23.73	05566	Apr 28	9;47	218.21
05580	May 12	19;43	52.13	05596	May 28	0;46	246.72
05610	Jun 11	3;30	80.21	05625	Jun 26	13;59	274.93
05639	Jul 10	12;31	108.22	05655	Jul 26	1;24	303.06
05668	Aug 8	23;32	136.41	05684	Aug 24	11;19	331.34
05698	Sep 7	12;58	165.00	05713	Sep 22	20;21	359.97
05728	Oct 7	4;53	194.13	05743	Oct 22	5;23	29.08
05757	Nov 5	22;52	223.85	05772	Nov 20	15;12	58.67
05787	Dec 5	17;56	254.01	05802	Dec 20	2;18	88.63
			1601				
05817	Jan 4	12;26	284.34	05831	Jan 18	14;46	118.70
05847	Feb 3	4;43	314.51	05861	Feb 17	4;23	148.63
05876	Mar 4	17;57	344.24	05890	Mar 18	18;54	178.20
05906	Apr 3	4;19	13.38	05920	Apr 17	10;08	207.31
05935	May 2	12;40	41.97	05950	May 17	1;50	235.99
05964	May 31	19;58	70.13	05979	Jun 15	17;28	264.35
05994	Jun 30	3;07	98.08	06009	Jul 15	8;13	292.58
06023	Jul 29	11;00	126.08	06038	Aug 13	21;32	320.90
06052	Aug 27	20;25	154.37	06068	Sep 12	9;24	349.49
06082	Sep 26	8;15	183.17	06097	Oct 11	20;21	18.50
06111	Oct 25	23;11	212.58	06127	Nov 10	7;02	47.98
06141	Nov 24	17;10	242.56	06156	Dec 9	17;52	77.84
06171	Dec 24	13;01	272.91				

Left page

NEW MOONS				FULL MOONS			
JD	DATE	TIME	LONG.	JD	DATE	TIME	LONG.
23+							
			1602	23+			
				06186	Jan 8	4;52	107.86
06201	Jan 23	8;35	303.28	06215	Feb 6	15;55	137.78
06231	Feb 22	1;56	333.32	06245	Mar 8	3;08	167.37
06260	Mar 23	16;15	2.80	06274	Apr 6	15;00	196.53
06290	Apr 22	3;42	31.68	06304	May 6	3;59	225.25
06319	May 21	12;52	60.03	06333	Jun 4	18;16	253.65
06348	Jun 19	20;31	88.06	06363	Jul 4	9;26	281.91
06378	Jul 19	3;25	115.98	06393	Aug 3	0;47	310.23
06407	Aug 17	10;32	144.07	06422	Sep 1	15;45	338.79
06436	Sep 15	18;58	172.57	06452	Oct 1	6;07	7.76
06466	Oct 15	5;48	201.63	06481	Oct 30	19;48	37.20
06495	Nov 13	19;47	231.29	06511	Nov 29	8;42	67.03
06525	Dec 13	12;49	261.45	06540	Dec 28	20;35	97.07
			1603				
06555	Jan 12	7;47	291.81	06570	Jan 27	7;22	127.04
06585	Feb 11	3;00	322.04	06599	Feb 25	17;16	156.73
06614	Mar 12	20;55	351.85	06629	Mar 27	2;50	185.96
06644	Apr 11	12;36	21.08	06658	Apr 25	12;45	214.72
06674	May 11	1;37	49.74	06687	May 24	23;37	243.11
06703	Jun 9	12;00	77.95	06717	Jun 23	11;49	271.30
06732	Jul 8	20;19	105.93	06747	Jul 23	1;29	299.50
06762	Aug 7	3;30	133.95	06776	Aug 21	16;36	327.95
06791	Sep 5	10;48	162.25	06806	Sep 20	8;56	356.82
06820	Oct 4	19;16	191.04	06836	Oct 20	1;52	26.21
06850	Nov 3	5;48	220.39	06865	Nov 18	18;20	56.04
06879	Dec 2	18;42	250.26	06895	Dec 18	9;17	86.14
			1604				
06909	Jan 1	9;51	280.44	06924	Jan 16	22;12	116.24
06939	Jan 31	2;45	310.66	06954	Feb 15	9;11	146.09
06968	Feb 29	20;39	340.63	06983	Mar 15	18;47	175.49
06998	Mar 30	14;32	10.14	07013	Apr 14	3;33	204.39
07028	Apr 29	7;08	39.11	07042	May 13	12;03	232.82
07057	May 28	21;26	67.58	07071	Jun 11	20;51	260.93
07087	Jun 27	9;08	95.72	07101	Jul 11	6;46	288.98
07116	Jul 26	18;43	123.78	07130	Aug 9	18;38	317.20
07146	Aug 25	3;15	152.01	07160	Sep 8	9;03	345.84
07175	Sep 23	11;48	180.63	07190	Oct 8	2;12	15.04
07204	Oct 22	21;05	209.76	07219	Nov 6	20;49	44.79
07234	Nov 21	7;28	239.39	07249	Dec 6	15;14	74.94
07263	Dec 20	19;07	269.37				
			1605				
				07279	Jan 5	7;58	105.21
07293	Jan 19	8;09	299.47	07308	Feb 3	22;20	135.29
07322	Feb 17	22;46	329.44	07338	Mar 5	10;21	164.96
07352	Mar 19	14;50	359.06	07367	Apr 3	20;20	194.08
07382	Apr 18	7;37	28.22	07397	May 3	4;41	222.65
07411	May 17	23;57	56.91	07426	Jun 1	11;59	250.80
07441	Jun 16	14;52	85.24	07455	Jun 30	19;10	278.76
07471	Jul 16	4;05	113.41	07485	Jul 30	3;26	306.77
07500	Aug 14	15;52	141.66	07514	Aug 28	13;59	335.11
07530	Sep 13	2;44	170.22	07544	Sep 27	3;38	3.99
07559	Oct 12	13;09	199.22	07573	Oct 26	20;21	33.48
07588	Nov 10	23;23	228.69	07603	Nov 25	15;09	63.52
07618	Dec 10	9;41	258.53	07633	Dec 25	10;28	93.85
			1606				
07647	Jan 8	20;22	288.54	07663	Jan 24	4;47	124.16
07677	Feb 7	7;53	318.48	07692	Feb 22	20;59	154.14
07706	Mar 8	20;31	348.13	07722	Mar 24	10;25	183.58
07736	Apr 7	10;14	17.35	07751	Apr 22	20;57	212.41
07766	May 7	0;46	46.12	07781	May 22	5;06	240.72
07795	Jun 5	15;40	74.54	07810	Jun 20	11;51	268.70
07825	Jul 5	6;40	102.79	07839	Jul 19	18;26	296.61
07854	Aug 3	21;30	131.09	07869	Aug 18	2;08	324.74
07884	Sep 2	11;52	159.64	07898	Sep 16	11;56	353.29
07914	Oct 2	1;22	188.59	07928	Oct 16	0;26	22.43
07943	Oct 31	13;45	217.98	07957	Nov 14	15;46	52.16
07973	Nov 30	1;08	247.75	07987	Dec 14	9;33	82.35
08002	Dec 29	11;54	277.74				
			1607				
				08017	Jan 13	4;53	112.73
08031	Jan 27	22;28	307.71	08047	Feb 12	0;14	142.96
08061	Feb 26	9;03	337.41	08076	Mar 13	17;49	172.74
08090	Mar 27	19;47	6.69	08106	Apr 12	8;21	201.92
08120	Apr 26	6;50	35.49	08135	May 11	19;37	230.50
08149	May 25	18;37	63.90	08165	Jun 10	4;20	258.63
08179	Jun 24	7;39	92.12	08194	Jul 9	11;39	286.57
08208	Jul 23	22;17	120.36	08223	Aug 7	18;46	314.59
08238	Aug 22	14;14	148.85	08253	Sep 6	2;36	342.92
08268	Sep 21	6;36	177.74	08282	Oct 5	11;52	11.75
08297	Oct 20	22;23	207.09	08311	Nov 3	23;05	41.14
08327	Nov 19	12;55	236.85	08341	Dec 3	12;45	71.05
08357	Dec 19	2;09	266.88				

Right page

NEW MOONS				FULL MOONS			
JD	DATE	TIME	LONG.	JD	DATE	TIME	LONG.
23+							
			1608	23+			
				08371	Jan 2	5;06	101.28
08386	Jan 17	14;10	296.95	08400	Jan 31	23;34	131.56
08416	Feb 16	1;04	326.78	08430	Mar 1	18;39	161.57
08445	Mar 16	10;50	356.19	08460	Mar 31	12;26	191.07
08474	Apr 14	19;45	25.08	08490	Apr 30	3;41	219.97
08504	May 14	4;27	53.51	08519	May 29	16;11	248.37
08533	Jun 12	13;59	81.65	08549	Jun 28	2;29	276.45
08563	Jul 12	1;20	109.75	08578	Jul 27	11;22	304.48
08592	Aug 10	15;03	138.05	08607	Aug 25	19;35	332.70
08622	Sep 9	6;56	166.76	08637	Sep 24	3;50	1.32
08652	Oct 9	0;10	195.98	08666	Oct 23	12;49	30.44
08681	Nov 7	17;45	225.70	08695	Nov 21	23;15	60.08
08711	Dec 7	10;45	255.80	08725	Dec 21	11;45	90.10
			1609				
08741	Jan 6	2;20	286.01	08755	Jan 20	2;32	120.27
08770	Feb 4	15;52	316.05	08784	Feb 18	19;05	150.32
08800	Mar 6	3;04	345.67	08814	Mar 20	12;20	179.97
08829	Apr 4	12;09	14.75	08844	Apr 19	5;10	209.12
08858	May 3	19;51	43.29	08873	May 18	20;50	237.77
08888	Jun 2	3;14	71.44	08903	Jun 17	10;57	266.07
08917	Jul 1	11;24	99.43	08932	Jul 16	23;22	294.21
08946	Jul 30	21;17	127.52	08962	Aug 15	10;10	322.44
08976	Aug 29	9;30	155.93	08991	Sep 13	19;47	350.95
09006	Sep 28	0;16	184.87	09021	Oct 13	4;57	19.91
09035	Oct 27	17;27	214.40	09050	Nov 11	14;28	49.35
09065	Nov 26	12;19	244.44	09080	Dec 11	0;58	79.21
09095	Dec 26	7;28	274.77				
			1610				
				09109	Jan 9	12;44	109.27
09125	Jan 25	1;05	305.05	09139	Feb 8	1;39	139.26
09154	Feb 23	15;51	334.96	09168	Mar 9	15;29	168.94
09184	Mar 25	3;29	4.31	09198	Apr 8	6;04	198.18
09213	Apr 23	12;36	33.07	09227	May 7	21;19	226.97
09242	May 22	20;11	61.35	09257	Jun 6	12;56	255.42
09272	Jun 21	3;14	89.34	09287	Jul 6	4;14	283.68
09301	Jul 20	10;37	117.29	09316	Aug 4	18;28	311.96
09330	Aug 18	19;11	145.45	09346	Sep 3	7;17	340.46
09360	Sep 17	5;46	174.05	09375	Oct 2	18;53	9.34
09389	Oct 16	19;11	203.24	09405	Nov 1	5;53	38.68
09419	Nov 15	11;47	233.04	09434	Nov 30	16;48	68.45
09449	Dec 15	6;59	263.29	09464	Dec 30	3;48	98.44
			1611				
09479	Jan 14	2;59	293.70	09493	Jan 28	14;45	128.43
09508	Feb 12	21;35	323.89	09523	Feb 27	1;39	158.13
09538	Mar 14	13;23	353.58	09552	Mar 28	12;51	187.41
09568	Apr 13	2;07	22.67	09582	Apr 27	0;56	216.25
09597	May 12	12;17	51.19	09611	May 26	14;20	244.72
09626	Jun 10	20;33	79.31	09641	Jun 25	5;00	273.00
09656	Jul 10	3;44	107.24	09670	Jul 24	20;22	301.28
09685	Aug 8	10;42	135.26	09700	Aug 23	11;45	329.75
09714	Sep 6	18;26	163.60	09730	Sep 22	2;41	358.60
09744	Oct 6	4;02	192.45	09759	Oct 21	17;01	27.90
09773	Nov 4	16;28	221.90	09789	Nov 20	6;36	57.63
09803	Dec 4	8;05	251.90	09818	Dec 19	19;10	87.64
			1612				
09833	Jan 3	2;13	282.21	09848	Jan 18	6;31	117.67
09862	Feb 1	21;24	312.52	09877	Feb 16	16;42	147.47
09892	Mar 2	15;59	342.49	09907	Mar 17	2;10	176.85
09922	Apr 1	8;46	11.93	09936	Apr 15	11;35	205.75
09951	Apr 30	23;03	40.78	09965	May 14	21;40	234.22
09981	May 30	10;42	69.13	09995	Jun 13	8;58	262.43
10010	Jun 28	20;00	97.17	10024	Jul 12	21;46	290.60
10040	Jul 28	3;43	125.16	10054	Aug 11	12;11	318.93
10069	Aug 26	10;56	153.35	10084	Sep 10	4;06	347.65
10098	Sep 24	18;51	181.97	10113	Oct 9	21;05	16.88
10128	Oct 24	4;26	211.13	10143	Nov 8	14;13	46.60
10157	Nov 22	16;13	240.83	10173	Dec 8	6;15	76.66
10187	Dec 22	6;16	270.92				
			1613				
				10202	Jan 6	20;20	106.81
10216	Jan 20	22;14	301.14	10232	Feb 5	8;14	136.78
10246	Feb 19	15;31	331.20	10261	Mar 6	18;21	166.34
10276	Mar 21	9;20	0.87	10291	Apr 5	3;17	195.40
10306	Apr 20	2;33	30.02	10320	May 4	11;39	223.96
10335	May 19	18;01	58.65	10349	Jun 2	20;02	252.14
10365	Jun 18	7;01	86.90	10379	Jul 2	5;09	280.16
10394	Jul 17	17;41	114.97	10408	Jul 31	15;52	308.29
10424	Aug 16	2;47	143.13	10438	Aug 30	5;01	336.75
10453	Sep 14	11;25	171.62	10467	Sep 28	20;59	5.75
10482	Oct 13	20;28	200.58	10497	Oct 28	15;13	35.34
10512	Nov 12	6;27	230.06	10527	Nov 27	10;09	65.40
10541	Dec 11	17;31	259.94	10557	Dec 27	4;01	95.68

NEW MOONS				FULL MOONS			
JD	DATE	TIME	LONG.	JD	DATE	TIME	LONG.
23+			1614	23+			
10571	Jan 10	5;45	290.02	10586	Jan 25	19;40	125.87
10600	Feb 8	19;23	320.03	10616	Feb 24	8;47	155.70
10630	Mar 10	10;30	349.76	10645	Mar 25	19;39	185.00
10660	Apr 9	2;45	19.06	10675	Apr 24	4;36	213.75
10689	May 8	19;14	47.89	10704	May 23	12;12	242.02
10719	Jun 7	10;52	76.32	10733	Jun 21	19;13	270.01
10749	Jul 7	0;59	104.53	10763	Jul 21	2;44	297.96
10778	Aug 5	13;36	132.76	10792	Aug 19	12;00	326.16
10808	Sep 4	1;07	161.21	10822	Sep 18	0;07	354.83
10837	Oct 3	11;59	190.08	10851	Oct 17	15;28	24.12
10866	Nov 1	22;31	219.42	10881	Nov 16	9;31	53.99
10896	Dec 1	8;55	249.16	10911	Dec 16	4;52	84.25
10925	Dec 30	19;24	279.15				
			1615				
				10940	Jan 14	23;54	114.62
10955	Jan 29	6;21	309.12	10970	Feb 13	17;15	144.75
10984	Feb 27	18;10	338.85	11000	Mar 15	8;02	174.39
11014	Mar 29	7;04	8.19	11029	Apr 13	19;54	203.43
11043	Apr 27	20;57	37.09	11059	May 13	5;04	231.90
11073	May 27	11;28	65.59	11088	Jun 11	12;19	259.96
11103	Jun 26	2;22	93.87	11117	Jul 10	18;49	287.87
11132	Jul 25	17;22	122.14	11147	Aug 9	1;51	315.89
11162	Aug 24	8;13	150.60	11176	Sep 7	10;34	344.28
11191	Sep 22	22;29	179.43	11205	Oct 6	21;46	13.21
11221	Oct 22	11;45	208.70	11235	Nov 5	11;47	42.74
11250	Nov 20	23;50	238.39	11265	Dec 5	4;27	72.79
11280	Dec 20	11;00	268.34				
			1616				
				11294	Jan 3	23;09	103.13
11309	Jan 18	21;37	298.34	11324	Feb 2	18;38	133.44
11339	Feb 17	8;03	328.15	11354	Mar 3	13;12	163.41
11368	Mar 17	18;29	357.56	11384	Apr 2	5;16	192.80
11398	Apr 16	5;04	26.49	11413	May 1	18;03	221.58
11427	May 15	16;09	55.00	11443	May 31	3;53	249.85
11457	Jun 14	4;18	83.24	11472	Jun 29	11;47	277.84
11486	Jul 13	18;02	111.43	11501	Jul 28	18;56	305.80
11516	Aug 12	9;26	139.82	11531	Aug 27	2;26	334.01
11546	Sep 11	1;55	168.57	11560	Sep 25	11;02	2.66
11575	Oct 10	18;23	197.78	11589	Oct 24	21;20	31.86
11605	Nov 9	9;52	227.44	11619	Nov 23	9;47	61.60
11634	Dec 8	23;58	257.44	11649	Dec 23	0;47	91.73
			1617				
11664	Jan 7	12;41	287.53	11678	Jan 21	18;12	122.02
11694	Feb 6	0;07	317.48	11708	Feb 20	13;01	152.13
11723	Mar 7	10;19	347.04	11738	Mar 22	7;28	181.81
11752	Apr 5	19;24	16.09	11767	Apr 20	23;56	210.92
11782	May 5	3;52	44.64	11797	May 20	13;44	239.47
11811	Jun 3	12;38	72.83	11827	Jun 19	1;04	267.65
11840	Jul 2	22;49	100.90	11856	Jul 18	10;40	295.68
11870	Aug 1	11;15	129.09	11885	Aug 16	19;17	323.82
11900	Aug 31	2;10	157.64	11915	Sep 15	3;38	352.31
11929	Sep 29	19;02	186.69	11944	Oct 14	12;21	21.26
11959	Oct 29	12;48	216.27	11973	Nov 12	22;06	50.75
11989	Nov 28	6;26	246.28	12003	Dec 12	9;33	80.65
12018	Dec 27	22;59	276.51				
			1618				
				12032	Jan 10	23;08	110.79
12048	Jan 26	13;40	306.66	12062	Feb 9	14;43	140.88
12078	Feb 25	2;00	336.45	12092	Mar 11	7;30	170.67
12107	Mar 26	11;56	5.71	12122	Apr 10	0;24	199.97
12136	Apr 24	20;03	34.41	12151	May 9	16;32	228.77
12166	May 24	3;18	62.66	12181	Jun 8	7;24	257.17
12195	Jun 22	10;48	90.66	12210	Jul 7	20;45	285.35
12224	Jul 21	19;39	118.67	12240	Aug 6	8;31	313.54
12254	Aug 20	6;39	146.94	12269	Sep 4	18;53	341.96
12283	Sep 18	20;12	175.68	12299	Oct 4	4;25	10.78
12313	Oct 18	12;20	205.01	12328	Nov 2	13;52	40.08
12343	Nov 17	6;38	234.90	12357	Dec 1	23;57	69.82
12373	Dec 17	2;00	265.17	12387	Dec 31	11;06	99.83
			1619				
12402	Jan 15	20;40	295.52	12416	Jan 29	23;21	129.86
12432	Feb 14	12;58	325.60	12446	Feb 28	12;31	159.65
12462	Mar 16	2;04	355.17	12476	Mar 30	2;26	189.02
12491	Apr 14	12;14	24.12	12505	Apr 28	17;05	217.93
12520	May 13	20;22	52.55	12535	May 28	8;23	246.46
12550	Jun 12	3;28	80.60	12564	Jun 26	23;53	274.76
12579	Jul 11	10;33	108.54	12594	Jul 26	14;50	303.03
12608	Aug 9	18;26	136.59	12624	Aug 25	4;35	331.46
12638	Sep 8	3;58	165.02	12653	Sep 23	17;01	0.23
12667	Oct 7	15;57	193.99	12683	Oct 23	4;32	29.43
12697	Nov 6	7;02	223.58	12712	Nov 21	15;40	59.08
12727	Dec 6	1;09	253.69	12742	Dec 21	2;44	89.04

NEW MOONS				FULL MOONS			
JD	DATE	TIME	LONG.	JD	DATE	TIME	LONG.
23+			1620	23+			
12756	Jan 4	21;02	284.08	12771	Jan 19	13;41	119.05
12786	Feb 3	16;33	314.39	12801	Feb 18	0;27	148.86
12816	Mar 4	9;47	344.29	12830	Mar 18	11;13	178.27
12845	Apr 2	23;57	13.59	12859	Apr 16	22;29	207.23
12875	May 2	11;14	42.30	12889	May 16	10;55	235.78
12904	May 31	20;19	70.53	12919	Jun 15	0;48	264.08
12934	Jun 30	3;57	98.51	12948	Jul 14	15;51	292.33
12963	Jul 29	10;58	126.47	12978	Aug 13	7;24	320.73
12992	Aug 27	18;16	154.68	13007	Sep 11	22;49	349.45
13022	Sep 26	2;55	183.34	13037	Oct 11	13;46	18.62
13051	Oct 25	13;56	212.58	13067	Nov 10	4;02	48.24
13081	Nov 24	4;01	242.40	13096	Dec 9	17;21	78.20
13110	Dec 23	21;01	272.62				
			1621				
				13126	Jan 8	5;25	108.27
13140	Jan 22	15;48	302.96	13155	Feb 6	16;06	138.18
13170	Feb 21	10;43	333.07	13185	Mar 8	1;42	167.71
13200	Mar 23	4;22	2.70	13214	Apr 6	10;49	196.75
13229	Apr 21	19;50	31.75	13243	May 5	20;14	225.33
13259	May 21	8;46	60.26	13273	Jun 4	6;39	253.58
13288	Jun 19	19;12	88.40	13302	Jul 3	18;31	281.71
13318	Jul 19	3;40	116.38	13332	Aug 2	8;05	309.96
13347	Aug 17	11;08	144.49	13361	Aug 31	23;22	338.53
13376	Sep 15	18;44	172.95	13391	Sep 30	16;06	7.58
13406	Oct 15	3;32	201.93	13421	Oct 30	9;34	37.16
13435	Nov 13	14;17	231.45	13451	Nov 29	2;33	67.16
13465	Dec 13	3;14	261.43	13480	Dec 28	17;51	97.34
			1622				
13494	Jan 11	18;13	291.62	13510	Jan 27	6;51	127.41
13524	Feb 10	10;44	321.75	13539	Feb 25	17;43	157.13
13554	Mar 12	4;10	351.55	13569	Mar 27	3;01	186.36
13583	Apr 10	21;36	20.86	13598	Apr 25	11;24	215.06
13613	May 10	13;56	49.66	13627	May 24	19;31	243.34
13643	Jun 9	4;12	78.02	13657	Jun 23	4;02	271.37
13672	Jul 8	16;04	106.15	13686	Jul 22	13;45	299.42
13702	Aug 7	2;00	134.28	13716	Aug 21	1;36	327.74
13731	Sep 5	10;58	162.65	13745	Sep 19	16;14	356.53
13760	Oct 4	19;56	191.46	13775	Oct 19	9;37	25.93
13790	Nov 3	5;36	220.77	13805	Nov 18	4;38	55.86
13819	Dec 2	16;13	250.55	13834	Dec 17	23;24	86.12
			1623				
13849	Jan 1	3;50	280.58	13864	Jan 16	16;20	116.39
13878	Jan 30	16;36	310.62	13894	Feb 15	6;42	146.38
13908	Mar 1	6;44	340.44	13923	Mar 16	18;34	175.87
13937	Mar 30	22;12	9.86	13953	Apr 15	4;18	204.80
13967	Apr 29	14;26	38.82	13982	May 14	12;22	233.21
13997	May 29	6;27	67.37	14011	Jun 12	19;27	261.27
14026	Jun 27	21;22	95.64	14041	Jul 12	2;30	289.20
14056	Jul 27	10;51	123.85	14070	Aug 10	10;45	317.27
14085	Aug 25	23;05	152.23	14099	Sep 8	21;22	345.76
14115	Sep 24	10;30	180.97	14129	Oct 8	11;11	14.82
14144	Oct 23	21;26	210.17	14159	Nov 7	4;07	44.49
14174	Nov 22	8;04	239.81	14188	Dec 6	23;07	74.66
14203	Dec 21	18;33	269.74				
			1624				
				14218	Jan 5	18;33	105.03
14233	Jan 20	5;10	299.74	14248	Feb 4	12;53	135.28
14262	Feb 18	16;20	329.56	14278	Mar 5	4;58	165.12
14292	Mar 19	4;26	359.02	14307	Apr 3	18;14	194.38
14321	Apr 17	17;33	28.04	14337	May 3	4;36	223.03
14351	May 17	7;32	56.64	14366	Jun 1	12;37	251.22
14380	Jun 15	22;07	84.96	14395	Jun 30	19;19	279.15
14410	Jul 15	13;05	113.21	14425	Jul 30	1;57	307.10
14440	Aug 14	4;12	141.59	14454	Aug 28	9;46	335.34
14469	Sep 12	19;04	170.30	14483	Sep 26	19;45	4.06
14499	Oct 12	9;11	199.45	14513	Oct 26	8;26	33.38
14528	Nov 10	22;07	229.03	14542	Nov 24	23;52	63.27
14558	Dec 10	9;53	258.93	14572	Dec 24	17;40	93.53
			1625				
14587	Jan 8	20;46	288.95	14602	Jan 23	12;52	123.88
14617	Feb 7	7;12	318.84	14632	Feb 22	8;00	153.99
14646	Mar 8	17;26	348.39	14662	Mar 24	1;22	183.60
14676	Apr 7	3;40	17.47	14691	Apr 22	15;44	212.59
14705	May 6	14;10	46.08	14721	May 22	2;55	241.02
14735	Jun 5	1;28	74.36	14750	Jun 20	11;39	269.08
14764	Jul 4	14;13	102.53	14779	Jul 19	19;06	297.03
14794	Aug 3	4;48	130.81	14809	Aug 18	2;25	325.14
14823	Sep 1	21;00	159.43	14838	Sep 16	10;32	353.63
14853	Oct 1	13;52	188.51	14867	Oct 15	20;04	22.65
14883	Oct 31	6;14	218.06	14897	Nov 14	7;29	52.21
14912	Nov 29	21;18	247.99	14926	Dec 13	21;11	82.23
14942	Dec 29	10;50	278.09				

NEW MOONS				FULL MOONS			
JD	DATE	TIME	LONG.	JD	DATE	TIME	LONG.
23+							
			1626	23+			
				14956	Jan 12	13;21	112.47
14971	Jan 27	22;55	308.12	14986	Feb 11	7;28	142.65
15001	Feb 26	9;38	337.83	15016	Mar 13	2;07	172.49
15030	Mar 27	19;04	7.05	15045	Apr 11	19;33	201.79
15060	Apr 26	3;32	35.75	15075	May 11	10;37	230.52
15089	May 25	11;49	64.02	15104	Jun 9	23;07	258.81
15118	Jun 23	21;00	92.08	15134	Jul 9	9;35	286.88
15148	Jul 23	8;08	120.18	15163	Aug 7	18;45	314.97
15177	Aug 21	21;50	148.58	15193	Sep 6	3;21	343.34
15207	Sep 20	13;57	177.45	15222	Oct 5	12;00	12.14
15237	Oct 20	7;36	206.87	15251	Nov 3	21;19	41.46
15267	Nov 19	1;40	236.77	15281	Dec 3	7;56	71.24
15296	Dec 18	19;03	266.98				
			1627				
				15310	Jan 1	20;24	101.31
15326	Jan 17	10;51	297.20	15340	Jan 31	10;52	131.43
15356	Feb 16	0;23	327.15	15370	Mar 2	2;55	161.31
15385	Mar 17	11;22	356.60	15399	Mar 31	19;36	190.77
15414	Apr 15	20;09	25.48	15429	Apr 30	11;59	219.73
15444	May 15	3;32	53.86	15459	May 30	3;26	248.24
15473	Jun 13	10;39	81.91	15488	Jun 28	17;36	276.48
15502	Jul 12	18;38	109.87	15518	Jul 28	6;18	304.66
15532	Aug 11	4;27	138.02	15547	Aug 26	17;33	333.01
15561	Sep 9	16;44	166.58	15577	Sep 25	3;41	1.70
15591	Oct 9	7;42	195.69	15606	Oct 24	13;19	30.86
15621	Nov 8	1;08	225.40	15635	Nov 22	23;10	60.47
15650	Dec 7	20;17	255.58	15665	Dec 22	9;49	90.42
			1628				
15680	Jan 6	15;37	285.96	15694	Jan 20	21;27	120.46
15710	Feb 5	9;17	316.17	15724	Feb 19	10;00	150.33
15739	Mar 5	23;58	345.94	15753	Mar 19	23;16	179.83
15769	Apr 4	11;26	15.11	15783	Apr 18	13;14	208.87
15798	May 3	20;21	43.70	15813	May 18	3;59	237.49
15828	Jun 2	3;45	71.85	15842	Jun 16	19;21	265.84
15857	Jul 1	10;41	99.79	15872	Jul 16	10;42	294.10
15886	Jul 30	18;05	127.78	15902	Aug 15	1;18	322.47
15916	Aug 29	2;44	156.05	15931	Sep 13	14;39	351.13
15945	Sep 27	13;28	184.82	15961	Oct 13	2;51	20.22
15975	Oct 27	3;02	214.19	15990	Nov 11	14;22	49.74
16004	Nov 25	19;46	244.14	16020	Dec 11	1;37	79.63
16034	Dec 25	15;01	274.47				
			1629				
				16049	Jan 9	12;41	109.66
16064	Jan 24	10;59	304.85	16078	Feb 7	23;27	139.56
16094	Feb 23	5;28	334.92	16108	Mar 9	9;57	169.11
16123	Mar 24	21;06	4.45	16137	Apr 7	20;37	198.19
16153	Apr 23	9;41	33.35	16167	May 7	8;06	226.84
16182	May 22	19;43	61.73	16196	Jun 5	21;02	255.18
16212	Jun 21	3;58	89.76	16226	Jul 5	11;26	283.41
16241	Jul 20	11;13	117.71	16256	Aug 4	2;52	311.73
16270	Aug 18	18;21	145.81	16285	Sep 2	18;35	340.34
16300	Sep 17	2;18	174.30	16315	Oct 2	10;05	9.37
16329	Oct 16	12;08	203.34	16345	Nov 1	1;01	38.87
16359	Nov 15	0;43	232.96	16374	Nov 30	15;05	68.77
16388	Dec 14	16;20	263.06	16404	Dec 30	3;57	98.84
			1630				
16418	Jan 13	10;20	293.39	16433	Jan 28	15;19	128.84
16448	Feb 12	5;15	323.60	16463	Feb 27	1;17	158.51
16477	Mar 13	23;30	353.40	16492	Mar 28	10;21	187.71
16507	Apr 12	16;00	22.65	16521	Apr 26	19;16	216.41
16537	May 12	6;09	51.34	16551	May 26	4;53	244.73
16566	Jun 10	17;49	79.59	16580	Jun 24	15;48	272.86
16596	Jul 10	3;15	107.61	16610	Jul 24	4;25	301.03
16625	Aug 8	11;13	135.67	16639	Aug 22	18;54	329.46
16654	Sep 6	18;47	164.00	16669	Sep 21	11;07	358.35
16684	Oct 6	3;02	192.80	16699	Oct 21	4;35	27.77
16713	Nov 4	12;53	222.15	16728	Nov 19	22;14	57.67
16743	Dec 4	0;47	251.98	16758	Dec 19	14;41	87.84
			1631				
16772	Jan 2	14;45	282.12	16788	Jan 18	4;57	118.00
16802	Feb 1	6;24	312.27	16817	Feb 16	16;48	147.87
16831	Mar 2	23;13	342.18	16847	Mar 18	2;41	177.26
16861	Apr 1	16;32	11.66	16876	Apr 16	11;16	206.13
16891	May 1	9;23	40.62	16905	May 15	19;16	234.52
16921	May 31	0;43	69.12	16935	Jun 14	3;20	262.60
16950	Jun 29	13;50	97.32	16964	Jul 13	12;14	290.60
16980	Jul 29	0;47	125.43	16993	Aug 11	22;53	318.79
17009	Aug 27	10;19	153.72	17023	Sep 10	12;07	347.39
17038	Sep 25	19;24	182.38	17053	Oct 10	4;18	16.58
17068	Oct 25	4;53	211.54	17082	Nov 8	22;53	46.35
17097	Nov 23	15;09	241.18	17112	Dec 8	18;11	76.54
17127	Dec 23	2;17	271.15				

NEW MOONS				FULL MOONS			
JD	DATE	TIME	LONG.	JD	DATE	TIME	LONG.
23+							
			1632	23+			
				17142	Jan 7	12;18	106.86
17156	Jan 21	14;22	301.20	17172	Feb 6	4;00	136.99
17186	Feb 20	3;34	331.09	17201	Mar 6	17;02	166.68
17215	Mar 20	18;06	0.63	17231	Apr 5	3;40	195.80
17245	Apr 19	9;47	29.73	17260	May 4	12;23	224.38
17275	May 19	1;51	58.40	17289	Jun 2	19;45	252.52
17304	Jun 17	17;20	86.74	17319	Jul 2	2;38	280.46
17334	Jul 17	7;36	114.96	17348	Jul 31	10;06	308.45
17363	Aug 15	20;37	143.28	17377	Aug 29	19;25	336.76
17393	Sep 14	8;39	171.90	17407	Sep 28	7;39	5.60
17422	Oct 13	20;04	200.96	17436	Oct 27	23;11	35.06
17452	Nov 12	7;04	230.48	17466	Nov 26	17;25	65.08
17481	Dec 11	17;44	260.34	17496	Dec 26	12;54	95.42
			1633				
17511	Jan 10	4;14	290.35	17526	Jan 25	7;57	125.76
17540	Feb 8	14;57	320.24	17556	Feb 24	1;14	155.77
17570	Mar 10	2;19	349.81	17585	Mar 25	15;52	185.24
17599	Apr 8	14;38	18.95	17615	Apr 24	3;34	214.10
17629	May 8	3;56	47.66	17644	May 23	12;35	242.43
17658	Jun 6	18;02	76.04	17673	Jun 21	19;47	270.42
17688	Jul 6	8;46	104.28	17703	Jul 21	2;19	298.34
17717	Aug 4	23;55	132.60	17732	Aug 19	9;28	326.45
17747	Sep 3	15;11	161.20	17761	Sep 17	18;22	354.99
17777	Oct 3	6;02	190.22	17791	Oct 17	5;45	24.10
17806	Nov 1	19;54	219.70	17820	Nov 15	19;54	53.80
17836	Dec 1	8;27	249.53	17850	Dec 15	12;36	83.96
17865	Dec 30	19;50	279.55				
			1634				
				17880	Jan 14	7;11	114.30
17895	Jan 29	6;25	309.51	17910	Feb 13	2;27	144.52
17924	Feb 27	16;34	339.18	17939	Mar 14	20;46	174.31
17954	Mar 29	2;33	8.40	17969	Apr 13	12;37	203.52
17983	Apr 27	12;36	37.14	17999	May 13	1;18	232.13
18012	May 26	23;11	65.49	18028	Jun 11	11;08	260.30
18042	Jun 25	10;58	93.65	18057	Jul 10	19;08	288.27
18072	Jul 25	0;33	121.86	18087	Aug 9	2;30	316.31
18101	Aug 23	16;05	150.34	18116	Sep 7	10;16	344.66
18131	Sep 22	8;58	179.27	18145	Oct 6	19;11	13.50
18161	Oct 22	2;00	208.69	18175	Nov 5	5;43	42.88
18190	Nov 20	18;03	238.54	18204	Dec 4	18;15	72.75
18220	Dec 20	8;32	268.63				
			1635				
				18234	Jan 3	9;08	102.93
18249	Jan 18	21;24	298.72	18264	Feb 2	2;14	133.14
18279	Feb 17	8;44	328.57	18293	Mar 3	20;37	163.11
18308	Mar 18	18;39	357.95	18323	Apr 2	14;39	192.59
18338	Apr 17	3;20	26.81	18353	May 2	6;52	221.51
18367	May 16	11;24	55.19	18382	May 31	20;35	249.94
18396	Jun 14	19;48	83.28	18412	Jun 30	8;03	278.07
18426	Jul 14	5;44	111.33	18441	Jul 29	17;55	306.15
18455	Aug 12	18;06	139.59	18471	Aug 28	2;55	334.42
18485	Sep 11	9;09	168.28	18500	Sep 26	11;41	3.08
18515	Oct 11	2;19	197.52	18529	Oct 25	20;46	32.23
18544	Nov 9	20;32	227.29	18559	Nov 24	6;46	61.87
18574	Dec 9	14;35	257.43	18588	Dec 23	18;15	91.85
			1636				
18604	Jan 8	7;24	287.70	18618	Jan 22	7;37	121.96
18633	Feb 6	22;09	317.78	18647	Feb 20	22;45	151.92
18663	Mar 7	10;20	347.42	18677	Mar 21	14;57	181.52
18692	Apr 5	20;01	16.50	18707	Apr 20	7;20	210.63
18722	May 5	3;50	45.02	18736	May 19	23;08	239.27
18751	Jun 3	10;48	73.15	18766	Jun 18	13;57	267.59
18780	Jul 2	18;07	101.10	18796	Jul 18	3;30	295.78
18810	Aug 1	2;53	129.15	18825	Aug 16	15;41	324.07
18839	Aug 30	13;54	157.54	18855	Sep 15	2;35	352.66
18869	Sep 29	3;36	186.45	18884	Oct 14	12;38	21.68
18898	Oct 28	19;58	215.96	18913	Nov 12	22;29	51.16
18928	Nov 27	14;32	246.01	18943	Dec 12	8;48	81.02
18958	Dec 27	10;05	276.36				
			1637				
				18972	Jan 10	19;55	111.05
18988	Jan 26	4;51	306.68	19002	Feb 9	7;53	140.99
19017	Feb 24	21;05	336.63	19031	Mar 10	20;32	170.60
19047	Mar 26	10;03	6.03	19061	Apr 9	9;51	199.77
19076	Apr 24	20;01	34.80	19090	May 8	23;56	228.50
19106	May 24	3;57	63.07	19120	Jun 7	14;51	256.90
19135	Jun 22	10;57	91.05	19150	Jul 7	6;15	285.16
19164	Jul 21	18;00	118.99	19179	Aug 5	21;27	313.49
19194	Aug 20	1;58	147.14	19209	Sep 4	11;42	342.06
19223	Sep 18	11;39	175.72	19239	Oct 4	0;45	11.02
19252	Oct 17	23;48	204.88	19268	Nov 2	12;49	40.43
19282	Nov 16	15;02	234.64	19298	Dec 2	0;22	70.24
19312	Dec 16	9;12	264.86	19327	Dec 31	11;37	100.25

NEW MOONS				FULL MOONS			
JD	DATE	TIME	LONG.	JD	DATE	TIME	LONG.
23+			**1638** 23+				
19342	Jan 15	5;03	295.26	19356	Jan 29	22;29	130.22
19372	Feb 14	0;26	325.47	19386	Feb 28	8;56	159.90
19401	Mar 15	17;29	355.20	19415	Mar 29	19;11	189.12
19431	Apr 14	7;29	24.32	19445	Apr 28	5;53	217.87
19460	May 13	18;39	52.87	19474	May 27	17;47	246.27
19490	Jun 12	3;41	81.00	19504	Jun 26	7;19	274.49
19519	Jul 11	11;22	108.95	19533	Jul 25	22;17	302.76
19548	Aug 9	18;33	136.98	19563	Aug 24	14;04	331.26
19578	Sep 8	2;05	165.33	19593	Sep 23	5;59	0.16
19607	Oct 7	10;59	194.18	19622	Oct 22	21;32	29.54
19636	Nov 5	22;13	223.60	19652	Nov 21	12;22	59.34
19666	Dec 5	12;21	253.55	19682	Dec 21	2;03	89.41
			1639				
19696	Jan 4	5;15	283.82	19711	Jan 19	14;13	119.47
19725	Feb 2	23;45	314.08	19741	Feb 18	0;46	149.27
19755	Mar 4	18;19	344.04	19770	Mar 19	10;01	178.62
19785	Apr 3	11;38	13.48	19799	Apr 17	18;41	207.47
19815	May 3	2;54	42.35	19829	May 17	3;38	235.88
19844	Jun 1	15;47	70.74	19858	Jun 15	13;38	264.03
19874	Jul 1	2;20	98.83	19888	Jul 15	1;16	292.14
19903	Jul 30	11;03	126.86	19917	Aug 13	14;47	320.45
19932	Aug 28	18;51	155.09	19947	Sep 12	6;16	349.17
19962	Sep 27	2;48	183.72	19976	Oct 11	23;25	18.41
19991	Oct 26	11;56	212.89	20006	Nov 10	17;23	48.19
20020	Nov 24	22;52	242.58	20036	Dec 10	10;50	78.32
20050	Dec 24	11;50	272.63				
			1640	20066	Jan 9	2;24	108.54
20080	Jan 23	2;34	302.80	20095	Feb 7	15;27	138.55
20109	Feb 21	18;39	332.80	20125	Mar 8	2;09	168.12
20139	Mar 22	11;33	2.41	20154	Apr 6	11;08	197.16
20169	Apr 21	4;32	31.53	20183	May 5	19;10	225.68
20198	May 20	20;38	60.17	20213	Jun 4	2;57	253.83
20228	Jun 19	10;55	88.46	20242	Jul 3	11;14	281.82
20257	Jul 18	23;02	116.60	20271	Aug 1	20;50	309.90
20287	Aug 17	9;21	144.82	20301	Aug 31	8;42	338.33
20316	Sep 15	18;46	173.35	20330	Sep 29	23;30	7.29
20346	Oct 15	4;12	202.35	20360	Oct 29	17;10	36.87
20375	Nov 13	14;13	231.84	20390	Nov 28	12;31	66.96
20405	Dec 13	1;00	261.73	20420	Dec 28	7;35	97.30
			1641				
20434	Jan 11	12;34	291.78	20450	Jan 27	0;38	127.54
20464	Feb 10	1;00	321.73	20479	Feb 25	14;58	157.41
20493	Mar 11	14;36	351.38	20509	Mar 27	2;39	186.74
20523	Apr 10	5;27	20.60	20538	Apr 25	12;09	215.49
20552	May 9	21;11	49.39	20567	May 24	19;59	243.75
20582	Jun 8	12;56	77.82	20597	Jun 23	2;55	271.73
20612	Jul 8	3;53	106.06	20626	Jul 22	9;55	299.66
20641	Aug 6	17;39	134.33	20655	Aug 20	18;11	327.83
20671	Sep 5	6;23	162.85	20685	Sep 19	4;55	356.46
20700	Oct 4	18;22	191.78	20714	Oct 18	18;53	25.70
20730	Nov 3	5;50	221.17	20744	Nov 17	11;59	55.55
20759	Dec 2	16;49	250.96	20774	Dec 17	7;07	85.82
			1642				
20789	Jan 1	3;25	280.95	20804	Jan 16	2;36	116.20
20818	Jan 30	13;54	310.90	20833	Feb 14	20;52	146.36
20848	Mar 1	0;43	340.58	20863	Mar 16	12;49	176.03
20877	Mar 30	12;15	9.85	20893	Apr 15	1;54	205.10
20907	Apr 29	0;46	38.67	20922	May 14	12;07	233.60
20936	May 28	14;15	67.12	20951	Jun 12	20;04	261.69
20966	Jun 27	4;33	95.37	20981	Jul 12	2;47	289.60
20995	Jul 26	19;32	123.64	21010	Aug 10	9;32	317.62
21025	Aug 25	10;58	152.14	21039	Sep 8	17;32	345.99
21055	Sep 24	2;22	181.03	21069	Oct 8	3;43	14.90
21084	Oct 23	17;06	210.38	21098	Nov 6	16;34	44.40
21114	Nov 22	6;34	240.14	21128	Dec 6	8;04	74.41
21143	Dec 21	18;39	270.13				
			1643	21158	Jan 5	1;47	104.71
21173	Jan 20	5;36	300.15	21187	Feb 3	20;46	134.99
21202	Feb 18	15;50	329.93	21217	Mar 5	15;37	164.95
21232	Mar 20	1;41	359.30	21247	Apr 4	8;44	194.37
21261	Apr 18	11;25	28.18	21276	May 3	22;58	223.20
21290	May 17	21;25	56.62	21306	Jun 2	10;07	251.51
21320	Jun 16	8;19	84.81	21335	Jul 1	18;56	279.53
21349	Jul 15	20;50	112.96	21365	Jul 31	2;34	307.52
21379	Aug 14	11;25	141.31	21394	Aug 29	10;11	335.75
21409	Sep 13	3;54	170.09	21423	Sep 27	18;36	4.41
21438	Oct 12	21;16	199.35	21453	Oct 27	4;25	33.61
21468	Nov 11	14;12	229.09	21482	Nov 25	15;58	63.33
21498	Dec 11	5;43	259.15	21512	Dec 25	5;38	93.41

NEW MOONS				FULL MOONS			
JD	DATE	TIME	LONG.	JD	DATE	TIME	LONG.
23+			**1644** 23+				
21527	Jan 9	19;29	289.29	21541	Jan 23	21;32	123.62
21557	Feb 8	7;34	319.24	21571	Feb 22	15;14	153.68
21586	Mar 8	18;04	348.80	21601	Mar 23	9;26	183.33
21616	Apr 7	3;09	17.83	21631	Apr 22	2;31	212.44
21645	May 6	11;14	46.35	21660	May 21	17;26	241.03
21674	Jun 4	19;09	74.50	21690	Jun 20	6;00	269.25
21704	Jul 4	4;03	102.52	21719	Jul 19	16;42	297.33
21733	Aug 2	15;03	130.66	21749	Aug 18	2;14	325.53
21763	Sep 1	4;47	159.17	21778	Sep 16	11;15	354.06
21792	Sep 30	21;08	188.22	21807	Oct 15	20;19	23.05
21822	Oct 30	15;10	217.83	21837	Nov 14	5;56	52.54
21852	Nov 29	9;39	247.89	21866	Dec 13	16;40	82.43
21882	Dec 29	3;21	278.17				
			1645	21896	Jan 12	5;01	112.50
21911	Jan 27	19;17	308.36	21925	Feb 10	19;07	142.52
21941	Feb 26	8;44	338.18	21955	Mar 12	10;37	172.24
21970	Mar 27	19;31	7.46	21985	Apr 11	2;44	201.49
22000	Apr 26	4;02	36.15	22014	May 10	18;41	230.28
22029	May 25	11;09	64.38	22044	Jun 9	9;56	258.68
22058	Jun 23	18;03	92.36	22074	Jul 9	0;12	286.90
22088	Jul 23	1;55	120.33	22103	Aug 7	13;15	315.15
22117	Aug 21	11;44	148.57	22133	Sep 6	1;01	343.64
22147	Sep 20	0;07	177.28	22162	Oct 5	11;41	12.53
22176	Oct 19	15;16	206.58	22191	Nov 3	21;47	41.87
22206	Nov 18	8;56	236.46	22221	Dec 3	7;57	71.63
22236	Dec 18	4;17	266.74				
			1646	22250	Jan 1	18;40	101.63
22265	Jan 16	23;44	297.12	22280	Jan 31	6;07	131.62
22295	Feb 15	17;23	327.25	22309	Mar 1	18;14	161.34
22325	Mar 17	7;57	356.85	22339	Mar 31	6;55	190.64
22354	Apr 15	19;14	25.83	22368	Apr 29	20;17	219.48
22384	May 15	3;57	54.26	22398	May 29	10;34	247.96
22413	Jun 13	11;14	82.31	22428	Jun 28	1;43	276.24
22442	Jul 12	18;08	110.24	22457	Jul 27	17;11	304.53
22472	Aug 11	1;36	138.30	22487	Aug 26	8;12	333.02
22501	Sep 9	10;25	166.71	22516	Sep 24	22;08	1.87
22530	Oct 8	21;20	195.66	22546	Oct 24	10;56	31.16
22560	Nov 7	11;02	225.21	22575	Nov 22	22;56	60.86
22590	Dec 7	3;51	255.28	22605	Dec 22	10;27	90.84
			1647				
22619	Jan 5	23;04	285.65	22634	Jan 20	21;31	120.85
22649	Feb 4	18;52	315.96	22664	Feb 19	8;02	150.64
22679	Mar 6	13;09	345.88	22693	Mar 20	18;06	180.00
22709	Apr 5	4;36	15.22	22723	Apr 19	4;13	208.88
22738	May 4	17;02	43.95	22752	May 18	15;10	237.36
22768	Jun 3	3;02	72.20	22782	Jun 17	3;40	265.60
22797	Jul 2	11;20	100.20	22811	Jul 16	17;53	293.82
22826	Jul 31	18;44	128.19	22841	Aug 15	9;26	322.22
22856	Aug 30	2;06	156.42	22871	Sep 14	1;33	350.99
22885	Sep 28	10;20	185.09	22900	Oct 13	17;37	20.23
22914	Oct 27	20;24	214.31	22930	Nov 12	9;08	49.92
22944	Nov 26	9;06	244.08	22959	Dec 11	23;39	79.95
22974	Dec 26	0;40	274.25				
			1648	22989	Jan 10	12;42	110.05
23003	Jan 24	18;25	304.54	23019	Feb 9	0;01	139.97
23033	Feb 23	12;57	334.62	23048	Mar 9	9;43	169.48
23063	Mar 24	6;51	4.24	23077	Apr 7	18;23	198.49
23092	Apr 22	23;04	33.31	23107	May 7	2;51	227.01
23122	May 22	13;07	61.85	23136	Jun 5	12;03	255.20
23152	Jun 21	0;50	90.02	23165	Jul 4	22;40	283.29
23181	Jul 20	10;30	118.07	23195	Aug 3	11;10	311.49
23210	Aug 18	18;48	146.22	23225	Sep 2	1;44	340.05
23240	Sep 17	2;44	174.72	23254	Oct 1	18;17	9.12
23269	Oct 16	11;21	203.70	23284	Oct 31	12;12	38.74
23298	Nov 14	21;27	233.22	23314	Nov 30	6;20	68.79
23328	Dec 14	9;26	263.17	23343	Dec 29	23;08	99.04
			1649				
23357	Jan 12	23;14	293.31	23373	Jan 28	13;31	129.16
23387	Feb 11	14;30	323.37	23403	Feb 27	1;16	158.90
23417	Mar 13	6;48	353.10	23432	Mar 28	10;53	188.12
23446	Apr 11	23;35	22.38	23461	Apr 26	19;08	216.79
23476	May 11	16;05	51.16	23491	May 26	2;48	245.04
23506	Jun 10	7;19	79.56	23520	Jun 24	10;37	273.04
23535	Jul 9	20;37	107.74	23549	Jul 23	19;23	301.05
23565	Aug 8	7;56	135.93	23579	Aug 22	6;00	329.33
23594	Sep 6	17;55	164.36	23608	Sep 20	19;21	358.10
23624	Oct 6	3;30	193.21	23638	Oct 20	11;47	27.47
23653	Nov 4	13;24	222.56	23668	Nov 19	6;40	57.41
23682	Dec 3	23;56	252.34	23698	Dec 19	2;16	87.71

Left Panel

NEW MOONS / FULL MOONS — 1650 (23+)

NEW MOONS JD	DATE	TIME	LONG.	FULL MOONS JD	DATE	TIME	LONG.
1650							
23712	Jan 2	11;06	282.36	23727	Jan 17	20;33	118.04
23741	Jan 31	22;56	312.36	23757	Feb 16	12;15	148.07
23771	Mar 2	11;40	342.10	23787	Mar 18	1;09	177.60
23801	Apr 1	1;36	11.44	23816	Apr 16	11;33	206.53
23830	Apr 30	16;41	40.33	23845	May 15	20;02	234.94
23860	May 30	8;20	68.86	23875	Jun 14	3;14	262.98
23889	Jun 28	23;44	97.14	23904	Jul 13	10;02	290.90
23919	Jul 28	14;13	125.40	23933	Aug 11	17;32	318.96
23949	Aug 27	3;40	153.83	23963	Sep 10	2;58	347.41
23978	Sep 25	16;17	182.64	23992	Oct 9	15;20	16.43
24008	Oct 25	4;16	211.90	24022	Nov 8	7;03	46.08
24037	Nov 23	15;43	241.60	24052	Dec 8	1;25	76.23
24067	Dec 23	2;36	271.56				
1651							
				24081	Jan 6	20;57	106.61
24096	Jan 21	13;04	301.54	24111	Feb 5	15;56	136.88
24125	Feb 19	23;29	331.32	24141	Mar 7	9;04	166.74
24155	Mar 21	10;22	0.71	24170	Apr 5	23;32	196.02
24184	Apr 19	22;05	29.64	24200	May 5	11;05	224.71
24214	May 19	10;50	58.18	24229	Jun 3	20;01	252.92
24244	Jun 18	0;34	86.46	24259	Jul 3	3;13	280.87
24273	Jul 17	15;11	114.70	24288	Aug 1	9;51	308.83
24303	Aug 16	6;32	143.10	24317	Aug 30	17;11	337.06
24332	Sep 14	22;15	171.86	24347	Sep 29	2;18	5.77
24362	Oct 14	13;43	201.09	24376	Oct 28	13;53	35.07
24392	Nov 13	4;09	230.75	24406	Nov 27	4;09	64.92
24421	Dec 12	17;07	260.71	24435	Dec 26	20;48	95.14
1652							
24451	Jan 11	4;39	290.76	24465	Jan 25	15;11	125.45
24480	Feb 9	15;09	320.64	24495	Feb 24	10;09	155.54
24510	Mar 10	0;59	350.15	24525	Mar 25	4;11	185.15
24539	Apr 8	10;30	19.18	24554	Apr 23	19;50	214.18
24568	May 7	20;04	47.73	24584	May 23	8;27	242.65
24598	Jun 6	6;13	75.96	24613	Jun 21	18;20	270.75
24627	Jul 5	17;41	104.07	24643	Jul 21	2;30	298.74
24657	Aug 4	7;10	132.32	24672	Aug 19	10;08	326.87
24686	Sep 2	22;52	160.94	24701	Sep 17	18;14	355.38
24716	Oct 2	16;09	190.04	24731	Oct 17	3;28	24.40
24746	Nov 1	9;44	219.66	24760	Nov 15	14;12	53.96
24776	Dec 1	2;17	249.67	24790	Dec 15	2;48	83.93
24805	Dec 30	17;07	279.83				
1653							
				24819	Jan 13	17;30	114.11
24835	Jan 29	6;04	309.89	24849	Feb 12	10;13	144.23
24864	Feb 27	17;16	339.61	24879	Mar 14	4;07	174.02
24894	Mar 29	2;52	8.81	24908	Apr 12	21;43	203.31
24923	Apr 27	11;11	37.48	24938	May 12	13;40	232.06
24952	May 26	18;52	65.72	24968	Jun 11	3;22	260.39
24982	Jun 25	2;59	93.73	24997	Jul 10	15;00	288.50
25011	Jul 24	12;43	121.78	25027	Aug 9	1;12	316.65
25041	Aug 23	1;03	150.13	25056	Sep 7	10;38	345.06
25070	Sep 21	16;15	178.98	25085	Oct 6	19;50	13.91
25100	Oct 21	9;44	208.41	25115	Nov 5	5;18	43.25
25130	Nov 20	4;21	238.35	25144	Dec 4	15;30	73.03
25159	Dec 19	22;45	268.60				
1654							
				25174	Jan 3	2;58	103.06
25189	Jan 18	15;46	298.88	25203	Feb 1	16;03	133.11
25219	Feb 17	6;31	328.87	25233	Mar 3	6;41	162.92
25248	Mar 18	18;33	358.34	25262	Apr 1	22;18	192.32
25278	Apr 17	3;59	27.23	25292	May 1	14;10	221.24
25307	May 16	11;32	55.59	25322	May 31	5;40	249.74
25336	Jun 14	18;16	83.61	25351	Jun 29	20;27	278.00
25366	Jul 14	1;28	111.55	25381	Jul 29	10;16	306.23
25395	Aug 12	10;12	139.66	25410	Aug 27	22;55	334.65
25424	Sep 10	21;18	168.19	25440	Sep 26	10;22	3.42
25454	Oct 10	11;09	197.28	25469	Oct 25	20;56	32.63
25484	Nov 9	3;44	226.97	25499	Nov 24	7;11	62.28
25513	Dec 8	22;29	257.15	25528	Dec 23	17;39	92.22
1655							
25543	Jan 7	18;09	287.53	25558	Jan 22	4;41	122.23
25573	Feb 6	12;54	317.79	25587	Feb 20	16;18	152.05
25603	Mar 8	5;03	347.60	25617	Mar 22	4;26	181.48
25632	Apr 6	17;51	16.80	25646	Apr 20	17;08	210.45
25662	May 6	3;39	45.41	25676	May 20	6;41	239.01
25691	Jun 4	11;27	73.57	25705	Jun 18	21;16	267.32
25720	Jul 3	18;24	101.50	25735	Jul 18	12;40	295.58
25750	Aug 2	1;31	129.49	25765	Aug 17	4;09	324.00
25779	Aug 31	9;38	157.76	25794	Sep 15	18;56	352.74
25808	Sep 29	19;30	186.51	25824	Oct 15	8;35	21.90
25838	Oct 29	7;49	215.85	25853	Nov 13	21;13	51.50
25867	Nov 27	23;08	245.75	25883	Dec 13	9;07	81.42
25897	Dec 27	17;17	276.03				

Right Panel

NEW MOONS JD	DATE	TIME	LONG.	FULL MOONS JD	DATE	TIME	LONG.
1656 (23+)							
				25912	Jan 11	20;28	111.46
25927	Jan 26	13;00	306.40	25942	Feb 10	7;11	141.34
25957	Feb 25	8;10	336.48	25971	Mar 10	17;16	170.86
25987	Mar 26	1;01	6.04	26001	Apr 9	3;01	199.88
26016	Apr 24	14;51	34.98	26030	May 8	13;11	228.46
26046	May 24	1;56	63.39	26060	Jun 7	0;37	256.73
26075	Jun 22	11;00	91.45	26089	Jul 6	13;52	284.91
26104	Jul 21	18;50	119.42	26119	Aug 5	4;49	313.22
26134	Aug 20	2;14	147.55	26148	Sep 3	20;53	341.86
26163	Sep 18	10;03	176.06	26178	Oct 3	13;17	10.94
26192	Oct 17	19;12	205.09	26208	Nov 2	5;25	40.52
26222	Nov 16	6;36	234.67	26237	Dec 1	20;44	70.47
26251	Dec 15	20;45	264.72	26267	Dec 31	10;43	100.60
1657							
26281	Jan 14	13;26	294.99	26296	Jan 29	22;56	130.63
26311	Feb 13	7;36	325.15	26326	Feb 28	9;18	160.30
26341	Mar 15	1;46	354.94	26355	Mar 29	18;12	189.46
26370	Apr 13	18;44	24.19	26385	Apr 28	2;27	218.12
26400	May 13	9;50	52.90	26414	May 27	10;58	246.39
26429	Jun 11	22;43	81.19	26443	Jun 25	20;38	274.46
26459	Jul 11	9;27	109.27	26473	Jul 25	8;05	302.59
26488	Aug 9	18;29	137.38	26502	Aug 23	21;37	330.99
26518	Sep 8	2;40	165.75	26532	Sep 22	13;20	359.88
26547	Oct 7	11;00	194.57	26562	Oct 22	6;52	29.32
26576	Nov 5	20;27	223.92	26592	Nov 21	1;18	59.27
26606	Dec 5	7;33	253.74	26621	Dec 20	19;08	89.50
1658							
26635	Jan 3	20;26	283.83	26651	Jan 19	10;54	119.72
26665	Feb 2	10;51	313.93	26680	Feb 17	23;55	149.62
26695	Mar 4	2;26	343.77	26710	Mar 19	10;25	179.03
26724	Apr 2	18;46	13.18	26739	Apr 17	19;06	207.87
26754	May 2	11;18	42.11	26769	May 17	2;49	236.23
26784	Jun 1	3;11	70.63	26798	Jun 15	10;20	264.28
26813	Jun 30	17;32	98.87	26827	Jul 14	18;27	292.26
26843	Jul 30	5;58	127.05	26857	Aug 13	4;01	320.41
26872	Aug 28	16;45	155.40	26886	Sep 11	15;57	348.98
26902	Sep 27	2;41	184.11	26916	Oct 11	6;56	18.13
26931	Oct 26	12;35	213.31	26946	Nov 10	0;52	47.89
26960	Nov 24	22;57	242.97	26975	Dec 9	20;30	78.11
26990	Dec 24	9;51	272.94				
1659							
				27005	Jan 8	15;45	108.48
27019	Jan 22	21;15	302.96	27035	Feb 7	8;50	138.66
27049	Feb 21	9;18	332.79	27064	Mar 8	23;04	168.38
27078	Mar 22	22;19	2.25	27094	Apr 7	10;34	197.51
27108	Apr 21	12;34	31.27	27123	May 6	19;50	226.09
27138	May 21	3;48	59.88	27153	Jun 5	3;31	254.24
27167	Jun 19	19;19	88.23	27182	Jul 4	10;21	282.17
27197	Jul 19	10;22	116.48	27211	Aug 2	17;22	310.15
27227	Aug 18	0;31	144.85	27241	Sep 1	1;45	338.44
27256	Sep 16	13;47	173.54	27270	Sep 30	12;37	7.25
27286	Oct 16	2;22	202.67	27300	Oct 30	2;45	36.67
27315	Nov 14	14;20	232.25	27329	Nov 28	19;59	66.66
27345	Dec 14	1;38	262.15	27359	Dec 28	15;10	96.99
1660							
27374	Jan 12	12;17	292.16	27389	Jan 27	10;34	127.34
27403	Feb 10	22;34	322.02	27419	Feb 26	4;41	157.37
27433	Mar 11	8;57	351.53	27448	Mar 26	20;28	186.87
27462	Apr 9	19;56	20.60	27478	Apr 25	9;23	215.76
27492	May 9	7;53	49.23	27507	May 24	19;31	244.11
27521	Jun 7	20;54	77.56	27537	Jun 23	3;28	272.13
27551	Jul 7	10;59	105.78	27566	Jul 22	10;16	300.07
27581	Aug 6	2;03	134.10	27595	Aug 20	17;12	328.18
27610	Sep 4	17;50	162.74	27625	Sep 19	1;26	356.72
27640	Oct 4	9;48	191.82	27654	Oct 18	11;51	25.81
27670	Nov 3	1;08	221.38	27684	Nov 17	0;51	55.48
27699	Dec 2	15;06	251.29	27713	Dec 16	16;21	85.59
1661							
27729	Jan 1	3;26	281.35	27743	Jan 15	9;54	115.89
27758	Jan 30	14;23	311.32	27773	Feb 14	4;35	146.07
27788	Mar 1	0;23	340.96	27802	Mar 15	23;06	175.86
27817	Mar 30	9;49	10.14	27832	Apr 14	15;57	205.09
27846	Apr 28	19;04	38.83	27862	May 14	6;02	233.74
27876	May 28	4;37	67.12	27891	Jun 12	17;13	261.96
27905	Jun 26	15;09	95.23	27921	Jul 12	2;11	289.96
27935	Jul 26	3;29	123.39	27950	Aug 10	10;05	318.03
27964	Aug 24	18;07	151.85	27979	Sep 8	18;01	346.40
27994	Sep 23	10;55	180.79	28009	Oct 8	2;48	15.25
28024	Oct 23	4;47	210.26	28038	Nov 6	12;53	44.64
28053	Nov 21	22;15	240.18	28068	Dec 6	0;34	74.49
28083	Dec 21	14;10	270.34				

Left Panel

NEW MOONS				FULL MOONS			
JD	DATE	TIME	LONG.	JD	DATE	TIME	LONG.
23+			**1662**	23+			
				28097	Jan 4	14;08	104.62
28113	Jan 20	4;07	300.48	28127	Feb 3	5;42	134.76
28142	Feb 18	16;08	330.34	28156	Mar 4	22;55	164.66
28172	Mar 20	2;23	359.72	28186	Apr 3	16;36	194.11
28201	Apr 18	11;07	28.55	28216	May 3	9;21	223.03
28230	May 17	18;50	56.91	28246	Jun 2	0;08	251.49
28260	Jun 16	2;26	84.96	28275	Jul 1	12;48	279.67
28289	Jul 15	11;08	112.95	28304	Jul 30	23;48	307.79
28318	Aug 13	22;03	141.16	28334	Aug 29	9;45	336.12
28348	Sep 12	11;52	169.82	28363	Sep 27	19;15	4.83
28378	Oct 12	4;28	199.05	28393	Oct 27	4;45	34.02
28407	Nov 10	22;52	228.85	28422	Nov 25	14;39	63.67
28437	Dec 10	17;42	259.04	28452	Dec 25	1;28	93.64
			1663				
28467	Jan 9	11;38	289.36	28481	Jan 23	13;36	123.69
28497	Feb 8	3;37	319.48	28511	Feb 22	3;16	153.58
28526	Mar 9	16;58	349.15	28540	Mar 23	18;11	183.10
28556	Apr 8	3;32	18.24	28570	Apr 22	9;44	212.15
28585	May 7	11;47	46.76	28600	May 22	1;16	240.77
28614	Jun 5	18;41	74.87	28629	Jun 20	16;23	269.09
28644	Jul 5	1;26	102.80	28659	Jul 20	6;48	297.32
28673	Aug 3	9;16	130.81	28688	Aug 18	20;15	325.67
28702	Sep 1	19;08	159.17	28718	Sep 17	8;34	354.33
28732	Oct 1	7;40	188.05	28747	Oct 16	19;48	23.42
28761	Oct 30	23;00	217.55	28777	Nov 15	6;21	52.95
28791	Nov 29	16;52	247.57	28806	Dec 14	16;47	82.83
28821	Dec 29	12;20	277.92				
			1664				
				28836	Jan 13	3;31	112.84
28851	Jan 28	7;47	308.27	28865	Feb 11	14;43	142.74
28881	Feb 27	1;21	338.27	28895	Mar 12	2;22	172.29
28910	Mar 27	15;46	7.70	28924	Apr 10	14;28	201.39
28940	Apr 26	2;53	36.50	28954	May 10	3;16	230.05
28969	May 25	11;29	64.79	28983	Jun 8	17;07	258.40
28998	Jun 23	18;42	92.77	29013	Jul 8	8;06	286.65
29028	Jul 23	1;38	120.71	29042	Aug 6	23;44	315.00
29057	Aug 21	9;13	148.86	29072	Sep 5	15;11	343.63
29086	Sep 19	18;13	177.43	29102	Oct 5	5;43	12.67
29116	Oct 19	5;19	206.56	29131	Nov 3	19;06	42.15
29145	Nov 17	19;10	236.27	29161	Dec 3	7;33	72.01
29175	Dec 17	11;59	266.45				
			1665				
				29190	Jan 1	19;17	102.05
29205	Jan 16	7;05	296.81	29220	Jan 31	6;18	132.01
29235	Feb 15	2;41	327.02	29249	Mar 1	16;33	161.67
29264	Mar 16	20;43	356.78	29279	Mar 31	2;10	190.84
29294	Apr 15	11;58	25.93	29308	Apr 29	11;45	219.53
29324	May 15	0;18	54.51	29337	May 28	22;12	247.86
29353	Jun 13	10;17	82.67	29367	Jun 27	10;20	276.02
29382	Jul 12	18;41	110.65	29397	Jul 27	0;25	304.25
29412	Aug 11	2;19	138.71	29426	Aug 25	16;06	332.77
29441	Sep 9	9;58	167.08	29456	Sep 24	8;38	1.71
29470	Oct 8	18;30	195.93	29486	Oct 24	1;15	31.14
29500	Nov 7	4;47	225.34	29515	Nov 22	17;19	61.01
29529	Dec 6	17;34	255.24	29545	Dec 22	8;13	91.14
			1666				
29559	Jan 5	9;00	285.44	29574	Jan 20	21;25	121.24
29589	Feb 4	2;26	315.66	29604	Feb 19	8;38	151.06
29618	Mar 5	20;33	345.59	29633	Mar 20	18;02	180.39
29648	Apr 4	14;02	15.02	29663	Apr 19	2;18	209.20
29678	May 4	5;59	43.90	29692	May 18	10;22	237.56
29707	Jun 2	19;58	72.32	29721	Jun 16	19;11	265.65
29737	Jul 2	7;49	100.45	29751	Jul 16	5;34	293.72
29766	Jul 31	17;46	128.54	29780	Aug 14	18;00	321.99
29796	Aug 30	2;27	156.82	29810	Sep 13	8;43	350.70
29825	Sep 28	10;48	185.49	29840	Oct 13	1;35	19.95
29854	Oct 27	19;46	214.67	29869	Nov 11	19;57	49.76
29884	Nov 26	6;07	244.35	29899	Dec 11	14;29	79.95
29913	Dec 25	18;08	274.37				
			1667				
				29929	Jan 10	7;32	110.23
29943	Jan 24	7;42	304.48	29958	Feb 8	21;58	140.28
29972	Feb 22	22;30	334.40	29988	Mar 10	9;35	169.87
30002	Mar 24	14;14	3.95	30017	Apr 8	18;56	198.90
30032	Apr 23	6;30	33.03	30047	May 8	2;54	227.40
30061	May 22	22;40	61.66	30076	Jun 6	10;18	255.53
30091	Jun 21	13;52	89.98	30105	Jul 5	17;55	283.49
30121	Jul 21	3;23	118.18	30135	Aug 4	2;37	311.54
30150	Aug 19	15;08	146.47	30164	Sep 2	13;17	339.94
30180	Sep 18	1;38	175.06	30194	Oct 2	2;46	8.87
30209	Oct 17	11;43	204.11	30223	Oct 31	19;25	38.43
30238	Nov 15	22;01	233.64	30253	Nov 30	14;33	68.52
30268	Dec 15	8;45	263.53	30283	Dec 30	10;20	98.89

Right Panel

NEW MOONS				FULL MOONS			
JD	DATE	TIME	LONG.	JD	DATE	TIME	LONG.
23+			**1668**	23+			
30297	Jan 13	19;52	293.56	30313	Jan 29	4;42	129.18
30327	Feb 12	7;25	323.46	30342	Feb 27	20;21	159.08
30356	Mar 12	19;38	353.03	30372	Mar 28	9;05	188.43
30386	Apr 11	8;57	22.17	30401	Apr 26	19;18	217.19
30415	May 10	23;29	50.89	30431	May 26	3;36	245.46
30445	Jun 9	14;48	79.29	30460	Jun 24	10;43	273.44
30475	Jul 9	6;09	107.56	30489	Jul 23	17;31	301.37
30504	Aug 7	20;54	135.88	30519	Aug 22	1;06	329.53
30534	Sep 6	10;50	164.46	30548	Sep 20	10;40	358.13
30564	Oct 6	0;02	193.46	30577	Oct 19	23;12	27.34
30593	Nov 4	12;35	222.92	30607	Nov 18	15;02	57.15
30623	Dec 4	0;25	252.75	30637	Dec 18	9;28	87.39
			1669				
30652	Jan 2	11;28	282.77	30667	Jan 17	4;57	117.77
30681	Jan 31	21;50	312.70	30696	Feb 15	23;48	147.94
30711	Mar 2	7;55	342.34	30726	Mar 17	16;44	177.64
30740	Mar 31	18;17	11.53	30756	Apr 16	7;02	206.74
30770	Apr 30	5;27	40.27	30785	May 15	18;28	235.27
30799	May 29	17;40	68.66	30815	Jun 14	3;23	263.38
30829	Jun 28	7;05	96.88	30844	Jul 13	10;39	291.32
30858	Jul 27	21;39	125.13	30873	Aug 11	17;27	319.35
30888	Aug 26	13;14	153.65	30903	Sep 10	1;01	347.73
30918	Sep 25	5;26	182.59	30932	Oct 9	10;23	16.62
30947	Oct 24	21;30	212.01	30961	Nov 7	22;10	46.09
30977	Nov 23	12;29	241.85	30991	Dec 7	12;29	76.07
31007	Dec 23	1;48	271.91				
			1670				
				31021	Jan 6	5;01	106.32
31036	Jan 21	13;26	301.95	31050	Feb 4	23;07	136.56
31065	Feb 19	23;46	331.72	31080	Mar 6	17;43	166.49
31095	Mar 21	9;16	1.05	31110	Apr 5	11;24	195.92
31124	Apr 19	18;20	29.88	31140	May 5	2;52	224.76
31154	May 19	3;26	58.27	31169	Jun 3	15;27	253.12
31183	Jun 17	13;12	86.40	31199	Jul 3	1;29	281.19
31213	Jul 17	0;26	114.50	31228	Aug 1	9;54	309.22
31242	Aug 15	13;52	142.82	31257	Aug 30	17;52	337.48
31272	Sep 14	5;47	171.59	31287	Sep 29	2;20	6.16
31301	Oct 13	23;29	200.89	31316	Oct 28	11;53	35.38
31331	Nov 12	17;35	230.70	31345	Nov 26	22;49	65.09
31361	Dec 12	10;35	260.83	31375	Dec 26	11;23	95.13
			1671				
31391	Jan 11	1;39	291.03	31405	Jan 25	1;49	125.28
31420	Feb 9	14;38	321.01	31434	Feb 23	18;04	155.26
31450	Mar 11	1;38	350.57	31464	Mar 25	11;26	184.85
31479	Apr 9	10;56	19.58	31494	Apr 24	4;36	213.95
31508	May 8	18;54	48.07	31523	May 23	20;20	242.55
31538	Jun 7	2;17	76.19	31553	Jun 22	10;04	270.81
31567	Jul 6	10;10	104.16	31582	Jul 21	21;57	298.94
31596	Aug 4	19;48	132.26	31612	Aug 20	8;33	327.20
31626	Sep 3	8;09	160.73	31641	Sep 18	18;27	355.78
31655	Oct 2	23;32	189.76	31671	Oct 18	4;08	24.83
31685	Nov 1	17;20	219.38	31700	Nov 16	13;57	54.34
31715	Dec 1	12;17	249.47	31730	Dec 16	0;19	84.22
31745	Dec 31	6;56	279.79				
			1672				
				31759	Jan 14	11;41	114.26
31775	Jan 30	0;03	310.03	31789	Feb 13	0;25	144.21
31804	Feb 28	14;45	339.89	31818	Mar 13	14;30	173.85
31834	Mar 29	2;36	9.19	31848	Apr 12	5;31	203.04
31863	Apr 27	11;48	37.89	31877	May 11	20;53	231.78
31892	May 26	19;08	66.12	31907	Jun 10	12;07	260.18
31922	Jun 25	1;44	94.07	31937	Jul 10	2;57	288.41
31951	Jul 24	8;51	122.02	31966	Aug 8	17;04	316.72
31980	Aug 22	17;38	150.23	31996	Sep 7	6;14	345.28
32010	Sep 21	4;51	178.91	32025	Oct 6	18;16	14.25
32039	Oct 20	18;52	208.19	32055	Nov 5	5;21	43.65
32069	Nov 19	11;37	238.04	32084	Dec 4	15;57	73.44
32099	Dec 19	6;29	268.31				
			1673				
				32114	Jan 3	2;32	103.43
32129	Jan 18	2;11	298.70	32143	Feb 1	13;24	133.39
32158	Feb 16	20;51	328.85	32173	Mar 3	0;38	163.06
32188	Mar 18	12;52	358.50	32202	Apr 1	12;13	192.29
32218	Apr 17	1;30	27.52	32232	May 1	0;20	221.07
32247	May 16	11;10	55.97	32261	May 30	13;23	249.48
32276	Jun 14	18;54	84.03	32291	Jun 29	3;41	277.73
32306	Jul 14	1;52	111.95	32320	Jul 28	19;06	306.02
32335	Aug 12	9;06	140.01	32350	Aug 27	10;56	334.55
32364	Sep 10	17;24	168.42	32380	Sep 26	2;17	3.48
32394	Oct 10	3;29	197.36	32409	Oct 25	16;33	32.84
32423	Nov 8	15;58	226.88	32439	Nov 24	5;41	62.61
32453	Dec 8	7;20	256.90	32468	Dec 23	17;54	92.63

Left Page

NEW MOONS				FULL MOONS			
JD	DATE	TIME	LONG.	JD	DATE	TIME	LONG.
23+							23+
1674							
32483	Jan 7	1;24	287.22	32498	Jan 22	5;18	122.66
32512	Feb 5	20;53	317.51	32527	Feb 20	15;49	152.43
32542	Mar 7	15;47	347.44	32557	Mar 22	1;30	181.76
32572	Apr 6	8;23	16.81	32586	Apr 20	10;45	210.58
32601	May 5	22;03	45.57	32615	May 19	20;25	238.99
32631	Jun 4	9;07	73.86	32645	Jun 18	7;25	267.16
32660	Jul 3	18;16	101.89	32674	Jul 17	20;25	295.33
32690	Aug 2	2;18	129.90	32704	Aug 16	11;25	323.72
32719	Aug 31	9;59	158.15	32734	Sep 15	3;48	352.51
32748	Sep 29	18;07	186.84	32763	Oct 14	20;43	21.80
32778	Oct 29	3;34	216.06	32793	Nov 13	13;24	51.56
32807	Nov 27	15;06	245.81	32823	Dec 13	5;11	81.65
32837	Dec 27	5;13	275.92				
1675							
32866	Jan 25	21;38	306.15	32852	Jan 11	19;24	111.81
32896	Feb 24	15;21	336.18	32882	Feb 10	7;35	141.76
32926	Mar 26	9;04	5.78	32911	Mar 11	17;43	171.27
32956	Apr 25	1;41	34.83	32941	Apr 10	2;16	200.24
32985	May 24	16;37	63.40	32970	May 9	10;07	228.72
33015	Jun 23	5;35	91.62	32999	Jun 7	18;16	256.86
33044	Jul 22	16;34	119.72	33029	Jul 7	3;40	284.89
33074	Aug 21	1;58	147.93	33058	Aug 5	14;59	313.06
33103	Sep 19	10;35	176.47	33088	Sep 4	4;35	341.59
33132	Oct 18	19;20	205.49	33117	Oct 3	20;33	10.65
33162	Nov 17	5;04	235.01	33147	Nov 2	14;28	40.29
33191	Dec 16	16;17	264.93	33177	Dec 2	9;18	70.39
1676							
				33207	Jan 1	3;27	100.69
33221	Jan 15	5;02	295.02	33236	Jan 30	19;20	130.87
33250	Feb 13	19;04	325.02	33266	Feb 29	8;16	160.65
33280	Mar 14	10;06	354.69	33295	Mar 29	18;33	189.87
33310	Apr 13	1;52	23.90	33325	Apr 28	2;57	218.53
33339	May 12	17;58	52.66	33354	May 27	10;24	246.76
33369	Jun 11	9;41	81.06	33383	Jun 25	17;43	274.74
33399	Jul 11	0;10	109.29	33413	Jul 25	1;44	302.73
33428	Aug 9	12;58	137.55	33442	Aug 23	11;19	330.97
33458	Sep 8	0;15	166.05	33471	Sep 21	23;21	359.70
33487	Oct 7	10;43	194.95	33501	Oct 21	14;31	29.03
33516	Nov 5	21;05	224.34	33531	Nov 20	8;41	58.96
33546	Dec 5	7;44	254.14	33561	Dec 20	4;31	89.27
1677							
33575	Jan 3	18;42	284.15	33590	Jan 18	23;52	119.65
33605	Feb 2	5;54	314.12	33620	Feb 17	16;56	149.73
33634	Mar 3	17;31	343.79	33650	Mar 19	7;02	179.28
33664	Apr 2	5;57	13.06	33679	Apr 17	18;21	208.24
33693	May 1	19;35	41.88	33709	May 17	3;26	236.65
33723	May 31	10;22	70.35	33738	Jun 15	11;00	264.70
33753	Jun 30	1;42	98.64	33767	Jul 14	17;49	292.62
33782	Jul 29	16;53	126.92	33797	Aug 13	0;54	320.68
33812	Aug 28	7;26	155.41	33826	Sep 11	9;25	349.10
33841	Sep 26	21;17	184.29	33855	Oct 10	20;28	18.09
33871	Oct 26	10;27	213.62	33885	Nov 9	10;45	47.69
33900	Nov 24	22;54	243.36	33915	Dec 9	4;03	77.80
33930	Dec 24	10;27	273.36				
1678							
				33944	Jan 7	23;12	108.17
33959	Jan 22	21;07	303.35	33974	Feb 6	18;28	138.44
33989	Feb 21	7;09	333.10	34004	Mar 8	12;23	168.33
34018	Mar 22	17;05	2.43	34034	Apr 7	3;58	197.64
34048	Apr 21	3;32	31.29	34063	May 6	16;45	226.36
34077	May 20	14;56	59.76	34093	Jun 5	2;50	254.60
34107	Jun 19	3;33	87.99	34122	Jul 4	10;51	282.58
34136	Jul 18	17;28	116.20	34151	Aug 2	17;48	310.56
34166	Aug 17	8;38	144.61	34181	Sep 1	0;58	338.80
34196	Sep 16	0;49	173.40	34210	Sep 30	9;28	7.51
34225	Oct 15	17;21	202.69	34239	Oct 29	20;07	36.79
34255	Nov 14	9;15	232.43	34269	Nov 28	9;14	66.60
34284	Dec 13	23;39	262.46	34299	Dec 28	0;40	96.77
1679							
34314	Jan 12	12;11	292.55	34328	Jan 26	17;57	127.03
34343	Feb 10	23;04	322.43	34358	Feb 25	12;15	157.08
34373	Mar 12	8;47	351.91	34388	Mar 27	6;23	186.68
34402	Apr 10	17;49	20.90	34417	Apr 25	22;58	215.72
34432	May 10	2;37	49.41	34447	May 25	12;59	244.24
34461	Jun 8	11;46	77.59	34477	Jun 24	0;15	272.40
34490	Jul 7	22;01	105.65	34506	Jul 23	9;27	300.42
34520	Aug 6	10;14	133.86	34535	Aug 21	17;40	328.59
34550	Sep 5	0;59	162.45	34565	Sep 20	1;59	357.13
34579	Oct 4	18;06	191.57	34594	Oct 19	11;07	26.17
34609	Nov 3	12;26	221.24	34623	Nov 17	21;28	55.73
34639	Dec 3	6;22	251.31	34653	Dec 17	9;13	85.68

Right Page

NEW MOONS				FULL MOONS			
JD	DATE	TIME	LONG.	JD	DATE	TIME	LONG.
23+							23+
1680							
34668	Jan 1	22;36	281.54	34682	Jan 15	22;36	115.80
34698	Jan 31	12;39	311.64	34712	Feb 14	13;46	145.84
34728	Mar 1	0;34	341.35	34742	Mar 15	6;26	175.56
34757	Mar 30	10;33	10.56	34771	Apr 13	23;37	204.81
34786	Apr 28	18;57	39.21	34801	May 13	16;02	233.57
34816	May 28	2;22	67.42	34831	Jun 12	6;46	261.93
34845	Jun 26	9;45	95.41	34860	Jul 11	19;37	290.10
34874	Jul 25	18;17	123.41	34890	Aug 10	6;58	318.30
34904	Aug 24	5;11	151.72	34919	Sep 8	17;23	346.77
34933	Sep 22	19;07	180.54	34949	Oct 8	3;22	15.67
34963	Oct 22	11;58	209.95	34978	Nov 6	13;16	45.05
34993	Nov 21	6;40	239.92	35007	Dec 5	23;25	74.83
35023	Dec 21	1;46	270.21				
1681							
				35037	Jan 4	10;14	104.84
35052	Jan 19	19;51	300.52	35066	Feb 2	22;08	134.83
35082	Feb 18	11;50	330.54	35096	Mar 4	11;19	164.56
35112	Mar 20	1;02	0.05	35126	Apr 3	1;38	193.88
35141	Apr 18	11;24	28.95	35155	May 2	16;36	222.75
35170	May 17	19;26	57.32	35185	Jun 1	7;46	251.23
35200	Jun 16	2;10	85.34	35214	Jun 30	22;48	279.50
35229	Jul 15	8;52	113.25	35244	Jul 30	13;25	307.77
35258	Aug 13	16;43	141.34	35274	Aug 29	3;20	336.25
35288	Sep 12	2;42	169.84	35303	Sep 27	16;14	5.09
35317	Oct 11	15;23	198.90	35333	Oct 27	4;02	34.38
35347	Nov 10	6;53	228.57	35362	Nov 25	15;01	64.08
35377	Dec 10	0;52	258.72	35392	Dec 25	1;39	94.04
1682							
35406	Jan 8	20;21	289.10	35421	Jan 23	12;19	124.03
35436	Feb 7	15;44	319.37	35450	Feb 21	23;13	153.80
35466	Mar 9	9;09	349.22	35480	Mar 23	10;22	183.16
35495	Apr 7	23;24	18.46	35509	Apr 21	21;52	212.06
35525	May 7	10;23	47.09	35539	May 21	10;08	240.55
35554	Jun 5	18;54	75.26	35568	Jun 19	23;37	268.81
35584	Jul 5	2;07	103.21	35598	Jul 19	14;29	297.06
35613	Aug 3	9;10	131.20	35628	Aug 18	6;20	325.50
35642	Sep 1	16;57	159.47	35657	Sep 16	22;17	354.30
35672	Oct 1	2;11	188.22	35687	Oct 16	13;26	23.55
35701	Oct 30	13;29	217.54	35717	Nov 15	3;24	53.22
35731	Nov 29	3;25	247.40	35746	Dec 14	16;14	83.19
35760	Dec 28	20;10	277.63				
1683							
				35776	Jan 13	4;06	113.25
35790	Jan 27	15;03	307.95	35805	Feb 11	15;00	143.13
35820	Feb 26	10;20	338.03	35835	Mar 13	0;55	172.62
35850	Mar 28	4;05	7.60	35864	Apr 11	10;05	201.60
35879	Apr 26	19;09	36.57	35893	May 10	19;10	230.11
35909	May 26	7;25	65.01	35923	Jun 9	5;11	258.31
35938	Jun 24	17;28	93.11	35952	Jul 8	17;00	286.44
35968	Jul 24	2;03	121.11	35982	Aug 7	7;01	314.72
35997	Aug 22	9;59	149.28	36011	Sep 5	22;55	343.37
36026	Sep 20	17;58	177.81	36041	Oct 5	15;53	12.50
36056	Oct 20	2;48	206.85	36071	Nov 4	9;02	42.13
36085	Nov 18	13;18	236.43	36101	Dec 4	1;35	72.15
36115	Dec 18	2;06	266.42				
1684							
				36130	Jan 2	16;48	102.34
36144	Jan 16	17;20	296.62	36160	Feb 1	6;04	132.40
36174	Feb 15	10;21	326.73	36189	Mar 1	17;07	162.08
36204	Mar 16	3;59	356.48	36219	Mar 31	2;13	191.23
36233	Apr 14	21;03	25.71	36248	Apr 29	10;06	219.85
36263	May 14	12;45	54.43	36277	May 28	17;48	248.07
36293	Jun 13	2;43	82.76	36307	Jun 27	2;20	276.09
36322	Jul 12	14;46	110.89	36336	Jul 26	12;33	304.17
36352	Aug 11	1;05	139.05	36366	Aug 25	0;59	332.55
36381	Sep 9	10;12	167.48	36395	Sep 23	15;52	1.42
36410	Oct 8	18;59	196.35	36425	Oct 23	9;03	30.87
36440	Nov 7	4;20	225.71	36455	Nov 22	3;48	60.84
36469	Dec 6	14;52	255.52	36484	Dec 21	22;40	91.13
1685							
36499	Jan 5	2;51	285.58	36514	Jan 20	15;55	121.40
36528	Feb 3	16;07	315.62	36544	Feb 19	6;20	151.36
36558	Mar 5	6;25	345.39	36573	Mar 20	17;46	180.77
36587	Apr 3	21;33	14.73	36603	Apr 19	2;52	209.62
36617	May 3	13;18	43.62	36632	May 18	10;33	237.96
36647	Jun 2	5;09	72.12	36661	Jun 16	17;44	265.99
36676	Jul 1	20;21	100.39	36691	Jul 16	1;15	293.93
36706	Jul 31	10;10	128.63	36720	Aug 14	9;56	322.06
36735	Aug 29	22;23	157.05	36749	Sep 12	20;42	350.60
36765	Sep 28	9;27	185.83	36779	Oct 12	10;20	19.71
36794	Oct 27	20;03	215.08	36809	Nov 11	3;11	49.45
36824	Nov 26	6;43	244.77	36838	Dec 10	22;31	79.67
36853	Dec 25	17;37	274.74				

Left Panel

NEW MOONS JD	DATE	TIME	LONG.	FULL MOONS JD	DATE	TIME	LONG.
23+			1686	23+			
				36868	Jan 9	18;25	110.07
36883	Jan 24	4;39	304.75	36898	Feb 8	12;46	140.29
36912	Feb 22	15;50	334.52	36928	Mar 10	4;19	170.05
36942	Mar 24	3;31	3.90	36957	Apr 8	16;52	199.21
36971	Apr 22	16;13	32.84	36987	May 8	2;55	227.79
37001	May 22	6;12	61.39	37016	Jun 6	11;05	255.94
37030	Jun 20	21;12	89.71	37045	Jul 5	18;10	283.88
37060	Jul 20	12;33	117.97	37075	Aug 4	1;01	311.87
37090	Aug 19	3;37	146.39	37104	Sep 2	8;46	340.15
37119	Sep 17	18;05	175.14	37133	Oct 1	18;31	8.92
37149	Oct 17	7;54	204.35	37163	Oct 31	7;14	38.31
37178	Nov 15	21;00	233.99	37192	Nov 29	23;10	68.27
37208	Dec 15	9;11	263.95	37222	Dec 29	17;34	98.57
			1687				
37237	Jan 13	20;19	293.98	37252	Jan 28	12;54	128.91
37267	Feb 12	6;31	323.82	37282	Feb 27	7;32	158.95
37296	Mar 13	16;14	353.29	37312	Mar 29	0;14	188.46
37326	Apr 12	2;06	22.28	37341	Apr 27	14;21	217.38
37355	May 11	12;42	50.84	37371	May 27	1;43	245.77
37385	Jun 10	0;29	79.11	37400	Jun 25	10;41	273.82
37414	Jul 9	13;37	107.29	37429	Jul 24	18;06	301.79
37444	Aug 8	4;11	135.60	37459	Aug 23	1;08	329.92
37473	Sep 6	20;03	164.25	37488	Sep 21	8;59	358.46
37503	Oct 6	12;45	193.39	37517	Oct 20	18;37	27.54
37533	Nov 5	5;24	223.01	37547	Nov 19	6;34	57.18
37562	Dec 4	20;53	253.00	37576	Dec 18	20;53	87.25
			1688				
37592	Jan 3	10;29	283.12	37606	Jan 17	13;14	117.50
37621	Feb 1	22;09	313.11	37636	Feb 16	6;58	147.63
37651	Mar 2	8;18	342.75	37666	Mar 17	1;08	177.39
37680	Mar 31	17;26	11.89	37695	Apr 15	18;29	206.62
37710	Apr 30	2;05	40.54	37725	May 15	9;47	235.31
37739	May 29	10;47	68.78	37754	Jun 13	22;24	263.58
37768	Jun 27	20;13	96.84	37784	Jul 13	8;37	291.63
37798	Jul 27	7;16	124.95	37813	Aug 11	17;20	319.74
37827	Aug 25	20;43	153.38	37843	Sep 10	1;41	348.14
37857	Sep 24	12;50	182.30	37872	Oct 9	10;32	17.01
37887	Oct 24	6;57	211.80	37901	Nov 7	20;24	46.41
37917	Nov 23	1;31	241.78	37931	Dec 7	7;29	76.25
37946	Dec 22	18;54	272.01				
			1689				
				37960	Jan 5	19;59	106.33
37976	Jan 21	10;09	302.20	37990	Feb 4	10;06	136.41
38005	Feb 19	23;05	332.09	38020	Mar 6	1;50	166.23
38035	Mar 21	9;54	1.47	38049	Apr 4	18;38	195.63
38064	Apr 19	18;53	30.30	38079	May 4	11;22	224.54
38094	May 19	2;33	58.64	38109	Jun 3	2;55	253.02
38123	Jun 17	9;41	86.66	38138	Jul 2	16;44	281.23
38152	Jul 16	17;25	114.61	38168	Aug 1	4;55	309.41
38182	Aug 15	2;59	142.78	38197	Aug 30	15;58	337.79
38211	Sep 13	15;24	171.39	38227	Sep 29	2;23	6.56
38241	Oct 13	6;58	200.60	38256	Oct 28	12;32	35.80
38271	Nov 12	1;03	230.40	38285	Nov 26	22;40	65.47
38300	Dec 11	20;16	260.61	38315	Dec 26	9;09	95.43
			1690				
38330	Jan 10	15;05	290.97	38344	Jan 24	20;22	125.44
38360	Feb 9	8;14	321.14	38374	Feb 23	8;42	155.26
38389	Mar 10	22;50	350.84	38403	Mar 24	22;12	184.70
38419	Apr 9	10;29	19.96	38433	Apr 23	12;37	213.70
38448	May 8	19;29	48.49	38463	May 23	3;30	242.28
38478	Jun 7	2;39	76.60	38492	Jun 21	18;31	270.59
38507	Jul 6	9;10	104.52	38522	Jul 21	9;26	298.83
38536	Aug 4	16;19	132.51	38551	Aug 19	23;56	327.24
38566	Sep 3	1;11	160.84	38581	Sep 18	13;38	355.97
38595	Oct 2	12;34	189.70	38611	Oct 18	2;17	25.14
38625	Nov 1	2;45	219.16	38640	Nov 16	13;52	54.73
38654	Nov 30	19;37	249.16	38670	Dec 16	0;45	84.63
38684	Dec 30	14;31	279.49				
			1691				
				38699	Jan 14	11;23	114.64
38714	Jan 29	10;08	309.83	38728	Feb 12	22;03	144.50
38744	Feb 28	4;40	339.85	38758	Mar 14	8;51	174.00
38773	Mar 29	20;30	9.32	38787	Apr 12	19;52	203.03
38803	Apr 28	9;00	38.17	38817	May 12	7;26	231.63
38832	May 27	18;35	66.48	38846	Jun 10	20;02	259.93
38862	Jun 26	2;19	94.48	38876	Jul 10	10;07	288.14
38891	Jul 25	9;22	122.43	38906	Aug 9	1;37	316.50
38920	Aug 23	16;47	150.59	38935	Sep 7	17;50	345.17
38950	Sep 22	1;20	179.16	38965	Oct 7	9;45	14.29
38979	Oct 21	11;38	208.28	38995	Nov 6	0;37	43.85
39009	Nov 20	0;15	237.96	39024	Dec 5	14;13	73.76
39038	Dec 19	15;36	268.07				

Right Panel

NEW MOONS JD	DATE	TIME	LONG.	FULL MOONS JD	DATE	TIME	LONG.
23+			1692	23+			
				39054	Jan 4	2;40	103.83
39068	Jan 18	9;28	298.38	39083	Feb 2	14;02	133.80
39098	Feb 17	4;38	328.56	39113	Mar 3	0;19	163.43
39127	Mar 17	23;12	358.32	39142	Apr 1	9;35	192.57
39157	Apr 16	15;33	27.50	39171	Apr 30	18;23	221.21
39187	May 16	5;08	56.11	39201	May 30	3;35	249.48
39216	Jun 14	16;13	84.32	39230	Jun 28	14;15	277.59
39246	Jul 14	1;32	112.34	39260	Jul 28	3;05	305.78
39275	Aug 12	9;50	140.43	39289	Aug 26	18;09	334.28
39304	Sep 10	17;52	168.84	39319	Sep 25	10;52	3.24
39334	Oct 10	2;21	197.71	39349	Oct 25	4;16	32.72
39363	Nov 8	12;03	227.10	39378	Nov 23	21;28	62.66
39392	Dec 7	23;41	256.97	39408	Dec 23	13;38	92.84
			1693				
39422	Jan 6	13;40	287.11	39438	Jan 22	4;00	122.99
39452	Feb 5	5;45	317.26	39467	Feb 20	16;07	152.82
39481	Mar 6	22;59	347.14	39497	Mar 22	2;00	182.16
39511	Apr 5	16;12	16.54	39526	Apr 20	10;12	210.95
39541	May 5	8;29	45.41	39555	May 19	17;42	239.27
39570	Jun 3	23;18	73.86	39585	Jun 18	1;32	267.31
39600	Jul 3	12;23	102.04	39614	Jul 17	10;44	295.33
39629	Aug 1	23;41	130.19	39643	Aug 15	21;59	323.57
39659	Aug 31	9;32	158.53	39673	Sep 14	11;42	352.25
39688	Sep 29	18;36	187.26	39703	Oct 14	3;54	21.50
39718	Oct 29	3;46	216.46	39732	Nov 12	22;10	51.31
39747	Nov 27	13;47	246.14	39762	Dec 12	17;21	81.54
39777	Dec 27	1;04	276.14				
			1694				
				39792	Jan 11	11;43	111.87
39806	Jan 25	13;37	306.19	39822	Feb 10	3;39	141.98
39836	Feb 24	3;12	336.06	39851	Mar 11	16;29	171.60
39865	Mar 25	17;39	5.53	39881	Apr 10	2;32	200.64
39895	Apr 24	8;50	34.54	39910	May 9	10;40	229.13
39925	May 24	0;31	63.14	39939	Jun 7	17;54	257.23
39954	Jun 22	16;06	91.47	39969	Jul 7	1;05	285.18
39984	Jul 22	6;46	119.72	39998	Aug 5	9;05	313.21
40013	Aug 20	20;00	148.08	40027	Sep 3	18;45	341.59
40043	Sep 19	7;51	176.75	40057	Oct 3	6;56	10.49
40072	Oct 18	18;52	205.86	40086	Nov 1	22;17	40.01
40102	Nov 17	5;40	235.43	40116	Dec 1	16;37	70.08
40131	Dec 16	16;34	265.34	40146	Dec 31	12;33	100.45
			1695				
40161	Jan 15	3;32	295.36	40176	Jan 30	7;54	130.78
40190	Feb 13	14;28	325.22	40206	Mar 1	0;52	160.73
40220	Mar 15	1;36	354.72	40235	Mar 30	14;49	190.10
40249	Apr 13	13;26	23.78	40265	Apr 29	1;57	218.88
40279	May 13	2;30	52.42	40294	May 28	10;55	247.16
40308	Jun 11	16;51	80.78	40323	Jun 26	18;25	275.15
40338	Jul 11	8;04	109.04	40353	Jul 26	1;18	303.09
40367	Aug 9	23;27	137.39	40382	Aug 24	8;33	331.25
40397	Sep 8	14;28	166.04	40411	Sep 22	17;16	359.84
40427	Oct 8	4;54	195.11	40441	Oct 22	4;30	29.01
40456	Nov 6	18;40	224.64	40470	Nov 20	18;54	58.78
40486	Dec 6	7;33	254.53	40500	Dec 20	12;13	88.98
			1696				
40515	Jan 4	19;18	284.57	40530	Jan 19	7;13	119.34
40545	Feb 3	5;53	314.51	40560	Feb 18	2;16	149.50
40574	Mar 3	15;37	344.11	40589	Mar 18	19;55	179.21
40604	Apr 2	1;06	13.24	40619	Apr 17	11;17	208.34
40633	May 1	11;01	41.92	40648	May 16	23;58	236.90
40662	May 30	21;55	70.24	40678	Jun 15	10;05	265.05
40692	Jun 29	10;12	98.41	40707	Jul 14	18;12	293.03
40722	Jul 29	0;00	126.64	40737	Aug 13	1;24	321.09
40751	Aug 27	15;20	155.16	40766	Sep 11	8;51	349.47
40781	Sep 26	7;55	184.13	40795	Oct 10	17;39	18.37
40811	Oct 26	1;00	213.62	40825	Nov 9	4;31	47.82
40840	Nov 24	17;28	243.53	40854	Dec 8	17;42	77.76
40870	Dec 24	8;14	273.66				
			1697				
				40884	Jan 7	9;00	107.96
40899	Jan 22	20;53	303.73	40914	Feb 6	1;58	138.14
40929	Feb 21	7;39	333.51	40943	Mar 7	19;49	168.03
40958	Mar 22	17;05	2.81	40973	Apr 6	13;33	197.44
40988	Apr 21	1;43	31.61	41003	May 6	5;52	226.13
41017	May 20	10;06	59.96	41032	Jun 4	19;50	254.71
41046	Jun 18	18;54	88.03	41062	Jul 4	7;15	282.82
41076	Jul 18	4;56	116.08	41091	Aug 2	16;44	310.90
41105	Aug 16	17;05	144.37	41121	Sep 1	1;20	339.20
41135	Sep 15	7;58	173.11	41150	Sep 30	10;04	7.92
41165	Oct 15	1;25	202.42	41179	Oct 29	19;34	37.15
41194	Nov 13	20;12	232.27	41209	Nov 28	6;09	66.87
41224	Dec 13	14;32	262.47	41238	Dec 27	17;54	96.89

NEW MOONS / FULL MOONS (left block)

NEW MOONS JD	DATE	TIME	LONG.	FULL MOONS JD	DATE	TIME	LONG.
23+				23+			
1698							
41254	Jan 12	7;02	292.73	41268	Jan 26	7;03	126.97
41283	Feb 10	21;07	322.75	41297	Feb 24	21;46	156.87
41313	Mar 12	8;52	352.32	41327	Mar 26	13;50	186.39
41342	Apr 10	18;35	21.33	41357	Apr 25	6;30	215.45
41372	May 10	2;41	49.80	41386	May 24	22;36	244.06
41401	Jun 8	9;51	77.90	41416	Jun 23	13;18	272.34
41430	Jul 7	17;02	105.84	41446	Jul 23	2;23	300.53
41460	Aug 6	1;31	133.89	41475	Aug 21	14;09	328.83
41489	Sep 4	12;26	162.33	41505	Sep 20	1;06	357.48
41519	Oct 4	2;32	191.32	41534	Oct 19	11;36	26.58
41548	Nov 2	19;36	220.92	41563	Nov 17	21;55	56.14
41578	Dec 2	14;34	251.04	41593	Dec 17	8;16	86.03
1699							
41608	Jan 1	9;52	281.39	41622	Jan 15	19;02	116.05
41638	Jan 31	4;01	311.66	41652	Feb 14	6;37	145.93
41667	Mar 1	19;55	341.56	41681	Mar 15	19;16	175.49
41697	Mar 31	8;58	10.88	41711	Apr 14	8;58	204.61
41726	Apr 29	19;08	39.60	41740	May 13	23;24	233.30
41756	May 29	3;00	67.84	41770	Jun 12	14;14	261.67
41785	Jun 27	9;39	95.79	41800	Jul 12	5;13	289.91
41814	Jul 26	16;20	123.73	41829	Aug 10	20;06	318.25
41844	Aug 25	0;16	151.91	41859	Sep 9	10;30	346.88
41873	Sep 23	10;25	180.57	41889	Oct 9	0;00	15.92
41902	Oct 22	23;16	209.82	41918	Nov 7	12;22	45.40
41932	Nov 21	14;54	239.65	41947	Dec 6	23;44	75.24
41962	Dec 21	8;55	269.89				
1700							
				41977	Jan 5	10;31	105.25
41992	Jan 20	4;20	300.26	42006	Feb 3	21;04	135.18
42021	Feb 18	23;34	330.43	42036	Mar 5	7;37	164.80
42051	Mar 20	16;48	0.10	42065	Apr 3	18;15	193.98
42081	Apr 19	6;53	29.16	42095	May 3	5;13	222.68
42110	May 18	17;46	57.65	42124	Jun 1	16;59	251.04
42140	Jun 17	2;16	85.73	42154	Jul 1	6;08	279.23
42169	Jul 16	9;34	113.67	42183	Jul 30	20;57	307.51
42198	Aug 14	16;47	141.73	42213	Aug 29	13;03	336.06
42228	Sep 13	0;49	170.15	42243	Sep 28	5;30	5.05
42257	Oct 12	10;17	199.08	42272	Oct 27	21;15	34.49
42286	Nov 10	21;46	228.57	42302	Nov 26	11;45	64.33
42316	Dec 10	11;45	258.55	42332	Dec 26	0;55	94.39
1701							
42346	Jan 9	4;22	288.82	42361	Jan 24	12;53	124.44
42375	Feb 7	22;57	319.06	42390	Feb 22	23;37	154.21
42405	Mar 9	17;53	348.98	42420	Mar 24	9;11	183.52
42435	Apr 8	11;20	18.37	42449	Apr 22	17;54	212.30
42465	May 8	2;12	47.16	42479	May 22	2;33	240.65
42494	Jun 6	14;27	75.49	42508	Jun 20	12;10	268.76
42524	Jul 6	0;37	103.55	42537	Jul 19	23;44	296.87
42553	Aug 4	9;27	131.60	42567	Aug 18	13;44	325.23
42582	Sep 2	17;43	159.90	42597	Sep 17	5;50	354.03
42612	Oct 2	2;04	188.61	42626	Oct 16	23;14	23.35
42641	Oct 31	11;13	217.84	42656	Nov 15	16;54	53.17
42670	Nov 29	21;54	247.56	42686	Dec 15	9;53	83.32
42700	Dec 29	10;40	277.62				
1702							
				42716	Jan 14	1;20	113.53
42730	Jan 28	1;38	307.78	42745	Feb 12	14;37	143.51
42759	Feb 26	18;12	337.75	42775	Mar 14	1;29	173.03
42789	Mar 28	11;18	7.31	42804	Apr 12	10;16	202.00
42819	Apr 27	3;56	36.35	42833	May 11	17;49	230.45
42848	May 26	19;25	64.92	42863	Jun 10	1;12	258.55
42878	Jun 25	9;25	93.18	42892	Jul 9	9;30	286.53
42907	Jul 24	21;43	121.33	42921	Aug 7	19;37	314.66
42937	Aug 23	8;27	149.61	42951	Sep 6	8;06	343.16
42966	Sep 21	18;03	178.21	42980	Oct 5	23;11	12.20
42996	Oct 21	3;18	207.27	43010	Nov 4	16;40	41.84
43025	Nov 19	12;59	236.80	43040	Dec 4	11;44	71.96
43054	Dec 18	23;40	266.72				
1703							
				43070	Jan 3	6;51	102.31
43084	Jan 17	11;32	296.77	43100	Feb 2	0;11	132.54
43114	Feb 16	0;26	326.70	43129	Mar 3	14;32	162.36
43143	Mar 17	14;12	356.30	43159	Apr 2	1;48	191.60
43173	Apr 16	4;44	25.44	43188	May 1	10;40	220.26
43202	May 15	19;58	54.15	43217	May 30	18;08	248.47
43232	Jun 14	11;35	82.55	43247	Jun 29	1;10	276.44
43262	Jul 14	2;51	110.80	43276	Jul 28	8;39	304.41
43291	Aug 12	17;00	139.12	43305	Aug 26	17;24	332.64
43321	Sep 11	5;45	167.70	43335	Sep 25	4;17	1.34
43350	Oct 10	17;23	196.67	43364	Oct 24	18;05	30.64
43380	Nov 9	4;30	226.11	43394	Nov 23	11;06	60.53
43409	Dec 8	15;30	255.94	43424	Dec 23	6;31	90.84

NEW MOONS / FULL MOONS (right block)

NEW MOONS JD	DATE	TIME	LONG.	FULL MOONS JD	DATE	TIME	LONG.
23+				23+			
1704							
43439	Jan 7	2;29	285.95	43454	Jan 22	2;26	121.22
43468	Feb 5	13;20	315.89	43483	Feb 20	20;42	151.34
43498	Mar 6	0;07	345.52	43513	Mar 21	12;06	180.93
43527	Apr 4	11;16	14.70	43543	Apr 20	0;30	209.91
43556	May 3	23;23	43.45	43572	May 19	10;24	238.34
43586	Jun 2	12;52	71.86	43601	Jun 17	18;32	266.41
43616	Jul 2	3;37	100.12	43631	Jul 17	1;38	294.34
43645	Jul 31	19;02	128.42	43660	Aug 15	8;38	322.41
43675	Aug 30	10;27	156.96	43689	Sep 13	16;34	350.83
43705	Sep 29	1;28	185.90	43719	Oct 13	2;32	19.79
43734	Oct 28	15;52	215.30	43748	Nov 11	15;23	49.35
43764	Nov 27	5;28	245.11	43778	Dec 11	7;21	79.42
43793	Dec 26	17;57	275.15				
1705							
				43808	Jan 10	1;39	109.75
43823	Jan 25	5;07	305.16	43837	Feb 8	20;45	140.01
43852	Feb 23	15;06	334.88	43867	Mar 10	15;06	169.89
43882	Mar 25	0;26	4.17	43897	Apr 9	7;34	199.21
43911	Apr 23	9;48	32.97	43926	May 8	21;33	227.97
43940	May 22	19;54	61.37	43956	Jun 7	8;54	256.25
43970	Jun 21	7;17	89.55	43985	Jul 6	17;58	284.27
43999	Jul 20	20;13	117.72	44015	Aug 5	1;36	312.28
44029	Aug 19	10;50	146.11	44044	Sep 3	8;56	340.54
44059	Sep 18	3;00	174.93	44073	Oct 2	17;06	9.26
44088	Oct 17	20;12	204.26	44103	Nov 1	3;00	38.53
44118	Nov 16	13;24	234.06	44132	Nov 30	15;05	68.31
44148	Dec 16	5;20	264.18	44162	Dec 30	5;20	98.45
1706							
44177	Jan 14	19;09	294.32	44191	Jan 28	21;24	128.65
44207	Feb 13	6;47	324.22	44221	Feb 27	14;41	158.64
44236	Mar 14	16;41	353.69	44251	Mar 29	8;23	188.20
44266	Apr 13	1;27	22.64	44281	Apr 28	1;25	217.24
44295	May 12	9;42	51.11	44310	May 27	16;33	245.79
44324	Jun 10	18;03	79.24	44340	Jun 26	5;15	274.00
44354	Jul 10	3;14	107.26	44369	Jul 25	15;44	302.08
44383	Aug 8	14;10	135.43	44399	Aug 24	0;50	330.30
44413	Sep 7	3;41	163.98	44428	Sep 22	9;37	358.88
44442	Oct 6	20;04	193.10	44457	Oct 21	18;53	27.94
44472	Nov 5	14;34	222.79	44487	Nov 20	5;03	57.51
44502	Dec 5	9;32	252.91	44516	Dec 19	16;14	87.45
1707							
44532	Jan 4	3;13	283.20	44546	Jan 18	4;35	117.52
44561	Feb 2	18;34	313.35	44575	Feb 16	18;17	147.48
44591	Mar 4	7;25	343.09	44605	Mar 18	9;27	177.13
44620	Apr 2	17;59	12.29	44635	Apr 17	1;40	206.32
44650	May 2	2;42	40.94	44664	May 16	18;00	235.07
44679	May 31	10;07	69.14	44694	Jun 15	9;24	263.44
44708	Jun 29	17;04	97.10	44723	Jul 14	23;22	291.65
44738	Jul 29	0;42	125.08	44753	Aug 13	11;55	319.91
44767	Aug 27	10;18	153.35	44782	Sep 11	23;29	348.45
44796	Sep 25	22;50	182.13	44812	Oct 11	10;27	17.41
44826	Oct 25	14;35	211.52	44841	Nov 9	21;03	46.84
44856	Nov 24	8;53	241.48	44871	Dec 9	7;28	76.65
44886	Dec 24	4;17	271.79				
1708							
				44900	Jan 7	17;59	106.64
44915	Jan 22	23;12	302.13	44930	Feb 6	4;59	136.58
44945	Feb 21	16;18	332.19	44959	Mar 6	16;52	166.24
44975	Mar 22	6;45	1.73	44989	Apr 5	5;46	195.49
45004	Apr 20	18;14	30.66	45018	May 4	19;35	224.29
45034	May 20	3;04	59.04	45048	Jun 3	10;03	252.73
45063	Jun 18	10;08	87.06	45078	Jul 3	0;54	280.99
45092	Jul 17	16;38	114.97	45107	Aug 1	15;57	309.28
45121	Aug 15	23;52	143.05	45137	Aug 31	6;52	337.82
45151	Sep 14	8;54	171.52	45166	Sep 29	21;10	6.74
45180	Oct 13	20;27	200.56	45196	Oct 29	10;25	36.11
45210	Nov 12	10;47	230.20	45225	Nov 27	22;28	65.86
45240	Dec 12	3;43	260.31	45255	Dec 27	9;36	95.85
1709							
45269	Jan 10	22;34	290.67	45284	Jan 25	20;13	125.83
45299	Feb 9	18;01	320.93	45314	Feb 24	6;36	155.57
45329	Mar 11	12;20	350.80	45343	Mar 25	16;57	184.88
45359	Apr 10	3;59	20.08	45373	Apr 24	3;26	213.71
45388	May 9	16;21	48.76	45402	May 23	14;28	242.14
45418	Jun 8	1;54	76.96	45432	Jun 22	2;40	270.35
45447	Jul 7	9;41	104.92	45461	Jul 21	16;35	298.55
45476	Aug 5	16;54	132.92	45491	Aug 20	8;12	327.00
45506	Sep 4	0;33	161.21	45521	Sep 19	0;49	355.84
45535	Oct 3	9;23	189.96	45550	Oct 18	17;19	25.16
45564	Nov 1	19;55	219.27	45580	Nov 17	8;46	54.91
45594	Dec 1	8;39	249.09	45609	Dec 16	22;48	84.95
45623	Dec 30	23;55	279.27				

NEW MOONS				FULL MOONS			
JD	DATE	TIME	LONG.	JD	DATE	TIME	LONG.
23+				**1710** 23+			
				45639	Jan 15	11;26	115.04
45653	Jan 29	17;31	309.53	45668	Feb 13	22;44	144.93
45683	Feb 28	12;18	339.57	45698	Mar 15	8;43	174.39
45713	Mar 30	6;29	9.14	45727	Apr 13	17;35	203.33
45742	Apr 28	22;36	38.13	45757	May 13	1;55	231.80
45772	May 28	12;05	66.61	45786	Jun 11	10;43	259.94
45801	Jun 26	23;15	94.75	45815	Jul 10	21;05	288.01
45831	Jul 26	8;47	122.79	45845	Aug 9	9;49	316.25
45860	Aug 24	17;26	151.00	45875	Sep 8	1;01	344.88
45890	Sep 23	1;51	179.57	45904	Oct 7	18;04	14.03
45919	Oct 22	10;42	208.63	45934	Nov 6	11;58	43.71
45948	Nov 20	20;39	238.20	45964	Dec 6	5;37	73.79
45978	Dec 20	8;21	268.16				
				1711			
				45993	Jan 4	22;05	104.04
46007	Jan 18	22;08	298.30	46023	Feb 3	12;33	134.14
46037	Feb 17	13;48	328.34	46053	Mar 5	0;32	163.84
46067	Mar 19	6;29	358.03	46082	Apr 3	10;09	192.99
46096	Apr 17	23;12	27.23	46111	May 2	18;02	221.59
46126	May 17	15;10	55.94	46141	Jun 1	1;12	249.77
46156	Jun 16	5;54	84.29	46170	Jun 30	8;48	277.76
46185	Jul 15	19;10	112.47	46199	Jul 29	17;52	305.79
46215	Aug 14	6;51	140.70	46229	Aug 28	5;08	334.14
46244	Sep 12	17;11	169.20	46258	Sep 26	18;59	2.98
46274	Oct 12	2;45	198.11	46288	Oct 26	11;26	32.42
46303	Nov 10	12;20	227.51	46318	Nov 25	6;00	62.40
46332	Dec 9	22;35	257.32	46348	Dec 25	1;27	92.72
				1712			
46362	Jan 8	9;51	287.35	46377	Jan 23	19;57	123.04
46391	Feb 6	22;08	317.34	46407	Feb 22	11;52	153.05
46421	Mar 7	11;14	347.04	46437	Mar 23	0;33	182.50
46451	Apr 6	1;05	16.32	46466	Apr 21	10;23	211.35
46480	May 5	15;42	45.14	46495	May 20	18;19	239.69
46510	Jun 4	7;01	73.60	46525	Jun 19	1;23	267.70
46539	Jul 3	22;31	101.88	46554	Jul 18	8;31	295.64
46569	Aug 2	13;25	130.17	46583	Aug 16	16;33	323.75
46599	Sep 1	3;07	158.67	46613	Sep 15	2;19	352.26
46628	Sep 30	15;32	187.52	46642	Oct 14	14;40	21.34
46658	Oct 30	3;07	216.83	46672	Nov 13	6;10	51.04
46687	Nov 28	14;20	246.56	46702	Dec 13	0;36	81.22
46717	Dec 28	1;26	276.54				
				1713			
				46731	Jan 11	20;34	111.62
46746	Jan 26	12;19	306.54	46761	Feb 10	15;51	141.88
46775	Feb 24	22;57	336.28	46791	Mar 12	8;41	171.68
46805	Mar 26	9;36	5.60	46820	Apr 10	22;27	200.87
46834	Apr 24	20;51	34.45	46850	May 10	9;27	229.48
46864	May 24	9;22	62.93	46879	Jun 8	18;20	257.64
46893	Jun 22	23;21	91.20	46909	Jul 8	1;52	285.60
46923	Jul 22	14;29	119.46	46938	Aug 6	8;52	313.59
46953	Aug 21	6;06	147.91	46967	Sep 4	16;18	341.87
46982	Sep 19	21;36	176.72	46997	Oct 4	1;14	10.64
47012	Oct 19	12;38	205.99	47026	Nov 2	12;40	39.99
47042	Nov 18	2;57	235.71	47056	Dec 2	3;09	69.90
47071	Dec 17	16;13	265.71	47085	Dec 31	20;23	100.16
				1714			
47101	Jan 16	4;06	295.77	47115	Jan 30	15;11	130.47
47130	Feb 14	14;34	325.62	47145	Mar 1	9;55	160.50
47159	Mar 15	23;58	355.05	47175	Mar 31	3;17	190.02
47189	Apr 14	9;00	23.99	47204	Apr 29	18;27	218.97
47218	May 13	18;24	52.49	47234	May 29	7;04	247.40
47248	Jun 12	4;53	80.69	47263	Jun 27	17;15	275.49
47277	Jul 11	16;53	108.82	47293	Jul 27	1;35	303.49
47307	Aug 10	6;38	137.11	47322	Aug 25	9;04	331.66
47336	Sep 8	22;10	165.77	47351	Sep 23	16;51	0.21
47366	Oct 8	15;19	194.93	47381	Oct 23	1;58	29.30
47396	Nov 7	8;47	224.61	47410	Nov 21	13;02	58.92
47426	Dec 7	1;44	254.67	47440	Dec 21	2;14	88.95
				1715			
47455	Jan 5	16;48	284.86	47469	Jan 19	17;19	119.14
47485	Feb 4	5;31	314.88	47499	Feb 18	9;52	149.20
47514	Mar 5	16;07	344.51	47529	Mar 20	3;14	178.91
47544	Apr 4	1;14	13.63	47558	Apr 18	20;32	208.12
47573	May 3	9;29	42.25	47588	May 18	12;37	236.83
47602	Jun 1	17;32	70.46	47618	Jun 17	2;36	265.15
47632	Jul 1	2;04	98.48	47647	Jul 16	14;14	293.26
47661	Jul 30	11;57	126.55	47677	Aug 15	0;04	321.43
47691	Aug 29	0;06	154.94	47706	Sep 13	9;07	349.87
47720	Sep 27	15;08	183.84	47735	Oct 12	18;17	18.79
47750	Oct 27	8;53	213.34	47765	Nov 11	4;08	48.20
47780	Nov 26	4;03	243.36	47794	Dec 10	14;53	78.04
47809	Dec 25	22;43	273.64				

NEW MOONS				FULL MOONS			
JD	DATE	TIME	LONG.	JD	DATE	TIME	LONG.
23+				**1716** 23+			
				47824	Jan 9	2;36	108.08
47839	Jan 24	15;22	303.89	47853	Feb 7	15;26	138.09
47869	Feb 23	5;26	333.80	47883	Mar 8	5;37	167.83
47898	Mar 23	17;01	3.20	47912	Apr 6	21;06	197.15
47928	Apr 22	2;29	32.02	47942	May 6	13;15	226.03
47957	May 21	10;20	60.35	47972	Jun 5	5;06	254.51
47986	Jun 19	17;18	88.36	48001	Jul 4	19;50	282.76
48016	Jul 19	0;24	116.30	48031	Aug 3	9;12	311.00
48045	Aug 17	8;52	144.43	48060	Sep 1	21;27	339.44
48074	Sep 15	19;53	173.01	48090	Oct 1	8;57	8.27
48104	Oct 15	10;08	202.18	48119	Oct 30	19;58	37.56
48134	Nov 14	3;24	231.96	48149	Nov 29	6;38	67.28
48163	Dec 13	22;32	262.19	48178	Dec 28	17;08	97.24
				1717			
48193	Jan 12	17;56	292.56	48208	Jan 27	3;46	127.22
48223	Feb 11	12;03	322.76	48237	Feb 25	14;59	156.98
48253	Mar 13	3;51	352.50	48267	Mar 27	3;06	186.34
48282	Apr 11	16;43	21.64	48296	Apr 25	16;11	215.26
48312	May 11	2;43	50.19	48326	May 25	6;06	243.79
48341	Jun 9	10;29	78.32	48355	Jun 23	20;38	272.08
48370	Jul 8	17;06	106.23	48385	Jul 23	11;38	300.33
48399	Aug 6	23;51	134.23	48415	Aug 22	2;50	328.77
48429	Sep 5	7;57	162.54	48444	Sep 20	17;47	357.58
48458	Oct 4	18;16	191.37	48474	Oct 20	7;54	26.82
48488	Nov 3	7;18	220.80	48503	Nov 18	20;48	56.48
48517	Dec 2	23;01	250.77	48533	Dec 18	8;30	86.43
				1718			
48547	Jan 1	17;00	281.07	48562	Jan 16	19;21	116.45
48577	Jan 31	12;16	311.39	48592	Feb 15	5;45	146.28
48607	Mar 2	7;15	341.42	48621	Mar 16	15;54	175.74
48637	Apr 1	0;16	10.91	48651	Apr 15	2;01	204.71
48666	Apr 30	14;12	39.79	48680	May 14	12;27	233.24
48696	May 30	1;02	68.15	48709	Jun 12	23;46	261.48
48725	Jun 28	9;35	96.17	48739	Jul 12	12;40	289.64
48754	Jul 27	17;01	124.14	48769	Aug 11	3;29	317.98
48784	Aug 26	0;29	152.31	48798	Sep 9	19;54	346.68
48813	Sep 24	8;48	180.90	48828	Oct 9	12;51	15.86
48842	Oct 23	18;33	210.02	48858	Nov 8	5;12	45.50
48872	Nov 22	6;12	239.67	48887	Dec 7	20;11	75.49
48901	Dec 21	20;10	269.74				
				1719			
				48917	Jan 6	9;37	105.60
48931	Jan 20	12;33	299.99	48946	Feb 4	21;35	135.59
48961	Feb 19	6;45	330.11	48976	Mar 6	8;06	165.21
48991	Mar 21	1;15	359.85	49005	Apr 4	17;18	194.32
49020	Apr 19	18;22	29.04	49035	May 4	1;36	222.92
49050	May 19	9;05	57.68	49064	Jun 2	9;50	251.14
49079	Jun 17	21;22	85.92	49093	Jul 1	19;09	279.18
49109	Jul 17	7;44	113.99	49123	Jul 31	6;33	307.32
49138	Aug 15	16;55	142.13	49152	Aug 29	20;34	335.79
49168	Sep 14	1;34	170.58	49182	Sep 28	12;56	4.76
49197	Oct 13	10;19	199.48	49212	Oct 28	6;45	34.28
49226	Nov 11	19;47	228.89	49242	Nov 27	0;52	64.27
49256	Dec 11	6;36	258.73	49271	Dec 26	18;13	94.51
				1720			
49285	Jan 9	19;16	288.82	49301	Jan 25	9;50	124.71
49315	Feb 8	9;53	318.90	49330	Feb 23	23;04	154.57
49345	Mar 9	1;55	348.71	49360	Mar 24	9;43	183.92
49374	Apr 7	18;28	18.07	49389	Apr 22	18;12	212.70
49404	May 7	10;41	46.93	49419	May 22	1;27	240.99
49434	Jun 6	1;59	75.38	49448	Jun 20	8;34	269.01
49463	Jul 5	16;04	103.60	49477	Jul 19	16;43	296.98
49493	Aug 4	4;41	131.80	49507	Aug 18	2;48	325.18
49522	Sep 2	15;54	160.22	49536	Sep 16	15;23	353.83
49552	Oct 2	2;01	189.00	49566	Oct 16	6;39	23.06
49581	Oct 31	11;43	218.25	49596	Nov 15	0;24	52.87
49610	Nov 29	21;43	247.94	49625	Dec 14	19;44	83.11
49640	Dec 29	8;30	277.93				
				1721			
				49655	Jan 13	15;00	113.49
49669	Jan 27	20;12	307.95	49685	Feb 12	8;22	143.64
49699	Feb 26	8;42	337.75	49714	Mar 13	22;37	173.31
49728	Mar 27	21;53	7.15	49744	Apr 12	9;41	202.36
49758	Apr 26	11;49	36.09	49773	May 11	18;21	230.86
49788	May 26	2;34	64.64	49803	Jun 10	1;39	258.95
49817	Jun 24	17;58	92.96	49832	Jul 9	8;36	286.88
49847	Jul 24	9;20	121.23	49861	Aug 7	16;06	314.91
49876	Aug 22	23;53	149.65	49891	Sep 6	0;58	343.26
49906	Sep 21	13;13	178.40	49920	Oct 5	12;01	12.14
49936	Oct 21	1;26	207.59	49950	Nov 4	1;59	41.62
49965	Nov 19	13;02	237.20	49979	Dec 3	19;07	71.66
49995	Dec 19	0;19	267.14				

NEW MOONS JD	DATE	TIME	LONG.	FULL MOONS JD	DATE	TIME	LONG.
23+			**1722**	23+			
				50009	Jan 2	14;34	102.02
50024	Jan 17	11;21	297.16	50039	Feb 1	10;24	132.36
50053	Feb 15	21;58	327.00	50069	Mar 3	4;31	162.34
50083	Mar 17	8;19	356.45	50098	Apr 1	19;44	191.75
50112	Apr 15	18;54	25.43	50128	May 1	7;58	220.55
50142	May 15	6;26	54.00	50157	May 30	17;47	248.84
50171	Jun 13	19;29	82.29	50187	Jun 29	1;54	276.85
50201	Jul 13	10;01	110.52	50216	Jul 28	9;07	304.81
50231	Aug 12	1;33	138.89	50245	Aug 26	16;19	332.98
50260	Sep 10	17;21	167.58	50275	Sep 25	0;30	1.57
50290	Oct 10	8;57	196.71	50304	Oct 24	10;41	30.72
50319	Nov 8	23;58	226.32	50333	Nov 22	23;41	60.44
50349	Dec 8	14;02	256.27	50363	Dec 22	15;37	90.60
			1723				
50379	Jan 7	2;44	286.36	50393	Jan 21	9;44	120.91
50408	Feb 5	13;52	316.31	50423	Feb 20	4;31	151.06
50437	Mar 6	23;36	345.89	50452	Mar 21	22;32	180.76
50467	Apr 5	8;31	14.98	50482	Apr 20	14;44	209.90
50496	May 4	17;24	43.60	50512	May 20	4;36	238.50
50526	Jun 3	3;03	71.86	50541	Jun 18	15;59	266.70
50555	Jul 2	14;05	99.97	50571	Jul 18	1;15	294.72
50585	Aug 1	2;53	128.17	50600	Aug 16	9;10	322.82
50614	Aug 30	17;35	156.67	50629	Sep 14	16;49	351.23
50644	Sep 29	10;05	185.66	50659	Oct 14	1;20	20.13
50674	Oct 29	3;47	215.19	50688	Nov 12	11;29	49.58
50703	Nov 27	21;29	245.16	50717	Dec 11	23;40	79.48
50733	Dec 27	13;47	275.37				
			1724				
				50747	Jan 10	13;48	109.64
50763	Jan 26	3;45	305.49	50777	Feb 9	5;30	139.76
50792	Feb 24	15;18	335.29	50806	Mar 9	22;18	169.58
50822	Mar 25	0;57	4.58	50836	Apr 8	15;31	198.96
50851	Apr 23	9;23	33.34	50866	May 8	8;10	227.82
50880	May 22	17;16	61.66	50895	Jun 6	23;14	256.25
50910	Jun 21	1;19	89.70	50925	Jul 6	12;06	284.43
50939	Jul 20	10;20	117.71	50954	Aug 4	22;54	312.57
50968	Aug 18	21;13	145.96	50984	Sep 3	8;26	340.92
50998	Sep 17	10;50	174.66	51013	Oct 2	17;40	9.68
51028	Oct 17	3;26	203.95	51043	Nov 1	3;21	38.94
51057	Nov 15	22;17	233.82	51072	Nov 30	13;46	68.65
51087	Dec 15	17;36	264.07	51102	Dec 30	1;01	98.66
			1725				
51117	Jan 14	11;30	294.38	51131	Jan 28	13;08	128.69
51147	Feb 13	2;53	324.45	51161	Feb 27	2;24	158.52
51176	Mar 14	15;37	354.04	51190	Mar 28	16;59	187.97
51206	Apr 13	1;58	23.06	51220	Apr 27	8;37	216.96
51235	May 12	10;26	51.54	51250	May 27	0;33	245.55
51264	Jun 10	17;38	79.62	51279	Jun 25	15;51	273.86
51294	Jul 10	0;27	107.55	51309	Jul 25	6;00	302.08
51323	Aug 8	8;05	135.58	51338	Aug 23	18;59	330.45
51352	Sep 6	17;44	163.97	51368	Sep 22	7;05	359.16
51382	Oct 6	6;24	192.92	51397	Oct 21	18;36	28.32
51411	Nov 4	22;20	222.50	51427	Nov 20	5;39	57.93
51441	Dec 4	16;49	252.60	51456	Dec 19	16;18	87.84
			1726				
51471	Jan 3	12;19	282.96	51486	Jan 18	2;48	117.84
51501	Feb 2	7;14	313.26	51515	Feb 16	13;32	147.68
51531	Mar 4	0;14	343.19	51545	Mar 18	0;56	177.16
51560	Apr 2	14;31	12.54	51574	Apr 16	13;14	206.21
51590	May 2	1;50	41.30	51604	May 16	2;29	234.83
51619	May 31	10;34	69.55	51633	Jun 14	16;33	263.17
51648	Jun 29	17;35	97.51	51663	Jul 14	7;18	291.41
51678	Jul 29	0;09	125.45	51692	Aug 12	22;32	319.77
51707	Aug 27	7;31	153.63	51722	Sep 11	13;55	348.45
51736	Sep 25	16;45	182.27	51752	Oct 11	4;49	17.57
51766	Oct 25	4;29	211.49	51781	Nov 9	18;39	47.13
51795	Nov 23	18;56	241.28	51811	Dec 9	7;07	77.03
51825	Dec 23	11;51	271.48				
			1727				
				51840	Jan 7	18;26	107.05
51855	Jan 22	6;33	301.82	51870	Feb 6	4;57	136.97
51885	Feb 21	1;45	331.97	51899	Mar 7	15;02	166.56
51914	Mar 22	19;49	1.66	51929	Apr 6	0;55	195.68
51944	Apr 21	11;17	30.76	51958	May 5	10;53	224.32
51973	May 20	23;34	59.29	51987	Jun 3	21;27	252.62
52003	Jun 19	9;09	87.41	52017	Jul 3	9;19	280.77
52032	Jul 18	17;04	115.38	52046	Aug 1	23;08	309.01
52062	Aug 17	0;31	143.46	52076	Aug 31	14;55	337.57
52091	Sep 15	8;28	171.90	52106	Sep 30	7;58	6.59
52120	Oct 14	17;35	200.84	52136	Oct 30	1;02	36.10
52150	Nov 13	4;20	230.32	52165	Nov 28	17;00	66.02
52179	Dec 12	17;07	260.25	52195	Dec 28	7;23	96.15
23+			**1728**	23+			
52209	Jan 11	8;14	290.45	52224	Jan 26	20;07	126.21
52239	Feb 10	1;29	320.63	52254	Feb 25	7;17	155.98
52268	Mar 10	19;49	350.51	52283	Mar 25	16;58	185.27
52298	Apr 9	13;37	19.88	52313	Apr 24	1;27	214.03
52328	May 9	5;29	48.71	52342	May 23	9;24	242.33
52357	Jun 7	18;56	77.07	52371	Jun 21	17;52	270.39
52387	Jul 7	6;15	105.18	52401	Jul 21	4;01	298.45
52416	Aug 5	16;05	133.28	52430	Aug 19	16;41	326.77
52446	Sep 4	1;07	161.62	52460	Sep 18	8;02	355.55
52475	Oct 3	9;58	190.38	52490	Oct 18	1;26	24.90
52504	Nov 1	19;10	219.62	52519	Nov 16	19;45	54.75
52534	Dec 1	5;20	249.34	52549	Dec 16	13;48	84.95
52563	Dec 30	17;01	279.36				
			1729				
				52579	Jan 15	6;29	115.22
52593	Jan 29	6;33	309.46	52608	Feb 13	20;58	145.24
52622	Feb 27	21;44	339.35	52638	Mar 15	8;49	174.78
52652	Mar 29	13;51	8.85	52667	Apr 13	18;10	203.75
52682	Apr 28	6;04	37.86	52697	May 13	1;45	232.18
52711	May 27	21;44	66.42	52726	Jun 11	8;40	260.25
52741	Jun 26	12;27	94.70	52755	Jul 10	16;06	288.20
52771	Jul 26	1;57	122.91	52785	Aug 9	1;06	316.29
52800	Aug 24	14;05	151.25	52814	Sep 7	12;25	344.76
52830	Sep 23	0;57	179.93	52844	Oct 7	2;26	13.78
52859	Oct 22	11;02	209.04	52873	Nov 5	19;07	43.40
52888	Nov 20	20;59	238.61	52903	Dec 5	13;55	73.52
52918	Dec 20	7;25	268.51				
			1730				
				52933	Jan 4	9;32	103.90
52947	Jan 18	18;37	298.54	52963	Feb 3	4;05	134.17
52977	Feb 17	6;34	328.42	52992	Mar 4	19;55	164.04
53006	Mar 18	19;09	357.95	53022	Apr 3	8;27	193.31
53036	Apr 17	8;22	27.01	53051	May 2	18;05	221.98
53065	May 16	22;27	55.66	53081	Jun 1	1;51	250.18
53095	Jun 15	13;25	84.02	53110	Jun 30	8;49	278.14
53125	Jul 15	4;54	112.29	53139	Jul 29	15;58	306.11
53154	Aug 13	20;06	140.66	53169	Aug 28	0;07	334.33
53184	Sep 12	10;19	169.31	53198	Sep 26	10;03	3.00
53213	Oct 11	23;21	198.37	53227	Oct 25	22;34	32.27
53243	Nov 10	11;29	227.87	53257	Nov 24	14;12	62.13
53272	Dec 9	23;05	257.73	53287	Dec 24	8;40	92.40
			1731				
53302	Jan 8	10;18	287.76	53317	Jan 23	4;33	122.78
53331	Feb 6	21;03	317.68	53346	Feb 21	23;40	152.92
53361	Mar 8	7;19	347.27	53376	Mar 23	16;18	182.54
53390	Apr 6	17;27	16.39	53406	Apr 22	5;55	211.55
53420	May 6	4;09	45.06	53435	May 21	16;48	240.01
53449	Jun 4	16;09	73.40	53465	Jun 20	1;41	268.10
53479	Jul 4	5;50	101.61	53494	Jul 19	9;18	296.05
53508	Aug 2	20;57	129.91	53523	Aug 17	16;30	324.13
53538	Sep 1	12;51	158.48	53553	Sep 16	0;11	352.57
53568	Oct 1	4;52	187.48	53582	Oct 15	9;22	21.52
53597	Oct 30	20;30	216.95	53611	Nov 13	20;59	51.04
53627	Nov 29	11;20	246.83	53641	Dec 13	11;29	81.06
53657	Dec 29	0;55	276.92				
			1732				
				53671	Jan 12	4;35	111.34
53686	Jan 27	12;52	306.95	53700	Feb 10	23;04	141.56
53715	Feb 25	23;09	336.67	53730	Mar 11	17;26	171.43
53745	Mar 26	8;12	5.93	53760	Apr 10	10;28	200.76
53774	Apr 24	16;47	34.68	53790	May 10	1;28	229.54
53804	May 24	1;44	63.01	53819	Jun 8	14;04	257.86
53833	Jun 22	11;51	91.13	53849	Jul 8	0;24	285.93
53862	Jul 21	23;37	119.26	53878	Aug 6	9;01	313.99
53892	Aug 20	13;22	147.63	53907	Sep 4	16;50	342.29
53922	Sep 19	5;08	176.44	53937	Oct 4	1;00	11.02
53951	Oct 18	22;33	205.80	53966	Nov 2	10;25	40.30
53981	Nov 17	16;41	235.67	53995	Dec 1	21;39	70.06
54011	Dec 17	10;04	265.85	54025	Dec 31	10;48	100.15
			1733				
54041	Jan 16	1;22	296.05	54055	Jan 30	1;37	130.29
54070	Feb 14	14;04	325.99	54084	Feb 28	17;42	160.22
54100	Mar 16	0;29	355.47	54114	Mar 30	10;32	189.73
54129	Apr 14	9;16	24.39	54144	Apr 29	3;23	218.75
54158	May 13	17;11	52.83	54173	May 28	19;15	247.32
54188	Jun 12	0;55	80.93	54203	Jun 27	9;17	275.56
54217	Jul 11	9;14	108.91	54232	Jul 26	21;12	303.71
54246	Aug 9	19;01	137.04	54262	Aug 25	7;28	331.98
54276	Sep 8	7;13	165.55	54291	Sep 23	16;59	0.61
54305	Oct 7	22;27	194.63	54321	Oct 23	2;36	29.71
54335	Nov 6	16;30	224.32	54350	Nov 21	12;49	59.30
54365	Dec 6	12;00	254.49	54379	Dec 20	23;42	89.24

Left Page

NEW MOONS				FULL MOONS			
JD	DATE	TIME	LONG.	JD	DATE	TIME	LONG.
23+			**1734**	**23+**			
54395	Jan 5	6;55	284.84	54409	Jan 19	11;18	119.28
54424	Feb 3	23;40	315.03	54438	Feb 17	23;46	149.18
54454	Mar 5	13;39	344.81	54468	Mar 19	13;23	178.74
54484	Apr 4	1;02	14.02	54498	Apr 18	4;15	207.86
54513	May 3	10;16	42.67	54527	May 17	19;55	236.56
54542	Jun 1	17;54	70.86	54557	Jun 16	11;31	264.94
54572	Jul 1	0;44	98.81	54587	Jul 16	2;20	293.17
54601	Jul 30	7;48	126.77	54616	Aug 14	16;03	321.49
54630	Aug 28	16;20	155.01	54646	Sep 13	4;50	350.09
54660	Sep 27	3;28	183.75	54675	Oct 12	16;54	19.12
54689	Oct 26	17;53	213.11	54705	Nov 11	4;24	48.61
54719	Nov 25	11;19	243.05	54734	Dec 10	15;25	78.46
54749	Dec 25	6;34	273.36				
			1735				
				54764	Jan 9	2;00	108.45
54779	Jan 24	1;57	303.72	54793	Feb 7	12;28	138.37
54808	Feb 22	19;59	333.80	54822	Mar 8	23;16	167.96
54838	Mar 24	11;37	3.37	54852	Apr 7	10;48	197.13
54868	Apr 23	0;20	32.33	54881	May 6	23;18	225.86
54897	May 22	10;13	60.74	54911	Jun 5	12;45	254.25
54926	Jun 20	17;56	88.78	54941	Jul 5	3;04	282.49
54956	Jul 20	0;36	116.69	54970	Aug 3	18;07	310.78
54985	Aug 18	7;29	144.77	55000	Sep 2	9;41	339.35
55014	Sep 16	15;46	173.23	55030	Oct 2	1;11	8.34
55044	Oct 16	2;18	202.25	55059	Oct 31	15;55	37.79
55073	Nov 14	15;28	231.85	55089	Nov 30	5;18	67.61
55103	Dec 14	7;14	261.93	55118	Dec 29	17;18	97.64
			1736				
55133	Jan 13	1;05	292.24	55148	Jan 28	4;10	127.63
55162	Feb 11	20;07	322.49	55177	Feb 26	14;20	157.34
55192	Mar 12	14;49	352.36	55207	Mar 27	0;05	186.62
55222	Apr 11	7;35	21.66	55236	Apr 25	9;43	215.40
55251	May 10	21;24	50.38	55265	May 24	19;39	243.77
55281	Jun 9	8;14	78.62	55295	Jun 23	6;35	271.92
55310	Jul 8	16;53	106.62	55324	Jul 22	19;17	300.08
55340	Aug 7	0;32	134.65	55354	Aug 21	10;08	328.50
55369	Sep 5	8;17	162.95	55384	Sep 20	2;51	357.36
55398	Oct 4	16;55	191.71	55413	Oct 19	20;20	26.73
55428	Nov 3	2;56	221.01	55443	Nov 18	13;13	56.56
55457	Dec 2	14;42	250.81	55473	Dec 18	4;38	86.66
			1737				
55487	Jan 1	4;35	280.93	55502	Jan 16	18;17	116.79
55516	Jan 30	20;42	311.13	55532	Feb 15	6;12	146.69
55546	Mar 1	14;27	341.11	55561	Mar 16	16;29	176.16
55576	Mar 31	8;30	10.67	55591	Apr 15	1;20	205.08
55606	Apr 30	1;17	39.67	55620	May 14	9;14	233.51
55635	May 29	15;54	68.17	55649	Jun 12	17;08	261.61
55665	Jun 28	4;15	96.35	55679	Jul 12	2;11	289.62
55694	Jul 27	14;53	124.45	55708	Aug 10	13;28	317.81
55724	Aug 26	0;26	152.70	55738	Sep 9	3;34	346.41
55753	Sep 24	9;32	181.32	55767	Oct 8	20;11	15.56
55782	Oct 23	18;41	210.41	55797	Nov 7	14;23	45.27
55812	Nov 22	4;26	239.99	55827	Dec 7	8;55	75.40
55841	Dec 21	15;20	269.93				
			1738				
				55857	Jan 6	2;32	105.69
55871	Jan 20	3;50	300.01	55886	Feb 4	18;14	135.84
55900	Feb 18	18;02	329.97	55916	Mar 6	7;22	165.57
55930	Mar 20	9;31	359.60	55945	Apr 4	17;48	194.74
55960	Apr 19	1;30	28.75	55975	May 4	2;01	223.33
55989	May 18	17;18	57.45	56004	Jun 2	9;00	251.50
56019	Jun 17	8;28	85.80	56033	Jul 1	15;56	279.45
56048	Jul 16	22;41	114.02	56063	Jul 31	0;00	307.45
56078	Aug 15	11;42	142.31	56092	Aug 29	10;07	335.76
56107	Sep 13	23;26	170.88	56121	Sep 27	22;49	4.58
56137	Oct 13	10;06	199.86	56151	Oct 27	14;18	33.99
56166	Nov 11	20;16	229.30	56181	Nov 26	8;15	63.96
56196	Dec 11	6;32	259.12	56211	Dec 26	3;45	94.28
			1739				
56225	Jan 9	17;20	289.13	56240	Jan 24	23;06	124.64
56255	Feb 8	4;48	319.08	56270	Feb 23	16;25	154.68
56284	Mar 9	16;50	348.71	56300	Mar 25	6;31	184.18
56314	Apr 8	5;26	17.92	56329	Apr 23	17;25	213.05
56343	May 7	18;46	46.67	56359	May 23	1;55	241.40
56373	Jun 6	9;07	75.10	56388	Jun 21	9;07	269.42
56403	Jul 6	0;21	103.37	56417	Jul 20	16;04	297.35
56432	Aug 4	15;53	131.69	56446	Aug 18	23;40	325.46
56462	Sep 3	6;53	160.24	56476	Sep 17	8;42	353.96
56491	Oct 2	20;47	189.18	56505	Oct 16	19;56	23.01
56521	Nov 1	9;36	218.56	56535	Nov 15	10;01	52.66
56550	Nov 30	21;38	248.33	56565	Dec 15	3;11	82.80
56580	Dec 30	9;09	278.34				

Right Page

NEW MOONS				FULL MOONS			
JD	DATE	TIME	LONG.	JD	DATE	TIME	LONG.
23+			**1740**	**23+**			
				56594	Jan 13	22;34	113.18
56609	Jan 28	20;08	308.33	56624	Feb 12	18;14	143.43
56639	Feb 27	6;29	338.04	56654	Mar 13	12;09	173.26
56668	Mar 27	16;23	7.31	56684	Apr 12	3;11	202.48
56698	Apr 26	2;26	36.09	56713	May 11	15;19	231.12
56727	May 25	13;27	64.51	56743	Jun 10	1;05	259.32
56757	Jun 24	2;06	92.72	56772	Jul 9	9;17	287.30
56786	Jul 23	16;30	120.95	56801	Aug 7	16;41	315.32
56816	Aug 22	8;11	149.42	56831	Sep 6	0;08	343.62
56846	Sep 21	0;25	178.27	56860	Oct 5	8;35	12.39
56875	Oct 20	16;34	207.61	56889	Nov 3	19;00	41.72
56905	Nov 19	8;09	237.39	56919	Dec 3	8;04	71.58
56934	Dec 18	22;37	267.46				
			1741				
				56948	Jan 1	23;56	101.79
56964	Jan 17	11;29	297.56	56978	Jan 31	17;45	132.05
56993	Feb 15	22;31	327.41	57008	Mar 2	12;09	162.05
57023	Mar 17	7;58	356.82	57038	Apr 1	5;47	191.56
57052	Apr 15	16;28	25.72	57067	Apr 30	21;44	220.52
57082	May 15	0;54	54.16	57097	May 30	11;31	248.98
57111	Jun 13	10;10	82.31	57126	Jun 28	23;00	277.13
57140	Jul 12	20;56	110.39	57156	Jul 28	8;31	305.18
57170	Aug 11	9;39	138.65	57185	Aug 26	16;47	333.39
57200	Sep 10	0;29	167.29	57215	Sep 25	0;50	1.98
57229	Oct 9	17;19	196.47	57244	Oct 24	9;42	31.07
57259	Nov 8	11;29	226.19	57273	Nov 22	20;06	60.68
57289	Dec 8	5;38	256.31	57303	Dec 22	8;20	90.68
			1742				
57318	Jan 6	22;14	286.56	57332	Jan 20	22;15	120.82
57348	Feb 5	12;17	316.63	57362	Feb 19	13;31	150.82
57377	Mar 6	23;42	346.28	57392	Mar 21	5;46	180.46
57407	Apr 5	9;04	15.39	57421	Apr 19	22;28	209.63
57436	May 4	17;10	43.97	57451	May 19	14;48	238.33
57466	Jun 3	0;45	72.15	57481	Jun 18	5;49	266.67
57495	Jul 2	8;35	100.14	57510	Jul 17	18;54	294.85
57524	Jul 31	17;29	128.18	57540	Aug 16	6;06	323.08
57554	Aug 30	4;23	156.53	57569	Sep 14	16;08	351.59
57583	Sep 28	18;08	185.40	57599	Oct 14	1;51	20.55
57613	Oct 28	11;00	214.89	57628	Nov 12	11;56	50.00
57643	Nov 27	6;08	244.92	57657	Dec 11	22;35	79.84
57673	Dec 27	1;42	275.25				
			1743				
				57687	Jan 10	9;48	109.86
57702	Jan 25	19;43	305.54	57716	Feb 8	21;38	139.82
57732	Feb 24	11;04	335.50	57746	Mar 10	10;23	169.48
57761	Mar 25	23;38	4.91	57776	Apr 9	0;21	198.72
57791	Apr 24	9;47	33.74	57805	May 8	15;25	227.53
57820	May 23	18;02	62.07	57835	Jun 7	7;00	255.99
57850	Jun 22	1;06	90.07	57864	Jul 6	22;16	284.26
57879	Jul 21	7;53	118.00	57894	Aug 5	12;40	312.54
57908	Aug 19	15;34	146.12	57924	Sep 4	2;08	341.05
57938	Sep 18	1;20	174.66	57953	Oct 3	14;50	9.95
57967	Oct 17	14;10	203.80	57983	Nov 2	2;54	39.31
57997	Nov 16	6;15	233.55	58012	Dec 1	14;21	69.08
58027	Dec 16	0;50	263.76	58042	Dec 31	1;11	99.06
			1744				
58056	Jan 14	20;21	294.14	58071	Jan 29	11;36	129.02
58086	Feb 13	15;10	324.35	58100	Feb 27	22;00	158.73
58116	Mar 14	8;01	354.12	58130	Mar 28	8;53	188.02
58145	Apr 12	22;08	23.29	58159	Apr 26	20;36	216.87
58175	May 12	9;19	51.88	58189	May 26	9;20	245.33
58204	Jun 10	17;59	80.02	58218	Jun 24	23;03	273.58
58234	Jul 10	1;02	107.96	58248	Jul 24	13;44	301.83
58263	Aug 8	7;43	135.95	58278	Aug 23	5;12	330.29
58292	Sep 6	15;17	164.27	58307	Sep 21	21;04	359.15
58322	Oct 6	0;44	193.08	58337	Oct 21	12;35	28.47
58351	Nov 4	12;40	222.49	58367	Nov 20	2;58	58.21
58381	Dec 4	3;12	252.42	58396	Dec 19	15;49	88.22
			1745				
58410	Jan 2	20;02	282.67	58426	Jan 18	3;15	118.26
58440	Feb 1	14;30	312.95	58455	Feb 16	13;38	148.08
58470	Mar 3	9;24	342.96	58484	Mar 17	23;23	177.50
58500	Apr 2	3;10	12.47	58514	Apr 16	8;48	206.42
58529	May 1	18;26	41.39	58543	May 15	18;16	234.89
58559	May 31	6;41	69.78	58573	Jun 14	4;24	263.07
58588	Jun 29	16;20	97.85	58602	Jul 13	16;00	291.18
58618	Jul 29	0;27	125.84	58632	Aug 12	5;45	319.48
58647	Aug 27	8;12	154.04	58661	Sep 10	21;45	348.18
58676	Sep 25	16;29	182.65	58691	Oct 10	15;14	17.40
58706	Oct 25	1;55	211.78	58721	Nov 9	8;51	47.11
58735	Nov 23	12;52	241.42	58751	Dec 9	1;19	77.18
58765	Dec 23	1;40	271.45				

NEW MOONS				FULL MOONS			
JD	DATE	TIME	LONG.	JD	DATE	TIME	LONG.
23+			1746	23+			
				58780	Jan 7	15;59	107.35
58794	Jan 21	16;33	301.63	58810	Feb 6	4;46	137.36
58824	Feb 20	9;23	331.69	58839	Mar 7	15;45	166.98
58854	Mar 22	3;13	1.38	58869	Apr 6	1;07	196.08
58883	Apr 20	20;35	30.56	58898	May 5	9;13	224.65
58913	May 20	12;14	59.22	58927	Jun 3	16;48	252.82
58943	Jun 19	1;42	87.50	58957	Jul 3	1;00	280.82
58972	Jul 18	13;13	115.61	58986	Aug 1	10;59	308.90
59001	Aug 16	23;25	143.80	59015	Aug 30	23;40	337.34
59031	Sep 15	8;54	172.30	59045	Sep 29	15;12	6.29
59060	Oct 14	18;12	201.25	59075	Oct 29	8;56	35.83
59090	Nov 13	3;45	230.69	59105	Nov 28	3;39	65.85
59119	Dec 12	14;06	260.52	59134	Dec 27	22;01	96.14
			1747				
59149	Jan 11	1;43	290.57	59164	Jan 26	14;51	126.39
59178	Feb 9	14;55	320.58	59194	Feb 25	5;18	156.29
59208	Mar 11	5;34	350.31	59223	Mar 26	16;57	185.66
59237	Apr 9	21;06	19.60	59253	Apr 25	2;03	214.44
59267	May 9	12;50	48.43	59282	May 24	9;24	242.72
59297	Jun 8	4;14	76.87	59311	Jun 22	16;07	270.71
59326	Jul 7	18;59	105.12	59340	Jul 21	23;27	298.66
59356	Aug 6	8;47	133.38	59370	Aug 20	8;27	326.83
59385	Sep 4	21;24	161.86	59399	Sep 18	19;52	355.45
59415	Oct 4	8;50	190.72	59429	Oct 18	10;02	24.65
59444	Nov 2	19;25	220.03	59459	Nov 17	2;55	54.44
59474	Dec 2	5;43	249.75	59488	Dec 16	21;54	84.67
59503	Dec 31	16;17	279.72				
			1748				
				59518	Jan 15	17;36	115.06
59533	Jan 30	3;21	309.71	59548	Feb 14	12;07	145.26
59562	Feb 28	14;55	339.46	59578	Mar 15	3;51	174.97
59592	Mar 29	2;57	8.79	59607	Apr 13	16;12	204.06
59621	Apr 27	15;35	37.67	59637	May 13	1;41	232.57
59651	May 27	5;09	66.16	59666	Jun 11	9;20	260.67
59680	Jun 25	19;50	94.44	59695	Jul 10	16;18	288.60
59710	Jul 25	11;20	122.72	59724	Aug 8	23;31	316.62
59740	Aug 24	2;53	151.20	59754	Sep 7	7;49	344.97
59769	Sep 22	17;39	180.02	59783	Oct 6	17;57	13.82
59799	Oct 22	7;17	209.27	59813	Nov 5	6;37	43.27
59828	Nov 20	19;56	238.95	59842	Dec 4	22;19	73.26
59858	Dec 20	7;51	268.93				
			1749				
				59872	Jan 3	16;45	103.58
59887	Jan 18	19;08	298.96	59902	Feb 2	12;28	133.91
59917	Feb 17	5;43	328.78	59932	Mar 4	7;21	163.91
59946	Mar 18	15;36	358.20	59961	Apr 2	23;47	193.35
59976	Apr 17	1;13	27.12	59991	May 2	13;14	222.18
60005	May 16	11;23	55.61	60021	Jun 1	0;04	250.51
60034	Jun 14	22;56	83.84	60050	Jun 30	8;59	278.54
60064	Jul 14	12;22	112.03	60079	Jul 29	16;46	306.53
60094	Aug 13	3;30	140.39	60109	Aug 28	0;13	334.72
60123	Sep 11	19;43	169.10	60138	Sep 26	8;12	3.32
60153	Oct 11	12;14	198.30	60167	Oct 25	17;38	32.46
60183	Nov 10	4;27	227.96	60197	Nov 24	5;24	62.14
60212	Dec 9	19;46	257.98	60226	Dec 23	19;53	92.25
			1750				
60242	Jan 8	9;37	288.12	60256	Jan 22	12;45	122.51
60271	Feb 6	21;34	318.09	60286	Feb 21	6;51	152.61
60301	Mar 8	7;38	347.67	60316	Mar 23	0;48	182.30
60330	Apr 6	16;19	16.73	60345	Apr 21	17;31	211.44
60360	May 6	0;28	45.30	60375	May 21	8;21	240.06
60389	Jun 4	9;02	73.50	60404	Jun 19	21;00	268.30
60418	Jul 3	18;49	101.57	60434	Jul 19	7;33	296.38
60448	Aug 2	6;27	129.72	60463	Aug 17	16;30	324.52
60477	Aug 31	20;15	158.20	60493	Sep 16	0;44	352.98
60507	Sep 30	12;16	187.19	60522	Oct 15	9;16	21.91
60537	Oct 30	6;06	216.74	60551	Nov 13	18;59	51.36
60567	Nov 29	0;40	246.77	60581	Dec 13	6;21	81.24
60596	Dec 28	18;23	277.03				
			1751				
				60610	Jan 11	19;23	111.34
60626	Jan 27	9;50	307.22	60640	Feb 10	9;50	141.40
60655	Feb 25	22;29	337.04	60670	Mar 12	1;23	171.16
60685	Mar 27	8;41	6.33	60699	Apr 10	17;39	200.47
60714	Apr 25	17;10	35.08	60729	May 10	10;05	229.31
60744	May 25	0;47	63.37	60759	Jun 9	1;48	257.76
60773	Jun 23	8;17	91.38	60788	Jul 8	15;57	285.99
60802	Jul 22	16;28	119.37	60818	Aug 7	4;13	314.19
60832	Aug 21	2;15	147.58	60847	Sep 5	14;57	342.61
60861	Sep 19	14;32	176.24	60877	Oct 5	0;59	11.42
60891	Oct 19	5;57	205.51	60906	Nov 3	11;04	40.72
60921	Nov 18	0;16	235.37	60935	Dec 2	21;34	70.45
60950	Dec 17	20;00	265.64				

NEW MOONS				FULL MOONS			
JD	DATE	TIME	LONG.	JD	DATE	TIME	LONG.
23+			1752	23+			
				60965	Jan 1	8;32	100.45
60980	Jan 16	15;04	296.01	60994	Jan 30	19;57	130.44
61010	Feb 15	7;50	326.12	61024	Feb 29	7;59	160.20
61039	Mar 15	21;42	355.74	61053	Mar 29	21;01	189.56
61069	Apr 14	8;54	24.77	61083	Apr 28	11;17	218.49
61098	May 13	17;56	53.25	61113	May 28	2;29	247.04
61128	Jun 12	1;25	81.34	61142	Jun 26	17;55	275.35
61157	Jul 11	8;12	109.27	61172	Jul 26	8;52	303.61
61186	Aug 9	15;18	137.28	61201	Aug 24	22;59	332.04
61215	Sep 7	23;56	165.65	61231	Sep 23	12;19	0.81
61245	Oct 7	11;13	194.57	61261	Oct 23	0;59	30.04
61275	Nov 6	1;47	224.10	61290	Nov 21	12;59	59.70
61304	Dec 5	19;20	254.17	61320	Dec 21	0;15	89.65
			1753				
61334	Jan 4	14;35	284.53	61349	Jan 19	10;50	119.65
61364	Feb 3	9;53	314.84	61378	Feb 17	21;04	149.46
61394	Mar 5	3;46	344.79	61408	Mar 19	7;25	178.88
61423	Apr 3	19;13	14.17	61437	Apr 17	18;24	207.85
61453	May 3	7;47	42.96	61467	May 17	6;21	236.40
61482	Jun 1	17;36	71.24	61496	Jun 15	19;22	264.69
61512	Jul 1	1;20	99.22	61526	Jul 15	9;30	292.90
61541	Jul 30	8;07	127.18	61556	Aug 14	0;41	321.27
61570	Aug 28	15;12	155.36	61585	Sep 12	16;37	349.99
61599	Sep 26	23;44	183.99	61615	Oct 12	8;42	19.18
61629	Oct 26	10;29	213.20	61645	Nov 11	0;02	48.82
61658	Nov 24	23;47	242.96	61674	Dec 10	13;53	78.78
61688	Dec 24	15;30	273.11				
			1754				
				61704	Jan 9	2;05	108.85
61718	Jan 23	9;09	303.40	61733	Feb 7	12;54	138.77
61748	Feb 22	3;50	333.52	61762	Mar 8	22;47	168.33
61777	Mar 23	22;11	3.21	61792	Apr 7	8;07	197.41
61807	Apr 22	14;43	32.33	61821	May 6	17;16	226.00
61837	May 22	4;26	60.89	61851	Jun 5	2;46	254.24
61866	Jun 20	15;19	89.06	61880	Jul 4	13;23	282.33
61896	Jul 20	0;10	117.06	61910	Aug 3	1;56	310.53
61925	Aug 18	8;06	145.18	61939	Sep 1	16;54	339.07
61954	Sep 16	16;12	173.64	61969	Oct 1	9;58	8.11
61984	Oct 16	1;11	202.60	61999	Oct 31	3;57	37.68
62013	Nov 14	11;28	232.08	62028	Nov 29	21;21	67.67
62042	Dec 13	23;19	261.99	62058	Dec 29	13;07	97.86
			1755				
62072	Jan 12	13;03	292.12	62088	Jan 28	2;53	127.97
62102	Feb 11	4;47	322.23	62117	Feb 26	14;42	157.74
62131	Mar 12	22;01	352.04	62147	Mar 28	0;43	187.03
62161	Apr 11	15;34	21.40	62176	Apr 26	9;13	215.76
62191	May 11	8;03	50.22	62205	May 25	16;47	244.04
62220	Jun 9	22;35	78.62	62235	Jun 24	0;23	272.05
62250	Jul 9	11;06	106.78	62264	Jul 23	9;16	300.07
62279	Aug 7	22;03	134.94	62293	Aug 21	20;31	328.34
62309	Sep 6	8;03	163.33	62323	Sep 20	10;43	357.10
62338	Oct 5	17;37	192.14	62353	Oct 20	3;36	26.44
62368	Nov 4	3;11	221.42	62382	Nov 18	22;10	56.32
62397	Dec 3	13;11	251.15	62412	Dec 18	17;00	86.57
			1756				
62427	Jan 2	0;06	281.14	62442	Jan 17	10;49	116.87
62456	Jan 31	12;23	311.16	62472	Feb 16	2;33	146.93
62486	Mar 1	2;07	340.99	62501	Mar 16	15;32	176.51
62515	Mar 30	16;59	10.41	62531	Apr 15	1;44	205.48
62545	Apr 29	8;25	39.37	62560	May 14	9;43	233.92
62574	May 28	23;51	67.92	62589	Jun 12	16;30	261.97
62604	Jun 27	14;55	96.21	62618	Jul 11	23;20	289.89
62634	Jul 27	5;20	124.45	62648	Aug 10	7;23	317.96
62663	Aug 25	18;48	152.87	62677	Sep 8	17;34	346.40
62693	Sep 24	7;05	181.61	62707	Oct 8	6;26	15.39
62722	Oct 23	18;19	210.79	62736	Nov 6	22;06	44.99
62752	Nov 22	4;54	240.41	62766	Dec 6	16;13	75.09
62781	Dec 21	15;23	270.33				
			1757				
				62796	Jan 5	11;48	105.47
62811	Jan 20	2;10	300.34	62826	Feb 4	7;08	135.77
62840	Feb 18	13;20	330.17	62856	Mar 6	0;20	165.68
62870	Mar 20	0;53	359.63	62885	Apr 4	14;17	194.99
62899	Apr 18	12;53	28.63	62915	May 4	1;01	223.68
62929	May 18	1;40	57.21	62944	Jun 2	9;24	251.90
62958	Jun 16	15;36	85.52	62973	Jul 1	16;33	279.86
62988	Jul 16	6;43	113.77	63002	Jul 30	23;34	307.82
63017	Aug 14	22;28	142.17	63032	Aug 29	7;20	336.04
63047	Sep 13	13;57	170.88	63061	Sep 27	16;34	4.71
63077	Oct 13	4;28	200.02	63091	Oct 27	3;59	33.95
63106	Nov 11	17;52	229.60	63120	Nov 25	18;12	63.76
63136	Dec 11	6;19	259.51	63150	Dec 25	11;21	93.99

Left panel

NEW MOONS				FULL MOONS			
JD	DATE	TIME	LONG.	JD	DATE	TIME	LONG.
23+				1758 23+			
63165	Jan 9	18;00	289.56	63180	Jan 24	6;35	124.34
63195	Feb 8	4;53	319.48	63210	Feb 23	2;00	154.48
63224	Mar 9	14;55	349.04	63239	Mar 24	19;39	184.12
63254	Apr 8	0;22	18.10	63269	Apr 23	10;30	213.16
63283	May 7	9;53	46.70	63298	May 22	22;31	241.65
63312	Jun 5	20;25	74.98	63328	Jun 21	8;19	269.77
63342	Jul 5	8;44	103.13	63357	Jul 20	16;40	297.75
63371	Aug 3	23;03	131.40	63387	Aug 19	0;18	325.87
63401	Sep 2	14;55	159.99	63416	Sep 17	8;04	354.32
63431	Oct 2	7;36	189.03	63445	Oct 16	16;49	23.27
63461	Nov 1	0;19	218.56	63475	Nov 15	3;26	52.78
63490	Nov 30	16;24	248.51	63504	Dec 14	16;34	82.75
63520	Dec 30	7;13	278.66				
			1759				
				63534	Jan 13	8;15	112.97
63549	Jan 28	20;11	308.73	63564	Feb 12	1;43	143.14
63579	Feb 27	7;05	338.46	63593	Mar 13	19;40	172.97
63608	Mar 28	16;13	7.69	63623	Apr 12	12;54	202.29
63638	Apr 27	0;19	36.40	63653	May 12	4;36	231.08
63667	May 26	8;21	64.69	63682	Jun 10	18;21	259.44
63696	Jun 24	17;17	92.75	63712	Jul 10	6;01	287.56
63726	Jul 24	3;51	120.84	63741	Aug 8	15;50	315.68
63755	Aug 22	16;32	149.18	63771	Sep 7	0;31	344.03
63785	Sep 21	7;33	177.98	63800	Oct 6	8;59	12.80
63815	Oct 21	0;43	207.34	63829	Nov 4	18;12	42.08
63844	Nov 19	19;18	237.24	63859	Dec 4	4;48	71.83
63874	Dec 19	13;50	267.48				
			1760				
				63888	Jan 2	17;01	101.88
63904	Jan 18	6;40	297.74	63918	Feb 1	6;40	131.97
63933	Feb 16	20;43	327.73	63947	Mar 1	21;27	161.83
63963	Mar 17	7;59	357.22	63977	Mar 31	13;08	191.28
63992	Apr 15	17;05	26.15	64007	Apr 30	5;20	220.26
64022	May 15	0;53	54.56	64036	May 29	21;22	248.81
64051	Jun 13	8;14	82.63	64066	Jun 28	12;23	277.10
64080	Jul 12	15;54	110.59	64096	Jul 28	1;43	305.30
64110	Aug 11	0;44	138.68	64125	Aug 26	13;22	333.65
64139	Sep 9	11;42	167.16	64154	Sep 24	23;55	2.33
64169	Oct 9	1;36	196.21	64184	Oct 24	10;09	31.48
64198	Nov 7	18;40	225.87	64213	Nov 22	20;36	61.10
64228	Dec 7	14;03	256.04	64243	Dec 22	7;25	91.03
			1761				
64258	Jan 6	9;47	286.42	64272	Jan 20	18;33	121.05
64288	Feb 5	3;51	316.67	64302	Feb 19	6;04	150.90
64317	Mar 6	19;08	346.49	64331	Mar 20	18;17	180.38
64347	Apr 5	7;32	15.73	64361	Apr 19	7;38	209.42
64376	May 4	17;29	44.38	64390	May 18	22;10	238.06
64406	Jun 3	1;35	72.58	64420	Jun 17	13;26	266.42
64435	Jul 2	8;34	100.53	64450	Jul 17	4;41	294.68
64464	Jul 31	15;23	128.48	64479	Aug 15	19;24	323.04
64493	Aug 29	23;09	156.71	64509	Sep 14	9;22	351.71
64523	Sep 28	9;05	185.42	64538	Oct 13	22;40	20.81
64552	Oct 27	22;04	214.74	64568	Nov 12	11;17	50.36
64582	Nov 26	14;17	244.64	64597	Dec 11	23;06	80.25
64612	Dec 26	8;53	274.93				
			1762				
				64627	Jan 10	10;03	110.27
64642	Jan 25	4;19	305.29	64656	Feb 8	20;19	140.15
64671	Feb 23	22;58	335.38	64686	Mar 10	6;21	169.70
64701	Mar 25	15;37	4.97	64715	Apr 8	16;43	198.80
64731	Apr 24	5;34	33.97	64745	May 8	3;52	227.46
64760	May 23	16;40	62.41	64774	Jun 6	16;06	255.79
64790	Jun 22	1;20	90.47	64804	Jul 6	5;33	283.99
64819	Jul 21	8;30	118.42	64833	Aug 4	20;13	312.28
64848	Aug 19	15;22	146.50	64863	Sep 3	11;59	340.87
64877	Sep 17	23;11	174.97	64893	Oct 3	4;21	9.91
64907	Oct 17	8;53	203.97	64922	Nov 1	20;28	39.44
64936	Nov 15	20;59	233.55	64952	Dec 1	11;23	69.34
64966	Dec 15	11;32	263.58	64982	Dec 31	0;32	99.42
			1763				
64996	Jan 14	4;13	293.85	65011	Jan 29	12;00	129.42
65025	Feb 12	22;21	324.03	65040	Feb 27	22;12	159.12
65055	Mar 14	16;52	353.88	65070	Mar 29	7;35	188.35
65085	Apr 13	10;19	23.20	65099	Apr 27	16;33	217.09
65115	May 13	1;26	51.95	65129	May 27	1;35	245.41
65144	Jun 11	13;42	80.24	65158	Jun 25	11;23	273.51
65173	Jul 10	23;31	108.29	65187	Jul 24	22;47	301.63
65203	Aug 9	7;55	136.35	65217	Aug 23	12;31	330.02
65232	Sep 7	16;00	164.69	65247	Sep 22	4;45	358.87
65262	Oct 7	0;39	193.47	65276	Oct 21	22;40	28.28
65291	Nov 5	10;23	222.79	65306	Nov 20	16;46	58.17
65320	Dec 4	21;29	252.57	65336	Dec 20	9;39	88.35

Right panel

NEW MOONS				FULL MOONS			
JD	DATE	TIME	LONG.	JD	DATE	TIME	LONG.
23+				1764 23+			
65350	Jan 3	10;13	282.64	65366	Jan 19	0;31	118.53
65380	Feb 2	0;48	312.76	65395	Feb 17	13;17	148.45
65409	Mar 2	17;08	342.67	65425	Mar 18	0;04	177.91
65439	Apr 1	10;27	12.17	65454	Apr 16	9;07	206.82
65469	May 1	3;24	41.17	65483	May 15	16;53	235.23
65498	May 30	18;53	69.70	65513	Jun 14	0;12	263.30
65528	Jun 29	8;25	97.92	65542	Jul 13	8;11	291.26
65557	Jul 28	20;14	126.07	65571	Aug 11	18;06	319.41
65587	Aug 27	6;51	154.38	65601	Sep 10	6;50	347.97
65616	Sep 25	16;49	183.05	65630	Oct 9	22;34	17.10
65646	Oct 25	2;34	212.20	65660	Nov 8	16;36	46.83
65675	Nov 23	12;27	241.79	65690	Dec 8	11;37	76.99
65704	Dec 22	22;54	271.72				
			1765				
				65720	Jan 7	6;12	107.32
65734	Jan 21	10;24	301.76	65749	Feb 5	23;07	137.51
65763	Feb 19	23;13	331.65	65779	Mar 7	13;29	167.28
65793	Mar 21	13;18	1.19	65809	Apr 6	0;57	196.46
65823	Apr 20	4;13	30.29	65838	May 5	9;49	225.07
65852	May 19	19;28	58.94	65867	Jun 3	16;57	253.22
65882	Jun 18	10;39	87.29	65896	Jul 2	23;33	281.15
65912	Jul 18	1;29	115.53	65926	Aug 1	6;51	309.13
65941	Aug 16	15;38	143.88	65955	Aug 30	15;55	337.42
65971	Sep 15	4;48	172.53	65985	Sep 29	3;28	6.20
66000	Oct 14	16;49	201.59	66014	Oct 28	17;50	35.59
66030	Nov 13	3;54	231.08	66044	Nov 27	10;52	65.54
66059	Dec 12	14;31	260.93	66074	Dec 27	5;56	95.85
			1766				
66089	Jan 11	1;09	290.94	66104	Jan 26	1;37	126.22
66118	Feb 9	12;01	320.85	66133	Feb 24	20;02	156.29
66147	Mar 10	23;10	350.43	66163	Mar 26	11;36	185.83
66177	Apr 9	10;38	19.56	66192	Apr 24	23;48	214.73
66206	May 8	22;41	48.25	66222	May 24	9;09	243.10
66236	Jun 7	11;47	76.61	66251	Jun 22	16;46	271.12
66266	Jul 7	2;14	104.84	66280	Jul 21	23;46	299.06
66295	Aug 5	17;49	133.18	66310	Aug 20	7;09	327.17
66325	Sep 4	9;45	161.79	66339	Sep 18	15;40	355.67
66355	Oct 4	1;06	190.80	66369	Oct 18	2;01	24.72
66384	Nov 2	15;21	220.26	66398	Nov 16	14;50	54.33
66414	Dec 2	4;28	250.10	66428	Dec 16	6;33	84.42
66443	Dec 31	16;39	280.13				
			1767				
				66458	Jan 15	0;50	114.75
66473	Jan 30	3;56	310.13	66487	Feb 13	20;17	144.99
66502	Feb 28	14;16	339.82	66517	Mar 15	14;53	174.82
66531	Mar 29	23;44	9.05	66547	Apr 14	7;04	204.07
66561	Apr 28	8;52	37.78	66576	May 13	20;24	232.74
66590	May 27	18;32	66.12	66606	Jun 12	7;14	260.97
66620	Jun 26	5;43	94.26	66635	Jul 11	16;16	288.98
66649	Jul 25	18;58	122.46	66665	Aug 10	0;17	317.04
66679	Aug 24	10;11	150.92	66694	Sep 8	8;03	345.37
66709	Sep 23	2;44	179.80	66723	Oct 7	16;21	14.16
66738	Oct 22	19;46	209.19	66753	Nov 6	2;04	43.48
66768	Nov 21	12;31	239.04	66782	Dec 5	13;56	73.30
66798	Dec 21	4;15	269.17				
			1768				
				66812	Jan 4	4;20	103.45
66827	Jan 19	18;17	299.32	66841	Feb 2	20;53	133.65
66857	Feb 18	6;11	329.20	66871	Mar 3	14;32	163.60
66886	Mar 18	16;00	358.61	66901	Apr 2	8;02	193.10
66916	Apr 17	0;20	27.48	66931	May 2	0;25	222.05
66945	May 16	8;05	55.87	66960	May 31	15;07	250.54
66974	Jun 14	16;17	83.97	66990	Jun 30	3;52	278.73
67004	Jul 14	1;50	112.00	67019	Jul 29	14;42	306.84
67033	Aug 12	13;23	140.21	67049	Aug 28	0;03	335.10
67063	Sep 11	3;16	168.83	67078	Sep 26	8;43	3.74
67092	Oct 10	19;33	198.00	67107	Oct 25	17;40	32.85
67122	Nov 9	13;46	227.74	67137	Nov 24	3;39	62.47
67152	Dec 9	8;43	257.91	67166	Dec 23	15;06	92.44
			1769				
67182	Jan 8	2;43	288.22	67196	Jan 22	3;58	122.53
67211	Feb 6	18;15	318.35	67225	Feb 20	18;00	152.47
67241	Mar 8	6;48	348.03	67255	Mar 22	8;58	182.04
67270	Apr 6	16;46	17.14	67285	Apr 21	0;41	211.15
67300	May 6	0;58	45.71	67314	May 20	16;42	239.83
67329	Jun 4	8;19	73.87	67344	Jun 19	8;17	268.19
67358	Jul 3	15;39	101.83	67373	Jul 18	22;36	296.41
67387	Aug 1	23;49	129.84	67403	Aug 17	11;16	324.70
67417	Aug 31	9;35	158.16	67432	Sep 15	22;32	353.28
67446	Sep 29	22;00	186.99	67462	Oct 15	9;06	22.29
67476	Oct 29	13;37	216.45	67491	Nov 13	19;38	51.78
67506	Nov 28	8;08	246.47	67521	Dec 13	6;24	81.64
67536	Dec 28	4;03	276.82				

NEW MOONS				FULL MOONS			
JD	DATE	TIME	LONG.	JD	DATE	TIME	LONG.
23+				1770 23+			
				67550	Jan 11	17;23	111.66
67565	Jan 26	23;11	307.16	67580	Feb 10	4;33	141.58
67595	Feb 25	15;53	337.16	67609	Mar 11	16;07	171.17
67625	Mar 27	5;37	6.60	67639	Apr 10	4;33	200.33
67654	Apr 25	16;37	35.45	67668	May 9	18;13	229.07
67684	May 25	1;29	63.78	67698	Jun 8	9;00	257.48
67713	Jun 23	8;53	91.79	67728	Jul 8	0;17	285.75
67742	Jul 22	15;40	119.72	67757	Aug 6	15;25	314.06
67771	Aug 20	22;52	147.83	67787	Sep 5	6;00	342.63
67801	Sep 19	7;41	176.35	67816	Oct 4	19;55	11.60
67830	Oct 18	19;09	205.45	67846	Nov 3	9;10	41.03
67860	Nov 17	9;51	235.16	67875	Dec 2	21;37	70.85
67890	Dec 17	3;26	265.34				
			1771				
				67905	Jan 1	9;06	100.87
67919	Jan 15	22;37	295.71	67934	Jan 30	19;39	130.82
67949	Feb 14	17;45	325.92	67964	Mar 1	5;35	160.50
67979	Mar 16	11;24	355.71	67993	Mar 30	15;28	189.73
68009	Apr 15	2;39	24.91	68023	Apr 29	1;55	218.50
68038	May 14	15;07	53.53	68052	May 28	13;20	246.90
68068	Jun 13	0;55	81.71	68082	Jun 27	2;00	275.11
68097	Jul 12	8;44	109.68	68111	Jul 26	16;00	303.33
68126	Aug 10	15;42	137.69	68141	Aug 25	7;20	331.80
68155	Sep 8	23;02	166.01	68170	Sep 23	23;42	0.69
68185	Oct 8	7;50	194.82	68200	Oct 23	16;21	30.07
68214	Nov 6	18;48	224.20	68230	Nov 22	8;14	59.89
68244	Dec 6	8;11	254.09	68259	Dec 21	22;29	89.97
			1772				
68273	Jan 4	23;48	284.30	68289	Jan 20	10;50	120.04
68303	Feb 3	17;10	314.53	68318	Feb 18	21;33	149.87
68333	Mar 4	11;27	344.50	68348	Mar 19	7;08	179.26
68363	Apr 3	5;26	14.00	68377	Apr 17	16;03	208.15
68392	May 2	21;43	42.94	68407	May 17	0;46	236.57
68422	Jun 1	11;23	71.38	68436	Jun 15	9;54	264.70
68451	Jun 30	22;23	99.50	68465	Jul 14	20;15	292.76
68481	Jul 30	7;29	127.54	68495	Aug 13	8;44	321.02
68510	Aug 28	15;46	155.77	68524	Sep 11	23;49	349.70
68540	Sep 27	0;15	184.40	68554	Oct 11	17;14	18.93
68569	Oct 26	9;35	213.55	68584	Nov 10	11;41	48.69
68598	Nov 24	20;05	243.19	68614	Dec 10	5;32	78.82
68628	Dec 24	7;58	273.18				
			1773				
				68643	Jan 8	21;35	109.05
68657	Jan 22	21;29	303.30	68673	Feb 7	11;25	139.10
68687	Feb 21	12;47	333.28	68702	Mar 8	23;06	168.74
68717	Mar 23	5;28	2.91	68732	Apr 7	8;50	197.83
68746	Apr 21	22;31	32.07	68761	May 6	17;00	226.39
68776	May 21	14;42	60.74	68791	Jun 5	0;17	254.53
68806	Jun 20	5;12	89.05	68820	Jul 4	7;40	282.50
68835	Jul 19	17;55	117.21	68849	Aug 2	16;25	310.54
68865	Aug 18	5;16	145.46	68879	Sep 1	3;41	338.92
68894	Sep 16	15;45	174.01	68908	Sep 30	18;01	7.85
68924	Oct 16	1;49	203.01	68938	Oct 30	11;10	37.38
68953	Nov 14	11;46	232.48	68968	Nov 29	6;01	67.42
68982	Dec 13	21;59	262.33	68998	Dec 29	1;06	97.74
			1774				
69012	Jan 12	8;53	292.34	69027	Jan 27	19;02	128.03
69041	Feb 10	20;51	322.28	69057	Feb 26	10;43	157.97
69071	Mar 12	10;05	351.94	69086	Mar 27	23;34	187.37
69101	Apr 11	0;21	21.16	69116	Apr 26	9;33	216.16
69130	May 10	15;14	49.94	69145	May 25	17;20	244.45
69160	Jun 9	6;19	78.36	69174	Jun 23	23;59	272.42
69189	Jul 8	21;21	106.62	69204	Jul 23	6;46	300.36
69219	Aug 7	12;01	134.92	69233	Aug 21	14;52	328.51
69249	Sep 6	1;58	163.48	69263	Sep 20	1;11	357.10
69278	Oct 5	14;51	192.42	69292	Oct 19	14;13	26.28
69308	Nov 4	2;38	221.79	69322	Nov 18	6;02	56.04
69337	Dec 3	13;36	251.56	69352	Dec 18	0;14	86.25
			1775				
69367	Jan 2	0;15	281.54	69381	Jan 16	19;49	116.63
69396	Jan 31	10;55	311.50	69411	Feb 15	15;02	146.84
69425	Mar 1	21;45	341.19	69441	Mar 17	8;05	176.58
69455	Mar 31	8;47	10.46	69470	Apr 15	21;52	205.71
69484	Apr 29	20;13	39.27	69500	May 15	8;29	234.25
69514	May 29	8;29	67.70	69529	Jun 13	16;48	262.37
69543	Jun 27	22;06	95.93	69559	Jul 13	0;00	290.31
69573	Jul 27	13;09	124.20	69588	Aug 11	7;09	318.34
69603	Aug 26	5;10	152.71	69617	Sep 9	15;08	346.69
69632	Sep 24	21;09	181.60	69647	Oct 9	0;36	15.54
69662	Oct 24	12;17	210.94	69676	Nov 7	12;12	44.96
69692	Nov 23	2;13	240.69	69706	Dec 7	2;28	74.90
69721	Dec 22	15;02	270.70				

NEW MOONS				FULL MOONS			
JD	DATE	TIME	LONG.	JD	DATE	TIME	LONG.
23+				1776 23+			
				69735	Jan 5	19;32	105.17
69751	Jan 21	2;49	300.75	69765	Feb 4	14;30	135.46
69780	Feb 19	13;33	330.57	69795	Mar 5	9;36	165.45
69809	Mar 19	23;14	359.95	69825	Apr 4	2;59	194.91
69839	Apr 18	8;13	28.83	69854	May 3	17;39	223.77
69868	May 17	17;15	57.26	69884	Jun 2	5;38	252.14
69898	Jun 16	3;22	85.42	69913	Jul 1	15;31	280.21
69927	Jul 15	15;25	113.56	69943	Jul 31	0;04	308.23
69957	Aug 14	5;41	141.89	69972	Aug 29	8;01	336.46
69986	Sep 12	21;48	170.62	70001	Sep 27	16;07	5.09
70016	Oct 12	14;55	199.85	70031	Oct 27	1;11	34.23
70046	Nov 11	8;10	229.58	70060	Nov 25	11;58	63.89
70076	Dec 11	0;43	259.66	70090	Dec 25	1;05	93.94
			1777				
70105	Jan 9	15;48	289.86	70119	Jan 23	16;32	124.14
70135	Feb 8	4;48	319.86	70149	Feb 22	9;34	154.18
70164	Mar 9	15;31	349.45	70179	Mar 24	3;02	183.83
70194	Apr 8	0;20	18.49	70208	Apr 22	19;51	212.96
70223	May 7	8;04	47.02	70238	May 22	11;21	241.59
70252	Jun 5	15;45	75.18	70268	Jun 21	1;06	269.87
70282	Jul 5	0;25	103.18	70297	Jul 20	13;00	298.00
70311	Aug 3	10;50	131.30	70326	Aug 18	23;12	326.21
70340	Sep 1	23;33	159.76	70356	Sep 17	8;20	354.72
70370	Oct 1	14;46	188.73	70385	Oct 16	17;15	23.69
70400	Oct 31	8;16	218.29	70415	Nov 15	2;49	53.15
70430	Nov 30	3;13	248.35	70444	Dec 14	13;35	83.02
70459	Dec 29	22;03	278.66				
			1778				
				70474	Jan 13	1;42	113.09
70489	Jan 28	15;01	308.91	70503	Feb 11	15;01	143.08
70519	Feb 27	5;02	338.77	70533	Mar 13	5;16	172.77
70548	Mar 28	16;06	8.08	70562	Apr 11	20;20	202.01
70578	Apr 27	0;56	36.81	70592	May 11	12;02	230.80
70607	May 26	8;29	65.08	70622	Jun 10	3;49	259.24
70636	Jun 24	15;39	93.08	70651	Jul 9	18;53	287.50
70665	Jul 23	23;14	121.04	70681	Aug 8	8;33	315.77
70695	Aug 22	8;06	149.22	70710	Sep 6	20;43	344.26
70724	Sep 20	19;10	177.86	70740	Oct 6	7;50	13.14
70754	Oct 20	9;14	207.10	70769	Nov 4	18;34	42.49
70784	Nov 19	2;31	236.93	70799	Dec 4	5;22	72.25
70813	Dec 18	22;03	267.21				
			1779				
				70828	Jan 2	16;18	102.25
70843	Jan 17	17;51	297.60	70858	Feb 1	3;17	132.22
70873	Feb 16	11;53	327.75	70887	Mar 2	14;24	161.92
70903	Mar 18	3;02	357.40	70917	Apr 1	2;04	191.21
70932	Apr 16	15;16	26.46	70946	Apr 30	14;48	220.05
70962	May 16	1;03	54.95	70976	May 30	4;49	248.54
70991	Jun 14	9;03	83.05	71005	Jun 28	19;50	276.83
71020	Jul 13	16;02	110.98	71035	Jul 28	11;09	305.11
71049	Aug 11	22;57	139.00	71065	Aug 27	2;12	333.58
71079	Sep 10	6;53	167.36	71094	Sep 25	16;44	2.43
71108	Oct 9	17;00	196.25	71124	Oct 25	6;38	31.73
71138	Nov 8	6;09	225.75	71153	Nov 23	19;46	61.46
71167	Dec 7	22;25	255.78	71183	Dec 23	7;53	91.45
			1780				
71197	Jan 6	16;58	286.11	71212	Jan 21	18;53	121.46
71227	Feb 5	12;14	316.40	71242	Feb 20	4;58	151.24
71257	Mar 6	6;38	346.36	71271	Mar 20	14;36	180.62
71286	Apr 4	23;04	15.76	71301	Apr 19	0;26	209.52
71316	May 4	12;52	44.58	71330	May 18	11;04	238.01
71345	Jun 2	23;55	72.90	71359	Jun 16	22;53	266.23
71375	Jul 2	8;40	100.92	71389	Jul 16	12;06	294.41
71404	Jul 31	16;00	128.90	71419	Aug 15	2;48	322.77
71433	Aug 29	23;07	157.10	71448	Sep 13	18;53	351.51
71463	Sep 28	7;13	185.74	71478	Oct 13	11;46	20.75
71492	Oct 27	17;11	214.93	71508	Nov 12	4;28	50.46
71522	Nov 26	5;26	244.66	71537	Dec 11	19;51	80.51
71551	Dec 25	19;57	274.78				
			1781				
				71567	Jan 10	9;15	110.63
71581	Jan 24	12;24	305.01	71596	Feb 8	20;42	140.57
71611	Feb 23	6;09	335.08	71626	Mar 10	6;40	170.11
71641	Mar 25	0;13	4.74	71655	Apr 8	15;41	199.15
71670	Apr 23	17;21	33.86	71685	May 8	0;13	227.71
71700	May 23	8;19	62.46	71714	Jun 6	8;52	255.89
71729	Jun 21	20;39	90.68	71743	Jul 5	18;21	283.93
71759	Jul 21	6;41	118.74	71773	Aug 4	5;36	312.08
71788	Aug 19	15;24	146.89	71802	Sep 2	19;24	340.59
71817	Sep 17	23;53	175.38	71832	Oct 2	11;53	9.62
71847	Oct 17	8;56	204.36	71862	Nov 1	6;13	39.23
71876	Nov 15	18;59	233.86	71892	Dec 1	0;47	69.29
71906	Dec 15	6;12	263.75	71921	Dec 30	18;00	99.54

NEW MOONS				FULL MOONS					NEW MOONS				FULL MOONS			
JD	DATE	TIME	LONG.	JD	DATE	TIME	LONG.		JD	DATE	TIME	LONG.	JD	DATE	TIME	LONG.
23+				**1782**	23+				23+				**1788**	23+		
71935	Jan 13	18;48	293.84	71951	Jan 29	9;01	129.70		74121	Jan 8	12;04	287.99	74136	Jan 23	2;11	122.85
71965	Feb 12	9;01	323.87	71980	Feb 27	21;43	159.50		74151	Feb 7	7;11	318.27	74165	Feb 21	13;02	152.65
71995	Mar 14	0;49	353.61	72010	Mar 29	8;16	188.78		74180	Mar 7	23;46	348.13	74195	Mar 22	0;07	182.06
72024	Apr 12	17;34	22.91	72039	Apr 27	17;00	217.50		74210	Apr 6	13;21	17.39	74224	Apr 20	11;58	211.02
72054	May 12	10;07	51.73	72069	May 27	0;29	245.76		74240	May 6	0;12	46.06	74254	May 20	1;04	239.59
72084	Jun 11	1;27	80.14	72098	Jun 25	7;34	273.74		74269	Jun 4	8;57	74.28	74283	Jun 18	15;29	267.91
72113	Jul 10	15;06	108.34	72127	Jul 24	15;26	301.72		74298	Jul 3	16;21	102.25	74313	Jul 18	6;42	296.17
72143	Aug 9	3;15	136.56	72157	Aug 23	1;19	329.95		74327	Aug 1	23;13	130.21	74342	Aug 16	22;04	324.57
72172	Sep 7	14;21	165.01	72186	Sep 21	14;08	358.67		74357	Aug 31	6;35	158.43	74372	Sep 15	13;07	353.29
72202	Oct 7	0;51	193.87	72216	Oct 21	6;05	27.99		74386	Sep 29	15;35	187.13	74402	Oct 15	3;39	22.46
72231	Nov 5	11;03	223.21	72246	Nov 20	0;23	57.88		74416	Oct 29	3;14	216.41	74431	Nov 13	17;28	52.08
72260	Dec 4	21;13	252.95	72275	Dec 19	19;39	88.15		74445	Nov 27	18;02	246.26	74461	Dec 13	6;19	82.03
									74475	Dec 27	11;35	276.52				
				1783									**1789**			
72290	Jan 3	7;45	282.94	72305	Jan 18	14;23	118.50		74490	Jan 11	17;57	112.07				
72319	Feb 1	19;02	312.92	72335	Feb 17	7;17	148.60		74505	Jan 26	6;36	306.85	74520	Feb 10	4;23	141.96
72349	Mar 3	7;25	342.66	72364	Mar 18	21;31	178.20		74535	Feb 25	1;28	336.94	74549	Mar 11	13;59	171.47
72378	Apr 1	20;54	12.01	72394	Apr 17	8;48	207.20		74564	Mar 26	18;52	6.55	74578	Apr 9	23;24	200.51
72408	May 1	11;14	40.91	72423	May 16	17;29	235.64		74594	Apr 25	9;56	35.57	74608	May 9	9;19	229.09
72438	May 31	2;03	69.42	72453	Jun 15	0;28	263.70		74623	May 24	22;18	64.04	74637	Jun 7	20;16	257.37
72467	Jun 29	17;04	97.71	72482	Jul 14	7;01	291.61		74653	Jun 23	8;09	92.15	74667	Jul 7	8;38	285.52
72497	Jul 29	8;02	125.98	72511	Aug 12	14;21	319.65		74682	Jul 22	16;08	120.13	74696	Aug 5	22;34	313.78
72526	Aug 27	22;36	154.44	72540	Sep 10	23;32	348.07		74711	Aug 20	23;21	148.24	74726	Sep 4	14;06	342.38
72556	Sep 26	12;20	183.26	72570	Oct 10	11;15	17.03		74741	Sep 19	6;59	176.72	74756	Oct 4	6;54	11.46
72586	Oct 26	0;56	212.53	72600	Nov 9	1;46	46.60		74770	Oct 18	16;05	205.72	74786	Nov 3	0;07	41.04
72615	Nov 24	12;29	242.19	72629	Dec 8	18;54	76.67		74800	Nov 17	3;15	235.27	74815	Dec 2	16;32	71.03
72644	Dec 23	23;22	272.14						74829	Dec 16	16;40	265.27				
				1784									**1790**			
				72659	Jan 7	13;58	107.03						74845	Jan 1	7;06	101.18
72674	Jan 22	9;59	302.13	72689	Feb 6	9;33	137.34		74859	Jan 15	8;06	295.48	74874	Jan 30	19;32	131.21
72703	Feb 20	20;37	331.94	72719	Mar 7	3;48	167.27		74889	Feb 14	1;06	325.61	74904	Mar 1	6;05	160.91
72733	Mar 21	7;19	1.34	72748	Apr 5	19;12	196.63		74918	Mar 15	18;56	355.41	74933	Mar 30	15;21	190.11
72762	Apr 19	18;14	30.28	72778	May 5	7;16	225.36		74948	Apr 14	12;30	24.71	74962	Apr 28	23;52	218.81
72792	May 19	5;45	58.79	72807	Jun 3	16;34	253.60		74978	May 14	4;33	53.49	74992	May 28	8;11	247.08
72821	Jun 17	18;26	87.05	72837	Jul 3	0;11	281.57		75007	Jun 12	18;13	81.83	75021	Jun 26	17;00	275.13
72851	Jul 17	8;41	115.26	72866	Aug 1	7;18	309.54		75037	Jul 12	5;24	109.93	75051	Jul 26	3;11	303.21
72881	Aug 16	0;23	143.67	72895	Aug 30	14;52	337.77		75066	Aug 10	14;49	138.04	75080	Aug 24	15;38	331.56
72910	Sep 14	16;43	172.43	72924	Sep 28	23;39	6.44		75095	Sep 8	23;31	166.42	75110	Sep 23	6;54	0.40
72940	Oct 14	8;39	201.64	72954	Oct 28	10;12	35.67		75125	Oct 8	8;25	195.24	75140	Oct 23	0;39	29.83
72969	Nov 12	23;29	231.29	72983	Nov 26	23;09	65.43		75154	Nov 6	18;07	224.57	75169	Nov 21	19;32	59.76
72999	Dec 12	13;03	261.26	73013	Dec 26	14;49	95.60		75184	Dec 6	4;48	254.35	75199	Dec 21	13;46	90.00
				1785									**1791**			
73029	Jan 11	1;26	291.34	73043	Jan 25	8;53	125.90		75213	Jan 4	16;39	284.39	75229	Jan 20	6;01	120.23
73058	Feb 9	12;40	321.27	73073	Feb 24	4;00	156.02		75243	Feb 3	5;53	314.44	75258	Feb 18	19;50	150.19
73087	Mar 10	22;43	350.81	73102	Mar 25	22;16	185.67		75272	Mar 4	20;40	344.27	75288	Mar 20	7;20	179.65
73117	Apr 9	7;47	19.84	73132	Apr 24	14;14	214.74		75302	Apr 3	12;46	13.70	75317	Apr 18	16;48	208.56
73146	May 8	16;26	48.39	73162	May 24	3;28	243.27		75332	May 3	5;19	42.67	75347	May 18	0;41	236.95
73176	Jun 7	1;41	76.60	73191	Jun 22	14;20	271.42		75361	Jun 1	21;14	71.20	75376	Jun 16	7;43	265.00
73205	Jul 6	12;32	104.69	73220	Jul 21	23;34	299.44		75391	Jul 1	11;46	99.46	75405	Jul 15	14;58	292.94
73235	Aug 5	1;38	132.92	73250	Aug 20	7;53	327.59		75421	Jul 31	0;46	127.66	75434	Aug 13	23;41	321.05
73264	Sep 3	16;59	161.50	73279	Sep 18	16;00	356.08		75450	Aug 29	12;34	156.04	75464	Sep 12	11;01	349.57
73294	Oct 3	9;53	190.57	73309	Oct 18	0;39	25.05		75479	Sep 27	23;35	184.77	75494	Oct 12	1;31	18.68
73324	Nov 2	3;25	220.15	73338	Nov 16	10;36	54.55		75509	Oct 27	10;09	213.97	75523	Nov 10	18;54	48.39
73353	Dec 1	20;40	250.16	73367	Dec 15	22;33	84.48		75538	Nov 25	20;29	243.61	75553	Dec 10	13;59	78.56
73383	Dec 31	12;44	280.37						75568	Dec 25	6;51	273.54				
				1786									**1792**			
				73397	Jan 14	12;46	114.63						75583	Jan 9	9;14	108.93
73413	Jan 30	2;53	310.48	73427	Feb 13	4;57	144.74		75597	Jan 23	17;39	303.54	75613	Feb 8	3;11	139.15
73442	Feb 28	14;40	340.23	73456	Mar 14	22;05	174.52		75627	Feb 22	5;16	333.36	75642	Mar 8	18;47	168.95
73472	Mar 30	0;13	9.46	73486	Apr 13	15;06	203.81		75656	Mar 22	17;57	2.83	75672	Apr 7	7;26	198.17
73501	Apr 28	8;12	38.14	73516	May 13	7;10	232.60		75686	Apr 21	7;36	31.85	75701	May 6	17;15	226.78
73530	May 27	15;37	66.39	73545	Jun 11	21;47	260.98		75715	May 20	21;58	60.46	75731	Jun 5	0;52	254.95
73559	Jun 25	23;32	94.41	73575	Jul 11	10;42	289.15		75745	Jun 19	12;45	88.79	75760	Jul 4	7;27	282.88
73589	Jul 25	8;55	122.45	73604	Aug 9	21;53	317.33		75775	Jul 19	3;47	117.04	75789	Aug 2	14;15	310.84
73618	Aug 23	20;26	150.75	73634	Sep 8	7;42	345.74		75804	Aug 17	18;45	145.43	75818	Aug 31	22;28	339.10
73648	Sep 22	10;28	179.53	73663	Oct 7	16;50	14.56		75834	Sep 16	9;14	174.14	75848	Sep 30	8;57	7.87
73678	Oct 22	3;00	208.89	73693	Nov 6	2;11	43.87		75863	Oct 15	22;44	203.28	75877	Oct 29	22;09	37.23
73707	Nov 20	21;33	238.80	73722	Dec 5	12;25	73.63		75893	Nov 14	11;03	232.85	75907	Nov 28	14;05	67.14
73737	Dec 20	16;49	269.08						75922	Dec 13	22;23	262.74	75937	Dec 28	8;19	97.43
				1787									**1793**			
				73751	Jan 3	23;52	103.65		75952	Jan 12	9;07	292.75	75967	Jan 27	3;48	127.78
73767	Jan 19	10;59	299.40	73781	Feb 2	12;29	133.68		75981	Feb 10	19;39	322.63	75996	Feb 25	22;50	157.87
73797	Feb 18	2;32	329.43	73811	Mar 4	2;02	163.46		76011	Mar 12	6;06	352.17	76026	Mar 27	15;41	187.44
73826	Mar 19	14;57	358.95	73840	Apr 2	16;25	192.84		76040	Apr 10	16;37	21.24	76056	Apr 26	5;20	216.38
73856	Apr 18	0;42	27.87	73870	May 2	7;34	221.76		76070	May 10	3;29	49.86	76085	May 25	15;51	244.78
73885	May 17	8;40	56.28	73899	May 31	23;13	250.30		76099	Jun 8	15;17	78.15	76115	Jun 24	0;10	272.82
73914	Jun 15	15;49	84.34	73929	Jun 30	14;44	278.60		76129	Jul 8	4;36	106.34	76144	Jul 23	7;28	300.77
73943	Jul 14	23;03	112.28	73959	Jul 30	5;17	306.86		76158	Aug 6	19;39	134.66	76173	Aug 21	14;49	328.89
73973	Aug 13	7;11	140.36	73988	Aug 28	18;24	335.27		76188	Sep 5	11;57	163.29	76202	Sep 19	23;03	357.40
74002	Sep 11	17;05	168.81	74018	Sep 27	6;14	4.03		76218	Oct 5	4;28	192.38	76232	Oct 19	8;46	26.44
74032	Oct 11	5;38	197.82	74047	Oct 26	17;20	33.24		76247	Nov 3	20;13	221.92	76261	Nov 17	20;33	56.03
74061	Nov 9	21;25	227.45	74077	Nov 25	4;17	62.89		76277	Dec 3	10;39	251.83	76291	Dec 17	10;50	86.08
74091	Dec 9	16;05	257.60	74106	Dec 24	15;15	92.84									

Left section

NEW MOONS				FULL MOONS			
JD	DATE	TIME	LONG.	JD	DATE	TIME	LONG.
23+			1794	23+			
76306	Jan 1	23;46	281.91	76321	Jan 16	3;43	116.35
76336	Jan 31	11;35	311.93	76350	Feb 14	22;21	146.54
76365	Mar 1	22;06	341.61	76380	Mar 16	17;05	176.36
76395	Mar 31	7;26	10.81	76410	Apr 15	10;09	205.63
76424	Apr 29	15;57	39.49	76440	May 15	0;39	234.33
76454	May 29	0;33	67.77	76469	Jun 13	12;38	262.59
76483	Jun 27	10;19	95.85	76498	Jul 12	22;41	290.64
76512	Jul 26	22;09	123.99	76528	Aug 11	7;31	318.74
76542	Aug 25	12;27	152.42	76557	Sep 9	15;50	347.11
76572	Sep 24	4;49	181.32	76587	Oct 9	0;19	15.92
76601	Oct 23	22;22	210.75	76616	Nov 7	9;40	45.25
76631	Nov 22	16;07	240.65	76645	Dec 6	20;37	75.05
76661	Dec 22	9;05	270.85				
			1795				
76691	Jan 21	0;22	301.04	76675	Jan 5	9;40	105.14
76720	Feb 19	13;20	330.95	76705	Feb 4	0;47	135.28
76749	Mar 20	23;50	0.37	76734	Mar 5	17;20	165.17
76779	Apr 19	8;20	29.22	76764	Apr 4	10;16	194.62
76808	May 18	15;44	57.59	76794	May 4	2;41	223.56
76837	Jun 16	23;08	85.65	76823	Jun 2	18;00	252.07
76867	Jul 16	7;36	113.64	76853	Jul 2	7;50	280.29
76896	Aug 14	17;57	141.81	76882	Jul 31	20;01	308.46
76926	Sep 13	6;44	170.40	76912	Aug 30	6;40	336.80
76955	Oct 12	22;08	199.55	76941	Sep 28	16;17	5.48
76985	Nov 11	15;56	229.29	76971	Oct 28	1;38	34.64
77015	Dec 11	11;11	259.48	77000	Nov 26	11;31	64.26
				77029	Dec 25	22;23	94.22
			1796				
77045	Jan 10	6;15	289.84	77059	Jan 24	10;22	124.27
77074	Feb 8	23;17	320.02	77088	Feb 22	23;17	154.14
77104	Mar 9	13;12	349.75	77118	Mar 23	12;59	183.64
77134	Apr 8	0;05	18.88	77148	Apr 22	3;27	212.69
77163	May 7	8;42	47.44	77177	May 21	18;40	241.32
77192	Jun 5	16;04	75.58	77207	Jun 20	10;15	269.67
77221	Jul 4	23;06	103.53	77237	Jul 20	1;25	297.93
77251	Aug 3	6;40	131.53	77266	Aug 18	15;28	326.29
77280	Sep 1	15;37	159.83	77296	Sep 17	4;10	354.94
77310	Oct 1	2;49	188.63	77325	Oct 16	15;51	24.02
77339	Oct 30	17;02	218.04	77355	Nov 15	3;04	53.55
77369	Nov 29	10;27	248.03	77384	Dec 14	14;10	83.43
77399	Dec 29	6;04	278.38				
			1797				
77429	Jan 28	1;52	308.74	77414	Jan 13	1;10	113.45
77458	Feb 26	19;47	338.77	77443	Feb 11	11;56	143.35
77488	Mar 28	10;48	8.25	77472	Mar 12	22;38	172.89
77517	Apr 26	22;52	37.13	77502	Apr 11	9;44	201.97
77547	May 26	8;32	65.48	77531	May 10	21;53	230.63
77576	Jun 24	16;29	93.50	77561	Jun 9	11;27	258.99
77605	Jul 23	23;32	121.44	77591	Jul 9	2;14	287.24
77635	Aug 22	6;36	149.56	77620	Aug 7	17;40	315.57
77664	Sep 20	14;45	178.08	77650	Sep 6	9;07	344.18
77694	Oct 20	1;04	207.15	77680	Oct 6	0;12	13.22
77723	Nov 18	14;21	236.81	77709	Nov 4	14;42	42.72
77753	Dec 18	6;38	266.94	77739	Dec 4	4;19	72.60
			1798				
77783	Jan 17	1;02	297.28	77768	Jan 2	16;41	102.66
77812	Feb 15	20;02	327.48	77798	Feb 1	3;41	132.63
77842	Mar 17	14;09	357.26	77827	Mar 2	13;30	162.27
77872	Apr 16	6;20	26.48	77856	Mar 31	22;43	191.46
77901	May 15	20;01	55.14	77886	Apr 30	8;04	220.17
77931	Jun 14	7;06	83.36	77915	May 29	18;12	248.51
77960	Jul 13	15;58	111.37	77945	Jun 28	5;39	276.66
77989	Aug 11	23;33	139.42	77974	Jul 27	18;43	304.85
78019	Sep 10	7;00	167.76	78004	Aug 26	9;31	333.31
78048	Oct 9	15;25	196.58	78034	Sep 25	1;56	2.22
78078	Nov 8	1;37	225.95	78063	Oct 24	19;20	31.66
78107	Dec 7	13;58	255.81	78093	Nov 23	12;34	61.55
				78123	Dec 23	4;20	91.69
			1799				
78137	Jan 6	4;23	285.96	78152	Jan 21	17;54	121.81
78166	Feb 4	20;30	316.13	78182	Feb 20	5;17	151.65
78196	Mar 6	13;47	346.04	78211	Mar 21	15;00	181.02
78226	Apr 5	7;24	15.51	78240	Apr 19	23;38	209.87
78256	May 5	0;12	44.46	78270	May 19	7;48	238.26
78285	Jun 3	15;05	72.94	78299	Jun 17	16;07	266.35
78315	Jul 3	3;33	101.11	78329	Jul 17	1;24	294.38
78344	Aug 1	13;53	129.21	78358	Aug 15	12;34	322.59
78373	Aug 30	23;00	157.49	78388	Sep 14	2;27	351.24
78403	Sep 29	7;55	186.16	78417	Oct 13	19;12	20.46
78432	Oct 28	17;22	215.33	78447	Nov 12	13;55	50.25
78462	Nov 27	3;40	244.98	78477	Dec 12	8;52	80.43
78491	Dec 26	14;57	274.96				

Right section

NEW MOONS				FULL MOONS			
JD	DATE	TIME	LONG.	JD	DATE	TIME	LONG.
23+			1800	23+			
				78507	Jan 11	2;21	110.73
78521	Jan 25	3;21	305.02	78536	Feb 9	17;25	140.82
78550	Feb 23	17;08	334.92	78566	Mar 11	6;00	170.47
78580	Mar 25	8;21	4.47	78595	Apr 9	16;19	199.57
78610	Apr 24	0;32	33.57	78625	May 9	0;47	228.12
78639	May 23	16;43	62.23	78654	Jun 7	8;01	256.25
78669	Jun 22	7;57	90.57	78683	Jul 6	14;57	284.19
78698	Jul 21	21;47	118.77	78712	Aug 4	22;45	312.21
78728	Aug 20	10;20	147.08	78742	Sep 3	8;40	340.55
78757	Sep 18	21;58	175.70	78771	Oct 2	21;37	9.44
78787	Oct 18	9;00	204.76	78801	Nov 1	13;46	38.94
78816	Nov 16	19;38	234.28	78831	Dec 1	8;17	68.98
78846	Dec 16	6;03	264.14	78861	Dec 31	3;42	99.32
			1801				
78875	Jan 14	16;34	294.14	78890	Jan 29	22;29	129.64
78905	Feb 13	3;36	324.03	78920	Feb 28	15;19	159.62
78934	Mar 14	15;30	353.61	78950	Mar 30	5;24	189.05
78964	Apr 13	4;23	22.76	78979	Apr 28	16;30	217.87
78993	May 12	18;09	51.48	79009	May 28	1;01	246.17
79023	Jun 11	8;34	79.86	79038	Jun 26	7;56	274.15
79052	Jul 10	23;27	108.11	79067	Jul 25	14;30	302.07
79082	Aug 9	14;36	136.44	79096	Aug 23	21;56	330.21
79112	Sep 8	5;38	165.05	79126	Sep 22	7;18	358.79
79141	Oct 7	19;58	194.07	79155	Oct 21	19;12	27.93
79171	Nov 6	9;10	223.53	79185	Nov 20	9;51	57.67
79200	Dec 5	21;09	253.34	79215	Dec 20	3;02	87.84
			1802				
79230	Jan 4	8;13	283.35	79244	Jan 18	22;00	118.20
79259	Feb 2	18;47	313.30	79274	Feb 17	17;23	148.40
79289	Mar 4	5;06	342.96	79304	Mar 19	11;24	178.17
79318	Apr 2	15;19	12.17	79334	Apr 18	2;37	207.33
79348	May 2	1;42	40.91	79363	May 17	14;34	235.91
79377	May 31	12;42	69.27	79392	Jun 15	23;51	264.05
79407	Jun 30	1;02	97.45	79422	Jul 15	7;34	292.01
79436	Jul 29	15;10	125.69	79451	Aug 13	14;52	320.06
79466	Aug 28	7;03	154.21	79480	Sep 11	22;43	348.43
79495	Sep 26	23;50	183.15	79510	Oct 11	7;47	17.29
79525	Oct 26	16;21	212.56	79539	Nov 9	18;34	46.69
79555	Nov 25	7;44	242.39	79569	Dec 9	7;34	76.59
79584	Dec 24	21;41	272.46				
			1803				
				79598	Jan 7	23;07	106.79
79614	Jan 23	10;11	302.53	79628	Feb 6	16;53	137.02
79643	Feb 21	21;18	332.35	79658	Mar 8	11;36	166.99
79673	Mar 23	7;03	1.72	79688	Apr 7	5;29	196.45
79702	Apr 21	15;41	30.56	79717	May 6	21;13	225.34
79731	May 20	23;55	58.94	79747	Jun 5	10;24	253.74
79761	Jun 19	8;47	87.05	79776	Jul 4	21;23	281.85
79790	Jul 18	19;23	115.12	79806	Aug 3	6;53	309.92
79820	Aug 17	8;25	143.42	79835	Sep 1	15;33	338.19
79849	Sep 15	23;55	172.14	79865	Oct 1	0;04	6.85
79879	Oct 15	17;11	201.40	79894	Oct 30	9;04	36.02
79909	Nov 14	11;12	231.17	79923	Nov 28	19;15	65.67
79939	Dec 14	4;52	261.31	79953	Dec 28	7;12	95.67
			1804				
79968	Jan 12	21;12	291.56	79982	Jan 26	21;12	125.80
79998	Feb 11	11;24	321.61	80012	Feb 25	12;55	155.78
80027	Mar 11	23;02	351.21	80042	Mar 26	5;30	185.37
80057	Apr 10	8;19	20.25	80071	Apr 24	22;02	214.48
80086	May 9	15;59	48.76	80101	May 24	13;48	243.10
80115	Jun 7	23;06	76.88	80131	Jun 23	4;24	271.41
80145	Jul 7	6;48	104.85	80160	Jul 22	17;32	299.59
80174	Aug 5	16;05	132.93	80190	Aug 21	5;08	327.87
80204	Sep 4	3;38	161.35	80219	Sep 19	15;26	356.44
80233	Oct 3	17;49	190.30	80249	Oct 19	1;04	25.46
80263	Nov 2	10;36	219.84	80278	Nov 17	10;48	54.95
80293	Dec 2	5;26	249.91	80307	Dec 16	21;14	84.82
			1805				
80323	Jan 1	0;57	280.26	80337	Jan 15	8;39	114.86
80352	Jan 30	19;13	310.55	80366	Feb 13	20;57	144.80
80382	Mar 1	10;44	340.47	80396	Mar 15	10;00	174.41
80411	Mar 30	22;59	9.80	80425	Apr 13	23;46	203.58
80441	Apr 29	8;30	38.55	80455	May 13	14;22	232.32
80470	May 28	16;16	66.81	80485	Jun 12	5;40	260.73
80499	Jun 26	23;17	94.79	80514	Jul 11	21;08	289.00
80529	Jul 26	6;29	122.74	80544	Aug 10	11;58	317.32
80558	Aug 24	14;40	150.91	80574	Sep 9	1;36	345.88
80588	Sep 23	0;43	179.52	80603	Oct 8	14;01	14.84
80617	Oct 22	13;26	208.72	80633	Nov 7	1;41	44.24
80647	Nov 21	5;22	238.51	80662	Dec 6	13;01	74.05
80677	Dec 21	0;07	268.77				